李新荣
回　嵘　著
赵　洋

Eco-physiology of
Biological Soil Crusts
in Desert Regions of China

中国荒漠生物土壤结皮生态生理学研究

高等教育出版社·北京
HIGHER EDUCATION PRESS　BEIJING

内容简介

生物土壤结皮是荒漠生态系统的重要组成成分，对荒漠系统生态和水文过程发挥着重要的影响，维持着荒漠生物地球化学循环和生态系统的健康和可持续发展。然而，生物土壤结皮的这些功能主要取决于组成其群落的生物体及多样性，它们对环境胁迫和气候变化具有高度的敏感性。本书全面阐述了这些结皮群落组成成分对非生物因子和生物因子等胁迫和干扰的生态生理响应，以及适应对策和机理，是正确认知和评估未来荒漠生态系统对全球变化响应和实现可持续生态系统管理的前提，是荒漠生态系统恢复与重建的重要科学依据。

本书可为生态学、土壤学、生理学、微生物学、水文学和全球变化研究者以及从事防沙治沙、荒漠生态恢复和生态系统管理的人员提供重要的参考。

图书在版编目（CIP）数据

中国荒漠生物土壤结皮生态生理学研究 / 李新荣，回嵘，赵洋著 . -- 北京：高等教育出版社，2016.12
ISBN 978-7-04-046979-0

Ⅰ. ①中… Ⅱ. ①李… ②回… ③赵… Ⅲ. ①荒漠－生态系统－土壤结皮－生理学－研究－中国 Ⅳ. ① P941.73 ② S152.4

中国版本图书馆 CIP 数据核字 (2016) 第 296644 号

策划编辑 关 焱 李冰祥 责任编辑 殷 鸽 关 焱 封面设计 刘晓翔
插图绘制 于 博 黄云燕 责任校对 刘春萍 责任印制 田 甜

出版发行	高等教育出版社	咨询电话	400-810-0598
社　　址	北京市西城区德外大街 4 号	网　　址	http://www.hep.edu.cn
邮政编码	100120		http://www.hep.com.cn
印　　刷	固安县铭成印刷有限公司	网上订购	http://www.hepmall.com.cn
			http://www.hepmall.com
开　　本	787mm × 1092mm 1/16		http://www.hepmall.cn
印　　张	28	版　　次	2016 年 12 月第 1 版
字　　数	550 千字	印　　次	2016 年 12 月第 1 次印刷
购书热线	010-58581118	定　　价	268.00 元

本书如有缺页、倒页、脱页等质量问题，请到所购图书销售部门联系调换

物 料 号 46979-00
审 图 号 GS（2016）2529 号
ZHONGGUO HUANGMO SHENGWU TURANG JIEPI SHENGTAI SHENGLIXUE YANJIU

序

生态文明建设是中国特色社会主义事业的重要内容，加大自然生态系统和环境保护力度是推进生态文明建设的重要举措。所谓自然生态系统是在一定时间和空间范围内，依靠自然调节能力维持的相对稳定的生态系统。我国北方广泛分布的荒漠、沙地和农牧交错带是承受风沙危害最严重的地区，是风沙治理、生态重建和恢复与生态文明建设的核心地区之一。生物土壤结皮作为干旱、半干旱荒漠自然生态系统中最重要的地表生物覆盖体，在维系荒漠自然生态系统稳定和持续发展中具有重要意义。

生物土壤结皮是由真菌界的地衣、细菌界的蓝细菌以及植物界的苔类、藓类和藻类为优势的微型生物与沙土颗粒缠绕在一起所形成的皮壳状结构。生物土壤结皮作为荒漠地区最重要的地表生物覆盖体，在遏制地表风蚀和水蚀，稳定地表环境，阻止就地起沙以切断沙尘暴沙源，促进荒漠和沙区土壤微生物及微小动物的繁衍、生存以改善土壤物质转化，如固氮中的硝化与反硝化作用、腐殖质的分解与合成等方面发挥着重要作用。同时，生物土壤结皮还能够保持土壤深层水分以利于相应数量维管植物的定居和生存，从而在维系荒漠自然生态系统的稳定和持续发展中具有重要意义。

近年来，生物土壤结皮在维系荒漠自然生态系统的稳定性和持续发展中的重要意义已越来越被学界所认可。然而，人类对于生物土壤结皮中微型生物的种类组成、结构、形成条件、演替规律及其生态功能机理还缺乏足够认识，这既是务必探明的科学问题，又是荒漠自然生态系统重建与恢复急需解决的实际问题，有待生物学、生态学等多学科的合作与综合研究。

《中国荒漠生物土壤结皮生态生理学研究》是继《荒漠生物土壤结皮生态与水文学研究》之后，李新荣研究员和他所在的中国科学院沙坡头沙漠试验研究站研究团队的又一力作。该书在上一本的基础之上，总结了研究团队自1999年至今十余年的研究成果。全书重点研究了生物土壤结皮群落中不同的微型生物类别、优势种分布格局对非生物因子(干旱、UV-B增强、增温、降水节律变化、沙埋和大气氮沉降等)、生物因子(维管植物覆盖度、土壤种子库和土壤微生物类群等)变化和干扰(生物入侵和火烧等)的响应，以及各种胁迫对特定生物土壤结皮群落组成、结构和功能的改变，突出回答了为什么生物土壤结皮能够发挥诸多复杂的生态和水文功能，以及其在干旱、半干旱荒漠生态重建和恢复中的重要作用。作者认为，未来环境变化，特别是气候变化会直接通过改变生物土壤结皮群落的组成(优势种的相互替代)和分布格局而影响其多样的生态功能，进一步导致荒漠生态系统功能变化的不确定性，进而提出生物土壤结皮维系荒漠自然生态系统中的沙地植被，特别是人工植被系统稳定性持续发展的观点。此外，作者通过十余年的长期定位研究，提出了利用以生物土壤结皮繁育拓殖为主的综合固沙模式，并在沙化土地恢复的实践中得到验证。

本书的出版必将在推动我国沙漠化治理和风沙区生态恢复与重建等相关科学领域的理论创新和实践应用方面有所启示和借鉴；为相关科学领域的科教人员及师生提供重要的参考。谨以此为序与读者共勉。

中国科学院院士，中国菌物学会名誉理事长
中国科学院中国孢子植物志编辑委员会主编
2016年8月16日于北京中关村

绪论

我国是一个受风沙危害和沙漠化十分严重的国家，风沙危害和沙漠化区域主要分布在东经75°~125°和北纬35°~50°范围，横跨半湿润、半干旱、干旱和极端干旱区等不同生物气候带。其中年降水量大于250 mm的东部沙地和农牧交错带以及年降水量小于200 mm的贺兰山以西的沙漠与绿洲、沙漠与荒漠草原过渡区是沙漠化和风沙危害最为严重的地区，也是进行无灌溉植被建设和构建国家北方生态屏障的关键区域。为了有效遏制风沙危害，防止沙化土地进一步扩张，国家先后在风沙危害区启动了“三北”防护林建设、退耕还林和京津风沙源治理等一批以人工植被建设作为主要生态修复措施的重大生态建设工程。近60年来，我国沙区植被建设取得了举世瞩目的成就，有效遏制了沙漠化的发展，促进了局地生境恢复。在充分肯定成绩之时，笔者发现实践中还存在许多问题。无论是在降水较多的东部沙区还是降水较少的贺兰山以西沙区，都不同程度地存在局地地下水下降以及固沙植被衰退和死亡的现象，直接影响了沙区的生态恢复和防风固沙效益的可持续性。实践中出现的这些问题从一定角度反映了理论研究的滞后，包括对许多科学问题没有给予应有的重视。例如，在进行植被恢复和流沙固定过程中，对固沙植被区形成的生物土壤结皮的生态功能没有给予应有的重视，或出现认识上的片面性，一些地方甚至认为破坏生物土壤结皮可以增加降水入渗，改变土壤水分状况。从恢复生态学和生态系统生态学的角度来看，这种认识恰似盲人摸象，存在着很大的局限性和片面性。

生物土壤结皮（biological soil crust, BSC），是指由隐花植物（非维管植物）如蓝藻、绿藻、地衣、藓类和土壤中微生物以及相关的其他生物体，通过菌丝体、假根和多聚糖分泌物等与土壤表层微小颗粒胶结形成的十分复杂的复合体，是干旱、半干旱荒漠地表景观的重要组成之一，其存在是该区域生态系统健康的重要标志之一。大多数荒漠生态系统是由非生物因素（abiotic factor）调控和胁迫的系统，特别是受水分的限制，地表不可能支撑大面积、相对均一且连续分布的高等植物群落，植物群落斑块状的分布为BSC的拓殖和覆盖提供了空间和适宜的生态位，使BSC的覆盖在干旱区占地表活体覆盖面积的40%以上，在有些地区甚至达到70%以上。仅从其分布的面积就可以看出它们的重要性。

BSC的演替规律是判断沙漠化发生、发展与逆转趋势的重要依据。以往判别沙漠化发生、发展和逆转趋势需要许多指标来综合判别，在实践应用中存在很多困难，而且需要训练有素的专业技术人员来完成，无论是理论研究还是在实践中会产生很多不确定性。例如，在利用大尺度研究中常用的植被指数和局地尺度的植被盖度作为沙漠化程度的判别指标时，会带来很多误差和误导。一个区域植被的盖度和植被指数有显著的季节差异，如何获取合理时段的数据本身就很困难。此外，在受风沙危害频繁的地区或沙漠化地区，植被盖度高低并不能说明生态恢复的好坏或沙漠化逆转的程度。在沙化草地，通常降水较多的年份会带来很高的植被盖度，可是这要看植被的基本组成是什么，在这种情况下往往是大量的一年生植物种群大爆发，表面上植被盖度很高，根据遥感图像或实地调查很容易误认为生态恢复明显，沙化得到遏制。但是当次年是一个干旱年时，这些一年生的草本植物不见了踪迹，地面风蚀依旧，沙化土地并未因一时的高植被盖度而得到修复。

BSC在沙化土地或流沙固定中存在有规律的发展和演替，当采取人为措施使流动沙丘表面达到相对稳定状态时，BSC在沙面就开始了拓殖、定居和发展。沙面得到固定，大量的降尘累积再经雨滴的打击，在沙面形成一层黏粒和粉粒含量较高的物理结皮，土壤微生物（如细菌）的增殖和蓝藻的拓殖使沙面在4～5年后形成了以蓝藻为优势的蓝藻结皮；大量的绿藻等旱生、超旱生的荒漠藻在BSC中逐渐占优势地位，形成荒漠藻结皮；此后，地表出现了大量的地衣结皮和地衣、蓝藻和荒漠藻的混生结皮；这些BSC的形成为藓类结皮逐渐在地势平缓或水分相对较好的局部开始大量拓殖创造了条件，50年后形成了高等植物和BSC隐花植物镶嵌分布的稳定格局特点。

在干扰程度较高的区域或常处于受干扰状态环境的BSC的演变阶段通常处在以蓝藻占优势地位的BSC状态，干扰会明显地减少地衣和藓类的物种多样性和多度；相对干燥、稳定的地表景观多为地衣为主的BSC覆盖；相对潮湿的土表或有利于集结雨水（包括凝结水）的微地形则有利于藓类结皮的拓殖和繁衍。由此可见，当一个固沙植被系统或沙化草地系统中BSC以蓝藻为优势，说明该系统处于不稳定或退化阶段，沙漠化发展趋势明显；当固沙植被系统或沙化草地系统中BSC以藓类和地衣为优势，说明该系统处于相对稳定阶段，土地沙化得到了遏制，生态处于恢复的趋势。这些简单易行的办法在过去并未得到应有的重视。

BSC抵御着风蚀和水蚀，维持了土壤的稳定性，对沙丘有持续的固定作用。大量的研究表明，BSC的存在增加了土壤的稳定性。风洞实验进一步验证了其抵抗风蚀的作用，BSC本身的形成机理就揭示了这一功能产生的机制。BSC中菌类和蓝藻等产生的菌丝体能够利用所分泌的多糖黏结小于0.25 mm的颗粒使之团聚成为稳定的大于0.25 mm的微团聚体，其过程是土壤微生物和藻类等把非结晶黏胶状的有机物密切地黏结在一起，而有机物又将矿物细粒进一步黏结，形成球状表面团聚。这样既借助于菌丝体将土壤细粒紧实地黏结，又通过微生物分泌物的黏结，促使土表的稳定性增强而避免风蚀和水蚀。BSC特殊的结构和其复杂的组成有效地缓解了雨滴对土表的溅击，控制了径流的发生和发展。这在土质结构紧密、容重大的黄土高原和钙含量高的荒漠灰钙土生境中尤为明显。此外，尽管BSC有阻止降水入渗的作用，但观测和模拟发现BSC存在的区域只有当昼夜降水强度达到40 mm时，才有地表产流，而在沙区这种降水事件十分罕见。BSC对缓解干旱、半干旱区脆弱生境土壤侵蚀的贡献得到了广泛的认可，除减少水土流失，还减少了直接关系到人类生存环境的沙尘暴的危害。BSC能够大量捕获大气降尘，为系统输入养分，促进沙区土壤成土过程，有效地改变了荒漠系统非生物因素的胁迫，为土壤生物繁衍创造了生境。BSC在沙面一旦拓殖发展，在没有人为干扰的情况下，就会发展到与其气候相关联的演替阶段，而且能够稳定持续，使沙子不再流动，也为其他物种在固沙区的拓殖、发展提供生境，因此，BSC也被称为荒漠/沙漠的“生态系统工程师”（ecosystem engineer）。

BSC维系了荒漠系统生物多样性。BSC为大量的沙区微小生物提供了生境，维持了荒漠系统生物多样性。研究表明，微小节肢动物的数量在有BSC存在的土表层最高。对位于腾格里沙漠东南缘的沙坡头地区4种不同生境的土壤动物多样性所做的比较研究认为，土壤动物的生物量和密度与BSC的发育阶段呈正相关关系。自然发育未受干扰的地衣、藓类结皮覆盖的土壤中，土壤动物的生物量和密度明显地高于BSC发育早期阶段土壤类型中的。BSC在土表的盖度、BSC的生物体（蓝藻、其他藻类、地衣和苔藓等）组成的差异与荒漠昆虫多样性和优势种组成密切相关。BSC盖度高且发育良好的植被区昆虫以草原和荒漠化草原指示种占优势，而BSC覆盖较低的植被区昆虫种类以流沙、荒漠指示种占优势地位。

应当指出，影响土壤动物的因素很多，如植被类型。BSC也在一定程度上影响着这些生境因子。例如，BSC通过对降水入渗的影响，对土壤水分起到再分配的作用。而土壤湿度直接影响着许多土壤微生物的活性与分布，间接地影响着土壤动物的食物链关系。BSC中的一些生物体及其分泌物是土壤动物的直接食物来源，如一些藻类所分泌的多聚糖物质，还有一些土壤动物直接采食BSC组成生物体。此外，藓类的孢蒴还是一些动物的食物来源。地衣和一些藻类对UV-B辐射的抵御作用、BSC对土壤温度的影响、对土壤养分（碳、氮等）的供给、对地上维管植物养分的供应、对地表稳定性的维持（抵御风蚀和水蚀作用）等，直接或间接地为土壤动物的繁衍和生存提供了适宜的生存环境，即BSC的存在很大程度上改善了相关生物的生境，维持了健康和可持续的荒漠系统。

土壤食物链的营养结构在土壤养分循环中十分重要。土壤初级生产者是BSC中的地衣、藓类、绿藻和蓝藻，这些生物体连同植物体一起既被土壤生物取食又被其分解。在分解过程中，早期增殖的酵母和细菌被线虫和原生动物（protozoa）所取食，而螨（mites）又控制着线虫的数量。真菌主导了后期的分解作用，而它则被线虫、跳虫（collembolan）和螨所取食。因此，

BSC是其他土壤食物链成分的重要食物来源。例如，细菌是蓝藻结皮的主要分解者和消费者，真菌比细菌更能克服难于分解的物质，放线菌则以蓝藻作为食物来源；原生动物包括变形虫、纤毛虫和鞭毛虫，它们和线虫在土壤食物链中以BSC中的蓝藻和其他某些藻类为食；微小节肢动物以某些藻类、真菌、蓝藻（尤其是具鞘微鞘藻）、细菌及其他无脊椎动物和植物碎片为食。

BSC“生态系统工程师”的作用连接了极端荒漠生境中非生物因素和生物因素，使极端风沙环境变得“生机勃勃”。此间BSC起着载体的作用，例如，大量的蚂蚁在有BSC覆盖的沙丘得到了繁衍，BSC不仅为它们创造了生存的环境，BSC特殊的结构确保了蚁穴不被沙埋，而且提供了重要的食物来源，如BSC生物体分泌的多聚糖。在没有降水的季节，BSC还可以利用它们对吸湿凝结水的“捕获能力”为这些微小的无脊椎动物提供珍贵的水分。因此，BSC的存在可以使原来单一的沙漠人工固沙系统转变为复杂的、多样物种参与的生态系统，为沙漠治理和沙区生态持续发展做出了贡献。

BSC促进了人工固沙系统的持续发展，使固沙成果得到巩固。BSC通过改变土壤性状而影响高等植物的萌发、定居和存活。BSC对土壤性质的改变包括改变土壤表面的粗糙度、土壤质地、温度、养分的有效性、有机质和水分等。就机理而言，BSC可以通过改变以下方面来影响高等植物的萌发、定居和存活：① 改变土壤表面的微地形。在热带荒漠，由于缺少土壤冻融，土壤表面粗糙度的改变较小，其中蓝藻结皮能使土壤表面光滑化，而地衣和藓类的生长增加了土表的粗糙度。但藓类的发育在温带荒漠特殊的地貌条件下，如在固定沙丘丘间低地，也使土表光滑化。相反，在寒冷荒漠地区，由于地表和各种BSC受到土壤侵蚀，土表的粗糙度明显增加，以至于土表高度能隆升15 cm。这些相对大的地形变化特征可捕获风和水带来的种子、有机质、土壤微粉粒和水分。② 通过影响种子的捕获间接地影响参与萌发与定居的种子数量。当土壤表面光滑时，其对种子的捕获（seed entrapment）量很低，而在那些因BSC存在而显著增加地表粗糙度的生境（已有BSC发育和冻胀丘），种子的捕获量则很高。种子在这些蓝藻形成的光滑BSC表面停留的能力很弱。进行干扰试验使这些BSC粗糙化，种子在土表停留的能力显著增强。与热带荒漠光滑土相比较，温带和寒区荒漠地表经常出现的是地衣－藓类结皮以及植物凋落物。冻胀丘土壤以有限的物理和化学板结为特征。这种粗糙的地表除了可捕获种子外，还可以捕获有机质、水分和土壤细颗粒，使土壤微生境肥力提高。③ 影响高等植物养分。BSC对土壤表面化学性质的改变与相关植物种子组织体中主要生命元素的含量变化密切相关。BSC的存在总的来说提高了植物对镁、钾、铜和锌的吸收，而同时减少了植物对铁的吸收。④ 影响种子本身的生物学特性。种子萌发对水分的需求存在差异。在空气干燥的荒漠，许多种子要求一定的植被覆盖和土壤来保持充分的湿度从而进行萌发。土表很小的裂缝和断裂微地形对小颗粒种子植物萌发已足够，但对大颗粒种子则需要额外土壤或凋落物的覆盖。一些缺少特殊穿透结构的植物种子，通常是一年生的，生长于土壤－土表干扰相对较大的地区，它们的萌发在那些凋落物较少、BSC完整稳定以及土表干扰较低的地区就会受到抑制。土壤的流动性也是制约种子萌发的一个重要因素。种子通常有其适合的埋藏深度，增深和变浅都会有碍于萌发。增加土壤氮肥能够刺激一些植物种的种子萌发。一些植物种子的萌发可能是因BSC的存在而被激发。此外，BSC能够阻止外来种的入侵，维持系统的稳定性，这对未来全球变化背景下荒漠生态系统生物安全有着十分重大的意义。

BSC为人工固沙系统提供了碳和氮，提高了系统的生产力。荒漠生态系统是一个土壤养分十分贫瘠的系统，BSC的存在对荒漠系统的能流和物流、养分循环产生了重要影响和贡献，有益于系统生物生产力的提高。

BSC中固氮的藻类主要由一些具异形细胞类（如鱼腥藻属、眉藻属、念珠藻属、裂须藻属和伪枝藻属）和一些非异形细胞类（如鞘丝藻属、微鞘藻属、颤藻属和单歧藻属）组成。地衣中具有固氮作用的主要是胶衣属（*Collema*）。以上所有的种类都可在我国沙漠BSC群落中见到。有研究表明，以地衣胶衣属为优势种的黑色BSC具有最高的固氮速率（13 kg hm^{-2} a^{-1}），以具鞘微鞘藻为优势种的浅色BSC固氮速率为1.4 kg hm^{-2} a^{-1}，而介于深色和浅色之间的是以葛仙米（*Pogostemon auricularius*）和伪枝藻（*Scytonema myochrous*）为优势种的BSC，其固氮速率为9 kg hm^{-2} a^{-1}。影响地衣和蓝藻结皮中单个种的固氮速率的主要因子是它们自身的生物学特性、土壤水分、温度和光照。另外，无论是自由生长的还是作为地衣中藻类的组成部分，它们能够固定大气中的氮，通常会有利于高等植物的生长。藓类的分解是土壤中营养元素的来源之一，特别是氮和磷，可以增加维管植物的存活率。由BSC固定的氮可以被周围的高等植物和另外的生物体如真菌、放线菌和细菌利用。大约70%由蓝藻和蓝藻–地衣固定的养分被立即释放到土壤环境中，而这些养分对相关的生物体包括高等植物、藓类和其他微生物群是有效的。研究证实，BSC的存在增加了200%的土壤环境氮含量，是荒漠土壤和植物的主要氮素来源之一。当然，BSC也可能通过反硝化作用和挥发过程造成氮的损失，尽管很少有具体的实验数据来支持这一过程。

BSC中的生物体通过光合作用、呼吸、分解和矿化作用对荒漠系统的碳循环起着直接或间接的作用，是干旱、半干旱地区荒漠系统碳的主要贡献者。它们通过调节分解和矿化率，进而调节着养分的有效性和初级生产力。因此，BSC中的微生物种群在荒漠生态系统的养分循环和能流中起着至关重要的作用。BSC通过其组成生物体中的蓝藻、地衣、绿藻和藓类的光合作用进行碳的固定，但许多因素影响着BSC对碳的贡献，如湿度决定着生物体光合作用的有效性，甚至在一些情况下如通过呼吸促使土壤碳的损失。即便如此，BSC对干旱荒漠系统碳循环的贡献也是不容置疑的。笔者在腾格里沙漠南缘的研究表明，以蓝藻为优势的BSC每年的固碳量是11.36 g C m^{-2}，而以地衣–藓类为优势的BSC每年的固碳量可达到26.75 g C m^{-2}。尽管这些数字是惊人的，但一直没有引起注意。

由以上5个方面不难看出BSC在干旱区生态恢复和沙化治理中的重要性和功能的多样性，因此，保护了我们脚下的BSC就是保护了土地不被沙化。这一理念在美国、以色列和澳大利亚的沙化土地治理和生态系统管理中已得到广泛的认可。认识到BSC的重要性，对我国沙区生态恢复和沙化土地治理将产生积极的影响，有利于形成全新的沙化治理理念，形成新的恢复和治理技术体系，进一步促进我国的防沙治沙事业。

有关BSC的功能（生态与水文功能）在笔者《荒漠生物土壤结皮生态与水文学研究》一书中有详尽的研究分析、实验验证和理论探讨，然而这些功能主要取决于形成BSC群落的种类组成、物种多样性和格局特点，这些种类组成的时空格局变化必然会导致BSC在荒漠/沙地生态系统中功能的改变，使生态恢复与重建的成效和持续性受到影响。因此，深入了解BSC群落中各功能群对环境中生物与非生物胁迫的响应，认知BSC群落组成物种多样性对干扰、气候变化的适应和响应规律，即对BSC生态生理学的研究是荒漠/沙地生态系统管理和BSC服务功能调控的理论依据和重要前提条件。

本书是沙坡头站研究组对国内荒漠BSC研究成果的总结和提升。全书共分5章，第1章论述了BSC研究进展以及中国沙区BSC群落种的多样性和空间分布格局规律，提出了BSC空间分布格局的理论假说；第2章阐述了影响BSC拓殖和发展的环境与生态因子，并重点介绍了影响BSC生态生理的非生物因子和生物因子；第3章集中阐述BSC对水分、光照、温度、UV-B辐射、CO_2浓度、风沙、N沉降、火烧、增温与降水减少等的生态生理响应；第4章集中阐述BSC与维管植物的关系，以及土壤微生物与BSC的相互作用；第5章从BSC的人工培养方法以及人工结皮的特征等方面对BSC人工培养及其在沙区的应用进行全面论述。从这本专著中，我们可以了解中国荒漠生物土壤结皮研究的历史和现状，当前研究的主要内容、前沿和热点，以及未来的主要研究方向。

在本书文字录入、准备、书稿校对等工作过程中，得到了研究组贾荣亮、李计红、王进、谭会娟、潘颜霞、刘艳梅、黄磊、苏延桂、陈应武、冯丽、张鹏、张亚峰、虎瑞等博士的大力支持，在此一并表示衷心感谢。

感谢中国科学院微生物研究所魏江春院士在百忙中为本书作序鼓励。

感谢中国科学院西北生态环境资源研究院、中国科学院沙坡头沙漠试验研究站和高等教育出版社的领导和编辑，没有他们的支持和努力，本书很难在较短的时间里问世。

本书中的研究得到了国家自然科学基金重点项目（项目号：41530746、40930636）、面上项目（项目号：41271061）和中国科学院寒区旱区环境与工程研究所“生物环保固沙技术在防沙治沙中的应用”项目（项目号：HHS-TSS-STS-1505）的支持。

生物土壤结皮的研究还存在许多未知的前沿科学问题，加之我们学识所限，书中遗漏和不妥之处在所难免，恳请读者不吝指正。

Introduction

Within China, a large area stretching from 75°–125° E and 35°–50° N is severely threatened by sand storms and desertification. This area spans semi-humid, semi-arid, arid, and extremely arid bioclimatic regions. The most vulnerable regions within the area are located in Northern China, and include the eastern desert and the transition region between agriculture and pasturage interlaced, where annual precipitation is more than 250 mm, and the transition regions between desert and oasis, desert and desert grassland, which are west of Helan Mountain and have an annual precipitation less than 200 mm. These extremely vulnerable regions make up the key target regions for construction of an ecological screen against desertification in the north of China through population without irrigation vegetation. To effectively contain the destruction caused by sand storms and prevent further desert expansion, the Chinese government has successively launched a batch of significant ecological restoration projects. These projects have focused on vegetative restoration, such as construction of the "Shelter Forest System Programme in Three-North Regions of China", "Grain for Green Project", and "Beijing-Tianjin Dust Storms Sources Control Project". Through the last 60 years, the Chinese government has implemented remarkable measures to curb desertification and promote restoration of vegetative ecosystems in desert regions. While fully recognizing these impressive achievements, it is also important to acknowledge that continued efforts are essential to the prevention of further desertification. Both the eastern desert regions, with their relatively high precipitation, and the desert region west of Helan Mountain, with its low precipitation, to different extents, suffer from declines in groundwater levels, and degeneration and death of sand fixing vegetation. These changes have directly affected the ecological restoration of these desert regions and the capacity to minimize the effects of wind and sustainably promote sand fixing. These problems, from a certain perspective, reflect a lack of theoretical knowledge, and need for more scientific research on desertification issues and restoration. For example, in the course of vegetation restoration for fixation of moving sand dune, few studies have investigated the ecological functions of biological soil crust (BSC), which are formed by sand fixing vegetation. The result is an overall poor understanding of how to best restore vegetation to promote BSC formation. Some managers even attempt to enhance permeation of precipitation and improve soil moisture status by destroying the BSC. With large knowledge gaps, restoration of these ecosystems is like "groping in the dark", which obviously has great limitations.

BSC make up a key component of the arid and semi-arid desert surface landscape and provide an important indicator of regional ecosystem health. They are made up of a complex composite of cryptogams (non-vascular plants), such as cyanobacteria, green algae, lichens, mosses, microorganisms in the soil, and other related organisms that bond with fine particles on the surface of soil via mycelia, rhizoids, and polysaccharides. Most desert ecosystems are controlled and limited by abiotic factors, such as moisture limitations, which prevent the event and continuous coverage with higher plant communities. The distribution of plant community patches within desert ecosystems provides space and appropriate ecological niches for BSC. The importance of BSC to arid ecosystems is evident in their immense distribution area. Within arid regions covered with living organisms, BSC account for more than 40% (and in some regions more than 70%) of the surface area.

The succession stages of BSC provide a useful basis for diagnosis of desertification trends. Before research increased our understanding of BSC, it was necessary to quantify a wide range of indicator variables in order to assess ecosystem desertification status and trends. Therefore, assessments of the status of desert ecosystems were impractical in application, requiring highly trained professional technicians, and possessing considerable uncertainty both for restoration and research applications. For example, using the vegetation index (commonly applied for large spatial scale quantification) and the vegetation coverage (used for small spatial scale quantification) to identify the degree of desertification introduces large errors and contributes to misleading conclusions. Vegetation coverage and index of a region vary independently with season. Therefore, data for these two indices is unreliable when collected over long time periods. Furthermore, vegetation coverage is a poor indicator of ecological restoration and/or desertification status in regions that are frequently hit by sand storms or experienced desertification. Vegetative indices may vary with environmental factors, for example, in desertificated grasslands, vegetation coverage is usually very high during high precipitation years, but low during low precipitation years. During high precipitation years, annual plant populations within desertificated grasslands spread like an outbreak. Remote sensing images or field investigations would quantify vegetation coverage as very high, and conclude that the ecosystem has been restored and desertification has been curbed. However, during a subsequent arid year, the same annual plants will disappear and the surface of the ecosystem will again be subject to wind erosion. Using only vegetative coverage as an assessment of ecosystem restoration, the temporary high vegetative coverage makes a non-restored ecosystem appear to be restored during a high precipitation year.

The development and evolution of BSC in desertificated or sand fixed areas follows regular trends. After restorative stabilization of moving sand dune surfaces, colonization of BSC will begin, and the crusts will settle and develop on the sand surface. Surface sand colonization will fix the sand allowing accumulation of dust. When rain falls on the accumulated dust, it forms a physical crust with a high content of clay and silt particles on the sand surface. Within four to five years, the sand surface will become colonized with bacteria, soil microbes, and cyanobacteria, forming a crust dominated by cyanobacteria. Considerable xeric and super-xeric desert algae, such as green algae, will then gradually become dominant in BSC, forming a desert algae crust. Finally, crusts will become populated with lichen, resulting in mixed BSC of lichen, cyanobacteria and desert algae. This mixed BSC provides conditions favorable for large amounts of moss to colonize the crust in relatively flat areas and areas with relatively high moisture content. Within 50 years, this BSC development process will give rise to a stable distribution pattern of higher plants and BSC cryptogams.

In highly disturbed regions, the diversity of BSC is reduced through reductions in abundance of lichen and moss, resulting in BSC dominated by cyanobacteria. In relatively dry and stable regions, BSC tends to be dominated by lichen. Whereas, in humid regions or small areas that collect rainfall or condensation, moss crusts tend to proliferate. Therefore, domination of sand-fixing vegetation system or desertificated grassland BSC with cyanobacteria suggests the system is at an unstable or degenerative stage, and desertification is occurring. Otherwise, when the BSC of sand-fixing vegetation system or desertificated grassland is dominated by moss and lichen, the system is likely at a relatively stable stage and desertification has been curbed allowing restoration of the ecosystem. These simple and easy indicators have not attracted due attention in past restoration efforts.

BSC resists wind and water erosion and maintains the stability of soil and fixed sand dunes. Several studies have shown BSC to enhance soil stability. Wind tunnel experiments have shown BSC to resist wind erosion. The mechanisms driving these functions stem from the processes leading to the formation of BSC. Fungi and cyanobacteria within BSC secrete saccharine, and mycelium (generated by fungi and cyanobacteria) binds particles smaller than 0.25 mm into stable micro-aggregates. Through this process, soil microbes, algae, etc., tightly bind to non-crystalline sticky organic matter and the organic matter further binds fine mineral particles to form spherical surface aggregates. Through these processes, mycelium tightly binds the soil particles, and secretions produced by the soil microbes strengthen the stability of the sand surface, stabilizing it against wind and water erosion. The special structure and complex composition of BSC can effectively attenuate the splash caused by rainfall and alleviate the erosion caused by runoff. Such effects are especially obvious in the Loess Plateau, where the soil structure is compact and dense, and within grey desert habitats, where calcium content is high. Despite the capacity of BSC to prevent permeation of precipitation, monitoring and simulation studies have shown that only when cumulative 24-hour precipitation reaches 40 mm and surface runoff occurring, which is rare in desert regions. Reductions to soil erosion within arid and semi-arid vulnerable habitats covered with BSC have been widely recognized. In addition to reducing water damage and soil loss, BSC can reduce the destructive impacts of sandstorms, which are directly caused by anthropogenic alterations to the environment. BSC can capture large amounts of dust, which infuses the system with nutrients and promotes the formation of soil in desert regions. These actions change the impacts of abiotic stressors and create a habitat for the reproduction of soil organisms within desert regions. In the absence of artificial disturbance, once the BSC has developed on the sand surface, it will progress to a stage compatible with the climate, and persist, acting to stop sands from moving, and provide a habitat for the colonization with and development of other species in the sand-fixing regions. Therefore, BSC is often considered the "ecosystem engineer" of the desert.

BSC is important to the maintenance of biological diversity within desert ecosystems. It provides habitat for huge communities of microorganisms. Research has shown that the greatest abundances of micro-arthropods on the earth's surface are found within BSC. A study comparing soil microorganism diversity among four different habitats within the Shapotou region at the southeastern edge of Tengger Desert, found both the biomass and density of soil organisms to positively correlate with the developmental stage of the BSC. That is, the biomass and density of soil organisms within lichen- and moss-dominated BSC, which had developed naturally with no disturbances, were clearly higher than biomass and densities within BSC at earlier developmental stages. The BSC coverage of a region and the composition of the BSC (*e.g.*, cyanobacteria, algae, lichen, and/or moss) are closely related to the diversity and dominant species of insects within a desert. When the soil surface around vegetation has a high coverage of later developmental stage BSC, the dominant insects tend to be indicator species of grasslands and desertificated grasslands; whereas vegetation with relatively low soil surface BSC coverage tends to be dominated with insects indicative of quicksand and deserts.

It should be pointed out that many habitat factors, including vegetation type, influence soil organism composition. BSC, to a certain extent, can indirectly influence soil organisms through impacts on habitat factors. For example, by influencing precipitation permeation, BSC can re-allocate soil moisture. Soil moisture directly influences the activity and distribution of many soil microbes and indirectly influences food chain relationships among soil organisms. Some BSC organisms and their secretions, for example the saccharine secreted by some BSC algae, are a food source for other soil organisms. Additionally, spores of some mosses are a food source for soil organisms. The lichen and some algae from BSC resistance to UV-B radiation, increasing soil temperature and supply of soil nutrients (*e.g.*, carbon and nitrogen), supply of nutrients to the vascular plants above the ground, and maintenance of earth's surface stability (resistance to wind and water corrosion) will directly or indirectly provide a suitable living environment for the reproduction and survival of soil animals. Namely, BSC plays important roles in maintaining habitats for desert organisms and the health and sustainability of desert ecosystems.

The soil food chain structure is very important to soil nutrient circulation. The primary nutrient producers in desert soils are lichen, moss, green algae and cyanobacteria, which are all found within the BSC. Soil organisms depend on these BSC organisms and plant materials for their nutritional supply. During process of decomposition, yeasts and bacteria at early developmental stages of BSC are consumed by nematodes and protozoa, while mites control the quantity of nematodes. Fungi become the dominant decomposers in later stage BSC, and are eaten by nematodes, collembolan, and mites. Therefore, the components of the BSC are important food sources for other organisms making up the soil food chain. For example, bacteria within algae crust are the dominant decomposer and consumer. Materials that are not decomposable by bacteria can often be decomposed by fungi. Actinomycetes consume cyanobacteria as a food source; Nematodes and protozoa, including amoebas, ciliates, and flagellates, feed on cyanobacteria and algae in BSC; micro-arthropods feed on algae, fungus, cyanobacteria (especially *Microcolus vaginatus*), bacteria, other invertebrate animals, and plant fragments.

The "ecosystem engineer" functions of BSC, it plays a role of a carrier and connects biotic and abiotic processes to facilitate life and diversity in extreme desert habitats. For example, many ant species reproduce in sand dunes covered with BSC. BSC provides a stable habitat with important food sources, such as the saccharine secreted by BSC organisms, which is secured from destruction through sand movement. The high capacity for BSC to capture water through absorption and condensation also provides precious moisture for tiny invertebrate animals during the dry season. Therefore, BSC can transform simple artificially produced sand-fixing desert ecosystems into complex, multiple species ecosystems that contribute to desertification control and the sustainable development of desert ecology.

BSC facilitates the sustainable development of artificial sand-fixing ecosystems and consolidates the benefits of sand fixing. By altering soil properties, such as surface roughness, texture, temperature, nutrient availability, organic matter content, and moisture, BSC can influence the seeds germination, settlement and survival of higher plants. The mechanisms through which BSC alters ecosystems include: i) In tropical deserts,where land surface disturbances through freezing and thawing do not occur, the soil surface can remain extremely smooth. For example, land surfaces of areas with algae crusted soil tend to be smooth, whereas the growth of lichen and moss in the BSC can introduce irregularities to the land surface. Notably, moss-dominated BSC in temperate deserts form a special exception, and can contribute to areas of smooth land surface in lowlands between sand-fixing dunes. However, in cold desert regions, the actions of soil erosion and diverse BSC, contribute to roughening of the land surface, and can raise the surface by as much as 15 cm. By increasing the roughness of the soil surface, BSC can facilitate entrapment of seeds, organic matter, fine soil particles, and moisture from wind and precipitation. ii) By influencing the entrapment of seeds, BSC can indirectly influence the quantity of seeds that successfully germinate and settle. Smooth land surfaces are poor at entrapping seeds. In habitats where the land surface is rough due to the influences of BSC and frost heaving, seed entrapment rates are very high. Cyanobacteria-dominated BSC does not introduce irregularities to the soil surface, which results in poor seed retention. Interfere test will roughen these BSCs and the seed entrapment on the earth's surface will be enhanced obviously. Tropical desert soils tend to be extremely smooth compared to temperate and cold desert soils, which are usually covered by lichen-moss crust and plant litter. Soils experiencing frost heaves are characterized by limited physical and chemical hardening. This kind of rough land surface can trap, not only seeds, but also organic matter, moisture and fine soil particles, and enhance the fertility of the soil micro-habitat. iii) BSC can influence the nutrients available to higher plants. Studies have shown the changes to soil chemical properties caused by BSC to correlate with changes in the biological element contained within the seed tissues of plants growing within the same area. Generally speaking, the existence of BSC can enhance the absorption of Mg, K, Cu and Zn, and reduce the absorption of Fe by plants. iv) BSC can influence the biological properties of seeds. The conditions which required to retain sufficient moisture for seed germination vary with desert type. In a desert with dry air, seeds often require specific vegetation coverage and soil to retain adequate moisture for germination. Tiny crevices and fractured micro-habitat within the land surface tend to trap enough moisture for germination of tiny seeds, but large seeds require extra soil and plant litter coverage. Plant seeds lacking a special penetration structure are usually annual species, and tend to be restricted to regions with relatively high land surface disturbance. Germination for these species would be inhibited in the regions with small amounts of plant litter, complete and stable BSC and relatively low land surface disturbance. Soil mobility is another important factor constraining seed germination. Most seeds are influenced by burying depth, with too deep or shallow depths hampering germination. Enhancement of the nitrogen fertility of soil can stimulate seed germination for some plants. Germination of some species' seeds can be stimulated through to the existence of BSC. Additionally, BSC can guard against the invasion of alien species, and thus maintain the stability of the system. This protection is becoming an increasingly significant advantage of BSC in desert ecosystems under the context of global changes.

BSC enhances the productivity of desert ecosystems through providing carbon and nitrogen to nutrient poor soils. BSC contributes to the flow of energy and materials and nutrient circulation within the desert ecosystem, which enhances the biological productivity of the system.

The nitrogen fixing cyanobacteria in BSC are mainly comprised of heterocyst cells (*e.g.*, *Anabaena, Calothrix, Nostoc, Schizothrix, Scytonema*) with a few non-heterocyst cells (*e.g.*, *Lyngbya, Microcolus,*

Oscillatoria, Tolypothrix). The primary nitrogen fixing lichen within BSC is *Collema*. All of the above algae and lichen species are found in BSC colonies across the deserts of China. Some studies have found black BSC, which is dominated by the lichen *Collema*, to fix nitrogen at the highest rate of all BSC types (13 kg hm^{-2} a^{-1}); light BSC, which is dominated by the algae *Microcolus vaginatus*, fixes nitrogen at a low rate of 1.4 kg hm^{-2} a^{-1}, whereas moderately dark BSC is dominated by the cyanobacteria of *Pogostemon auricularius* and *Scytonema myochrous* and can fix nitrogen at a moderate rate of 9 kg hm^{-2} a^{-1}. The nitrogen fixation rate of the individual lichen and blue-algae crust species are mainly influenced by their own biological properties or abiotic properties, such as soil moisture, temperature, and sunshine. Additionally, whether the BSC nitrogen fixing species grow freely or are part of algae or lichen, they can fix atmospheric nitrogen, which is usually beneficial to growth of the surrounding higher plants. Moss decomposition provides an additional nutrient, especially nitrogen and phosphorus, source for soil, and therefore, further enhancement of vascular plant survival. The nitrogen fixed by BSC can be utilized by surrounding higher plants and other living bodies such as the fungus *Actinomycetes* and bacteria. More than 70% of the nutrients fixed by cyanobacteria and lichen are immediately released into the soil environment, and available to higher plants, moss, and micro-organisms. Studies have shown BSC, which can increase soil nitrogen content by 200%, to be the main source of nitrogen for desert soil and plants. However, BSC may also cause nitrogen loss through denitrification and volatilization, but limited empirical data describing these processes exist.

The organisms within BSC have direct and indirect effects on the carbon cycle of desert ecosystems through photosynthesis, breathing, decomposition, and mineralization, which are the main processes contributing carbon to arid and semi-arid desert ecosystems. Through regulating decomposition and mineralization rates, BSC further regulates primary productivity and the nutrient assimilation efficiency. Therefore, microbe colonies in BSC play vital roles in the nutrient circulation and energy flow within desert ecosystems. BSC can fix carbon via photosynthesis by cyanobacteria, lichen, green algae, and moss. However, many factors influence the carbon contribution of BSC; for example, humidity can influence the efficiency of photosynthesis within organisms, and, under some conditions, it can even lead to loss of soil carbon through breathing. Regardless, the contributions of BSC to carbon circulation within arid desert ecosystems are well established. Studies carried out within the southern edge of Tengger Desert found cyanobacteria-dominated BSC to fix carbon at an annual rate of up to 11.36 g C m^{-2}, while lichen-moss-dominated BSC carbon fixation rate can be as high as 26.75 g C m^{-2}. Though these dates are amazing, this contribution of BSC to the carbon cycle has not always been accepted.

The above five mechanisms for ecosystem alteration by BSC show the importance and functional diversity of BSCs in ecological restoration and sand fixing of arid regions. Therefore, we can say that protecting the BSC is equal to protecting land from desertification. This idea has been extensively recognized in the restoration of desertificated land and ecosystem management practices of the US, Israel and Australia. Recognizing the importance of BSC in management plans will improve ecological restoration of desert regions and desertificated lands in China. Although this is a completely new restoration approach for China, it will provide a scientifically-based system of restoration that will improve desertification prevention and control in China.

Detailed research, experimental verification, and theoretical explanations of the ecological and hydrological functions of BSC are provided in our book *Eco-hydrology of Biological Soil Crusts in Desert Regions of China* (*by* Higher Education Press, 2012). However, these functions mainly depend on the composition, diversity, and patterns of the BSC. Temporal and spatial variability in these populations can influence the functions of BCS in desert/sandy ecosystems and affect the effectiveness and sustainability of ecological restoration and reconstruction. Therefore, the theoretical basis and essential precondition to management of desert/sandy ecosystems through regulation of BCS service functions is a thorough understanding of responses of individual BSC organisms to biotic and abiotic environmental changes and the adaptability and responses of the BSC population composition and diversity to disturbances and climate change. This knowledge requires research on eco-physiology of BSC.

This book is a collective work from the research group of the Shapotou Station. It represents a summary and exhibition of studies on BSC in China. The book consists of 5 chapters. Chapter 1 discusses the progress and development of research at the station in terms of community composition, diversity, and distribution patterns of BSC in desert areas of China and proposes theoretical hypotheses of BSC spatial distribution patterns. Chapter 2 describes environmental and ecological factors affecting colonization and development of BSC, as well as factors influencing the eco-physiology of BSC. Chapter 3 elaborates on the eco-physiological responses of desert BSC to hydrology, light, temperature, UV-B radiation, CO_2 concentrations, sand activity, nitrogen deposition, fire, warming, and decreasing precipitation. Chapter 4 examines the relationship between BSC and vascular plants, and the interactions of BSC with soil microbes. Chapter 5 presents a thorough discussion on artificial cultivation and application of BSC in desert zones and their practical application . From this book, the readers can better understand the history and current status of BSC, as well as the current research frontiers, hot spots, and the main study directions of BSC in the desert regions of China.

Many members of our research group provided support in the contents, preparation and proofreading of this book, including Dr. Rongliang Jia, Dr. Jihong Li, Dr. Jin Wang, Dr. Huijuan Tan, Dr. Yanxia Pan, Dr. Yanmei Liu, Dr. Lei Huang, Dr. Yangui Su, Dr. Yingwu Chen, Dr. Li Feng, Dr. Peng Zhang, Dr. Yafeng Zhang, Dr.

Rui Hu and many more. We would like to express our great appreciation to all contributors.

We thank academician Jiangchun Wei from Institute of Microbiology Chinese Academy of Sciences (CAS) for encouragement and writing the preface of this book.

We would also like to acknowledge the support from the leaders of the Northwest Institute of Eco-Environment and Resources, CAS, Shapotou Desert Research and Experiment Station, CAS, and the help from editors at the Division of Natural Science Academic Publishing, Higher Education Press. This book would not have been published in such a short time without their support and help.

The research presented in this book has been supported by the key program of National Natural Science Foundation of China (41530746 and 40930636), the general program of National Natural Science Foundation of China (41271061) and the "Application of Biological Environment Protection and Sand-fixing Technologies in Desertification Prevention and Control" program of the Cold and Arid Regions Environmental and Engineering Research Institute, Chinese Academy of Sciences (HHS-TSS-STS-1505).

Despite more than 10 years of BSC research, many important questions about BSC remain unanswered. Hence, this book is bound to include mistakes or omissions. Suggestions for improvement are welcome and appreciated.

目录

Contents

第1章　中国沙区BSC群落的组成、物种多样性和格局特点

我国沙区约占国土陆地面积的16.8%，横跨半湿润、半干旱、干旱和极端干旱等不同生物气候带（图1-1）（朱震达和刘恕，1989；CCICCD，2002）。由于自然环境严酷，且易受气候变化和不合理的人类活动的长期影响，沙区存在植被退化和风沙危害加剧等一系列问题，严重威胁和制约着国家生态安全和区域可持续发展（沈培平等，2006）。自20世纪70年代以来，国家实施了一系列以植被建设为主要措施的重大生态工程，有效地促进了沙区生态的恢复与重建。2015年年底，第5次全国荒漠化和沙化土地监测结果显示，截至2014年，全国荒漠化土地面积261.16×10^4 km^2，占国土面积27.20%，相比于2009年净减少12120 km^2，年均减少2424 km^2；沙化土地面积172.12×10^4 km^2，占国土面积17.93%，相比于2009年净减少9902 km^2，年均减少1980 km^2。有明显沙化趋势的土地面积30.03×10^4 km^2，占国土面积3.12%。实际有效治理的沙化土地面积20.37×10^4 km^2，占沙化土地面积11.80%。监测结果显示，相比于2009年，荒漠化和沙化程度呈现出由极重度向轻度转变的良好趋势。沙区植被状况进一步好转，区域风沙天气明显减少，北京地区平均每年出现2次，较上一个监测期减少63.00%（寇江泽，2015）。

虽然荒漠化和沙化土地面积整体呈缩小趋势，但是治理和保护的任务依然艰巨。现今荒漠化和沙化土地面积分别占国土面积的1/4以上和1/6以上，与20世纪50年代初相比，我国沙化土地增加了8×10^4 km^{2}以上，相当于增加了两个腾格里沙漠的面积。目前有明显沙化趋势的30.03×10^4 km^{2}土地，如果保护利用不当，极易成为新的沙化土地，尤其是沙区无序开发建设现象严重，开垦、超载放牧、水资源过度利用、内陆湖泊面积萎缩、河流断流现象和地下水位逐年下降等问题仍然突出。

除以上不合理的人为干扰外，大面积和高密度的植被建设也导致了一些沙区不同程度的地下水位下降和植被退化，使生态恢复效果出现不可持续的现象，甚至出现了新的土地沙化。相关理论研究，特别是对土壤生态过程、水文过程及相互作用的机理研究滞后，缺乏对实践有效指导是造成这种现状的重要原因（李新荣等，2013，2014；Li *et al.*，2014a）。因此，科学、全面认知沙区土壤生态水文过程和作用机理，并探索适宜的恢复对策和途径是国

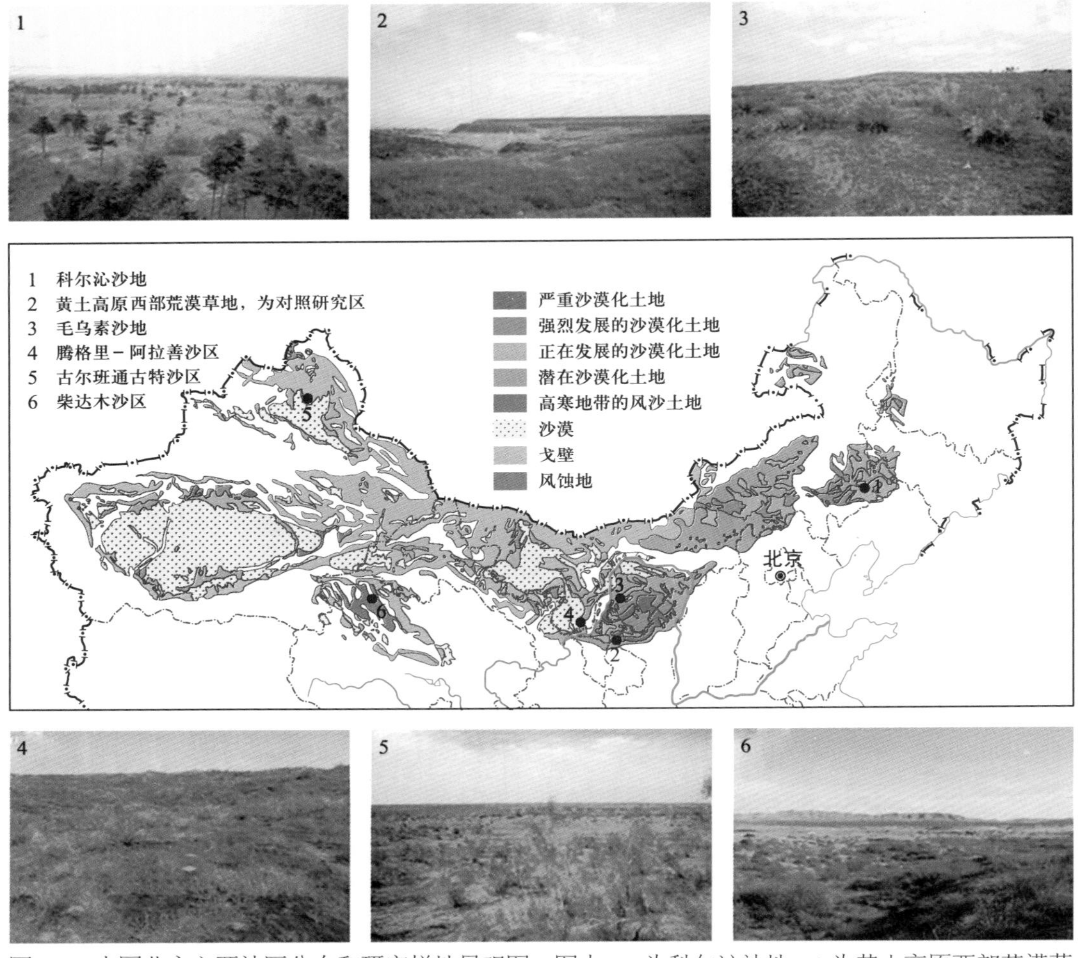

图 1-1　中国北方主要沙区分布和研究样地景观图。图中，1 为科尔沁沙地；2 为黄土高原西部荒漠草地，为对照研究区；3 为毛乌素沙地；4 为腾格里－阿拉善沙区；5 为古尔班通古特沙区；6 为柴达木沙区

Figure 1-1　The distribution of sites in different desert regions of Northern China. 1. Horqin Sandy Land; 2. western Loess Plateau; 3. Mu Us Sandland; 4. Tengger-Alxa Desert; 5. Gurbantunggut Desert; 6. Qaidam Desert

家在防沙治沙和沙区生态恢复与重建中的重大科技需求，也是构建我国北方生态屏障所面临的关键科学问题之一。

尽管20世纪90年代兴起的生态水文学为客观、全面诠释沙区植被与土壤系统相互作用与反馈机理提供了新的理念和途径，并将理解生物类群与土壤水文过程之间的相互作用及功能关系作为沙区生态水文学研究的核心内容之一。但是，相关研究主要集中在不同时空尺度上维管植物（vascular plant）群落组成、结构对土壤水文过程的响应和反馈上（王新平等，2004；Owens *et al.*，2006;刘俊杉等，2010；Wang *et al.*，2012；李柳等，2013），而忽视了生物土壤结皮（biological soil crust，BSC）的存在及其在生态恢复与重建中的重要作用（Berdugo *et al.*，2014），以致在以水分为主要限制因子的沙区生态系统中对生态恢复机理的认识忽略了一个关键的生物调节者（李新荣，2012）。此外，BSC是生物群落中重要的组成部分，在植被格局和过程中发挥着关键作用，然而传统的生态过程和格局研究及治沙工程很少考虑BSC的拓殖和演替（王涛，2011）。

受土壤水分的限制，沙区固定沙丘表面不可能支撑大面积、相对均一且连续分布的维管植物群落，在维管植物群落空间分布的斑块之间的“裸地”常被藻类、地衣、藓类等隐花植物（cryptogam）和土壤中微生物以及相关的其他生物体与土壤表层颗粒胶结形成的BSC所覆盖。在干旱沙区，BSC的分布面积往往大于维管植物，占地表活体覆盖度的40%以上（Belnap and Lange，2003；李新荣，2012）。作为沙区土壤表面特殊的生物活动层，BSC扮演了“生态系统工程师”（ecosystem engineer）的重要角色，在联系地表生物与非生物成分中起着无法替代的作用，已成为表征沙区生态系统健康的主要指标之一（李新荣等，2009）。因此，考虑BSC的重要作用是全面认知沙区生态格局与过程机理的必要前提，是干旱区恢复生态学发展的新趋势和新生长点之一（Li *et al.*，2013；李新荣等，2014），是揭示人为促进沙区生态重建和恢复机理以及沙地生态系统恢复可持续性及管理的重要基础。

然而，BSC无论在沙区生态系统生物地球化学循环，还是在土壤物理化学过程、生物过程和生态与水文过程以及景观过程中所发挥的独特作用均取决于其组成、结构和功能的多样性和独特性，这些组成BSC的生物体已被认为是干旱、半干旱生态系统中重要的物种组成。因此，组成BSC群落的物种多样性及其时空格局的变化均会导致BSC在沙地生态系统中功能的改变，这些功能的改变为沙地生态重建与恢复和生态系统管理带来很多不确定性。那么，我们的问题是：BSC群落组成的物种多样性在时空分布和格局上有什么规律？决定BSC种组成及其丰富度的关键因素是什么，在不同的空间尺度上是否亦存在差异？同样，BSC在不同的尺度上所发挥的功能也是否存在差异？这些问题的回答是深入了解BSC在沙地生态恢复中的功能和作用及探讨恢复可持续性的重要基础。

1.1　国内外研究进展及发展动态分析

有关BSC的研究在20世纪30年代就有报道，国内最早的研究来自50年代腾格里沙漠沙坡头地区（李新荣，2012），而大量研究始于20世纪的90年代初期，并受到前所未有的重视（Li *et al.*，2004；Bowker，2007），研究的重点主要聚焦其形成、功能和作用（李新荣等，2000；Belnap and Lange，2003；李新荣等，2009）。关于BSC在沙区土壤生态过程、水文过程中的作用已开展了大量研究，涉及土壤生态过程（土壤微生物、土壤微小动物群落组成、结构和功能动态变化）和水文过程（包括入渗、产流、凝结水捕获、蒸发、土壤水重新分配等）的各个方面（李新荣，2012）。在BSC影响土壤生态过程方面，国内外的研究得到的普遍的结论是，BSC的存在和发展提高了沙区土壤细菌（Zhang *et al.*，2012）、真菌（Grishkan and Kidron，2013）、原生动物如线虫（Liu *et al.*，2011）、跳虫、螨（李新荣，2012）和微小节肢动物如蚂蚁（Li *et al.*，2011，2014a）、土壤昆虫（李新荣等，2008；Chen and Li，2012）的多样性和土壤酶活性（Miralles *et al.*，2012；Liu *et al.*，2014）。在BSC影响沙区土壤水文过程方面，相关研究主要涉及BSC对降水入渗、产流、蒸发和凝结水捕获4个过程的影响。虽然越来越多的研究认为，BSC的形成和发展改变了降水在沙区土壤中的重新分配，导致水分有效性在土壤垂直方向上的“浅层化”和水平方向上的“异质性”（李新荣，2012），但BSC对前三个过程的影响仍然存在很大的争议：有的研究认为BSC的存在增加了入渗（Eldridge，1993；Maestre *et al.*，2002），从而减少地表径流（Perez，1997）；有的研究则持相反的观点（West，1990；Bisdom，1993；Kidron and Yair，1997；Eldridge *et al.*，2000；Wang *et al.*，2007；Li *et al.*，2010a），甚至少数研究认为它们之间不存在任何影响（Eldridge and Greene，1994）。关于BSC对地表蒸发的影响也仍存在着促进（West，1990；陈荷生，1992）和抑制（Brotherson and Rushforth，1983）两种截然不同的观点。大量来自我国腾格里沙漠沙坡头地区的研究认为，BSC对地表蒸发的影响表现出明显的阶段性，且主要受控于降水强度和BSC发育阶段（Liu *et al.*，2007），当降水较大时，演替后期阶段的藓类结皮在降水初期增加了水分蒸发，而在降水后期则减少了水分的蒸发（Li *et al.*，2010a）。BSC对凝结水捕获的研究结果在不同区域间较为一致，即随着BSC由初期阶段的蓝藻结皮向后期阶段的地衣结皮、藓类结皮演替，表层土壤对凝结水的吸附量呈增加趋势（Kidron *et al.*，2003，2009；Liu *et al.*，2006；Pan *et al.*，2008），但是关于这些增加的水分是否参与不同降水量沙区的土壤水文过程以及贡献率如何，还没有确切的答案。造成上述影响机理不清楚的原因可归纳为以下几点：①地域差异。BSC外部形态在不同气候带沙区存在很大差异使得生态过程、水文过程复杂化（Belnap and Lange，2003）。②研究手段限制。虽然在早期研究中将土壤中细菌、真菌、线虫、跳虫、螨和蚂蚁等异养生物体作为BSC的重要组成之一，但由于在野外对它们进行分类和鉴定较为困难，绝大多数工作集中在自养生物体方面，近期的研究也仅仅是附带提及它们的作用（Büdel，2005），至今BSC对这些土壤生物的影响多为定性描述；而研究水文过程中所采用的去除/覆盖BSC、入渗测量仪、盖度面积的技术处理及利用除藻剂去除蓝藻干扰等方法都存在很大局限性。③未考虑BSC覆盖下的土壤理化性质，尤其是土壤质地（Williams *et al.*，1999；Eldridge，2000）的影响。如何区分BSC层和其下的土壤性质对入渗、径流等水文过程的贡献也是目前研究的难点。④忽略了最为关键的一个科学问题，即BSC的多样功能（multifunctionality）主要取决于BSC本身的特征种的组成、格局和功能的多样性。

对上述问题认识的不足，不仅制约了对沙区土壤生态水文过程机理的深入理解，也影响了对沙区生态恢复机理及其可持续性的进一步认知。目前，就机理解释而言，比较一致的观点主要有两个：①直接作用。BSC生物体直接参与了土壤生物的食物链、土壤水分受体和供体（传递网络）的构成，由不同BSC生物体提供的食物、截留水量差异来直接驱动土壤生态、水文过程（李新荣等，2008；李新荣，2012）。②间接作用。BSC通过捕获降尘（Li *et al.*，2006）、改变表层土壤稳定性（Eldridge and Greene，1994）、孔隙度（王新平等，2011）、容重（Li *et al.*，2004）、团粒结构和土层厚度（Greene and Tongway，1989；Greene and Chartres，1990；Wang *et al.*，2007）、地表辐射反射和温度（Zhang *et al.*，2012）、pH、盐分含量、碳酸钙累计、微量元素含量和多聚糖含量（Li *et al.*，2004；Guo *et al.*，2008），通过光合作用（Zaady *et al.*，2000；Brostoff *et al.*，2002，2005；Belnap and Lange，2003；Housman *et al.*，2006；Su *et al.*，2011；Li *et al.*，2012）、呼吸作用（Zhao *et al.*，2013，2014，2016）、固氮作用（Belnap，2002；张鹏等，2011）、矿化作用（Delgado-Baquerizo *et al.*，2013；虎瑞等，2014）和分解作用（李康宁，2010）影响土壤养分含量等土壤理化、水文特性及微地形地貌特征（Li *et al.*，2005，2008），间接作用于土壤生物群落结构、功能及表土层降水入渗、径流、蒸发和凝结水的捕获（李新荣等，2009；李新荣，2012）。然而，这些机理的提出多基于间接实验或普通生态学和水文学知识的推断，缺乏直接实验和观测数据的支撑（李新荣，2012）。例如，不同BSC生物体提供的食物在数量和质量上有哪些差异？不同土壤生物体如何影响土壤各种功能性酶活性？不同BSC生物体通过光合、固氮、分解和矿化作用究竟能为土壤增加多少碳和氮？BSC中菌丝体、假根、叶鞘和胞外分泌物如何调节土壤水分动态变化及其量化？……笔者仍然不能确切回答。机理认知不清也在一定程度上限制了对BSC影响土壤生态水文过程的深入了解和整个沙区生态恢复的理论探讨。Li等（2002）和Wang等（2007）根据在沙坡头地区的研究结果提出了BSC在土壤水文过程中，特别是对降水入渗的影响取决于降水强度、区域的降水量和BSC层下土壤基质的理化性质以及隐花植物组成差异的综合评价观点。这个观点虽然较好地解释了国际上来自不同研究区的长期争论，但仍未得到足够的数据支撑，特别是缺乏来自涵盖我国不同降水梯度、土壤质地和干扰强度沙区的相关研究数据验证。

值得注意的是，BSC对沙区土壤生态过程和水文过程的影响不是孤立存在的。虽然沙区土壤生态过程和土壤水文过程间可以产生直接联系，但作为沙区生态系统重要组成部分的BSC在这两个过程相互作用中起到了关键的联结和调节作用。来自以色列内盖夫沙漠不同降水梯度（86 ~ 160 mm）的研究发现，BSC的存在颠覆了一个普遍的观念——较多降水导致更深的水分入渗，并影响了深层土壤生态过程（Yair *et al.*，2011）。尽管土壤质地相似，但BSC的存在改变了降水的再分配：降水多的沙区表土层发育较厚的BSC吸收和存储了大量的水分，从而限制了降水入渗，而在降水少的地区分布的较薄的BSC只能截留少量的降水，导致地表径流的产生以及在径流区水分入渗到深层土壤，从而产生了径流发生频率、径流量随着年均降水量的增加反而减少的“非常”现象。我国腾格里沙漠东南缘人工固沙植被区土壤生态与水文过程的长期定位监测也印证了BSC在沙区土壤生态－水文过程中的重要调节作用，即植被建设后BSC的形成和演替通过对土壤生态水文过程的反馈作用，最终主导了植被的演变（Li *et al.*，2004，2007），从理论上诠释了BSC对年降水量小于200 mm沙区生态水文过程的重要影响在于促使了沙地土壤有效水分含量的浅层化，这一影响深刻地改变了沙地原来的水循环，影响了沙地植被的组成和格局，较好地揭示了该地区固沙植被演变的基本规律，即向特定生物气候区地带性植被的演替（Li *et al.*，2014a）。这也从侧面佐证了不考虑BSC的调节作用，超过土壤水分承载力的不合理植被建设将导致前面所提到的地下水位下降和土壤生境恶化的后果（Li *et al.*，2014b）。由此可见，沙区土壤生态和水文过程紧密相连，BSC可通过改变自身以及土壤生物学属性（物种组成、物种多样性、结构等）、化学属性（pH、盐分、养分等）和物理属性（粗糙度、孔隙度、稳定性、温度等）直接和间接地调节两个过程（李新荣，2012）。那么，对于横跨我国半湿润、半干旱和极端干旱沙区的BSC在各自土壤生态水文过程中扮演的角色也是如此吗？目前还无法得知。有关考虑BSC参与的沙区土壤生态与水文过程互馈作用研究也仅局限于局地尺度，较大尺度的跨区域联网研究还很少，因而限制了相关理论假说的普适性检验（Li *et al.*，2013）。总之，对BSC在沙区生态恢复与重建实践中的作用和功能以及机理的了解，必须建立在对BSC组成、结构、空间格局规律认识的基础之上。

为什么在不同生物气候带的沙区，BSC在组成和结构上存在很大差异？并不是所有沙区生境中都有BSC的存在。对其形成条件、分布与演替特征的研究是深入了解BSC在沙区地表过程中功能与地位的重要基础（West，1990；李新荣等，2009）。有研究发现，BSC的形成、分布和演替是其对外界环境因子和干扰因素综合适应的结果（Eldridge and Tozer，1997；Bowker，2007），并存在尺度效应（程军回和张元明，2010）：①在区域尺度上，BSC的分布与年降水量、凝结水量和土壤含水量呈正相关趋势，温度对BSC分布的影响因组成BSC的物种而异；②在局地尺度上，BSC中隐花植物的组成、物种多样性、时空分布由土壤质地、理化和生物学属性决定（Li *et al.*，2010a），而稳定的土壤表层（Maestre *et al.*，2005）、细沙

(0.01 ~ 0.05 mm)(Duan *et al.*, 2004)和微生物的活动(Maestre *et al.*, 2005)是BSC形成和发育的前提条件；③在小微尺度上，复杂的微地貌过程造成的土壤水分、养分资源分异是维持BSC多样性的关键因素(Li *et al.*, 2010b；吴永胜等，2010)；④适度干扰，如风蚀(Wang *et al.*, 2007；Jia *et al.*, 2012；Pu *et al.*, 2015)、水蚀(Belnap and Lange, 2003)、火烧(Bowker *et al.*, 2004)、沙埋(Jia *et al.*, 2008)、动物活动(Chen and Li, 2012)和车辆碾压(Belnap and Lange, 2003)等对BSC分布和生态功能无明显影响，但随着干扰程度的增加，BSC由演替后期类型向早期类型演变，导致其结构和功能的退化与丧失(李新荣，2012)。

土壤生态水文过程也是影响BSC的形成、分布与演替的重要因素，但过去一直被忽视。一方面，有研究发现土壤微生物、线虫等原生动物呼吸和活动所产生的分泌物是BSC形成及发展的重要条件之一(Maestre *et al.*, 2005)，但对于土壤生物群落组成、结构和功能(活动、酶活性)的变化是否会影响BSC形成和演替，目前还不清楚。另一方面，沙区土壤各个水文过程(降水入渗、径流、蒸发和凝结水捕获)都会影响到BSC对水分或养分资源的获取(Li *et al.*, 2008)，并影响到BSC生物体的生理生长、存活与演替(李新荣等，2009；李新荣，2012)。通常而言，具有一定持水能力的土壤表层是形成和维系BSC生存的关键，极端干燥的气候条件下，由于水分的限制，不利于藓类和一些中生或中旱生的隐花植物参与BSC生物体的组成，而多以蓝藻、绿藻类为主。随着降雨量的增加，藓类、地衣的多样性和盖度也随之增加，并出现其混生结皮。与中国其他沙区相比，毛乌素沙地与科尔沁沙地的物种多样性高，并且藓类植物的盖度也较高，而古尔班通古特沙漠的固定沙丘上藓类植物的物种多样性相对低，这可能是由于古尔班通古特沙漠的年降雨量仅为80 mm，而年蒸发量却高达2800 mm。土壤水分的快速蒸发对BSC有抑制作用：在降水量相同的情况下，蒸发较快的沙区地表湿润时间过短，不利于BSC生物生长，往往分布演替早期类型的蓝藻结皮或裸沙；而蒸发相对较慢的沙表面，降水后湿润持续时间较长，有利于演替后期的地衣结皮和藓类结皮的形成和分布(Kidron *et al.*, 2009)。特别地，由于BSC的特殊地表生境和结构特点，其生物组成、盖度和生态功能对未来降水格局(Cable and Huxman, 2004；Kidron *et al.*, 2012；Zelikova *et al.*, 2012)、温度(赵允格等，2010；Escolar *et al.*, 2012；Zelikova *et al.*, 2012)和人类活动的干扰方式及强度(Jia *et al.*, 2008, 2012；李新荣，2012)的变化表现出快速而敏感的反应。不难想象，降水、温度等气象条件及人类活动的干扰方式、强度的变化又可通过影响土壤生态和水文过程而间接影响到BSC的结构和功能。未来BSC的这种不确定性也势必反馈作用于沙区土壤生态水文过程，进而引起沙区生态系统功能的不确定性。因此，基于全球气候变化和人类活动干扰加剧情景下BSC的组成、结构和功能演变及其对沙区土壤生态水文过程的反馈机理研究，对判别沙区生态系统是否稳定和健康具有重要的理论指导意义。

目前，对于沙区土壤生态过程、水文过程以及它们联合起来怎样影响BSC，机理如何，特别是在全球气候变化及人类活动导致的干扰加剧的情况下，上述影响及机理又将如何变化等一系列问题还缺乏确切了解。尽管如此，一些学者试图利用模型来描述BSC分布与土壤生态因子间的关系。Bowker等（2006）在科罗拉多高原的研究发现，BSC分布显著受到土壤质地和养分的影响，并提出了不同尺度上地衣和藓类结皮分布的等级概念模型。Read等（2008）利用促进回归树（boosted regression tree，BRT）模拟分析了影响BSC分布的关键环境因子及其相互作用，提出了不同土壤质地和斑块面积BSC分布的模型。Li等（2010b）基于腾格里沙漠中微空间尺度上BSC分布格局与土壤环境因子的关系，提出了BSC分布与土壤质地、微地形和降尘关系间的概念模型，揭示了它们之间的相互作用机理。不难看出，这些模型多属定性描述，对影响BSC分布的关键因子尚缺乏定量的表达，且没有考虑区域土壤生态水文过程。如何从机理上解释BSC的分布格局，成为未来研究BSC的重点，也是全面认知BSC在沙区生态水文过程中功能的重要前提。

1.2 科学意义

在近期召开的几次BSC研究领域最具影响力的国际研讨会［2010年在德国召开的以“生态系统中的生物土壤结皮——多样性、生态和管理”为主题的第一届生物土壤结皮国际学术研讨会（Biocrust 1）、2012年在奥地利召开的以“生物土壤结皮和生态地貌过程”为主题的欧洲地球科学联合会2012年会员大会分会场研讨会、2013年在西班牙召开的以“生物土壤结皮改变世界”为主题的第二届生物土壤结皮国际学术研讨会（Biocrust 2）和2016年在美国召开的以“生理、系统分类、遗传和基因组、恢复、个体、群落、生态系统和景观生态学”为主题的第三届生物土壤结皮国际学术研讨会（Biocrust 3）］都强调了BSC对干旱区包括沙区生态系统的影响以及BSC在生态重建与恢复和生态系统管理中的重要性，认为BSC组成和格局的变化直接影响着其生态服务功能的改变，尤其是在全球变化的背景下，事关干旱区/沙区生态恢复与重建的可持续性；相关研究是目前和未来一段时期内的重点研究方向，并指出跨区域、多尺度的长期定位监测是推动相关研究的最有效技术手段之一。然而，国际上大量的研究多集中在寒区荒漠和热带荒漠，其降水大多分布在冬、春季节。除古尔班通古特沙漠外，我国沙区降水主要集中在较热的夏季和秋季（Reynaud and Lumpkin，1988），就研究区域而言，本书的研究是对全球BSC相关研究的重要补充。这是因为我国沙区分布范围广，从西部极端干旱区（降水量<100 mm）到东部半干旱区（降水量介于300 ~ 450 mm），

生态梯度变化明显，为BSC的大量拓殖和繁衍提供了多样的生态位，BSC结构和种类组成复杂多样，几乎涵盖了所有BSC发育阶段，这也为综合比较和全面探讨BSC对气候、土壤、植被和人类活动干扰等变化的响应和反馈机理提供了得天独厚的天然实验条件和研究平台。此外，20世纪80年代以来，我国在沙区相继建立了以监测和研究环境及生态系统变化为目的的生态监测和研究网络（如中国生态系统研究网络），为跨区域联合研究提供了便利，也为深入开展多尺度BSC对沙区生态恢复与重建的生态水文机理研究提供了良好研究背景和基础数据支撑。为此，本章我们以中国主要沙区的BSC为研究对象，分析了中国主要沙区BSC种类组成、空间分布格局的规律及其机理，提出BSC主要类型/演替阶段随降水（土壤水）、土壤特性（土壤质地组成）、植被特征（多年生植物与一年生植物盖度变化）和干扰梯度变化而相互转化的理论假说。

1.3　中国沙区BSC群落物种多样性、空间分布格局及规律

我们选取了科尔沁沙地（样地1）、毛乌素沙地（样地3）、腾格里－阿拉善沙区（样地4）、古尔班通古特沙漠（样地5）和柴达木沙区（样地6）作为研究样地，这些样地反映了从东到西降水从450 mm至80～110 mm的变化梯度。此外，BSC发育良好的黄土高原西部荒漠草地区作为对照样地（样地2），见图1-1。共357个调查样方被分别设置在以上不同沙区。各沙区样地自然概况、气候条件和样方分布与取样见表1-1。

在以上6个研究样地中分别设置综合植被调查样方数量为60、62、60、65、60和50个，样方大小为100 m^2，样品采集于2009—2012年的秋季。以上样方在不同沙区的分布主要取决于BSC优势种如地衣、藓类以及BSC的盖度。植被调查采用5 m × 5 m的样框在每个样方中随机进行，对植被总盖度、一年生和多年生植物的盖度分别进行了记录和估计。BSC的盖度采用点针法进行估算（李新荣，2012）。BSC和表层土（0~10 cm）采样也在每个样方中进行，由于各沙区深层沙土（超过10 cm）质地相对均一，故仅对含BSC的表层土进行了采样。在样品采集前先用蒸馏水把土表喷湿，使BSC优势种在野外便于识别，然后鉴定样方中的BSC组成种，如蓝藻、绿藻、地衣和藓类等。用每个样方中的物种丰富度来表示不同沙区BSC群落组成的多样性，用蓝藻、绿藻、地衣和藓类的总叶绿素含量来表示BSC的生物量；测定BSC覆盖下的土壤质地组成、pH、全氮、全磷、全钾、土壤有机碳、碳酸钙、总盐和表土层水分含量。

表 1-1　中国主要沙区（研究区）气候、土壤和植被总体特征与研究样地设置和取样说明

Table 1-1　General properties,climatic and vegetation features of study sites and sampling in Chinese deserts

总体特征	样地					
	科尔沁沙地（样地 1）	黄土高原西部荒漠草原（样地 2）	毛乌素沙地（样地 3）	腾格里－阿拉善沙区（样地 4）	古尔班通古特沙漠（样地 5）	柴达木沙区（样地 6）
面积 /($\times 10^4$ km^2)	10.56	>10	3.2	4.27	4.88	3.49
年降水量 /mm	450	430	335	186	110	80
降水主要月份	5—9	5—9	5—9	5—9	1—3	5—9
平均温度 /℃	5.5	12.3	6.2	10.5	8.3	4.6
1 月均温 /℃	−17.3	−5.6	−12.3	−6.9	−9.3	−10.4
7 月均温 /℃	24.3	23.6	24.0	24.3	21.5	17.3
年均蒸发量 /mm	2100	2300	2300	2600	2700	2900
植被组成（优势种，在每一样地调查样方中出现频率大于 5%）	*Agropyron cristatum* *Artemisia halodendron* *Artemisia frigida* *Atraphaxis bracteata* *Caragana microphylla* *Cleistogenes squarrosa* *Enneapogon borealis* *Phragmites communis* *Pinus sylvestris* var. *Mongolica* *Poa pratensis* *Lespedeza davurica* *Leymus secalinus* *Salsola pestifer* *Sonchus brachyotus* *Stipa glareosa* *Suaeda glauca* *Setaria viridis*	*Agropyron cristatum* *Allium polyrhizum* *Artemisia gmelinii* *Artemisia intramongolica* var. *microphylla* *Bothriochloa ischaemum* *Caragana stenophylla* *Ceratoides latens* *Ephedra rhytidosperma* *Eragrostis poaeoides* *Kochia scoparia* *Hippophae rhamnoides* *Leymus secalinus* *Plantago asiatica* *Reaumuria songarica* *Salsola passerina* *Stipa breviflora* *Stipa bungeana* *Poa annua*	*Artemisia ordosica* *Artemisia frigida* *Astragalus melilotoides* *Bassia dasyphylla* *Caragana korshinskii* *Caragana microphylla* *Chenopodium aristatum* *Corispermum mongolicum* *Echinops gmelini* *Euphorbia humifusa* *Hedysarum laeve* *Inula salsoloides* *Lespedeza davurica* *Oxytropis psammocharis* *Psammochloa villosa* *Salsola ruthenica* *Setaria viridis* *Stipa breviflora* *Scorzonera divaricata*	*Agriophyllum squarrosum* *Allium mongolicum* *Artemisia capillaris* *Artemisia ordosica* *Ammopiptanthus* *mongolicus* *Bassia dasyphylla* *Caragana korshinskii* *Cleistogenes songorica* *Chloris virgata* *Corispermum patelliforme* *Echinops gmelini* *Eragrostis poaeoides* *Reaumuria songarica* *Stipa breviflora* *Stipa gobica* *Salsola passerina* *Stipa glareosa*	*Artemisia arenaria* *Aristida pennata* *Agriophyllum squarrosum* *Allium* sp. *Bassia dasyphylla* *Ceratocarpus arenarius* *Corispermum lehmannianum* *Echinops gmelini* *Eremurus anisopteris* *Haloxylon persicum* *Lappula rupestris* *Salsola collina* *Torularia torulosa*	*Aneurolepidium dasystachys* *Artemisia capillaris* *Bassia dasyphylla* *Calligonum zaidamense* *Carex przewalskii* *Eragrostis poaeoides* *Haloxylon ammodendron* *Leymus secalinus* *Lycium ruthenicum* *Nitraria tangutorum* *Reaumuria songarica* *Salsola ruthenica* *Stipa glareosa* *Stipa gobica* *Phragmites communis* *Poa annua*

总体特征	样地					
	科尔沁沙地（样地 1）	黄土高原西部荒漠草原（样地 2）	毛乌素沙地（样地 3）	腾格里－阿拉善沙区（样地 4）	古尔班通古特沙漠（样地 5）	柴达木沙区（样地 6）
取样数量和取样时间	18 (2009) 22 (2010) 10 (2011) 10 (2012)	18 (2009) 22 (2010) 10 (2011) 12 (2012)	20 (2009) 20 (2010) 10 (2011) 10 (2012)	20 (2009) 25 (2010) 10 (2011) 10 (2012)	20 (2009) 10 (2010) 15 (2011) 15 (2012)	10 (2009) 15 (2010) 10 (2011) 15 (2012)

表 1-2　中国沙区（研究区）BSC 群落隐花植物种的丰富度、生物量和不同维管植物盖度

Table 1-2　Species richness of cryptogams,biomass of BSC and covers of different vascular plants of the study sites (mean ± SE) in northern Chinese deserts

样地（样方数量）	种丰富度（每个样方种的数量）				BSC 生物量 /(mg cm^{-2})	维管植物盖度 /%		
	蓝藻	其他藻类	地衣	藓类		一年生植物	多年生植物	总的植被
1 (60)	3.3 ± 1.4a	6.1 ± 3.2a	1.5 ± 1.1a	11.9 ± 8.5a	5.8 ± 1.9a	28.8 ± 6.9a	43.6 ± 11.7a	46.7 ± 9.9a
2 (62)	3.5 ± 1.9a	7.6 ± 5.1b	7.3 ± 4.7b	5.4 ± 3.2b	5.4 ± 2.4b	34.5 ± 8.7b	40.8 ± 8.9b	47.5 ± 9.3a
3 (60)	4.6 ± 1.8b	4.6 ± 1.9c	1.9 ± 1.2c	7.4 ± 4.1c	4.7 ± 2.3c	42.6 ± 12.7c	38.2 ± 17.5c	54.7 ± 8.1b
4 (65)	8.1 ± 3.5c	9.8 ± 5.5d	2.4 ± 2.1d	2.5 ± 2.4d	3.3 ± 1.8d	43.5 ± 9.6d	32.3 ± 7.5d	48.9 ± 5.1c
5 (60)	9.7 ± 5.8d	14.9 ± 5.4e	4.2 ± 2.1e	2.0 ± 1.3e	1.9 ± 0.8e	49.8 ± 14.5e	27.5 ± 11.5e	53.7 ± 9.3b
6 (50)	12.7 ± 5.3e	9.3 ± 4.0d	2.1 ± 1.2d	1.2 ± 0.8f	0.8 ± 0.8f	64.1 ± 4.9f	21.6 ± 9.2f	65.7 ± 7.1f

注：不同小写字母表示样地间差异显著 ($p < 0.01$)，字母相同表示无显著差异；样地 1~6 含义见图 1-1。表 1-3 同。

不同样地之间的植被总盖度、一年生植物和多年生植物盖度以及测定的土壤属性参数和BSC的盖度、物种丰富度、生物量用方差分析（ANOVA）的Tukey' s test进行分析，在此之前检验待测参数是否符合正态分布；BSC中物种丰富度、分布格局沿降水梯度的变化与年降水量、植被盖度和测定的土壤属性参数之间的关系则利用线性逐步回归进行了分析；BSC物种丰富度和生物量作为环境条件的参数，通过比较多种排序（除趋势对应分析（DCA）、典范对应分析（CCA）和主成分分析（PCA））的结果，我们利用冗余分析（RDA）进行了排序和环境解释（Jongman *et al.*，1995；Ter Braak and Smilauer，2002）。

1.3.1 不同沙区BSC群落组成种的物种丰富度和生物量

以上6个研究区含黄土高原西部荒漠草地对照区，BSC群落组成种存在很大差异。藓类种丰富度最高的记录出现在样地1，即科尔沁沙地具有最丰富的藓类分布，在60个调查样方中达到37个种（参见本章末附表1-1）；最多的地衣种类出现在黄土高原西部封育未干扰的样地2，在62个调查样方中发现有17种地衣；而在古尔班通古特沙漠（样地5）和柴达木沙区（样地6）记录（已鉴定的）的BSC蓝藻种类分别达33和30种之多。东部沙地BSC具有较高的藓类多样性，而西部沙地具有较高的蓝藻种类多样性，黄土高原区BSC具有较高的地衣种多样性。BSC种类组成的多样性使BSC对沙区降水梯度的变化有着不同的响应：随着从东至西沙区降水量的递减，BSC群落中蓝藻和其他荒漠藻的多样性增加，各样地间（除东部样地1和对照样地2外）存在显著的差异（$p<0.01$），藓类种的丰富度却出现了明显的下降趋势（表1-2），且各样地之间差异显著（$p<0.01$），而地衣种的丰富度随降水梯度的递减无明显增加或减少的趋势。

BSC生物量与降水量密切相关，在降水量较高的东部沙地（样地1）和对照（样地2）都具有较高的生物量，而降水量较小的西部沙区BSC的生物量较小。这种BSC生物量的空间分布格局可能与BSC中藓类所占的比例相关，降水量高的沙区BSC中藓类所占的比例较大（图1-2和表1-2，藓类盖度在东部沙地和对照样地明显高于西部沙区BSC中藓类的盖度）。在区域尺度上（针对每一个沙区或每一个研究区），除了多年生植物的盖度外，BSC生物量的大小主要取决于表层土壤质地中粉粒的百分含量（样地1）和黏粒的百分含量（样地2），而在样地3至样地6中则主要取决于表土层的含水量（图1-3）。在景观尺度上（所有研究样地或中国主要沙区），BSC生物量的分布格局主要取决于沙地表土层含水量，而非其他所测定的土壤属性参数（图1-4）。也就是说，降水量决定了BSC的生物量大小。如图1-4所示，BSC生物量与多年生植物盖度、表土层含水量和土壤有机碳密切相关（r分别为0.72、0.61和0.50；参见本章末附表1-2）。

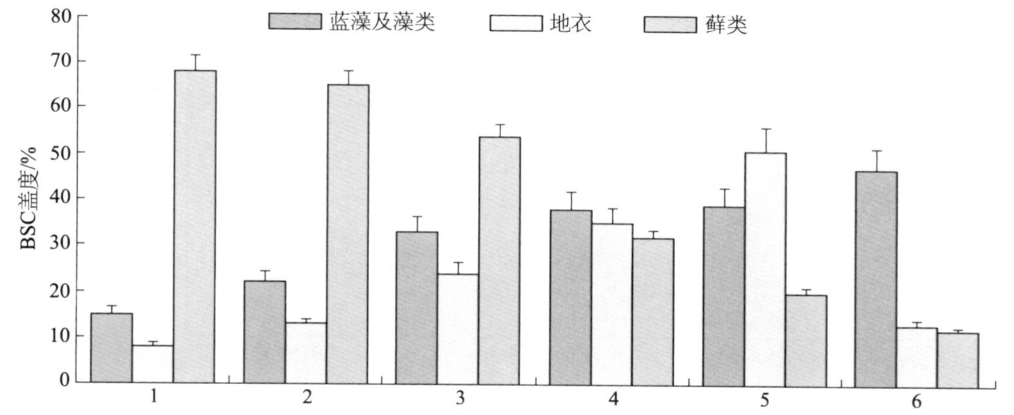

图 1-2　不同沙区（6 个样地）中不同类型 BSC 的盖度（平均值 ± 标准误）。每个样地的调查样方数见表 1-1，样地 1 ~ 6 见图 1-1

Figure 1-2　The covers of different BSC(mean ± SE). Investigative number of plots for each site see Table 1-1, and names of site 1–6 see Figure 1-1. in six study sites in deserts of China

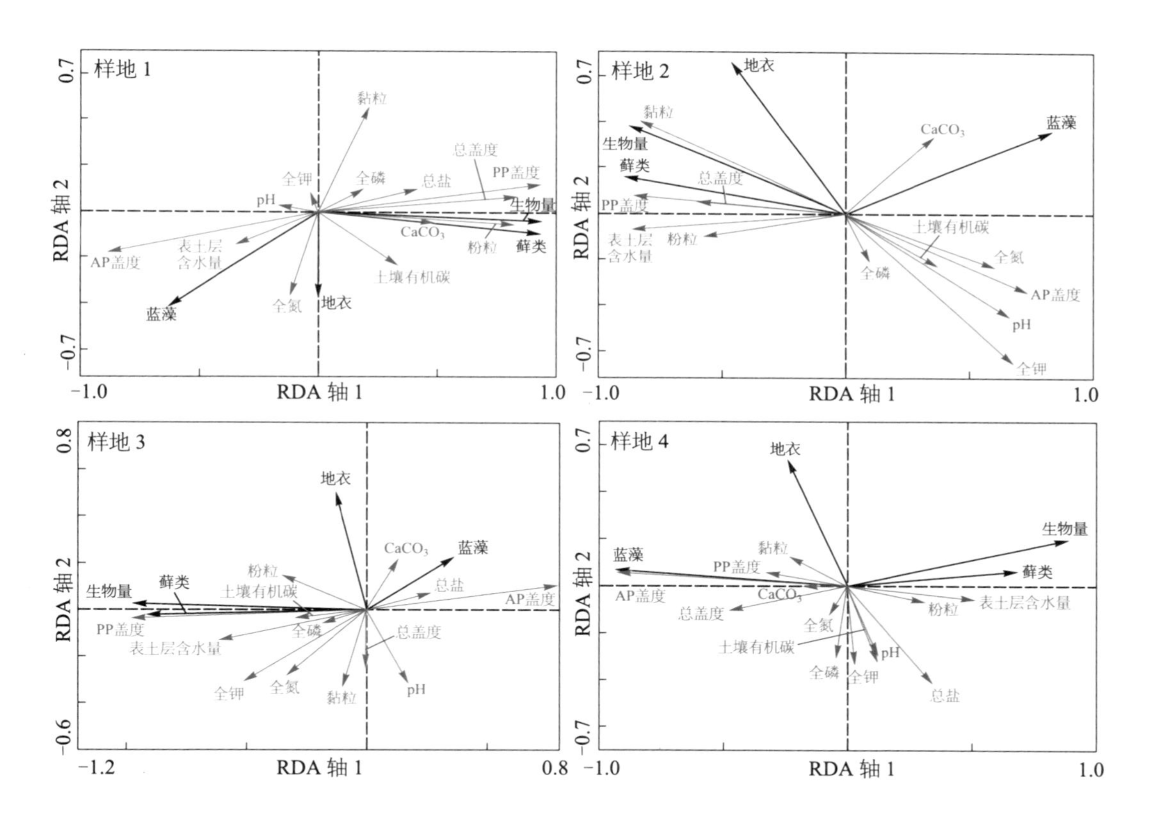

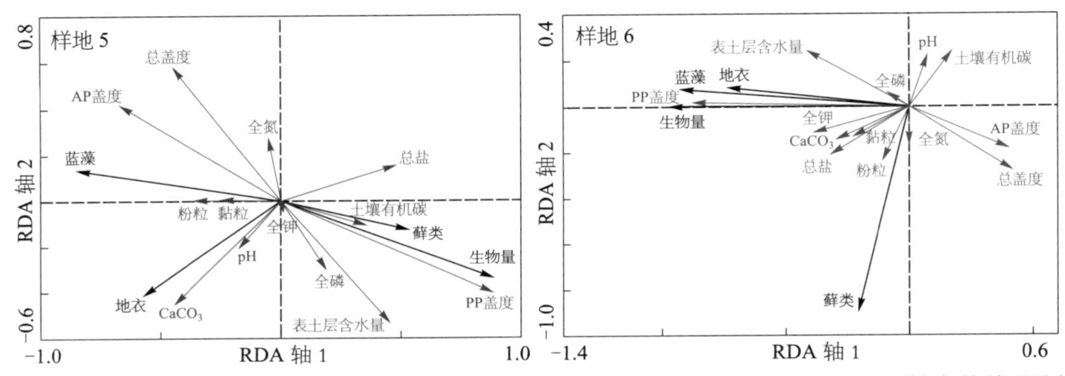

图 1-3　区域尺度上（每一个沙区 / 研究区）BSC 组成种的丰富度、生物量和土壤属性参数以及维管植物盖度的 RDA 排序图。图中 AP 为一年生草本，PP 为多年生草本；样地 1 ~ 6 见图 1-1

Figure 1-3　Biplot of species richness, biomass of BSC and soil property parameters, as well as covers of vascular plants in different study sites (regional scale) by RDA ordination. AP: annual plant, PP: perennial plant. Names of site 1−6 see Figure 1-1

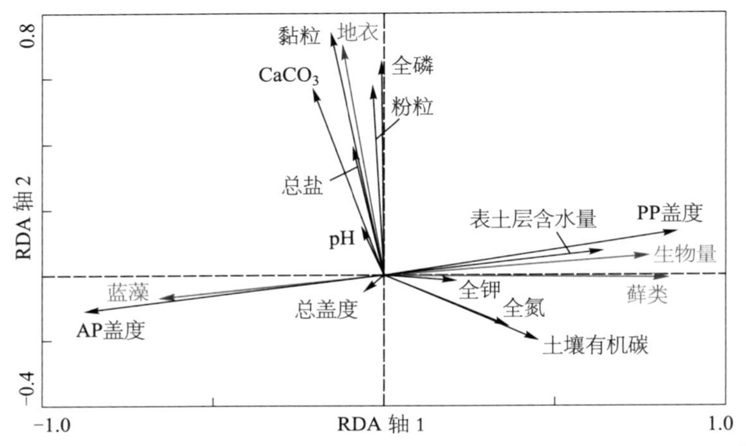

图 1-4　景观尺度上（所有研究样地 / 中国主要沙区）BSC 组成种的物种丰富度、生物量和土壤属性参数，以及维管植物盖度的 RDA 排序图。图中 AP 为一年生草本，PP 为多年生草本；样地 1 ~ 6 见图 1-1

Figure 1-4　Biplot of species richness, biomass of BSC and soil property parameters, as well as covers of different vascular plants in China's deserts (landscape scale) by RDA ordination. AP: annual plant, PP: perennial plant. Names of site 1−6 see Figure 1-1

1.3.2　沿降水梯度植被盖度和土壤属性的变化

如表 1-2 所示，多年生植物和一年生植物的盖度随降水梯度呈相反的变化趋势，在东部沙地有较高的多年生植物的盖度，而干旱的西部沙地中一年生植物的盖度相对较高。但是，

表 1-3　中国主要沙区调查样地 BSC 覆盖和表土层土壤属性

Table 1-3　Topsoil properties of the investigated sites (mean ± SE) in northern Chinese deserts covered by different BSC

样地（样方数量）	粉粒 /%	黏粒 /%	pH	土壤有机碳 /(g kg^{-1})	全氮 /(g kg^{-1})
1 (60)	7.27 ± 5.22a	0.68 ± 0.42 a	7.44 ± 0.49a	7.58 ± 4.63a	0.80 ± 0.39a
2 (62)	59.23 ± 4.44b	12.49 ± 2.16b	8.64 ± 0.15b	5.04 ± 0.16b	0.61 ± 0.12b
3 (60)	0.45 ± 0.26c	0.49 ± 0.09c	7.69 ± 0.41c	10.59 ± 4.93c	0.73 ± 0.29c
4 (65)	20.14 ± 3.77d	2.64 ± 1.06d	7.86 ± 0.19c	4.86 ± 0.77d	0.43 ± 0.15d
5 (60)	16.64 ± 3.64e	2.39 ± 1.96d	8.48 ± 0.51d	2.38 ± 0.62e	0.50 ± 0.66e
6 (50)	2.37 ± 0.37f	2.44 ± 1.25d	8.76 ± 0.35d	0.41 ± 0.17f	0.37 ± 0.10f

样地（样方数量）	全磷 /(g kg^{-1})	全钾 /(g kg^{-1})	碳酸钙 /(g kg^{-1})	总盐 /(g kg^{-1})	表土层含水量 /%
1 (60)	0.18 ± 0.03a	2.44 ± 0.38a	0.35 ± 0.10a	0.23 ± 0.07a	14.27 ± 1.27a
2 (62)	1.02 ± 0.06b	1.78 ± 0.12b	0.91 ± 0.06b	1.46 ± 0.18b	9.35 ± 0.93b
3 (60)	0.37 ± 0.13c	1.33 ± 0.39c	0.33 ± 0.07a	0.28 ± 0.08a	6.60 ± 2.15c
4 (65)	0.36 ± 0.07c	1.93 ± 0.15d	0.25 ± 0.08c	0.85 ± 0.27c	2.08 ± 0.88d
5 (60)	0.38 ± 0.09c	1.99 ± 0.35d	0.95 ± 0.16b	0.29 ± 0.08a	2.02 ± 0.54d
6 (50)	0.16 ± 0.07d	2.08 ± 0.65e	0.89 ± 0.74d	1.74 ± 0.28d	2.31 ± 0.46e

较高的植被总盖度却出现在西部沙地，那里蓝藻和其他荒漠藻具有较高的丰富度。我们也发现一年生草本植物对植被的总盖度有很大的贡献（表1-2），这也与我们在雨季末进行调查有关。一年生植物和多年生植物的盖度在不同的样地间均呈显著的差异（$p<0.01$），而植被的总盖度则在样地1与2之间，样地3与5之间无显著的差异（$p>0.01$）。

除了表土层含水量（从东至西下降）外，其他所测定的土壤属性参数随降水梯度无显著的变化趋势（表1-3）。相对于其他沙区，黄土高原西部荒漠草地（样地2）表土层具有较高的黏粒和粉粒含量，而东部沙地（样地1和样地3）比西部沙地（样地4～6）有较高的土壤有机碳和全氮含量。总之，随着降水梯度的变化，BSC覆盖下的表层土壤（0～10 cm）的理化性状没有明显的变化，但有机碳和全氮含量在6个研究样地差异显著（$p<0.01$，表1-3）。

1.3.3 不同尺度上BSC群落物种丰富度和空间分布格局对气候和土壤属性变化的响应

总的来看，BSC群落中蓝藻、绿藻、地衣和藓类对气候变化和土壤属性的变化在不同尺度上有着不同的响应。在区域尺度上，即在给定的降水条件下，BSC物种丰富度及分布格局对所测的土壤属性参数均存在不同的响应。当年降水量较高时（样地1～3），藓类的丰富度和盖度最高，而蓝藻和其他藻类的物种丰富度和盖度很低。后者与多年生植物盖度呈负相关关系，而与一年生植物盖度呈正相关关系（图1-2和表1-2）。除此之外，藻类和其他所测土壤属性参数之间没有明显的关联；在科尔沁沙地（样地1），就地衣在BSC中的物种多样性而言，其仅与土壤全氮含量相关，而藓类与多年生植物的盖度和土壤粉粒百分含量呈显著的正相关关系（图1-3和表1-4）；由于地表没有受到干扰，在黄土高原西部的荒漠草地，表土层土壤质地组成较细（黏粒和粉粒含量较高），蓝藻和其他藻与土壤中的碳酸钙含量呈正相关，而与表土层含水量呈负相关关系，尽管高的总盐含量和全钾含量对地衣的拓殖发育有很大的影响，但在黄土高原研究区（样地2），地衣多样性与绝大多数所测定的土壤属性参数无显著的相关关系，然而，藓类多样性与土壤黏粒含量和多年生植物盖度呈高度正相关关系（图1-3和表1-4）；在毛乌素沙地（样地3），藻类种的丰富度仍然与表土含水量呈负相关关系（表1-4），地衣种丰富度却与土壤碳酸钙含量和粉粒百分含量呈正相关，与土壤pH呈负相关，藓类多样性与多年生植物盖度和土壤有机碳呈正相关；在年降水量小于200 mm的沙区（样地4～6），BSC中藻类多样性与植被盖度密切相关：在腾格里－阿拉善沙区（样地4）和古尔班通古特沙漠（样地5），其与一年生植物的盖度呈正相关关系，而在柴达木沙区（样地6）与多年生植物盖度呈正相关关系。地衣多样性在腾格里－阿拉善沙区和柴达木沙区与所测定的土壤属性参数有显著相关关系，但在古尔班通古特沙漠与土壤碳酸钙含量密切相关。藓类多样性在古尔班通古特沙漠主要取决于多年生植物的盖度，在柴达木沙区藓类种类分布很少（附表1-1），但与土壤总盐含量有着密切的关联（图1-3和表1-4）。

在景观尺度上，蓝藻和其他藻类在BSC群落中的多样性与一年生植物的盖度呈正相关（r=0.64，附表1-2，图1-4和表1-4），而与表土层含水量呈负相关关系（r=−0.74）。土壤属性似乎对地衣种多样性的影响远远大于降水差异或变化的影响，质地组成较细的土壤有利于维持较高多样性的地衣种类的共存。此外，较高的碳酸钙积累、全磷和总盐含量也有利于地衣种类多样性共存。然而，全氮、全钾、土壤有机碳和植被总盖度的增加与地衣多样性呈负相关关系（图1-4）。对BSC中的藓类而言，其多样性主要依赖于表层土壤含水量，较高的降水有利于维持BSC中的藓类组成的多样性。此外，藓类多样性还与多年生植物盖度、土壤有机碳含量、全氮含量呈正相关。BSC的生物量也和藓类多样性（r=0.66）、多年生植物盖度（r=0.71）和表层土壤含水量（r=0.63）有较高的相关关系（附表1-2）。

表 1-4　BSC 群落物种丰富度、生物量和所测定参数之间关系在区域尺度和景观尺度上的逐步回归结果

Table 1-4　Results of stepwise progress analysis of the relationship between crustal species richness,biomass and measured parameters at the regional and landscape scales

	方程	R^2	p
区域尺度			
样地 1	蓝藻和其他藻类 =2.918−1.247 PP cover	0.429	<0.0001
	地衣 =0.272+0.166 N	0.039	0.131
	藓类 =0.054+1.154 PP cover+0.429 Silt−0.929 AP cover	0.804	<0.0001
	生物量 =0.96−0.65 AP cover+0.395 PP cover+0.096 Silt	0.875	<0.0001
样地 2	蓝藻和其他藻类 =10.534−9.661 moisture+2.842 $CaCO_3$	0.658	<0.0001
	地衣 =1.131−4.225 Salt−1.356 N	0.603	<0.0001
	藓类 = −5.525+1.575 PP cover+3.226 Clay	0.800	<0.0001
	生物量 = −1.189+1.167 Clay+0.680 PP cover−1.924 K	0.916	<0.0001
样地 3	蓝藻和其他藻类 =1.248−0.35 moisture−0.401 Clay	0.238	<0.0001
	地衣 =0.947−0.442 pH+0.561 $CaCO_3$	0.099	0.049
	藓类 = −1.446+1.318 PP cover+0.197 SOC	0.807	<0.0001
	生物量 = −1.056+1.123 PP cover+0.345 $CaCO_3$−0.205 Salt	0.875	<0.0001
样地 4	蓝藻和其他藻类 = −0.741+3.386 AP cover	0.792	<0.0001
	地衣 =0.214−0.323 Salt	0.082	0.073
	藓类 =2.534−1.277 AP cover	0.407	<0.0001
	生物量 =3.547−1.911 AP cover	0.662	<0.0001
样地 5	蓝藻和其他藻类 =1.726−0.595 PP cover+0.278 AP cover	0.632	<0.0001
	地衣 =0.728+0.791 $CaCO_3$−0.669 K+0.178 N	0.381	<0.0001
	藓类 = −0.162+0.429 PP cover+0.095 N	0.320	<0.0001
	生物量 = −1.077+0.952 PP cover	0.845	<0.0001
样地 6	蓝藻和其他藻类 =2.903+0.662 PP cover+0.28 K−1.332 Total cover	0.845	<0.0001
	地衣 =3.828+0.634 PP cover−2.406 AP cover	0.497	0.003
	藓类 =0.431+0.197 Salt−0.239 pH	0.187	0.172
	生物量 =3.928+2.201 PP cover−3.414 Total cover+0.569 K	0.847	<0.0001
景观尺度			
所有样地	蓝藻和其他藻类 = −0.256+0.943 AP cover−0.21 moisture+0.139 $CaCO_3$−0.14 Salt	0.501	<0.0001
	地衣 =0.601+0.071Clay+0.297$CaCO_3$−0.248 Salt+0.385 P−0.538 K+0.135 Silt	0.457	<0.0001
	藓类 = −0.03−1.133 AP cover+0.678 PP cover−0.125 Clay+0.886 Topsoil water+0.23 N	0.711	<0.0001
	生物量 = −1.083+1.026 PP cover+0.273 moisture−0.153 SOC	0.604	<0.0001

注：N: 全氮，Silt：粉粒，moisture：表土层水含量，K：全钾，P：全磷，$CaCO_3$：碳酸钙，SOC：土壤有机碳，Salt：总盐，Total cover：总盖度，Clay：黏粒，PP cover：多年生植物盖度，AP cover：一年生植物盖度；样地 1 ~ 6 含义见图 1-1。

1.4 中国沙区BSC空间分布格局的理论假说

植被通过覆盖地表提供遮阴、延长地表湿润时间、减小地表径流而影响着土壤水分含量和提高激活BSC生物活性的水分有效性，进而间接地影响着BSC群落的物种组成及其丰富度和格局特征（Housman *et al.*，2006；Li *et al.*，2012；Kidron and Benenson，2014）。一般在多年生植物冠幅之下比一年生植物冠幅下或冠幅间的空地有较高的土壤含水量、表土层厚度和有机质积累，主要是多年生植物冠幅对降水的再分配作用和凋落物及降尘在冠幅下的积累和沉积作用所造成的（Shachak and Lovett，1998；Li *et al.*，2008）。

BSC中藓类所占BSC总盖度的百分比、藓类种的丰富度与多年生植物盖度和土壤有机质呈正相关关系，可以理解为：与一年生草本的覆盖相比，多年生植物，特别是灌木的冠幅更能增加对大气降尘的捕获（Danin and Ganor，1991）、延长降水后冠幅以内土表的湿润时间（Wang *et al.*，2011）以及形成“沃岛效应”，有利于养分和凋落物的积累，尤其是在风蚀环境条件下，其作用更加明显（Garner and Steinberger，1989）。这使得多年生植物斑块下的藓类得到良好的发育生境，且藓类下面的土层比藻类和地衣覆盖下的土层都厚（Li *et al.*，2010b）。藓类在BSC组成中其多样性和盖度与一年生植物盖度的负相关关系可以解释为：藓类结皮在未受到干扰时，所形成的BSC很紧实，在多风的沙区其存在减少了一年生植物的种子进入土壤种子库的机会（Li *et al.*，2005；Su *et al.*，2011），使有机会在藓类结皮上萌发和定居的一年生植物种群数量减少。

蓝藻和其他藻类与一年生植物呈正相关关系，这主要是因为蓝藻和其他藻类主要分布在多年生植物冠幅之间的“裸地”（Li *et al.*，2005），由于受风蚀干扰，蓝藻和藻类结皮很难保持完整和紧实状态，一年生植物种子容易进入土壤，同样风蚀活动也限制了“裸地”上地衣和藓类在BSC组成中所占的比例（Jia *et al.*，2012），此外，蓝藻结皮的氮固定也为这些一年生植物的生存提供了宝贵的养分需求（Su *et al.*，2011）。

无论是在区域尺度上还是在景观尺度上，地衣均与植被（一年生植物和多年生植物）的盖度、表层土壤的含水量无显著的关联（图1-3和图1-4），其主要原因是地衣对地表的干扰相对敏感，其发展和存在需要一个稳定的表层土壤环境，通常稳定的土表质地组成中具有较高的黏粒和粉粒百分含量，如我们的对照研究区——黄土高原西部封育的荒漠草地（样地2）就是这样的生境（表1-4）。针对中国北方主要沙区（研究区），从景观尺度上看，土壤属性即土壤质地中较高的黏粒和粉粒百分含量、碳酸钙累积都有利于地衣种类多样性和较高的盖度（图1-4和附表1-2）。这些研究结论也支持了一些国际同行的研究（Kleiner and Harper，1977；Flechtner，2007）。土表较长时间的湿润和较多的降水似乎不利于地衣的生存、多样种类的共存和较高的盖度（图1-2），其原因之一是降水和阳光暴晒的频繁交替往往对地衣菌体（thalli）相关器官有较大的损害（Schuster *et al.*，1985；Lange and Green，1996；Green *et al.*，2011；Li，2012）。

相反，藓类分布及多样种类的共存对土壤湿度有很大的依赖性，这已被许多不同的生物气候带研究所证实（Kidron *et al.*，2010；Li，2012）。在本研究中，我们所选的样地（含对照）基本反映了降水从东至西减少的变化规律（图1-1和表1-3），BSC中藓类的多样性、所占比例和盖度也是按此梯度递减的。由于藓类比蓝藻和其他隐花植物有较高的叶绿素含量（Li *et al.*，2010a），因此是BSC群落生物量的主要贡献者。这一发现意味着BSC在全球变化背景下容易发生组成、结构和功能等多层次的改变，甚至导致从一种演替阶段向另一种演替阶段发展，如蓝藻为优势的BSC和地衣为优势的BSC相互转变，或蓝藻为优势的BSC和藓类为优势的BSC相互转变，以及地衣BSC与藓类BSC之间的相互替代。

尤其是，当区域降水增加时，有利于藓类为优势的BSC发展；相反，在降水量减少且地表有较大干扰时，有利于以蓝藻为优势的BSC发展。当地表干扰较小或土表相对稳定时，则有利于以地衣为优势的BSC发展。事实证明，BSC对干扰的响应是十分敏感的（Belnap and Lange，2003；Csotonyi and Addicott，2004；Dougill and Thomas，2004；Eldridge *et al.*，2006）。于是，我们就在区域尺度上提出了以下BSC物种多样性、分布格局及其相互转变的理论模型。这种相互转变或相互替代的关系主要取决于气候（降水）、土壤（土壤质地）、植被覆盖（一年生植物和多年生植物的盖度）和干扰（放牧强度强弱、人为交通碾压、地表破坏、踩踏和采集发菜等）。

如图1-5所示，在不考虑干扰的情况下，当全球变化正如许多气候模型预测的那样，即干旱、半干旱区降水增加时，地衣结皮将会被藓类结皮取代，BSC群落组成中藓类将具有更高的多样性，且无论在盖度还是生物量方面，藓类将占据绝对的优势，BSC的多功能将主要受到多样的藓类所调控；当降水量在一定的时间尺度上变异不大时，或在给定的降水条件下，表土层稳定且具有较高黏粒、粉粒含量时，地表将支撑各类地衣的发展，被以多样地衣为主的BSC所占据；当表层土壤质地相对较粗或土壤质地由细变粗时，地表多支撑多样蓝藻拓殖所形成的BSC。另外，当发生干扰或强干旱胁迫导致多年生植物盖度下降时，演替后期阶段的藓类结皮将被演替初期阶段的蓝藻结皮取代，但是适当的干扰、适中的植被覆盖和降水变化范围较小均有利于多样BSC组成种共存的混生结皮（mixed crusts）发展。

图1-6、图1-7和图1-8分别为BSC群落中常见藻类、地衣和藓类。

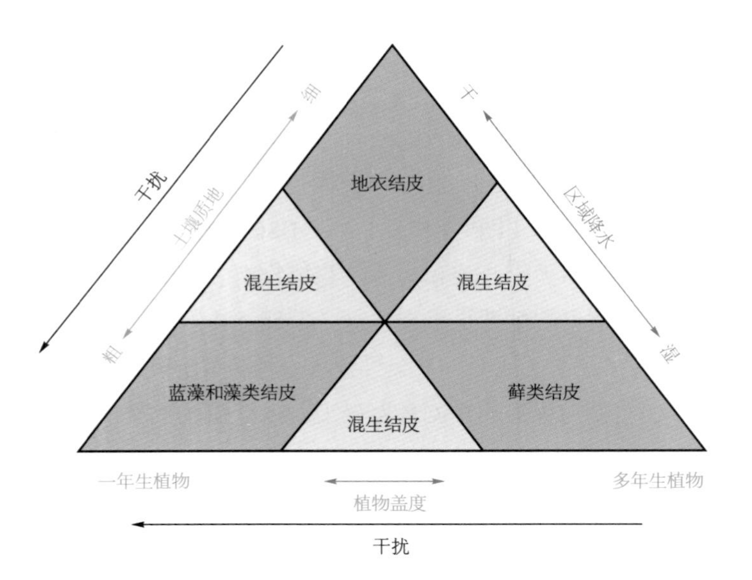

图 1-5　区域尺度上 BSC 组成种分布格局的变化框图。当地表受干扰较小或无干扰时，BSC 群落中往往形成地衣或藓类，或两者为优势的混生种类组成的分布格局。当区域降水增加时，地衣在群落中的优势地位将被藓类所取代，形成以多样藓类为优势的藓类结皮；反之，则有利于地衣为优势的 BSC 存在。当干扰增加时，藓类和地衣种的丰富度均会降低，最终有利于 BSC 中蓝藻和其他荒漠藻类繁衍，形成以蓝藻为优势的 BSC；当干扰去除后，多年生植物盖度的增加有利于藓类的发育和种的丰富度增加，最终蓝藻为优势的 BSC 被藓类为优势的 BSC 所取代。从土壤质地的变化来看，去除干扰，表土层得到长时间稳定，增加了质地组成中的黏粒和粉粒百分含量，即质地由粗变细，地衣种的丰富度将增加，蓝藻结皮将被稳定的地衣结皮所替代

Figure 1-5　The diagram of distribution pattern variations of cryptogams in BSC at a regional scale. When soil surface was subjected to no disturbance or slight disturbance, BSC community were mainly composed of lichens, or mosses, or lichen-moss predominant mixed crusts. Once precipitation enhanced, lichens's dominance within BSC community will be replaced by mosses, and moss-dominated crust community consisting of various mosses will be formed. Otherwise, lichens will be the dominant component of BSC community. When disturbance intensity increased, species richness of mosses and lichens decreased; this would result in a condition in favour of cyanobacteria and other desert algae, and helps to form a cyanobacteria-dominated crust community. When disturbance was removed, coverage increased of perennial plants species are beneficial for developing and increasing richness of mosses. From the perspective of the change of soil texture, removal of disturbance will make surface soil layer stabilize at a long time and increase the clay and silt content of surface soil. Namely, soil texture from coarse to fine, lichens species richness will increase and cyanobacteria crust will be replaced by a stable lichen crust.

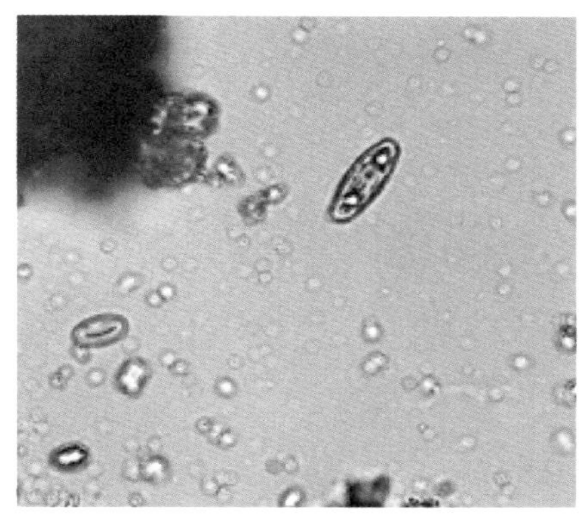

舟形藻属 *Navicula* sp. (400×)

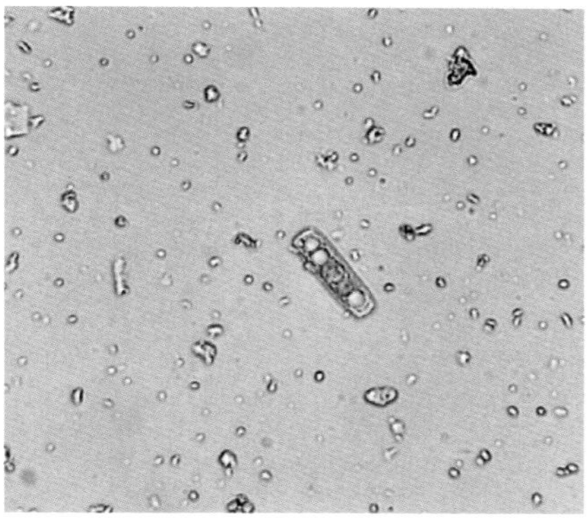

舟形藻属 *Navicula* sp. (400×)

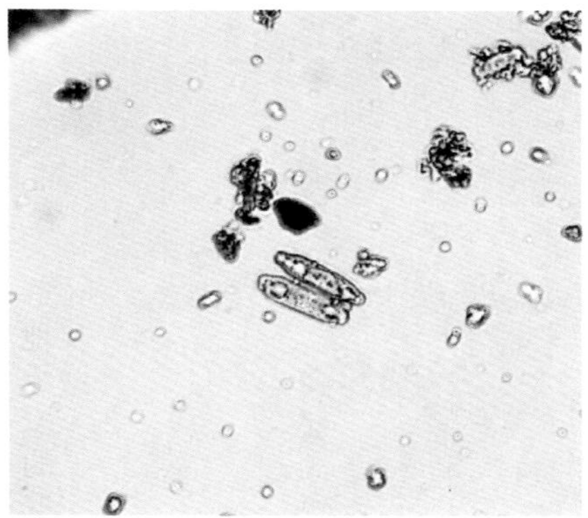

舟形藻属 *Navicula* sp. (400×)

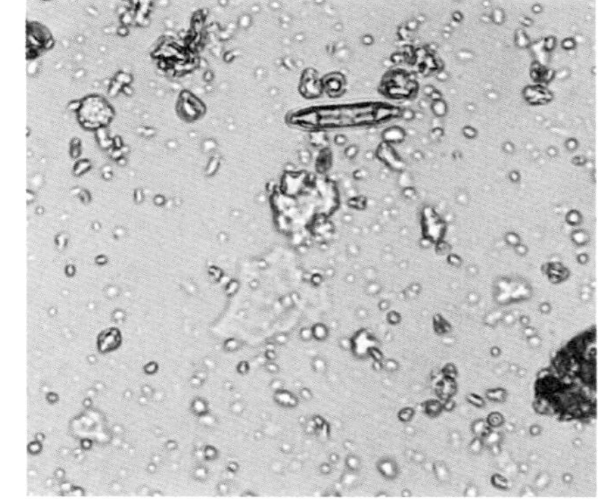

舟形藻属 *Navicula* sp. (400×)

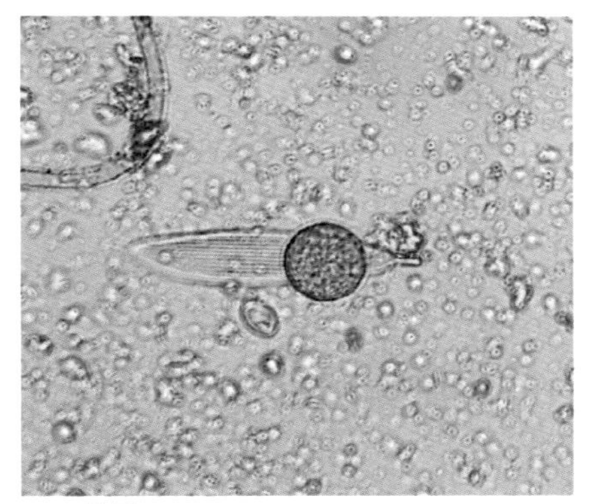

羽纹藻属 *Pinnularia* sp. (400×)

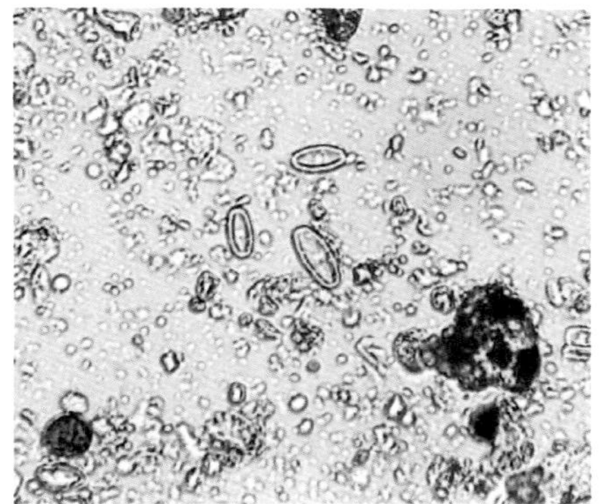

辐节藻属 *Stauroneis* sp. (400×)

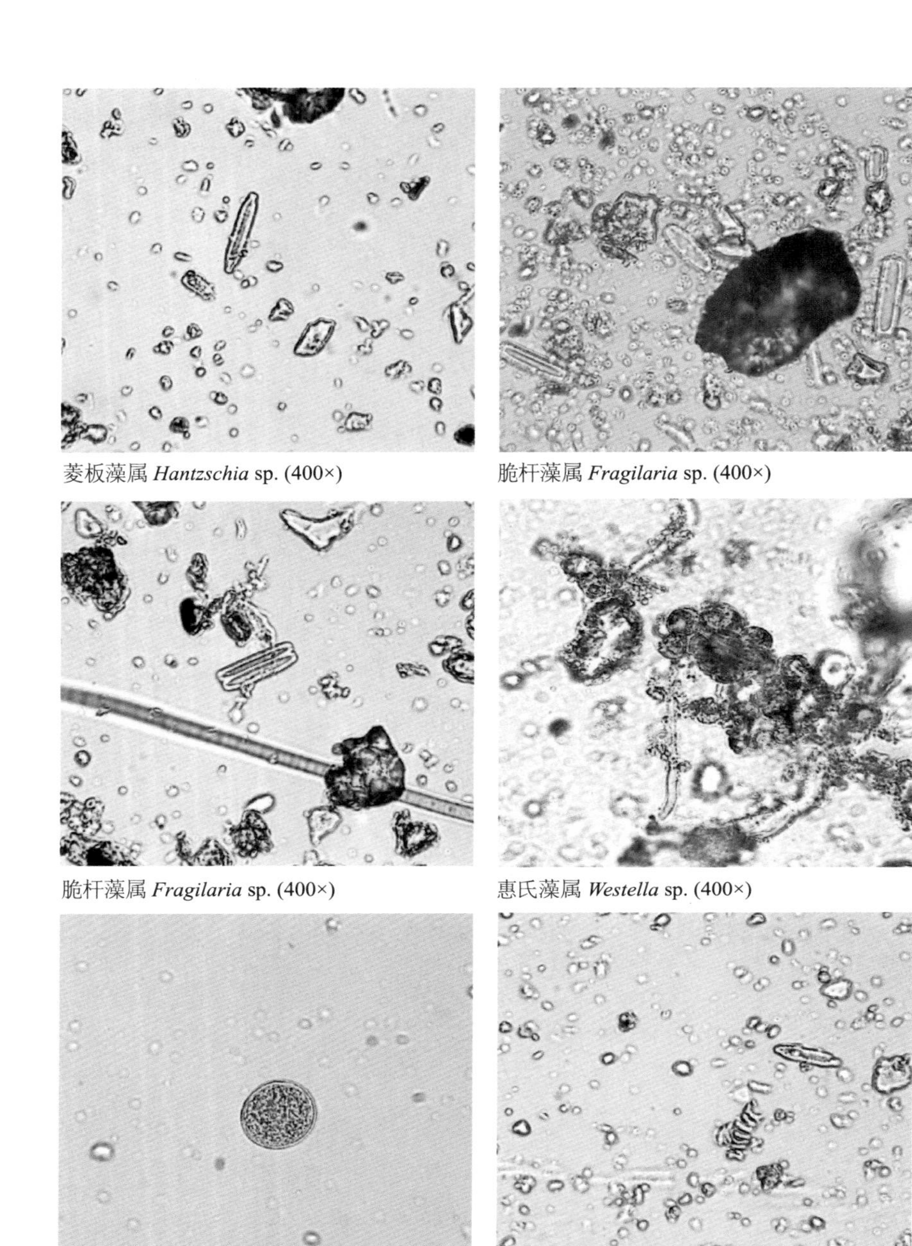

菱板藻属 *Hantzschia* sp. (400×)

脆杆藻属 *Fragilaria* sp. (400×)

脆杆藻属 *Fragilaria* sp. (400×)

惠氏藻属 *Westella* sp. (400×)

多芒藻属 *Golenkinia* sp. (400×)

栅藻属 *Scenedesmus* sp. (400×)

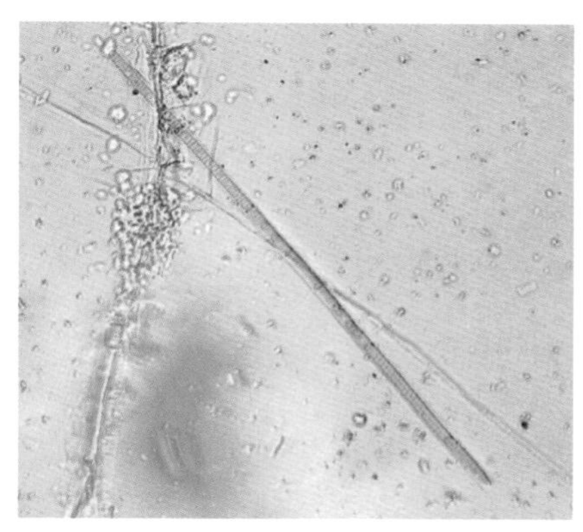

鱼腥藻属 *Anabaena* sp. (400×)

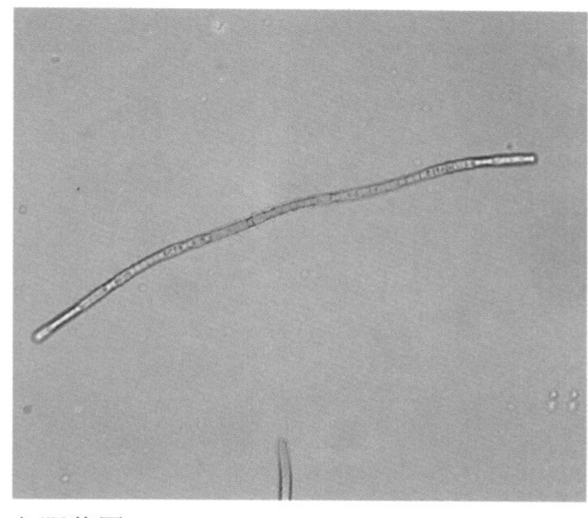

鱼腥藻属 *Anabaena* sp. (400×)

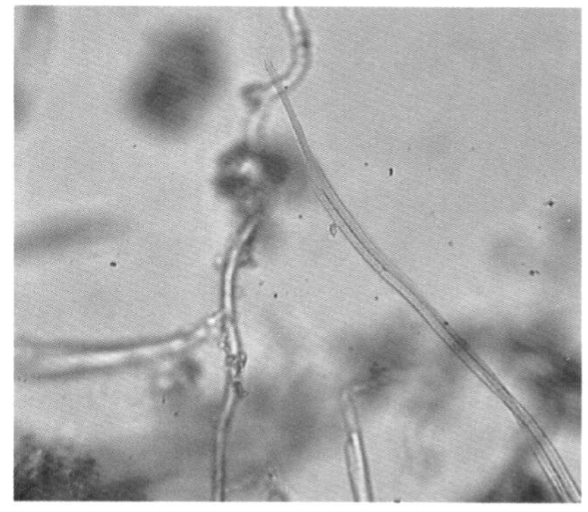

瘦鞘丝藻属 *Leptolyngbya* sp. (400×)

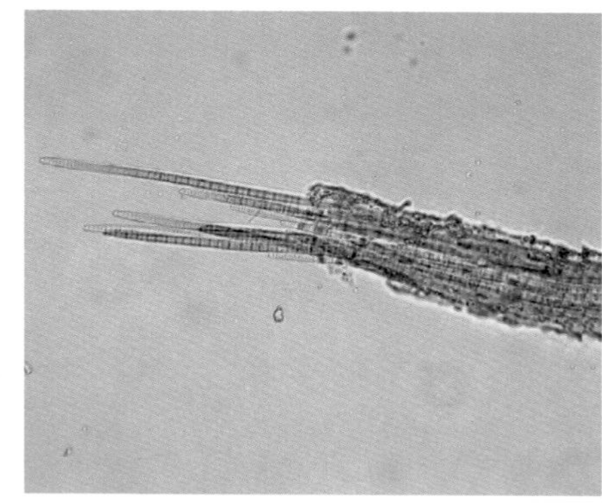

微鞘藻属 *Microcoleus* sp. (400×)

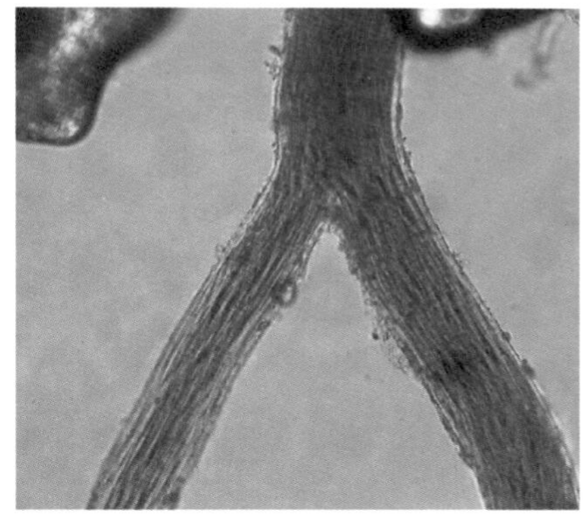

微鞘藻属 *Microcoleus* sp. (400×)

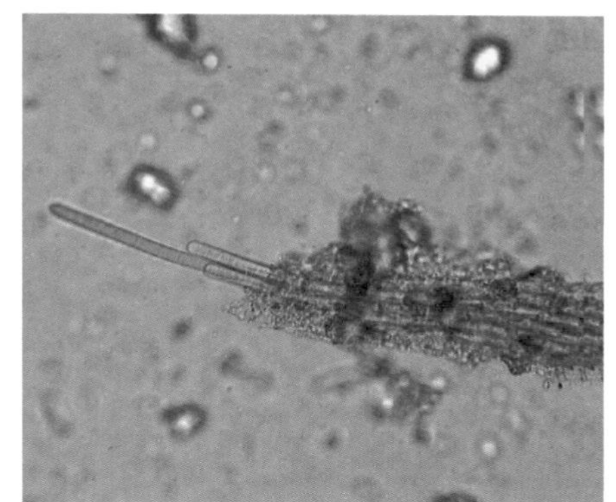

微鞘藻属 *Microcoleus* sp. (400×)

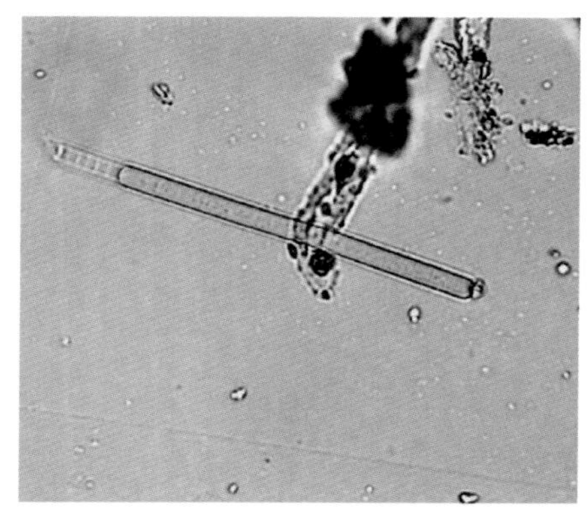

席藻属 *Phormidium* sp. (400×)

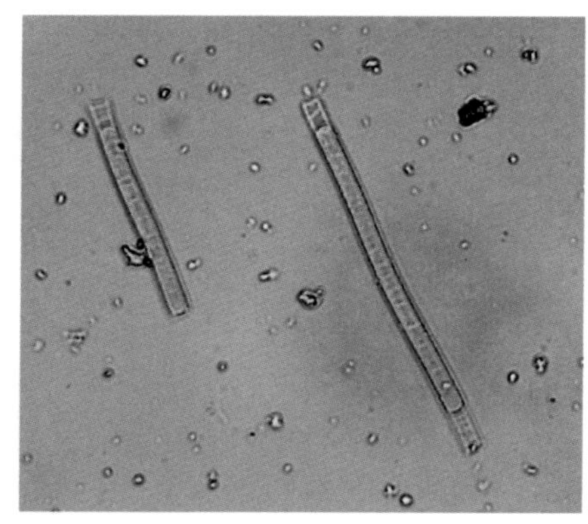

席藻属 *Phormidium* sp. (400×)

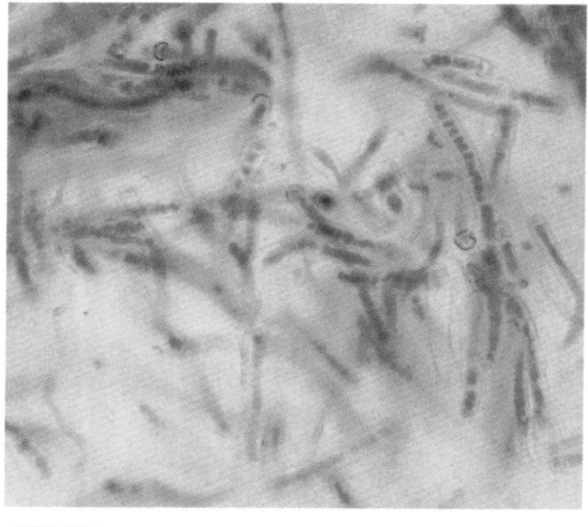

眉藻属 *Calothrix* sp. (400×)

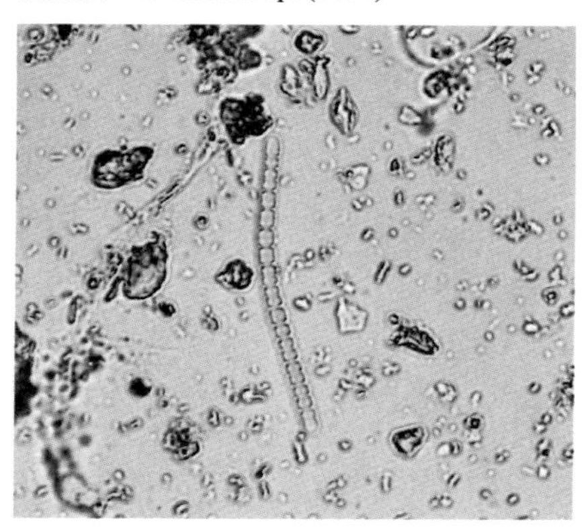

束丝藻属 *Aphanizomenon* sp. (400×)

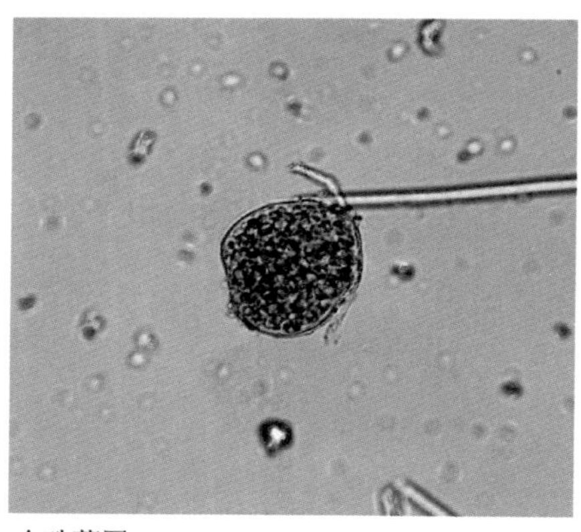

念珠藻属 *Nostoc* sp. (400×)

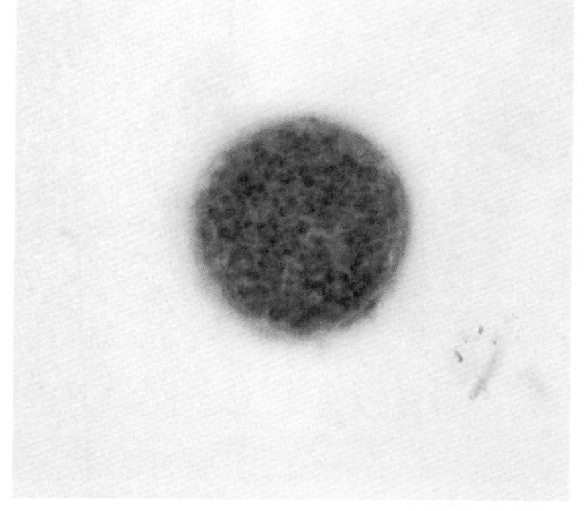

念珠藻属 *Nostoc* sp. (400×)

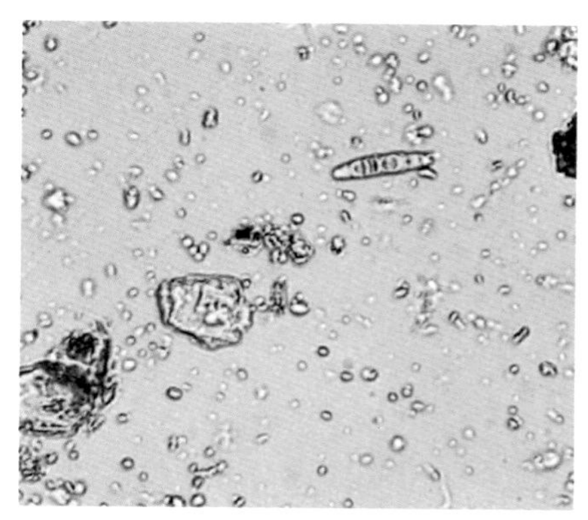

筒孢藻属 *Cylindrospermum* sp. (400×)

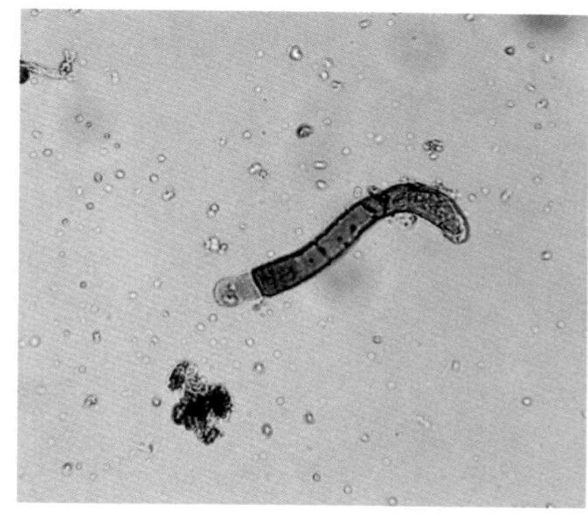

伪枝藻属 *Stigonema* sp. (400×)

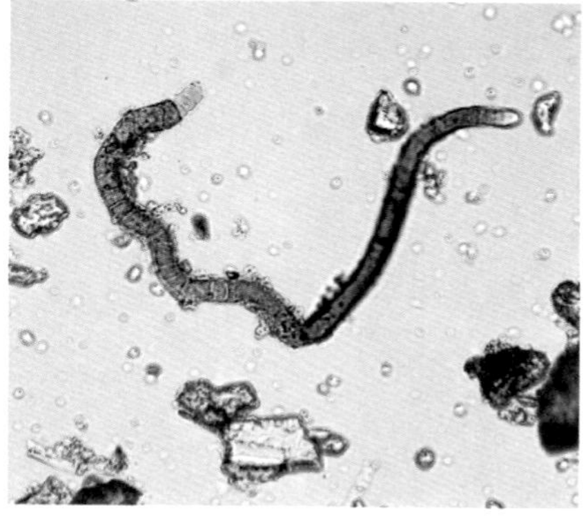

伪枝藻属 *Stigonema* sp. (400×)

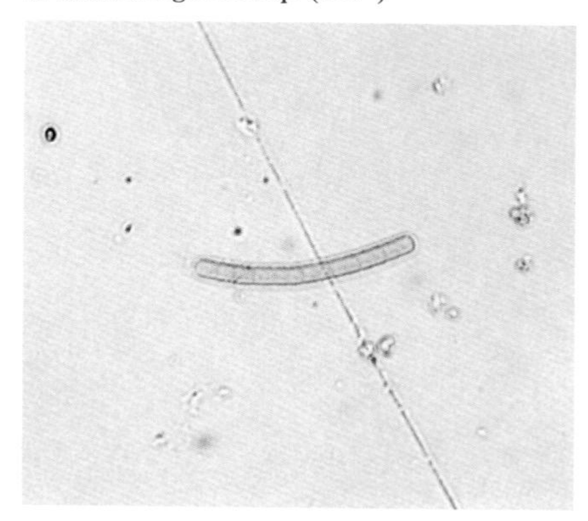

鞘丝藻属 *Lyngbya* sp. (400×)

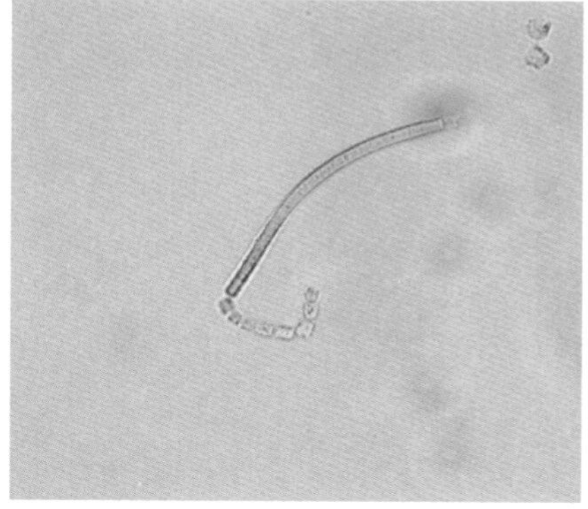

鞘丝藻属 *Lyngbya* sp. (400×)

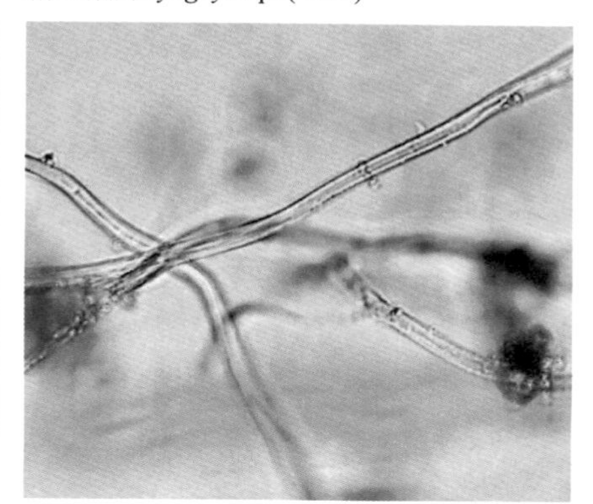

鞘丝藻属 *Lyngbya* sp. (400×)

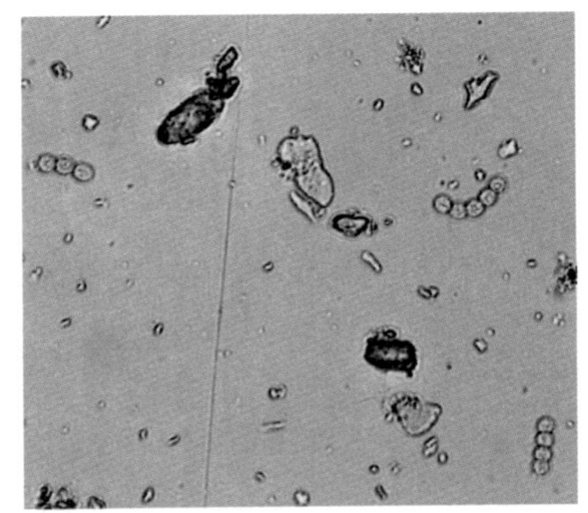

色球藻属 *Chroococcus* sp. (400×)

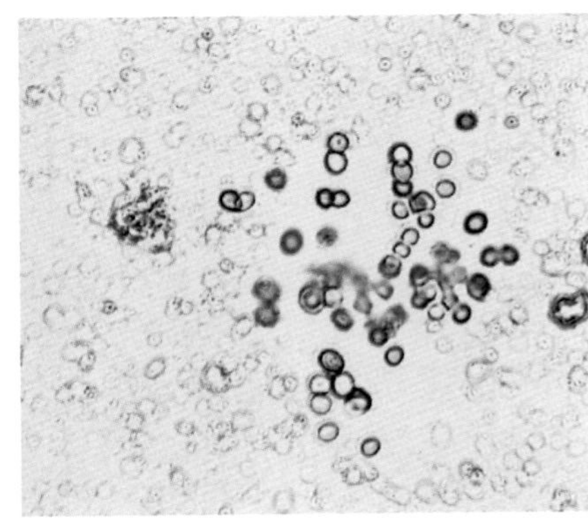

色球藻属 *Chroococcus* sp. (400×)

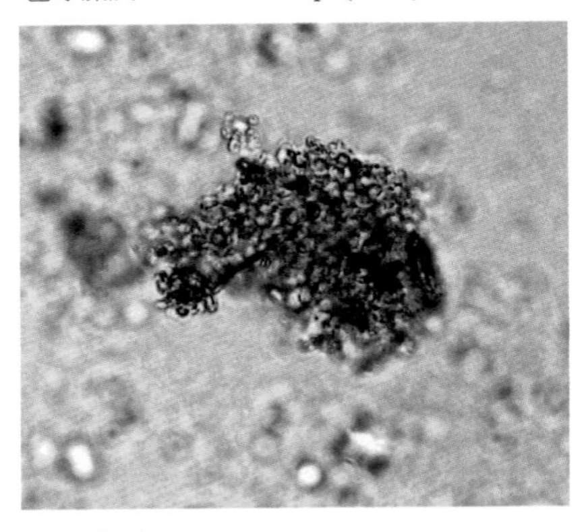

粘球藻属 *Gloeocapsa* sp. (400×)

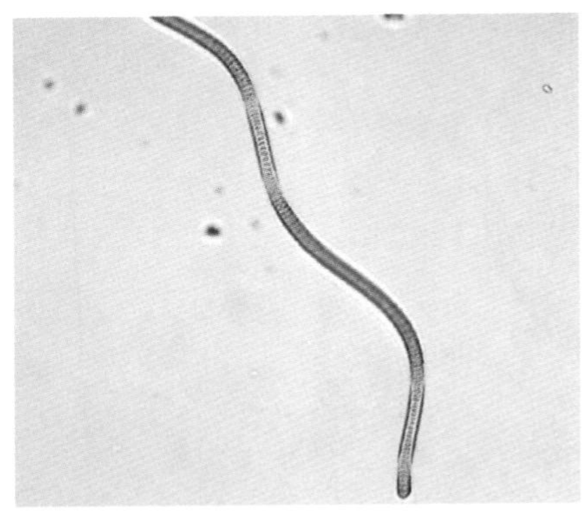

螺旋藻属 *Spirulina* sp. (400×)

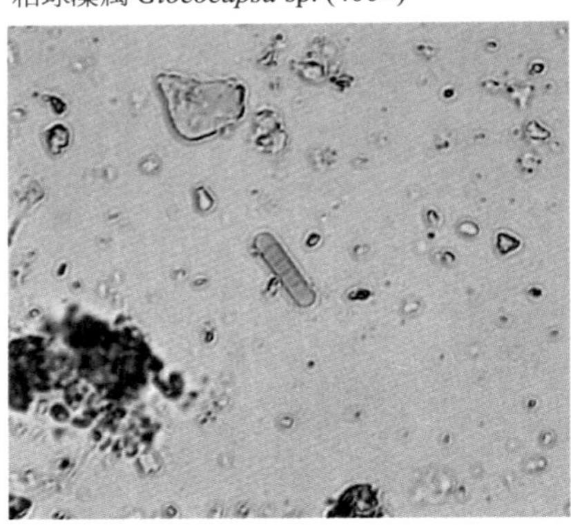

颤藻属 *Oscillatoria* sp. (400×)

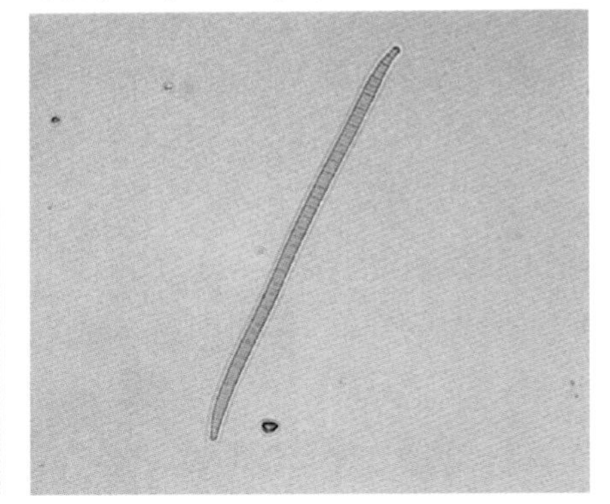

颤藻属 *Oscillatoria* sp. (400×)

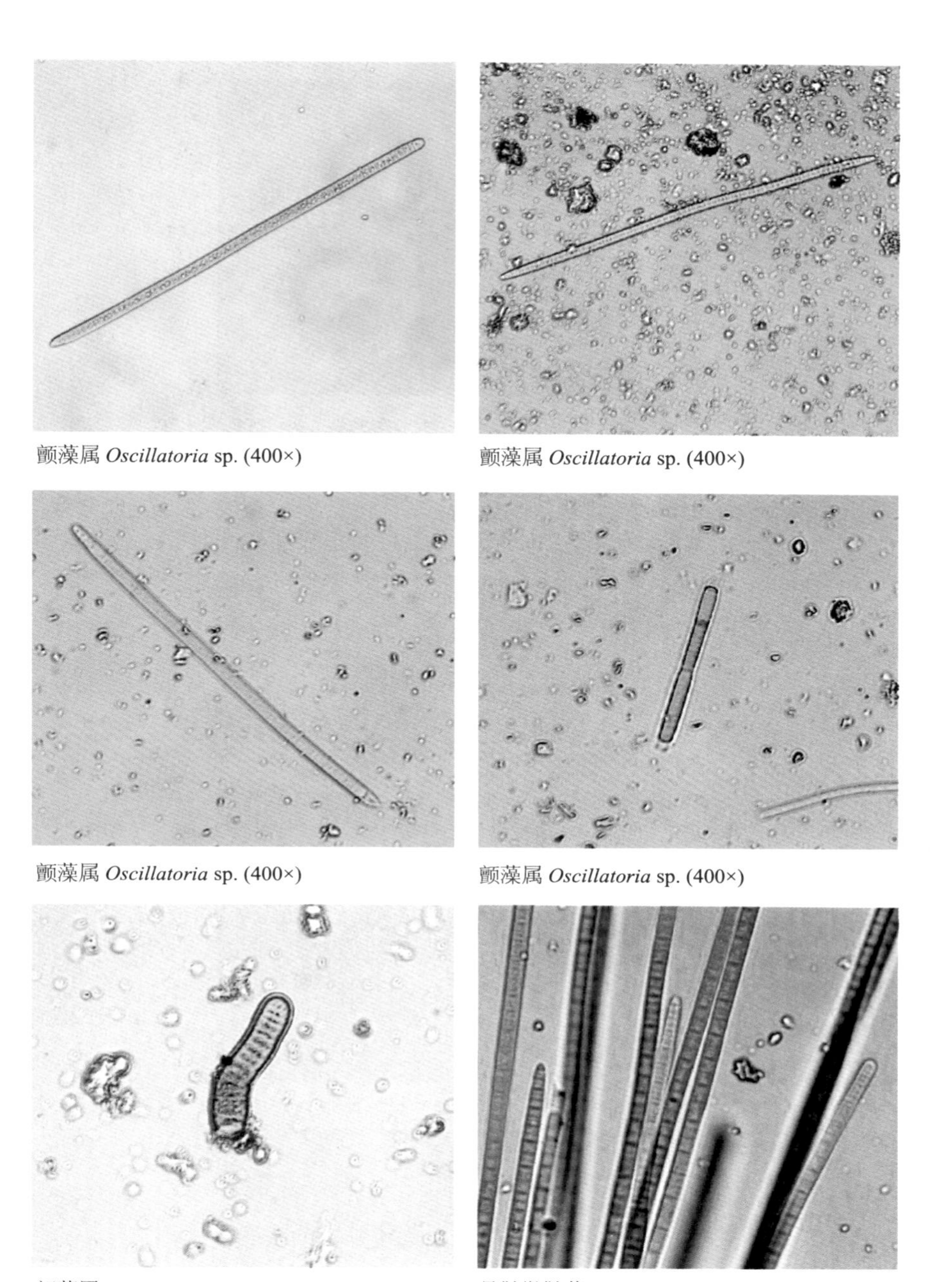

颤藻属 *Oscillatoria* sp. (400×)

颤藻属 *Oscillatoria* sp. (400×)

颤藻属 *Oscillatoria* sp. (400×)

颤藻属 *Oscillatoria* sp. (400×)

颤藻属 *Oscillatoria* sp. (400×)

具鞘微鞘藻 *Microcoleus vaginatus*

图 1-6　BSC 群落中常见藻类（含蓝藻）属的显微结构

Figure 1-6　Microscopic structure of the algae and cyanobacteria genus in BSC communities

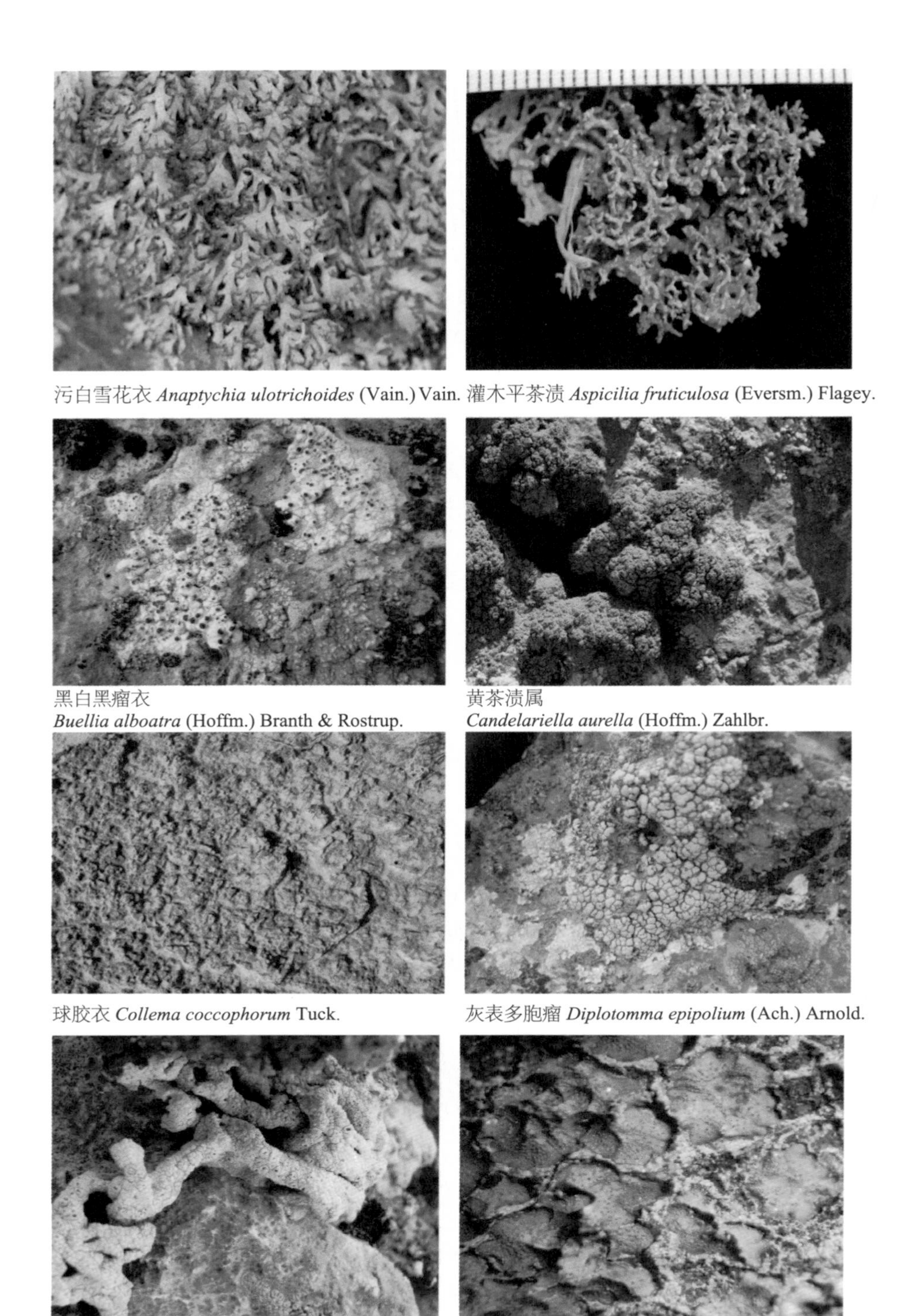

污白雪花衣 *Anaptychia ulotrichoides* (Vain.) Vain.

灌木平茶渍 *Aspicilia fruticulosa* (Eversm.) Flagey.

黑白黑瘤衣
Buellia alboatra (Hoffm.) Branth & Rostrup.

黄茶渍属
Candelariella aurella (Hoffm.) Zahlbr.

球胶衣 *Collema coccophorum* Tuck.

灰表多胞瘤 *Diplotomma epipolium* (Ach.) Arnold.

藓生双缘衣 *Diploschistes muscorum* (Scop.) R. Sant.

石果衣属 *Endocarpon adsurgens* Vain.

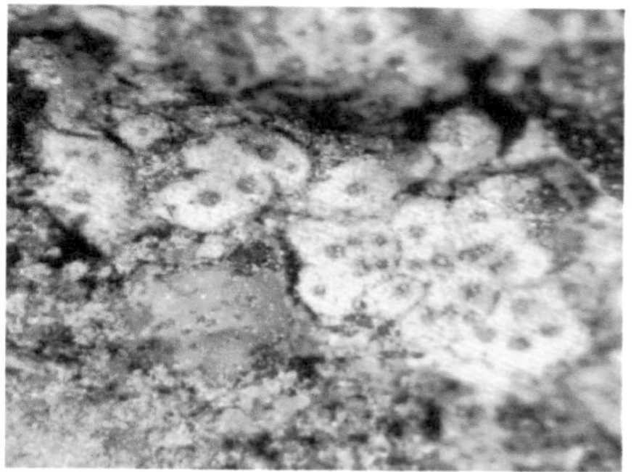

石果衣属 *Endocarpon crystallinum* Wei & Yang.

中华石果衣原种 *Endocarpon sinense* H. Magn.

石果衣属
Endocarpon lepidallum Nyl.

黑边石果衣
Endocarpon nigromarginatum H. Harada.

荒漠拟橙衣 *Fulgensia desertorum* (Tomin) Poelt

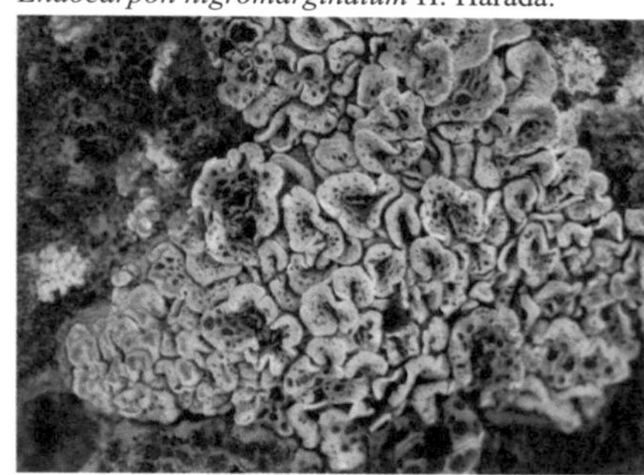

糙聚盘衣 *Glypholecia scabra* (Pers.) Müll. Arg.

碎茶渍 *Lecanora argopholis* (Ach.) Ach.

绳鲚茶渍 *Lecanora garovaglii* (Koerb.) Zahlbr.

小网衣属
Lecidella stigmatea (Ach.) Hertel & Leuckert.

裂片茶渍衣属
Lobothallia praeradiosa (Nyl.) Hafellner.

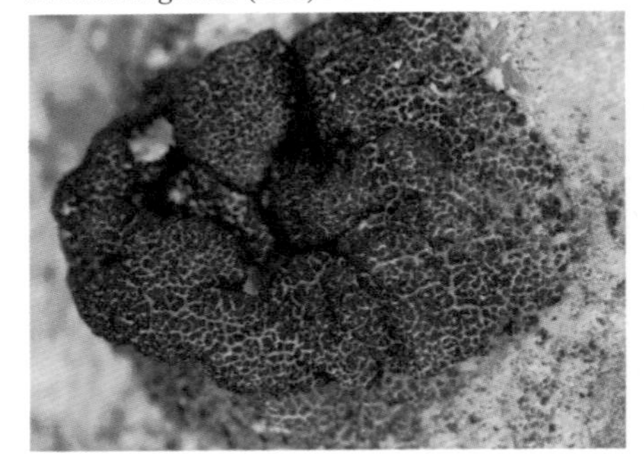

盾衣属 *Peltula cylindrical*

鳞网衣属 *Psora crenata* (Taylor.) Reinke.

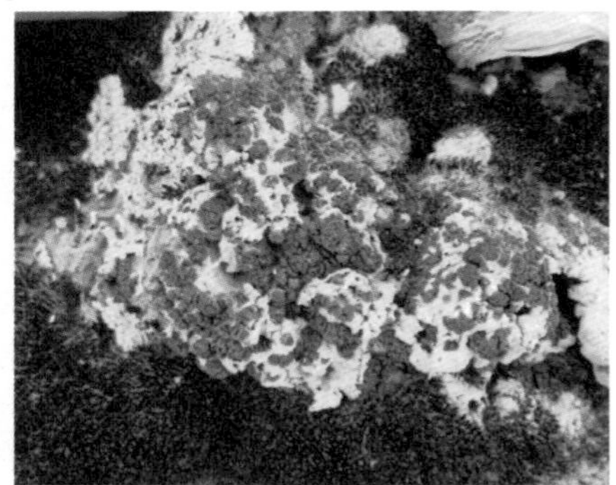

红鳞网衣 *Psora decipiens* (Hedw.) Hoffm.

条斑鳞茶渍 *Squamarina lentigera* (Weber) Poelt

暗色泡磷衣 *Toninia tristis* (Th. Fr.) Th. Fr.

淡腹黄梅
Xanthoparmelia Mexicana (Gyeln.) Hale.

旱黄梅
Xanthoparmelia camtschadalis (Ach.) Hale.

丽石黄衣 *Xanthoria elegans* (Link.) Th. Fr.

椰子黑盘衣 *Pyxine cocoes* (Sw.) Nyl.

图 1-7　BSC 群落中常见的地衣
Figure 1-7　The common lichens in BSC communities

真藓 *Bryum argenteum* Hedw.

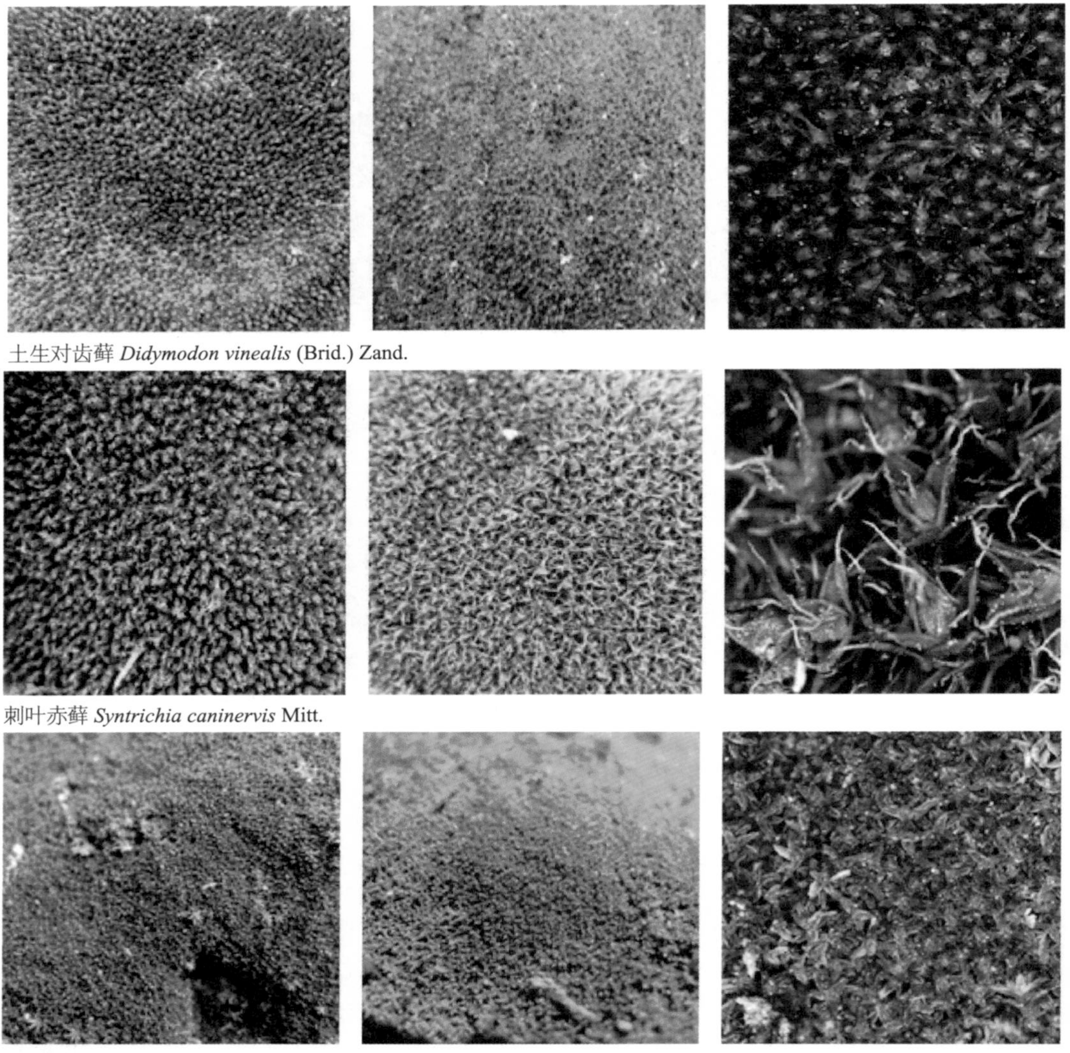

土生对齿藓 *Didymodon vinealis* (Brid.) Zand.

刺叶赤藓 *Syntrichia caninervis* Mitt.

净口藓 *Gymnostomum calcareum* Nees. et Hornsch.

阔叶紫萼藓 *Grimmia laevigata* (Brid.) Brid.

卷叶墙藓 *Tortula atrovirens*

缨齿藓菱形变种 *Jaffueliobryum wrightii* (Sull.) Thér. var. *rhombicum* X L Bai & Sarula

图 1-8　BSC 群落中常见的藓类。以上照片从左到右，依次为干燥、湿润和放大 10 倍处理

Figure 1-8　The common mosses in BSC communities. From left to right,moss under natural dry,wetting and the microscopic structure with enlarging 10 times

附表 1-1　中国主要沙区研究样地 BSC 群落中主要隐花植物种的分布

Table S1-1　Record of cryptogamic species in the crustal communities in the different deserts, sandland and western Loess Plateau of China

研究区样地	蓝藻和其他藻类	地衣	藓类
科尔沁沙地 (样地 1)	*Anabaena azotica* Ley (+++) *Asterocapsa purpurea* (Jao) Chu (+) *Characium angustum* A. Bruan (+) *Chlamydomonas* sp. (+) *Chlorella vulgaris* Beij. (++) *Chlorococcum humicola* (Näg.) Rab. (+) *Microcystis densa* G. S. West (++) *Nostoc* sp. (+++) *Pediastrum boryanum* (Turp.) Menegh. (+) *Synechocystis aquatilis* Sauv. (+) *Synechocystis crassa* Woronichin (+)	*Acarospora schleicheri* (Ach.) Massal. (+) *Fulgensia bracteata* (Hoffm.) Räsänen (+++) *Glypholecia scabra*(Pers.) Müll. Arg. (+) *Leptogium lichenoides* (l.) Zahlbr. (++) *Peltigera leucophlebia* (Nyl.) Gyeln. (++)	*Abietinella abietina* (Hedw.) Fleisch. (++) *Amblystegiun serpens* (Hedw.) B.S.G. var. *serpens* (+) *Anoectemgiun aestivum* (Hedw.) Mitt. (+) *An. stracheyanum* Mitt. (++) *Barbula unguiculata* var. *unguiculata* Hedw. (++) *Bra chythecium albicans* (Hedw.) B.S.G. (+) Bra. *plumosum* (Hedw.) B.S.G. var. *nitdum* Tak. (+) *Bra. perminusculum C.* Mull. (+) *Bryoerythrophyllum recurvirostre*(Hedw.) Chen (+) *Bryum arcticum* (R. Brown) B.S.G. (+) *Bryum alpinum* Huds ex With. (+) *Bryum argenteum* Hedw. (+++) *Bryum dichotomum* Hedw. (+) *Bryum kunzei* Hoppe et Hornsh (++) *Bryum lonchocaulon* C. Müll. (+) *Bryum pseudotriquetrum* (Hedw.) Gaertn. (+) *Bryum pallescens* Schleich ex Schwägr. (++) *Bryum thomsonii* Mitt. (++) *Bryum uliginosum* (Brid.) B.S.G. (+++) *Campylium chrysophyllum*(Brid.) Lange (+) *Ca. hispidulum* (Brid.) Mitt. (+) *Cratoneuron filicinum* (Hedw.) Spruc. (+) *Crossidium crassinerve* (De Not.) Jur. (+) *Desmatodon leucostoma* (R. Br.) Berggr. (+) *Didymodon vinealis* (Brid.) Zand. (++) *Drepanocladus aduncus* (Hedw.) Warnst. (++) *Entodon caliginosus* (Mitt.) Jaeg. (+) *E. concinnus* (De Not.) Par. (++)

续表

研究区样地	蓝藻和其他藻类	地衣	藓类
科尔沁沙地 (样地 1)			*Homomallium incurvatum* (Brid.) Loesk. (++) *Lindbergia brachyptera* (Mitt.) Kindb. (+) *Phascum cuspidatum* Schreb. ex Hedw. (++) *Pterygoneurum subsessile* (Brid.) Jour. (++) *Pottia intermedia* (Turn.) Fuernr. (+) *Pylaisiella brotheri* (Besch.) Iwats. et Nog. (+) *Rauiella fujisana* (Par.) Reim. (+) *Tortula mucronifolia* Schwägr. (++) *Weissia controversa* Hedw. (++)
黄土高原西部 荒漠草地 (样地 2)	*Anabaena azotica* Ley (+++) *Bracteococcus* sp. (+) *Microcoleus vaginatus* (Vauch.) Gom. (++) *Moncilia* sp. (++) *Myrmecia* sp. (++) *Nostoc* sp. (+++) *Hydrocoleum* sp. (+) *Lyngbya digueti* Gom. (++) *Oscillatoria nigra* Vauch. (+++) *Phormidium mucicola* Naum. (++) *Synechococcus aeruginosus* Näg. (+) *Tolypothrix* sp. (++)	*Collema coccophorum* Tuck. (+++) *C. tenax* (Sw.) Ach. Em. Degel. (++) *Diploschistes muscorum* (Scop.) R.Samt. (++) *Endocarpon aridum* P. M. McCarthy (+++) *E. pallidum* Ach. (+) *E. pusillum* Hedw. (+) *E. rogersii* P. M. McCarthy (+) *E. rosettum* (+) *E. simplicatum* (Nyl.) Nyl. (++) *Fulgensia bracteata* (Hoffm.) Räsänen (+++) *Gyalidea asteriscus* (Anzi) Aptroot & Lücking ssp. *Gracilispora* J. Yang & J. C. Wei (+) *Heppia lutosa* (Ach.) Nyl. (++) *Buellia mongolica* H. Magn. (+) *Placidium rufescens* (Ach.) A. Massal. (+) *Psora decipiens* (Hedw.) Hoffm. (+++) *Peltigera collina* (Ach.) Schrader (+) *Toninia sedifolia* (Scop.) Timdal. (++)	*Aloina rigida* Var. *rigida* (Hedw.) Limpr. (++) *Barbula fallax* Hedw. (++) *Bryum algovicum* Sendt. (+) *Br. argenteum* Hedw. (++) *Br. caespiticium* L. ex Hedw. (++) *Br. funkii* Schwägr (+) *Crossidium crassinerve* (De Not.) Jur. (+) *Didymodon constrictus* (Mitt.) Saito (+++) *D. nigrescens* (Mitt.) Saito (+++) *Gymnostomum calcareum* Nees. et Hornsch. (++) *Plagiobryum zierii* (Hedw.) Lindb. (+) *Pterygoneurum subsessile* (Brid.) Jour. (+)

续表

研究区样地	蓝藻和其他藻类	地衣	藓类
毛乌素沙地 (样地3)	*Anabaena azotica* Ley (++) *Chroococcus minutus* (Kuetz.) Näg. (+++) *Diatoma vulgare* Bory (++) *Gloeocapsa arenaria* (Has.) Rebenh. (+) *Gl. aeruginosa* (Carm.) Kuetz. (+) *Gl. atrata* (Turp.) Kuetz. (+) *Gomphonema constrictum* Ehr. (+) *Lyngbya digueti* Gom. (++) *Microcoleus vaginatus* (Vauch.) Gom. (++) *Penium cruciferum* (DeBary) Wittr. (++) *Penium* sp. (+) *Scytonema incrasstum* Jao (++) *Sc. stuposum* (Kuetz.) Born. (++) *Sc. javanicum* (Kuetz.) Born. (++) *Synechocystis crassa* Woronichin (++)	*Collema coccophorum* Tuck. (+++) *C. tenax* (Sw.) Ach. Em. Degel. (+++) *Diploschistes muscorum* (Scop.)R. Sant (++) *Psora decipiens* (Hedw.) Hoffm. (+++)	*Aloina rigida* var. *rigida* (Hedw.) Limpr. (+) *A. brevirostris* (Hook. & Grev.) Kindb. (++) *A. cornifolia* Delgad. (+) *A. obliquifolia* (C. Müll.) Broth. (+) *Barbula indica* (Hook.) Spreng. (+) *Ba. unguiculata* var. *unguiculata* Hedw. (+) *Bryoerythrophyllum recurvirostrum* (Hedw.) Chen (+) *Bryum argenteum* Hedw. (+++) *Bryum caespiticium* L. ex Hedw. (++) *Ceratodon purpureus* var. *purpureus* (Hedw.) Brid. (++) *Crossidium chloronotos* (Brid.) Limpr. (+) *Cr. crassinerve* (De Not.) Jur. (++) *Cr. squamigerum* (Viv.) Jur. (++) *Dicranella varia* (Hedw.) Schimp. (++) *Didymodon acutus* (Brid.) Saito (+) *Did. constrictus* (Mitt.) Saito (++) *Did. icmadophilus* (Schimp. ex C. Müll.) Saito (++) *Did. nigrescens* (Mitt.) Saito (+++) *Did. perobtusus* Broth. (++) *Did. rigidulus* Hedw. (+) *Funaria hygrometrica* Hedw. (++) *Gymnostomum calcareum* Nee. et Hornsch. (+) *Hilpertia velenovskyi* (Schiffn.) Zand. (++) *Microbryum rectum* (With.) Zand. (++) *Pterygoneurum subsessile* (Brid.) Jour. (+) *P. ovatum* (Hedw.) Dix. (++) *Tortula atrovirens* (Sm.) Lindb. (+++) *T. cernua* (Hueb.) Lindb.(++) *T. desertorum* Broth. (++) *T. randii* (Kenn.) Zand. (+) *Weissia controversa* Hedw. (+)

续表

研究区样地	蓝藻和其他藻类	地衣	藓类
腾格里 - 阿拉善沙区（样地 4）	*Anabaena azotica* Ley (++)	*Collema coccophorum* Tuck. (+++)	*Aloina brevirostris*(Hook. & Grev.) Kindb.(+)
	Chlamydomonas sp. (+)	*C. tenax* (Sw.) Ach. Em. Degel. (++)	*A.obliquifolia* (C. Müll.) Broth. (+)
	Chlorella vulgaris Beij. (+)	*Diploschistes muscorum* (Scop.) Hoffm. (+++)	*A. rigida* (Hedw.) Limpr var. *rigida*. (+)
	Chlorococcum humicola (Näg.) Rab. (++)	*Endocarpon aridum* P. M. McCarthy (++)	*Barbula ditrichoides* Broth. (+++)
	Chroococcus epiphyticus Jao (+)	*E. crystallinum* Wei & Yang (+)	*Bryoerythrophyllum recurvirostrum* (Hedw.)Chen (+)
	Cyambella sp. (+)	*E. pallidum* Ach. (+)	*Bryum argenteum* Hedw. (+++)
	Desmococcus olivaceus (Pers. ex Ach.) Laundon (+)	*E. pusillum* Hedw. (+)	*Didymodon constrictus* (Mitt.) Saito (++)
	Diatoma vulgare var. *Ovalis* (Frick.) Hust. (++)	*E. rogersii* P. M. McCarthy (+)	*D. nigrescens* (Mitt.) Saito (+++)
	Euglena sp1. (++)	*E. rosettum Amar* Singh & Upreti (+)	*D. perobtusus* Broth. (++)
	Euglena sp2. (++)	*E. simplicatum* (Nyl.) Nyl. (++)	*D. tectorum* (C. Mull.) Saito (++)
	Fragilaria intermedia Grun. (+)	*E. sinense* H. Magn. (+)	*Pterygoneurum subsessile* (Brid.) Jour. (+)
	Gloeocapsa sp. (+)	*Fulgensia bracteata* (Hoffm.) Räsänen (+++)	*Tortula bidentata* Bai. X. L.(+++)
	Gomphonema constrictum Ehr. (+)	*Heppia lutosa* (Ach.) Nyl. (++)	*T. desertorum* Broth. (++)
	Hantzschia amphioxys (Ehr.) Grun. (+)	*Buellia mongolica* H. Magn. (+)	
	Lyngbya crytoraginatus Schk. (+++)	*Placidium rufescens* (Ach.) A. Massal. (+)	
	Microcoleus vaginatus (Vauch.) Gom. (+++)	*Toninia sedifolia* (Scop.) Timdal. (++)	
	Navicula cryptocephala Kuetz. (+)		
	Nostoc flagelliforme Born. et Flah. (++)		
	Nostoc sp. (+)		
	No. commune (L.) Vauch. (+)		
	Palmellococcus miniatus Kuetz. Chod (++)		
	Pinnularia borealis Ehr. (++)		
	Phormidium tenue (Menegh.) Gom. (++)		
	Synechocystis pevalekii Ercegoa (+)		
	Scytonema javanicum(Kuetz.) Born. (++)		

续表

研究区样地	蓝藻和其他藻类	地衣	藓类
古尔班通古特沙漠（样地 5）	*Anabaena azotica* Ley (+++)	*Acarospora strigata* (Nyl.) Jatta. (++)	*Bryum argenteum* Hedw. (+++)
	Aphanocapsa delicatissima W. et G.S.West (+)	*Caloplaca songoricum* A. Abbas. (++)	*B. capillare* L. ex Hedw. (++)
	Calothrix sp. (++)	*Candeleriella aurella* (Hoffm.) Zahlbr. (+++)	*Crossidium chloronotos* (+)
	Chroococcus minutus (Kuetz.) Näg. (++)	*Catapyrenium* sp. (++)	*Grimmia anodon* B. S. G. (+)
	Ch. turgidus Näg. (+++)	*Collema tenax* (Sw.) Ach. Em. Degel. (+++)	*G. pulvinata* (Hedw.) Sm. (+)
	Ch. westii (W. West) Boye-Petersen (+)	*Co. tenax* var. *corallinum* (Massal.) Degel. (++)	*Tortula desertorum* Broth. (++)
	Clastidium sp. (++)	*Dimelaena oreina* (Ach.) Nonnon (+)	*T. muralis* Hedw. (+)
	Gloeocapsa sp. (+)	*Diploschistes muscorum* (Scop.) R. Sant (+++)	
	Gomphosphaeria sp. (++)	*Fulgensia bracteata* (Hoffm.) Räsänen. (++)	
	Homoeothrix juliana (Menegh.) Kirchn. (+)	*Lecanora argopholis* (Ach.) Ach. (++)	
	Hydrocoleum sp. (+)	*Lecidea* sp. (+)	
	Lyngbya martensiana Meneghini (+)	*Psora decipiens* (Hedw.) Hoffm. (+++)	
	L. gracilis Rabenh. (++)	*Xanthoria elegans* (Link.) Th. Fr. (++)	
	Microcoleus vaginatus (Vauch.) Gom. (+++)		
	Microcoleus paludosus (Kuetz.) Gom. (+++)		
	Microcystis sp. (+)		
	Myxosarcina sp. (+)		
	Nodularia spumigena Mertens (++)		
	Nostoc sp. (++)		
	Oscillatoria agardhii Gom. (++)		
	O. cortiana Menegh. (+)		
	O. formosa Bory (+)		
	O. limosa Ag. (++)		
	O. tenuis Ag. (++)		
	Phormidium foveolarum (Mont.) Gom. (+)		
	Porphyrosiphon sp. (+)		

续表

研究区样地	蓝藻和其他藻类	地衣	藓类
古尔班通古特沙漠（样地 5)	*Raphidiopsis* sp. (+) *Spirulina jenneri* (Stiz.) Geiller (++) *Symploca* sp. (+) *Synechococcus* sp. (+) *Synechococcus aeruginosus* Näg. (++) *Synechocystis parvus* Migula (++) *Xenococcus lyngbyae* Jao (+++)		
柴达木沙区（样地 6)	*Anabaena oscillarioides* var. *minor* Jao et Lee (++) *A. variabilis* (++) *Calothrix parietina* (Näg.) Thuret (++) *Chlorococcum humicola* (Näg.) Rab. (+) *Chroococcus epiphyticus* Jao (+++) *Chroococcus* sp.(++) *Gloeocapsa arenaria* (Has.) Rebenh. (++) *G. punctata* Näg. (++) *Homoeothrix juliana* (Menegh.) Kirchn. (++) *Hydrocoleum coccineus* Gom. (+) *Lyngbya digueti* Gom. (+++) *Microcoleus tenerrimus* Gom. (++) *Mi. vaginatus*(Vauch.) Gom. (+++) *Myxosarcina chroococcoides* Printz (+++) *My. concinna* Printz (++) *Navicula cryptocephala* Kuetz. (+) *Nodularia harveyana* var. *sphaerocarpa* (Born. et Flah.) Elenk. (+) *Nostoc flagelliforme* Borm. (+++) *Nos. humifusum* Carm.(+++) *Nos. piscinale* Kuetz. (+) *Nos. sphaericum* Vauch. (+)	*Collema coccophorum* Web. (+++) *C. tenax* (Sw.) Ach.Em. Degel. (+) *Diploschistes muscorum* (Scop.) R. Sant (++) *Psora decipiens* (Hedw.) Hoffm. (++) *Endocarpon* sp. (+) *Toninia sedifolia* (Scop.) Timdal. (+)	*Bryum argenteum* Hedw. (+++) *B. caespiticium* L. ex Hedw. (+) *Didymodon constrictus* (Mitt.) Saito (++) *Pterygoneurum subsessile* (Brid.) Jour. (++)

续表

研究区样地	蓝藻和其他藻类	地衣	藓类
柴达木沙区（样地 6）	*Nos. spongaeforme* Ag. (+) *Phormidium foveolarum* (Mont.) Gom. (+++) *Porphyrosiphon* sp. (+) *Schizothrix undulatus* Gom. (+++) *Scytonema javanicum* (Kuetz.) Born. (+) *Symploca* sp. (++) *Synechococcus aeruginosus* Näg. (+) *Synechocystis crassa* Woronichin (++) *Tolypothrix* sp. (++)		

注： +++ 为优势种，++ 为常见种，+ 为少有种。

附表 1-2　BSC 的物种丰富度、生物量在景观尺度上与环境因子之间的 Pearson 相关性分析

Table S1-2　Correlation matrix of species richness,biomass BSC and environment factors （for all sites,at landscape scale）,the correlation coefficient （r） were calculated using Pearson correlation analysis method which were performed by CANOCO version 4.5

	蓝藻和其他藻类	地衣	藓类	生物量	粉粒	黏粒	pH	土壤有机碳	全氮	全磷	全钾	碳酸钙	总盐	湿度	一年生植物盖度	多年生植物盖度	植被总盖度
蓝藻和其他藻类	1.00																
地衣	0.17	1.00															
藓类	−0.65	−0.13	1.00														
生物量	−0.52	−0.01	0.66	1.00													
粉粒	−0.08	0.51	−0.01	−0.07	1.00												
黏粒	−0.01	0.53	−0.20	−0.03	0.75	1.00											
pH	0.04	0.11	−0.07	−0.02	0.14	0.19	1.00										
土壤有机碳	−0.37	−0.19	0.42	0.50	−0.17	−0.26	0.22	1.00									
全氮	−0.24	−0.15	0.33	0.23	−0.22	−0.23	0.19	0.53	1.00								
全磷	−0.10	0.45	−0.04	0.02	0.35	0.63	0.42	0.24	0.12	1.00							
全钾	−0.13	−0.04	0.19	0.16	0.36	0.15	0.71	0.17	0.11	0.15	1.00						
碳酸钙	0.21	0.43	−0.20	0.01	0.28	0.50	0.40	−0.55	−0.17	0.28	0.29	1.00					
总盐	−0.03	0.27	−0.21	0.11	0.51	0.74	0.30	−0.09	−0.06	0.53	0.16	0.40	1.00				
湿度	−0.51	−0.04	0.63	0.61	−0.17	−0.18	0.19	0.53	0.56	0.10	0.27	−0.12	−0.08	1.00			
一年生植物盖度	0.64	0.03	−0.73	−0.63	−0.23	−0.02	0.04	−0.28	−0.14	−0.03	−0.27	0.15	0.08	−0.47	1.00		
多年生植物盖度	−0.74	−0.01	0.71	0.72	0.16	0.04	−0.07	0.33	0.17	0.13	0.24	−0.19	0.02	0.41	−0.75	1.00	
植被总盖度	0.14	−0.02	0.01	−0.06	−0.24	−0.06	−0.08	−0.19	−0.05	−0.12	−0.18	0.19	−0.05	−0.28	0.27	−0.18	1.00

参考文献

陈荷生. 沙坡头地区生物结皮的水文物理特点及环境意义. 干旱区研究,1992,9(1): 31−38.

程军回,张元明. 影响生物土壤结皮分布的环境因子. 生态学杂志,2010,29(1): 133−141.

虎瑞,王新平,潘颜霞,张亚峰,张珂,张浩. 沙坡头地区藓类结皮土壤净氮矿化作用对水热因子的响应. 应用生态学报,2014,25(2): 394−400.

寇江泽. 荒漠化沙化土地连续十年缩减. 人民日报,2015年12月30日 16 版.

李康宁. 浑善达克沙地生物土壤结皮中降解纤维素菌的分离、鉴定及降解活性分析. 内蒙古农业大学硕士学位论文,2010.

李柳,李小雁,马育军,赵国琴,吴华武. 油蒿灌木的生态水文过程研究综述. 干旱气象,2013,31(4): 814−819.

李新荣. 荒漠生物土壤结皮生态与水文学研究. 北京: 高等教育出版社,2012.

李新荣,陈应武,贾荣亮. 生物土壤结皮:荒漠昆虫多样性的重要食物链组成. 中国沙漠,2008,28(2): 245−248.

李新荣,张元明,赵允格. 生物土壤结皮研究: 进展、前沿与展望. 地球科学进展,2009,24(1): 11−24.

李新荣,张志山,黄磊,王新平. 我国沙区人工植被系统生态−水文过程和互馈机理研究评述. 科学通报,2013,58(5): 397−410.

李新荣,赵洋,回嵘,苏洁琼,高艳红. 中国干旱区恢复生态学研究进展及趋势评述. 地理科学进展,2014,33(11): 1435−1443.

刘俊杉,徐霞,张勇,田玉强,高琼. 长期降雨波动对半干旱灌木群落生物量和土壤水分动态的效应. 中国科学（生命科学）, 2010,40(2): 166−174.

沈培平,岳耀杰,王静爱,吕红峰,易湘生. 基于生态安全条件的沙区土地结构优化与高效利用. 干旱区研究,2006,23(3): 433−438.

王涛. 中国风沙防治工程. 北京:科学出版社,2011.

王新平,康尔泗,张景光,李新荣. 荒漠地区主要固沙灌木的降水截留特征. 冰川冻土,2004,26(1): 89−94.

王新平,李新荣,潘颜霞,王正宁,仝桂静,Mele G,Tedeschi A. 我国温带荒漠生物土壤结皮孔隙结构分布特征. 中国沙漠, 2011,31(1): 58−62.

吴永胜,哈斯,李双权,刘怀泉,贾振杰. 毛乌素沙地南缘生物土壤结皮的发育特征. 水土保持学报,2010,24(5): 258−261.

赵允格,许明祥,Belnap J. 生物结皮光合作用对光温水的响应及其对结皮空间分布格局的解译——以黄土丘陵区为例. 生态学报,2010,30(17): 4668−4675.

朱震达,刘恕. 中国的荒漠化及其治理. 北京: 科学出版社,1989.

Belnap J,Lange OL. *Biological Soil Crusts: Structure,Function,and Management*. Berlin: Springer-Verlag,2003.

Belnap J. Nitrogen fixation in biological soil crusts from southeast Utah,USA. *Biology and Fertility of Soils*,2002,35: 128-135.

Berdugo M,Soliveres S,Maestre FT. Vascular plants and biocrusts modulate how abiotic factors affect wetting and drying events in drylands. *Ecosystems*,2014,17: 1242-1256.

Bisdom EBA,Dekker LW,Schoute JFT. Water repellency of sieve fractions from sandy soils and relationships with organic material and soil structure. *Geoderma*,1993,56: 105-118.

Bowker MA. Biological soil crust rehabilitation in theory and practice: An underexploited opportunity. Restoration Ecology,2007,15: 13-23.

Bowker MA,Belnap J,Davidson DW,Goldstein H. Correlates of biological soil crust abundance across a continuum of spatial scales: Support for a hierarchical conceptual model. *Journal of Applied Ecology*,2006,43: 152−163.

Bowker MA,Belnap J,Rosentreter R,Graham B. Wildfire-resistant biological soil crusts and fire-induced loss of soil stability in Palouse Prairies,USA. *Applied Soil Ecology*,2004,26: 41−52.

Brostoff,WN,Sharifi MR,Rundel PW. Photosynthesis of cryptobiotic crusts in a seasonally inundated system of pans and dunes at Edwards Air Force Base,western Mojave Desert,California: Laboratory studies. *Flora*,2002,197: 143−151.

Brotherson JD,Rushforth SR. Influence of cryptogamic crusts on moisture relationships of soils in Navajo National Monument,Arizona. *The Great Basin Naturalist*,1983: 73−78.

Büdel B. Microorganisms of biological crusts on soil surface. In: Buscot F,Varma A (eds.),*Microorganisms in Soils: Roles in Genesis*. Berlin: Springer-Verlag,2005,307−323.

Cable JM,Huxman TE. Precipitation pulse size effects on Sonoran Desert soil microbial crusts. *Oecologia*,2004,141: 317−324.

Chen YW,Li XR. Spatiotemporal distribution of nests and influence of ant (*Formica cunicularia* Lat.) activity on soil property and seed bank after revegetation in the Tengger Desert. *Arid Land Research and Management*,2012,26: 365−378.

CCICDD(China National Committee for Implementation of the United Nations Convention to Combat Desertification). *China County Paper to Combat Desertification*. Beijing: China Forestry Publishing House,2002.

Csotonyi JT,Addicott JF. Influence of trampling-induced micro-topography on growth of the soil crust bryophyte *Ceratodon purpureus* in Jasper National Park. *Canadian Journal of Botonia*,2004,82: 1382−1392.

Danin A,Ganor E. Trapping of airborne dust by mosses in the Negev Desert,Israel. *Earth Surface Processes and Landforms*,1991,16: 153-162.

Delgado-Baquerizo M,Maestre FT,Gallardo A. Biological soil crusts increase the resistance of soil nitrogen dynamics to changes in temperatures in a semi-arid ecosystem. *Plant and Soil*,2013,366: 35–47.

Dougill AJ,Thomas AD. Kalahari sand soils: Spatial heterogeneity,biological soil crusts and land degradation. *Land Degradation and Development*,2004,15: 233–242.

Duan ZH,Xiao XH,Li XR,Dong ZB. Evolution of soil properties on stabilized sands in the Tengger Desert,China. *Geomorphology*,2004,59: 237–246.

Eldridge DJ. Cryptogam cover and soil surface condition: Effects on hydrology in a semi-arid woodland. *Arid Soil Research and Rehabilitation*,1993,7: 203–217.

Eldridge DJ,Greene RSB. Microbiotic soil crusts: A view of their roles in soil and ecological processes in the rangelands of Australia. *Australian Journal of Soil Research*,1994,32: 389–415.

Eldridge DJ,Tozer ME. Environmental factors relating to the distribution of terricolous bryophytes and lichens in semi-arid eastern Australia. *Bryologist*,1997,100: 28–39.

Eldridge DJ,Freudenberger D,Koen TB. Diversity and abundance of biological soil crust taxa in relation to fine and coarse-scale disturbance in a grassy eucalypt woodland in eastern Australia. *Plant and Soil*,2006,281: 255–268.

Eldridge DJ,Zaady E,Shachak M. Infiltration through three contrasting biological soil crusts in patterned landscapes in the Negev,Israel. *Catena*,2000,40: 323–336.

Escolar C,Martínez I,Bowker MA,Maestre FT. Warming reduces the growth and diversity of biological soil crusts in a semi-arid environment: Implications for ecosystem structure and functioning. *Philosophical Transactions The Royal Society Biological Sciences*,2012,367: 3087–3099.

Flechtner VR. North American desert microbiotic soil crust communities: Diversity despite challenge. In: Seckbach J (ed.). *Algae and Cyanobacteria in Extreme Environments*. Berlin: Springer-Verlag,2007,537–551.

Garner W,Steinberger YA. A proposed mechanism for the formation of "Fertile Islands" in the desert ecosystem. *Journal of Arid Environments*,1989,16: 257–262.

Green TGA,Sancho LG,Pintado A. Ecophysiology of desiccation rehydration cycles in mosses and lichens. In: Lüttge U (ed.). *Plant Desiccation Tolerance,Ecological Studies*. Berlin: Springer-Verlag,2011,89–120.

Greene RSB,Tongway DJ. The significance of (surface) physical and chemical properties in determining soil surface condition of red earths in rangelands. *Soil Research*,1989,27: 213–225.

Greene RSB,Chartres CJ,Hodgkinson KC. The effects of fire on the soil in a degraded semiarid woodland. I. Cryptogam cover and physical and micromorphological properties. *Soil Research*,1990,28: 755–777.

Grishkan I,Kidron GJ. Biocrust-inhabiting cultured microfungi along a dune catena in the western Negev Desert,Israel. *European Journal of Soil Biology*,2013,56: 107–114.

Guo YR,Zhao HL,Zuo XA,Drake S,Zhao XY. Biological soil crust development and its top soil properties in the process of dune stabilization,Inner Mongolia,China. *Environmental Geology*,2008,54: 653–662.

Housman DC,Powers HH,Collins AD,Belnap J. Carbon and nitrogen fixation differ between successional stages of biological soil crusts in the Colorado Plateau and Chihuahuan Desert. *Journal of Arid Environments*,2006,66: 620–634.

Jia RL,Li XR,Liu LC,Gao YH,Li XJ. Responses of biological soil crusts to sand burial in a revegetated area of the Tengger Desert,Northern China. *Soil Biology and Biochemistry*,2008,40: 2827–2834.

Jia RL,Li XR,Liu LC,Gao YH,Zhang XT. Differential wind tolerance of soil crust mosses explains their micro-distribution in nature. *Soil Biology and Biochemistry*,2012,45,31–39.

Jongman RHG,ter Braak CJF,Van Tongeren OFR. *Data Analysis in Community and Landscape Ecology*. Cambridge: Cambridge University Press,1995.

Kidron GJ,Benenson I. Biocrusts serve as biomarker for the upper 30 cm soil water content. *Journal of Hydrology*,2014,509: 398–405.

Kidron GJ,Yair A. Rainfall-runoff relationship over encrusted dune surfaces,Nizzana western Negev,Israel. *Earth Surface Processes Land Forms*,1997,22: 1169–1184.

Kidron GJ,Barinova S,Vonshak A. The effects of heavy winter rains and rare summer rains on biological soil crusts in the Negev Desert. *Catena*,2012,95: 6–11.

Kidron GJ,Yair A,Vonshak A,Abeliovich A. Microbiotic crust control of runoff generation on sand dunes in the Negev Desert. *Water Resources Research*,2003,39: 1108–1112.

Kidron GJ,Vonshak A,Abeliovich A. Microbiotic crusts as biomarkers for surface stability and wetness duration in the Negev Desert. *Earth Surface Processes and Landforms*,2009,34: 1594–1604.

Kleiner E,Harper KT. Soil properties in relation to cryptogamic ground cover in Canyonlands National Park. *Journal of Rangeland Management*,1977,30: 202-205.

Lange OL,Green A. High thallus water content severely limits photosynthetic carbon gain of central European epilithic lichens under natural conditions. *Oecologia*,1996,108: 13–20.

Li XJ,Li XR,Song WM,Gao YP,Zheng JG,Jia RL. Effects of crust and shrub patches on runoff,sedimentation,and related nutrient (C,N) redistribution in the desertified steppe zone of the Tengger Desert,Northern China. *Geomorphlogy*,2008,96: 221–232.

Li XR,Wang XP,Li T,Zhang JG. Microbiotic soil crust and its effect on vegetation and habitat of artificially stabilized desert dunes in Tengger Desert,North China. *Biology and Fertility of Soils*,2002,35: 147–154.

Li XR,Xiao HL,Zhang JG,Wang XP. Long-term ecosystem effects of sand-binding vegetation in Shapotou region of Tengger Desert,Northern China. *Restoration Ecology*,2004,2: 376–390.

Li XR,Jia XH,Long LQ,Zerbe S. Effects of biological soil crusts on seed bank,germination and establishment of two desert annual plants. *Plant and Soil*,2005,277: 379–389.

Li XR,Xiao HL,He MZ,Zhang JG. Sand barriers of straw checkerboards for habitat restoration in extremely arid desert regions. *Ecological Engineering*,2006,28: 149–157.

Li XR,He MZ,Duan ZH,Xiao HL,Jia XH. Recovery of topsoil physicochemical properties in revegetated sites in the sand-burial ecosystems of the Tengger Desert,Northern China. *Geomorphology*,2007,88: 254–265.

Li XR,Tian F,Jia RL,Zhang ZS,Liu LC. Do biological soil crusts determine vegetation changes in sandy deserts? Implications for managing artificial vegetation. *Hydrological Process*,2010a,24: 3621–3630.

Li XR,He MZ,Zerbe S,Li XJ,Liu LC. Micro-geomorphology determines community structure of biological soil crusts at small scales. *Earth Surface Processes and Landform*,2010b,35: 932–940.

Li XR,Jia RL,Chen YW,Huang L,Zhang P. Association of ant nests with successional stages of biological soil crusts in the Tengger Desert,Northern China. *Applied Soil Ecology*,2011,47: 59–66.

Li XR,Zhang P,Su YG,Jia RL. Carbon fixation by biological soil crusts following stabilization of sand dune in arid desert regions of China: A four-year field study. *Catena*,2012,97: 119–126.

Li XR,Zhang ZS,Huang L,Wang XP. Review of the ecohydrological processes and feedback mechanisms controlling sand-binding vegetation systems in sandy desert regions of China. *Chinese Science Bulletin*,2013,58: 1–14.

Li XR,Gao YH,Su JQ,Jia RL,Zhang ZS. Ants mediate soil water in arid desert ecosystems: Mitigating rainfall interception induced by biological soil crusts? *Applied Soil Ecology*,2014a,78: 57–64.

Li XR,Zhang ZS,Tan HJ,Gao YH,Liu LC,Wang XP. Ecological restoration and recovery in the wind-blown sand hazard areas of Northern China: Relationship between soil water and carrying capacity for vegetation in the Tengger Desert. *Science in China (Life Sciences)*,2014b,57: 539–548.

Liu LC,Li SZ,Duan ZH,Wang T,Zhang ZS,Li XR. Effects of microbiotic crusts on dew deposition in the restored vegetation area at Shapotou,Northwest China. *Journal of Hydrology*,2006,328: 331–337.

Liu LC,Song YX,Gao YH,Wang T,Li XR. Effects of microbiotic crusts on evaporation from the revegetated area in a Chinese desert. *Australian Journal of Soil Research*,2007,45: 422–427.

Liu YM,Li XR,Jia RL,Huang L,Zhou YY,Gao YH. Effects of biological soil crusts on soil nematode communities following dune stabilization in the Tengger Desert,Northern China. *Applied Soil Ecology*,2011,49: 118–124.

Liu YM,Yang HY,Li XR,Xing ZS. Effects of biological soil crusts on soil enzyme activities in revegetated areas of the Tengger Desert. *Applied Soil Ecology*,2014,80: 6–14.

Maestre FT,Escudero A,Martinez I,Guerrero C,Rubio A. Does spatial pattern matter to ecosystem functioning? Insights from biological soil crusts. *Functional Ecology*,2005,19: 566–573.

Maestre FT,Huesca MT,Zaady E,Bautistaa S,Cortinaa J. Infiltration,penetration resistance and microphytic crust composition in contrasted microsites within a Mediterranean semi-arid steppe. *Soil Biology and Biochemistry*,2002,34: 895–898.

Miralles I,Domingo F,Cantón Y,Trasar-Cepedac C,Carmen Leirósd M,Gil-Sotres F. Hydrolase enzyme activities in a successional gradient of biological soil crusts in arid and semi-arid zones. *Soil Biology and Biochemistry*,2012,53:124–132.

Owens MK,Lyons RK,Alejandro CL. Rainfall partitioning within semiarid juniper communities: Effects of event size and canopy cover. *Hydrological Process*,2006,20: 3179–3189.

Pan YX,Wang XP,Jia RL,Chen YW,He MZ. Spatial variability of surface soil moisture content in a re-vegetated desert area in Shapotou,Northern China. *Journal of Arid Environments*,2008,72: 1675–1683.

Perez FL. Microbiotic crusts in the high equatorial Andes,and their influence on Paramo soils. *Catena*,1997,31: 173–198.

Pu CF,Zhao Y,Hill RL,Zhao CL,Yang YS,Zhang P,Wu SF. Wind erosion prevention characteristics and key influencing factors of bryophytic soil crusts. *Plant and Soil*,2015,397: 163–174.

Read CF,Duncan DH,Vesk PA,Elith J. Biological soil crust distribution is related to patterns of fragmentation and landuse in a dryland agricultural landscape of southern Australia. *Landscape Ecology*,2008,23: 1093–1105.

Reynaud PA,Lumpkin TA. Microalgae of the Lanzhou (China) cryptogamic crust. *Arid Land Research and Management*,1988,2:

145–155.

Schuster G,Ott S,Jahns HM. Artificial cultures of lichens in natural environment. *Lichenologist*,1985,17: 247–253.

Shachak M,Lovett GM. Atmospheric deposition to a desert ecosystem and its implications for management. *Ecological Application*,1998,8: 455–463.

Su YG,Zhao X,Li AX,Li XR,Huang G. Nitrogen fixation in biological soil crusts from the Tengger Desert,Northern China. *European Journal of Soil Biology*,2011,47: 182–187.

Ter Braak CJF,Šmilauer P. Canoco for Windows version 4.5. Biometrics–Plant Research International,Wageningen 1.The Netherlands,2002.

Wang XP,Li XR,Xiao HL. Effects of surface characteristics on infiltration pattern in an arid shrub desert. *Hydrological Processes*,2007,21: 72–79.

Wang XP,Wang ZN,Berndrtsson R,Zhang YF,Pan YX. Desert shrub stemflow and its significance in soil moisture replenishment. *Hydrology and Earth System Sciences*,2011,15: 561–567.

Wang XP,Zhang YF,Hu R,Pan YX,Berndtsson R. Canopy storage capacity of xerophytic shrubs in Northwestern China. *Journal of Hydrology*,2012,454-455: 152–159.

West NE. Structure and function of microphytic soil crusts in wild-land ecosystems of arid to semi-arid regions. *Advances in Ecological Research*,1990,20: 179–223.

Williams JD,Dobrowolski JP,West NE. Microbiotic crust influence on unsaturated hydraulic conductivity. *Arid Soil Research and Rehabilitation*,1999,13: 145–154.

Yair A,Almog R,Veste M. Differential hydrological response of biological topsoil crusts along a rainfall gradient in a sandy arid area: Northern Negev desert,Israel. *Catena*,2011,87: 326–333.

Zaady E,Kuhn U,Wilske B,Sandoval-Sotob L,Kesselmeier J. Patterns of CO_2 exchange in biological soil crusts of successional age. *Soil Biology and Biochemistry*,2000,32: 959–966.

Zelikova TJ,Housman DC,Grote EE,Neher DA,Belnap J. Warming and increased precipitation frequency on the Colorado Plateau: Implications for biological soil crusts and soil processes. *Plant and Soil*,2012,335: 265–282.

Zhang W,Zhang GS,Liu GX,Chen T, Zhang M, Dyson PJ, An L. Bacterial diversity and distribution in the southeast edge of the Tengger Desert and their correlation with soil enzyme activities. *Journal of Environmental Sciences*,2012,24: 2004–2011.

Zhang YF,Wang XP,Pan YX,Hu R. Diurnal relationship between the surface albedo and surface temperature in revegetated desert ecosystems,Northwestern China. *Arid Land Research and Management*,2012,26: 32–43.

Zhao Y,Li XR,Zhang ZS,Hu YG,Chen YL. Biological soil crusts influence carbon release responses following rainfall in a temperate desert,Northern China. *Ecological Research*,2014,29: 889–896.

Zhao Y,Li XR,Zhang ZS,Jia RL,Hu YG,Zhang P. The effects of extreme rainfall events on carbon release from biological soil crusts covered soil in fixed sand dunes in the Tengger Desert,Northern China. *Sciences in Cold and Arid Regions*,2013,5: 191–196.

Zhao Y,Zhang ZS,Hu YG,Chen YL. The seasonal and successional variations of carbon release from biological soil crust-covered soil. *Journal of Arid Environments*,2016,127: 148–153.

第 2 章　荒漠 BSC 与生存环境

2.1　影响 BSC 拓殖和发展的环境与生态因子

2.1.1　环境的基本要素及分类

环境是指生物有机体生活空间的外界自然条件的总和，包括生物有机体以外所有的环境要素，具有综合性和可调剂性（蒋高明，2004）。环境既包括空气、水、土地、植物、动物等物质因素，也包括观念、制度、行为准则等非物质因素；既包括非生命体形式，也包括生命体形式。

构成环境的各要素称为环境因子（environmental factor）。其中对生物生长、发育、生殖、行为和分布等生命活动有直接或间接影响的环境因子称为生态因子（ecological factor）。所有的生态因子构成生物的生态环境（ecological environment）。

环境是一个十分庞大和复杂的体系，分类的标准和依据各不相同。依据环境范围大小可将生物的环境分为大环境和小环境。大环境是指具有不同气候和植被特点的地理区域。小环境又称微环境，是相对自然环境、区域环境等大环境单元而言的，是对生物有着直接影响的邻近环境，如植物根际环境、叶表面的温湿度和气流变化等。依据人类影响程度可将环境分为自然环境和人工环境。自然环境包括非生物环境（无机环境）和生物环境（有机环境）两部分。其中，生物环境指影响环境的其他植物、动物、微生物及其群体。人工环境有广义和狭义之分。广义的人工环境包括所有的栽培植物及其所需的环境；狭义的人工环境指的是人工控制下的植物环境，如人工温室等。依据环境空间尺度可将环境分为地球环境、区域环境、群落环境、种群环境和植物个体环境。植物生态生理学研究的尺度一般是植物个体环境，即接近植物个体表面或表面不同部位的环境（蒋高明，2004）。

2.1.2 生态因子及其作用特征

生态因子是指环境中对生物的生长、发育、生殖、行为和分布有直接或间接影响的外界环境要素。根据不同的研究目的和划分标准，可以划分出多种生态因子类型。

根据生态因子是否具有生物成分，可将生态因子划分为非生物因子和生物因子（Molles, 2000）。非生物因子即非生命物质，如水分、温度、光照、大气和土壤等；生物因子包括植物、动物和微生物及土壤微小动物，它们在生态系统中维持着复杂的种内和种间关系。根据生态因子的组成性质，可将生态因子划分为：①气候因子，如水分、温度、光照、大气等；②土壤因子，土壤的物理化学特性、土壤肥力等；③生物因子，与生物发生相互关系的植物、动物和微生物等；④地理因子，如高原、山地、平原、低地、坡度、坡向、洪积扇等，它们的存在使各种因子及其影响产生差异或进行重新分配，如光照、温度、湿度、土壤的分布会因不同的地理位置而不同；⑤人为因子，是对生物产生影响的人类活动，包括人类的垦殖、放牧和采伐、交通、环境污染等。在各种生态因子中，并非所有的因子都为生物的生长所必需。我们把生物生长所必需的因子称为生存条件，生物缺少它们就不能生长。对于绿色植物来说，所必需的因子是氧气、二氧化碳、光照、水分、温度和无机盐。

生态因子作用的一般特征：①综合性。生态环境中一种生态因子变化，都必将引起其他因子不同程度的响应。任何一种生物学过程都是诸多因子共同作用的结果，即环境因子是综合在一起起作用的，每一因子的作用依赖于其他因子的作用。例如，一个地区的湿润程度不只取决于降水量这一个因素，而是诸多因素相互作用的综合效应。湿润程度既取决于水分收入（降水），又取决于水分支出（蒸发、蒸腾、径流和渗漏等）。②主导因子。在众多生态因子中，必有一种或少数几种生态因子是主要的，称为主导因子。主导因子起主要作用，影响其他因子。例如，光合作用时，光强是主导因子，温度和CO_2为次要因子。生态因子的主次不是一成不变的，在一定条件下可以发生转换，处于不同时期和条件下的生物对生态因子的要求和反应不同，某种特定条件下的主导因子在另一条件下会降为次要因子。③直接作用和间接作用。根据生态因子与生物的作用方式可将生态因子分为直接作用和间接作用，区分其作用方式对认识生物的生长、发育、繁殖及分布都很重要。例如，环境中的地理因子——坡向、坡度、海拔及经纬度等对生物的作用是直接的，但它们能够影响光照、温度、降水等因子的分布格局，因而对生物产生的作用又是间接的，而这些地方的光照、温度、降水对生物生长和分布产生直接作用。④阶段性。由于生物生长发育不同阶段对生态因子的需求不同，因此生态因子对生物的作用也具有阶段性，这种阶段性是由生态环境的规律性变化造成的。例如，光照在植物的春化阶段并不起作用，但光照长短在光周期阶段则至关重要。⑤不可替代性和补偿作用。环境中各种生态因子对生物的作用虽然不尽相同，但都各具有重要性，尤其是起主导作用的因子，如果缺少，便会影响生物的正常生长发育，甚至造成其生病或死亡。所以从总体上说生态因子是不可替代的，但是从局部来说是能补偿的。例如，在某一由多个生态因子综合作用的过程中，某因子量上的不足可以由其他因子来补偿，以获得相似的生态效应。以植物的光合作用为例，如果光照不足，可以增加CO_2量来补偿。生态因子的补偿作用只能在一定范围内作部分补偿，而不能以一个因子来替代另一个因子，且因子之间的补偿作用也不是经常存在的（黄璐琦和王康才，2012）。

环境中的各种因子影响着生物的生长和发育等生命活动，同时生物对环境也产生一系列响应和适应机制。近年来，国内外学者在环境因子对生物的影响方面做了大量研究，开展了有关干旱、低温、高温、光照强度、UV-B辐射、CO_2浓度、土壤营养等环境因子对植物生长发育、光合作用、呼吸作用、抗氧化酶系统等的形态学、生理学和分子生物学方面的研究（蒋高明，1997；刘国顺等，2007；郑有飞等，2007；孙敬松等，2010；Li *et al.*，2012；武立权等，2012；Martins *et al.*，2014），但对BSC的主要生物体——非维管植物（no vascular plant）或隐花植物（cryptogam）如蓝藻、绿藻、地衣和藓类的研究和了解却很少。

2.1.3 植物与生态因子之间的相互作用

植物与生态因子之间的关系主要表现为作用、适应和反作用三种形式。如对于温度，各物种反应不同，有些物种能适应的温度却可能使另一些物种死亡。一般来说，同种植物在不同发育阶段的适应性也不相同。随着环境的变化，植物的适应性也会发生改变。一个物种可能通过特有的生理过程适应一个新的环境，当新旧环境差别太显著时，可能需要更长时期的适应过程。植物不仅对环境具有适应能力，而且对环境也有巨大的改造作用，即生态反作用，因而有些生物被称为“生态系统工程师”，如沙地系统中的蚂蚁和BSC中的生物群落（Li *et al.*，2012）。从短时间尺度上看，植物与环境的关系以适应为主，反作用为辅；但从较长的进化尺度看，植物与环境的关系则以反作用为主。例如，我国在20世纪50年代开始的风沙危害治理中，利用植物固沙的实践就很好地证实了植物与生态因子之间的这种关系。起初在固沙区人工种植的固沙植被对流沙的土壤物理、化学及生物学性状没有显著的影响，旱生灌木从生态生理学特征上对沙区夏季高温、沙层含水量低以及强风沙流、频繁沙埋等胁迫存在一个适应的过程，表现在生长状况受到抑制，种植的成活率和保存率相对低下，也不能完成自我更新。但5~6年后，人工种植的灌木对环境特别是土壤的影响发生了很大变化，随着沙面物理环境的稳定，沙面BSC的出现，表层土壤的理化性状和生物学性状发生了显著变化，有利于其他微小生物和动物的繁衍及其他植物种的定居，而人工种植的灌木也出现了自我更新现象（Li *et al.*，2004）。

但是，适应是相对的，任何植物对环境因子的适应性都有一定的界限范围，对某环境因子能够忍耐的最小剂量为下限临界点，忍耐的最大剂量为上限临界点，最适宜于植物生命活动的为最适点，这三点合称植物对环境因子响应的“三基点”。一般来说，植物在某一生态因子维度上的分布呈正态曲线（图2-1），该曲线可以称为资源利用曲线（resource utilization curve）。植物适应的上限和下限之间的环境范围就是植物的适应范围，又称为植物的生态幅（ecological amplitude），在该生态幅之间的环境区域就是该植物的分布区。

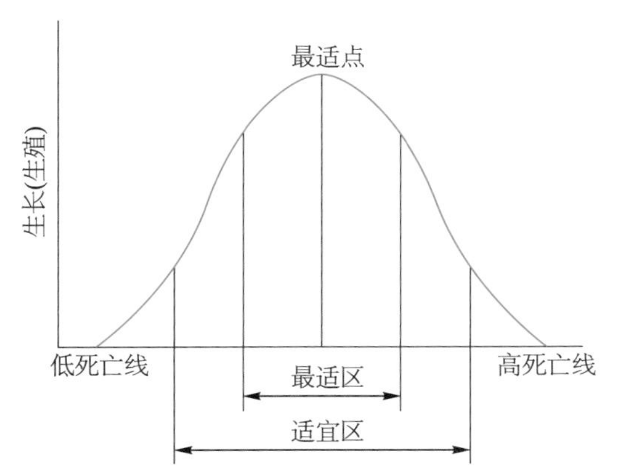

图 2-1　植物适应环境“三基点”示意图（据孙儒泳等，2002）
Figure 2-1　The “three basis points” sketch map of plants adapting environment

生态位（ecological niche）是一个物种所处的环境以及其自身生活习性的总称（蒋高明，2004）。某种植物在某一生态因子梯度上的生态幅实际上也就是该植物的生态位。不同的物种占据不同的生态位，两个或两个以上生态位相似的物种生活于同一空间时分享或竞争共同资源的现象称为生态位重叠（niche overlap）。生态位重叠的两个物种因竞争排斥原理而难以长期共存，除非空间和资源十分丰富。某一物种在无任何竞争物种存在时所占据的最大空间，称为基础生态位（fundamental niche）。基础生态位只是一个理论上的生态位，以假定一个物种种群单独存在，无其他任何竞争环境资源的别的物种干扰为前提。但实际上在此生态位边界以内总是有别的竞争物种存在并与之竞争资源，当在群落中有竞争对象存在时，其实际栖息的空间要小得多，称为实际生态位（realized niche）。

当环境中某种生存条件出现异常，便会抑制植物生命活动或威胁植物生存，这种现象称为环境胁迫（environmental stress）。在生态学上，环境胁迫是指一种显著偏离于植物适宜生活需求的环境条件，也称为环境应力。动物啮食、寄生、风害、火灾和土壤侵蚀等现象可以部分或全部破坏植物生命活动的产物，被称为干扰（disturbance）（刘鸿雁，2005）。环境变化引起的植物在功能性水平上产生的变化和反应，尽管在开始时是可逆的，但后来可变为永久性的。

胁迫因子是指对植物胁迫刺激的生态因子，胁迫反应或胁迫状态是指植物对胁迫刺激的反应或适应的即时状态。当胁迫超过自身调节能力的极限时，潜在损伤就发展成慢性病或不可逆伤害，这一极限即为阈值（threshold value）（Larcher，1997）。例如，干旱引发的胁迫对于植物个体来说，其危害程度不仅依赖于其干旱抗性，也依赖于其环境条件。萎蔫系数就是一个阈值，是指植物根系不能迅速吸取到能满足蒸腾需要的水分，植物开始出现萎蔫时的土壤含水量。其他还有许多如温度、辐射、光照甚至人为胁迫造成的阈值。可见，胁迫因子是非常复杂的，既有非生物的温度、水分、辐射等因子，又有生物的微生物、动物、植物因子以及人类的影响，胁迫常常是多种因子同时起作用，相互影响，相互制约。

2.2 影响 BSC 生态生理的因子

2.2.1 非生物因子

（1）水分

BSC组成的主要生物体如藻类、地衣和藓类均为变水植物，只有在湿润的环境中才有生理代谢活性，尤其对碳和氮的固定过程。因此，水分被认为是影响BSC生态生理功能最主要的生态因子。干旱沙区水分的主要形态有气态和液态。气态水主要是指空气中的水汽，一般用相对湿度来表示空气中水汽的含量；液态水主要包括雨、雾、露等，其中降雨和降雪是最主要的降水形式，对植物生长起重要作用。几乎所有的关于BSC生态生理功能的研究都涉及水分的影响，尤其是它们本身的含水量。

国内外一些学者研究了BSC在脱水-复水过程中的光合速率变化趋势，并找出了相应的水分补偿点和饱和点（Sonesson *et al.*，1992；Palmqvist，2000；Li *et al.*，2014）。一般而言，很少的水分就能刺激BSC进行光合作用。如地衣 *Collema tenax* 的水分补偿点仅为0.22 mm的液态水，而绿藻只需利用露水和大气中的水汽就能进行光合作用，并达到补偿点（Lange *et al.*，1998；Lange，2003）。目前为止，水分对BSC的CO_2交换影响的研究均表明，BSC干燥后复水首先是CO_2的大爆发（Lange *et al.*，1994；Deltoro *et al.*，1998；Zotz *et al.*，2000），接着才是进行光合作用，这些最终影响到BSC在复水过程中的碳平衡。值得注意的是，BSC类生物在干燥复水后能够迅速恢复光合作用，相对地，维管植物则需要较长时间的恢复（Lange *et al.*，1998；杜晓濛等，2014）。光合速率与叶绿素荧光的研究表明，干燥程度和时间也是影响BSC生态生理特征的重要原因。一般而言，干燥程度越大、时间越长，BSC生物光合作用恢复时间越长（Jeffrise *et al.*，1993a，1993b），这也与干旱区降水格局的变化将引起BSC结构和组成的改变，从而导致其光合固碳能力变化的研究结果相互印证（Belnap *et al.*，2004；Sponseller，2007）。但如果含水量过高又会显著抑制BSC的净光合速率，目前对这一现象的解释是较高的水分阻碍了CO_2的通路，进而减少了外界供给BSC光合有机体碳同化所需要的CO_2。这样的解释已经被CO_2浓度倍增的实验所证明，CO_2浓度的增加能够补偿多余水分对气体流通的阻碍，并显著提高BSC的净光合速率（Lange *et al.*，1997）。对藓类植物水分含量与光合作用关系的研究表明，藓类植物净光合速率随组织含水量的增加而提高，但过剩的水分含量会增加植物体对CO_2吸收的阻力，从而引起光合速率下降（吴玉环等，2001）。李刚等（2014）通过对降雪处理下两种BSC光合和呼吸作用的测定，证实了冬季BSC净光合速率、呼吸速率受空气温度、辐射强度及土壤水分的影响，其中水分是关键影响因子。

（2）光照

光照是十分复杂而重要的生态因子，它是仅次于水分对BSC生态生理特征产生重要影响的环境因子。这主要是因为光照不仅是BSC进行光合作用的必要条件，直接影响到BSC的光合能力，这可以用BSC的光反应曲线来很好地表述；而且光照对BSC的组成和色素含量有重要影响，能调节BSC生长和发育的过程（Green *et al.*，1997；Zaady *et al.*，2000）。BSC光合作用一般对光照的要求不太高（Tretiach and Pecchiari，1995；Green *et al.*，1997），这是因为这些变水植物的光合作用一般在雨后或凌晨光照低的情况下进行（Lange *et al.*，1998；Palmqvist，2000）。如果光照太强将刺激BSC产生大量的保护性色素，这将影响到BSC的组成（Campbell *et al.*，1996；Deltoro *et al.*，1998）。因此，在高光照的荒漠地区，BSC为适应环境形成不同的功能型（Bowker *et al.*，2002；Belnap *et al.*，2004）：第一类是以*Microcoleus vaginatus*为主的浅色的蓝藻结皮，它们定居在地表以下几毫米处，只有在条件适宜时将光合器官伸出地表进行光合作用（Garcia and Pringault, 2001）；第二类是以*Scytonema myochrous*和*Nostoc commune*为主的深色BSC，它们具有高的保护色素含量，能够避免高光照对BSC的损伤（Dodds，1989）；第三类是以地衣和藓类为主的BSC，它们主要通过复水修复损坏的光合器官而适应环境（Garcia and Pringault，2001；Belnap *et al.*，2004）。

大量研究表明，BSC光合作用对光照的响应都受到BSC水分的影响（Lange, 2003）。一般而言，BSC光合作用光照饱和点随着水分的增加而增加，地衣结皮在最佳水分状态下时，甚至在2000 μmol m^{-2} s^{-1}时仍未出现光抑制现象（Lange *et al.*，1998）；而BSC光合作用光照补偿点却似乎不受水分条件的影响，随着BSC生物的不同而有较大的浮动，并表现出喜阳植物的特征（Lange *et al.*，1997，1998）。然而光合速率与叶绿素荧光的研究表明，藓类结皮即使生活在易受光直射的开阔生境中，通常要求较低的光照就能进行光合作用，表现出喜阴的光合特征，其光合作用所需的饱和光照强度远远低于高等植物所需要的（Alpert and Oechel，1987；Marschall and Proctor，2004）。这其实是藓类结皮对环境的长期适应和演化的结果，因为强光照往往伴随着高温，两个生态因子的综合作用可使藓类植物很快地失去水分而变干燥。因此，藓类植物的光合作用及其他代谢活动一般仅仅在弱光或中等光照强度和较低的温度范围内进行（吴玉环等，2001），这也与第1章中我们对BSC中藓类多样性及分布格局的描述相吻合，即藓类结皮与多年生植物，尤其是灌木的盖度密切相关，因为植物冠幅为其种类多样性共存提供了遮阴等适宜的微生境。

（3）温度

温度是影响植物分布的主要条件之一，但它常和水分结合在一起决定植物的分布界限；温度也是影响植物生长发育最重要的因素之一，制约着植物的萌发、生长发育速度以及体内的一切生理生化变化。每种BSC的生长都有一个温度范围。生长的最低温度、最适温度和最高温度通常称为生长温度的“三基点”。超过两个极限范围，植物生理活动就会停止，甚至死亡。

BSC各生理功能和新陈代谢的速率与温度变化密切相关（Davey and Rothery, 1996）。研究表明，夏季温度升高改变了藻类、藓类及地衣结皮的生物多样性和某些生理功能（Bowker *et al.*，2002）。在亚北极和北极地区，地衣的种类由于温度升高而减少，维管植物由于温度升高而多样性增加（Cornelissen *et al.*，2001）。BSC光合作用的温度适宜范围很广，从2 ℃（Lange *et al.*，1998）至45 ℃（San José and Bravo，1991）BSC都能保持较高的净光合速率。但也有例外，以色列沙漠地区的BSC在35 ℃时由于强烈的呼吸速率而使净光合速率降低（Levi *et al.*，1981）。

温度与BSC光合碳固定的关系研究都侧重于做出相应的光合速率反应曲线（Sonesson *et al.*，1992；Lange *et al.*，1998；Palmqvist，2000）。在此基础上，BSC温度-光合关系研究注重温度的敏感性，并与相应的呼吸作用的温度敏感性进行对比，从而预测BSC在温度变化（主要是全球变化背景下的温度升高）情况下的碳“源-汇”关系（Lange *et al.*，1994；Davey，1997）。研究表明，温度对呼吸速率的刺激作用要大于光合作用，随着温度的增加，光合型植物群落将会变成异养型。这种温度对CO_2气体交换的影响作用，对于经历昼夜和季节温度浮动的地区碳循环研究有重要启示作用（Hancke and Glud，2004）。藓类结皮对气候变暖的响应是不可预料的（房世波等，2008），现在多数研究认为，气候变暖会使非维管植物的数量减少，这主要由于气候变暖提高了植物营养的可用性而导致维管植物的生长和郁闭度增加（Chapin *et al.*，1995）。然而，短期增温试验模拟的结果往往是藓类植物生物量变化不大或生物量增加。而在亚北极区的一项关于塔藓的研究表明，在开顶式气室中提高气温时塔藓的生长受抑制；但在该区域自然环境温度梯度中却随温度的增加而增长（Deslippe *et al.*，2005）。很多研究表明，温度变化能引起藓类暗呼吸速率的强烈变化，CO_2交换速率也明显变化，总光合作用和电子传递速率呈显著线性相关关系，说明藓类的暗呼吸和CO_2交换之间的关系较复杂，且与其他较高等的植物不同（Green *et al.*，1998；吴玉环等，2001）。

（4）UV-B辐射

人类活动导致大量氯氟烃和氮氧化物等气体进入平流层，导致臭氧浓度日益下降（McKenzie *et al.*，2007），而且这种趋势还将加剧，2010—2019年期间，平流层臭氧水平下降将更为严重（Valkama *et al.*，2003）。臭氧层的主要功能是吸收太阳辐射中的紫外线。紫外线是指波长为200～380 nm的太阳光，包括3类：UV-A波长为315～380 nm；UV-B波长为280～315 nm；UV-C波长为200～280 nm。臭氧层对UV-C的吸收能力很强，即使臭氧层减少90%，也不会有UV-C到达地面；对UV-A的吸收能力很弱，但UV-A对生物基本无影响；而臭氧层变薄导致到达地面的UV-B辐射增强，这一全球变化已成为国际社会广泛关注的重大环境问题（郑有飞等，2007）。UV-B增强给人类健康和生态环境带来诸多影响和危害（李豪杰等，2004）。高剂量的UV-B辐射对植物的生长发育、光合作用、物质代谢、抗氧化酶系统及细胞膜都产生影响（D' surney *et al.*，1993；黄梅玲等，2010；张红霞等，2010）。室内外试验表明，大多数植物在UV-B辐射胁迫下，表现出植株矮化、节间缩短、叶面积减少、光合速率降低、生产力下降等特征（Allen *et al.*，1998；安黎哲等，2001）。

近年来，国内外很多学者从不同角度对组成BSC的不同种类隐花植物进行了大量的研究（梁少民等，2005；张丙昌等，2007）。UV-B辐射能抑制或阻止隐花植物的一系列生态、生理和生化过程（Newsham *et al.*，2002；肖媛等，2010），如影响它们的生物量、光合作用和相关色素、抗氧化酶含量以及破坏蛋白质和脱氧核糖核酸（DNA）等。Suresh等（1998）利用5 W m^{-2}的UV-B辐射强度处理蓝藻种*Nostoc* sp.、*Anabaena* sp.和*Scytonema* sp.，发现UV-B阻碍了3种蓝藻的生长，使其生长速率分别降低了28%、30%和36%，干重分别降低了12%、27%和10%，叶绿素、蛋白质、淀粉含量与对照相比也有不同程度的降低，并且藻细胞存活率随着UV-B辐射强度的升高而降低，在高强度UV-B照射下，*Nostoc* sp.的存活率仅为40%。

也有少数研究报道了UV-B辐射对隐花植物生物量影响不大。Gehrke（1999）通过室内模拟亚北极地区UV-B辐射增强30%的胁迫条件，探讨其对长毛砂藓和塔藓生物量的影响。试验发现塔藓生物量持续降低，对UV-B辐射增强反应敏感，而长毛砂藓生物量基本没有受到影响，抵抗UV-B辐射的能力较强。此外，UV-B辐射通过不同途径影响隐花植物的光合作用。苏延桂等（2011）对沙坡头地区不同年代的人工植被区藻类结皮进行研究，结果表明，紫外辐射增强通过降低荒漠藻结皮的光合色素含量，减少了荒漠藻结皮的净光合速率，从而对其生产力产生影响。Gehrke（1999）对锈色泥炭藓和大金发藓的研究指出，增强UV-B辐射使锈色泥炭藓和大金发藓叶绿素a（Chl-a）浓度降低。

在应对UV-B辐射的胁迫方面，大多数隐花植物都发展了完整的保护和修复机制。绝大多数藻类植物本身含有或者经UV-B辐射诱导后合成的保护色素，以屏蔽UV-B的伤害（Bassman，2004）。这些植物保护色素大部分具有很强的UV-B吸收能力，主要有两类：一类是伪枝藻素（scytonemin），是分布在细胞外的一类黄褐色脂溶性二聚体色素，能够起到屏蔽UV-B进入细胞的作用；另一类是分布于细胞内的海胆酮（echinenone）、类胡萝卜素（carotenoid）、叶黄素（lutein）等色素，它们使细胞能抵御UV-B产生的单线态氧对细胞产生的损伤。Monika等（1997）研究发现UV-B辐射处理使*Nostoc commune*的类胡萝卜素含量增高，经过短时间UV-B照射，类胡萝卜素与Chl-a的比值上升，类胡萝卜素含量的升高是藻类植物对UV-B辐射的响应机制。而大多数藓类和地衣体内具有高含量的酚类化合物，这些化合物具有过滤UV-B辐射，减少UV-B辐射对自身的伤害，保护植物体的作用（Gauslaa and Mcevoy，2005）。其中，类黄酮被认为是主要的植物保护色素，其在植物表层的累积可减少UV-B辐射的透过率（Lindroth and Hofmann，2000）。从藓类和地衣植物中已发现并分离出黄酮类化合物如反式－桂皮酸、阿魏酸等UV-B吸收化合物（娄红祥，2006），这些化合物使得植物在生理生化层面具有保护自身不受UV-B辐射伤害的有效防御机制。Hall等（2002）通过室内和室外试验对UV-B辐射处理前后石蕊属地衣林鹿蕊的酚类含量的变化进行研究比较，发现高的UV-B辐射胁迫处理下，林鹿蕊产生较高的酚类化合物。Dietz等（2000）也发现在UV-B辐射处理下地衣表面色素含量的变化最为明显，要高于地衣表面的厚度和反射系数的变化，地衣色素含量的多少对其能通过多少光合有效辐射起重要作用。

尽管不同隐花植物对UV-B的敏感性存在差异，但高强度的UV-B辐射使大多数隐花植物细胞产生大量的活性氧（reactive oxygen species，ROS）。大量的活性氧自由基在体内的生成和滞留是UV-B辐射对隐花植物产生伤害的主要原因之一（Shiu and Lee，2005）。而隐花植物消除ROS的抗氧化系统在受到胁迫时会进化出一套清除ROS的保护网络，抗氧化酶活性会发生相应改变以适应和抵抗辐射胁迫造成的伤害（Collen and Davison，1999）。He和Häder（2002）通过观察UV-B辐射条件下鱼腥藻ROS含量的变化指出，UV-B辐射诱导鱼腥藻ROS产生使自由基消除系统失去平衡，进而影响一系列抗氧化酶含量的变化。UV-B对隐花植物蛋白质含量的影响还存在很多争议。Araoz等（1998）和Sauter等（1993）认为，UV-B辐射可抑制蛋白质合成、导致蛋白质降解、使蛋白质含量减少。而一些学者（Tevini *et al.*，1981）则认为UV-B辐射可以使蛋白质含量升高，可能是因为UV-B辐射导致植物体内类黄酮含量增高，而芳香族氨基酸是合成类黄酮的前体物质。DNA是UV-B辐射伤害植物的主要位点之一，植物经过UV-B诱导后，引起DNA分子中的嘧啶环发生一系列反应。高强度的UV-B辐射使DNA中相邻嘧啶环结合形成嘧啶二聚体，包括环丁烷嘧啶二聚体（CPDs）、6,4-光产物和Dewar异构体（Rajeshwar and Donat，2002）。最常见的是形成CPDs，CPDs会阻碍DNA聚合酶和RNA聚合酶Ⅱ在DNA双螺旋上的推进，进而阻止DNA的复制和转录（Doanhue *et al.*，1994）。UV-B诱导产生CPDs的多少随辐射强度和持续时间而上升（Tevini *et al.*，1981）。Hall等（2003）研究发现，增强UV-B辐射胁迫会使脱水地衣的光合作用减弱、色素系统受到伤害，且UV-B辐射在DNA中形成CPDs，地衣只有再水化后才能修复UV-B对DNA造成的损伤。脂肪酸是构成植物细胞膜的主要成分，其中不饱和脂肪酸能够保持细胞膜的相对流动性，以保证细胞的正常生理功能。UV-B辐射能导致隐花植物细胞的不饱和脂肪酸发生氧化，使细胞膜结构发生变化，导致物质的泄漏，从而影响植物一系列的生态生理指标（Rai *et al.*，1995）。为了应对UV-B辐射胁迫带来的损伤，大多数隐花植物通过改变自身脂肪酸组分比来提高抗UV-B能力。

（5）CO_2浓度

CO_2是大气中最主要的温室气体之一，人类活动导致大气圈中CO_2浓度正在逐年增加。CO_2是植物光合作用的底物，环境中CO_2浓度升高将对植物生长发育及生物量产生重要影响。首先，随着CO_2浓度的增加，植物光合作用增加，进而增加植物生物量，使进入土壤的凋落物数量增加，给土壤中的微生物提供更多的营养和能量（蒋高明等，1997）。一般认为C_3植物在CO_2加倍下光合能力提高10%～50%，C_4植物提高的程度<10%（Cure and Acock，1986）。其次，CO_2浓度升高将造成植物保卫细胞收缩，气孔关闭，从而使细胞内氧分压降低，进而呼吸作用降低（Stulen and Herton，1993）。再次，CO_2浓度升高对植物化学成分的影响主要表现在对非结构性碳水化合物、关键酶、蛋白质及化学组分方面的改变。

CO_2浓度升高影响BSC的光合作用，且光合作用的变化与其他环境因子密切相关。苏延桂等（2010）以腾格里沙漠东南缘沙坡头地区不同年代建植的人工植被区及相邻自然植被区藻类结皮为对象，研究了藻类结皮光合速率与BSC含水量、温度和CO_2浓度之间的关系，结果表明，CO_2浓度升高时，藻类结皮的净光合速率在高含水量时显著增加。在较大的降水事件后，藻类结皮的净光合速率首先处于一个较低的量值，随水分的减少，净光合速率逐渐上升，直至水分成为藻类结皮的限制因子，净光合速率再次下降。有研究表明，BSC光合速率与大气中CO_2的关系与BSC生物体的物种组成有关（Lange *et al.*，1997）。由于BSC中存在固氮酶，CO_2浓度升高可以提高固氮菌的光合作用，可能导致氮固定的增加（Deslippe *et al.*，2005）。氮素会影响BSC的光合作用、有机质的分解、同化产物的分配等过程。还有研究表明，BSC的分布和盖度与土壤中的碳、氮含量没有直接关系，但与土壤呼吸呈正相关关系（程军回和张元明，2010）。

（6）风沙

风沙活动直接影响着风沙地区植物资源的可持续利用与发展。在我国北方，风沙活动十分猖獗，尤其是在沙漠环境和干旱、半干旱地区，风沙活动十分频繁（史培军等，2000）。以往的研究主要是围绕风沙与其他逆境综合条件（降水、温度、湿度、养分等因子）进行的，研究内容涉及植物的生理生化、物质代谢及生态适应性等。大多数研究表明，风沙胁迫可使植物的净光合速率（P_n）、气孔导度（C_s）、叶温（T_i）、叶片水势（W_p）降低，使蒸腾速率（T_r）升高；且风速越大，吹风间隔越短，这些参数变幅越大；风沙能降低植物的水分利用率，进而增加植物的干燥作用；同时使脯氨酸含量增加（于云江等，2002）。

有关风沙对BSC影响的研究几乎全部集中在BSC可以大幅度提高土壤起动风速、减小地表风蚀（Belnap and Gillette，1997；Williams *et al.*，1999；王雪芹等，2004；李晓丽和申向东，2006）及BSC破损所造成的土壤细粒和有机质损失（West，1990）等方面，而关于风沙对BSC生态生理学特征影响的研究较少（Jia *et al.*，2008）。吴永胜等（2012）通过野外观测测定了固定沙丘纵断面表面气流和BSC的发育特征，结果表明，BSC的厚度、抗剪强度和细颗粒物含量均从沙丘底部向中上部呈减少的趋势，表现出“风速越高区域BSC的发育程度越低，风速越低区域BSC的发育程度越高”的特点。风的存在一方面加快了空气的流通，提高了土壤和植物水分蒸散速率（Grace and Russell，1977；Chintakovid *et al.*，2002；于云江等，2003），使干旱区植物较早进入干燥状态，而无风环境也不利于隐花植物进行光合作用（Kitaya *et al.*，2003）；另一方面，强风还可作为一个机械胁迫，影响植物叶片的形态和解剖结构，进而影响生理功能（Biddington，1986；Jaffe and Forbes，1993）。不难想象，上述两个方面都会对BSC生物产生影响，因为BSC光合作用产量大小不仅与受水分含量影响的净光合速率有关，还与BSC自身的水分保持能力有关（Lange，2003），而风沙作用的存在会通过改变BSC保水能力影响BSC生物的光合作用产量。

（7）干扰

BSC是干旱荒漠地区中比较脆弱和敏感的生态系统组成成分或构建者，其分布受干扰的影响波动较大，BSC对来自地表的人、动物及机械重压和火烧的干扰尤其敏感（参见第1章）。长期过度的干扰，将导致土壤水、氮和其他营养物质在时空上的异质性增强，土壤理化性质发生改变，诱发沙漠化的发生和发展（Manzano and Navar，2000）。影响BSC分布的干扰主要有放牧干扰、火烧干扰和机械性碾压。

干扰主要通过采食、践踏和排泄等方式对BSC产生影响，其中主要表现在践踏作用。受践踏后，土壤中真菌、地衣和藓类的种类和数量明显减少（Williams *et al.*，2008）；放牧干扰影响了土壤的机械组成，导致土壤中粗沙比例增加，钙含量下降，土壤有机质和pH发生变化（王雪芹等，2007）。放牧干扰使BSC在小尺度上的盖度和分布迅速减少，Pierre和Charles（1999）在撒哈拉沙漠的研究表明，BSC的盖度从原来没有放牧干扰的51.5%分别下降到适度放牧的18.8%和重度放牧的14.0%。BSC盖度的降低必然导致土壤侵蚀加剧，进而影响土壤的理化性质。

火烧通常被看作许多干旱、半干旱景观中一种常见的干扰形式。由于BSC主要分布在土壤表面1～3 mm，火烧使BSC生物体完全毁灭，并且火烧后地表覆盖率极低，导致土壤养分流失严重。因此，火烧干扰比放牧干扰对BSC的破坏性强。火烧干扰影响植被种群的形成过程、控制群落的组成和外貌（邱杨，1998）。火烧干扰后土壤的属性发生改变，这是由于燃烧向土壤中施加了热量、残留了灰烬，并且改变了原始土壤基质的状况（姜勇等，2003）。另外，火烧强度影响着物种的组成和多样性。低强度的火烧干扰对藓类和地衣结皮盖度没有显著影响，而高强度火烧显著降低了BSC的盖度和固氮速率（Belnap and Lange，2003）。但也有研究表明，火烧后不久，土壤表层（0～30 cm）微生物数量低于未烧地，在火烧次年，火烧地微生物数量超过未烧地（周道玮等，1999）。因此，从短期来看，火烧对土壤结构有不利影响，对BSC的盖度和物种种类产生负面作用；但从长期来看，火烧可以提高土壤结构的通透性，改变土壤有机质的组成，对土壤微生物数量和活性有一定的刺激作用，在一定程度上，火烧干扰对土壤有积极作用。

BSC对机械性碾压非常敏感，如施工车辆碾压、军事演习等对BSC有很大的影响，导致土壤理化性质改变。机械干扰后土壤容易被侵蚀，明显降低BSC的生物多样性和土壤稳定性。车辆碾压使地表敏感性提高13.72%～33.99%，即使恢复一年，其敏感性依然高达11.4%～29.50%（梁少民等，2005）。机械性碾压对BSC氮循环产生显著影响，降低BSC的硝酸酶活性和固氮量（Belnap *et al.*，1994），且冷沙漠（cold desert）比热沙漠（hot desert）对机械性碾压的敏感性更强，这是由于两种沙漠中不同特性的蓝藻造成的。冷沙漠的蓝藻系是大型、高度易变的非异形胞类型，固氮能力差，环境易受干扰。而热沙漠的蓝藻系是小型、相对稳定的有异形胞的类型，固氮能力强（Belnap，2002）。机械性碾压导致BSC土壤的破碎化，加速了风蚀和水蚀，进而降低了土壤肥力，影响BSC的生长和恢复。

对BSC在干扰后恢复能力的评价通常采用可视性指标，即地表盖度或生物量，有时也采用BSC中各组分所占的比例（Anderson *et al.*，1982）。BSC在受到干扰后的完全恢复是一个漫长的过程。通常，BSC完全恢复到可见的程度至少需要1～5年，恢复成较厚的BSC可能需要50年，恢复藓类和地衣结皮则需要250年时间（Belnap and Gardner，1993）。

2.2.2 生物因子

（1）维管植物

BSC盖度与维管植物生长之间是复杂且相互影响的关系（Bowker，2007）。BSC的分布与维管植物之间的关系受维管植物种类及影响途径的影响（Maestre and Cortina，2002）。研究表明，BSC 的分布与维管植物的出现存在互利关系。Harper 和 Pendleton（1993） 认为，BSC的存在可以促进维管植物对一些所需营养元素的吸收，从而有利于维管植物的生长。对干旱区草原的研究发现，荒漠植物形成的密集草丛减少了沉积物的流失，产生了明显的“沃岛效应”，使有机质和总稳定度分别提高了4.8倍和1.3倍；同时，维管植物过滤和吸收了部分到达地表的太阳辐射，提供了一个温度较低、土壤含水量较高的微环境（Bochet *et al.*，1999）。养分的富集和适宜的微环境加速了 BSC 的形成和发育。在刺槐（*Robinia pseudoacacia*）、兴安悬钩子（*Rubus chamaemorus*）、无芒雀麦（*Bromus inermis*）和长柔毛野豌豆（*Vicia villosa*）等维管植物之间，BSC的盖度和物种多样性较高（Neher *et al.*，2003；Thompson *et al.*，2006）。然而，Eldridge等（2006）的研究表明，BSC的分布与维管植物之间存在负相关关系。与有维管植物覆盖的地方相比，没有维管植物覆盖的地方BSC中的物种数和盖度显著增加（Eldridge and Tozer，1997）。随着植被的不断恢复，植物盖度和生物量的增加使得资源（水分、光照、养分等）竞争加剧，从而加剧了水分、光照、养分限制等胁迫对植物和BSC的影响。Schofield（1985）指出，在北美因过度放牧造成的草类植物大幅度降低导致地表苔藓类覆盖的增加。这可能是因为维管植物的存在对土壤中有限的养分竞争，限制了BSC形成和分布所需的资源，并且维管植物凋落物的增加，减少了BSC潜在的分布空间（Martinez *et al.*，2006；Briggs and Morgan，2008）。凋落物盖度在40%以内时，BSC盖度保持在20%左右；当凋落物盖度增加到40%～60%时，BSC盖度降低至10%；当凋落物盖度增加到60%～100%时，BSC盖度仍维持在10%。

（2）土壤动物

土壤动物是荒漠生态系统的重要成员，在生态系统中与微生物协同作用，对生态系统中物质转化、凋落物分解、土壤发育和成熟等过程起极其重要的作用，在一定程度上维持着整个生态系统的健康和生态服务功能的发挥（Williams *et al.*，1995；Whitford，2002）。土壤动物多样性的组成在一定程度上反映了其生境的变化特征。陈进福等（2006）利用样筐和网捕法收集了腾格里沙漠东南缘沙坡头地区的昆虫，调查发现，BSC盖度高且发育良好的植被区中昆虫以草原和荒漠化草原指示种占优势，而BSC盖度较低的植被区中昆虫种类以流沙、荒漠指示种占优势地位。

由于 BSC 的存在增加了土壤表层温度， 在一定程度上缓解了土壤温度的时空变异（Belnap and Lange，2003）；增加了土壤表面对凝结水的捕获，为许多爬行昆虫和动物提供水分（刘立超等，2005）；提高了土壤表层生物体抵御紫外线辐射的能力，使土壤中的生物免受紫外线辐射的危害（Garcia-Pichel and Castenholz，1993）。BSC作为荒漠生态系统的重要组成者，显著增加了土壤动物多样性和物种丰富度，这可能是由于BSC稳定了土表、改善了植被系统中的土壤环境，为荒漠土壤动物生存或完成生活史提供了适宜的生境，改善了它们生存的严酷生境，并作为土壤动物的部分食物来源，参与食物链的构成、物质循环和能量流动（陈进福等,2006）。BSC的存在明显增加了表层土壤微小节肢动物的数量（Steinberger，1989）。对位于腾格里沙漠东南缘沙坡头地区的4种不同生境的土壤动物多样性所做的比较认为，土壤动物的生物量和密度与BSC的发育阶段呈正相关关系（李新荣等，2009）。李新荣等（2008）通过实验发现，尖尾东鳖甲能取食藓类结皮，并对地衣和藻类也可能有一定的取食作用，同时解剖了几种拟步甲昆虫的成虫，在其消化道内均发现有BSC生物的存在。陈应武等（2007）的研究发现，掘穴蚁筑丘的生物干扰主要表现在改变土壤水分、土壤质地结构和理化性质，这些干扰导致土壤异质性增强，改善了土壤的环境状况。掘穴蚁的活动使得地表出现丘状流沙的聚集，导致地下出现巢穴，孔隙度增大，土壤容重下降。蚁丘流沙的疏松结构有利于降水下渗，同时有机物被埋在地下增加了土壤的保水能力（Wang *et al.*，1996）；此外，流沙的覆盖截断了土壤毛细管的提水作用，导致土壤水分蒸发量的降低（肖洪浪等，2003）。这些因素都导致蚁巢土壤含水量高于邻近土壤。蚂蚁的活动形成的蚁丘覆盖了表层BSC和有机质，导致有机物（BSC和凋落物）的腐烂分解；蚂蚁粪便和排泄物在蚁巢的聚集以及蚂蚁聚集食物（动物尸体和蚜虫蜜露等）均导致了土壤养分和无机盐在蚁巢的富集。

（3）土壤微生物和线虫

土壤微生物是形成BSC的主导因素之一，其数量和种群分布受到所处生态环境的影响。土壤微生物在土壤的物质转化和能量流动中起着重要的作用，参与土壤中有机物质的分解、土壤腐殖质的形成和分解过程以及土壤养分的转化和循环等（牛新胜等，2009）。在干旱、半干旱地区，许多土壤微生物通过自身的生理代谢活动及数量变化改变表层土壤的理化性质，进而影响地表植物的分布。

Belnap和Lange（2003）指出，BSC可提高土壤微生物的数量。边丹丹等（2011）研究了黄土丘陵区不同植被BSC及其下层土壤中的土壤微生物数量，发现土壤微生物数量与BSC的覆盖程度有关，有BSC的土壤微生物数量高于无BSC的土壤微生物数量，其中细菌数量以中等厚度BSC最高，放线菌和真菌数量以薄BSC最高。吴楠等（2005）的研究表明，土壤微生物的数量变化特征与BSC的类型、分布以及土壤有机质含量等因子有一定的相关性。

土壤微生物量是指土壤中除植物根茬等残体和大于$5\times10^3\ \mu m^3$的土壤动物以外，体积小于$5\times10^3\ \mu m^3$的具有生命活动的生物总量，其主要类群为细菌、真菌、放线菌、藻类和原生动物等，是土壤有机质中最活跃和最易变化的部分。土壤微生物量主要包括微生物碳、微生物氮和微生物磷（Zhong and Cai，2007；Kaschuk *et al.*，2010）。土壤酶是一种生物催化剂，主要来源于土壤微生物和植物根系的分泌物及动植物残体的分解，包括氧化还原酶类、水解酶类、裂合酶类和转移酶类等。陈政等（2009）的报道显示，铜尾矿废弃地的BSC可提高表层尾矿中的土壤微生物量。陈祝春（1991）研究了沙坡头地区BSC形成过程中土壤微生物和土壤酶活性的变化，表明流动沙丘上建立人工植被后，土壤中有机质含量、微生物数量均增加，土壤酶活性也相应增强，从而加速了土壤中的生物化学转化过程，促进BSC的形成。刘艳梅等（2014）以腾格里沙漠东南缘1956年、1964年、1981年和1987年人工固沙区不同BSC覆盖的沙丘土壤为研究对象，以流沙区为对照，探讨了BSC对土壤微生物生物量碳和氮的影响。研究表明，与流沙区相比，54龄、46龄、29龄和23龄人工固沙植被区BSC可显著提高土壤微生物生物量碳和氮含量，且固沙年限与土壤微生物生物量碳和氮含量呈显著正相关关系。

线虫是土壤中最丰富的后生动物，广泛存在于各种生境，在土壤生态系统腐屑食物网中占有重要地位（肖能文等，2011）。线虫通过与微生物的相互作用不仅能维持土壤生态系统稳定，促进物质循环和能量流动，而且能敏感地指示土壤的恢复程度，是衡量沙区生态恢复与健康的重要生物学属性。Belnap和Lange（2003）报道了BSC能为土壤线虫的生存提供食物来源和适宜的居住场所。此外，Darby等（2007）的研究表明，土壤线虫群落受BSC发育阶段的影响，即相对于早期阶段的BSC，发育晚期的BSC下土壤线虫群落更成熟、更复杂。刘艳梅等（2013）调查了腾格里沙漠东南缘的人工植被固沙区藻类结皮和藓类结皮下的土壤线虫，发现演替后期的藓类结皮比演替早期的藻类结皮中土壤线虫多度和丰富度更高，更有利于土壤线虫的生存，且随着土层厚度的增加，土壤线虫多度和丰富度降低，说明BSC的存在有利于浅层土壤线虫多样性的维持。这可能是因为相对于演替早期的藻类结皮，演替晚期的藓类结皮由于生长慢、BSC层较厚，能为土壤线虫提供更多、更丰富的食物来源、更适宜的土壤温度、更高的土壤湿度和有机质含量以及更稳定的食物网结构。

参考文献

安黎哲, 冯虎元, 王勋陵. 增强的紫外线–B辐射对几种作物和品种生长的影响. 生态学报, 2001, 21(2): 249–253.

边丹丹, 廖超英, 孙长忠, 李晓明, 许永霞, 唐海滨. 黄土丘陵区土壤生物结皮对土壤微生物分布特征的影响. 干旱地区农业研究, 2011, 29(4): 109–114.

陈进福, 李新荣, 陈应武, 苏延桂. 生物土壤结皮对荒漠昆虫多样性的影响. 中国沙漠, 2006, 26(6): 986–996.

陈应武, 李新荣, 苏延桂, 窦彩虹, 贾晓红, 张志山. 腾格里沙漠人工植被区掘穴蚁(*Formica cunicularia*)的生态功能. 生态学报, 2007, 27(4): 1508–1514.

陈政, 阳贵德, 孙庆业. 生物结皮对铜尾矿废弃地土壤微生物量及酶活性的影响. 应用生态学报, 2009, 20(9): 2193–2198.

陈祝春. 沙丘结皮层形成过程的土壤微生物和土壤酶活性. 环境科学, 1991, 12(1): 19–23.

程军回, 张元明. 影响生物土壤结皮分布的环境因子. 生态学杂志, 2010, 29(1): 133–141.

杜晓濛, 李菁, 田向荣, 李朝阳, 李鹄鸣. 尖叶拟船叶藓光系统Ⅱ光合荧光特性、活性氧代谢与耐脱水生理生态适应的关系. 生态学报, 2014, 34(23): 6807–6816.

房世波, 冯凌, 刘华杰. 生物土壤结皮对全球气候变化的响应. 生态学报, 2008, 28(7): 3312–3321.

黄璐琦, 王康才. 药用植物生理生态学. 北京: 中国中医药出版社, 2012.

黄梅玲, 江洪, 金清, 余树全. UV–B辐射胁迫下不同起源时期的3种木本植物幼苗的生长及光合特性. 生态学报, 2010, 30(8): 1998–2009.

姜勇, 诸葛玉平, 梁超, 张旭东. 火烧对土壤性质的影响. 土壤学报, 2003, 34(1): 65–69.

蒋高明, 韩兴国, 林光辉. 大气CO_2浓度升高对植物的直接影响——国外十余年来模拟实验研究之主要手段及基本结论. 植物生态学报, 1997, 21(6): 489–502.

蒋高明. 植物生理生态学. 北京: 高等教育出版社, 2004.

李刚, 刘立超, 高艳红, 赵杰才, 杨昊天. 降雪对生物土壤结皮光合及呼吸作用的影响. 中国沙漠, 2014, 34(4): 998–1006.

李豪杰, 郭世昌, 陈辉, 陈宗瑜, 常有礼, 张秀年, 温永琴, 周平. 昆明地区春季紫外辐射监测及变化研究. 云南大学学报(自然科学版), 2004, 26(6): 509–515.

李晓丽, 申向东. 结皮土壤的抗风蚀性分析. 干旱区资源与环境, 2006, 20(2): 203–207.

李新荣, 陈应武, 贾荣亮. 生物土壤结皮：荒漠昆虫食物链的重要构建者. 中国沙漠, 2008, 28(2): 245–248.

李新荣, 张元明, 赵允格. 生物土壤结皮研究：进展、前沿与展望. 地球科学进展, 2009, 24(1): 11–24.

梁少民, 吴楠, 王红玲, 聂华丽, 张元明. 干扰对生物土壤结皮及其理化性质的影响. 干旱区地理, 2005, 28(6): 818–823.

刘国顺, 乔新荣, 王芳, 杨超, 郭桥燕, 云菲. 光照强度对烤烟光合特性及其生长和品质的影响. 西北植物学报, 2007, 27(9): 1833–1837.

刘鸿雁. 植物学. 北京: 北京大学出版社, 2005.

刘立超, 李守中, 宋耀选, 张志山, 李新荣. 沙坡头人工植被区微生物结皮对地表蒸发影响的试验研究. 中国沙漠, 2005, 25(2): 191–195.

刘艳梅, 李新荣, 赵昕, 张鹏, 回嵘. 生物土壤结皮对荒漠土壤线虫群落的影响. 生态学报, 2013, 33(9): 2816–2824.

刘艳梅, 杨航宇, 李新荣. 生物土壤结皮对荒漠区土壤微生物生物量的影响. 土壤学报, 2014, 51(2): 394–401.

娄红祥. 苔藓植物化学与生物学. 北京: 北京科学技术出版社, 2006.

牛新胜, 张宏彦, 王立刚. 玉米秸秆覆盖冬小麦免耕播种对土壤微生物量碳的影响. 中国土壤与肥料, 2009, 46(1): 64–68.

邱杨. 森林植被的自然火干扰. 生态学杂志, 1998, 17(1): 54–60.

史培军, 张宏, 王平, 周武光. 我国沙区防沙治沙的区域模式. 自然灾害学报, 2000, 9(3): 1–7.

苏延桂, 李新荣, 陈应武, 崔艳, 鲁艳. 温度和CO_2浓度升高对荒漠藻结皮光合作用的影响. 应用生态学报, 2010, 21(9): 2217–2222.

苏延桂, 李新荣, 赵昕, 王正宁. 紫外辐射增强对不同发育阶段荒漠藻结皮光合作用的影响. 中国沙漠, 2011, 31(4): 889–893.

孙儒泳, 李庆芬, 牛翠娟, 娄安如. 基础生态学. 北京: 高等教育出版社, 2002.

孙敬松, 周广胜, 韩广轩. 太阳辐射对玉米农田土壤呼吸作用的影响. 生态学报, 2010, 30(21): 5925–5932.

王雪芹, 张元明, 张伟民, 韩致文. 古尔班通古特沙漠生物结皮对地表风蚀作用影响的风洞实验. 冰川冻土, 2004, 26(5): 632–638.

王雪芹, 张元明, 蒋进, 杨维康, 陈明, 张继凯, 陈均杰, 宋春武. 放牧对古尔班通古特沙漠南部沙垄地表性质的影响. 地理学报, 2007, 62(7): 698–706.

吴楠, 潘伯荣, 张元明, 王红玲, 梁少民, 聂华丽. 古尔班通古特沙漠生物结皮中土壤微生物垂直分布特征. 应用与环境生物学, 2005, 11(3): 349–353.

吴永胜, 哈斯, 屈志强. 影响生物土壤结皮在沙丘不同地貌部位分布的风因子讨论. 中国沙漠, 2012, 32(4): 980–984.

吴玉环, 黄国宏, 高谦, 曹同. 苔藓植物对环境变化的响应及适应性研究进展. 应用生态学报, 2001, 12(6): 943–946.
武立权, 尤翠翠, 柯建, 黄义德. 高温对水稻黄叶突变体剑叶光合特性和叶绿体超微结构的影响. 西北植物学报, 2012, 32(11): 2264–2269.
肖洪浪, 李新荣, 段争虎, 李涛, 李守中. 流沙固定过程中土壤–植被系统演变. 中国沙漠, 2003, 23(6): 605–611.
肖能文, 谢德燕, 王学霞, 闫春红, 胡理乐, 李俊生. 大庆油田石油开采对土壤线虫群落的影响. 生态学报, 2011, 31(13): 3736–3744.
肖媛, 王高鸿, 刘永定. UV–B辐射对雨生红球藻光合特性和虾青素含量的影响及其响应. 水生生物学报, 2010, 34(6): 1077–1082.
于云江, 史培军, 贺丽萍, 刘家琼. 风沙流对植物生长影响的研究. 地球科学进展, 2002, 17(2): 262–267.
于云江, 史培军, 鲁春霞, 刘家琼. 不同风沙条件对几种植物生态生理特征的影响. 植物生态学报, 2003, 27(1): 53–58.
张丙昌, 赵建成, 张元明, 李文斋, 郑云普. 不同生态因子对生物结皮中土生绿球藻生长的影响. 干旱区研究, 2007, 24(5): 641–646.
张红霞, 吴能表, 胡丽涛, 洪鸿. 不同强度UV–B辐射胁迫对蚕豆幼苗生长及叶绿素荧光特性的影响. 西南师范大学学报(自然科学版), 2010, 35(1): 105–110.
郑有飞, 刘建军, 王艳娜, 吴荣军. 增强UV–B辐射与其他因子复合作用对植物生长的影响研究. 西北植物学报, 2007, 27(8): 1702–1712.
周道玮, 岳秀泉, 孙刚, 李月胜. 草原火烧后土壤微生物的变化. 东北师大学报(自然科学版), 1999, 49(1): 118–124.
Allen DJ, Nogues S, Baker NR. Ozone depletion and increased UV–B radiation: Is there a real threat to photosynthesis? *Journal of Experimental Botany*, 1998, 49: 1775–1788.
Alpert P, Oechel WC. Comparative patterns of net photosynthesis in an assemblage of mosses with contrasting microdistributions. *The American Journal of Botany*, 1987, 74: 1787–1796.
Anderson DC, Harper KT, Rushforth SR. Recovery of cryptogamic soil crusts from grazing on Utah winter ranges. *Journal of Range Management*, 1982, 35: 355–359.
Araoz R, Lebert M, Häder DP. Translation activity under ultraviolet radiation and temperature stress in the cyanobacterium *Nostoc* sp. *Journal of Photochemistry and Photobiology B: Biology*, 1998, 47: 115–120.
Bassman JH. Ecosystem consequences of enhanced solar ultraviolet radiation: Secondary plant metabolites as mediators of multiple trophic interactions in terrestrial plant communities. *Photochemistry and Photobiology*, 2004, 79: 382–398.
Belnap J. Impacts of off–road vehicles on nitrogen cycles in biological soil crusts: Resistance in different U.S. deserts. *Journal of Arid Environments*, 2002, 52: 155–165.
Belnap J, Gardner JS. Soil microstructure in soils of the Colorado Plateau: The role of the cyanobacterium *Microcoleus vaginatus*. *Great Basin Naturalists*, 1993, 53: 40–47.
Belnap J, Gillette DA. Disturbance of biological soil crusts: Impacts on potential wind erodibility of sandy desert soils in southeastern Utah. *Land Degradation and Development*, 1997, 8: 355–362.
Belnap J, Lange OL. *Biological Soil Crusts: Structure, Function, and Management*. Berlin: Springer–Verlag, 2003.
Belnap J, Harper KT, Warren SD. Surface disturbance of cryptobiotic soil crusts: Nitrogenase activity, chlorophyll content and chlorophyll degradation. *Arid Soil Research and Rehabilitation*, 1994, 8: 1–8.
Belnap J, Phillips SL, Miller ME. Response of desert biological soil crusts to alterations in precipitation frequency. *Oecologia*, 2004, 141: 306–316.
Biddington NL. The effects of mechanically–induced stress in plants—a review. *Plant Growth Regulation*, 1986, 4: 103–123.
Bochet E, Rubio JL, Poesen J. Modified topsoil islands within patchy Mediterranean vegetation in SE Spain. *Catena*, 1999, 38: 23–44.
Bowker MA, Reed SC, Belnap J, Phillips SL. Temporal variation in community composition, pigmentation, and F_v/F_m of desert cyanobacterial soil crusts. *Microbial Ecology*, 2002, 43: 13–25.
Bowker MA. Biological soil crust rehabilitation in theory and practice: An underexploited opportunity. *Restoration Ecology*, 2007, 15: 13–23.
Briggs A, Morgan JW. Morphological diversity and abundance of biological soil crusts differ in relation to landscape setting and vegetation type. *Australian Journal of Botany*, 2008, 56: 246–253.
Campbell D. Complementary chromatic adaptation alters photosynthetic strategies in the cyanobacterium Calothrix. *Microbiology*, 1996, 142: 1255–1263.
Chapin FS, Shaver GR, Giblin AE, Nadelhoffer KJ, Laundre JA. Responses of arctic tundra to experimental and observed changes in climate. *Ecology*, 1995, 76: 694–711.
Chintakovid W, Kubota C, Bostick WM, Kozai T. Effect of air current speed on evapotranspiration rate of transplant canopy under artificial light. *Japanese Society of High Technology in Agriculture*, 2002, 14: 25–31.

Collen J, Davison IR. Reactive oxygen metabolism in intertidal *Fucus* spp. (Phaeophyceae). *Journal of Phycology*, 1999, 35: 62–69.

Cornelissen JHC, Callaghan TV, Alatalo JM, Michelsen A, Graglia E, Hartley AE, Hik DS, Hobbie SE, Press MC, Robinson CH, Henry GHR, Shaver GR, Phoenix GK, Jones DG, Jones DG, Jonasson S, Chapin III FS, Molau U, Neill C, Lee JA, Melillo JM, Sveinbjörnsson B, Aerts R. Global change and arctic ecosystems: Is lichen decline a function of increases in vascular plant biomass? *Journal of Ecology*, 2001, 89: 984–994.

Cure JD, Acock B. Crop response to carbon dioxide doubling: A literature survey. *Agriculture and Forest Meteorology*, 1986, 38: 127–145.

D' surney SJ, Tschaplinski TJ, Edwards NT, Shugart LR. Biological responses of two soybean cultivars exposed to enhanced UVB radiation. *Environmental and Experimental Botany*, 1993, 33: 347–356.

Darby BJ, Neher DA, Belnap J. Soil nematode communities are ecologically more mature beneath late–than early–successional stage biological soil crusts. *Applied Soil Ecology*, 2007, 35: 203–212.

Davey MC. Effects of continuous and repeated dehydration on carbon fixation by bryophytes from the maritime Antarctic. *Oecologia*, 1997, 110: 25–31.

Davey MC, Rothery P. Seasonal variation in respiratory and photosynthetic parameters in three mosses from the maritime antarctic. *Annals of Botany*, 1996, 78: 719–728.

Deltoro VI, Calatayud A, Gimeno C, Abadía A, Barreno E. Changes in chlorophyll a fluorescence, photosynthetic CO_2 assimilation and xanthophyll cycle interconversions during dehydration in desiccation–tolerant and intolerant liverworts. *Planta*, 1998, 207: 224–228.

Deslippe JR, Egger KN, Henry GHR. Impacts of warming and fertilization on nitrogen–fixing microbial communities in the Canadian High Arctic. *FEMS Microbiology Ecology*, 2005, 53: 41–50.

Dietz S, Büdel B, Lange OL, Bilger W. Transmittance of light through the cortex of lichens from contrasting habitats. *Bibliotheca Lichenologica*, 2000, 75: 171–182.

Doanhue BA, Taylor JS, Reines D. Transcript cleavage by RNA polymerase Ⅱ arrested by a cyclobutane dimmer in the DNA template. *Proceedings of the National Academy of Sciences of the United States of America*, 1994, 91: 8502–8506.

Dodds WK. Microscale vertical profiles of N_2 fixation, photosynthesis, O_2, chlorophyll *a* and light in a cyanobacterial assemblage. *Applied and Environmental Microbiology*, 1989, 55: 882–886.

Eldridge DJ, Tozer ME. Environmental factors relating to the distribution of terricolous bryophytes and lichens in semi–arid eastern Australia. *Bryologist*, 1997, 100: 28–39.

Garcia PF, Pringault O. Cyanobacteria track water in desert soils. *Nature*, 2001, 413: 380–381.

Garcia–Pichel F, Castenholz RW. Occurrence of UV–absorbing, mycosporine–like compounds among cyanobacterial isolates and an estimate for their screening capacity. *Applied Environmental Microbiology*, 1993, 59: 163–169.

Gauslaa Y, Mcevoy M. Seasonal changes in solar radiation drive acclimation of the sun–screening compound parietin in the lichen *Xanthoria parietina*. *Basic and Applied Ecology*, 2005, 6: 75–82.

Gehrke C. Impacts of enhanced ultraviolet–b radiation on mosses in a subarctic heath ecosystem. *Ecology*, 1999, 80: 1844–1851.

Grace J, Russell G. The effect of wind on grasses: Ⅲ. Influence of continuous drought or wind on anatomy and water relations in *Festuca arundinacea* Schreb. *Journal of Experimental Botany*, 1977, 28: 268–278.

Green TGA, Büdel B, Mayer A, Zellner H, Lange OL. Temperate rainforest lichens in New Zealand: Light response of photosynthesis. *New Zealand Journal of Botany*, 1997, 35: 493–504.

Green TGA, Schroeter B, Kappen L, Seppelt R, Maseyk K. An assessment of the relationship between chlorophyll fluorescence and CO_2 gas exchange from field measurements on a moss and lichen. *Planta*, 1998, 206: 611–618.

Hall RSB. Effects of increased UV-B radiation on the lichen *Cladonia arbuscula* ssp. *Mitis:* UV-absorbing pigments and dNA damage. Lund: Lund University, 2002.

Hall RSB, Paulsson M, Duncan K, Tobin A, Widell S, Bornman J. Water-and temperature–dependence of DNA damage and repair in the fruticose lichen *Cladonia arbuscula* ssp. *mitis* exposed to UV–B radiation. *Physiologia Plantarum*, 2003, 118: 371–379.

Hancke K, Glud RN. Temperature effects on respiration and photosynthesis in three diatom–dominated benthic communities. *Aquatic Microbial Ecology*, 2004, 37: 265–281.

Harper KT, Pendleton RL. Cyanobacteria and cyanolichens: Can they enhance availability of essential minerals for higher plants? *Great Basin Naturalist*, 1993, 53: 59–72.

He YY, Häder DP. Involvement of reactive oxygen species in the UV–B damage to the cyanobacterium Anabaena sp. *Journal of Photochemistry and Photobiology B: Biology*, 2002, 66: 73–80.

Jaffe MJ, Forbes S. Thigmomorphogenesis: The effect of mechanical perturbation on plants. *Plant Growth Regulation*, 1993, 12: 313–324.

Jeffrise DL, Link SO, Klopatek JM. CO_2 fluxes of cryptogamic crusts I. Response to resaturation. *New Phytologist*, 1993a, 125:

163–173.

Jeffrise DL, Link SO, Klopatek JM. CO_2 fluxes of cryptogamic crusts Ⅱ. Response to dehydration. *New Phytologist*, 1993b, 125: 391–396.

Jia RL, Li XR, Liu LC, Gao YH, Li XJ. Responses of biological soil crusts to sand burial in a revegetated area of the Tengger Desert, Northern China. *Soil Biology and Biochemistry*, 2008, 40: 2827–2834.

Kaschuk G, Alberton O, Hungria M. Three decades of soil microbial biomass studies in Brazilian ecosystems: Lessons learned about soil quality and indications for improving sustainability. *Soil Biology and Biochemistry*, 2010, 42: 1–13.

Kitaya Y, Tsuruyama J, Shibuya T, Yoshida M, Kiyota M. Effects of air current speed on gas exchange in plant leaves and plant canopies. *Advances in Space Research*, 2003, 31(1): 177–182.

Lange OL. Photosynthesis of soil–crust biota as dependent on environmental factors. In: Belnap J, Lange OL (eds.). *Biological Soil Crusts: Structure, Function, and Management.* Berlin: Springer–Verlag, 2003.

Lange OL, Belnap J, Reichenberger H. Photosynthesis of the cyanobacterial soil–crust lichen *Collema tenax* from arid lands in southern Utah, USA: Role of water content on light and temperature responses of CO_2 exchange. *Functional Ecology*, 1998, 12: 195–202.

Lange OL, Belnap J, Reichenberger H, Meyer A. Photosynthesis of green algal soil crust lichens from arid lands in southern Utah, USA: Role of water content on light and temperature responses of CO_2 exchange. *Flora*, 1997, 192: 1–15.

Lange OL, Meyer A, Zellner H, Heber U. Photosynthesis and water relations of lichen soil crusts: Field measurements in the coastal fog zone of the Namib Desert. *Functional Ecology*, 1994, 8: 253–264.

Larcher W. 植物生态生理学. 翟志席, 郭玉海, 马永泽, 等译. 北京: 中国农业大学出版社, 1997.

Levi Y, Berner T, Cohen Y. CO_2 exchange and growth rate of the loess soil crusts algae in the Negev Desert of Israel. In: Hillel I(ed). *Developments in Arid Zone Ecology and Environmental Quality*. Philadelphia: Balaban ISS, 1981: 43–48.

Li JH, Li XR, Chen CY. Degradation and reorganization of thylakoid proteins complexes are involved in the rapid photosynthetic changes of desert moss *Bryum argenteum* in response to dehydration and rehydration. *The Bryologist*, 2014, 117: 110–118.

Li XM, Zhang LH, Li YY, Ma LJ, Bu N, Ma CY. Changes in photosynthesis, antioxidant enzymes and lipid peroxidation in soybean seedlings exposed to UV–B radiation and/or Cd. *Plant and Soil*, 2012, 352: 377–387.

Li XR, Xiao HL, Zhang JG, Wang XP. Long–term ecosystem effects of sand–binding vegetation in Shapotou region of Tengger Desert, Northern China. *Restoration Ecology*, 2004, 2: 376–390.

Li XR, Zhang P, Su YG, Jia RL. Carbon fixation by biological soil crusts following stabilization of sand dune in arid desert regions of China: A four–year field study. *Catena*, 2012, 97: 119–126.

Li XR, Gao YH, Su JQ, Jia RL, Zhang ZS. Ants mediate soil water in arid desert ecosystems: Mitigating rainfall interception induced by biological soil crusts? *Applied Soil Ecology*, 2014, 78: 57–64.

Lindroth RL, Hofmann RW. Population differences in *Trifolium repens* L. response to ultraviolet–B radiation: Foliar chemistry and consequences for two *Lepidopteran* herbivores. *Oecologia*, 2000, 122: 20–28.

Maestre FT, Cortina J. Spatial patterns of surface soil properties and vegetation in a Mediterranean semi–arid steppe. *Plant and Soil*, 2002, 241: 279–291.

Manzano MG, Navar J. Processes of desertification by goats over–grazing in the *Tamaulipan thornscrub* (matorral) in northeastern Maxico. *Journal of Arid Environments*, 2000, 44: 1–17.

Marschall M, Proctor MCF. Are bryophytes shade plants? Photosynthetic light responses and proportions of chlorophyll *a*, chlorophyll *b* and total carotenoids. *Annals of Botany*, 2004, 94: 593–603.

Martinez I, Escudero A, Maestre FT, de la Cruz A, Guerrero C, Rubio A. Small–scale patterns of abundance of mosses and lichens forming biological soil crusts in two semi–arid gypsum environments. *Australian Journal of Botany*, 2006, 54: 339–348.

Martins LD, Tomaz MA, Lidon FC, DaMatta FM, Ramalho JC. Combined effects of elevated [CO_2] and high temperature on leaf mineral balance in *Coffea* spp. plants. *Climate Change*, 2014, 126: 365–379.

McKenzie RL, Aucamp PJ, Bais AF, Björn LO, Ilyas M. Changes in biologically–active ultraviolet radiation reaching the earth's surface. *Photochemical and Photobiological Science*, 2007, 6: 218–231.

Molles M C. Ecology: *Concepts and Applications*. Beijing: Science Press, 2000.

Monika ES, Wolfgang B, Ssiegfried S. UV–B–induced synthesis of photoprotective pigments and extracellular polysaccharides in the terrestrial cyanobacterium *Nostoc commune*. *Journal of Bacteriology*, 1997, 179: 1940–1945.

Neher DA, Walters T, Tramer E, Weicht TR, Veluci RM, Saiya–Cork K, Will–Wolf S, Toppin J, Traub J, Johansen JR. Biological soil crust and plant communities in a sand savanna of northwestern Ohio. *Journal of the Torrey Botanical Society*, 2003, 130: 244–252.

Newsham KK, Hodgson DA, Murray AWA, Peat HJ, Smith RIL. Response of two Antarctic bryophytes to stratospheric ozone depletion. *Global Change Biology*, 2002, 8: 972–983.

Palmqvist K. Carbon economy in lichens. *New Phytologist*, 2000, 148: 11–36.

Pierre H, Charles L. Effects of livestock grazing on physical and chemical properties of sandy soils in Sahelian range lands. *Journal of Arid Environments*, 1999, 41: 231–245.

Rai LC, Tyagi B, Mallick N, Rai PK. Interactive effects of UV–B and copper on photosynthetic activity of the cyanobacterium *Anabaena doliolum*. *Environmental and Experimental Botany*, 1995, 35: 177–185.

Rajeshwar PS, Donat PH. UV–induced DNA damage and repair: A review. *Photochemistry and Photobiological Sciences*, 2002, 1: 225–236.

San José JJ, Bravo CR. CO_2 exchange in soil algal crusts occurring in the *Trachypogon savannas* of the Orinoco Llanos, Venezuela. *Plant and Soil*, 1991, 135: 233–244.

Sauter M, Seagull RW, Kende H. Internodal elongation and orientation of cellulose microfibrils and microtubules in deepwater rice. *Planta*, 1993, 190: 354–362.

Schofield WB. *Introduction of Bryology*. New York: Macmillan, 1985.

Shiu CT, Lee TM. Ultraviolet–B–induced oxidative stress and responses of the ascorbate–glutathione cycle in amarine macroalga *Ulvafasciata*. *Journal of Experimental Botany*, 2005, 56: 2851–2865.

Sonesson M, Gehrke C, Tjus M. CO_2 environment, microclimate and photosynthetic characteristics of the moss *Hylocomium splendens* in a subarctic habitat. *Oecologia*, 1992, 92: 23–29.

Sponseller RA. Precipitation pulses and soil CO_2 flux in a Sonoran Desert ecosystem. *Global Change Biology*, 2007, 13: 426–436.

Steinberger Y. Energy and protein budgets of the desert isopod *Hemilepistrus reaumuri*. *Acta Oecologia,* 1989, 10: 117–134.

Stulen I, Herton J. Root growth and function under atmospheric CO_2 enrichment. *Vegetatio*, 1993, 104/105: 99–115.

Suresh BG, Joshi PC, Viswanathan PN. UV–B induced reduction in biomass and overall productivity of cyanobacteria. *Biochemical and Biophysical Research Communications*, 1998, 244: 138–142.

Tevini M, Iwanzik W, Thoma U. Some effects of enhanced UV–B irradiation the growth and composition of plants. *Planta*, 1981, 153: 388–394.

Thompson WA, Eldridge DJ, Boner SP. Structure of biological soil crust communities in *Callitris glaucophylla* woodlands of New South Wales, Australia. *Journal of Vegetation Science*, 2006, 17: 271–280.

Tretiach M, Pecchiari M. Gas exchange rates and chlorophyll content of epi– and endolithic lichens from the Trieste Karst (NE Italy). *New Phytologist*, 1995, 130: 585–592.

Valkama E, Kivimäenpää M, Hartikainen H, Wulff A. The combined effects of enhanced UV–B radiation and selenium on growth, chlorophyll fluorescence and ultrastructure in strawberry (*Frigaria ananassa*) and barley (*Hordeum vulgare*) treated in the field. *Agricultural and Forest Meteorology* , 2003, 120: 267– 278.

Wang D, Lowery B, McSweeney K, Norman JM. Spatial and temporal patterns of ant burrow openings as affected by soil properties and agricultural practices. *Pedobiologia*, 1996, 40: 201–211.

West NE. Structure and function of microphytic soil crusts in wildland ecosystems of arid to semi–arid regions. *Advances in Ecological Research*, 1990, 20: 179–223.

Whitford W. *Ecology of Desert Systems*. San Diego: Academic Press, 2002, 343.

Williams JD, Dobrowolski JP, West NE. Microbiotic crust influence on unsaturated hydraulic conductivity. *Arid Soil Research and Rehabilitation*, 1999, 13: 145–154.

Williams JD, Dobrowolski JP, West NE, Gillette DA. Microphytic soil crust influence on wind erosion. A*merican Society of Agricultural Engineers*, 1995, 38: 131–137.

Williams WJ, Eldridge DJ, Alchin BM. Grazing and drought reduce cyanobacterial soil crusts in an Australian *Acacia* woodland. *Journal of Arid Environments*, 2008, 72: 1064–1075.

Zaady E, Kuhn U, Wilske B, Sandoval–Soto L, Kesselmeier J. Patterns of CO_2 exchange in biological soil crusts of successional age. *Soil Biology and Biochemistry*, 2000, 32: 959–966.

Zhong WH, Cai ZC. Long–term effects of inorganic fertilizers on microbial biomass and community functional diversity in a paddy soil derived from quaternary red clay. *Applied Soil Ecology*, 2007, 36: 84–91.

Zotz G, Schweikert A, Jetz W, Westerman H. Water relations and carbon gain are closely related to cushion size in the moss *Grimmia pulvinata*. *New Phytologist,* 2000, 148: 59–67.

第3章 荒漠BSC对非生物因子的生态生理响应

正如第2章所讲，影响荒漠生态系统生物体的非生物环境因子很多，而对BSC群落的主要构建者蓝藻、绿藻、地衣和藓类而言，对其影响最大的莫过于水分。不同于非维管植物（no-vascular plant），BSC群落中的这些组成成分常处于“干燥”状态，它们遇水而“活”，包括在有限的降水、吸湿凝结水和降雪事件后诱发生物活性。但是不同的湿润过程、湿润强度和湿润持续时间以及频繁的干湿交替也对这些BSC组成造成了很大的“伤害”。此外，光照、高温、UV-B辐射的增强、CO_2浓度升高和大气氮沉降的增加等全球变化诱发的潜在胁迫以及风沙活动与火烧干扰等，均在不同程度上抑制着BSC的拓殖和发展，直接影响着BSC多元生态功能的发挥。同时，BSC群落中不同类群的组分对这些胁迫有着特殊的响应，长期以来相关研究没有得到应有的重视，妨碍了我们从荒漠植物空间分布格局（维管植物斑块与BSC斑块镶嵌分布）及其动态这一“视角”探讨荒漠/沙地生态过程及其机理。

3.1 BSC对水分的生态生理响应

水分是荒漠生态系统的主要限制因子。尽管BSC群落中优势种如蓝藻、绿藻、地衣和藓类都可以归类为变水植物，对水分的需求远远低于维管植物，但是土壤水分状况，尤其是表土层的湿度对其存在和发展也起着十分重要的作用，甚至决定着不同种类、不同生活型和功能群在BSC群落中的分布格局（见第1章）。在小微尺度上，微小的地表土壤水分差异，如地表小土堆（soil mound）的存在造成的湿度差异（土堆四个方向的湿度均存在较大差异），使BSC隐花植物的分布也不相同（李新荣，2012）。不同的隐花植物对表层土壤水分的变化本身存在着不同的生态生理响应，对这些响应机理的研究是全面揭示荒漠/沙地生态系统植被格局和过程的理论基础和重要前提。因为因水分差异和变化导致的BSC群落组成、结构和功能的变化，直接关系到其在荒漠/沙地系统中功能的发挥，为生态系统健康评估和生态系统管理带来许多不确定性。

3.1.1 荒漠藓类耐干燥的生理生化机制研究

苔藓植物是一类以孢子繁殖、由水生向陆生过渡的高等植物，是物种数仅次于种子植物的高等植物，有15000（Gradstein *et al.*，2001）~25000（Crum，2001）种，在自然界的不同生境中广泛分布，如沙漠、苔原以及一些微生境，如叶片、树干、岩石和土壤等，但参与BSC形成的仅是那些分布在地表与土壤颗粒发生胶结的种类（李新荣，2012）。尽管苔藓普遍被认为具有耐干燥能力（Proctor *et al.*，2007b），但实际上对苔藓耐干燥机制的研究目前仅局限于几个种。Wood（2007）曾鉴定了一部分苔藓包括158种藓、51种苔和1种角苔具有耐干燥能力，而这些仅占苔藓总数的大约1%。其中墙藓（*Tortula muralis*）一直被作为模式耐干燥苔藓，从生理、生化、结构及分子方面已有深入研究（Oliver and Bewley，1997；Wood and Oliver，2004）。图3-1集中解释了目前对苔藓耐干燥的生理生化调控的认识。

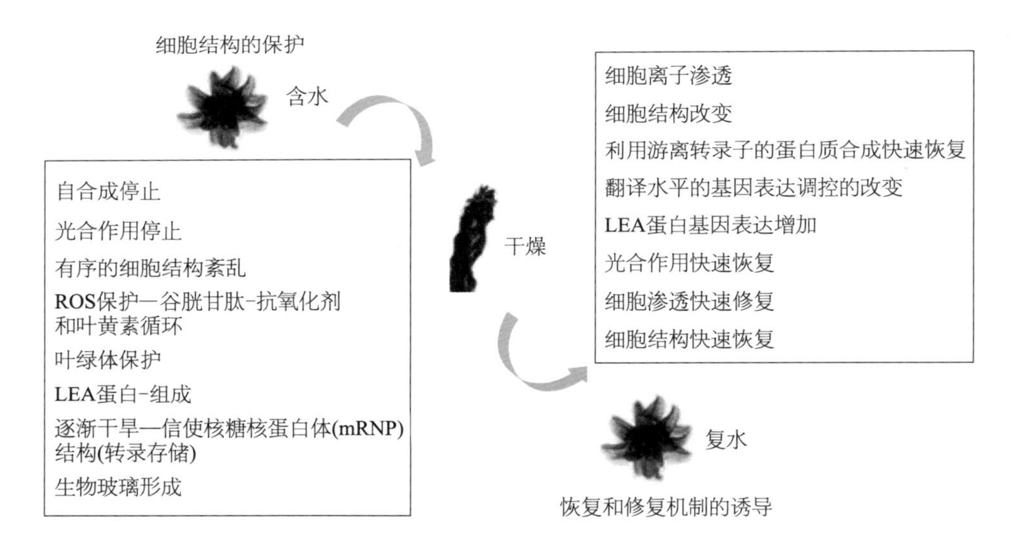

图3-1 苔藓耐干燥的生理生化调控模式（Goffinet and Show，2008）
Figure 3-1 Summary depletion of the current view of desiccation tolerance in bryophytes from a biochemical and molecular perspective（Goffinet and Show，2008）

目前，藓类耐干燥的组成型保护机制理论大多来源于对其干燥后所保留物质成分的分析，包括：① 糖类，主要是蔗糖。一直以来，可溶性糖被认为与植物耐干燥能力紧密相关（Walters *et al.*，2002）。种子、花粉和大多数耐干燥植物在干燥后积累可溶性糖。蔗糖占墙藓配子体干重的 10%，而且这个值在干燥和复水过程中都不会发生变化（Bewley *et al.*，1979）。目前认为在干燥过程中，蔗糖可以使组织玻璃化或者维持大分子物质的氢键结构稳定性，但苔藓在干燥过程中的玻璃化还没有被证明。② 保护蛋白，主要是胚胎发育晚期富集蛋白（late embryogenesis abundant protein，LEA）。尽管其功能还不十分明确，但其被认为在种子和营养组织中参与耐干燥能力的建立（Kermode and Finch-Savage，2002）。在拟南芥［*Arabidopsis thaliana*（L.）Heynh］脱落酸（abscisic acid，ABA）合成和 ABA 不敏感突变体中，由于缺乏 LEAs 蛋白导致这些植物不能忍耐干燥（Meurs *et al.*，1992）。大麦（*Hordeum vulgare*）的一个 LEAs 蛋白 HVA1 超表达后，对水分胁迫的抗性增强（Xu *et al.*，1996）。在目前所有研究的耐干燥植物的营养器官中都存在 LEAs，其中 LEAs 家族的第二个成员被命名为脱水素（dehydrin），与水分胁迫反应紧密相关，参与滞留离子、非折叠蛋白的复性、膜的保护和结合到 DNA 上稳定染色体（Cuming，1999）。LEAs 家族的第三个成员能形成细胞骨架微丝参与细胞质的有序干燥和稳定膜结构（Wise and Tunnacliffe，2004）。另外，在被子植物中证明与耐干燥相关的保护蛋白还有低分子量的热激蛋白，在脱水过程中有助于蛋白折叠和保持活性结构，然而其是否作用在苔藓中还没有被证明（Bartels and Sunkar，2005）。③ 抗氧化物质和清除 ROS 的酶。苔藓在脱水过程中由于光合受到抑制产生活性氧。积累的活性氧不仅损伤光系统Ⅱ，还能通过巯基氧化反应使蛋白质变性、色素解体、膜质过氧化和游离脂肪酸增加（Apel and Hurt，2004；Smirnoff，2005）。在脱水过程中墙藓能通过抑制脂氧合酶的活性以阻止膜脂肪酸的过氧化反应。墙藓配子体细胞质中的油滴作为一种过氧化反应的缓冲剂也参与这种保护机制。对于中度耐干燥藓（*Atrichum androgynum*），其在脱水过程中膜磷脂过氧化反应明显增强（Guschina *et al.*，2002）。另外，墙藓在干燥过程中通过消耗谷胱甘肽（glutathione，GSH）以抵御氧化物的损伤，还有的藓类在干燥过程中其 GSH 没有发生变化，而其抗坏血酸含量下降。

与干燥过程中细胞保护机制同等重要的是苔藓复水后的快速修复机制。很多研究发现，苔藓在复水后细胞会遭到破坏，但这种破坏是干燥造成的损伤还是复水后水流瞬间进入细胞造成的损伤还需进一步研究。其中，细胞质中的两亲性物质在组织含水量增加时会进入细胞膜，导致细胞膜的完整性遭到破坏，尤其在干燥敏感组织复水过程中。但是 Buitink 等（2000）发现，这些两亲性物质在组织含水量高时进入膜中会破坏细胞膜的完整性，在含水量低时进入细胞膜却会增加膜的稳定性。耐干燥机理中很重要的一部分是阻止或限制干燥细胞复水后造成的损伤（Osborne *et al.*，2002）。种子依靠缓慢的复水速度和二次复水过程缓解复水带来的损伤，而苔藓的复水几乎在瞬时发生，其可溶性物质会发生泄漏，但这种现象在耐干燥苔藓中只是很短暂的，说明耐干燥苔藓能很快修复这种损伤并使这种泄露保持在一定的限度。复水后可溶性物质的泄露程度与复水前的干燥速度相关，干燥越快，复水后可溶性物质的泄露程度越严重（Oliver *et al.*，1998）。然而，苔藓能在 24 h 内恢复正常，干燥敏感型藓却不能恢复正常生理活性（Krochko *et al.*，1978）。耐干燥苔藓在复水后能很快恢复代谢，墙藓配子体能在复水 2 h 后即可恢复蛋白合成，如果是快速干燥，也可在 3 ~ 4 h 后恢复蛋白合成（Gwózdz *et al.*，1974）。复水后 2 h 内的蛋白合成不同于干燥前的模式，墙藓在复水后比干燥前湿润时增加了 74 个新合成的蛋白。苔藓进化出一种适应机制识别长期和短期的干燥，只有在长期干燥后才会启动相应的应对机制，是苔藓在可变的周围环境中保存能量的一种方式（Oliver，1991）。在基因表达上，复水后墙藓的 mRNA 含量不发生变化，对蛋白的调控发生在翻译水平而非转录水平。在被子植物上，干燥和复水引发的基因调控主要在转录水平。在苔藓中，复水后需要合成的蛋白质已经转录成 mRNA。在遭遇快速脱水后，由于转录不能及时完成，在复水后 1 h mRNA 会重新聚集以补充这种缺失（Velten and Oliver，2001）。如果是缓慢干燥，在干燥过程中已经聚集了复水后需要合成的蛋白的 mRNA（Wood and Oliver，1999），它们与蛋白形成稳定的聚合形式，作为信使核糖核蛋白颗粒（messenger ribonucleoprotein particle，mRNP）存在于细胞中。一旦复水这些 mRNPs 立即脱离蛋白，开始启动翻译程序。mRNPs 的存在也进一步说明了组成型保护机制的存在，而且快速脱水后缓慢的恢复现象也说明 mRNPs 没有被储存，从而导致了复水后的重新转录和缓慢恢复。另外，耐干燥苔藓在复水后光合作用能快速恢复，尤其是光系统Ⅱ（Proctor，2001）；耐干燥苔藓能集中和有效地保护叶绿体，尤其是参与产生 ATP 的器官和组织。

BSC 中的藓类植物均具有较强的耐干燥性，对藓类植物的耐干燥生理响应和机理研究始于 20 世纪 70 年代（Bewley，1979）。分布在阴冷潮湿环境的藓类对水分缺失极为敏感，而分布在岩石表面和沙丘的一些藓类不仅能忍耐干燥而且它们生活史中的大部分时间都处于干燥状态，也使这些植物能更好地忍耐极端胁迫环境，如丛本藓（*Anoectangium compactum*）在收藏 19 年后仍能保持生命力。土生墙藓（*Tortula ruralis*）能很好地适应干旱环境，经受几个月甚至几十年的干旱仍能保持生命力。在长期的进化过程中，荒漠生境中的藓类植物形成了许多特殊的形态结构和生理特征以适应常年干燥的荒漠环境（张元明和曹同，2002）。在缺水条件下，藓类植物的颜色和形态结构通常会发生变化。例如，刺叶赤藓（*Syntrichia caninervis* Mitt.）失水后，叶片皱缩，直立贴茎，外表呈黑色；复水后叶片会迅速展开，中肋明显，外表变成黄绿色（魏美丽和张元明，2010）；为了应对干燥环境，很多藓类植物在长期进化的过程中，形成了一系列形态特殊的结构。例如，土生墙藓遇到干旱时叶缘会内卷或背卷；大帽藓科（Encalyptaceae）的有些种类还会出现卷缩或扭转（杨武，2008）。另外，许多藓类植物的叶片发育出疣、乳状突起或毛状的叶尖，能够反射光线，减少水分蒸发，如紫萼藓科（Grimmiaceae）和一些真藓属（*Bryum*）的种类（Schuster，1983）。垫丛状紫萼藓（*Grimmia pulvinata*）中具毛状叶尖的个体比不具此类结构的少 30% 的水分丧失。

在我国北方广袤的干旱、半干旱地区广泛分布着藓类植物，尤其是耐干燥藓类。目前对其耐干燥的生态生理方面已有诸多研究，但主要的研究力量和研究对象都在国外，国内与之相关的研究不仅较少，而且主要集中在对耐干燥藓类种类的鉴定、形态结构特征的解剖分析以及对其功能和作用的认识阶段。对国内耐干燥藓类的研究有助于拓宽研究范围，揭示其在不同生态环境的适应机制，进而深入了解 BSC 生态与水文功能的多样性形成，为荒漠 / 沙地生态系统的恢复和建设服务。

腾格里沙漠东南缘的沙坡头地区（37° 32′ ~ 37° 26′ N，105° 02′~ 104° 30′ E）属于荒漠化草原向草原化荒漠的过渡区，海拔 1500 m，以高大、密集的格状新月形沙丘链连绵分布而著称。土壤基质为疏松、贫瘠的流沙，沙层稳定含水量仅为 2% ~ 3%。该地区年均降水量为 186 mm，且主要集中在 5—9 月，年蒸发量为 2800 mm，全年日照时数为 3264.7 h，平均气温 9.6 ℃，其中 1 月平均温度为 −6.9 ℃，7 月平均温度为 24.3 ℃，年均风速 2.9 m s^{-1}。天然植被以细枝岩黄耆（*Hedysarum scoparium*）和沙蓬（*Agriophyllum squarrosum*）等为主，盖度 1% 左右（Li *et al.*，2003；李新荣，2012）。为了确保包兰铁路沙漠地段的畅通无阻，自 1956 年起建立了“以固为主、固阻结合”的植被固沙防护体系，随后又经历了多次大规模扩建。随着固沙植被的建立和沙面的固定，相继形成了荒漠藻类结皮、地衣结皮、藓类结皮以及目前的高等植物和隐花植物镶嵌分布的格局。该区的生态环境得到了大幅改善，原有的流沙区演变成了一个复杂的人工 – 天然荒漠植被景观系统（李新荣，2012）。

采集沙坡头人工固沙植被区发育良好、成片分布的真藓（*Bryum argenteum*）（2012 年 7 月）。在样品采集前，用蒸馏水湿润 BSC 表面，以保证样品采集的易操作性和样品的完整性。采集样品时，使用 PVC 管整齐地将真藓结皮取出，放入直径相同的培养皿中，带回实验室，用蒸馏水完全湿润。为了调查脱水时间对真藓表型、叶绿体结构和光合特性的影响，蒸馏水完全湿润 48 h 的样品在室温环境自然脱水 2 h、3 天、7 天和 1 年。复水过程使用氯霉素水溶液（chloramphenicol solution，3 mmol L^{-1}）以抑制蛋白合成，蒸馏水作为对照。为了调查不同脱水速度对真藓结构的影响，将真藓配子体用蒸馏水湿润 48 h（完全复水方式），后用去离子水冲洗 3 次，装入尼龙网，放置在装有活性硅胶的大玻璃瓶内快速脱水，每隔 5 min 称重，计时 2 h。脱水 12 h 后用封口袋密封低温储藏备用。脱水及称重条件为光照 30 μmol m^{-2} s^{-1}，温度为 25 ± 1 ℃。取部分快速脱水样品再复水，室温下复水 48 h。

（1）短期脱水和复水对真藓的影响

不同水分条件对真藓表型的影响：为了了解 BSC 层真藓在水分充足条件下的生长状况，将野外同一样地采集的真藓样品 1 份用蒸馏水完全湿润（图 3-2a），1 份带回实验室湿润后，每天上、下午分别浇水 1 次，在实验室条件下生长 2 周（图 3-2b）。相比野外生长的样品，在充足的水分条件下，真藓生长良好，配子体显著伸长，体积增大。

(a) (b)

图 3-2　不同水分条件下的真藓表型。（a）野外环境；（b）室内培养

Figure 3-2　Morphological changes of *B. argenteum* in different water environment.（a）desert environment；（b）room environment

短期脱水对真藓表型和含水量的影响：野外采集的真藓样品被蒸馏水完全湿润并在实验室环境培养 3 天后，配子体呈鲜绿色，叶片张开，含水量为 1565%（相对干重）。放置 2 h 后，真藓配子体呈圆形，叶片围合，颜色依然很绿，但含水量已迅速下降为 75%（图 3-3a）。同样的环境放置 3 天后，真藓颜色变为暗绿，配子体稍有皱缩，含水量仅 25%（图 3-3b）。继续放置 1 周后，真藓样品变得暗黑，体积明显变小，含水量已降到 5%（图 3-3c）。在野外环境中，真藓含水量鲜有机会达到完全饱和，在模拟一次短暂降雨后，真藓含水量为 75% 左右，与夏季在实验室放置 2 h 后的含水量相当。所以，在脱水实验中，以 75% 的含水量为真藓的脱水初期（图 3-3a）。真藓连续 1 周脱水后，其含水量变化如图 3-4 所示。

图 3-3　短期脱水后的真藓表型。(a) 75% 含水量；(b) 25% 含水量；(c) 5% 含水量
Figure 3-3　Morphological changes in different dehydrated *B. argenteum*. (a) The 75% water content sample；(b) The 25% water content sample；(c) The 5% water content sample

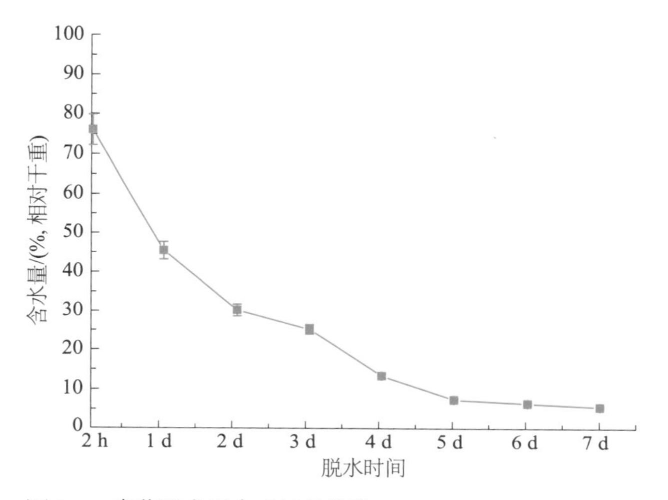

图 3-4　真藓脱水后含水量的变化
Figure 3-4　The water content curve of dehydrated *B. argenteum*

短期脱水对真藓叶绿素和光合参数的影响：在真藓脱水过程中，其表型发生了明显的变化，颜色随含水量的下降而变黑变暗，而总叶绿素 Chl (a+b) 表现出先上升后下降的趋势 (图 3-5a)。相比脱水 2 h 的样品，当脱水 3 天后，Chl (a+b) 增加了 42.57%，而继续脱水 1 周（7 天）后，其 Chl (a+b) 又比脱水 2 h 的样品减少了 30.24%。叶绿素 a 和 b 的比值 Chl-a/Chl-b 也呈现出相似的变化规律 (图 3-5a)，在脱水 3 天后，Chl-a/Chl-b 相比脱水 2 h 的样品增加了 30.24%，随着含水量的继续下降，在脱水 1 周时，Chl-a/Chl-b 又比脱水 2 h 的样品下降了 6.15%。Chl (a+b) 和 Chl-a/Chl-b 分别在脱水 2 h 和 1 天后出现了急速上升的趋势，但在随后的脱水过程中又逐渐降低，到脱水 1 周后，其数值相比脱水 2 h 的样品无显

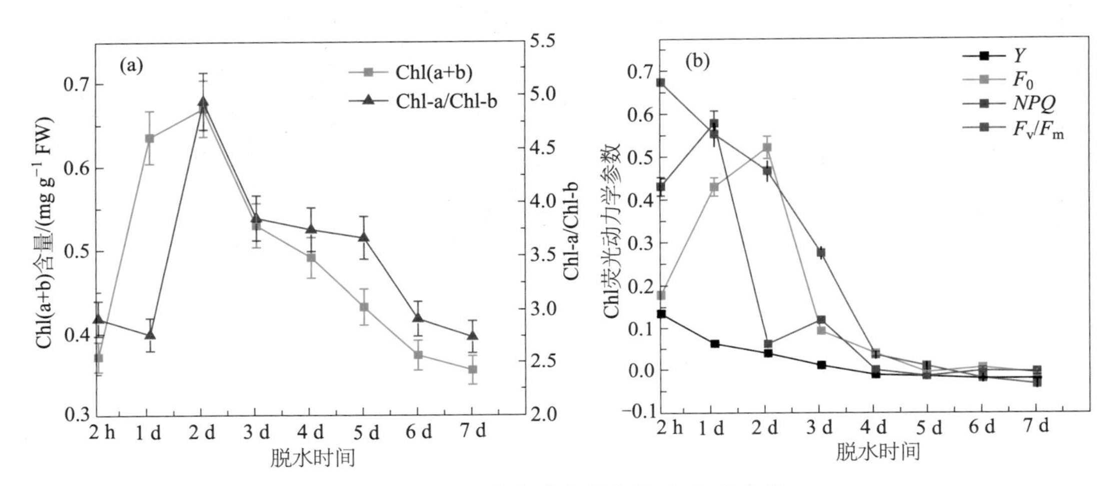

图 3-5　真藓短期脱水后叶绿素含量（a）和荧光动力学参数（b）的变化

Figure 3-5　The changes of Chl（a+b）contents and Chl-a/Chl-b value（a）and Chl fluorescence parameters（b）in dehydrated *B. argenteum*

著性差异。这些数据显示真藓在几乎完全脱水后 Chl（a+b）和 Chl-a/Chl-b 基本保持不变。

分别测定了 Chl-a 荧光动力学参数 F_m、Y、F_0 和 F_v/F_m，结果表明，随着脱水时间的延长，F_v/F_m 和 Y 都逐渐降低，F_0 和 NPQ 则是先升高后降低（$p<0.05$，图 3-5b）。与脱水 2 h 相比，真藓在脱水 3 天后，F_v/F_m、Y、F_0 和 NPQ 分别减少到 58.97%、81.69%、45.01% 和 70%。当持续脱水 1 周后，F_v/F_m 和 Y 都基本降为 0，只有 NPQ 还保持较低的数值。这些结果说明，随着脱水时间的延长，真藓光合系统关闭，只有 NPQ 显示存在光保护机制，多余的光能被耗散。

短期脱水对真藓类囊体膜蛋白的影响：短期的自然脱水尽管保留了叶绿素，但当含水量降为 5% 左右时，光合系统会逐渐关闭。期间的类囊体膜蛋白复合物变化通过蓝绿温和聚丙烯酰胺凝胶电泳分析显示，类囊体膜蛋白复合物主要被分离成 4 条带——PS Ⅱ超级复合物、PS Ⅰ与 LHC Ⅰ复合物的二聚体、PS Ⅱ复合物的单体以及 LHC Ⅱ的三聚体（Guo *et al.*, 2005b）。随着脱水时间的延长，真藓的 4 个复合物条带含量都逐渐降低，降低的幅度与水分含量呈正比（图 3-6a）。

进一步运用十二烷基硫酸钠－聚丙烯酰胺凝胶电泳（SDS-PAGE 电泳）分析类囊体膜蛋白复合物的组成。结果显示，脱水 3 天后，类囊体膜 PS Ⅱ反应中心蛋白亚基 CP47、CP43 和 D1 蛋白含量均低于脱水 2 h 的样品，同时 PS Ⅱ复合物的外周蛋白 LHC Ⅱ和光系统Ⅰ反应中心 PsaA/B 蛋白含量也低于脱水 2 h 样品。当脱水 1 周后，类囊体膜蛋白复合物的降低幅度更加明显（图 3-6b）。这些图说明随着脱水时间的延长，真藓类囊体膜蛋白发生了部分降解。

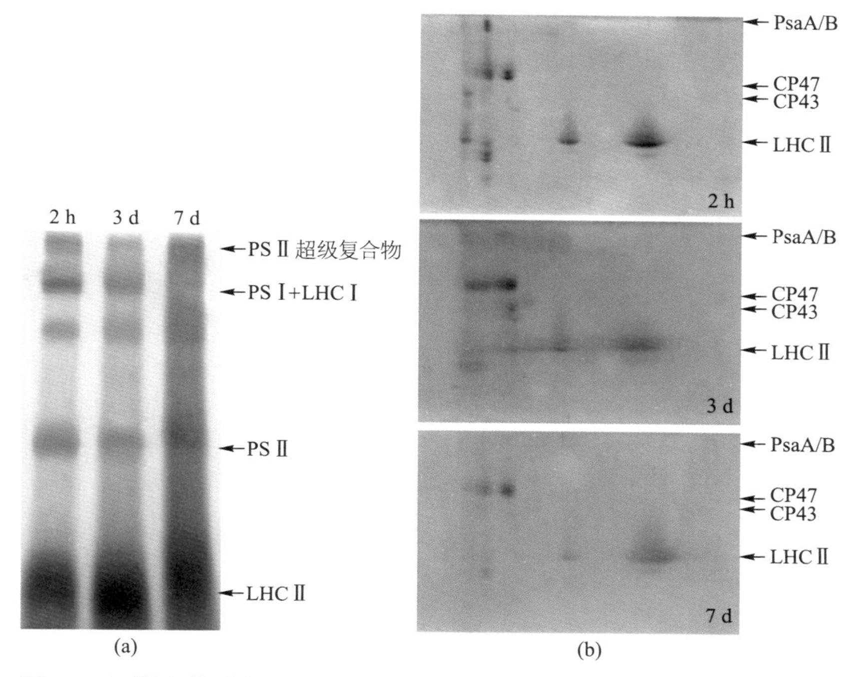

图 3-6　短期自然脱水对类囊体膜蛋白的影响：(a) 蓝绿温和胶电泳分析类囊体膜蛋白复合物；(b) SDS-PAGE 电泳分离类囊体膜蛋白复合物组成

Figure 3-6　Changes in thylakoid membrane proteins complexes of *B. argenteum* during dehydration. (a) Blue native (BN) gel analysis of thylakoid membrane protein complexes; (b) SDS-urea-PAGE separation of protein complexes in the thylakoid membrane

短期脱水对真藓叶绿体结构的影响：很多研究显示，在脱水过程中，真藓的叶绿体保持完整。通过透射电镜分析真藓的叶绿体：在脱水的不同阶段，真藓的叶绿体膜没有遭到明显损伤，但随着脱水时间的延长，叶绿体有些皱缩。相比脱水 2 h 样品，脱水 3 天后，细胞出现质壁分离，小液泡增多，叶绿体中的基粒明显减少，类囊体结构变得松散和无序，淀粉粒变小，嗜锇颗粒数量增加（图 3-7a、b）。继续脱水 1 周后，叶绿体中外形稍显不规则的圆形类囊体，基粒减少非常明显，淀粉粒更少，而嗜锇颗粒数量大量增加（图 3-7c）。这些明显的变化说明真藓在短期的自然脱水过程中，尽管叶绿体膜基本保持完好，但叶绿体中的成分，包括基粒、类囊体、淀粉粒和嗜锇颗粒，都发生了明显的变化。

短期脱水后复水对真藓类囊体膜蛋白的影响：上面的分析结果显示真藓的类囊体结构和类囊体膜蛋白在脱水过程中发生了部分降解。通过进一步的复水实验，蓝绿温和胶电泳分析显示，类囊体膜蛋白复合物依然被分成主要的 4 条带，包括 PS Ⅱ超级复合物、PS Ⅰ与

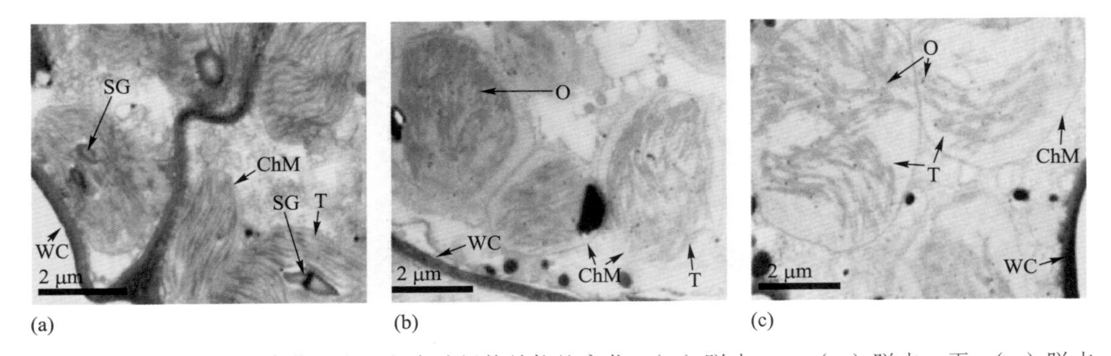

(a) (b) (c)

图 3-7 透射电镜分析真藓脱水过程中叶绿体结构的变化 :(a) 脱水 2 h ;(b) 脱水 3 天 ;(c) 脱水 1 周。ChM : 叶绿体膜，CW : 细胞壁，O : 嗜锇颗粒，SG : 淀粉粒，T : 类囊体

Figure 3-7 Transmission electron microscopy (TEM) micrographs showing chloroplast ultrastructure in a *B. argenteum*. (a) 2 h-dehydrated sample ; (b) 3 d-dehydrated sample ; (c) 7 d-dehydrated sample.ChM : chloroplast membrane ; CW : cell wall ; O : osmiophilic granule ; SG : starch grain ; T : thylakoid

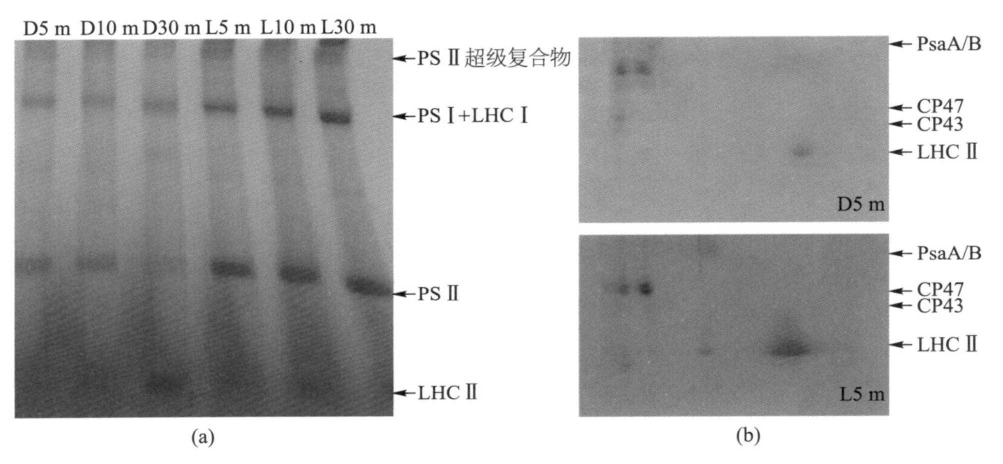

(a) (b)

图 3-8 短期自然脱水后复水对类囊体膜蛋白的影响 :(a) 蓝绿温和胶电泳分析类囊体膜蛋白复合物 ;(b) SDS-PAGE 电泳分离类囊体膜蛋白复合物组成。D5 m、D10 m 和 D30 m 分别表示在暗处复水 5 min、10 min 和 30 min ; L5 m、L10 m 和 L30 m 分别表示在光下复水 5 min、10min 和 30min

Figure 3-8 The changes in thylakoid membrane proteins of *B. argenteum* after rehydration in light and dark. (a) Blue native (BN) gel analysis of thylakoid membrane protein complexes ; (b) SDS-urea-PAGE separation of protein complexes in the thylakoid membrane.Line D5 m, D10 m, D30 m represent in dark rehydration for 5 min, 10 min, 30 min ; line L5 m, L10 m, L30 m represent in light rehydration for 5 min, 10 min, and 30 min

LHC Ⅰ 复合物的二聚体、PS Ⅱ 复合物的单体以及 LHC Ⅱ 的三聚体。但在光下复水时，类囊体膜蛋白复合物的量随着复水时间的延长而逐渐增加，而在暗处复水时类囊体膜蛋白复合物的量明显低于光下复水初期的量（图 3-8a）。二向的 SDS-PAGE 电泳分析也进一步说明在

复水 5 min 时，暗处复水的真藓类囊体膜 PS Ⅱ反应中心蛋白亚基 CP47、CP43 和 D1 蛋白，PS Ⅱ复合物的外周蛋白 LHC Ⅱ和光系统 Ⅰ 反应中心 PsaA/B 蛋白的量均低于在光下复水的量（图 3-8b）。这些结果进一步说明真藓在脱水后存在部分类囊体膜蛋白的降解和复水后的重新组装。

短期脱水后复水对真藓光合参数的影响：随干燥时间的延长，真藓 Chl-a 荧光动力学参数 F_m、Y、F_0 和 F_v/F_m 逐渐降低直至光合系统关闭，类囊体膜蛋白也发生了部分的降解。当复水时，类囊体膜蛋白会开始重新组装。在光下复水时，叶绿素荧光参数 Y（图 3-9a）、F_0（图 3-9b）、NPQ（图 3-9c）和 F_v/F_m（图 3-9d）随复水时间的延长都逐渐恢复到脱水前（pre-dehydration）状态。复水 5 min 后真藓的 F_v/F_m 达到正常值的 91.82%，F_0 和 NPQ 在 10 min 后也恢复到正常值，Y 值在 1 h 后达到最大值。这些结果说明真藓在光下复水后其光合能力能完全恢复。

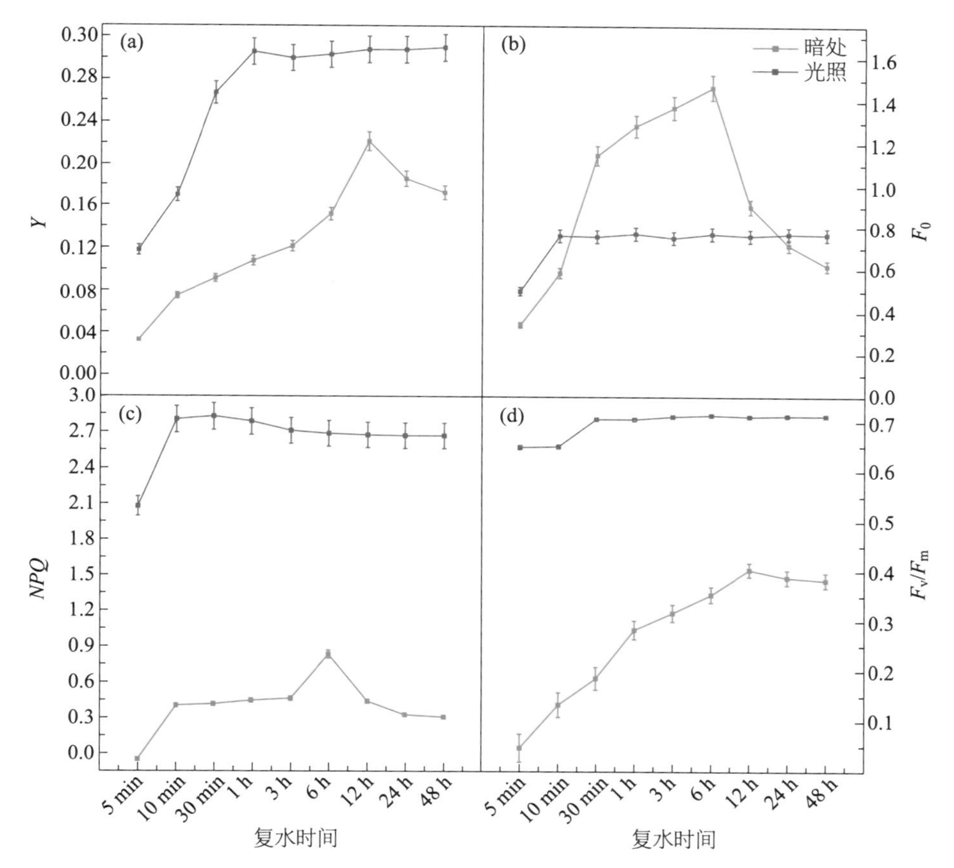

图 3-9　真藓短期脱水后复水叶绿素 a 荧光参数的变化

Figure 3-9　The changes of Y（a）、F_0（b）、NPQ（c）and F_v/F_m（d）of *B.argenteum* after rehydration in light and dark

然而，当在暗处复水时，F_v/F_m（图 3-9d）在 5 min 后逐渐增加，但随着复水时间的延长，这些值保持在一个稳定的状态，大约是正常值的 60%，然后不再上升。这些数据说明真藓短期干燥后在暗处复水不能完全恢复光合能力。

为了进一步说明类囊体膜蛋白降解和重新组装对叶绿素 a 荧光动力学参数的影响，在复水过程中使用氯霉素溶液抑制叶绿体编码蛋白的合成。结果显示，在暗处复水时，5 min 后 F_v/F_m 开始上升，但达到正常值的大约 60% 时，基本不再上升（图 3-10）。而在光下复水时，F_v/F_m 在 5 min 内即达到正常值的 70%，在 30 min 后又降到正常值的 50%，随复水时间的延长，又逐步降低（图 3-10）。这个结果说明在复水过程中，光合的恢复可能受类囊体膜蛋白的影响。

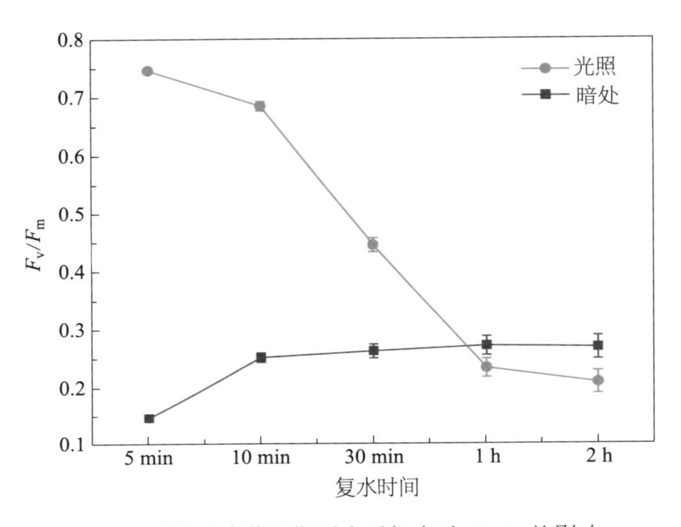

图 3-10　氯霉素对真藓短期脱水后复水时 F_v/F_m 的影响

Figure 3-10　The changes of F_v/F_m of *B. argenteum* after rehydration with chloramphenicol treatments

（2）长期干燥对真藓生态生理的影响

沙漠环境干燥少雨，尤其是冬春季，真藓会遭遇更长的干燥期。在长期的干燥过程中，真藓会有怎样的变化？前面的分析表明，真藓在短期的脱水干燥后细胞膜和叶绿体保持完整，叶绿素滞留，但叶绿体内部结构有部分损伤，类囊体膜蛋白部分降解。目前，对苔藓植物能忍受的干燥时间的研究相对较少，大多研究集中在几小时或几天的短期脱水。尽管在野外环境中鲜有长达 1 年或数年干燥，但对于研究苔藓植物的耐干燥能力而言，干燥时间是反映其耐干燥能力的重要因素。有文献报道 *Tortula caninervis* 和 *T. ruralis* 在储藏 3 年后依然能恢复生理活性（Oliver *et al.*，1998），*Grimmia laevigata* 作为标本放置 10 年后依然能在复水后变绿（Keever，1957）。然而 *T. norvegica* 在放置 12 个月后遭受严重损伤，不能恢复生理活性（Oliver *et al.*，1998）。事实上，很多资料都在陈述苔藓能耐干燥，但具体的耐干燥时间仅能查到有限的几篇报道，也仅仅集中在几种苔藓长期干燥后能否恢复生理活性方面，但长期干燥造成的变化却鲜有研究（Proctor *et al.*，2007a，b）。

长期干燥对真藓含水量和表型的影响：与前面短期干燥实验同时采集真藓样品，将 3 份样品带回实验室用蒸馏水充分湿润后，在实验室环境下放置 1 年。用同样的方法将样品在实验室放置 1 周，比较短期和长期干燥对真藓的影响。干燥 1 年后，真藓的含水量降为 <2%，从表型上真藓变得更黑更小，只在土壤表面看到极薄的一层颗粒状分布。干燥 1 周的真藓尽管颜色、大小会有变化，但能看到土壤表面有很明显的一层片状的分布，含水量在 5% 左右。

长期干燥后复水对真藓光合参数的影响：首先分析了干燥 1 年后真藓的叶绿素含量，结果发现与干燥 1 周的相比，总叶绿素含量只是轻微减少，变化不显著。对干燥 1 年的真藓使用蒸馏水复水，30 min 后 F_v/F_m 开始增加，说明在 30 min 后光合系统开始启动。干燥 1 周的真藓在复水 5 min 后 F_v/F_m 就达到了正常值的 90%。尽管光合系统启动较慢，但在复水 6 h 后，干燥 1 年的真藓的 F_v/F_m 也恢复到了正常值（图 3-11）。这些分析说明，真藓能忍受 1 年的干燥而恢复光合系统。同时又对干燥 1 周和 1 年的真藓类囊体膜蛋白进行了比较，发现类囊体膜蛋白有轻微的减少。

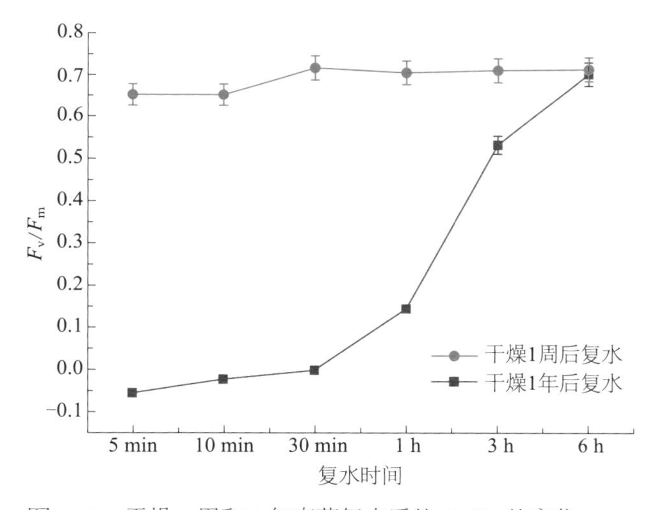

图 3-11　干燥 1 周和 1 年真藓复水后的 F_v/F_m 的变化

Figure 3-11　The changes of F_v/F_m of dried *B. argenteum* after rehydration after desiccated for 1 week and 1 year

长期干燥后复水对真藓微表型的影响：既然真藓能忍受长达 1 年的干燥而恢复光合能力，我们通过透射电镜对其叶绿体结构进行了分析。相比较干燥 1 周的样品，真藓在干燥 1 年后，其细胞壁变薄，叶绿体皱缩，类囊体减少和排列紊乱。最明显的一个现象是在细胞膜周围出现了很多黑色的质体小球（plastoglobule）（图 3-12）。

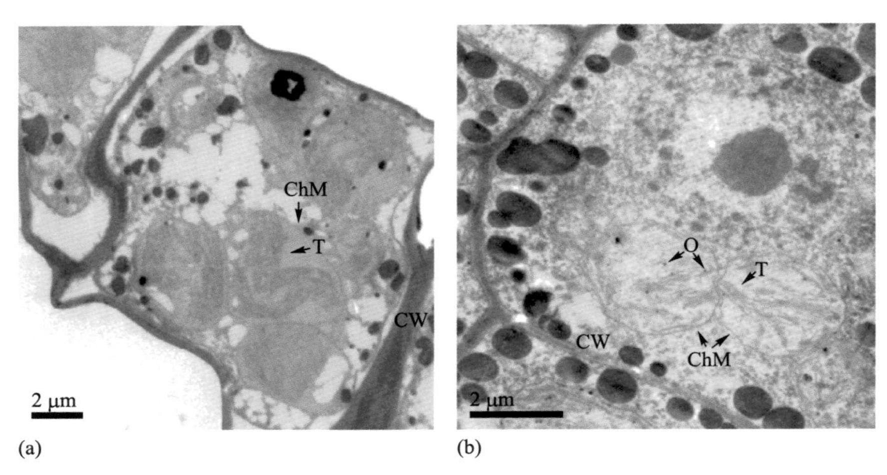

图 3-12　干燥 1 周（a）和 1 年（b）的真藓透射电镜图。ChM：叶绿体膜；CW：细胞壁；O：嗜锇颗粒；T：类囊体

Figure 3-12　Transmission electron microscopy（TEM）micrographs showing cell ultrastructure in *B.argenteum* after desiccated for 1 week（a）and 1 year（b）.ChM：chloroplast membrane；CW：cell wall；O：osmiophilic granule；T：thylakoid

通过扫描电镜分析发现，干燥 1 周的真藓样品（图 3-13a，c，e），叶状体完整，细胞排列整齐，细胞外形完整，细胞间隙明显。干燥 1 年的样品（图 3-13b，d，f），叶状体依然完整，但明显地出现了皱缩，细胞排列紊乱，细胞外形有些不太规则，细胞壁变薄，细胞间隙消失。

长期干燥后复水对真藓质膜结构的影响：图 3-12 和图 3-13 的电镜图中显示真藓在经历 1 年的干燥后，细胞壁变薄，细胞外形发生变化，细胞膜变得模糊。利用傅里叶变换红外光谱法（FTIR）研究膜结构在医学上比较广泛，而植物组织膜结构对环境胁迫和非生物胁迫的响应是近几年国内外研究热点。

① 真藓的红外指纹图谱：红外光谱常被用来区别和鉴定不同物种及化合物，每种物质包括活体生物都会显示独特的红外吸收峰，类似于指纹对于个体的独特性，因此称为指纹图谱。图 3-14 是真藓、小立碗藓（*Physcomitrella patens*）和土生对齿藓的红外指纹图谱。从图谱中峰的位置显示他们同属于藓类植物，从峰形显示它们三者的膜结构中含有不同的成分，真藓与土生对齿藓更为接近，它们明显比小立碗藓的很多成分含量高。

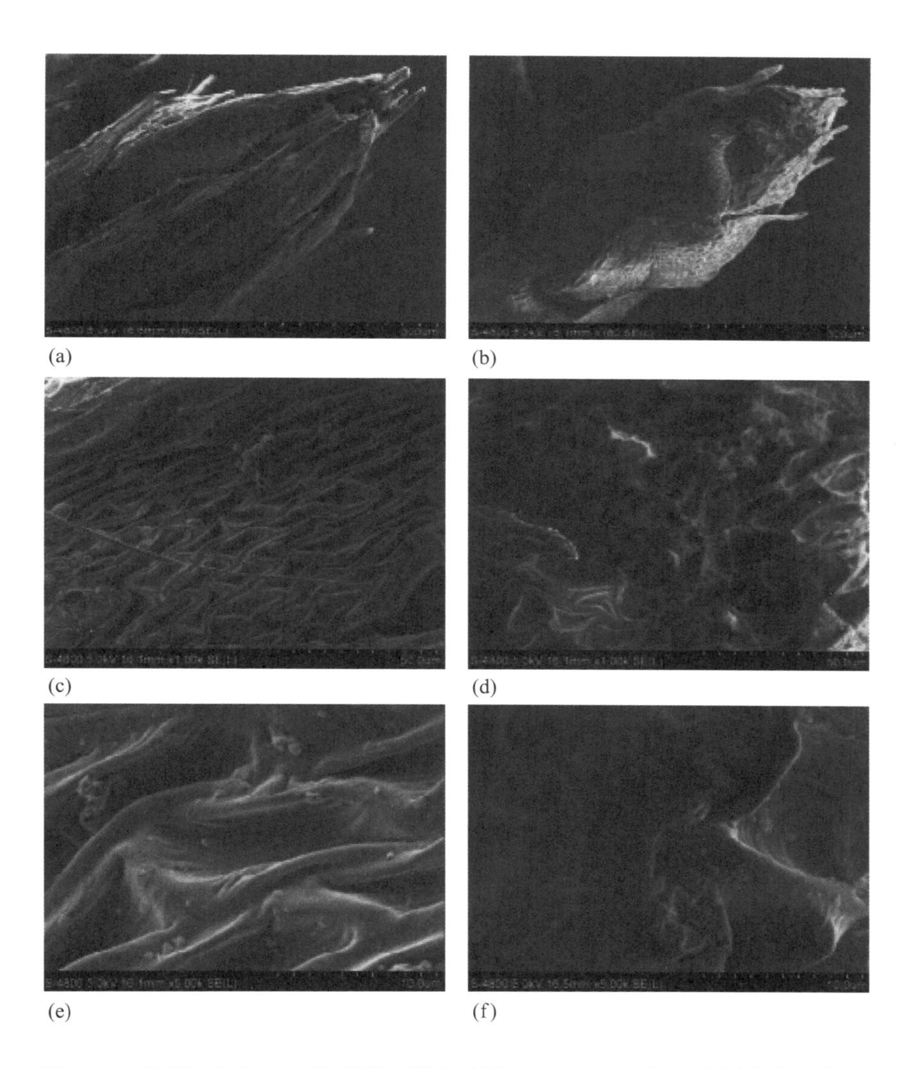

图 3-13　干燥 1 周和 1 年的真藓扫描电镜图 :（a）、（c）、（e）分别代表干燥一周后放大 180 倍、1000 倍、5000 倍 ;（b）、（d）、（f）分别代表干燥 1 年后放大 180 倍、1000 倍、5000 倍
Figure 3-13　Scanning electron microscopy showing the microscopic morphology of *B. argenteum*. Desiccated for 1 week ×180（a）, ×1000（c）, ×5000（e）; desiccated for a year ×180（b）, ×1000（d）, × 5000（f）

图 3-15 是干燥 1 周和 1 年真藓样品的红外指纹图谱。真藓红外光谱在 3400 cm^{-1} 附近的一个明显的吸收峰为—OH 和—NH 化学键伸缩振动谱带，表明组织内部含有大量蛋白质和多糖类。2925 cm^{-1} 附近为 C—H 伸缩振动谱带，主要由碳氢化合物引起。C—H 变角振动吸收位于 1200 ~ 1500 cm^{-1}，往往与其他吸收峰重叠。蛋白吸收峰存在于 1658 cm^{-1}（C=O，酰胺 I 带）、1553 cm^{-1}（N—H，酰胺Ⅱ带）和 1260 cm^{-1}（C—N，酰胺Ⅲ带）三个区域。1744 cm^{-1}

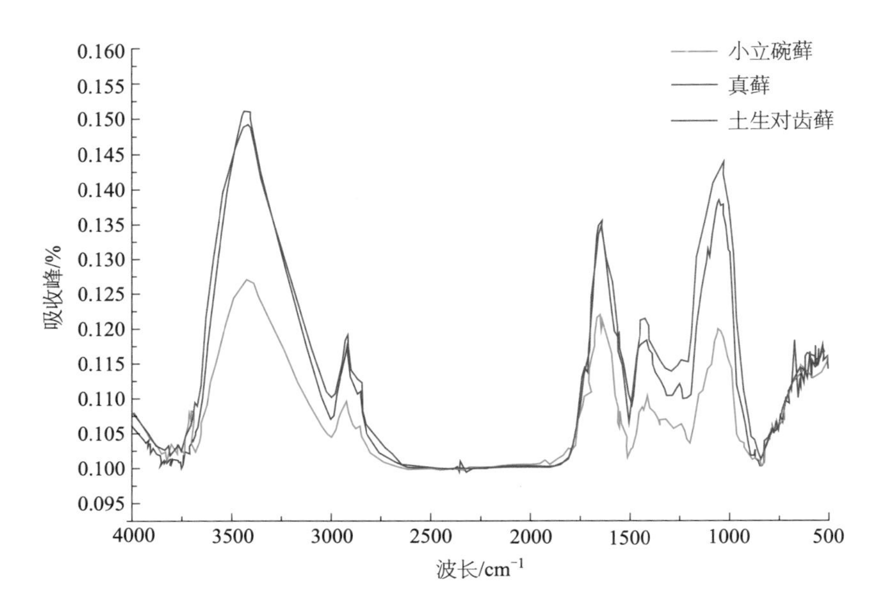

图 3-14　真藓、小立碗藓和土生对齿藓在 4500 ~ 500 cm^{-1} 红外吸收指纹图谱

Figure 3-14　Absorption FTIR spectra in the 4500−500 cm^{-1} region in *B. argenteum*, *P. patens* and *D. vinealis*

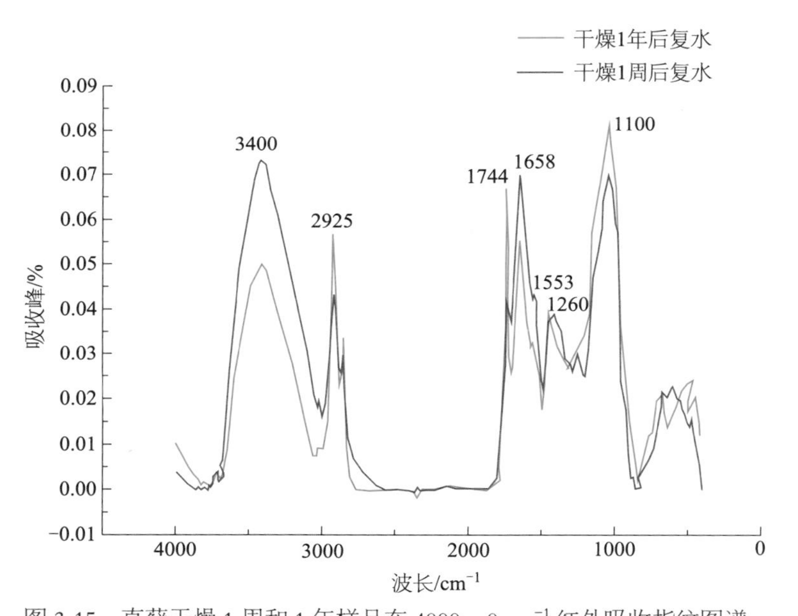

图 3-15　真藓干燥 1 周和 1 年样品在 4000 ~ 0 cm^{-1} 红外吸收指纹图谱

Figure 3-15　Absorption FTIR spectra in the 4000−0 cm^{-1} region in *B. argenteum* after desiccated for 1 week and 1 year

附近为 COOR 特征峰，主要是细胞壁胶质和细胞膜质中化合物。指纹区 1100 cm^{-1} 附近吸收峰主要由 C—H 弯曲或者 C—O 或者 C—C 伸缩引起，主要为碳氢化合物。

② 长期干燥对真藓膜透性的影响：真藓的原始图谱（图 3-14，图 3-15）中很多吸收峰重叠在一起，不易辨别。为了分辨出一些被掩盖的小峰，二阶导数技术被用来增加波峰灵敏度。对吸收峰明显的波段进行了二阶导数谱分析（图 3-16）。

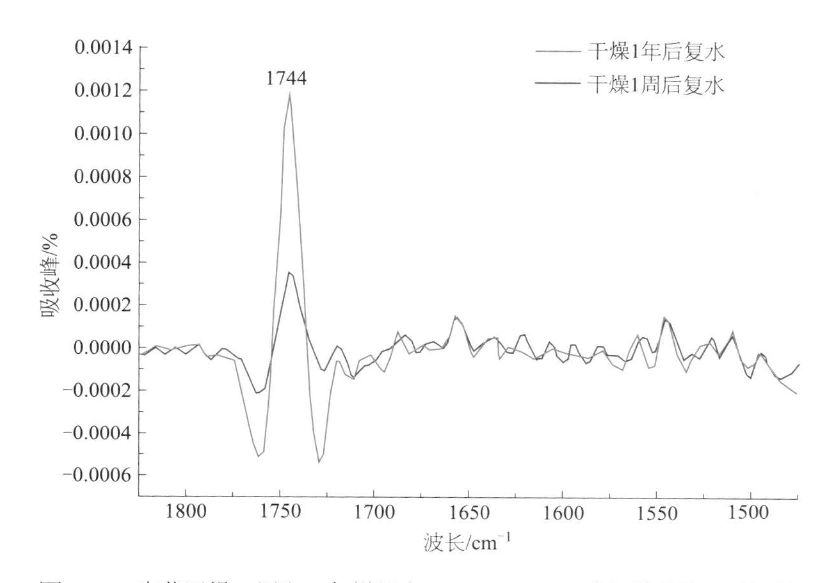

图 3-16　真藓干燥 1 周和 1 年样品在 1800 ~ 1500 cm^{-1} 红外吸收二阶导数图谱

Figure 3-16　Second-derivative transportation FTIR spectra in the 1800−1500 cm^{-1} region in *B. argenteum* after desiccated for 1 week and 1 year

在图 3-16 中，位于 1780 ~ 1720 cm^{-1} 处可以观察到质膜界面磷脂分子构象变化，即羰基伸缩振动。1744 cm^{-1} 和 1728 cm^{-1} 分别为磷酸甘油基上 SN-1 和 SN-2 位点上酯酰键 C=O 伸缩振动，SN-1 位 C=O 靠近疏水环境，SN-2 位 C=O 靠近极性亲水区域。从图 3-16 可以看出，相比较干燥 1 周的样品，真藓在干燥 1 年后 1728 cm^{-1} 处峰值减小，表明 SN-2 酯酰键 C=O 伸缩振动减弱，即靠近极性亲水区 C=O 减少。与此结果一致的是真藓干燥 1 年后 1744 cm^{-1} 处峰值大幅度增加，表明 SN-1 酯酰键 C=O 的伸缩振动加强，靠近疏水区的 C=O 增多，也就进一步说明甘油基骨架取向发生变化，脂分子极性区构象发生变化。细胞膜是双层磷脂膜结构，朝外的是亲水区，疏水区在中间，红外图谱说明长期的干燥造成细胞膜磷脂结构发生变化，使真藓细胞膜表面的亲水区遭到破坏，细胞膜透性增强。很多研究表明，1744 cm^{-1} 酯类吸收峰可以作为细胞膜破损程度表征，吸收峰值越大，说明酯类含量增多。酯类吸收特征峰（1744 cm^{-1} 或者 1728 cm^{-1}）比值也可用来反映细胞膜破损程度，变化越大，说明细胞破损程度越高，细胞膜透性越大。真藓干燥 1 周时，1744 cm^{-1}/1728 cm^{-1} 为 −3.7，而干燥 1 年后的比值为 −2.36，长期干燥的比值远远大于短期干燥，也进一步说明长期干燥对膜结构造成了损伤。

③ 长期干燥对真藓糖类化合物的影响：由于在指纹区内，碳氢化合物种类繁多且吸收峰大多重叠，在红外光谱上难以区分，较少用于比较含量变化。二阶导数增加了光谱敏感度，可以分辨出微小峰位。对真藓长、短期干燥的光谱作二阶导数，得到图 3-17。在 1200 ~ 500 cm^{-1} 存在 1052 cm^{-1}、1101 cm^{-1}、1113 cm^{-1}、1126 cm^{-1} 等多个吸收峰。按照 Yang 和 Yen（2002）的方法选择 1101 cm^{-1} 区域为植物胁迫糖类特征峰。可以看出，真藓在干燥 1 年后 1101 cm^{-1} 吸收峰明显增加，说明长期干燥使糖类物质含量增加。另外，真藓不同干燥时间的红外图谱存在不同的吸收峰，说明两者存在不同的糖类化合物的组成。

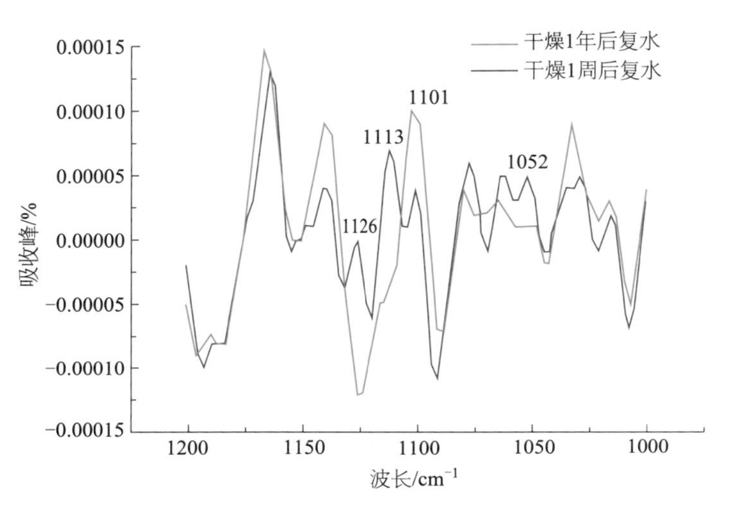

图 3-17 真藓干燥 1 周和 1 年样品在 1200 ~ 1000 cm^{-1} 红外吸收二阶导数图谱

Figure 3-17 The second-derivative transportation FTIR spectra in the 1200−1000 cm^{-1} region in *B. argenteum* after desiccated for 1 week and 1 year

④ 长期干燥对膜蛋白二级结构的影响：1700 ~ 1600 cm^{-1} 区域内红外图谱为蛋白质酰胺 I 带（图 3-18），存在有多个吸收峰。对苔藓叶片膜蛋白二级结构进行指认，确定 1658 cm^{-1} 左右处为 α- 螺旋，1638 cm^{-1}、1675 cm^{-1} 和 1688 cm^{-1} 为 β−折叠或者转角，1641 cm^{-1} 为无规则卷曲（Surewicz *et al.*，1993）。从图 3-18 中可以看出，真藓干燥 1 年后，1658 cm^{-1} 的 α−螺旋峰值增高，而且向波数增大的方向移动，即“蓝移”。1638 cm^{-1}、1675 cm^{-1} 和 1688 cm^{-1} 的 β−折叠或者转角也都出现了不同程度的“蓝移”。蓝移的出现代表了需要的能量高，表示蛋白结构趋于不稳定。

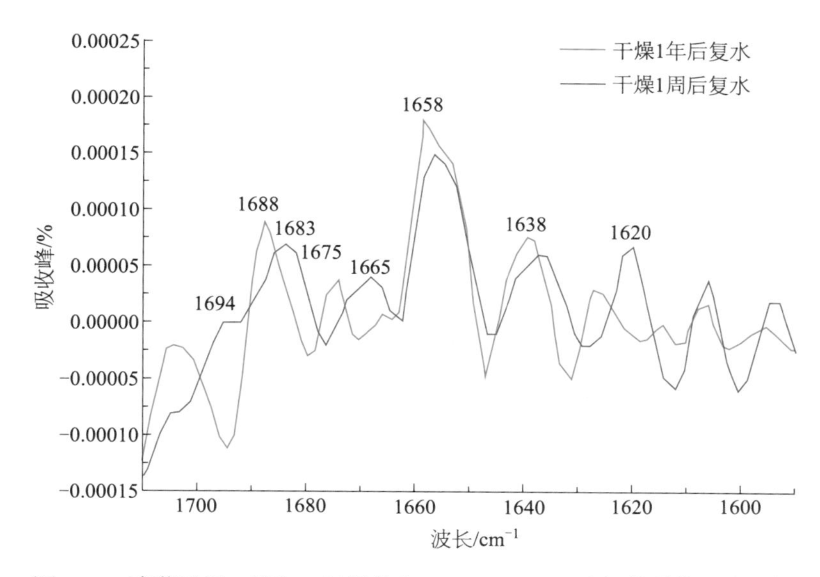

图 3-18　真藓干燥 1 周和 1 年样品在 1700 ~ 1600 cm^{-1} 红外吸收二阶导数图谱

Figure 3-18　The second-derivative transportation FTIR spectra in the 1700−1600 cm^{-1} region in *B. argenteum* after desiccated for 1 week and 1 year

（3）“干燥 – 复水”交替对真藓的影响

在开放的沙漠环境中，由于降水少而不规律，藓类作为 BSC 组成的一部分，必然遭受时间不等的“脱水 – 干燥 – 复水”的循环。

多次干燥 – 复水对真藓光合参数的影响：将野外采集的真藓样品置于实验室环境，充分湿润后放置 1 周，当含水量降至 5% 以下时，用蒸馏水充分湿润，自然干燥后再湿润，共经历 5 次脱水 – 干燥 – 复水处理。对照为采集后带回实验室湿润后直接自然干燥的样品。

对以上处理的样品进行了叶绿素含量分析，结果发现，对照真藓的叶绿素仅比处理的低了 1.5%，没有表现出明显的差异。同时，对这些干燥的样品复水后发现光合参数也没有很大的变化，对照和处理的 F_v/F_m 在 5 min 后都达到了干燥前的 90%，只是处理的真藓在 10 min 后 F_v/F_m 达到了干燥前的 93.5%，比对照的稍微高一些，对照在 10 min 时为 91.7%，但两者在 30 min 后都基本恢复了干燥前的 F_v/F_m（图 3-19a）。其他光合参数 Y（图 3-19b）、NPQ（图 3-19c）和 F_0（图 3-19d）在处理和对照样品之间差异同样很小。

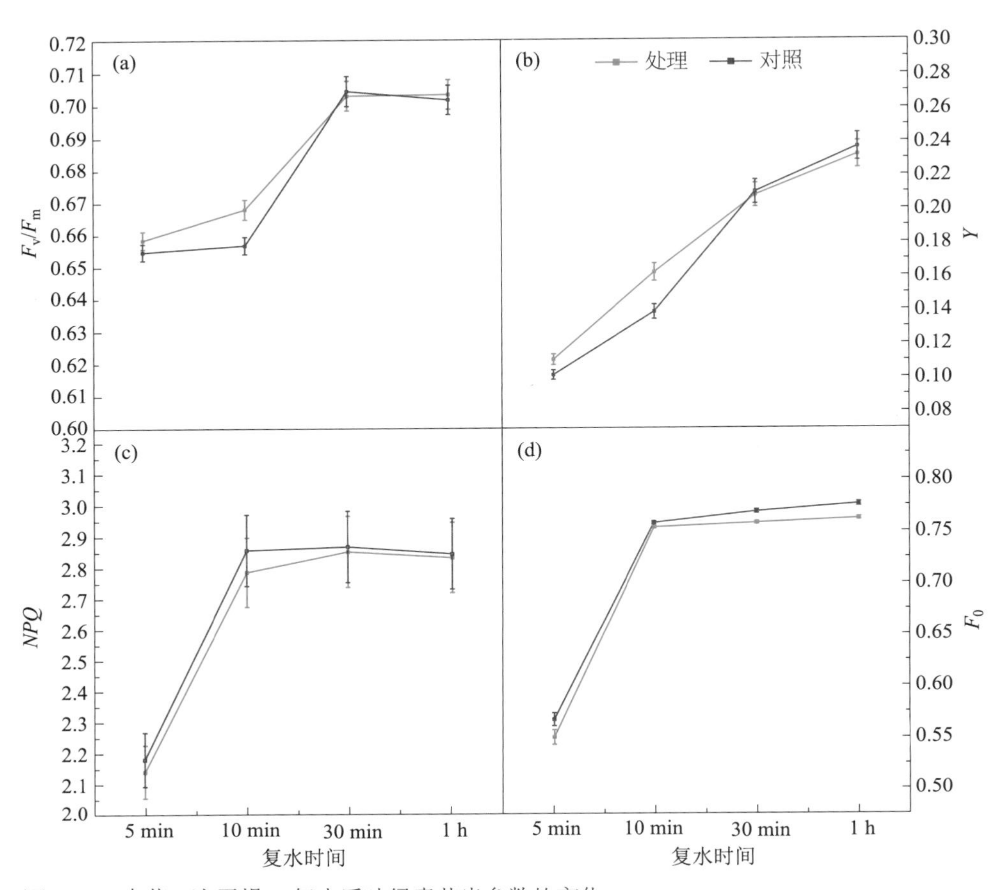

图 3-19　真藓 5 次干燥 – 复水后叶绿素荧光参数的变化
Figure 3-19　The changes of F_v/F_m（a），Y（b），NPQ（c）and F_0（d）of rehydration *B. argenteum* after 5 times desiccation

多次干燥 – 复水对真藓微表型的影响：通过扫描电镜发现经过多次干燥 – 复水处理的真藓样品外形依然保持完整。相比对照真藓（图 3-13），处理的真藓的叶状体稍微张开，皱缩程度不严重（图 3-20a），双层细胞壁保护的细胞排列整齐（图 3-20b）。最明显的不同是处理的真藓表面分布有更多的蜡质层（图 3-20c，d）。透射电镜则显示处理后的真藓细胞膜完整，细胞壁更厚一些，叶绿体外形完整，只是其中的嗜锇颗粒少一些（图 3-21）。

多次干燥 – 复水对真藓膜结构的影响：对处理的真藓膜结构通过傅里叶变换显微红外光谱法分析后发现，相比对照的真藓，其膜中的脂类物质和碳水化合物有不同程度的增加，而蛋白结构并没有太大的变化（图 3-22）。

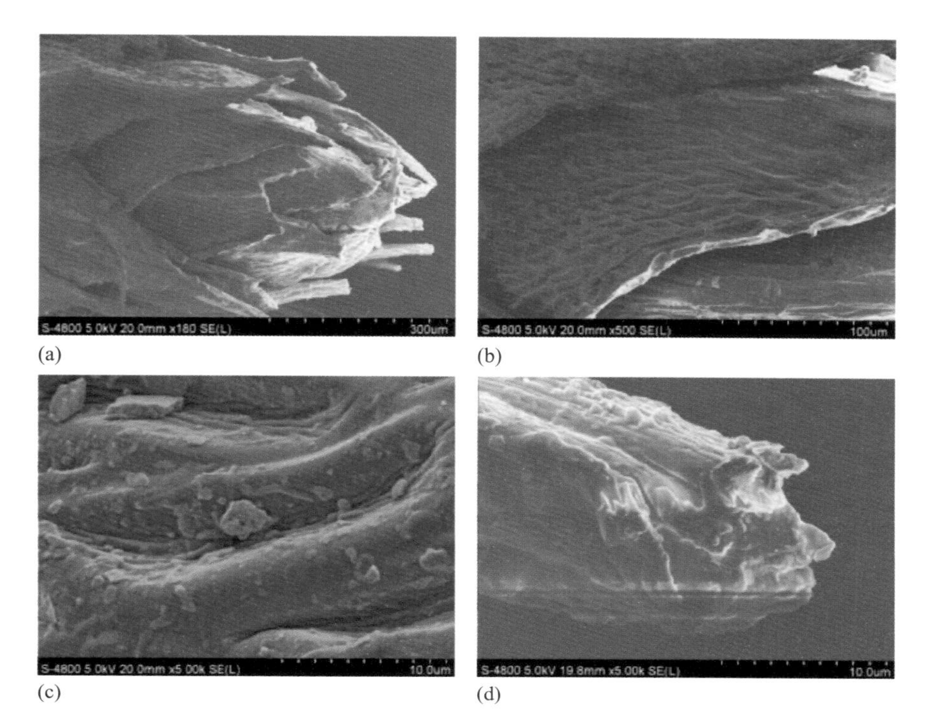

图 3-20　处理的真藓扫描电镜图：(a) 放大 180 倍；(b) 放大 1000 倍；(c) 放大 5000 倍；(d) 放大 5000 倍
Figure 3-20　Scanning electron microscopy showing the microscopic morphology of *B. argenteum*. Desiccated for many time ×180 (a), ×1000 (b), ×5000 (c), ×5000 (d)

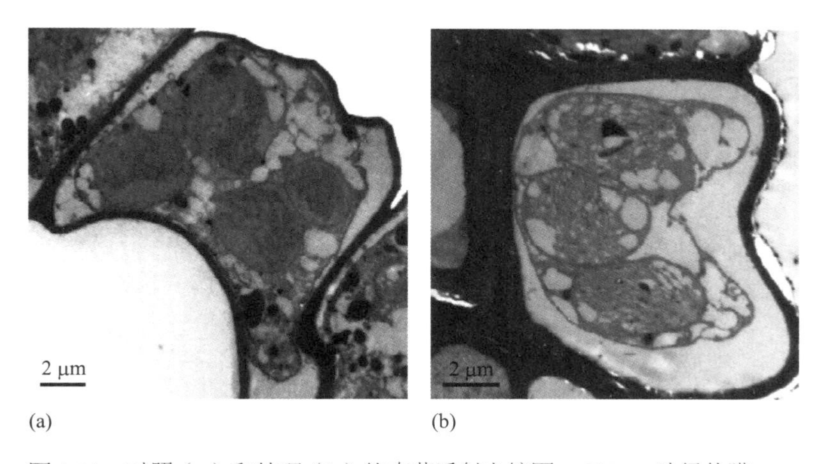

图 3-21　对照（a）和处理（b）的真藓透射电镜图。ChM：叶绿体膜；CW：细胞壁；O：嗜锇颗粒；T：类囊体
Figure 3-21　Transmission electron microscopy (TEM) micrographs showing cell ultrastructure in *B. argenteum* after desiccated for one time (a) and many times (b). ChM: chloroplast membrane; CW: cell wall; O: osmiophilic granule; T: thylakoid

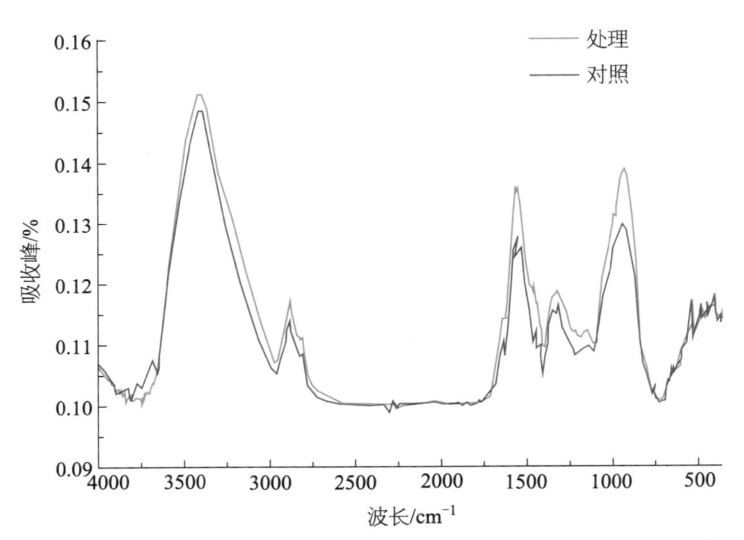

图 3-22　处理和对照的真藓样品在 4000 ~ 0 cm^{-1} 红外吸收指纹图谱

Figure 3-22　Absorption FTIR spectra in the 4000−0 cm^{-1} region in *B. argenteum* after desiccated for one time and many times

（4）脱水速度对真藓表型和结构的影响

完全耐干燥（fully desiccation-tolerant）和修饰性耐干燥（modified desiccation-tolerant）植物具有不同的保护机制响应脱水胁迫。完全耐干燥植物能忍受快速失去所有的细胞水，而修饰性耐干燥植物需要慢速脱水诱导和建立相关保护机制（Oliver *et al.*，1998；Tuba *et al.*，1998）。在修饰性耐干燥植物响应慢速脱水时，其细胞壁往往折叠使脱水危害最小化（Farrant，2000），但并非所有耐干燥植物都具有此功能（Cooper and Farrant，2002）。Cooper 等（1999）比较了不同脱水速度对修饰性耐干燥植物 *Craterostigma wilmsii* 可溶性蔗糖含量的影响发现，快速脱水显著降低了蔗糖积累。快速脱水使 *C.wilmsii* 叶绿体含量降低，光合系统元件受损。而对于完全耐干燥的土生墙藓，色素并没有显著变化且细胞结构完整，光合系统元件结构完整（Ingram and Bartels，1996；Proctor and Smirnoff，2000）。

根据对沙坡头地区 BSC 和地面温度的研究，固沙区内 BSC 表面温度都是在 6：00 开始上升，13：00 达到最大值，其中 1956 年、1964 年和 1987 年固沙区最大值分别为 51.6 ℃、45.4 ℃和 44.6 ℃。随着固沙年限的推移，BSC 表面温度的变化幅度增大，即 1956 年固沙区内 BSC 表面温度日变化幅度最大，达 47.5 ℃；1964 年固沙区次之，为 39.6 ℃；1987 年固沙区最小，为 37.4 ℃（石莎等，2004）。总体表现为长势好的 BSC 的表面温度日变化幅度大于长势差的 BSC 的，因为日出后，发育厚且藓类植物长势好的 BSC（厚度 1.5 ~ 2.2 cm）吸热多，增温快，且下垫面反射率下降幅度大，导致其表面温度高于藓类植物长势不好的 BSC（厚度 0.4 ~ 0.8 cm）；直到日落后，由于藓类植物长势好的 BSC 散热快，所以表面温度低于长势不好的 BSC。所以，不论生长年限，沙坡头地区固沙区的藓类结皮要忍受大约 39.6 ~ 47.5 ℃的最高日变化温度，对于单细胞的藓类植物，意味着必须要忍受快速脱水的胁迫（石莎等，2004）。尽管藓类被认为能忍耐快速脱水，但对脱水速度的研究也仅集中在有限的几个物种中，真藓在快速脱水中发生的变化也鲜有报道。

脱水速度对真藓含水量和表型的影响：为了比较自然脱水和干热条件对真藓的影响，将同时采集的真藓配子体离体材料 1 份放入硅胶瓶快速脱水；1 份放入硅胶瓶中再置于烘箱，温度设为 60 ℃，使样品同时遭受快速加高温脱水；另外 1 份在室内自然脱水。自然脱水的样品放置在 7 月的室内环境，空气湿度为 33%，放置 2 h 后，含水量下降为 59%，4 h 后下降为 42%，6 h 后下降为 30% 左右，8 h 后基本稳定在 32% 左右。快速脱水处理的样品每 10 min 测一次含水量，结果发现在 10 min 内含水量已经下降到 65%，20 min 已经降为 37%，然后缓慢下降，到 1 h 以后恒定为 1.5%。快速加高温脱水的样品在 10 min 后含水量下降为 50% 左右，20 min 后含水量为 25%，30 min 后达到 1.5% 的恒定含水量。从表型上观察，快速脱水的样品依然保持鲜绿，快速加高温脱水的样品和自然脱水的样品都有些暗绿。

脱水速度对真藓光合色素的影响：比较了不同处理对真藓光合色素含量的影响，结果发现不同处理样本之间没有显著差异，只是快速脱水处理样品的总叶绿素含量比其他稍高 2%，但差异不显著，表明脱水速度对真藓细胞光合色素含量无显著影响。

脱水速度对真藓微表型的影响：通过透射电镜发现快速脱水的真藓叶绿体中保留有较大的淀粉粒，胞膜和叶绿体膜都保持完整（图 3-23a）。增加了高温处理后，淀粉粒依然明显存在，叶绿体膜保持完整，只是胞膜出现了部分破损（图 3-23b）。

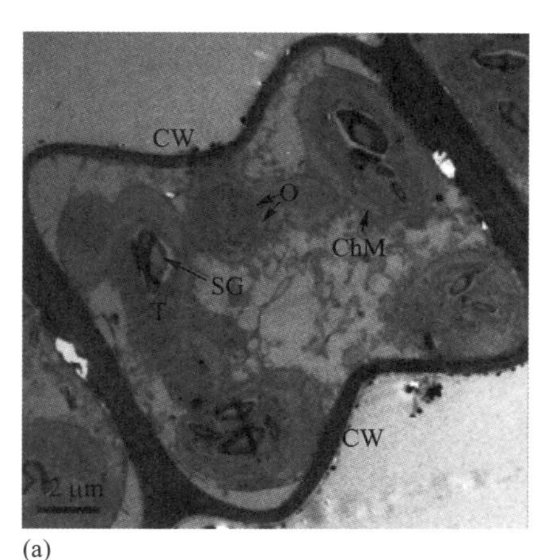

(a)

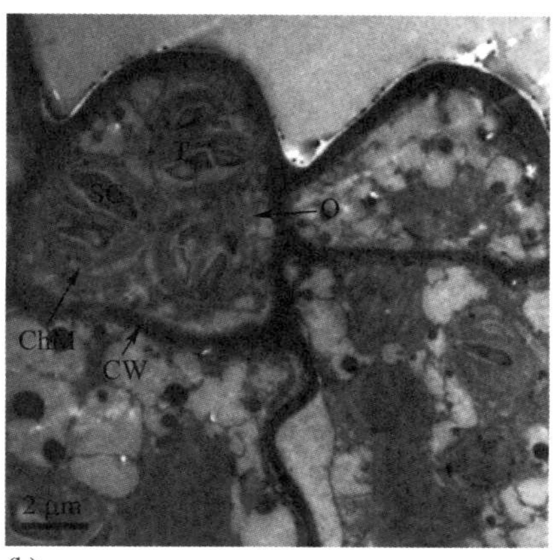

(b)

图 3-23　真藓快速脱水（a）和高温快速脱水（b）的透射电镜图。ChM：叶绿体膜；CW：细胞壁；O：嗜锇颗粒；T：类囊体

Figure 3-23　Transmission electron microscopy（TEM）micrographs showing cell ultrastructure in *B. argenteum* after fast desiccation（a）and high temperature fast combined desiccation（b）.ChM：chloroplast membrane；CW：cell wall；O：osmiophilic granule；T：thylakoid

通过扫描电镜发现，真藓在快速脱水后，外形完整，叶状体干瘪皱缩，细胞排列依然整齐有序（图 3-24a，c，e）。而经过快速加高温共同处理后，外形还基本完整，叶状体干缩稍卷曲，细胞结构排列紊乱，部分细胞有破损（图 3-24b，d，f）。

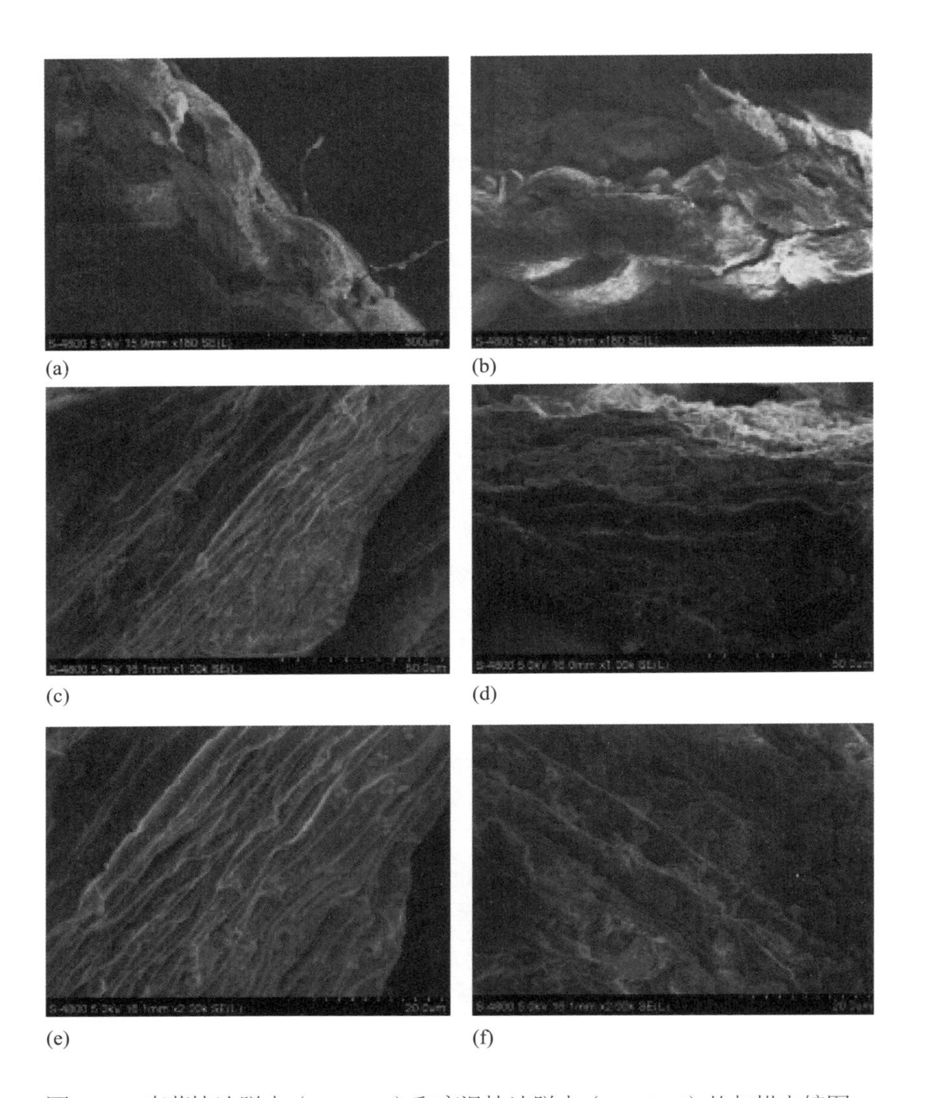

图 3-24 真藓快速脱水（a，c，e）和高温快速脱水（b，d，f）的扫描电镜图

Figure 3-24 Scanning electron microscopy（SEM）micrographs showing cell ultrastructure in *B. argenteum* after fast desiccation（a,c,e）and high temperature fast combined desiccation（b,d,f）

脱水速度对真藓光合参数的影响：图 3-23 反映出高温快速脱水对真藓膜结构造成了损伤，但叶绿体依然完整。进一步分析了这些脱水的样品在光下复水后的光合反应，结果表明，自然脱水的样品在复水 5 min 后，F_v/F_m 达到干燥前的 91.82%，30 min 后达到正常值。快速脱水的样品 30 min 后 F_v/F_m 开始增加，一直到 12 h 后恢复正常。而快速加高温脱水的样品在复水 1 h 后 F_v/F_m 开始增加，3 h 后达到了干燥前水平的 25%，然后缓慢上升，到 48 h 后基本恢复正常（图 3-25）。

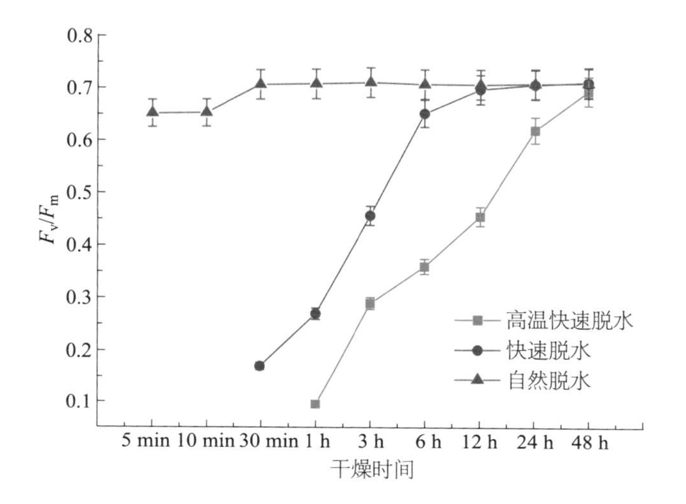

图 3-25　真藓快速脱水、高温快速脱水后复水的 F_v/F_m 的变化

Figure 3-25　The changes of F_v/F_m of *B. argenteum* after fast dried and high temperature fastdried following rehydration，natural dried as control

脱水速度对真藓膜结构的影响：真藓经历快速脱水和高温快速脱水之后，尽管从微表型上有不同程度的损伤，但在复水之后，光合系统都能够逐渐恢复。从红外光谱上显示，真藓在经历高温和快速脱水的双重胁迫后膜中各种化合物成分吸收峰最高（图 3-26、图 3-27 和图 3-28）。这说明真藓在快速脱水过程中膜中的水分散失很快，因为这三种处理最后恒定的总含水量都基本一致，差异不明显，所以说明在快速脱水过程中，通过快速失去膜中的水分稳定膜结构是耐干燥的机理之一。另外，可能真藓在应对干燥胁迫时，快速脱水是最明显的表型，而脱水后处于干燥状态的真藓能够忍受更加不良的环境胁迫，如高温等，所以比较了快速脱水和高温快速脱水的处理，高温只是略微增加了膜中各种化合物的含量。

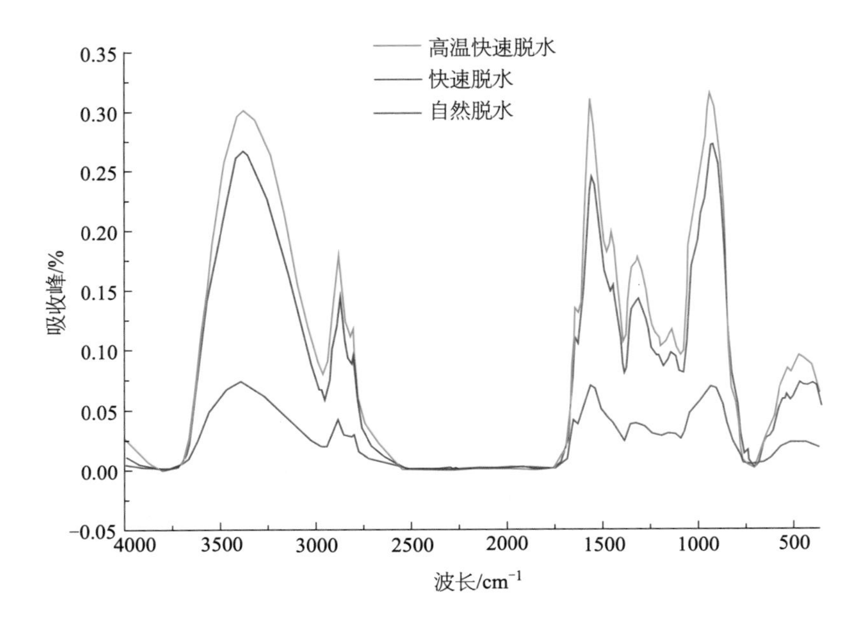

图 3-26 真藓快速脱水、高温快速脱水后在 4000 ~ 500 cm^{-1} 红外吸收指纹图谱

Figure 3-26 Absorption FTIR spectra in the 4000−500 cm^{-1} region of *B. argenteum* after fast dried and high temperature fast dried

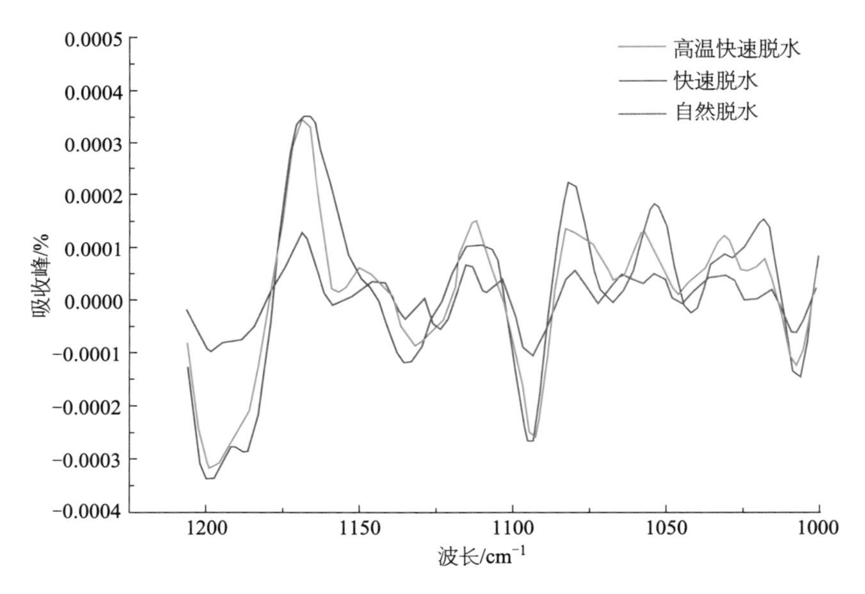

图 3-27 真藓快速脱水、高温快速脱水后在 1200 ~ 1000 cm^{-1} 红外吸收指纹图谱

Figure 3-27 Absorption FTIR spectra in the 1200−1000 cm^{-1} region of *B. argenteum* after fast dried and high temperature fast dried

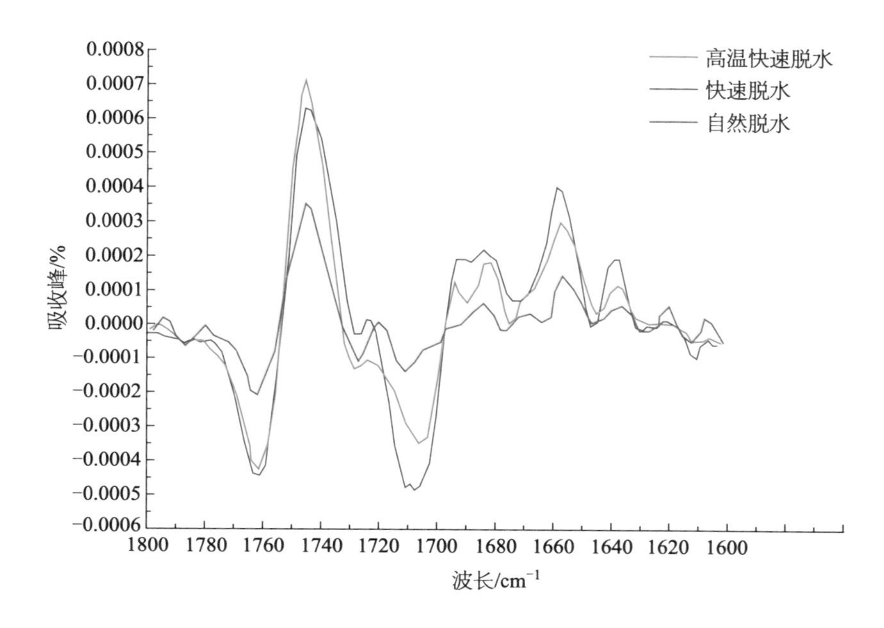

图 3-28 真藓快速脱水、高温快速脱水后在 1800 ~ 1600 cm^{-1} 红外吸收指纹图谱

Figure 3-28 Absorption FTIR spectra in the 1800−1600 cm^{-1} region of *B. argenteum* after fast dried and high temperature fast dried

3.1.2 干旱胁迫下 BSC 真藓和土生对齿藓信号转导物质的响应及活性氧清除机制

（1）BSC真藓和土生对齿藓对干旱胁迫的信号转导物质响应

真藓和土生对齿藓形态结构相对简单，但其面临的环境因子非常复杂，如高光强、高温、干旱、高盐碱、有机营养不平衡等。当这些环境因子持续时间过长或程度过剩，就会对植物造成胁迫（Bohnert and Sheveleva，1998）。植物经过漫长的进化过程，形成多种适应机制来抵御各种胁迫。植物胁迫应答的第一步是感受胁迫，然后通过信号转导途径传播胁迫信号。胁迫信号传播的最终结果是引起各种生理应答反应，如气孔关闭、基因表达以及相应的细胞和分子过程的改变（Knight and Knight，2001）。植物对环境胁迫的感受及其信号转导是通过多途径进行的，而且这些途径之间又是纵横交错的（Bohnert and Sheveleva，1998）。参与植物胁迫信号转导途径的主要有一氧化氮（NO）、脱落酸（ABA）、Ca^{2+} 等。目前，对藓类植物的研究主要集中在 ABA 或其他植物激素对藓类原丝体发育的影响，对其抗逆信号转导的分子机制研究还处于起步阶段（Takezawa *et al.*，2011）。

干旱胁迫是土生对齿藓和真藓面临的主要胁迫因子之一，在长期的进化过程中产生的抵抗干旱胁迫的信号转导机制可能与其他高等植物不一样。因此，研究 BSC 中土生对齿藓和真藓在水分胁迫后各种信号转导相关物质的响应，对探讨两者的抗旱机理，理解它们在 BSC 发展和发挥生态功能中的地位有重要的意义。

对采自腾格里沙漠东南缘沙坡头地区的土生对齿藓和真藓给予定量水分湿润，保持在 25 ℃下 16 h 光照、8 h 黑暗的昼夜循环，光照强度为 150 mmol m^{-2} s^{-1} 和相对湿度 60% 的培养间中培养，经室内培养一段时间，长势基本一致后，采取断水逐渐自然干旱的胁迫处理。分别取断水后 1 h（对照，测得植株含水量为 75%），24 h（中度干旱胁迫，测得植株含水量为 30%）和 48 h（重度干旱胁迫，测得植株含水量为 10%）的土生对齿藓和真藓地上部分为试验材料。为避免瞬时水分的影响，提取样本时用液氮研磨，测定各项信号转导指标。每项指标做 3 次重复。

逐渐干旱胁迫对土生对齿藓和真藓 Ca^{2+} 含量的影响：从图 3-29 可以得出，随着干旱胁迫程度的增加，土生对齿藓 Ca^{2+} 含量显著增加，在水分含量为 10% 时达到最大值 315.705 μmol g^{-1} FW，为对照 106.571 μmol g^{-1} FW 的 3 倍；真藓中 Ca^{2+} 含量先减小后增加，在水分含量为 30% 时含量最低，为 9.853 μmol g^{-1} FW，10% 时含量最高，为 23.278 μmol g^{-1} FW，但是变化不太明显，没有显著性差异。

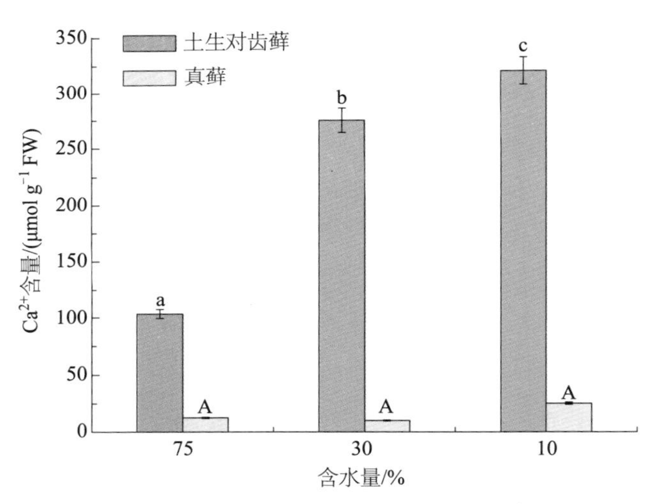

图 3-29 逐渐干旱胁迫对土生对齿藓和真藓 Ca^{2+} 含量的影响。图中数据为平均值 ± 标准误（n=3）；不同字母表示差异显著（p<0.05），图 3-30 ~ 图 3-44 同

Figure 3-29 Effects of gradual drought stress on content of Ca^{2+} of *Didymodon vinealis* and *Bryum argenteum*. Means ± SE（n=3）; different letters indicates significant difference（p<0.05）, it is the same as Figure 3-30−Figure 3-44

对土生对齿藓和真藓 K^+ 含量的影响：图 3-30 显示，随着干旱胁迫程度增加，土生对齿藓的 K^+ 含量从对照到水分含量为 30% 时变化不明显，从 2.62 mg g^{-1} FW 降到 2.54 mg g^{-1} FW，但在水分含量为 10% 时达 2.97 mg g^{-1} FW，显著增加。真藓的 K^+ 含量随水分含量的减少而逐渐增加，变化显著，分别为 2.96 mg g^{-1} FW、3.46 mg g^{-1} FW 和 3.76 mg g^{-1} FW。

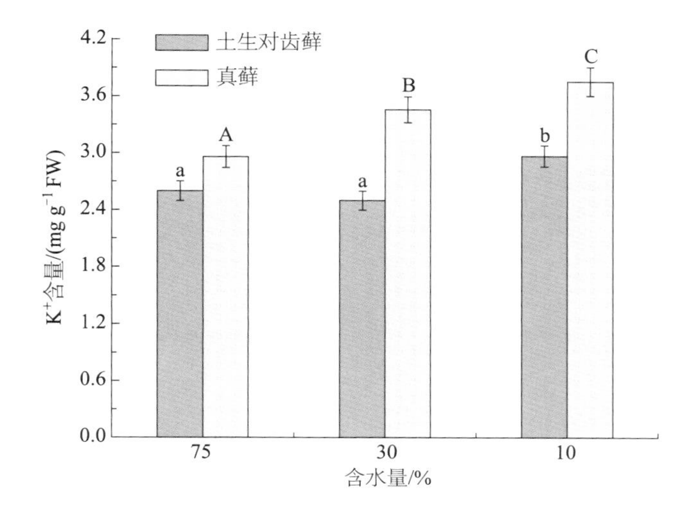

图 3-30 逐渐干旱胁迫对土生对齿藓和真藓 K^+ 含量的影响

Figure 3-30 Effects of gradual drought stress on content of K^+ of *D. vinealis* and *B. argenteum*

对土生对齿藓和真藓 NO 含量和一氧化氮合酶（NOS 酶）活性的影响：图 3-31 表明，在逐渐干旱胁迫下，土生对齿藓的 NO 含量表现出显著减少的趋势：对照为 0.089 μmol mg^{-1}pro，中度干旱胁迫处理为 0.081 μmol mg^{-1}pro，较对照减少了 9%，重度干旱胁迫处理为 0.009 μmol mg^{-1}pro，较对照减少了近 90%。真藓的 NO 含量则表现出先增加后减小的趋势，在水分含量为 30% 时达到最大值 0. 057 μmol mg^{-1}pro，为对照 0.024 μmol mg^{-1}pro 的 2 倍多，之后出现大幅度的降低，在重度干旱胁迫减少到 0.003 μmol mg^{-1}pro，为对照的 5.3%。

图 3-32 显示，随干旱胁迫程度增加，土生对齿藓的 NOS 酶活性表现出显著降低的趋势，中度干旱胁迫处理时降低到 0. 106 U mg^{-1}pro，为对照 0.255 U mg^{-1}pro 的 42%，重度干旱胁迫处理降低到 0. 043 U mg^{-1}pro，为对照的 17%。真藓的 NOS 酶则表现出先升高后降低的趋势，中度干旱胁迫时达到最大值 0. 062 U mg^{-1}pro，为对照 0.042 U mg^{-1}pro 的 1.5 倍，之后 NOS 酶活性降低到对照水平。土生对齿藓和真藓的 NOS 酶变化和 NO 含量变化有一定的相似性。

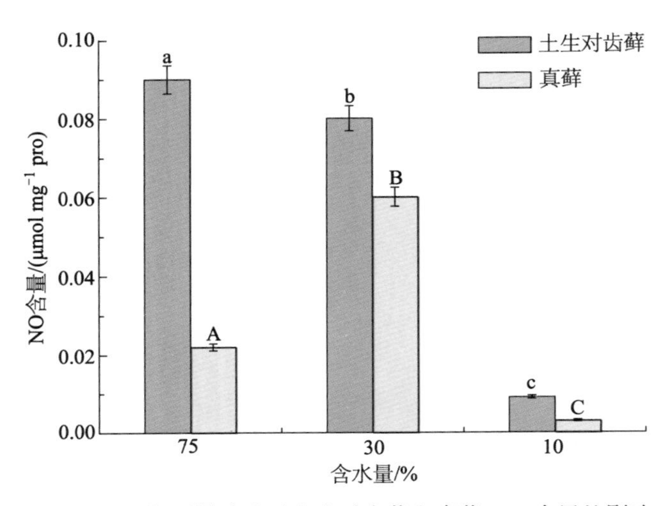

图 3-31　逐渐干旱胁迫对土生对齿藓和真藓 NO 含量的影响

Figure 3-31　Effects of gradual drought stress on content of NO of *D. vinealis* and *B. argenteum*

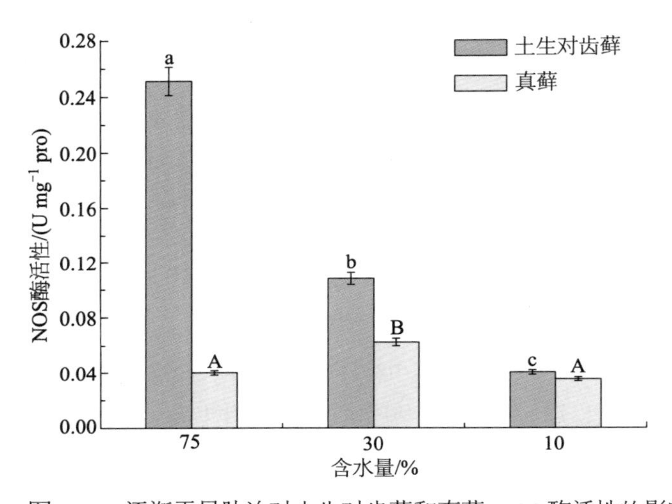

图 3-32　逐渐干旱胁迫对土生对齿藓和真藓 NOS 酶活性的影响

Figure 3-32　Effects of gradual drought stress on NOS activity of *D. vinealis* and *B. argenteum*

对土生对齿藓和真藓 ABA 含量的影响：图 3-33 显示，在逐渐干旱胁迫下，土生对齿藓 ABA 含量表现出显著减少的趋势，中度干旱胁迫处理达 1.527 mg g^{-1} FW，为对照 1.796 mg g^{-1} FW 的 85%，减少的幅度较小；重度干旱胁迫处理为 0.739 mg g^{-1} FW，为对照的 41%，减少得十分显著。在逐渐干旱胁迫下，真藓的 ABA 含量同样表现出逐渐减少的趋势，中度干旱胁迫为 1.537 mg g^{-1} FW，为对照 1.678 mg g^{-1} FW 的 92%，重度干旱胁迫达

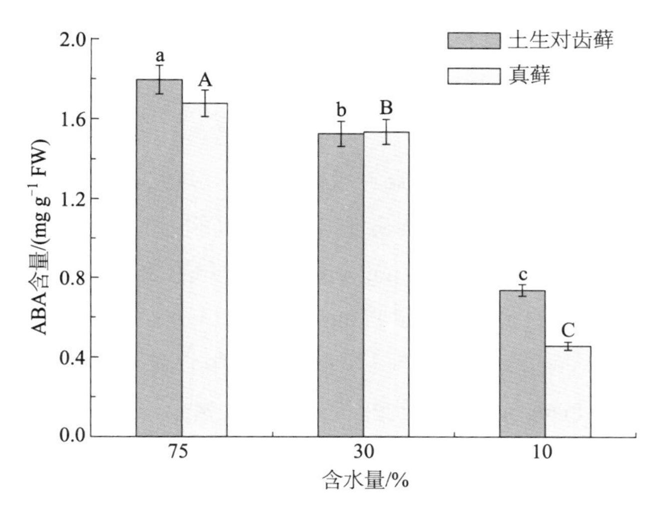

图 3-33　逐渐干旱胁迫对土生对齿藓和真藓 ABA 含量的影响

Figure 3-33　Effects of gradual drought stress on content of ABA of *D. vinealis* and *B. argenteum*

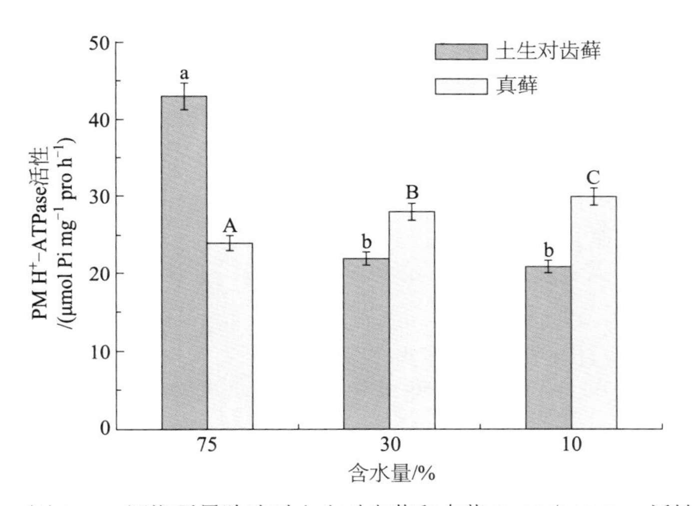

图 3-34　逐渐干旱胁迫对土生对齿藓和真藓 PM H^+-ATPase 活性的影响

Figure 3-34　Effects of gradual drought stress on PM H^+-ATPase activity of *D. vinealis* and *B. argenteum*

0.458 mg g^{-1} FW，为对照的 27%。

对土生对齿藓和真藓质膜 H^+-ATP 酶（PM H^+-ATPase）活性的影响：图 3-34 显示，在逐渐干旱胁迫下，土生对齿藓的 PM H^+-ATPase 活性表现出降低的趋势，当其水分含量减少到 30% 时，PM H^+-ATPase 活性显著降低到 22.00 μmol Pi mg^{-1}pro h^{-1}，较对照 43.35 μmol Pi mg^{-1}pro h^{-1} 降低了近 50%，而之后的重度胁迫处理，其活性略有降低，但没有显著性变化。真藓的 PM H^+-ATPase 活性表现出逐渐升高的趋势，中度干旱胁迫为 28.00 μmol

Pi mg^{-1}pro h^{-1}，较对照 24.30 μmol Pi mg^{-1}pro h^{-1} 升高了 15%，重度干旱胁迫为 30.13 μmol Pi mg^{-1}pro h^{-1}，较对照升高了 24%，变化显著。

对土生对齿藓和真藓液泡膜 H^+-ATP 酶（TP H^+-ATPase）活性的影响：图 3-35 显示，在逐渐干旱胁迫下，土生对齿藓和真藓 TP H^+-ATPase 活性均表现出显著升高的趋势。土生对齿藓在中度干旱胁迫下的 TP H^+-ATPase 活性为 28.00 μmol Pi mg^{-1}pro h^{-1}，较对照 25.82 μmol Pi mg^{-1}pro h^{-1} 升高了 8%，重度干旱胁迫时为 31.40 μmol Pi mg^{-1}pro h^{-1}，较对照升高了 22%。真藓的 TP H^+-ATPase 活性，在中度干旱胁迫处理和重度干旱胁迫处理中分别为 25.00 和 32.74 μmol Pi mg^{-1}pro h^{-1}，较对照 18.87 μmol Pi mg^{-1}pro h^{-1} 分别升高了 32% 和 74%，显著增加。

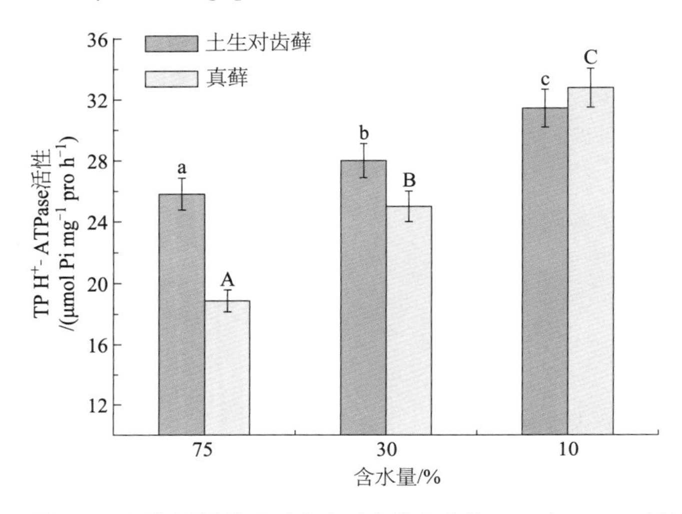

图 3-35　逐渐干旱胁迫对土生对齿藓和真藓 TP H^+-ATPase 活性的影响

Figure 3-35　Effects of gradual drought stress on TP H^+-ATPase activity of *D. vinealis* and *B. argenteum*

综上所述，土生对齿藓和真藓的 PM H^+-ATPase、TP H^+-ATPase 可以不同程度地提高两者的 K^+ 含量，增加渗透压，引发抗逆反应。在本研究中两种藓类植物的 ABA 含量都减少，可能两者均不是通过 ABA 达到抗旱目的，其机理有待进一步研究。Ca^{2+} 参与了土生对齿藓的抗旱信号转导过程，但对真藓作用不明显。NO 对土生对齿藓的抗旱作用不明显，但能提高真藓的抗旱性。上述结果进一步证实，K^+、Ca^{2+}、NO 等分子在极端干旱环境下对土生对齿藓和真藓抗旱信号转导具有调控作用（Zhao *et al.*，2015），在干旱时期对土生对齿藓和真藓起到保护作用，即缓解干旱胁迫对土生对齿藓和真藓正常生长的抑制作用，提高两者抵抗干旱胁迫的能力，明显增加了其成活指数。

（2）逐渐干旱胁迫下BSC中真藓和土生对齿藓的活性氧清除机制

干旱胁迫与植物膜脂过氧化及保护酶系统关系的研究已受到普遍重视（王军辉等，2006）。干旱胁迫可以加剧植物体内活性氧（reactive oxygen species，ROS）的产生，ROS包括超氧化物阴离子（O_2^-）、羟自由基（OH^-）、过氧化氢（H_2O_2）和单线态氧（O_2）等。这些物质十分活跃，被认为是植物代谢过程中的毒副产品，它们能够氧化植物体内膜系统，导致生物膜脂过氧化、蛋白质变性、DNA链断裂（Hasegawa *et al.*，2000；Mehdy，1994）以及光合作用受阻等多种有害的细胞学效应。ROS一旦超出了植物的清除能力会造成植物体氧化损伤（Sakaki *et al.*，1983），严重时导致植物细胞死亡（Mudd，1996）。为了适应外界的逆境条件，植物在长期进化中已形成了清除ROS自由基、抑制膜脂过氧化作用的酶促和非酶促系统，前者主要包括超氧化物歧化酶（SOD）、过氧化物酶（POD）、过氧化氢酶（CAT）、谷胱甘肽还原酶（GR）和抗坏血酸过氧化物酶（APX）等酶类，后者主要是维生素C（Vc）、维生素E（Ve）、类胡萝卜素（Car）和谷胱甘肽（GSH）等小分子抗氧化物质（许长城和邹琦，1993；陈少裕，1989）。

真藓和土生对齿藓是沙坡头地区BSC藓类植物重要组成成分。因此，本试验以真藓与土生对齿藓为试验材料，研究了逐渐干旱胁迫下真藓和土生对齿藓活性氧清除机制，探讨两者的抗旱机理差异，有助于理解两者在BSC生态恢复过程中的重要作用。

藓类植物过氧化氢、丙二醛和可溶性蛋白含量的变化：H_2O_2是植物体内主要的ROS之一，其含量变化在一定程度上可以反映ROS的产生情况。如图3-36所示，逐渐干旱胁迫下，真藓的H_2O_2含量有少量增加，但没有显著差异且含量较低。土生对齿藓的H_2O_2含量随干旱胁迫加剧而显著增加，尤其是在重度干旱胁迫下，H_2O_2含量较对照增加了近87%。

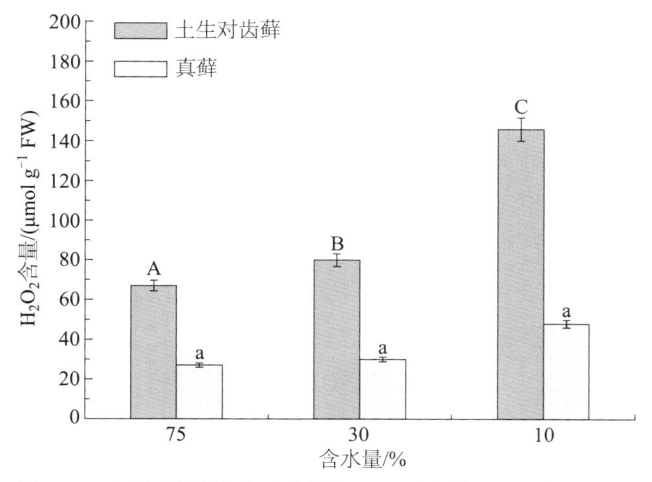

图3-36　逐渐干旱胁迫对真藓和土生对齿藓H_2O_2含量的影响

Figure 3-36　Effects of gradual drought stress on H_2O_2 contents in *B. argenteum* and *D. vinealis*

植物在逆境胁迫下，细胞原生质膜中的不饱和脂肪酸受 ROS 攻击易发生膜脂过氧化，形成脂氢过氧化物（ROOH）。ROOH 可以分解产生丙二醛（MDA），因此，MDA 的含量可以反映膜脂过氧化作用的强弱（Chen *et al.*，1991）。从图 3-37 可以看出，逐渐干旱胁迫下真藓和土生对齿藓的 MDA 含量变化不一致。随着干旱胁迫的程度增加，真藓的 MDA 含量降低得十分显著，中度干旱胁迫较对照降低了 34%，重度干旱胁迫较对照降低了近 80%。而逐渐干旱胁迫下，土生对齿藓的 MDA 含量增加较平缓，中度干旱胁迫较对照没有明显变化，重度干旱胁迫较对照有显著增加，但较中度干旱胁迫没有明显变化。

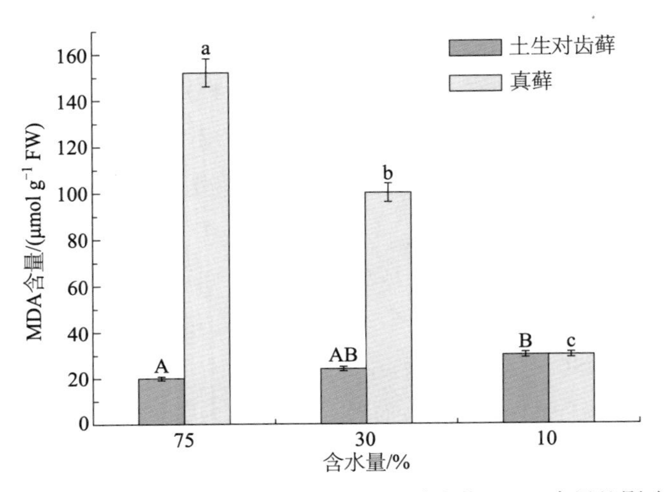

图 3-37 逐渐干旱胁迫对真藓和土生对齿藓 MDA 含量的影响
Figure 3-37 Effects of gradual drought stress on MDA contents in *B. argenteum* and *D. vinealis*

当植物遭受干旱胁迫时，很容易导致蛋白质氧化损伤。ROS 能够氧化很多种蛋白质氨基酸残基（如精氨酸、组氨酸、赖氨酸、脯氨酸、苏氨酸和色氨酸等）形成羰基化蛋白质，从而导致可溶性蛋白含量减少。因此，检测可溶性蛋白含量可用来评价蛋白质氧化损伤情况（伏毅等，2010）。逐渐干旱胁迫下，真藓和土生对齿藓的可溶性蛋白含量都有所下降（图 3-38）。随着干旱胁迫程度的增加，真藓的可溶性蛋白含量有略微降低，但无显著差异。土生对齿藓的可溶性蛋白含量在中度干旱胁迫下显著降低，降低了近 75%，重度干旱胁迫较对照显著降低，降低了近 80%，但重度干旱胁迫较中度干旱胁迫无显著差异。

藓类植物抗氧化系统酶活性变化：① SOD 活性的变化。SOD 是膜脂过氧化防御系统的主要保护酶，能歧化 O_2^- 成 H_2O_2，是清除活性氧的第一道防线（Neil *et al.*，2002）。较高的 SOD 活性是植物抵抗逆境胁迫的生理基础。从图 3-39 可以看出，逐渐干旱胁迫下，真藓和土生对齿藓的 SOD 活性虽然有所变化，但均没有显著差异。② POD 活性的变

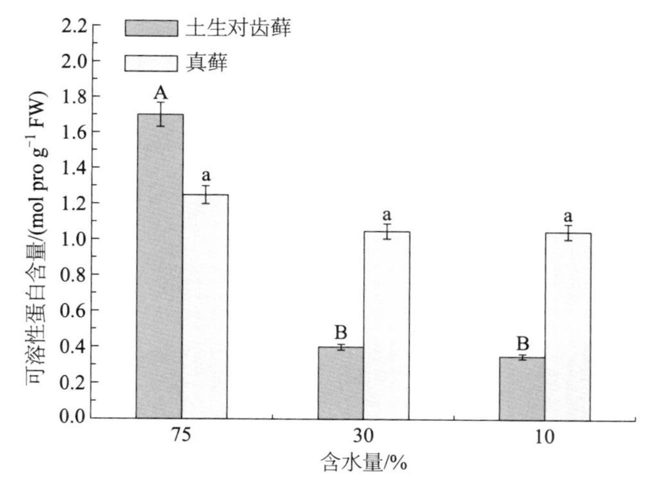

图 3-38 逐渐干旱胁迫对真藓和土生对齿藓可溶性蛋白含量的影响

Figure 3-38 Effects of gradual drought stress on soluble protein contents in *B. argenteum* and *D. vinealis*

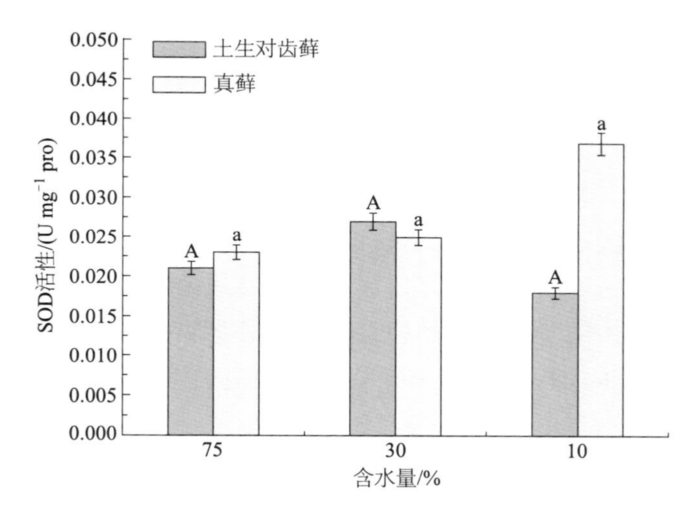

图 3-39 逐渐干旱胁迫对真藓和土生对齿藓 SOD 活性的影响

Figure 3-39 Effects of gradual drought stress on SOD activities of *B. argenteum* and *D. vinealis*

化。POD 是植物体内普遍存在的一种酶，参与植物的生长发育以及多种生理生化代谢过程和对逆境的适应调节机制。POD 可清除多种类型的活性氧，在植物体内清除活性氧过程中起着重要作用。在逐渐干旱胁迫下，真藓和土生对齿藓的 POD 活性变化不一致（图 3-40）。真藓 POD 活性随着干旱胁迫加剧而下降，中度干旱胁迫较对照有所降低，但没有显著差异，而重度干旱胁迫较对照有显著降低，降低了近 90%。土生对齿藓 POD 活性随着干旱胁迫程度加剧先增加后减少，中度干旱胁迫较对照有显著增加，增加了近 3 倍；重度干旱胁

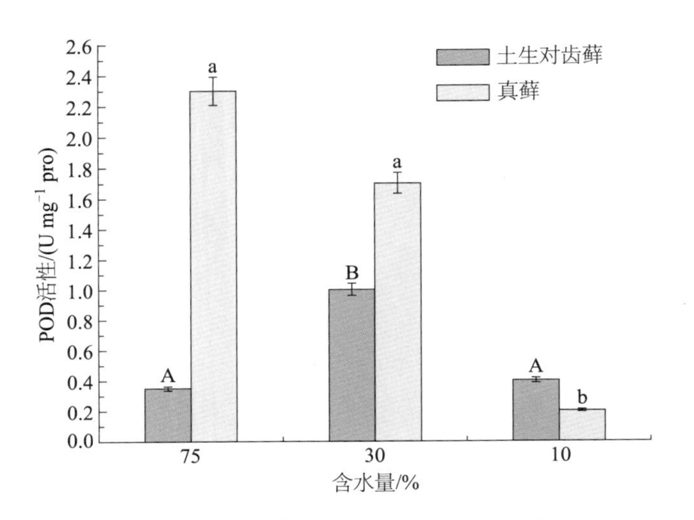

图 3-40 逐渐干旱对真藓和土生对齿藓 POD 活性的影响

Figure 3-40 Effects of gradual drought stress on POD activities of *B. argenteum* and *D. vinealis*

迫较中度干旱胁迫有显著降低，降低了近 60%。③ CAT 活性的变化。CAT 不仅是植物中清除 H_2O_2 的关键酶，而且是植物耐受胁迫所必需的保护酶（Willekens *et al.*，1994），它催化 H_2O_2 生成 H_2O，此反应不需要另外的动力。逐渐干旱胁迫下，真藓的 CAT 活性显著增加，中度干旱胁迫较对照增加了 37%，重度干旱胁迫较对照增加了 73%（图 3-41）。而土生对齿藓的 CAT 活性在逐渐干旱胁迫下的变化没有显著差异。④ APX 活性的变化。APX 是植物 AsA-GSH 氧化还原途径的重要组分之一。APX 利用抗坏血酸（AsA）为电子供体将 H_2O_2 转化为 H_2O，是植物体中清除 H_2O_2 的关键酶。随着干旱胁迫的程度增加，真藓和土生对齿藓的 APX 活性都会降低（图 3-42）。真藓的 APX 活性随着干旱胁迫程度增加而显著下降，中度干旱胁迫较对照下降了 29%，重度干旱胁迫较对照下降了 58%。土生对齿藓的 APX 活性在中度干旱胁迫下较对照有显著下降，降低了约 55%；重度干旱胁下 APX 活性较对照显著下降，降低了约 69%，但重度干旱胁迫较中度干旱胁迫没有显著变化。⑤ GR 活性和 GSH 含量的变化。GR 通过参与 AsA-GSH 循环而在细胞活性氧的清除中起重要作用。在氧化胁迫反应中，它对于保护细胞内的谷胱甘肽大部分处于还原状态起着关键的作用（谷胱甘肽完成它的生理功能必须保持还原的状态）。由图 3-43 可以看出，随着干旱胁迫程度的增加，真藓和土生对齿藓的 GR 活性变化完全相反。逐渐干旱胁迫下，真藓的 GR 活性有所增加。中度干旱胁迫较对照有显著增加，重度干旱胁迫较对照也有显著增加，但重度干旱胁迫较中度干旱胁迫变化不大。然而，逐渐干旱胁迫下，土生对齿藓的 GR 活性显著下降，中度干旱胁迫较对照降低了近 25%，重度干旱胁迫较对照降低了近 63%。GSH 是普遍存在于植物组织中的小分子量抗氧化物质，在减缓氧化胁迫、抵御逆境伤害方面具有非常重要

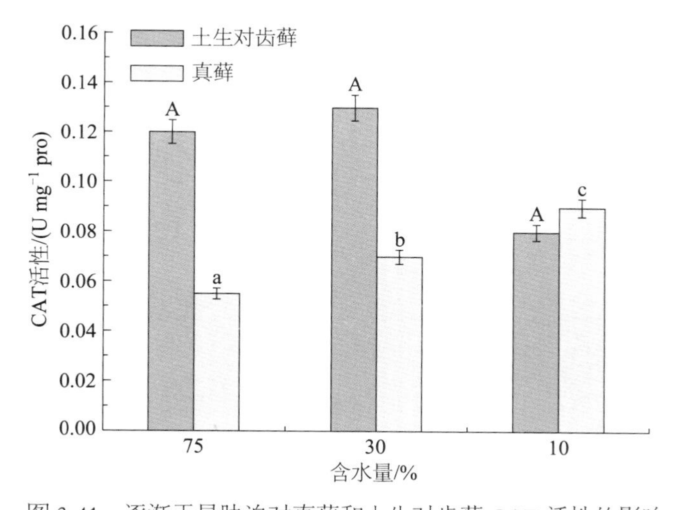

图 3-41　逐渐干旱胁迫对真藓和土生对齿藓 CAT 活性的影响

Figure 3-41　Effects of gradual drought stress on CAT activities of *B. argenteum* and *D. vinealis*

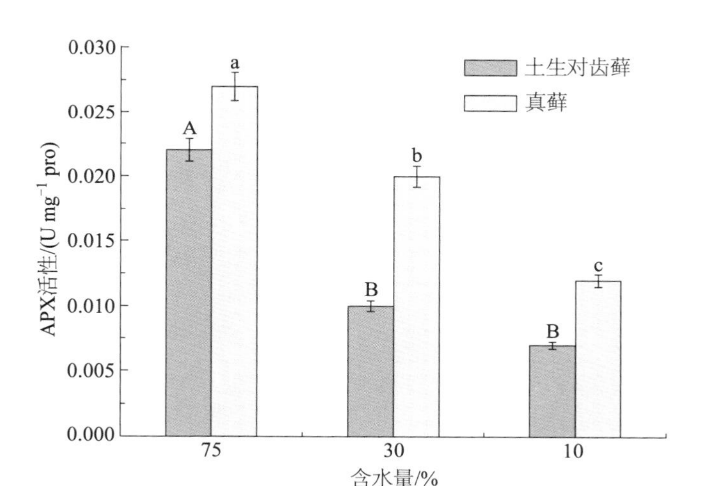

图 3-42　逐渐干旱胁迫对真藓和土生对齿藓 APX 活性的影响

Figure 3-42　Effects of gradual drought stress on APX activities of *B. argenteum* and *D. vinealis*

的作用（Noctor and Foyer，1998）。它不但可直接同活性氧自由基反应将其还原，还可与 APX、GR 等一起形成 AsA-GSH 循环，参与清除 H_2O_2（Liu *et al.*，2007）。随着干旱胁迫程度的增加，真藓和土生对齿藓的 GSH 含量变化并不一致（图 3-44）。随着干旱胁迫加剧，真藓的 GSH 含量先增加后减少，中度干旱胁迫较对照有显著增加，增加了近 1 倍，重度干旱胁迫较中度干旱胁迫有显著降低，降低了近 18%。而土生对齿藓的 GSH 含量随着干旱胁迫加剧而显著增加，中度干旱胁迫较对照增加了 58%，重度干旱胁迫较对照增加了近 80%。

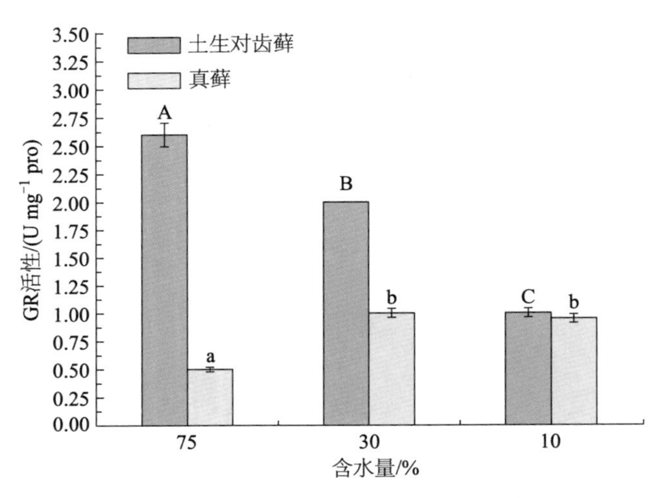

图 3-43　逐渐干旱胁迫对真藓和土生对齿藓 GR 活性的影响

Figure 3-43　Effects of gradual drought stress on GR activities of *B. argenteum* and *D. vinealis*

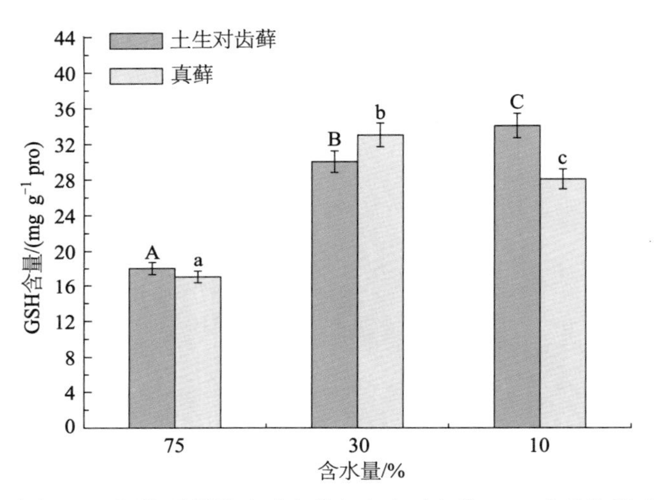

图 3-44　逐渐干旱胁迫对真藓和土生对齿藓 GSH 含量的影响

Figure 3-44　Effects of gradual drought stress on GSH contents in *B. argenteum* and *D. vinealis*

本试验表明，真藓和土生对齿藓的活性氧清除系统在逐渐干旱胁迫下有着不同的响应机制。逐渐干旱胁迫下，真藓的 MDA 含量下降，H_2O_2 含量略微升高，但可溶性蛋白含量变化不大。真藓的抗氧化系统在逐渐干旱胁迫下的变化主要为：CAT 活性增加，GR 活性和 GSH 含量均先增加后降低，SOD 活性变化不大，而 APX 和 POD 活性下降。这表明逐渐干旱胁迫下真藓的 ROS 增加不是很剧烈，而且 CAT 在清除 ROS 过程中起着重要的作用，SOD 则起着辅助作用，而 APX 和 POD 作用不明显。GR 和 GSH 在中度干旱胁迫下清除 ROS 有着重要作用，但在重度干旱胁迫下作用不明显。逐渐干旱胁迫下，土生对齿藓的 MDA 和 H_2O_2 含量均显著增加，而可溶性蛋白含量显著下降。其抗氧化系统中除了 GSH 含量上升，POD 活性先升后降外，CAT 和 SOD 活性均变化不大，而 APX 和 GR 活性却下降。这表明逐渐干旱胁迫下，土生对齿藓的 ROS 显著增加且对其造成氧化胁迫，GSH 在清除 ROS 过程中起主要作用，SOD 和 CAT 起辅助作用，而 APX 和 GR 的作用不明显。POD 则在中度干旱胁迫下清除 ROS，在重度干旱胁迫下作用不明显。

3.1.3 极端降雨事件对不同类型 BSC 覆盖土壤碳释放的影响

气候变化对陆地生态系统的影响引起了科学界和公众的广泛关注，因为全球气候变化潜在地影响到生态系统功能及其生态服务功能的维持（Fay *et al.*，2008；Bennett *et al.*，2005）。极端降雨事件指与历史同期相比出现较少的小概率降雨事件，它具有危害性高、突发性强等特征。对极端降雨事件的划分多采用世界气象组织（World Meteorological Organization，WMO）的定义标准。极端降雨事件的增加是气候变化的一个重要表现（Christensen and Hewitson，2007）。大量事实证明，在全球、地区及小尺度上的极端降雨事件呈现增加趋势。例如，20 世纪全球陆地（除了南极洲）降雨量增加了 9 mm，但是在地区尺度上，北美洲的部分地区、欧洲和南非的降雨天数增加了；在欧洲、非洲、澳大利亚和地中海地区，干旱时期的降雨频率增加了；与此同时，一些地区的极大降雨事件增加了（New *et al.*，2001）。21 世纪，极端降雨事件的出现频次还会继续增加（Groisman *et al.*，2005；Alley *et al.*，2007；王兴梅等，2011）。对 BSC 碳循环方面的研究多集中在碳交换（Castillo-Monroy *et al.*，2011）、CO_2 年释放量（Thomas *et al.*，2008）及人工模拟降雨对 CO_2 年释放的影响（Thomas and Hoon，2010）等方面，而对极端降雨条件下 BSC 覆盖土壤碳释放的研究鲜见报道。因此，对极端降雨条件下 BSC 覆盖土壤碳释放的研究，尤其是在自然降雨条件下进行研究，将为探索极端降雨对 BSC 覆盖的土壤碳释放的影响和全球变化背景下 BSC 对极端降雨的响应特点提供可靠的数据支持。

本研究以处于腾格里沙漠东南缘的沙坡头人工固沙植被区不同类型 BSC 覆盖的土壤为研究对象，通过对极端降雨量和降雨强度下 BSC 覆盖土壤的碳释放和土壤含水量的监测，研究不同类型 BSC 覆盖土壤的碳释放对不同数量和强度的降雨的响应差异及规律。

采用WMO的极端气候事件定义标准进行极端降雨事件的划分。WMO规定：如果某个（些）气候要素的时、日、月、年值达到25年以上一遇，或者与其对应的多年平均值（一般应达30年左右）的“差”超过其二倍的均方差时，这个（些）气候要素值就属于“异常”气候值。出现“异常”气候值的事件就是“极端气候事件”。降雨是表征气候属性的一个最基本要素，因此将符合上述标准的降雨事件归属于“极端降雨事件”（Christensen and Hewitson，2007）。卫伟等（2007）根据WMO的标准确定的黄土丘陵沟壑区（年均降雨量427 mm）的极端降雨事件为降雨总量大于40.1 mm或降雨强度大于0.55 mm min^{-1}的降雨事件。根据WMO的标准，本试验区降雨总量大于30.0 mm或降雨强度大于0.15 mm min^{-1}的降雨事件为极端降雨事件。如表3-1所示，试验期间，降雨事件1和降雨事件2可以定义为极端降雨事件，降雨事件3为普通降雨事件。

表3-1 试验期间降雨事件及其特征

Table 3-1 The quantity，duration and intensity of three rainfall events in the experimental period

序号	时间（年－月－日）	降雨量/mm	降雨历时/min	降雨强度/（mm min^{-1}）
1	2011-08-10	44.7	1 170	0.04
2	2011-08-23	8.3	15	0.55
3	2011-05-09	16.3	870	0.02

试验于2011年5—8月进行，采用Li-6400-09土壤呼吸室（LI-COR，Inc.，美国）测定BSC覆盖土壤呼吸速率。降雨事件1结束后立即开始观测，之后每日9：00观测一次，直到呼吸速率恢复到降雨前水平时停止观测；降雨事件2，分别于降雨结束后0.5 h、2 h、6 h、9 h、14 h、26 h、30 h、34 h和38 h进行观测；降雨事件3，分别采用上述的测定方法进行测定。在试验开始前，尽量去除BSC表面的沙粒和杂物，保持表面清洁，每种类型的BSC设3次重复。

（1）试验期间BSC样品含水量和空气温度变化

试验期间（2011年5—8月），极端降雨事件结束后平均气温为19～33 ℃，普通降雨事件结束后平均气温为13～27 ℃。采用称重法测定0～10 cm土壤含水量。极端降雨事件结束后，三类BSC覆盖土壤含水量为0.5%～25%，普通降雨事件结束后三类BSC覆盖土壤含水量为0.5%～18%。无论是极端降雨事件还是普通降雨事件后，藻类结皮和混生结皮覆盖土壤的平均含水量均高于藓类结皮覆盖土壤，这是由于藓类结皮相对于藻类结皮土壤具有较高的生物量，水分散失快于藻类结皮和混生结皮；由于样品表层水分的不断散失，试验后期，三类BSC覆盖土壤的含水量差异不显著（图3-45）。

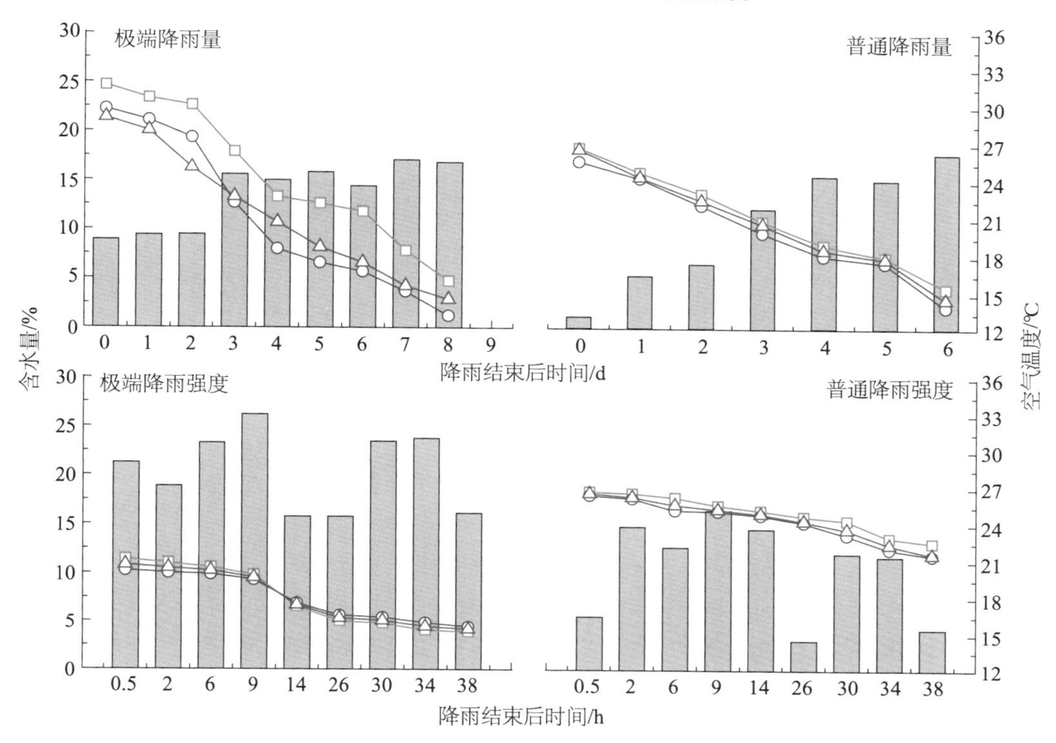

图 3-45 试验期间 BSC 含水量及空气温度的变化。折线表示 BSC 含水量、直方表示空气温度
Figure 3-45 The air temperature (histograms) and the water content (lines) of the BSC samples after rainfall

（2）极端降雨和普通降雨对不同类型BSC覆盖土壤碳释放的影响

降雨前，藻类结皮、藓类结皮和混生结皮覆盖土壤呼吸速率分别为 0.17 ~ 0.22 μmol m^{-2} s^{-1}、0.28 ~ 0.36 μmol m^{-2} s^{-1} 和 0.22 ~ 0.27 μmol m^{-2} s^{-1}（图 3-46）。在极端降雨事件和普通降雨事件中，BSC 类型和降雨后时间均显著影响 BSC 覆盖土壤碳释放（表 3-2）。

极端降雨量条件下，藓类结皮覆盖土壤的碳释放过程呈单峰曲线，而藻类结皮和混生结皮覆盖土壤的变化趋势基本一致，呈双峰曲线（图 3-47）。降雨结束初期，藓类结皮、藻类结皮和混生结皮覆盖土壤呼吸速率较低，分别为 0.83 μmol m^{-2} s^{-1}、0.10 μmol m^{-2} s^{-1} 和 0.12 μmol m^{-2} s^{-1}，最大呼吸速率均出现在降雨结束后第 4 天，分别为 1.61 μmol m^{-2} s^{-1}、0.83 μmol m^{-2} s^{-1} 和 1.31 μmol m^{-2} s^{-1}。普通降雨量条件下，不同类型 BSC 碳释放过程均表现双峰曲线的变化趋势，最大呼吸速率均出现在雨后第 4 天，分别为 1.29 μmol m^{-2} s^{-1}、0.74 μmol m^{-2} s^{-1} 和 1.19 μmol m^{-2} s^{-1}。无论是极端降雨量还是普通降雨量条件下，相同

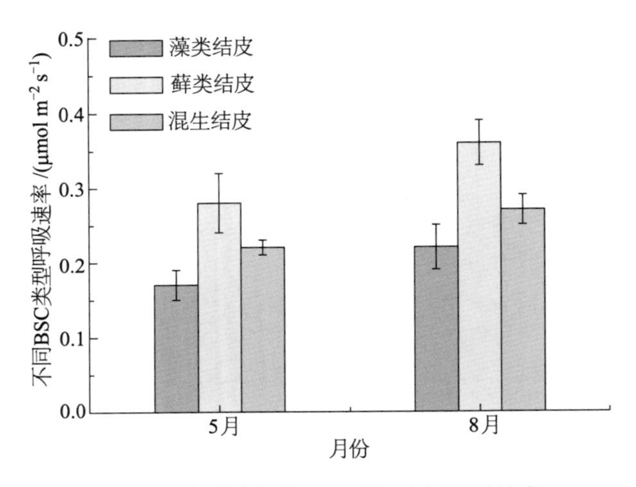

图 3-46 降雨前不同类型 BSC 覆盖土壤呼吸速率

Figure 3-46 The respiration rate of the soil covered by different types of BSC before rainfall

表 3-2 BSC 类型和降雨后时间对 BSC 覆盖土壤碳释放影响的二维方差分析

Table 3-2 Two way-ANOVA analysis on effects of BSC types and time after rainfall on carbon release of BSC covered soil

方差来源	F	p	方差来源	F	p
极端降雨量			普通降雨量		
校正模型	17.7	0.000	校正模型	11.0	0.000
结皮类型	102	0.000	结皮类型	4.44	0.018
降雨后时间	23.7	0.000	降雨后时间	25.6	0.000
结皮类型 × 降雨后时间	2.77	0.004	结皮类型 × 降雨后时间	4.77	0.000
R^2=0.895（adjust R^2=0.844）			R^2=0.839（adjust R^2=0.763）		
极端降雨强度			普通降雨强度		
校正模型	10.6	0.000	校正模型	60.3	0.000
结皮类型	66.7	0.000	结皮类型	164	0.000
降雨后时间	16.2	0.000	降雨后时间	119	0.000
结皮类型 × 降雨后时间	2.05	0.026	结皮类型 × 降雨后时间	15.8	0.000
R^2=0.851（adjust R^2=0.771）			R^2=0.967（adjust R^2=0.951）		

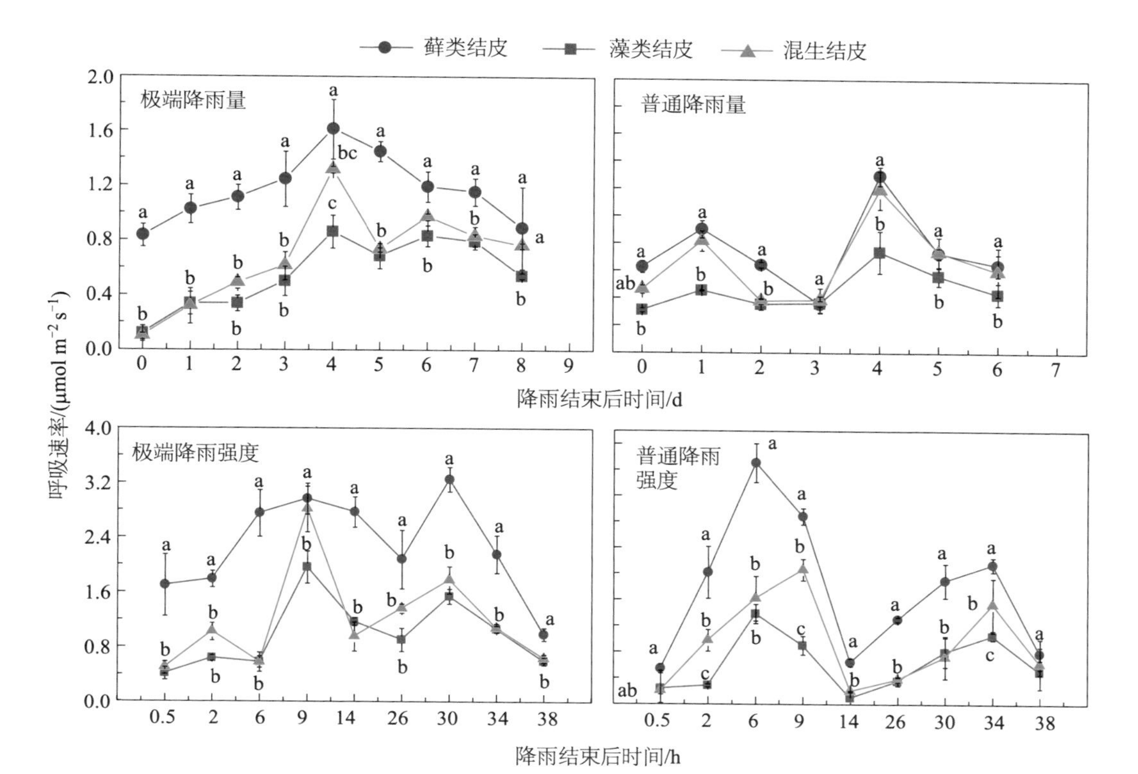

图 3-47 降雨量和降雨强度对不同 BSC 呼吸速率（mean ± SE）的影响

Figure 3-47 The effect of rainfall quantity and rainfall intensity on the respiration rate (mean ± SE) of different types of BSC

时间点上，藓类结皮覆盖土壤的碳释放量始终显著高于藻类结皮和混生结皮覆盖土壤。极端降雨结束初期的藓类结皮覆盖土壤呼吸速率及其最高值均高于普通降雨；而藻类结皮和混生结皮覆盖土壤，在极端降雨结束初期表现出极低的呼吸值。

极端降雨强度条件下，藓类结皮和藻类结皮、混生结皮覆盖土壤碳释放过程分别呈双峰和三峰曲线（图 3-47）。藓类结皮、藻类结皮和混生结皮覆盖土壤呼吸速率最大值分别出现在降雨后 30 h 和 9 h，分别为 3.25 μmol m^{-2} s^{-1} 和 1.53 μmol m^{-2} s^{-1}、1.77 μmol m^{-2} s^{-1}。普通降雨强度下，不同类型 BSC 覆盖土壤碳释放过程表现出双峰曲线变化趋势，藓类结皮和藻类结皮、混生结皮覆盖土壤呼吸速率最高值分别出现在降雨结束后 6 h 和 9 h，分别为 3.71 μmol m^{-2} s^{-1}、1.38 μmol m^{-2} s^{-1} 和 2.06 μmol m^{-2} s^{-1}。对相同时间点不同类型 BSC 碳释放量进行方差分析发现，在整个碳释放过程中，藓类结皮覆盖土壤的呼吸速率始终高于或显著高于藻类结皮和混生结皮覆盖土壤。极端降雨和普通降雨强度条件下，藓类结皮覆盖土壤呼吸速率最大值出现的时间不同，极端降雨后的最大值出现在降雨结束后的后期，而普通降雨出现在降雨结束后的初期；在极端降雨结束初期，藻类结皮和混生结皮覆盖土壤呼吸速率较低。

表 3-3 描述了极端降雨和普通降雨结束后土壤含水量和不同类型 BSC 覆盖土壤碳释放间的关系。不论是极端降雨条件下还是普通降雨条件下，土壤含水量和不同类型 BSC 覆盖土壤碳释放间的拟合方程均有统计学意义，能够很好地反映两者间的关系。

表 3-3　不同类型 BSC 覆盖土壤碳释放与土壤水分的关系

Table 3-3　The relationships between the moisture content and the respiration rate of soil covered by different type of BSC

降雨类型	结皮类型	回归方程	R^2	p
极端降雨量	藻类	$Rs=0.273+0.086W-0.004W^2$	0.901	0.001
	藓类	$Rs=0.804+0.115W-0.005W^2$	0.768	0.013
	混生	$Rs=0.582+0.082W-0.005W^2$	0.735	0.019
普通降雨量	藻类	$Rs=2.408-0.118W+0.009W^2$	0.625	0.049
	藓类	$Rs=0.652+0.027W-0.001W^2$	0.498	0.050
	混生	$Rs=-1.18+1.14W-0.007W^2$	0.997	0.000
极端降雨强度	藻类	$Rs=-2.331+1.07W-0.072W^2$	0.543	0.046
	藓类	$Rs=-8.415+3.091W-0.024W^2$	0.553	0.041
	混生	$Rs=-4.083+1.596W-0.106W^2$	0.545	0.047
普通降雨强度	藻类	$Rs=4.001-4.083W-0.012W^2$	0.640	0.038
	藓类	$Rs=-4.515+2.03W-0.068W^2$	0.590	0.049
	混生	$Rs=-3.838-0.647W-0.022W^2$	0.490	0.050

在极端降雨量和极端降雨强度条件下，大量水分聚集在土壤表面，土壤的通透性变差，藻类结皮和混生结皮的呼吸速率受到了明显抑制，随着土壤含水量的下降，呼吸速率逐渐增大；而这两种情况对藓类结皮的呼吸速率影响不明显。Li 等（2003）通过腾格里沙漠 50 多年的定位监测，研究了沙丘经沙障固定表面、种植旱生灌木形成人工固沙植被后 BSC 的拓殖和演替过程，发现 BSC 拓殖和演替的过程为物理结皮—藻类结皮—混生结皮—藓类结皮。藻类结皮和藓类结皮对水分的拦截能力不同。藻类在生长季节能够分泌大量胞外多聚糖，随着这种多糖物质浓度的提高，土壤持水性能增大，土壤水分蒸发速率降低，水分在土壤中的运动速率降低，这使得降雨过后水分大量集中在土壤表层。藓类植物叶子的季节

变化也会影响土壤水文特性，许多藓类植物长有一定长度的尖梢，可以把水分引入植物体中间去，水分通过植物体向土壤深层入渗，降雨过后水分不会大量集中在土壤表层（Li *et al.*，2010a）。水分取代了土壤中 CO_2 占据的位置的同时也使土壤的通透性变差，CO_2 在土壤中的扩散阻力因此增大，是导致土壤 CO_2 排放量减少的主要原因（Kursar，1989）。张志山等（2007）和何明珠等（2006）对沙坡头人工固沙植被区不同类型 BSC 蒸散特性的研究表明，藓类结皮覆盖土壤的水分蒸发量和蒸发速率显著高于藻类结皮覆盖土壤的；不同类型 BSC 覆盖土壤在蒸发的开始阶段，蒸发速率显著高于蒸发的后期阶段。极端降雨结束后，土壤表层含水量很高，蒸发量较大，土壤表层含水量会迅速下降。此后，水分对呼吸的抑制作用得到缓解，呼吸速率逐渐增加。藓类结皮覆盖土壤的水分蒸发量和蒸发速率显著高于藻类结皮覆盖土壤的，藓类结皮覆盖土壤的表层含水量的下降速率远高于藻类结皮和混生结皮覆盖土壤的，进而有效减缓了水分对呼吸的抑制，这可以解释极端降雨对藓类结皮覆盖土壤的呼吸影响不明显。

综上所述，极端降雨（降雨量和降雨强度）短期内会明显抑制藻类结皮和混生结皮覆盖土壤的呼吸速率，而对藓类结皮覆盖土壤的影响不明显；处于演替高级阶段的藓类结皮能够很好地应对短期的极端降雨事件。本研究仅针对极端降雨对 BSC 覆盖土壤碳释放过程的短期影响，对 BSC 覆盖土壤碳释放过程的长期影响将是今后研究的重点。

3.1.4 持续湿润时间对 BSC 固氮活性的影响

BSC 中的隐花植物属变水（poikilohydric）植物，自身对水分的调节能力极差，无法利用深层土壤水分（Nash，1996）。而 BSC 仅在湿润时才能进行新陈代谢和氮固定，因此其生物活性主要与降雨量、降雨频度和降雨的持续时间密切相关（Belnap *et al.*，2004），特别是降雨事件发生后其湿润的持续时间将直接影响其固氮速率，被认为是影响其固氮功能的关键因子之一（Belnap，2002）。国外关于 BSC 水分含量与固氮速率关系的研究已有很多，但由于大部分结果是在实验室人工控制条件下得出的，与少量的野外试验结果相差较大，甚至相反（Belnap and Lange，2003），说明生态因子与环境因子对其固氮活性的影响极其复杂，这可能与研究的区域条件特别是降雨发生后的环境条件、实验时间尺度等密切相关。而目前我国来自野外研究的实验数据非常稀少，很难将实验室得出的研究结果外推到区域尺度和自然环境水平上。

以腾格里沙漠东南缘沙坡头地区人工固沙植被区和相邻天然植被区发育良好的两类典型 BSC（藻类和藓类为优势种）为研究对象，野外采集原状 BSC 土壤样品，在一次较大的自然降雨事件发生后，利用开顶式生长室（open-top growth chamber，OTC）连续测定了两类 BSC 固氮活性的变化，研究持续湿润时间、含水量和温度变化对其固氮活性的影响，分析并阐述我国温带荒漠区 BSC 对荒漠生态系统的氮贡献及对人工植被系统演替的指示意义。

2011 年 5 月 7—9 日有降雨，降雨量 17 mm，10 日天气转晴，早 8：00 时将摆放在实验观测场的 BSC 样品用电子天平（TD31001）进行称重后布置在不同规格的 OTC（大 OTC：边长 1.3 m、高 2 m 八棱柱；小 OTC：边长 1 m、高 1.5 m 长方体）内中心位置及自然状态下（对照，CK）。两个地区（沙坡头 1964 年建植人工固沙植被区和相邻天然植被区），两种 BSC 类型（藻类结皮和藓类结皮），3 个温度梯度［大 OTC、小 OTC 和对照（CK）］，各 3 个重复。将样品放置 2 h 后用红外温度计（IRT）测量样品表面温度，之后将有机玻璃罩（直径 12 cm，高 15 cm，下端开口，上端密封，中间留一圆孔用橡皮塞密封）罩在 BSC 样品上，下端插入沙子中，创造一个密闭环境。然后用注射器向培养器中注入乙炔气体，使容器内乙炔体积分数为 10%。

在自然状态下培养 4 h，每隔 1 h 记录大、小 OTC 内及室外的空气温度。培养结束后用注射器收集培养器顶部气体 5 mL。收气结束后即刻去除收气罩，并用红外温度计测量样品表面温度。将收集的气体带回实验室后即刻用气相色谱仪（Agilent GC6820，美国）测定乙烯生成量。固氮活性结果以乙烯生成速率（nmol C_2H_4 m^{-2} h^{-1}）表示。10 日晚至 11 日下午有降雨，降雨量 2 mm。12—19 日天气晴好，无降雨发生，从 12 日开始每日（12—15 日）和隔日（17 日和 19 日）于相同时间重复上述实验过程，一直到样品质量恒定，结束实验。

（1）实验期温度与样品含水量变化

实验期（2011 年 5 月 10 日、12—15 日、17 日和 19 日）全天气温为 12 ~ 28.5 ℃，平均为 18.6 ℃。培养期（10：00—14：00）平均气温为 19.1 ~ 34.5 ℃，平均为 27.4 ℃，其中小 OTC 气温显著高于大 OTC 和 CK，平均值分别为 29.8 ℃、26.4 ℃和 26 ℃。培养期样品表面温度（用培养时加罩前和除罩后样品表面温度的平均值表示）为 19.2 ~ 45.1 ℃，平均为 28.7 ℃；藻类结皮的表面温度为 19.1 ~ 45.9 ℃，平均为 29.1 ℃，由于受天气、太阳辐射及 OTC 本身局部遮光的影响，不同温度处理下的样品表面温度不一致，总体表现为 CK>小 OTC>大 OTC；藓类结皮的表面温度为 19.0 ~ 44.3 ℃，平均为 28.3 ℃，不同温度处理下的样品表面温度表现为小 OTC>CK> 大 OTC。

实验期两类 BSC 含水量为 0.2% ~ 15.2%，降雨后第 1 天和第 3 天（第 2 天有降雨，降雨量 2 mm）藻类结皮样品平均含水量高于藓类结皮，分别为 14.1% 和 13.3%，这是由于藓类结皮相对于藻类结皮具有较高的生物量，水分散失快于藻类结皮，含水量相对较低；其后由于样品表层水分散失殆尽，两类 BSC 的含水量差异不大，约为 0.1% ~ 0.2%。不同温度处理下两类 BSC 样品平均含水量表现一致，都表现为 CK>小 OTC>大 OTC，这是因为大、小 OTC 内相对较高的气温导致样品水分散失快于 CK（表 3-4，图 3-48）。

表 3-4　实验期温度与样品含水量的变化，平均值（标准差）

Table 3-4　Variation of temperature and samples water content in experimental period,mean(SD)

	BSC 类型	温度处理	降雨后天数 /d						
			1	3	4	5	6	8	10
全天气温 /℃			16.1 (3.6)	12.0 (2.5)	13.9 (2.6)	20.7 (3.1)	22.0 (2.8)	28.5(1.9)	17.3(1.1)
培养期气温 /℃		大 OTC	29.6(0.4)a	23.3(0.6)a	23.0(0.7)a	29.0(1.1)a	29.6(0.5)a	32.3(1.0)a	18.3(0.8)a
		小 OTC	33.4(0.8)b	26.3(0.9)b	23.7(0.8)b	31.9(1.9)b	34.4(0.4)b	36.9(1.2)b	21.7(0.4)b
		CK	28.2 (0.3)c	21.6(0.4)c	20.9(0.6)c	28.1(1.4)a	31.6(0.2)c	34.2(0.8)c	17.3(1.1)a
		平均	30.4	23.7	22.5	29.7	31.9	34.5	19.1
样品表面温度 /℃	藻类结皮	大 OTC	30.3(0.5)a	18.6(0.23)a	18.3(0.17)a	32.2(0.8)a	34.5(0.4)a	44.9(0.3)a	18.7(0.23)a
		小 OTC	31.9(0.5)b	21.9(0.14)b	19.2(0.11)b	30.8(0.3)a	37.7(0.1)b	44.1(0.5)a	19.3(0.10)b
		CK	32.6(0.2)b	19.5(0.04)c	18.8(0.04)c	31.0(0.8)a	37.8(0.4)b	48.7(0.5)b	19.3(0.04)b
		平均	31.6	20	18.8	31.3	36.7	45.9	19.1
	藓类结皮	大 OTC	29.1 (0.5)a	18.8 (0.1)a	18.5 (0.1)a	27.5(0.4)a	32.9(0.2)a	42.2(0.3)a	18.9(0.1)a
		小 OTC	32.0(0.4)b	22.2(0.2)b	19.5(0.2)b	30.6(0.6)b	37.0(0.4)b	44.4(1.0)ab	19.6(0.2)b
		CK	33.4(0.2)c	19.6 (0.1)c	18.9 (0.1)c	29.0(0.2)b	35.4(0.2)c	46.4(0.2)b	19.4(0.1)b
		平均	31.5	20.2	19.0	29.0	35.1	44.3	19.3
		总平均	31.6	20.1	18.9	30.2	35.9	45.1	19.2
样品含水量 /%	藻类结皮	大 OTC	14.6(0.3)a	12.9(0.2)a	8.8(0.2)a	5.5(0.1)a	3.5(0.1)a	2.1(0.02)a	0.4(0.00)a
		小 OTC	15.0(0.3)a	13.2(0.3)ab	9.5(0.2)ab	6.7(0.2)b	4.1(0.2)b	2.2(0.05)a	0.4(0.02)a
		CK	15.2(0.2)a	14.0 (0.2)b	9.9(0.1)b	6.4(0.2)b	3.6(0.1)ab	2.2(0.04)a	0.4(0.02)a
		平均	14.9	13.4	9.4	6.2	3.7	2.2	0.4
	藓类结皮	大 OTC	13.4(0.2)a	12.0(0.2)a	8.6(0.3)a	5.7(0.3)a	3.7(0.1)a	2.0(0.03)a	0.2(0.01)a
		小 OTC	13.8(0.4)ab	12.1(0.4)a	8.8(0.4)a	6.3(0.4)ab	3.7(0.2)ab	2.0(0.06)a	0.2(0.02)a
		CK	14.8(0.3)b	13.7(0.3)b	10.2(0.2)b	7.1(0.2)b	4.3(0.1)b	2.2(0.02)a	0.4(0.02)a
		平均	14	12.6	9.2	6.4	3.9	2.1	0.3
		总平均	14.5	13.0	9.3	6.3	3.8	2.2	0.4

注：数据来自沙坡头沙漠试验研究站气候观测场；不同字母表示培养期气温、样品表面温度、样品含水量在不同温度处理下差异显著（$p<0.05$）。

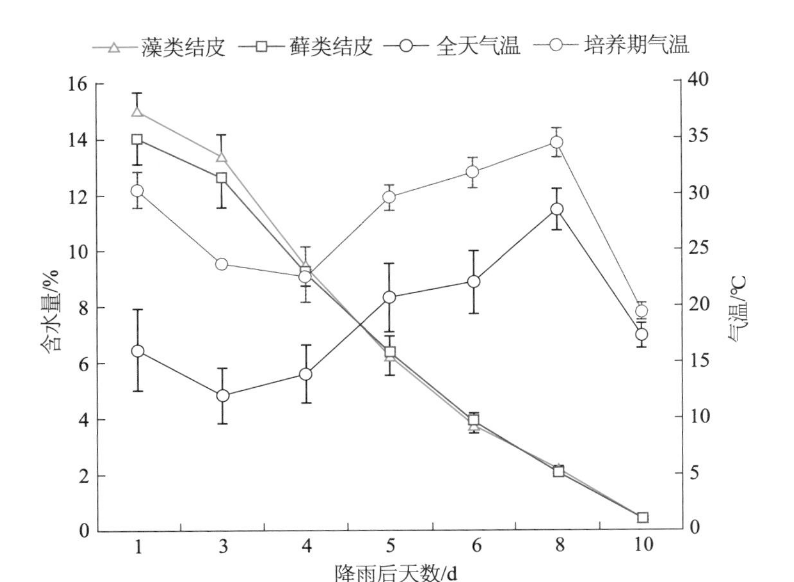

图 3-48 实验期间样品含水量（mean ± SD）及气温（mean ± SD）的变化

Figure 3-48 Variation of samples water content and temperature in experimental period

（2）持续湿润时间对BSC固氮活性的影响

实验期藻类结皮固氮活性在 $1.3 \times 10^4 \sim 12.8 \times 10^4$ nmol C_2H_4 m^{-2} h^{-1} 变化，平均为 5.3×10^4 C_2H_4 m^{-2} h^{-1}。自然植被区和人工植被区藻类结皮固氮活性的变化区间分别为 $1.7 \times 10^4 \sim 12.8 \times 10^4$ C_2H_4 m^{-2} h^{-1} 和 $1.3 \times 10^4 \sim 12.2 \times 10^4$ C_2H_4 m^{-2} h^{-1}，平均为 4.9×10^4 C_2H_4 m^{-2} h^{-1} 和 5.7×10^4 C_2H_4 m^{-2} h^{-1}，人工植被区藻类结皮固氮活性高于自然植被区。不同温度处理下自然植被区藻类结皮平均固氮活性表现为小 OTC>大 OTC>CK，人工植被区藻类结皮平均固氮活性则表现为小 OTC>CK>大 OTC。实验期藓类结皮固氮活性在 $0.2 \times 10^4 \sim 2.3 \times 10^4$ C_2H_4 m^{-2} h^{-1} 变化，平均为 0.9×10^4 C_2H_4 m^{-2} h^{-1}，显著低于藻类结皮的平均固氮活性。自然植被区和人工植被区藓类结皮固氮活性的变化区间分别为 $0.2 \times 10^4 \sim 2.3 \times 10^4$ C_2H_4 m^{-2} h^{-1} 和 $0.2 \times 10^4 \sim 1.7 \times 10^4$ C_2H_4 m^{-2} h^{-1}，平均为 1.0×10^4 C_2H_4 m^{-2} h^{-1} 和 0.8×10^4 C_2H_4 m^{-2} h^{-1}，自然植被区藓类结皮固氮活性略高于人工植被区。不同温度处理下自然植被区和人工植被区藓类结皮固氮活性差异不大，仅为 $0.1 \sim 0.2 \times 10^4$ C_2H_4 m^{-2} h^{-1}（表 3-5）。

表 3-5　实验期不同植被区、不同类型 BSC 固氮活性在不同温度处理下的变化。平均值（标准差），单位为 10^4 nmol C_2H_4 m^{-2} h^{-1}

Table 3-5　Variation of nitrogenase activity of different BSC types and vegetation areas at different temperature treatments in experimental period,mean（SD），$\times 10^4$ nmol C_2H_4 m^{-2} h^{-1}

BSC 类型	植被区	温度处理	降雨后天数 /d						
			1	3	4	5	6	8	10
藻类结皮	自然植被区	大 OTC	2.6(0.1)a	4.7(1.8)a	6.1(0.2)a	6.5(1.8)a	6.3(0.2)a	4.1(1.73)a	1.9(0.7)a
		小 OTC	1.9(0.5)a	4.2(1.1)ab	12.8(2.2)b	9.1(2.1)b	7.7(1.0)b	4.7(0.54)a	1.7(0.8)a
		CK	1.7(0.8)a	2.4(1.1)b	7.4(2.6)a	6.3(1.7)a	5.8(0.9)a	2.7(0.01)b	1.7(0.2)a
		平均	2.1	3.8	8.8	7.3	6.6	3.8	1.8
	人工植被区	大 OTC	2.3(0.1)a	3.9(0.4)a	10.0(2.6)a	6.2(0.4)a	7.1(2.5)a	4.6(1.2)a	1.9(0.3)a
		小 OTC	1.6(0.4)b	4.0(1.2)a	12.1(2.0)a	10.6(1.4)b	8.6(1.2)a	4.7(0.8)a	1.3(0.2)b
		CK	1.3(0.3)b	3.3(1.1)b	12.2(3.1)a	10.4(0.5)b	7.6(1.0)a	3.4(0.2)a	1.8(0.5)a
		平均	1.7	3.7	11.4	9.1	7.8	4.2	1.7
		总平均	1.9	3.8	10.3	8.2	7.2	4.1	1.7
藓类结皮	自然植被区	大 OTC	0.5(0.1)a	0.9(0.2)a	1.1(0.3)a	0.9(0.2)a	1.6(0.4)a	0.6(0.1)a	0.7(0.3)a
		小 OTC	0.3(0.1)b	0.7(0.4)a	2.3(0.5)b	1.6(0.6)a	1.4(0.1)a	0.6(0.2)a	0.6(0.2)a
		CK	0.2(0.1)b	0.6(0.1)a	1.7(0.2)ab	1.5(0.3)a	1.3(0.1)a	0.9(0.1)a	0.7(0.3)a
		平均	0.3	0.7	1.7	1.3	1.4	0.7	0.7
	人工植被区	大 OTC	0.3(0.08)a	0.7(0.18)a	1.4(0.26)a	0.9(0.28)a	0.9(0.15)a	0.7(0.21)a	0.7(0.09)a
		小 OTC	0.2(0.04)a	0.5(0.21)a	1.7(0.15)a	0.9(0.22)a	1.1(0.34)a	0.6(0.13)a	0.5(0.11)b
		CK	0.2(0.03)a	0.6(0.07)a	1.5(0.25)a	1.3(0.23)a	1.2(0.46)a	0.7(0.11)a	0.6(0.07)ab
		平均	0.2	0.6	1.5	1.0	1.1	0.7	0.6
		总平均	0.3	0.7	1.6	1.2	1.3	0.7	0.7

注：不同字母表示相同植被区、同一类型 BSC 固氮活性在不同温度处理下差异显著（$p<0.05$）。

降雨后藻类结皮和藓类结皮固氮活性随水分散失都表现出相同的变化趋势，但整个实验期藻类结皮的固氮活性都要显著高于藓类结皮（图 3-49）。降雨后第 1 天两类 BSC 的固氮活性都较低，从第 3 天开始，随着水分散失，固氮活性逐渐增强，至第 4 天迅速达到最大值（藻类结皮和藓类结皮平均固氮活性分别为 10.3×10^4 nmol C_2H_4 m^{-2} h^{-1} 和 1.6×10^4 nmol

$C_2H_4\ m^{-2}\ h^{-1}$），是第 1 天固氮活性的 5 倍多。此后，由于气温升高（图 3-48），水分散失加快，两类 BSC 固氮活性缓慢下降（第 5、6 天），但都显著高于降雨后第 1、3、8 和第 10 天的固氮活性（图 3-49）。至第 6 天后，由于样品水分已散失殆尽，两类 BSC 的固氮活性急剧下降，至第 10 天，两类 BSC 的含水量仅为 0.3 % 左右，其固氮活性也下降到 1.7×10^4 nmol $C_2H_4\ m^{-2}\ h^{-1}$ 和 0.6×10^4 nmol $C_2H_4\ m^{-2}\ h^{-1}$，仅为最大固氮活性时的 1/6 和 1/3。

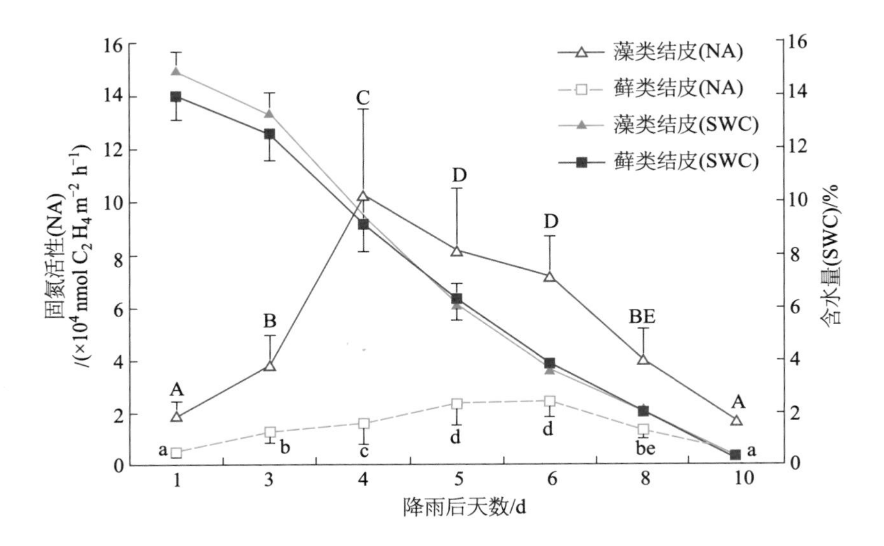

图 3-49　实验期 BSC 固氮活性（mean ± SD）的变化

Figure 3-49　Variation of nitrogenase activity of different BSC in experimental period

（3）BSC 固氮活性对水分和温度的响应

整个实验期藻类结皮和藓类结皮固氮活性与样品含水量之间均呈显著的二次函数关系，随着水分含量增加，固氮活性呈现先上升后降低的趋势，差异显著（图 3-50，$p<0.001$）。藻类结皮样品含水量在 6.7% 时的固氮活性（平均为 8.5×10^4 nmol $C_2H_4\ m^{-2}\ h^{-1}$）显著高于较高含水量（14.1 %）和较低含水量（1.3%）时的固氮活性（分别为 2.8×10^4 nmol $C_2H_4\ m^{-2}\ h^{-1}$ 和 2.9×10^4 nmol $C_2H_4\ m^{-2}\ h^{-1}$）。藓类结皮固氮活性在样品含水量为 6.8% 时（平均为 1.3×10^4 nmol $C_2H_4\ m^{-2}\ h^{-1}$）显著高于较高含水量（13.4%）和较低含水量（1.3%）时的固氮活性（分别为 0.5×10^4 nmol $C_2H_4\ m^{-2}\ h^{-1}$ 和 0.6×10^4 nmol $C_2H_4\ m^{-2}\ h^{-1}$）（表 3-5，图 3-49 和图 3-50）。

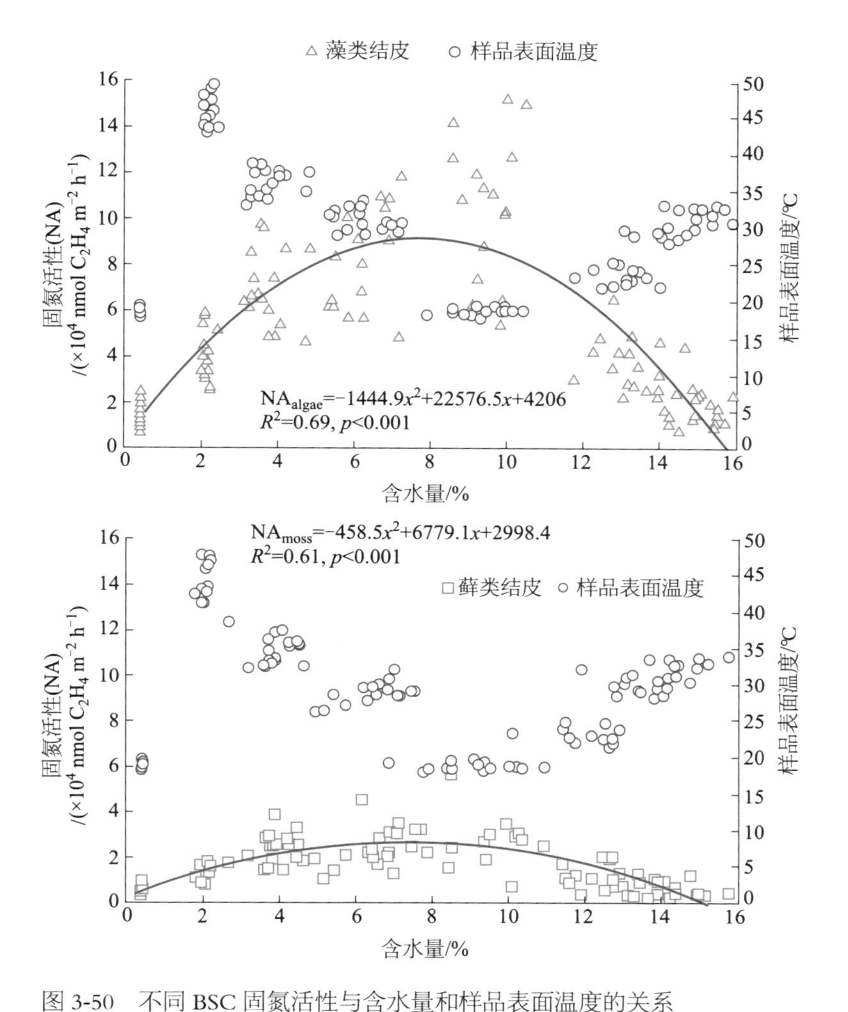

图 3-50 不同 BSC 固氮活性与含水量和样品表面温度的关系

Figure 3-50 The relationship between nitrogenase activity of different BSCs and samples water content and surface temperature

藻类结皮具有较高固氮活性时的平均样品表面温度为 28.9 ℃，具有较低固氮活性时的平均表面温度分别为 25.8 ℃（样品含水量较高）和 32.5 ℃（样品含水量较低）。藓类结皮具有较高固氮活性时的平均样品表面温度为 27.7 ℃，而较低固氮活性时的平均表面温度分别为 25.8 ℃（样品含水量较高）和 31.8 ℃（样品含水量较低）（表 3-4 和表 3-5；图 3-49 和图 3-50）。

综上所述，实验期藻类结皮和藓类结皮的平均固氮活性分别为 5.3×10^4 nmol C_2H_4 m^{-2} h^{-1} 和 0.9×10^4 nmol C_2H_4 m^{-2} h^{-1}，藻类结皮对系统的氮贡献显著高于藓类结皮；在经历 31 天持

续干旱，降雨发生后两类 BSC 固氮活性均表现为第 1 天较低，之后迅速增加，至第 4 天达到最大值，而后持续下降，第 6 天后迅速下降，至第 10 天降至最低。实验前样品所经历的持续干旱时间决定了降雨发生后其到达最大固氮速率的时间；实验期两类 BSC 固氮活性与样品含水量之间均呈显著的二次函数关系，其固氮活性随样品含水量的增加呈先上升后下降的趋势，藻类结皮的固氮活性显著高于藓类结皮；OTC 装置模拟增温效果显著，但短期模拟增温并不能显著提高两类 BSC 的固氮活性，增温主要通过加速 BSC 水分散失来影响其固氮活性。

3.1.5 不同频率干湿交替对两类 BSC 斑块土壤呼吸的影响

由降水、蒸散以及排水等过程所驱动的土壤水分的变化导致土壤常常经历干湿交替（de Oliveira *et al.*，2005）。大量研究表明，风干土壤再湿润会引起短暂的土壤碳释放脉冲效应，且与恒湿条件相比，经历了干湿交替的土壤累积碳释放量显著增加（Wu and Brookes，2005；Beare *et al.*，2009；Butterly *et al.*，2010），然而也有研究发现相反的结果（Franzluebbers *et al.*，1994；Mikha *et al.*，2005；Muhr *et al.*，2010），这可能是由于土壤质地、有机质含量以及植被覆盖等均会影响土壤呼吸对干湿交替的响应程度（Franzluebbers *et al.*，2000；Fierer and Schimel，2002）。荒漠区土壤长期处于干旱状态，土壤呼吸微弱，降水所引起的土壤再湿润可能会加速土壤碳周转速率，是该区土壤碳循环的重要驱动因子，且土壤再湿润所导致的碳释放脉冲可能在很大程度上决定着长时间尺度上的土壤碳释放量。

目前，国内外关于 BSC 在土壤水文过程、土壤生物地球化学循环过程和干旱区景观过程中作用的研究已得到了重视（李新荣，2012），特别是在腾格里沙漠东南缘沙坡头地区，有许多学者发表相关研究（李守中等，2005；黄磊等，2012；Zhao *et al.*，2014）。然而，BSC 斑块土壤碳释放对不同降水格局所引起的不同频率干湿交替如何响应，还未见报道。本研究以该区最常见的两类 BSC（藓类结皮和藻类结皮）所覆盖的土壤为研究对象，揭示了土壤质地、有机质含量对土壤碳释放在干湿交替过程中响应的影响，为探讨固沙植被区在演替过程中其碳释放对降水格局改变的响应提供依据。

腾格里沙漠东南缘沙坡头地区主要土壤类型为灰钙土和风沙土。藓类结皮和藻类结皮盖度分别达到 31% 和 42%。0 ~ 10 cm 土壤理化性质见表 3-6。

表 3-6　土壤理化性质（平均值 ± 标准差）

Table 3-6　Soil properties of the study site（mean ± SD）

土壤类型	黏粒含量 /%	粉粒含量 /%	砂粒含量 /%	容重 /（g cm^{-2}）	有机碳 /（g kg^{-1}）
藓类结皮斑块土壤	7.28 ± 0.52a	20.36 ± 2.83a	72.36 ± 1.08a	1.34 ± 0.08a	3.44 ± 0.27a
藻类结皮斑块土壤	6.65 ± 0.13a	13.56 ± 1.68b	79.79 ± 1.49b	1.42 ± 0.11a	2.43 ± 0.21b

注：不同字母表示不同 BSC 斑块土壤之间差异显著，p<0.05。

2013 年 6 月，距离上一次降雨事件 30 天时（降低上一次降雨对实验的影响），在实验区选择灌丛间地势平坦的开阔地，分别在藓类和藻类结皮斑块按照随机原则采集原状土样品各 6 个。将样品置于智能生化培养箱（SHP-250，上海鸿都电子科技有限公司）中，在 25 ℃条件下进行恒温暗培养，预培养 3 天（使土壤达到稳定状态）后，开始实验。根据沙坡头沙漠试验研究站多年平均降水数据显示，雨季（7—9 月）单次降雨量以 <10 mm 的降水出现频次最多，降雨周期约为 6 ~ 15 天，因此，本实验分别模拟了相同总降雨量（20 mm）、不同降雨频率（10 天和 20 天）条件下的多重干湿交替过程。分别记作 10 天循环和 20 天循环，10 天循环每次降雨 5 mm，20 天循环每次降雨 10 mm。每个处理设 3 个重复。模拟降雨采用喷壶喷洒的方式，向土壤表面均匀喷洒蒸馏水，第 1 次加水当天记作第 0 天。10 天循环分别于第 0 天、第 10 天、第 20 天和第 30 天采用相同方法向土壤表面喷洒等量蒸馏水，共 4 轮循环，第 0 天加水后至第 10 天加水前记作循环 1，依此类推，分别记作循环 2、循环 3 和循环 4；20 天循环分别于第 0 天和第 20 天采用相同方法向土壤表面喷洒等量蒸馏水，共 2 轮干湿循环，记作循环 1 和循环 2。实验共计 43 天。

土壤呼吸速率采用 Li-6400-09 土壤呼吸室（LI-COR，Inc.，美国）进行测定。分别于加水后 0.5 h、2 h、5 h、12 h、24 h 进行测定，之后每隔 24 h 测定一次，直到下一次加水后进行新一轮测定。同时采用 1/100 电子天平称量样品质量，实验结束后用烘干法（105 ℃条件下烘干 24 h）测定并计算土壤质量含水量。

（1）干湿交替对土壤呼吸速率的影响

不同频率干湿交替条件下，两种 BSC 斑块土壤呼吸速率均在模拟降雨后迅速升高，在 0.5 ~ 2 h 达到最大值，随后逐渐下降，并在 2 ~ 4 天内恢复到降雨前水平并趋于稳定（图 3-51）。其中，10 天干湿循环条件下藓类结皮斑块土壤 4 个循环呼吸速率峰值分别为 6.17 μmol m^{-2} s^{-1}、4.17 μmol m^{-2} s^{-1}、3.17 μmol m^{-2} s^{-1} 和 4.42 μmol m^{-2} s^{-1}，分别为模拟降雨

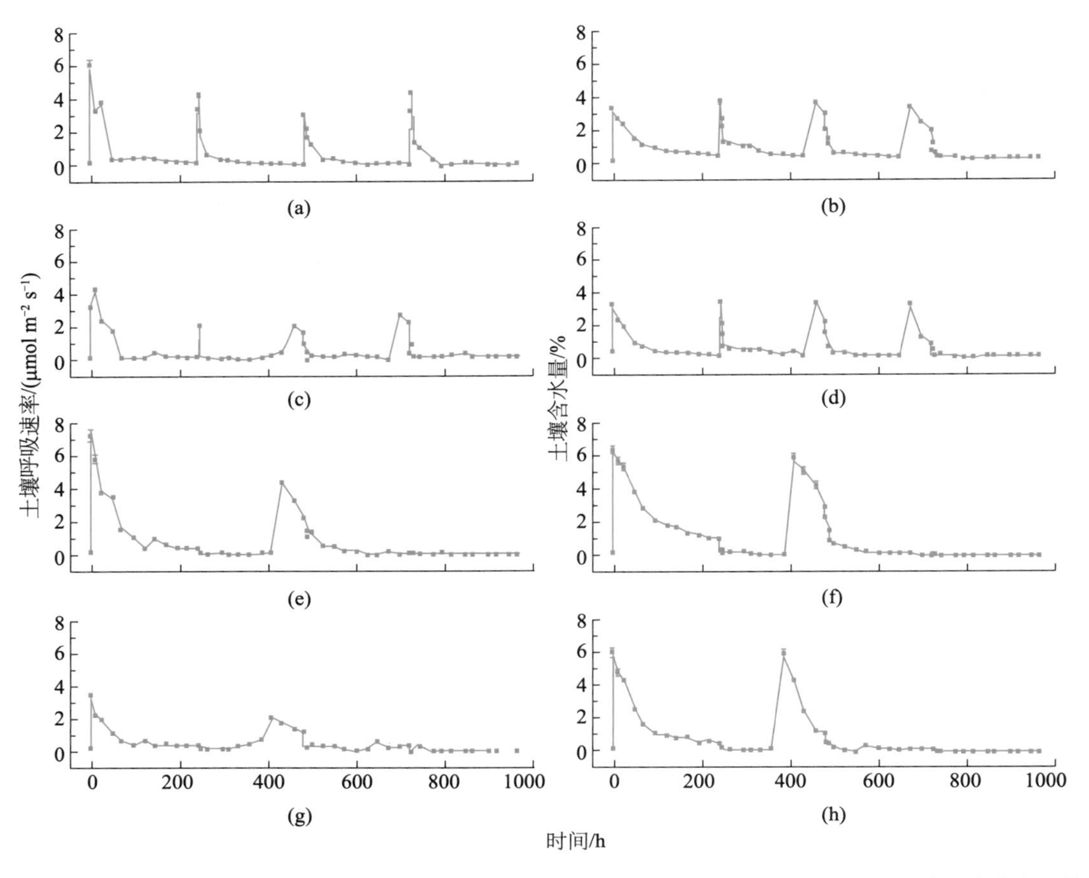

图 3-51　干湿交替条件下土壤呼吸速率和土壤含水量随时间的变化特征。（a）和（c）分别为藓类和藻类结皮斑块土壤在 10 天循环条件下的土壤呼吸速率；（b）和（d）分别为藓类和藻类结皮斑块土壤在 10 天循环条件下的土壤含水量；（e）和（g）分别为藓类和藻类结皮斑块土壤在 20 天循环条件下的土壤呼吸速率；（f）和（h）分别为藓类和藻类结皮斑块土壤在 20 天循环条件下的土壤含水量

Figure 3-51　Dynamics of soil respiration rate and soil moisture under repeated drying and rewetting cycles. Respiration rate of mosses（a）and algae（c）under 10 days repeated drying and rewetting cycles; Soil moisture of mosses（b）and algae（d）under 10 days repeated drying and rewetting cycles; Respiration rate of mosses（e）and algae（g）under 20 days repeated drying and rewetting cycles; Soil moisture of mosses（f）and algae（h）under 20 days repeated drying and rewetting cycles

前（0.13 μmol m^{-2} s^{-1}）的 47 倍、32 倍、24 倍和 34 倍；藻类结皮斑块土壤 4 个循环呼吸速率峰值分别为 3.34 μmol m^{-2} s^{-1}、2.18 μmol m^{-2} s^{-1}、1.77 μmol m^{-2} s^{-1} 和 2.56 μmol m^{-2} s^{-1}，分别为模拟降雨前（0.15 μmol m^{-2} s^{-1}）的 22 倍、15 倍、12 倍和 17 倍。20 天循环条件下藓类结皮斑块土壤 2 个循环呼吸速率峰值分别为 7.29 μmol m^{-2} s^{-1} 和 4.40 μmol m^{-2} s^{-1}，分别为模拟降雨前的 56 倍和 34 倍；藻类结皮斑块土壤 2 个循环呼吸速率峰值分别为 3.33 μmol m^{-2} s^{-1} 和 2.01 μmol m^{-2} s^{-1}，分别为模拟降雨前的 22 倍和 13 倍。随着干湿交

替次数的增加，两种 BSC 斑块土壤呼吸速率峰值均呈现减小的趋势。相同处理条件下，藻类结皮斑块土壤呼吸速率峰值均显著低于藓类结皮斑块土壤（$p<0.05$）。

两种 BSC 斑块土壤在各循环的平均呼吸速率均显著高于模拟降雨前呼吸速率（$p<0.05$）。其中，10 天循环条件下藓类结皮斑块土壤平均呼吸速率分别为 0.89 μmol m^{-2} s^{-1}、0.29 μmol m^{-2} s^{-1}、0.34 μmol m^{-2} s^{-1} 和 0.28 μmol m^{-2} s^{-1}，分别为模拟降雨前的 6.8 倍、2.2 倍、2.6 倍和 2.2 倍；藻类结皮斑块土壤平均呼吸速率分别为 0.49 μmol m^{-2} s^{-1}、0.25 μmol m^{-2} s^{-1}、0.24 μmol m^{-2} s^{-1} 和 0.23 μmol m^{-2} s^{-1}，分别为模拟降雨前的 3.3 倍、1.7 倍、1.6 倍和 1.5 倍。20 天循环条件下藓类结皮斑块土壤平均呼吸速率分别为 0.80 μmol m^{-2} s^{-1} 和 0.45 μmol m^{-2} s^{-1}，分别为模拟降雨前的 6.2 倍和 3.5 倍；藻类结皮斑块土壤平均呼吸速率分别为 0.45 μmol m^{-2} s^{-1} 和 0.28 μmol m^{-2} s^{-1}，分别为模拟降雨前的 3.0 倍和 1.9 倍。随着干湿交替次数的增加，两种 BSC 斑块土壤平均呼吸速率均显示减小的趋势。相同处理条件下，藓类结皮斑块土壤平均呼吸速率均高于藻类结皮斑块土壤（图 3-52）。

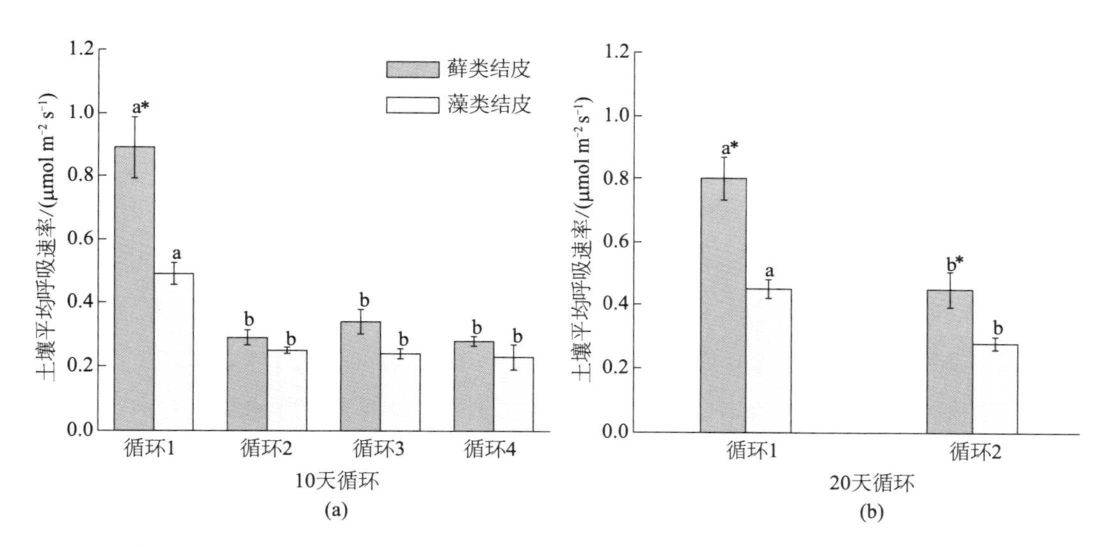

图 3-52　各干湿循环周期两种 BSC 斑块土壤平均呼吸速率：（a）10 天循环条件；（b）20 天循环条件。* 表示不同结皮斑块土壤类型之间差异显著（$p<0.05$）

Figure 3-52　Average respiration rates of two BSC in each drying and rewetting cycle：（a）10 d cycle；（b）20 d cycle. * indicates significant difference among soil types（$p<0.05$）

（2）干湿交替对土壤碳释放量的影响

随着干湿交替次数的增加，两种 BSC 斑块土壤累积碳释放量均显示减小的趋势。其中，10 天循环条件下，藓类结皮斑块土壤各干湿循环累积土壤碳释放量分别为 9.24 g C m^{-2}、2.97 g C m^{-2}、3.57 g C m^{-2} 和 2.93 g C m^{-2}，循环 2、循环 3、循环 4 分别较循环 1 下降了

68%、61% 和 68%；藻类结皮斑块土壤各干湿循环累积土壤碳释放量分别为 5.04 g C m^{-2}、2.62 g C m^{-2}、2.49 g C m^{-2} 和 2.44 g C m^{-2}，循环 2、循环 3、循环 4 分别较循环 1 下降了 48%、51% 和 52%。20 天循环条件下，藓类结皮斑块土壤累积土壤碳释放量分别为 16.60 g C m^{-2} 和 9.38 g C m^{-2}，循环 2 较循环 1 下降了 43%；藻类结皮斑块土壤累积土壤碳释放量分别为 9.25 g C m^{-2} 和 5.87 g C m^{-2}，循环 2 较循环 1 下降了 37%。相同处理条件下，藓类结皮斑块土壤累积碳释放量均高于藻类结皮斑块土壤（图 3-53）。

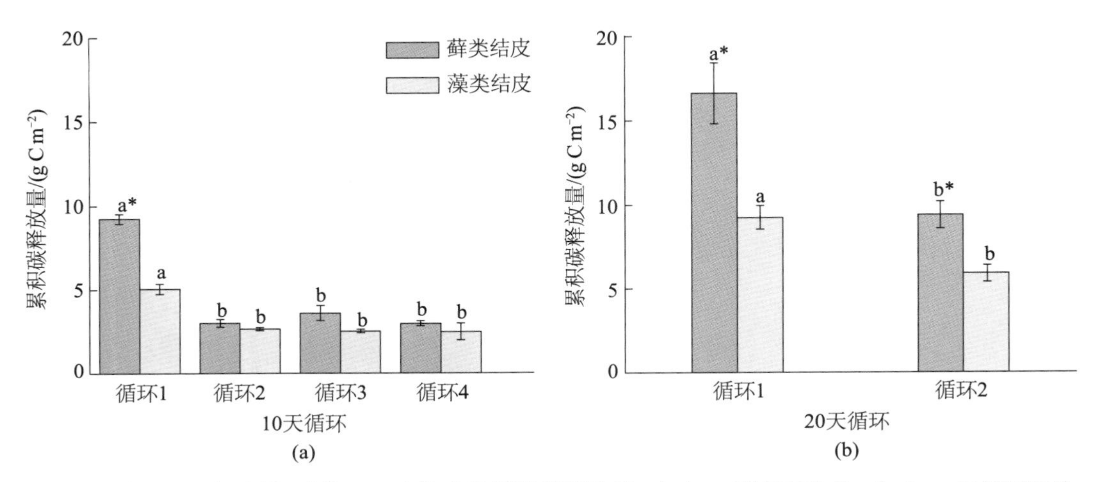

图 3-53 各干湿循环周期两种 BSC 斑块土壤累积碳释放量。（a）10 天循环条件；（b）20 天循环条件
Figure 3-53 Accumulated carbon release of two BSC in each drying and rewetting cycle.（a）10 d cycle；（b）20 d cycle

（3）干湿交替频率对土壤呼吸的影响

实验过程中，10 天循环条件下藓类结皮斑块土壤累积碳释放量为 18.71 g C m^{-2}，藻类结皮斑块土壤累积碳释放量为 12.59 g C m^{-2}；20 天循环条件下藓类结皮斑块土壤累积碳释放量为 25. 98 g C m^{-2}，藻类结皮斑块土壤累积碳释放量为 15.12 g C m^{-2}。20 天循环条件下的两类 BSC 斑块土壤累积碳释放量均比 10 天循环条件下土壤累积碳释放量高，即在一段既定时间内，相同的总降雨量，不同干湿交替频率，土壤所释放的碳不同，且单次降雨量较高、降雨间隔较长的条件下所释放的碳更多。藓类结皮斑块土壤在 20 天循环条件下的累积碳释放量比 10 天循环条件下增加了 39%，藻类结皮斑块土壤在 20 天循环条件下的累积碳释放量比 10 天循环条件下增加了 21%（图 3-54）。

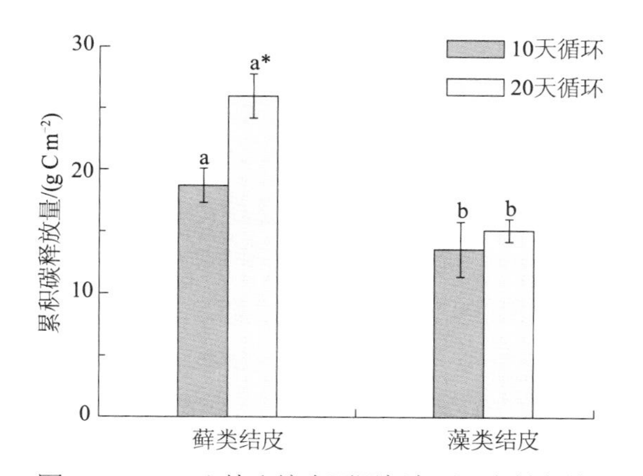

图 3-54　BSC 斑块土壤在不同频率干湿交替条件下的累积碳释放量

Figure 3-54　Accumulated carbon release of two BSC with different drying and rewetting frequencies

荒漠区土壤长期处于干旱状态，土壤呼吸作用微弱，土壤再湿润改变了这种状况，加速了土壤有机碳的转化，使土壤碳释放出现短暂的脉冲现象。然而，随着干湿交替的反复发生，这种脉冲逐渐减小，但减小的程度则依干旱期的不同而异，即随着干旱期的延长，再湿润后所引起的呼吸脉冲减小的程度越小。实验结果表明，在相对较短的时间尺度内（40 天），相同降雨量，高频次的小降雨事件所引起的碳释放量低于低频次的大降雨事件。这就意味着，全球变化所引起的降水格局的改变，即更多的干旱和强降水事件，可能会进一步影响温带荒漠土壤碳释放动态及土壤有机碳储量。此外，相同处理条件下，藓类结皮斑块土壤所释放的 CO_2 量均高于藻类结皮斑块，即处于固沙植被演替后期的藓类结皮斑块对土壤再湿润的响应程度更大。而随着干湿交替频率的降低，藓类结皮斑块土壤累积碳释放量的增加幅度比藻类结皮斑块土壤的更大，即藓类结皮斑块土壤对于干湿交替频率的变化更为敏感。

3.1.6　冬季降雪对 BSC 光合作用及呼吸作用和生理生化的影响

（1）BSC光合作用及呼吸作用对冬季降雪的响应

由于干旱、高温等严酷的环境限制了维管植物的发育以及光合固碳能力，BSC 的碳固定就成为荒漠生态系统碳的重要来源。因此，对 BSC 碳固定的研究是对目前进行的相关陆地生态系统碳循环研究的一个重要补充，可为客观评价荒漠地区在陆地系统碳循环中的地位提供科学依据（李新荣，2012）。目前，对BSC光合固碳的研究主要集中于美国、澳大利亚、以色列荒漠地区和欧洲部分温带草原以及南北两极和亚北极地（Harley *et al.*，1989；Davey，1997；Zotz，2000）。中国对BSC的光合作用和碳循环的研究报道很少。

目前，在评估荒漠结皮固碳能力时，一般都是基于在较高温度下开展的模拟试验，冬季BSC的碳固定/释放作用往往被忽略，并假设冬季BSC与大气之间的碳交换为0（Kappen，1993；Li *et al.*，2012）。然而，这种假设一方面缺少观测证据的支持，另一方面也忽视了BSC极强的环境适应能力。目前，有关冬季BSC固碳过程影响的研究还非常少见，尤其是冬季降雪对BSC固碳过程的研究更是少之又少。总体来说，冬季降雪对BSC碳循环的影响可能体现在三个方面：① 可能是荒漠地区水分的重要来源之一；② 改变了地表BSC的光照条件和空气通透性；③ 积雪覆盖的保温效应，为雪盖下层BSC生物体的生理活性提供适宜的温度条件。我们以宁夏沙坡头地区两种典型的BSC为研究对象，通过自然降雪结合增倍添加降雪的方式，采用气体交换法研究了冬季降雪对BSC光合和呼吸作用的影响，通过本研究，可以为深入研究冬季降雪对荒漠地区BSC碳循环的影响提供理论依据和基础数据，是有关研究的重要补充。

将PVC管采集到的藻类结皮和藓类结皮的原状结皮土壤样品放置野外培养（BSC采样时间为2012年7月）。降雪前，选取两种BSC类型（藓类结皮和藻类结皮）、5个降雪处理（无降雪、正常降雪、2倍降雪、3倍降雪及5倍降雪），各4个重复，无降雪BSC作为对照，其中3组样品用于光合作用以及呼吸作用试验，另外1组样品用于测量土壤含水量。具体方法：除无降雪处理的BSC使用直径为10 cm的塑料圆盘遮盖外，其余不同处理的BSC均接受自然降雪，此外，使用直径为10 cm的塑料圆盘接受自然降雪，然后向2倍降雪处理的BSC添加1份塑料圆盘所接受的降雪，向3倍处理的BSC添加2份塑料圆盘所接受的降雪，向5倍降雪处理的BSC添加4份塑料圆盘所接受到的降雪。

2012年冬季降雪分别发生在12月20日以及12月23日（降水量分别为0.1 mm和0.3 mm）。净光合速率（P_n，μmol m^{-2} s^{-1}）、呼吸速率（R_d，μmol m^{-2} s^{-1}）分别在降雪前、降雪后0～12天以及降雪后第30天进行测量。测定时间为每日11：00（此时光合、呼吸作用最为活跃）。采用土壤呼吸测定系统（Li-840A，LI-COR Inc.，美国）分别测量BSC的光合作用与呼吸作用。在光照条件下测量净光合速率，之后用黑布完全遮住圆柱形叶室，测量其暗呼吸作用。

试验期间环境因子状况：试验期间，空气温度一直保持在较低水平，从2012年12月19日到23日，温度呈明显下降趋势，最低温度达−12.46 ℃。此后28日及29日，温度再次急剧下降，降温原因为天气转阴，但未能形成有效降水；12月29日以后，温度大幅度增高，31日以后平均温度一直保持在−3～7 ℃。光合有效辐射（PAR）变化幅度较大，从12

月20日至24日，光合有效辐射大幅度增高，12月28日光合有效辐射达到试验期间最低值214.3 μmol m^{-2} s^{-1}，原因是12月24日至28日期间，天气转阴使得光照减弱，之后光合有效辐射从12月29日至2013年1月1日期间保持较高水平，其中最高值为758.9 μmol m^{-2} s^{-1}（图3-55）。试验期间两种BSC土壤含水量的变化趋势基本一致，不同降雪处理的BSC的含水量随着降雪量的增加而增加，两种BSC的土壤水分在2013年1月1日基本恢复至降雪前土壤含水量水平，原因是2012年12月29日至2013年1月1日期间温度和辐射一直保持较高水平（图3-55）。

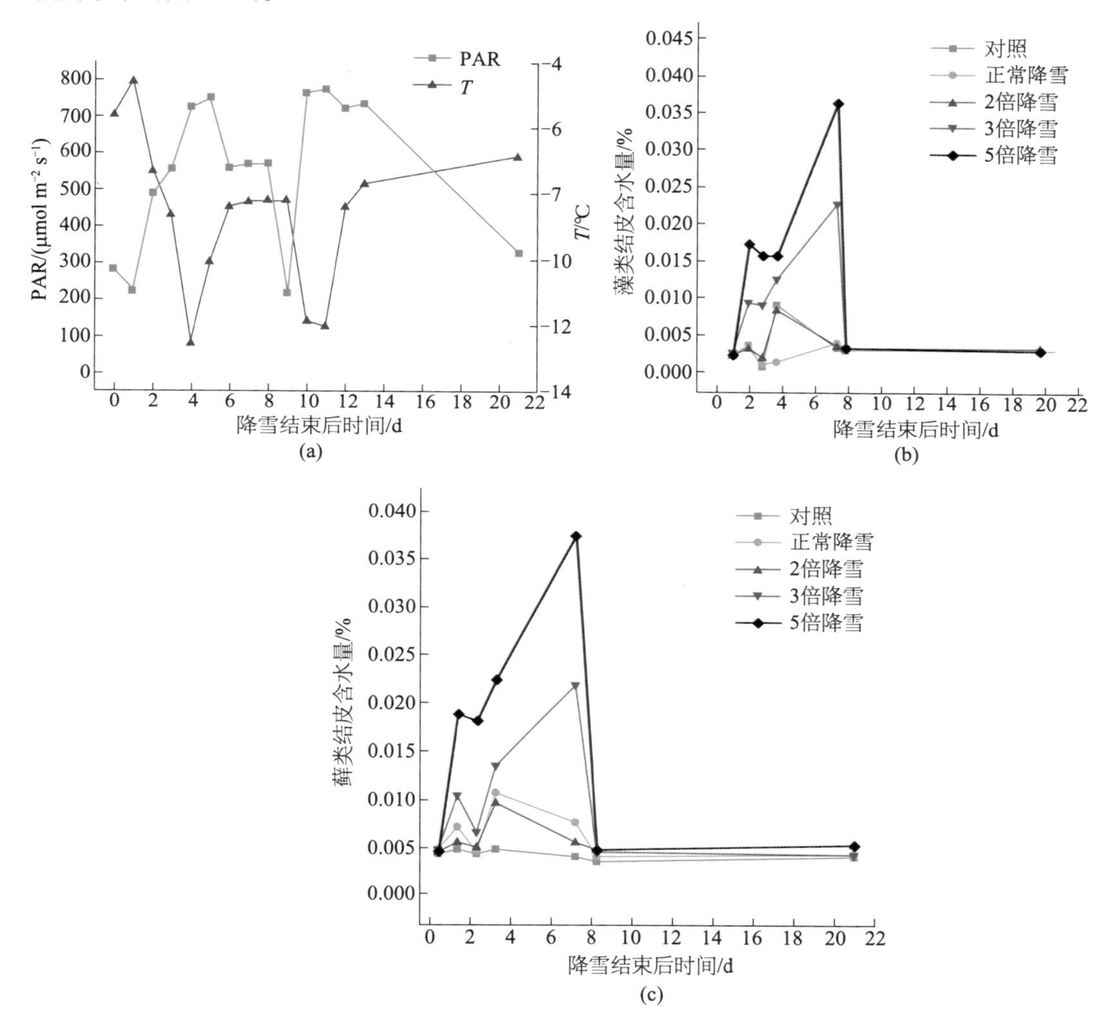

图3-55　试验期间环境因子变化

Figure 3-55　Variation of environment factors during the experiment

降雪后 BSC 的光合作用与呼吸作用变化：不同降雪处理下两种 BSC 的净光合作用表现出相似的变化特征，明显分为 3 个阶段（图 3-56）。2012 年 12 月 20 日至 23 日为第 1 阶段，20 日第 1 次降雪后不同处理下两种 BSC 的净光合速率均先显著增高（$p_{藓}<0.05$，$p_{藻}<0.05$）然后又极显著降低（$p_{藓}<0.001$，$p_{藻}<0.001$）；其中不同降雪量处理的藓类结皮的净光合速率的大小顺序为 2 倍降雪 >3 倍降雪 > 正常降雪 > 对照 >5 倍降雪；不同降雪量处理的藻类结皮的净光合速率顺序为 2 倍降雪 >3 倍降雪 >5 倍降雪 > 正常降雪 > 对照。2012 年 12 月 23 日至 28 日为第 2 阶段，23 日第 2 次降雪后，不同处理下两种 BSC 的净光合速率均先极显著增高（$p_{藓}<0.001$，$p_{藻}<0.001$），然后显著降低（$p_{藓}<0.05$，$p_{藻}<0.05$）；两种 BSC 的净光合速率顺序为正常降雪 >2 倍降雪 > 对照 >3 倍降雪 >5 倍降雪。2012 年 12 月 28 日至 2013 年 1 月 19 日为第 3 个阶段，两种 BSC 的光合速率表现出先增高后降低的趋势，光合速率顺序为 5 倍降雪 >3 倍降雪 >2 倍降雪 > 正常降雪 > 对照，即随着降雪量的增加，两种 BSC 的净光合速率显著增高（$p_{藓}<0.001$，$p_{藻}<0.001$）。

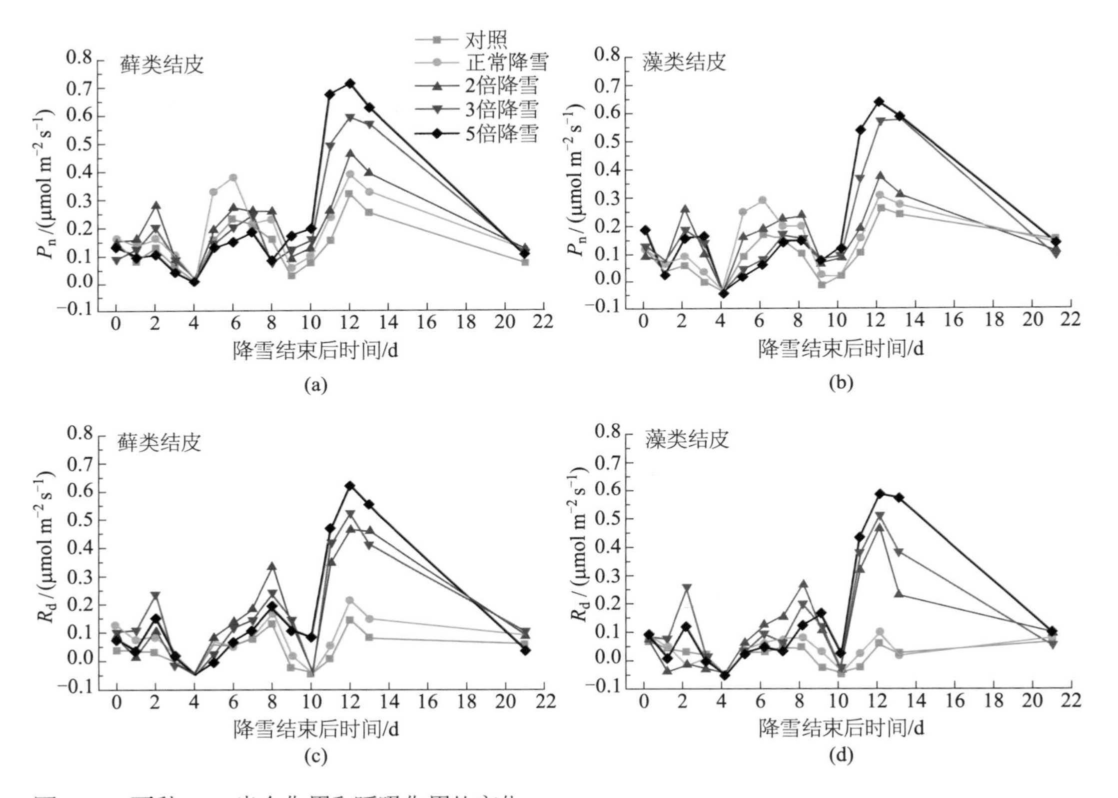

图 3-56　两种 BSC 光合作用和呼吸作用的变化

Figure 3-56　Variation of photosynthesis and respiration of two kinds of BSC

与 BSC 净光合速率类似，不同降雪处理的两种 BSC 的呼吸速率也表现为 3 个阶段（图 3-56）。2012 年 12 月 20 日至 23 日为第 1 阶段，2012 年 12 月 20 日降雪后，两种 BSC 呼吸速率先增高后显著降低（$p_{藓}<0.05$，$p_{藻}<0.05$），藓类结皮呼吸速率的顺序为 3 倍降雪 >5 倍降雪 >2 倍降雪 > 正常降雪 > 对照，藻类结皮呼吸速率的顺序为 3 倍降雪 >5 倍降雪 > 对照 >2 倍降雪 > 正常降雪。2012 年 12 月 23 日至 29 日为第 2 阶段，不同处理下两种 BSC 呼吸速率先显著增高（$p_{藓}<0.05$，$p_{藻}<0.05$）后显著降低（$p_{藓}<0.05$，$p_{藻}<0.05$），呼吸速率顺序均为 2 倍降雪 >3 倍降雪 >5 倍降雪 > 正常降雪 > 对照。2012 年 12 月 29 日到 2013 年 1 月 19 日为第 3 阶段，不同降雪处理的 BSC 呼吸速率先增高后降低，呼吸速率顺序为 5 倍降雪 >3 倍降雪 >2 倍降雪 > 正常降雪 > 对照，即随着降雪量的增加，两种 BSC 的呼吸速率显著增高（$p_{藓}<0.05$，$p_{藻}<0.05$）。

（2）降雪对环境因子的改变及对BSC碳交换的影响

降雪后气温与 BSC 光合作用及呼吸作用的关系：对降雪后试验期间的气温与 BSC 的光合作用与呼吸作用的相关分析表明（图 3-57），藓类结皮和藻类结皮的净光合速率均与试验期间的气温呈显著正相关（$R_{藓}=0.323$，$p<0.001$；$R_{藻}=0.305$，$p<0.001$）；藓类结皮和藻类结皮的呼吸速率均与试验期间的气温呈显著正相关（$R_{藓}=0.388$，$p<0.001$；$R_{藻}=0.379$，$p<0.001$）。

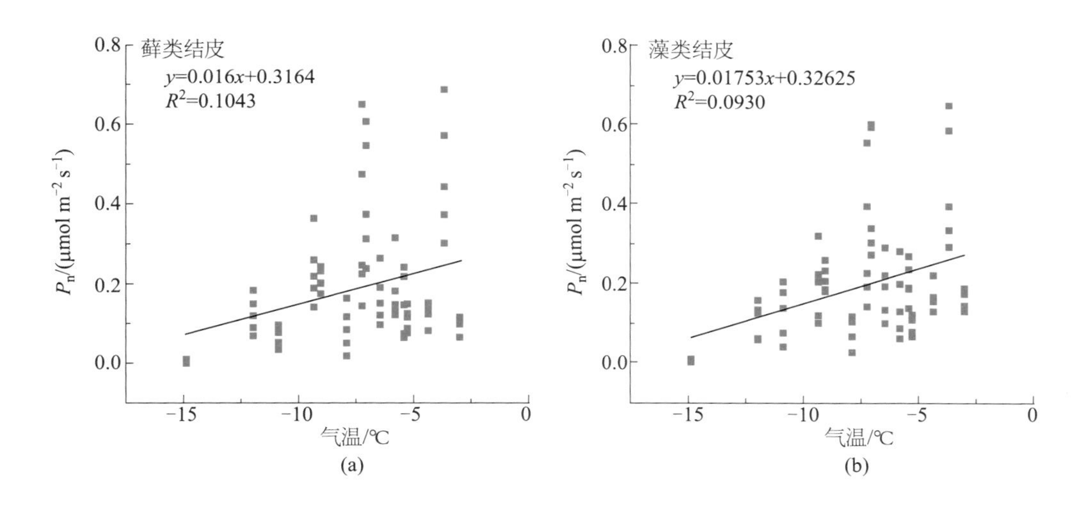

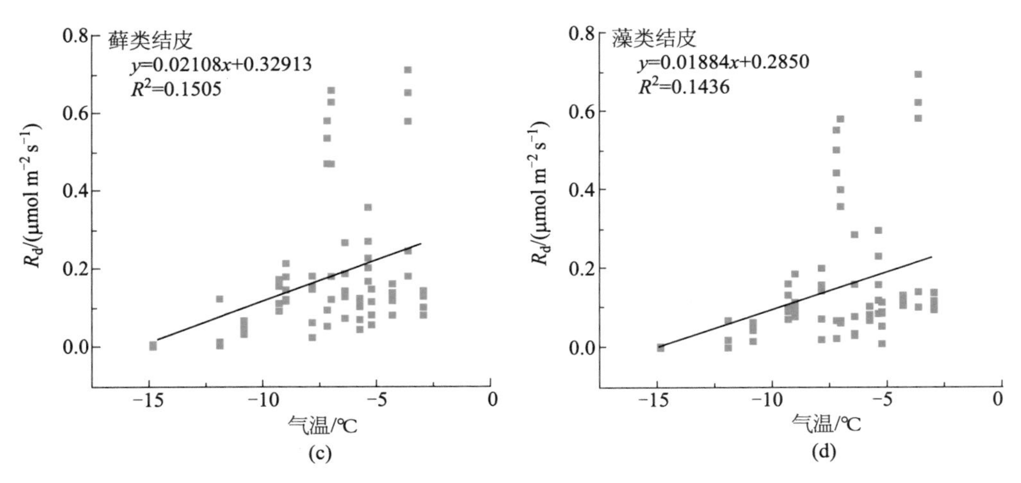

图 3-57　降雪后气温与两种 BSC 光合作用和呼吸作用的关系
Figure 3-57　Relationship between air temperature and photosynthesis activity and respiration activity of BSC

降雪后光合有效辐射与 BSC 光合作用的关系：对降雪后试验期间的光合有效辐射与 BSC 的光合作用的相关分析表明（图 3-58），藓类结皮和藻类结皮的净光合速率均与试验期间的光合有效辐射呈显著正相关（$R_{藓}$=0.390，p<0.001；$R_{藻}$=0.317，p<0.001）。

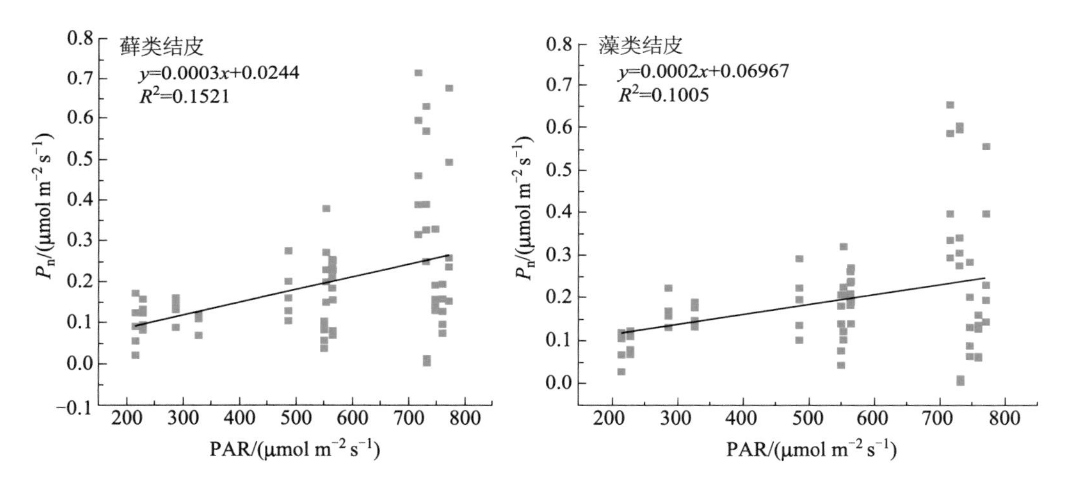

图 3-58　降雪后光合有效辐射与两种 BSC 光合作用的关系
Figure 3-58　Relationship between PAR and photosynthesis activity of BSC

降雪后土壤含水量与 BSC 光合作用及呼吸作用的关系：对降雪后试验期间的土壤含水量与 BSC 的光合作用与呼吸作用的相关分析表明（图 3-59），藓类结皮和藻类结皮的净光合速率均与试验期间的土壤含水量呈显著正相关（$R_{藓}=0.340$，$p<0.05$；$R_{藻}=0.282$，$p<0.05$）；藓类结皮和藻类结皮的呼吸速率均与试验期间的土壤含水量呈显著正相关（$R_{藓}=0.355$，$p<0.05$；$R_{藻}=0.424$，$p<0.05$）。

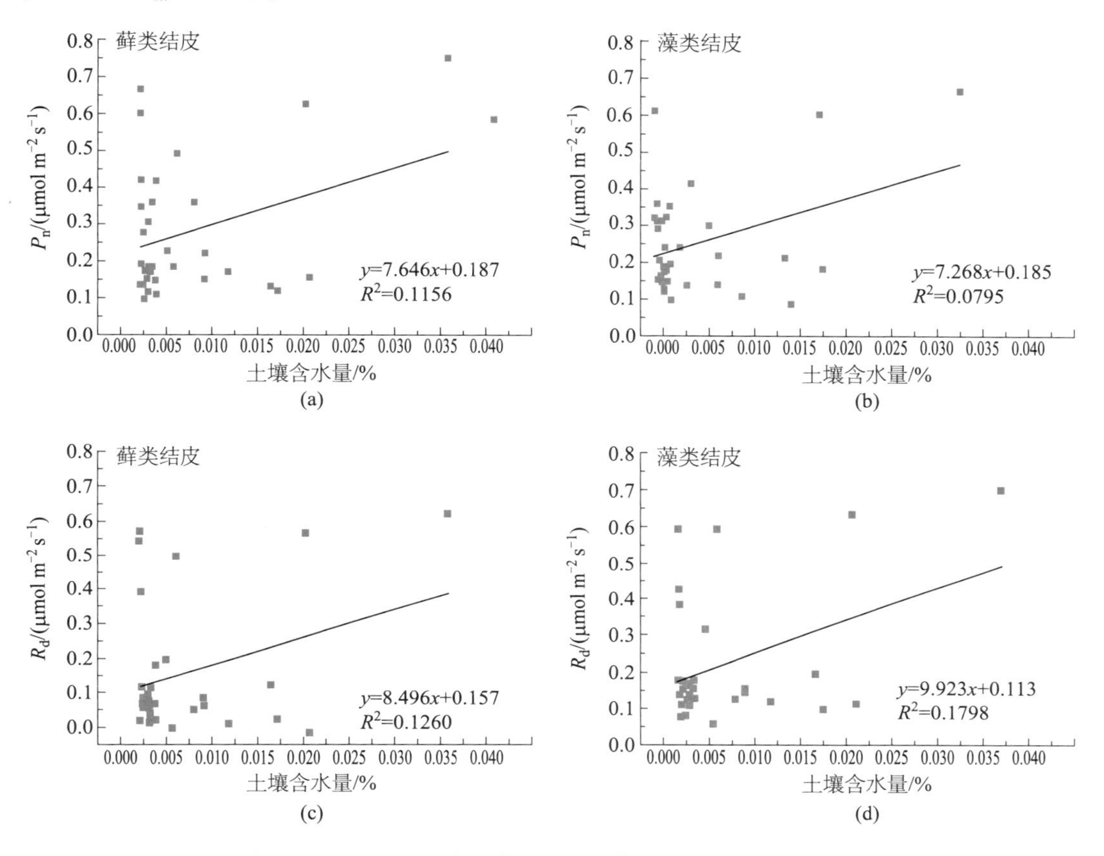

图 3-59　降雪后土壤含水量与两种 BSC 光合作用和呼吸作用的关系

Figure 3-59　Relationship between soil water content and photosynthesis activity and respiration activity of BSC

降雪量与 BSC 光合作用及呼吸作用的关系：对降雪量与实验期间两种 BSC 的光合作用和呼吸作用的相关分析表明（图 3-60），藓类结皮和藻类结皮的净光合速率与降雪量呈相关（$R_{藓}=0.171$，$p>0.05$；$R_{藻}=0.226$，$p<0.05$）；藓类结皮和藻类结皮的呼吸速率与降雪量呈显著正相关（$R_{藓}=0.294$，$p<0.05$；$R_{藻}=0.330$，$p<0.05$）。

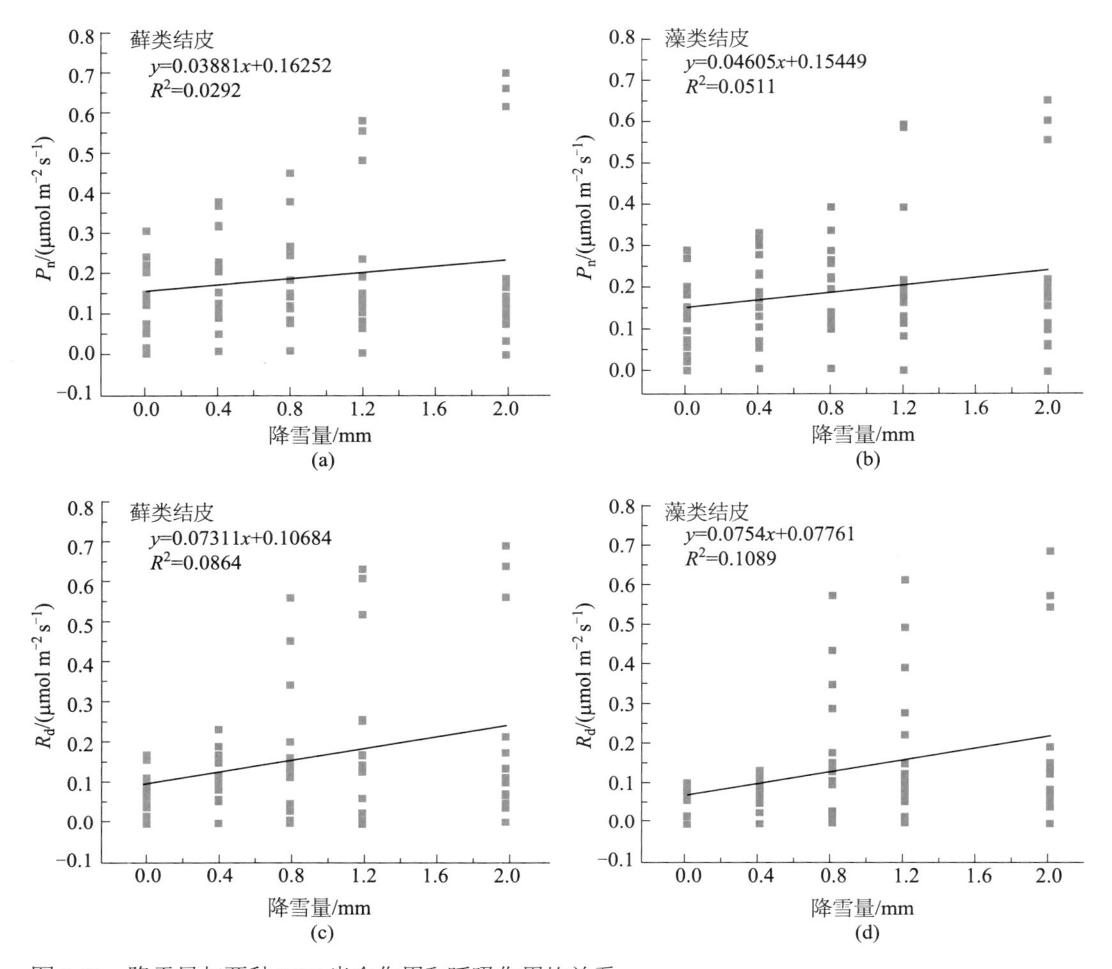

图 3-60　降雪量与两种 BSC 光合作用和呼吸作用的关系

Figure 3-60　Relationship between the amount of snow and photosynthesis activity and respiration activity of BSC

关键因子探讨：通过逐步回归分析的方法，对 BSC 光合作用与呼吸作用的影响因子进行计算发现，向回归方程引入自变量的过程中所引入的变量均为土壤含水量，即两种 BSC 光合作用与呼吸作用的关键影响因素均为土壤水分。藓类结皮光合作用和呼吸作用（y）引入土壤含水量（x）时的回归方程分别为 $y=0.187+7.646x$（$p<0.001$），$y=0.185+7.268x$（$p<0.001$）；藻类结皮光合作用和呼吸作用（y）引入土壤含水量（x）时的回归方程分别为 $y=0.157+8.496x$（$p<0.001$），$y=0.113+9.923x$（$p<0.05$）。

当 BSC 经过充分的水化作用恢复生理功能时，温度就成为限制其光合作用的重要因素。0 ~ 50 ℃下都可观察到其光合作用（贾子毅等，2011），但是对于冬季低温状态下 BSC

光合作用研究比较少。测量发现，冬季低温状态下藓类结皮和藻类结皮均具有微弱的光合作用能力，降雪前藓类结皮最大净光合速率均值约为 0.160 ± 0.032 μmol m^{-2} s^{-1}，藻类结皮最大净光合速率均值约为 0.17 ± 0.023 μmol m^{-2} s^{-1}。而前期研究表明，BSC 的净光合速率为 0.11 ~ 11.5 μmol m^{-2} s^{-1}（Lange，2003）；在降雪前，藓类结皮最大呼吸速率均值约为 0.21 ± 0.027 μmol m^{-2} s^{-1}，藻类结皮最大呼吸速率均值约为 0.14 ± 0.040 μmol m^{-2} s^{-1}。可见，冬季 BSC 的光合作用与呼吸作用不容忽视。

其次，光合作用对于环境变化的反应通常是具有"弹性"的，这种"弹性"其中也包括降雪后 BSC 光合作用发生的变化（Levitt，1972；Salisbury and Ross，1992）。一般情况下，维管植物需要较多的水并且必须要在特定的季节才能进行正常的新陈代谢活动，而 BSC 只需要很少的水分就可以启动正常的新陈代谢（贾子毅等，2011）。研究表明，有些种类的 BSC 在 0.2 ~ 0.3 mm 的降水条件下就能保持最大的光合能力（Lange *et al.*，1997）。不难想象，降雪会对 BSC 的光合作用产生 3 种效果：为 BSC 提供水分，影响 BSC 的光照条件及空气通透性，影响 BSC 的温度。这 3 种效果对于 BSC 的光合作用和呼吸作用都会产生影响，但是因为随着时间和外界条件（温度、辐射等）的变化，覆盖在 BSC 上的雪会融化，所以对于 BSC 的光合作用具有阶段性，在不同的阶段，不同降雪的处理对于 BSC 的影响作用不同，但总体上，随着降雪的融化，降雪处理会增加 BSC 的光合能力，我们的测定结果也证明了这一点。

降雪后 BSC 的光合作用受到时间、温度、辐射和土壤含水量以及降雪量的影响。首先，从时间变化上来看，两次降雪后两种 BSC 的光合作用都经历一个先增高再降低的过程，这表明降雪确实对于 BSC 的光合作用产生影响。主要原因则是降雪为 BSC 提供光合作用所需要的水分和温度。研究表明，在前两个阶段中，两种 BSC 的净光合速率并不随着降雪量的增加而增加，这主要是因为降雪后产生的第 2 种效果造成的，即降雪影响了 BSC 的光照条件，进而使得降雪量较多的 BSC 的光合作用反而小于降雪量较少的 BSC。最后一个阶段中，随着降雪量的增加，两种 BSC 的光合作用不断增加，这则是由于降雪融化后，为 BSC 提供了水分。

对于大多数生态系统，温度和水分都是影响土壤呼吸的重要因子，土壤温度是土壤呼吸作用的主导因子（Schimel *et al.*，2001；王凤玉等，2003）。许多研究表明，在土壤含水量充足并且不成为限制因素的情况下，土壤呼吸与土壤温度呈正相关，但是在干旱、半干旱地区土壤含水量成为限制因子的条件下，含水量和温度共同起作用（Reiners，1968；Chapman，1979）。这是降雪为 BSC 提供了呼吸作用所需要的水分以及对于温度的改变引起的。但是在前两个阶段中，两种 BSC 的呼吸速率变化并不与降雪量呈正相关，我们认为这是由降雪产生的第 2 种效果——降雪影响 BSC 的空气通透性引起的。一方面，积雪覆盖影响了 BSC 中的气体交换，另一方面由于低温导致 BSC 中的水分结冰，也会影响 BSC 的气体交换，这两种共同作用使得 BSC 产生的 CO_2 量减少。所以在 BSC 呼吸作用的第 1 阶段中，由于降雪量较少，正常降雪和 2 倍降雪对 BSC 影响效果较低，不妨将其忽略，可以发现 3 倍降雪的 BSC 呼吸速率反而高于 5 倍降雪的 BSC 的呼吸速率。在 BSC 呼吸作用的第 2 阶段中，则是 2 倍降雪 >3 倍降雪 >5 倍降雪。第 3 阶段中，我们发现不同处理的 BSC 的呼吸速率与降雪量呈正相关，这是因为温度、辐射等连续升高，降雪全部融化，为 BSC 的呼吸作用提供了水分及温度条件。这与我们前面的推断是一致的。

最后，通过相关性分析，我们发现温度、水分及光照和 BSC 的光合作用均呈显著正相关，降雪量与 BSC 光合作用呈正相关；温度、水分和 BSC 的呼吸作用均呈显著正相关，降雪量与 BSC 的呼吸作用呈正相关。通过逐步回归分析法发现，水分是 BSC 的光合作用和呼吸作用的关键影响因子，这与赵允格等（2010）的研究结果一致。构成 BSC 的藻类、地衣及藓类植物多为变水植物，其重要特征之一就是植物体内的含水量会随环境含水量的变化而变化，而且其生物体许多生理代谢过程，包括光合作用和呼吸作用均与植物体含水量密切相关，只有含水量达到一定水平，才有光合作用和呼吸作用等生理活动的可能（赵允格等，2010）。本试验的研究结果也证实了这一点。

综上所述，我们认为冬季 BSC 的光合作用与呼吸作用是不可忽略的，是全球碳循环的重要组成部分。此外，降雪对于 BSC 的光合作用和呼吸作用都有促进作用，降雪后两种 BSC 的净光合速率和呼吸速率都显著增高，并且最终的结果显示，降雪量与 BSC 的光合作用与呼吸作用呈显著正相关。但是由于试验采用的是气体交换的方法来测量 BSC 的呼吸作用，所测量的呼吸速率并不是真正意义上的 BSC 的呼吸速率，所以在试验方法上还必须进行改进。在冬季低温状态下，BSC 光合作用与呼吸作用均可以发生，但是由于此次数据量较少，很难对 BSC 呼吸作用的温度敏感性进行研究，研究 BSC 呼吸对低温的敏感性以及相关的关键生态过程和研究冬季降雪下 BSC 光合作用的临界温度将是今后该领域研究的重点。

（3）冬季降雪对荒漠地区藓类结皮中真藓生理生化的影响

水分是 BSC 发育的重要限制因子，而冬季降雪可能是结皮中藓类植物的重要水分来源之一。藓类植物属于典型的变水植物，冬季降雪能够引起其体内水分的变化，并可能会引起其生理生化特性的变化。国内外做了降雪低温对结皮层生物体光合作用和呼吸作用方面的研究。早期的研究认为，地衣能够在极低温度下维持净光合速率，甚至当叶状体温度降低至 −11.5 ℃和 −22 ℃时，地中海地区的一些地衣也能够产生净光合速率（Kappen and Lange，1972）。也有研究表明，降雪和低温能够延长 BSC 光合活性时间（Su *et al.*，2013）。降雪可能对真藓结皮的生理产生影响，但目前关于降雪对结皮层生物体生理生化特性影响的研究相对较少（李刚等，2014）。我们以沙坡头人工植被区的真藓结皮为研究对象，研究降雪影响下结皮层真藓植物的光合色素含量、渗透调节物质、可溶性蛋白含量及丙二醛（MDA）含量的变化特征，为研究藓类植物对降雪的生理响应机制提供理论依据和数据支撑，对维持荒漠生态系统的稳定性具有重要意义。

在 2014 年 2 月的一次降雪期间（9 h 内降雪量达 2.3 mm），将采集的真藓结皮样品放置在空旷平地，做 4 个降雪处理，分别为无降雪（0S）、0.5 倍降雪（0.5S）、1 倍降雪（S）、2 倍降雪（2S），将无降雪处理作为对照，每个处理 3 个重复。无降雪处理的样品用直径为 10 cm 的透光塑料圆盘遮盖；其余不同处理的样品均接受自然降雪，将 0.5 倍降雪处理样品去除一半的降雪量。此外，使用直径为 10 cm 的透光塑料圆盘接受自然降雪，然后向 2 倍降雪处理的样品添加 1 份圆盘所接受的降雪量。经降雪处理后的结皮，待雪层融化后，将结皮取下带到实验室内并放置于 −84 ℃冰箱备用。

降雪影响下光合色素含量的变化特征：真藓光合色素的含量随降雪量的变化趋势较为一致，随降雪量的增加呈上升趋势（图 3-61）。在无降雪时，Chl-a、Chl-b、总 Chl 以及类

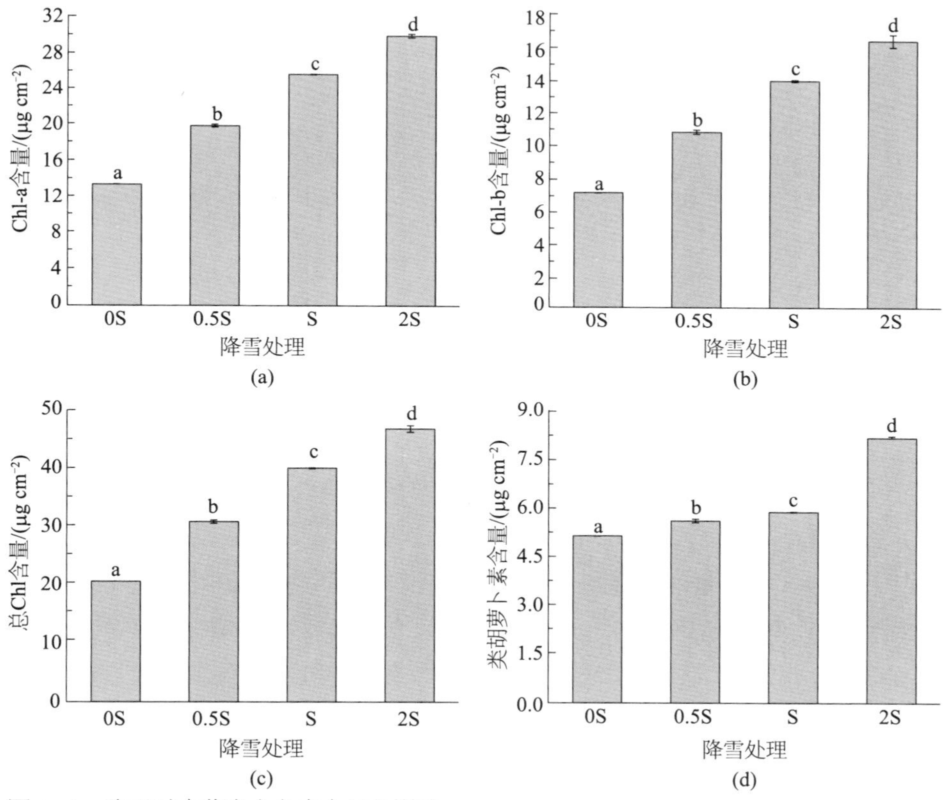

图 3-61　降雪对真藓光合色素含量的影响

Figure 3-61　The effects of snow on photosynthetic pigments concentration of *B. argenteum*

胡萝卜素含量最低，在降雪量达到2倍正常降雪量时，色素含量达到最大值，分别比无降雪处理时高出122.93%、133.11%、126.47%、58.53%，并且两者之间存在显著性差异（$p<0.01$）。方差分析也表明真藓色素含量在不同处理之间均存在显著性差异（组内、组间$p<0.01$）。

降雪影响下真藓渗透调节物质含量的变化特征：真藓可溶性糖含量随着降雪量的增加呈降低趋势，0.5倍降雪处理与对照相比，可溶性糖含量降低了4.52%，差异显著（图3-62，p=0.049）；1倍降雪和2倍降雪与对照相比，可溶性糖含量分别降低了14.43%和27.29%，2组均表现为极显著差异（$p<0.01$）。真藓游离脯氨酸含量变化随降雪量的增加也呈现降低趋势，由图3-62可看出，脯氨酸的降幅较可溶性糖大，0.5倍降雪处理、1倍降雪处理、2倍降雪处理与对照相比，脯氨酸含量分别降低了8.31%、31.96%、43.47%，三组比较均表现为极显著差异（$p<0.01$）。

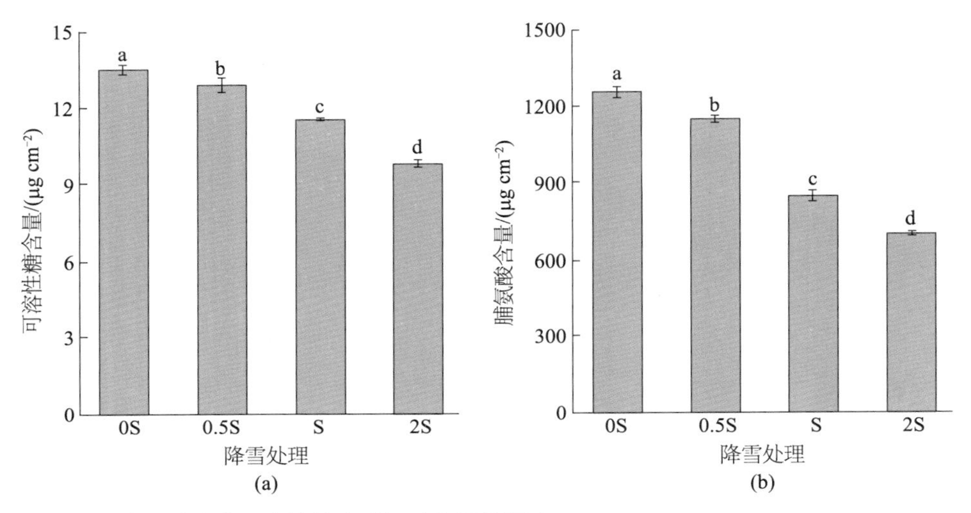

图3-62 降雪对真藓可溶性糖和脯氨酸含量的影响

Figure 3-62 The effects of snow on soluble sugar content and proline content of *B. argenteum*

降雪影响下可溶性蛋白含量的变化特征：真藓的可溶性蛋白含量随着降雪量的增加而显著增加。与对照相比，0.5倍降雪处理、1倍降雪处理、2倍降雪处理的真藓可溶性蛋白含量分别增加了6.65%、29.75%、41.34%。方差分析表明，各处理组与对照组相比具有显著差异，各处理组间差异同样显著（图3-63，$p<0.05$）。

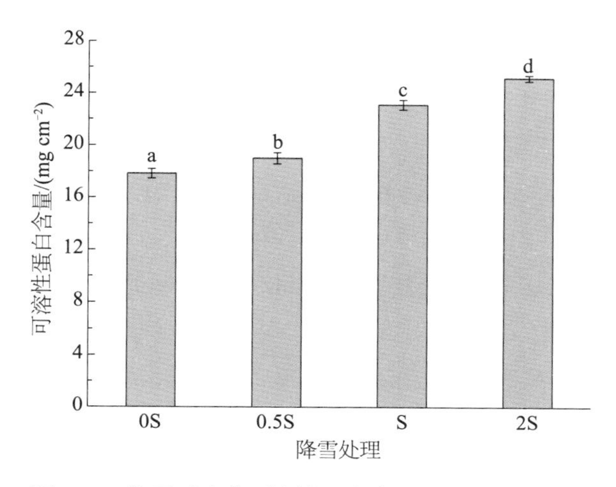

图 3-63　降雪对真藓可溶性蛋白含量的影响
Figure 3-63　The effects of snow on soluble protein content of *B. argenteum*

降雪影响下 MDA 含量的变化特征：真藓 MDA 含量随降雪量的增加而降低，与对照相比，0.5 倍降雪处理、1 倍降雪处理、2 倍降雪处理的真藓 MDA 含量分别降低了 2.95%、6.54%、12.58%。且方差分析表明，0.5 倍降雪处理与对照相比，差异不显著（图 3-64，$p>0.05$）；1 倍降雪处理与对照相比差异显著（$p<0.05$）；2 倍降雪处理与对照相比差异极显著（$p<0.01$）。1 倍降雪与 0.5 倍降雪处理间却表现为差异不显著（$p>0.05$）。

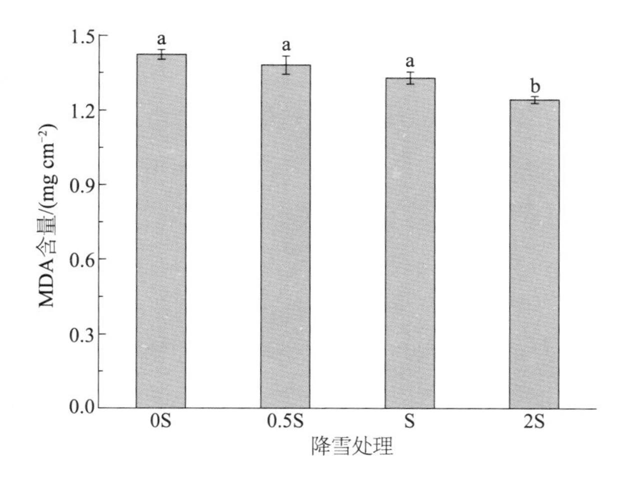

图 3-64　降雪对真藓 MDA 含量的影响
Figure 3-64　The effects of snow on malonaldehyde content of *B. argenteum*

BSC是荒漠景观重要组成部分，冬季降雪对BSC生物体生理生化特性的影响经常受到忽视。我们的研究表明，冬季降雪会引起结皮层生物体光合色素含量、渗透调节物质含量、可溶性蛋白含量以及MDA含量等的改变以提高其对环境的适应能力。

叶绿素含量的高低反映了结皮生物体光合作用能力的强弱。一般认为，结皮层生物体的光合速率与其叶绿素含量呈线性关系（Tuba *et al.*，1996；Palmqvist，2000）。换言之，叶绿素含量与光合速率密切相关，叶绿素含量越高，藓类植物的光捕获能力越强。叶绿素a（Chl-a）是光能的捕获者，也是叶绿体膜内光传导者；叶绿素b（Chl-b）也是光能捕获者（包维楷和冷俐，2005）。李亚敏和肖红利（2011）研究了Cd-Cu复合胁迫对两种藓类植物生理特性的影响，其中两种藓类的叶绿素含量都随着重金属浓度的升高而降低，呈明显抑制效应，光合作用受到破坏。回嵘等（2012）研究了UV-B辐射对真藓结皮生理特性的影响，结果表明，辐射也会引起叶绿素含量的降低。魏美丽和张元明（2010）研究了脱水过程中刺叶赤藓光合色素含量及叶绿体结构的变化，结果表明，脱水时叶绿素含量快速升高后再降低，湿水后叶绿素含量又升高。这说明相对重金属与辐射胁迫而言，水分胁迫对光合色素的影响是可恢复的。本文研究表明，Chl-a、Chl-b的含量均随着降雪增加呈现上升的趋势。这是因为在冬季无降雪时，植物受低温胁迫影响表现为低光合能力。降雪过后，积雪融化能够为藓类植物提供所需的水分，使得真藓植株含水量迅速增加，光合色素含量随之增加。从图3-61可以看出Chl-a的增幅较Chl-b大，这说明降雪不仅能够提升真藓的光捕获能力，也能增强其光传导能力。总之，在不考虑其他因素影响时，一定的降雪范围内，随着降雪量的增多，光合色素含量也越多，光就越容易为植物所利用，真藓植物光合能力就越强。

藓类植物不仅可以通过外部形态调节，还可通过渗透调节物质来适应外部环境变化。渗透调节物质包括可溶性糖、脯氨酸和可溶性蛋白。当藓类植物受到环境胁迫失水时，通常会通过渗透调节作用来增加含水量，可溶性糖对渗透调节有着不可替代的作用；脯氨酸通常以游离态的形式存在于植物体内，也是重要的渗透调节物质之一。张显强等（2004）认为鳞叶藓可溶性糖含量随着干旱强度变大而显著升高，而游离脯氨酸含量的升高幅度较微弱，说明在干旱环境下对鳞叶藓起调节作用的物质主要是可溶性糖，而脯氨酸几乎无影响。冬季无降雪时，藓类植物由于受低温和干旱胁迫影响，表现为高的可溶性糖与脯氨酸含量，以及低的可溶性蛋白含量，通过降低代谢活性以适应低温环境；而冬季降雪则对藓类植物的生理代谢过程具有改善作用。本研究中，降雪量小的时候可溶性糖与脯氨酸含量高，无降雪条件下可溶性糖与脯氨酸含量最高，然后随着降雪量增加呈显著下降趋势。这是因为冬季降雪是荒漠地区的主要水分来源之一，能够为真藓提供其生理生化活性所需的水分，随着含水量的增加，两者含量均会有所下降。这与陈文佳等（2013）的水分含量多时会降低可溶性糖与脯氨酸含量的研究结果相一致。可溶性蛋白与植物新陈代谢息息相关，含量越高新陈代谢能力越强。卜楠等（2012）研究了不同水分条件下沙漠豆生理指标的变化，结果表明在干旱胁迫下，可溶性蛋白含量先下降后上升但低于含水量较高的对照。而本研究结果表明随着降雪量增加，可溶性蛋白含量呈显著上升趋势。降雪量增加，积雪融化所提供的水分量就会增多，进而激发真藓的生理活性，例如可溶性蛋白含量升高，即BSC中真藓植物的新陈代谢能力有所提升。

当植物面临水分、温度胁迫时，细胞膜系统会受到破坏，主要表现为MDA含量的增加。MDA是生物膜质过氧化强度和酶系统受伤害程度的重要指标，MDA积累越多，组织受伤害越严重。国春晖（2014）研究了三种藓类植物旱后复水过程中的生理特性，结果表明，随着干旱胁迫加剧，三种藓类MDA含量均迅速增多，复水后，山墙藓的MDA含量在急剧下降后直接趋于平稳，具有较强的保水能力。在冬季无降雪时，真藓植物由于受低温干旱胁迫的影响，在体内积累了大量的MDA，而积雪融化会使胁迫得到缓解。本研究中，随着降雪量的增加，MDA含量也呈现急剧下降的趋势，说明真藓在此过程中也有较强的保水能力。这可能是因为降雪量增加为植株提供更多水分，会使得真藓植物处于复水状态，干旱胁迫减缓，膜脂过氧化过程减弱，MDA含量就会降低。

综上所述，降雪对真藓的生理生化活性具有一定的促进作用。冬季降雪是荒漠地区水分的重要来源之一，且降雪覆盖具有一定的保温效应，在冬季严寒条件下为真藓提供适宜的水分条件和温度条件。冬季降雪通过缓解干旱胁迫改善了真藓的生理特性，促进了其光合作用，而且冬季低温也起到了一定的促进作用。降雪提高了低温下真藓的叶绿素含量和可溶性蛋白含量，进而促进其光合作用能力与新陈代谢能力。与此同时，降低了可溶性糖含量、脯氨酸含量以及MDA含量，进而影响其渗透调节作用以及膜脂过氧化程度。本研究为探讨藓类植物对降雪的生理响应机制提供理论依据和数据支撑，对维持荒漠生态系统的稳定性具有重要意义。

3.2 BSC对光照的生态生理响应

3.2.1 光照强度对BSC固氮活性的影响

BSC中包含多种固氮藻类，如*Anabaena azotica*、*Lyngbya crytoraginatus*、*Nostoc* *flagelliforme*、*Oscillatoria pseudogeminata* 等，它们的固氮活性在 5 个数量级之间变化（Belnap and Lange，2003）。Hartley 和 Schlesinger（2002）报道了奇瓦瓦沙漠 BSC 的固氮活性为 $1.0 \times 10 \sim 2.0 \times 10^5$ nmol m^{-2} h^{-1}；Zaady 等（1998）的研究指出，内盖夫沙漠 BSC 的固氮活性为 $3.4 \times 10^5 \sim 1.8 \times 10^6$ nmol m^{-2} h^{-1}。这些研究表明，BSC 是干旱、半干旱区除大气氮沉降和豆科植物固氮外，土壤氮素输入的又一个重要贡献者。对 BSC 固氮活性的研究只集中在少数区域的几个荒漠生态系统中（Zaady *et al.*，1998；Hartley and Schlesinger，2002；Aranibar *et al.*，2003）。我国荒漠面积占国土面积的 20%，BSC 在我国荒漠区发育良好、类型多样（Li *et al.*，2003）。20 世纪末，我国科研工作者从多个层面研究并报道了 BSC 的生理特征和生态功能，然而，对其固氮活性的研究还较为少见。

本文以腾格里沙漠东南缘沙坡头地区发育良好的藻类结皮、藓类结皮和地衣结皮为研究对象，利用乙炔还原法，通过室内大型人工气候室模拟不同光强，研究三种类型 BSC 的固氮活性。本项研究有助于评价 BSC 在温带荒漠生态系统氮循环中的地位和作用，为荒漠生态系统的管理和可持续发展提供科学依据。

在研究区选择 6 个典型沙丘作为 6 个重复，2008 年 8 月初，在每一个沙丘上，利用样线法从丘顶至丘间低地依次采集足够试验用的藻类结皮、地衣结皮和藓类结皮。样品采集点距离灌丛 3 m 以外。本实验利用乙炔还原法（ARA）测定 BSC 的固氮活性（Belnap，2003）。选择 60 mL 透明可密封的塑料罐作为培养器。先将样品置于培养器中，然后在结皮表面喷洒蒸馏水，密封培养器。用注射器抽取培养器中 20% 的空气（12 mL），再向培养器中注入等体积的 C_2H_2 气体。将培养器置于人工气候室中（Thermoline L+M，澳大利亚），设定气候室温度和光强（通过调节样品与高压钠灯的距离使到达培养器表面的光强为试验所需，利用 Li-6400 光量子探头测定培养器表面光强）。样品培养一定时间后，利用 5 mL 注射器抽取培养器中的混合气体，通过排水法将气体收集于密封的透明玻璃小瓶中。在气体收集后的 4 h 内，利用气相色谱仪（Agilent GC6820，美国）测定样品和标气（C_2H_2 体积浓度为 10%）中 C_2H_2 峰面积。试验中每测定 5 个样品后测定 1 次标气，并按下述公式计算 BSC 的固氮活性：

$$[C_2H_2] = \left(\frac{A_{sample}}{A_{standard}}\right) \times 10\% \times V_{sample} \times (1/22.4) \times \left(\frac{1}{T_c}\right) \times \left(\frac{1}{S}\right) \times \left(\frac{273}{273+T_c}\right) \times \left(\frac{P}{760}\right) \tag{3-1}$$

式中，$[C_2H_2]$ 为 C_2H_2 气体的生成速率（mmol m^{-2} h^{-1}）；A_{sample} 为样品中 C_2H_2 的峰面积；$A_{standard}$ 为标气中 C_2H_2 的峰面积；10% 为 C_2H_2 标气浓度；V_{sample} 为培养器体积（L）；T_c 为培养时间；S 为样品的垂直投影面积（m^2）；T_t 为向培养器中注入 C_2H_2 气体时的环境温度

（℃）；P 为向培养器中注入 C_2H_2 气体时的大气压（mmHg）。

试验 1：为了研究 BSC 固氮活性的时间效应，将藻类结皮、地衣结皮和藓类结皮充分湿润后在室温环境中培养 10 min、25 min、40 min、60 min、90 min、120 min 和 180 min，测定 C_2H_2 浓度；

试验 2：BSC 湿润后才有固氮作用，设置 3 个光强梯度（0 μmol m^{-2} h^{-1} PPFD、400 μmol m^{-2} h^{-1} PPFD 和 1000 μmol m^{-2} h^{-1} PPFD），温度设定为 24 ℃，在每个光强梯度下设置不同的水分处理（1 mm、3 mm 和 5 mm），研究 BSC 固氮活性在不同模拟降雨后对光照的响应。本试验中的所有样品在人工气候室中培养 3 h，测定 C_2H_2 浓度。以上 2 个试验，每个处理设置 6 个重复。

（1）三种类型BSC固氮活性的时间效应

三种类型 BSC 的固氮活性随时间变化的趋势不一致（图 3-65）。藻类结皮在 10 min 内的平均固氮活性为 60.8 mmol m^{-2} h^{-1}，显著高于在 40 min 内的固氮活性（图 3-65，$p<0.05$），前者是后者的 3.3 倍。藻类结皮的固氮活性在 90 min 内和 180 min 内无显著差异（15.6 ~ 16.2 mmol m^{-2} h^{-1}）（图 3-65，$p>0.05$）。藓类结皮和地衣结皮固氮活性随培养时间的延长而增加，但无显著差异（图 3-65，$p>0.05$）。

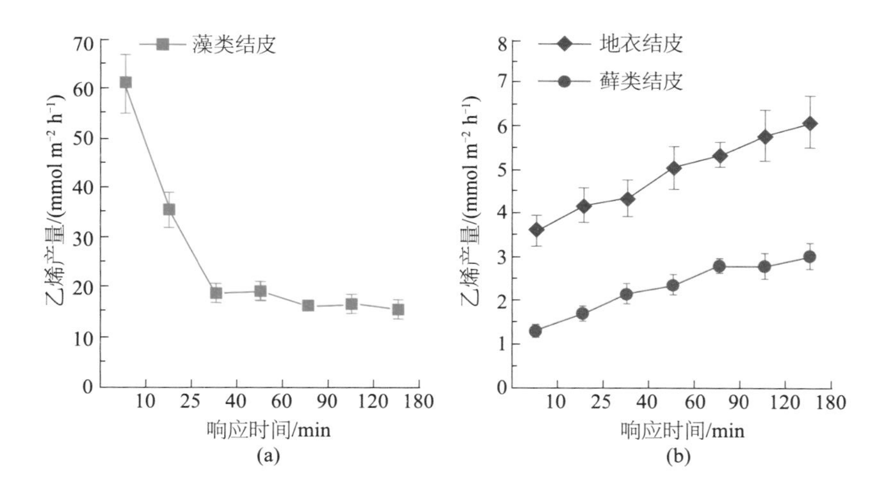

图 3-65　藻类结皮、地衣结皮和藓类结皮固氮活性的时间效应

Figure 3-65　Nitrogenase activity of algal crusts，lichen crusts and moss crusts with the time elapse

（2）光照强度对BSC固氮活性的影响

BSC 固氮活性与光照强度关系的研究结果表明，在无光照、400 μmol m^{-2} h^{-1} PPFD 和 1000 μmol m^{-2} h^{-1} PPFD 光强下，藻类结皮、藓类结皮和地衣结皮的固氮活性均无显著差异（图 3-66，$p>0.05$）。

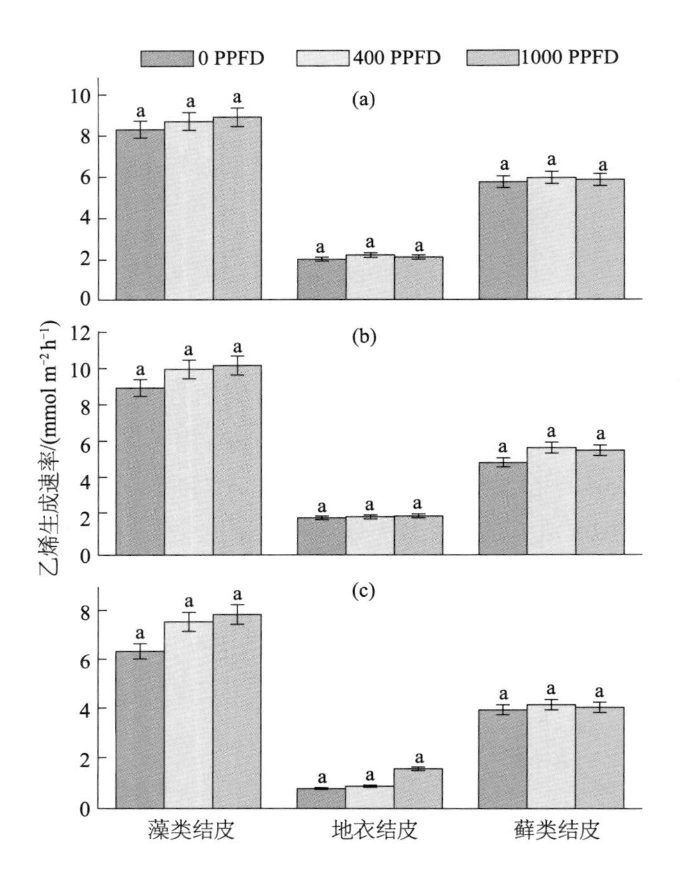

图 3-66　不同光强下三种 BSC（藻类结皮、藓类结皮和地衣结皮）的固氮活性（mean ± SE）（mmol m^{-2} h^{-1}）。（a）5 mm 模拟降雨量；（b）3 mm 模拟降雨量；（c）1 mm 模拟降雨量。不同小写字母表示光强对结皮固氮活性在 p=0.05 水平上影响显著

Figure 3-66　Nitrogenase activity（mean ± SE）（mmol m^{-2} h^{-1}）of BSC（algal crusts，moss crusts and lichen crusts）under three light intensities.（a）5 mm precipitation；（b）3 mm precipitation；（c）1 mm precipitation.Different lowercase letters indicate significant difference in nitrogenase activity among temperatures at the level of p=0.05

BSC 固氮活性与光强的关系研究非常有限，并且，由于现有研究所采用的光强单位不一致，导致许多研究结果之间可比性较低（Belnap，2003）。John（1977）测定黑暗环境中蓝绿藻结皮的固氮活性，发现结皮的固氮活性在培养 4 ~ 6 h 后降为 0。由于光合作用为

BSC 固氮作用提供能量，在黑暗的环境中，BSC 无光合作用；对土壤表层 0～3 cm 有机质含量的分析表明，藻类结皮、地衣结皮和藓类结皮下层土壤的有机质含量依次为 4.27 g kg^{-1}、6.81 g kg^{-1} 和 12.54 g kg^{-1}，这说明 BSC 层及下层土壤中的有机质能为固氮微生物的生理活动提供能量。我们对 BSC 在无光照、400 μmol m^{-2} h^{-1} PPFD 和 1000 μmol m^{-2} h^{-1} PPFD 光强下固氮活性的研究表明，三种 BSC 的固氮活性在黑暗和有光照的情况下无显著差异（图 3-66，$p>0.05$），这说明 BSC 固氮活性在我们测定的 3 h 中未受到显著影响，即 BSC 有机体已储存了足够结皮维持 3 h 固氮活性所需的能量。

3.2.2 BSC 改变了沙地生态系统的能流交换

地表反照率（surface albedo）是地表辐射平衡中的重要参数，表征地表对太阳短波辐射（0.3～3 μm）的吸收和反射能力，对某特定地表而言，定义为总反射辐射通量与入射辐射通量之比。

以腾格里沙漠沙坡头流沙区和 1964 年人工固沙植被区为研究区域。在 2009 年 4—8 月期间，选择晴天，分别运用日本 EKO 公司生产的 MR-32 型辐射仪（光谱范围 0.3～2.8 μm，灵敏度 7 mV kW^{-1} m^{-2}）和 Minolta/Land 公司生产的 Cyclops Compac3 型红外测温仪（测温范围：−50～500 ℃）对流动沙丘和藓类结皮的地面反照率和地表温度（0 cm 地温）进行同步测量。白天的测量时段为北京时间 8：30—17：30（当地时间为 7：30—16：30），每隔 0.5 h 测量一次。辐射仪的半球感应器水平朝下可测得地表反射辐射通量 $R\uparrow$，朝上测得太阳入射辐射通量 $Q\downarrow$，两者之比即为反照率值 $\alpha=R\uparrow/Q\downarrow$。太阳入射辐射和地表反射辐射数值（辐射仪输出毫伏值）连续读取 3 组，用平均值代表该时刻的实测值。为减少太阳辐射变动造成的影响，每组测量在 1 min 内完成。红外测温仪保持在距地表 1 m 高度处。此外，用铝盒每间隔 1 h 对 0～2 cm 的表层土壤进行取样，取得 3 个重复，根据土壤水分烘干称重法规程称量土样湿重，然后在 105 ℃下烘 24 h 后称取土样干重，计算土壤的重量含水量（Zhang *et al.*，2012）。

（1）流动沙丘和藓类结皮两种下垫面的反照率对比

当地表处于干燥状态时，流动沙丘和藓类结皮两种下垫面的地表反照率在晴天的平均日变化曲线均呈“U”型（图 3-67）。早晨和傍晚的地表反照率较高，对应较低的太阳高度角；地表反照率在当地时间正午最小，对应的太阳高度角最大。沙丘反照率的晴天日平均变化范围为 0.238～0.327，藓类结皮反照率的晴天日平均变化范围为 0.208～0.256。通过显著性检验，发现流动沙丘的反照率（0.266）显著（$p<0.05$）大于藓类结皮的反照率（0.226）。这说明，1964 年建植的人工固沙植被区地表经过 50 多年的演变，特别是 BSC 的发育使其反照率发生了很大改变。地表反照率的变化必然导致地表能量分配格局的改变，反照率较低的藓类土壤 BSC 地表要比反照率较高的流动沙丘地表吸收更多的太阳辐射能。由于荒漠地表干燥，实际蒸发很小，吸收的太阳辐射热量主要用于加热大气和地表土壤，即地表能量平衡中以感热和地热流量为主要因子。在这种情况下，地表反照率是调控地表温度的主要因子，导致植被区的地表温度要高于流沙区（这一现象已经得到大量研究支持）。地表反照率对地表温度的调控反过来会影响生态系统的一系列生物物理过程，如种子萌发、根系生长、植物发育以及微生物活性。此外，反照率与地表特征的紧密关系也意味着反照率可以作为监测地表覆盖变化的重要参数。

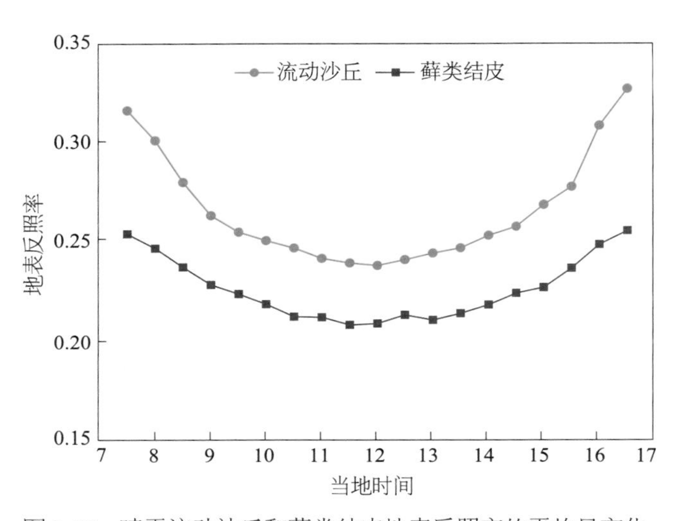

图 3-67 晴天流动沙丘和藓类结皮地表反照率的平均日变化

Figure 3-67 Diurnal variations of the averaged albedo for sand dune and moss-crusted soil under clear days

为了量化地表反照率和地表温度日变化的相关性，分上午和下午两个时段分别进行分析。图 3-68 反映了流动沙丘和藓类结皮这两种地表的反照率与地表温度之间的相关性，拟合方程及其 R^2 值和 p 值见表 3-7。可知，在上午和下午这两个时段，地表反照率和地表温度之间都高度相关，存在显著的线性关系。但是，拟合曲线的斜率在这两个时段的差别很大，意味着地表温度随地表反照率的变化速度在上、下午不同。随着地表反照率的增加，地表温度在上午随地表反照率的变化速度要明显快于下午。

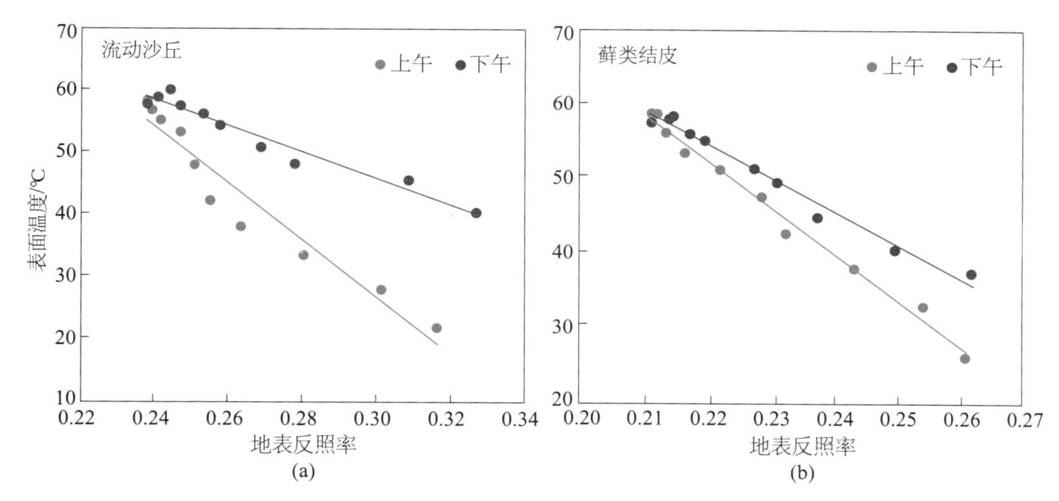

图 3-68　流动沙丘（a）和藓类结皮（b）的地表温度与地表反照率之间的相关性

Figure 3-68　Statistical relationship between surface temperature and surface albedo of sand dune (a) and moss-crusted soil (b)

表 3-7　地表反照率与地表温度关系的拟合方程

Table 3-7　Fitted formulas relating surface temperature to surface albedo

不同地表	上午 / 下午	拟合方程	R^2	p
流动沙丘	上午	$T_s = -454.5\alpha+163.4$	0.94	<0.001
	下午	$T_s = -211.3\alpha+109.3$	0.93	<0.001
藓类结皮	上午	$T_s = -626.6\alpha+190.7$	0.99	<0.001
	下午	$T_s = -450.0\alpha+154.6$	0.98	<0.001

注：T_s 表示地表温度；α 表示地表反照率。

根据以上结果，我们构造了一个用地表温度来间接计算地表反照率的经验模型，该模型基于算术平均法并融入了时间步长，如下：

$$\bar{\alpha}=\frac{\sum_{i=1}^{m}\alpha\left(T_s\right)_{AM,\ i}\cdot\Delta t_i+\sum_{i=m+1}^{n}\alpha\left(T_s\right)_{PM,\ i}\cdot\Delta t_i}{\sum_{i=1}^{n}\Delta t_i} \tag{3-2}$$

式中，$\bar{\alpha}$ 为地表反照率日平均值；$\alpha(T_s)_{AM,i}$ 为上午地表反照率关于地表温度的函数；$\alpha(T_s)_{PM,i}$ 为下午地表反照率关于地表温度的函数；T_s 为地表温度，单位为℃；m 为上午观测的总次数；n 为一天的观测总次数；Δt_i 为第 $i-1$ 次观测到第 i 次观测的时间步长，单位为 min。

对于流动沙丘和藓类结皮这两种地表，将表 3-7 中地表温度和地表反照率之间的拟合方程分别代入方程，就可得到分别适合两种地表的经验方程。

对于流动沙丘，其地表反照率的日平均值为

$$\overline{\alpha_{sd}} = \frac{\sum_{i=1}^{m} \frac{163.4-T_{s,i}}{454.5} \cdot \Delta t_i + \sum_{i=m+1}^{n} \frac{109.3-T_{s,i}}{211.3} \cdot \Delta t_i}{\sum_{i=1}^{n} \Delta t_i}$$

对于藓类结皮，其地表反照率的日平均值为

$$\overline{\alpha_{bsc}} = \frac{\sum_{i=1}^{m} \frac{190.7-T_{s,i}}{626.9} \cdot \Delta t_i + \sum_{i=m+1}^{n} \frac{154.6-T_{s,i}}{450} \cdot \Delta t_i}{\sum_{i=1}^{n} \Delta t_i}$$

以上方程适合于测量日晴朗无云、地表裸露干燥、蒸发很小的情况。此外，如果不对时间段进行划分来讨论反照率和地表温度的相关性，而是基于整个白天这样一个时间尺度，由于下午的地表温度整体上大于上午，而上、下午的地表反照率基本相等，在两者回归关系图上将出现随地表反照率的增加，与之对应的上午和下午地表温度的数据点越来越分散的情况。这样，用地表反照率来反映地表温度变化的准确性将大大降低（Zhang *et al.*, 2012）。

（2）两种地表的反照率与土壤含水量关系

当地表处于干燥状态，土壤湿度对地表反照率的影响微乎其微，地表反照率主要受太阳高度角的影响。但是，当地表处于湿润状态时，地表反照率同时受太阳高度角和土壤湿度两因素的作用。因此，要分析地表反照率随土壤湿度的变化关系，须首先排除太阳高度角的影响，即对地表反照率进行归一化处理。将太阳高度角对地表反照率的效应排除后，我们发现，藓类结皮和流动沙丘两种地表的归一化反照率均随土壤含水量的增加而递减（图 3-69），呈显著负相关（$p<0.05$）。该曲线包含两层意思：地表反照率随土壤含水量的增加而下降；下降的程度逐渐变缓。这是由于水的反射率很小，而透射率较大，包裹在土壤颗粒外围的水分增加了对太阳光的吸收路径，因而较高的土壤含水量对应较低的反照

率。此外，土壤含水量增加后，水分首先被土壤颗粒表面吸附，然后才用来填充土壤中的小孔或大孔隙。一旦土壤水分包裹了绝大多数土壤颗粒的表面，用来填充大孔隙而增加的水分对反射率的影响较小，因而随着土壤含水量的增加，地表反照率下降趋势减缓（Zhang *et al.*，2014）。

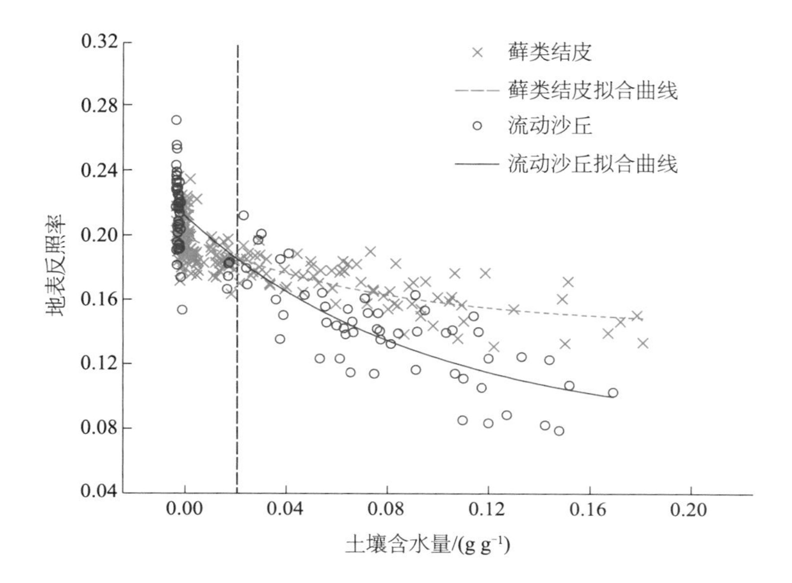

图 3-69　藓类结皮和流动沙丘的地表反照率随土壤含水量变化对比图

Figure 3-69　Comparison of the relationships between surface albedo and soil moisture for moss-crusted soil and sand dune

图 3-69 中两条拟合曲线存在一个交点，此处土壤含水量为 0.024 g g^{-1}。当土壤含水量小于 0.024 g g^{-1} 时，同等土壤含水量条件下，流动沙丘的地表反照率要大于藓类结皮的；而当土壤含水量大于该值时，流动沙丘的地表反照率反过来要小于藓类结皮且其随土壤含水量增加而减小的速度比藓类结皮快。说明流动沙丘地表反照率对土壤含水量变化的响应要比 BSC 敏感。该研究结果可用于通过土壤含水量估算地表反照率。此外，还可用于间接对比说明不同地表的蒸发状况。由于地表湿润时流动沙丘地表的反照率小于藓类结皮的反照率，在这种情况下，流动沙丘地表要比藓类结皮地表吸收更多的太阳辐射用于地表蒸发。因而，从地表反照率这一角度讲，与流动沙丘相比，在地表湿润状况下藓类结皮对地表蒸发具有一定的抑制作用。

3.3 BSC 对温度的生态生理响应

3.3.1 温度对 BSC 斑块土壤净氮矿化作用的影响

我们以腾格里沙漠东南缘一碗泉天然植被区（37° 25′ N，104° 36′ E）发育良好的两种 BSC 斑块（藓类结皮和藻类 – 地衣混生结皮）为研究对象，该区域内大量分布着藓类结皮和藻类–地衣结皮，具体土壤理化性质见表 3-8。运用原状土培养法，研究温度变化对两种 BSC 斑块土壤净氮矿化作用的影响，以期揭示不同 BSC 斑块土壤氮素矿化特征，认识荒漠生态系统氮循环机制，为荒漠生态系统可持续发展和科学管理提供理论依据。

表 3-8 研究区土壤理化性质

Table 3-8 The description of physico-chemical properties of soils in the research areas

土壤性质	藓类结皮土壤	藻类 – 地衣结皮土壤
pH	8.65 ± 0.015	8.79 ± 0.069
有机质 /%	0.18 ± 0.014	0.15 ± 0.010
全氮 /%	0.04 ± 0.002	0.03 ± 0.001
C/N	5.1 ± 0.167	5.6 ± 0.040
砂粒 /%	94.76 ± 1.461	95.97 ± 1.929
粉粒 + 黏粒 /%	5.24 ± 1.461	4.03 ± 1.929

注：数值为平均值 ± 标准误（n=3）。

2010 年 6 月，在试验区内选择发育良好的藓类结皮和藻类 – 地衣结皮斑块，分别设置小样方各 6 个（0.5 m × 0.5 m），样方间相隔 2 ~ 3 m，每个样方内随机设 6 个采样点。样品采集时先去除表层凋落物，用 PVC 管（内径为 5 cm，高 10 cm）在每个样点上取两个土柱，一个土柱上下封口后带回实验室进行培养试验，将另外一个土柱装入封口袋中进行 NH_4^+–N 和 NO_3^-–N 初始值测定。将带回实验室的 PVC 原状土柱用称重法将水分调节一致后（田间持水量的 85%），置于恒温培养箱中，分别在 −10 ℃、5 ℃、15 ℃、25 ℃、35 ℃和 40 ℃温度下进行培养，每个处理设 3 个重复。为保持土壤样品的湿度和空气流通，将 PVC 管的两端用保鲜膜封住，因为研究表明保鲜膜具有适度的透气性，并能减少水分的传递过程（Zhou and Ouyang，2001）。培养 14 天后取出样品，分析其硝态氮（NO_3^-–N）和铵态氮

（NH_4^+–N）含量，并计算在一定时间内累积的矿化氮含量和净矿化速率。

（1）不同培养温度对BSC斑块土壤净硝化速率的影响

如图 3-70 所示，温度对土壤净硝化速率具有显著影响（$p<0.05$）。−10 ~ 15 ℃培养时，藓类结皮、藻类 – 地衣结皮斑块土壤净硝化速率对温度的反应都很微弱，差异不显著。随着温度的升高，在 25 ~ 40 ℃培养下，特别是 25 ℃培养时，藓类结皮斑块土壤净硝化速率明显增加，达到显著水平（$p<0.05$）。此外，当培养温度从 25 ℃升高到 40 ℃时，藻类–地衣结皮斑块土壤净硝化速率增加了 145%，明显高于藓类结皮斑块土壤（34%）。

在 −10 ℃培养时，土壤净硝化作用均以固持态为主。在高温培养下（25 ~ 40 ℃），特别是 25 ℃培养时，藓类结皮斑块土壤表现出较强的硝化能力（$p<0.05$），净硝化速率是藻类 – 地衣结皮斑块土壤的 2.4 倍。40 ℃培养时，藓类结皮斑块土壤的净硝化速率比藻类 – 地衣结皮高 32%，呈显著差异（$p<0.05$）。

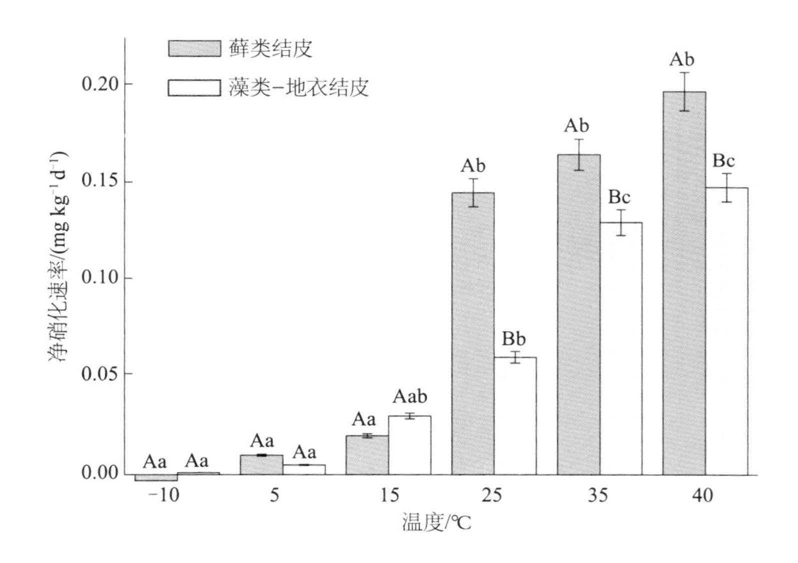

图 3-70　不同培养温度对土壤净硝化速率的影响。数值为平均值 ± 标准误。不同小写字母表示同一类型 BSC 斑块净硝化速率在不同温度处理下差异显著，$p<0.05$；不同大写字母表示不同类型 BSC 斑块净硝化速率在相同温度处理下差异显著，$p<0.05$。图 3-71 和图 3-72 同

Figure 3-70　The effects of temperatures on net soil nitrification rate（mean ± SE）.Different lowercase letters dewote significant differences among temperature treatments for the same BSC type,$p<0.05$；different capital letters denote significant differences among BSC types under the same temperature treatment, $p<0.05$. It's the same as Figure 3-71 and Figure 3-72

（2）不同培养温度对BSC斑块土壤净氮矿化速率的影响

从图 3-71 可知，低温培养（−10～15 ℃）不利于土壤氮的转化，净氮矿化速率均较低，且各温度间差异不显著（$p>0.05$）。25 ℃培养时，两种结皮斑块土壤净氮矿化速率明显升高（$p<0.05$），比 15 ℃培养时分别增加了 659% 和 490%。此外，高温培养下（25～40 ℃），藓类结皮和藻类－地衣结皮斑块土壤平均净氮矿化速率分别是低温培养下（−10～15 ℃）的 23 和 40 倍。

相同温度培养时，藓类结皮斑块土壤净氮矿化速率大于藻类－地衣结皮斑块土壤。在低温培养时（−10～15 ℃），两种结皮斑块土壤净氮矿化速率均处于较低水平，且差异不显著（$p>0.05$）。随着培养温度的升高（25～40 ℃），藓类结皮斑块土壤净氮矿化速率明显高于藻类－地衣结皮斑块土壤，其平均净氮矿化速率是藻类－地衣结皮斑块土壤的1.16倍。

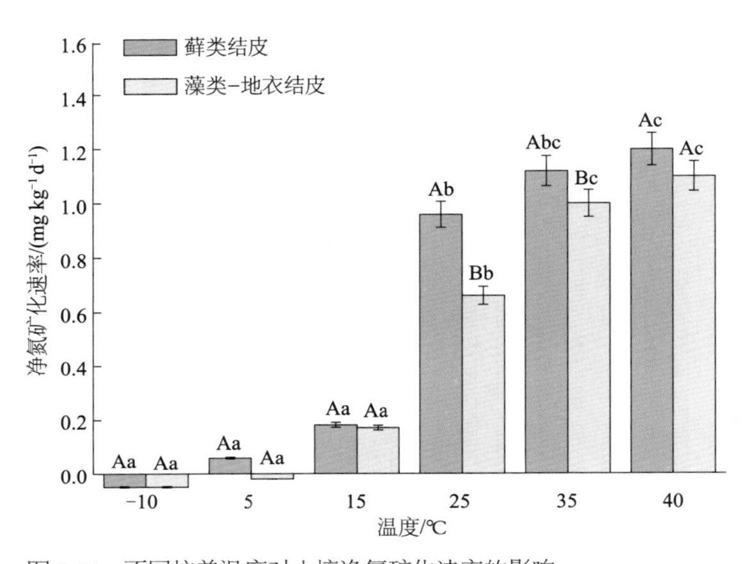

图 3-71　不同培养温度对土壤净氮矿化速率的影响

Figure 3-71　The effects of temperatures on net N mineralization rate（mean ± SE）

（3）不同培养温度对BSC斑块土壤无机氮积累的影响

温度对两种结皮斑块土壤无机氮积累的影响与净氮矿化速率相似，　低温培养时（−10～15 ℃）以固持态为主，无机氮含量均较低。随着温度的升高，无机氮含量明显增加（图3-72）。40 ℃培养时，两种结皮斑块土壤无机氮含量均最高。并且，藓类结皮斑块的土壤在高温培养下（40 ℃）无机氮的含量显著高于25 ℃培养（$p<0.05$），而与其在35 ℃培养时的含量无明显差异。藻类－地衣结皮斑块土壤无机氮含量虽然在较高温度培养下（25～40 ℃）

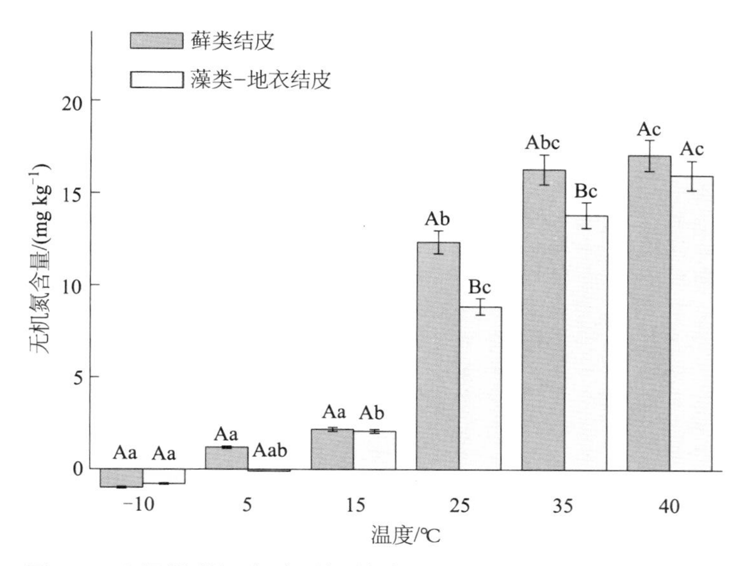

图 3-72　不同培养温度对土壤无机氮积累的影响

Figure 3-72　The effects of temperatures on inorganic N accumulation（mean ± SE）

明显升高，但各温度间的差异均不显著。此外，从25 ℃到40 ℃，藻类–地衣结皮和藓类结皮斑块土壤，无机氮含量分别增加了67%和36%。

两种结皮斑块土壤无机氮积累对温度的响应也不相同（图3-72）。5 ℃培养时，藓类结皮斑块土壤无机氮开始积累，而藻类–地衣结皮斑块土壤则以固持态为主。−10 ℃和15 ℃培养下，两种结皮斑块土壤无机氮含量均处于较低水平，差异不显著（$p>0.05$）。25 ℃和35 ℃时，藓类结皮斑块土壤无机氮含量明显高于藻类–地衣结皮斑块土壤，分别是后者的1.34倍和1.11倍，差异显著（$p<0.05$）。

（4）不同BSC斑块土壤净硝化速率和净氮矿化速率温度敏感性的比较

指数回归方程$y=ae^{bx}$能够很好地描述土壤氮转化速率（硝化和矿化）与温度之间的相关关系（图3-73）。本试验中，藓类结皮和藻类–地衣结皮斑块土壤的净氮转化速率能够与温度变量很好地拟合，两种结皮斑块的土壤净氮转化速率对温度的敏感性可以用温度敏感系数Q_{10}表示。利用函数拟合得到的温度响应系数b进行计算，得知两种结皮斑块土壤净氮转化速率的Q_{10}值变动范围为2.46 ~ 3.33。藓类结皮斑块土壤净硝化速率和净氮矿化速率对温度的敏感性均高于藻类–地衣结皮斑块土壤。整体来看，无论何种土壤，其硝化过程比总矿化过程对温度的变化敏感。

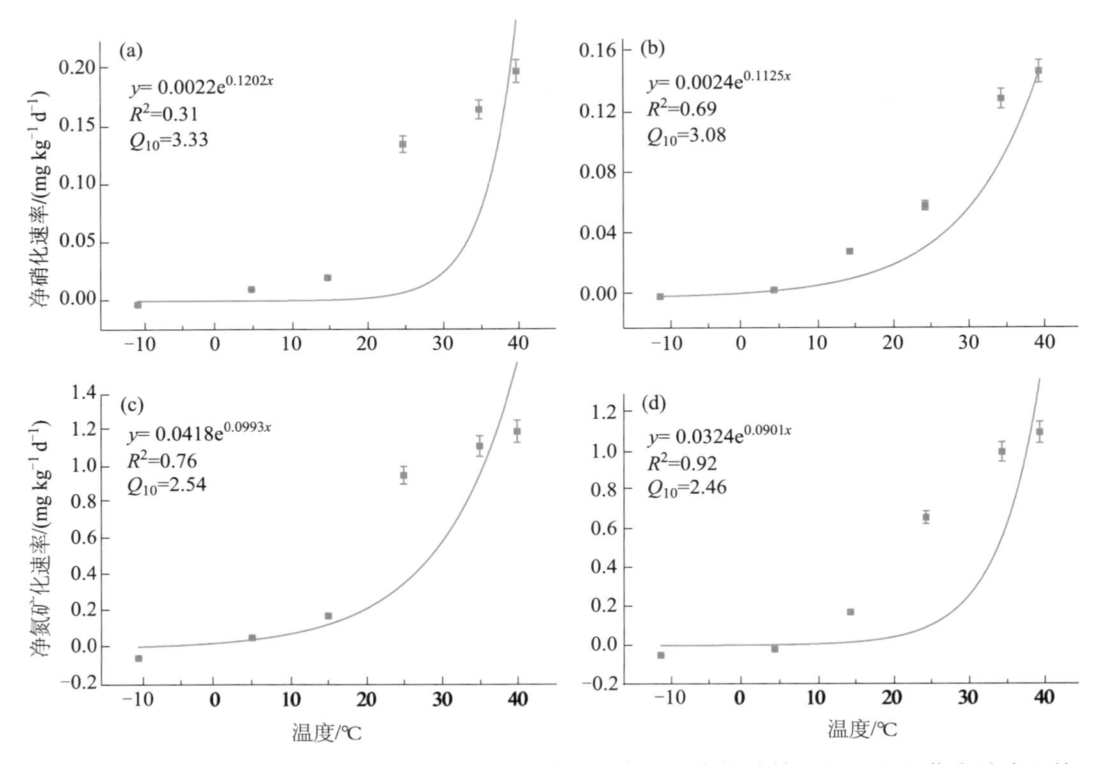

图 3-73　不同 BSC 斑块土壤净硝化速率和净氮矿化速率的温度敏感性。(a)、(c)藓类结皮斑块；(b)、(d)藻类–地衣结皮斑块。数值为平均值 ± 标准误

Figure 3-73　Temperature sensitivity of net nitrification and N mineralization rates (mean ± SE) under different biological soil crusts. (a)、(c) moss crusts；(b)、(d) algae-lichen crusts

综上所述，温度对土壤氮转化速率具有显著影响。低温培养下，两种结皮斑块土壤氮转化速率（硝化和矿化）均较低，以固持态为主。高温有利于土壤氮转化，特别是当温度达到25 ℃时，氮转化速率明显升高；相同温度条件下，藓类结皮促进了土壤氮转化，净硝化速率、净氮矿化速率以及无机氮积累均明显高于藻类–地衣结皮斑块土壤；不同温度条件下，两种结皮斑块土壤氮转化速率（硝化和矿化）遵循Q_{10}方程，在2.46 ~ 3.33范围内变动。藓类结皮斑块土壤氮转化对温度的响应比藻类–地衣结皮斑块土壤敏感。此外，就整个土壤氮矿化过程而言，硝化过程对温度的响应更为敏感。

3.3.2　模拟增温对荒漠 BSC– 土壤系统 CO_2、CH_4 和 N_2O 通量的影响

目前，众多国内外学者已开展了生态系统CO_2、CH_4和N_2O通量的多种监测研究（Keller

et al.，1986；Christensen *et al.*，1996；Du *et al.*，2006），并通过各种模拟增温试验，如被动式增温、OTC、红外线辐射器等（Aronson and McNulty，2009）模拟气候变暖，研究增温条件下不同生态系统与大气CO_2、CH_4和N_2O之间交换的变化规律（Rustad and Fernandez，1998；Chimner and Welker，2005；Oberbauer *et al.*，2007）。然而，大多数研究集中在冻原、高寒草甸和温带草原等地区（Welker *et al.*，2004；Wan *et al.*，2005；Xia *et al.*，2009；Lin *et al.*，2011），而对荒漠地区生态系统CO_2、CH_4和N_2O通量的相关报道并不多见，增温条件下的长期监测试验更是十分缺乏。

关于自然状态和气候变暖背景下，BSC覆盖的荒漠土壤与大气CO_2、CH_4和N_2O之间交换的研究报道相当匮乏。本文拟采用OTC增温方式来模拟气候变暖，通过对腾格里沙漠藓类、藻类以及两者混生的三种典型的BSC所覆盖的土壤温室气体通量的监测，试图弄清该系统在自然状态下的CO_2、CH_4和N_2O通量大小，增温条件下CO_2、CH_4和N_2O通量的变化，CO_2、CH_4和N_2O通量与土壤温度和土壤湿度之间的关系，为准确地评估荒漠系统温室气体通量对气候变暖的响应与反馈的方向和程度提供科学依据。

在一碗泉天然植被区（2012年6月初），用直径为20 cm的PVC管分别采集由藻类、藓类以及两者混生的三种不同类型的BSC覆盖的原状土壤样品（直径 × 深度=20 cm × 20 cm），每种BSC类型样品采集6份，共18个原状土样带回沙坡头沙漠试验研究站开展增温试验。藓类结皮、藻类结皮和混生结皮的盖度分别为95.1%、90.3%和91.3%，其中，混生结皮中藓类结皮、藻类结皮和地衣结皮的盖度分别为57.5%、30.8%和2.92%。每种BSC类型分增温和不增温两种处理，每种处理3个重复。2010年在沙坡头沙漠试验研究站水分平衡观测场西边建立了3个OTC用于增温试验，每个OTC为边长1.3 m、高2 m的八棱柱（张鹏等，2012）。各BSC类型的3个样品埋放在OTC中作为增温处理（W），另外3个样品埋放于OTC外作为不增温处理（NW），埋放时确保原状土壤样品与地表齐平。安置完毕后，将直径为25 cm带有水槽的底座安装于地表，使原状土芯处于底座的中心位置。增温处理和不增温处理均安装气象站（HOBO U30，Onset Computer Corporation，美国）1套，每0.5 h自动测定并储存5 cm深度的土壤温度和10 cm深度的土壤湿度数据。

每月中下旬用静态箱法采集气体样品，采样时间为9：00—11：00。采集气体样品时，将顶箱（高度 × 直径=40 cm × 25 cm）合扣在预先安装在土壤中的底座上，底座预先在水槽中装有约1/3深的水，用来密封顶箱。顶箱内装有风扇，使箱内气体混合均匀。在密封后的0 min、10 min、20 min和30 min，用50 mL的注射器通过三通阀采集静态箱内的气体25 mL。用手持式温度记录仪（JM624，今明仪器有限公司，中国）测定采样开始和结束时静态箱内的温度。所采集的气体样品带回室内，在气相色谱仪（Agilent GC6820，Agilent Technologies，Palo Alto，美国）上分析气体中的CO_2、CH_4和N_2O浓度。所有气体样品均在采集后24 h内完成分析。

（1）OTC的增温效应

如图3-74所示，OTC显著增加了5 cm深处的土壤温度。增温处理下，5 cm深处土壤的年均温度为13.5 ℃，比不增温处理增加了1.64 ℃（$p<0.001$，图3-74）。非生长季（11月—次年3月）和生长季（4—10月），非生长季的增温幅度明显高于生长季。

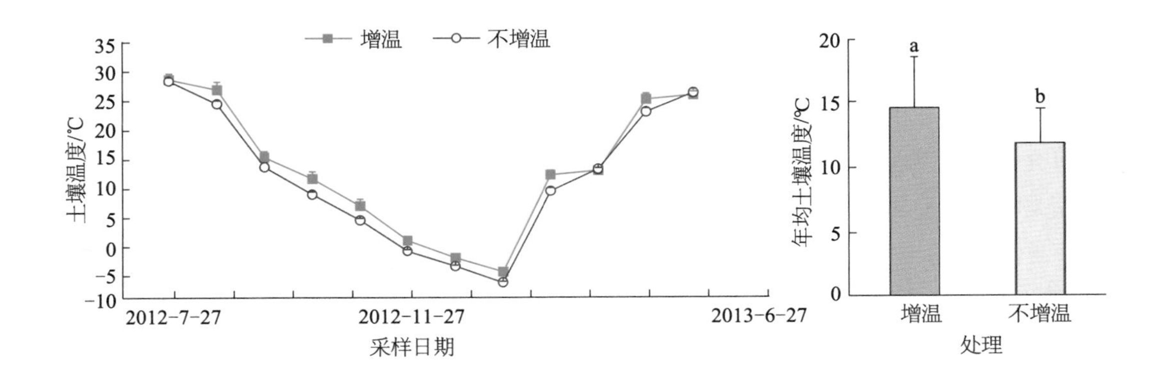

图3-74　增温对5 cm深度土壤温度的影响。图中数据为平均值 ± 标准误。不同字母表示差异显著，$p<0.05$

Figure 3-74　Effects of warming on soil temperature at 5 cm depth. Mean ± SE. Different letters indicate significant differences at 0.05 level

（2）CO_2通量

增温和BSC类型对CO_2通量没有直接的显著影响，但采样日期对CO_2通量的影响达到极显著水平；另外，BSC类型与采样日期两者的互作，以及增温、BSC类型和采样日期三者的互作对CO_2通量的影响均达到极显著水平（表3-9）。增温显著减少了藓类结皮2012年9月和12月的CO_2通量（图3-75a），显著增加了藻类结皮2012年8月（图3-75b）和混生结皮2013年3月的CO_2通量（图3-75c）。尽管增温对三种BSC类型年均CO_2通量的影响均不显著，但增温对不同BSC类型的影响不一致。不增温处理下藓类结皮年均CO_2通量为50.3 mg m^{-2} h^{-1}，增温使其减少了13.5%，生长季和非生长季分别减少3.90%和46.1%。不增温处理下藻类结皮年均CO_2通量为43.1 mg m^{-2} h^{-1}，增温使其减少了18.9%，生长季和非生长季分别减少了21.1%和12.2%。相反，增温处理下混生结皮的年均CO_2通量却增加了23.0%，生长季和非生长季分别增加了16.0%和38.6%。

表 3-9 CO_2、CH_4 和 N_2O 通量的多因素方差分析

Table 3-9 Multivariate analysis of CO_2, CH_4 and N_2O fluxes

因素	CO_2		CH_4		N_2O	
	F	p	F	p	F	p
增温（W）	0.0411	0.522	4.818	0.159	0.842	0.360
BSC 类型（T）	2.364	0.0988	1.265	0.285	0.287	0.751
采样日期（D）	25.879	<0.001	5.001	<0.001	1.120	0.350
T×D	5.124	<0.001	2.421	<0.001	0.658	0.874
W×D	1.683	0.083	2.113	0.023	0.825	0.615
W×T	1.815	0.167	1.864	0.159	0.637	0.530
W×T×D	2.407	0.001	2.137	0.004	1.480	0.090

（3）CH_4通量

CH_4通量受采样日期、结皮类型和采样日期互作、增温和采样日期互作以及采样日期、增温和结皮类型三者互作的显著影响，增温和结皮类型分别对CH_4通量的影响不显著（表3-9）。三种BSC类型的CH_4通量绝大多数情况下表现为负值（图3-76a，b，c），说明大部分情况下荒漠BSC–土壤系统是CH_4的汇。不增温处理下，藓类结皮、藻类结皮和混生结皮的年均CH_4通量分别为−0.038 μg m^{-2} h^{-1}、−0.011 μg m^{-2} h^{-1}、−0.056 μg m^{-2} h^{-1}；增温增加了藓类结皮和藻类结皮年均CH_4的吸收通量，分别增加了0.33倍和4.78倍。与之相反，增温使混生结皮的年均CH_4吸收通量减少35%。增温显著增加了藓类结皮2012年10月（图3-76a）和藻类结皮2013年5月（图3-76b）的CH_4吸收通量，但增温和不增温两种处理下，三种BSC类型年均CH_4通量差异都不显著。

（4）N_2O通量

采样日期、增温和结皮类型对N_2O通量的影响均不显著，三者之间的互作效应没有达到显著水平（表3-9）。三种BSC类型的N_2O通量并没有表现出明显的季节变化规律。不增温处理下藓类结皮、藻类结皮和混生结皮的年均N_2O通量分别为−4.42 μg m^{-2} h^{-1}、−2.12 μg m^{-2} h^{-1}和−3.12 μg m^{-2} h^{-1}，表明荒漠BSC–土壤系统在全年水平上表现为N_2O的汇。增温显著增加了藓类结皮2012年9月（图3-77a）、藻类结皮2012年7月（图3-77b）以及混生结皮2012年9月（图3-77c）的N_2O吸收通量，但两种处理下的三种BSC类型年均N_2O通量差

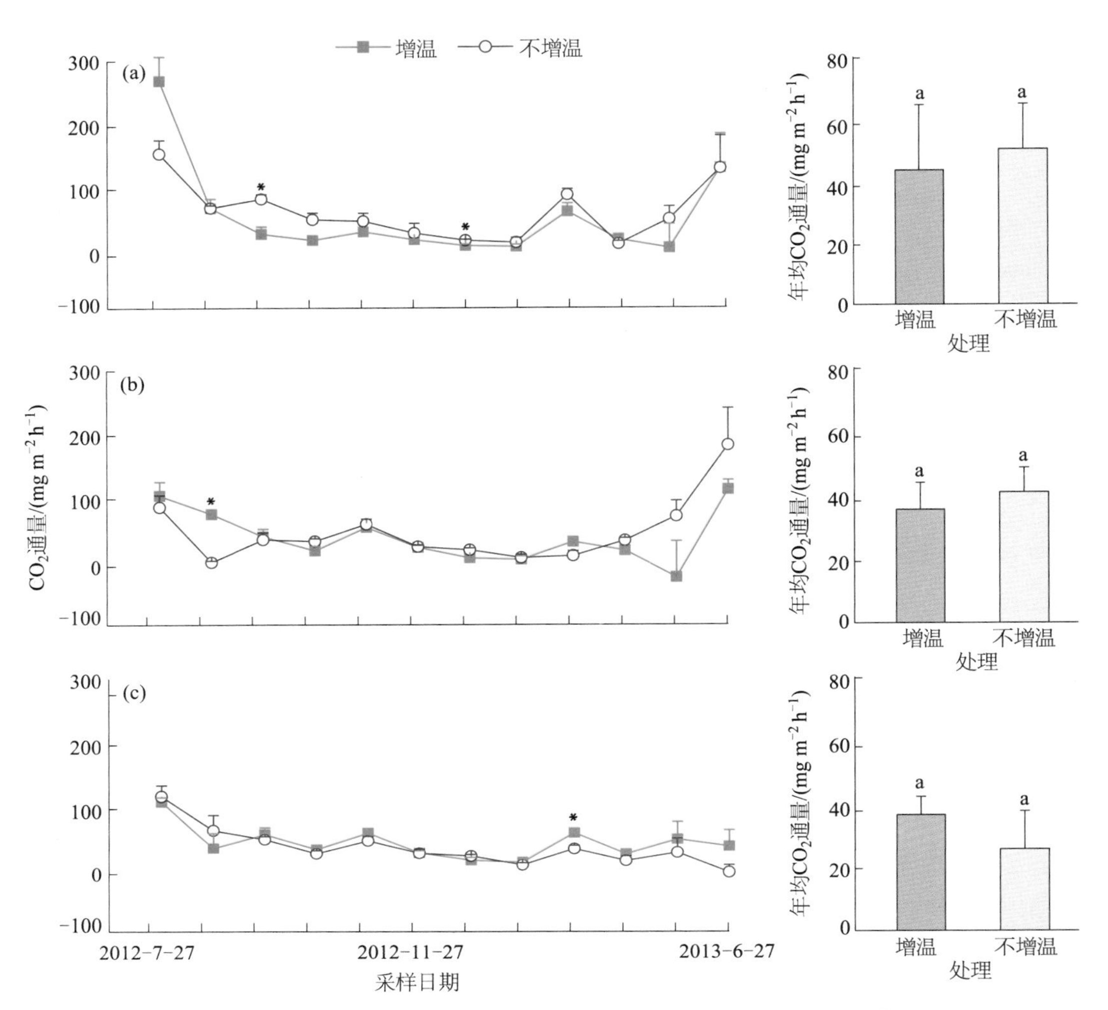

图 3-75　增温和不增温处理下不同BSC类型的CO_2通量：(a)藓类结皮；(b)藻类结皮；(c)混生结皮。平均值 ± 标准误；* 表示处理间差异显著，$p<0.05$；相同字母表示处理间差异不显著，$p>0.05$

Figure 3-75　CO_2 fluxes of various BSC types under warming and non-warming treatments：(a) moss crusts；(b) algae crusts；(c) mixed crusts.Mean ± SE；* indicates significant differences between treatments，$p<0.05$；The same letters indicate no significant differences between treatments，$p>0.05$

异均不显著。增温都促进了三种BSC类型对N_2O的吸收，藓类结皮、藻类结皮以及混生结皮的年均N_2O吸收通量分别增加了0.54倍、2.34倍和2.27倍。

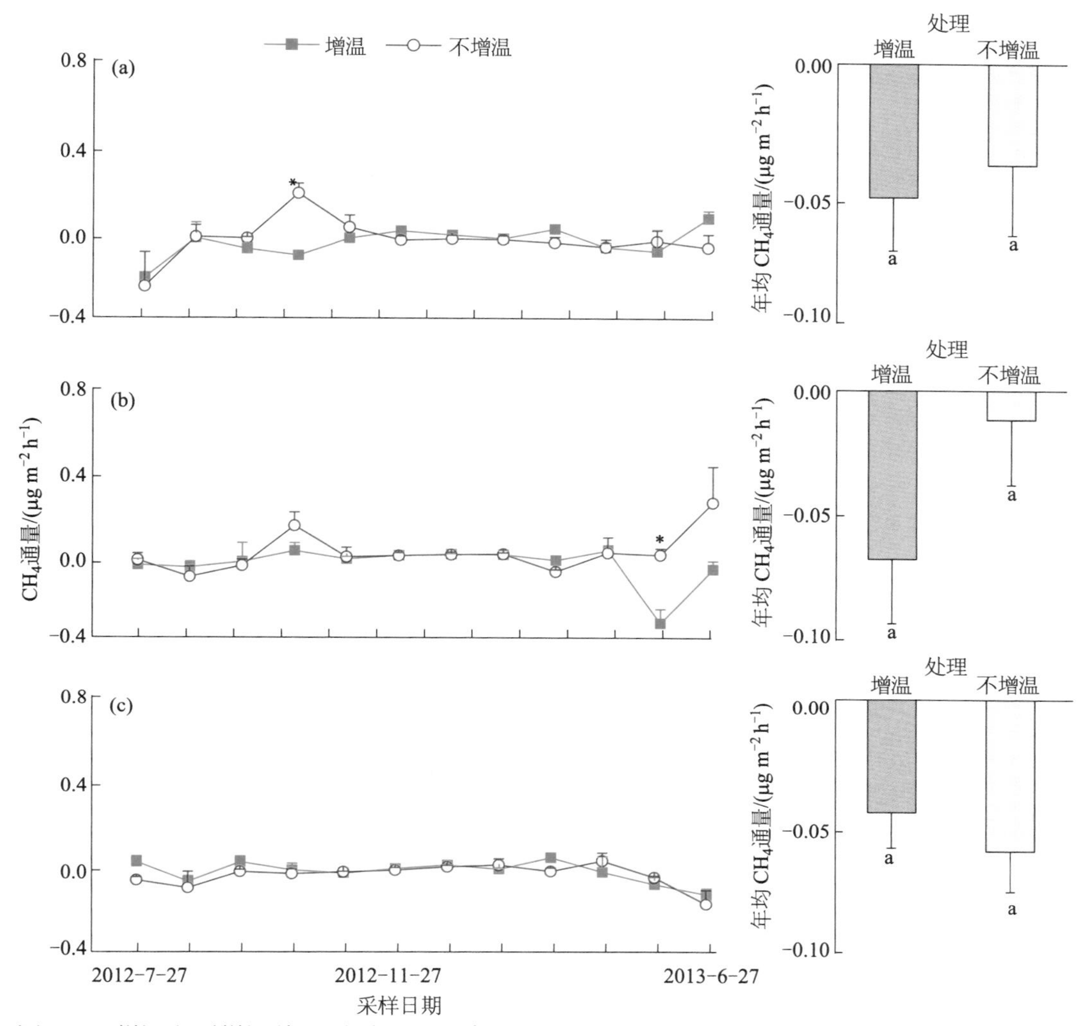

图 3-76　增温和不增温处理下不同 BSC 类型的 CH_4 通量变化：(a) 藓类结皮；(b) 藻类结皮；(c) 混生结皮。平均值 ± 标准误；* 表示处理间差异显著，$p<0.05$；相同字母表示处理间差异不显著，$p>0.05$

Figure 3-76　CH_4 fluxes of various BSC types under warming and non-warming treatments：(a) moss crusts；(b) algae crusts；(c) mixed crusts.Mean ± SE；* indicates significant differences between treatments ($p<0.05$)；The same letters indicate no significant differences between treatments ($p>0.05$)

（5）温室气体的全球增温潜能（GWP）

根据IPCC（2007）报道的CH_4和N_2O全球平均100年GWP值（CH_4: 25，N_2O: 298），将三种BSC类型全年累积CH_4和N_2O通量折合为CO_2当量。结果表明，增温对荒漠BSC-土壤系统的GWP影响根据不同类型的BSC而异，增温处理下藓类结皮和藻类结皮的CO_2当量

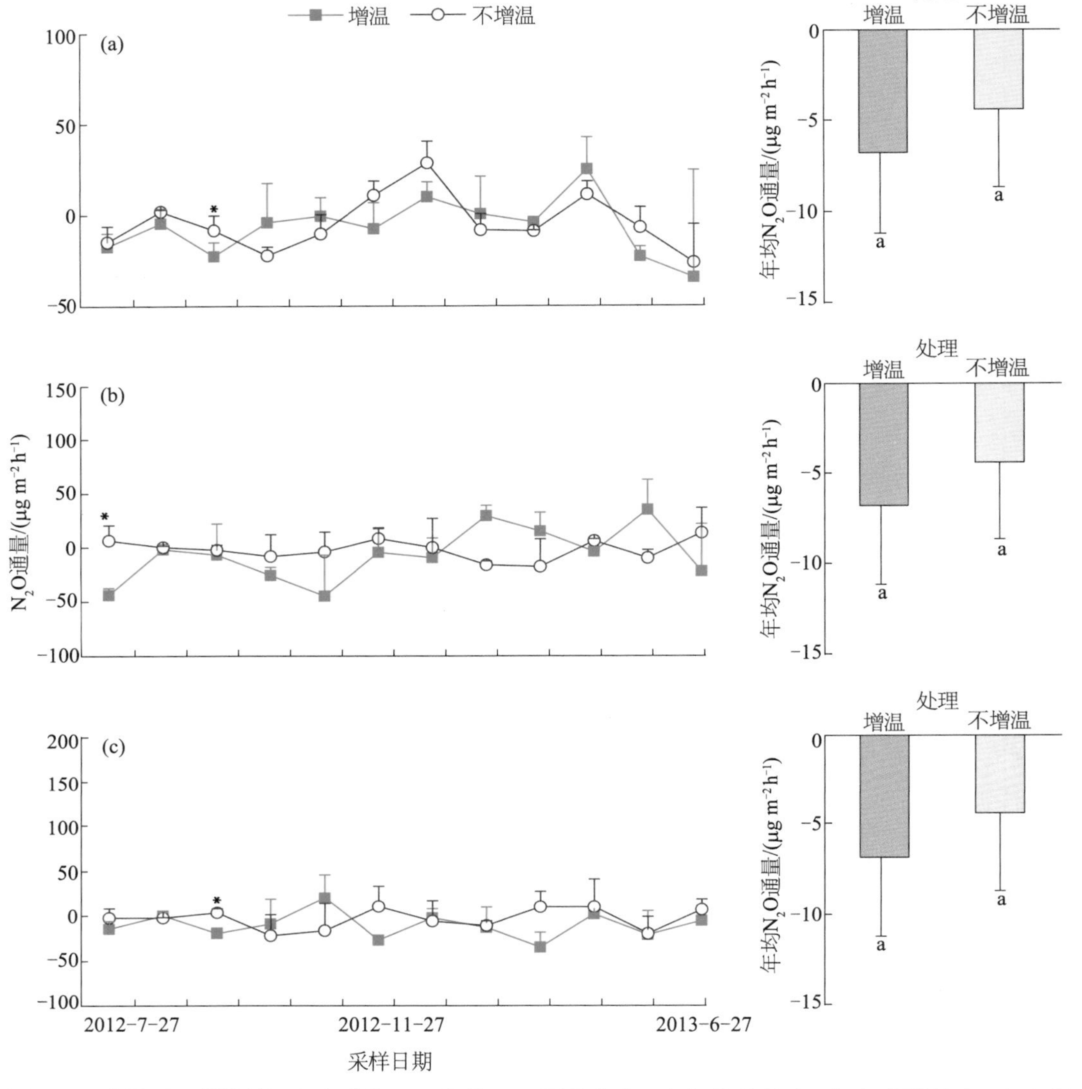

图3-77　增温和不增温处理下不同BSC类型N_2O通量变化：(a)藓类结皮；(b)藻类结皮；(c)混生结皮。平均值 ± 标准误；* 表示处理间差异显著，$p<0.05$；相同字母表示处理间差异不显著，$p>0.05$

Figure 3-77　N_2O fluxes of various BSC types under warming and non-warming treatments：(a) moss crusts；(b) algae crusts；(c) mixed crusts.Mean ± SE；* indicates significant differences between treatments，$p<0.05$；The same letters indicate no significant differences between treatments，$p>0.05$

分别减少15.3%和22.6%，而混生结皮增加了17.1%。但增温对三种BSC类型年平均GWP的影响不显著（表3-10）。

表 3-10 增温（W）和不增温（NW）处理下 CO_2、CH_4 和 N_2O 累积通量和全球增温潜能（GWP）

Table 3-10 Cumulative CO_2, CH_4 and N_2O emission and global warming potentials (GWP) (mean ± SD) under warming (W) and non-warming (NW) treatments

BSC 类型	处理	CO_2 累积通量 / ($g\ m^{-2}$)	CH_4 累积通量 / ($g\ m^{-2}$)	N_2O 累积通量 / ($g\ m^{-2}$)	GWP/ ($g\ m^{-2}$)
藓类结皮	W	380 ± 185.9	$-4.36\times10^{-4}\pm1.74\times10^{-4}$	−0.060 ± 0.039	363 ± 197
	NW	440 ± 110.8	$-3.29\times10^{-4}\pm2.31\times10^{-4}$	−0.387 ± 0.037	429 ± 122
藻类结皮	W	306 ± 99.8	$-5.72\times10^{-4}\pm2.17\times10^{-4}$	−0.062 ± 0.061	288 ± 118
	NW	377 ± 121.2	$-9.91\times10^{-4}\pm2.22\times10^{-4}$	−0.019 ± 0.024	372 ± 128
混生结皮	W	337 ± 65.0	$-3.65\times10^{-4}\pm1.17\times10^{-4}$	−0.089 ± 0.036	311 ± 76
	NW	273 ± 78.1	$-4.93\times10^{-4}\pm1.30\times10^{-4}$	−0.027 ± 0.029	265 ± 87

（6）温室气体通量与土壤温度和土壤湿度的关系

三种BSC类型的CO_2通量与5 cm深处的土壤温度均呈显著性指数正相关关系（图3-78a，b，c），土壤温度分别解释了藓类结皮、藻类结皮和混生结皮56.9%、29.2%和18.5%的CO_2通量变化。指数方程$y=ae^{bx}$中，b值表征了生态系统呼吸对温度的敏感性，藓类结皮CO_2通量的温度敏感性最大，藻类结皮次之，混生结皮最小。相反，藓类结皮和混生结皮CH_4通量与5 cm深处的土壤温度呈显著的线性负相关关系（图3-78d，f），说明这两种结皮随着土壤温度的增加，其CH_4吸收通量增加；土壤温度分别解释了藓类结皮和混生结皮5.40%和31.2%的CH_4通量变化。而藻类结皮CH_4通量与5 cm深处的土壤温度没有显著相关关系（图3-78e）。同样，三种BSC类型的N_2O通量与5 cm深处的土壤温度均没有显著的相关关系（图3-78g，h，i）。

土壤湿度与三种BSC类型的CO_2通量均呈显著的线性正相关关系（图3-79a，b，c），分别解释了藓类结皮、藻类结皮和混生结皮CO_2通量变化的30.9%、25.0%和13.5%。土壤湿度与藓类结皮和混生结皮的CH_4通量均呈显著性负相关关系，并分别解释了CH_4通量变化的4.4%和23.6%，与藻类结皮的CH_4通量不相关（图3-79d，e，f）。土壤湿度与藓类结皮N_2O通量呈显著线性负相关关系（图3-79g），并解释了5.9%的N_2O通量变化，藻类结皮与混生结皮N_2O通量与土壤湿度无显著相关关系（图3-79h，i）。土壤湿度与CO_2通量之间线性回归方程的斜率可以用来表征生态系统呼吸对湿度的敏感性，斜率越大表示敏感性越强，从图3-79可以看出，三种BSC类型CO_2通量的温度敏感性为藓类结皮>藻类结皮>混生结皮。

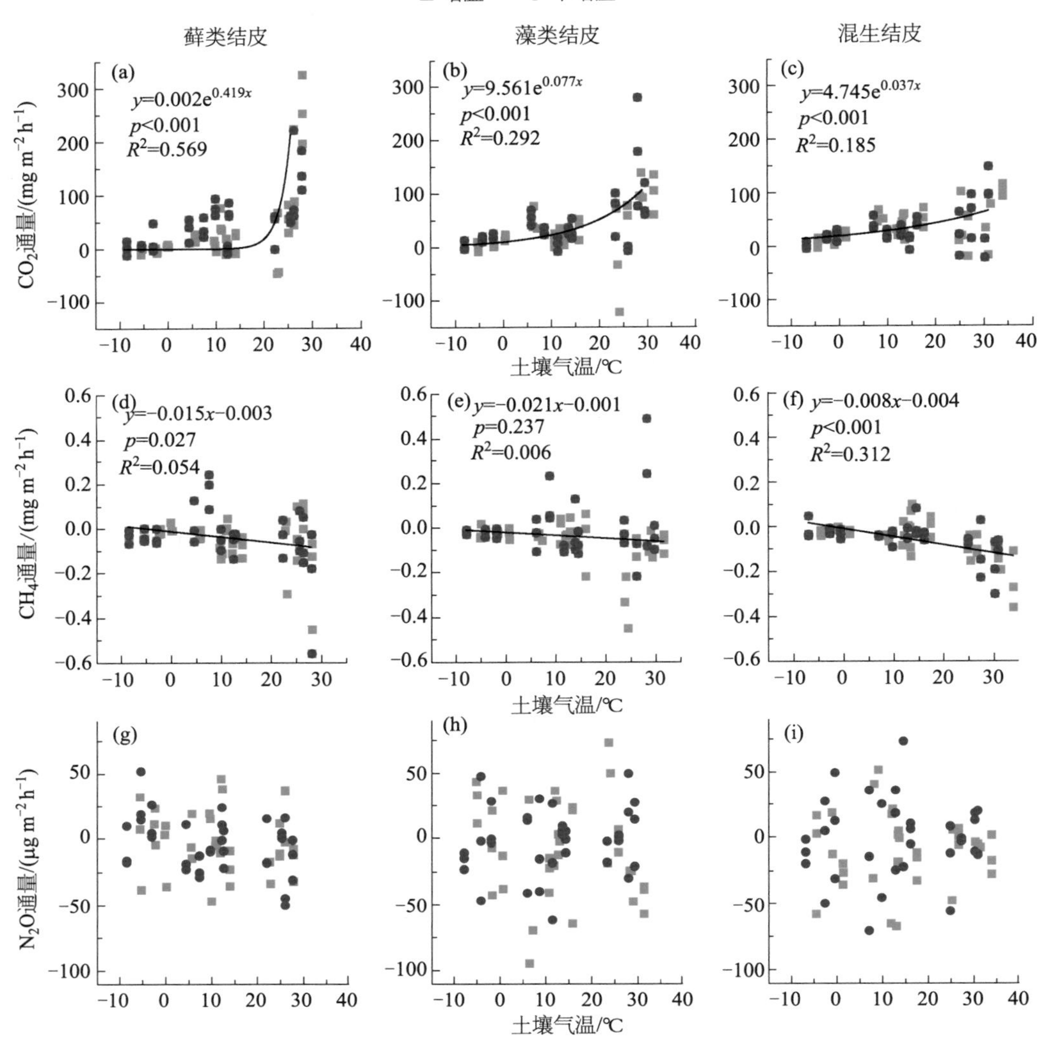

图 3-78　CO_2、CH_4 和 N_2O 通量与 5 cm 深处的土壤温度的关系：(a)、(d)、(g) 藓类结皮；(b)、(e)、(h) 藻类结皮；(c)、(f)、(i) 混生结皮

Figure 3-78　Relationships of CO_2, CH_4 and N_2O fluxes with soil temperature at 5 cm depth：(a)、(d)、(g) moss crusts；(b)、(e)、(h) algae crusts；(c)、(f)、(i) mixed crusts

增温和不增温两种处理下，三种温室气体通量差异与其5 cm深处的土壤温度差异的相关关系分析结果表明，藓类结皮CO_2和CH_4通量差异与温度差异呈显著线性负相关关系（表3-11）。表明随着温差的增加，藓类结皮的CO_2通量增加幅度和CH_4吸收通量的减小幅度逐渐减小，其他两种BSC类型的CO_2和CH_4通量变化程度与温度差异不相关；温度差异与藻

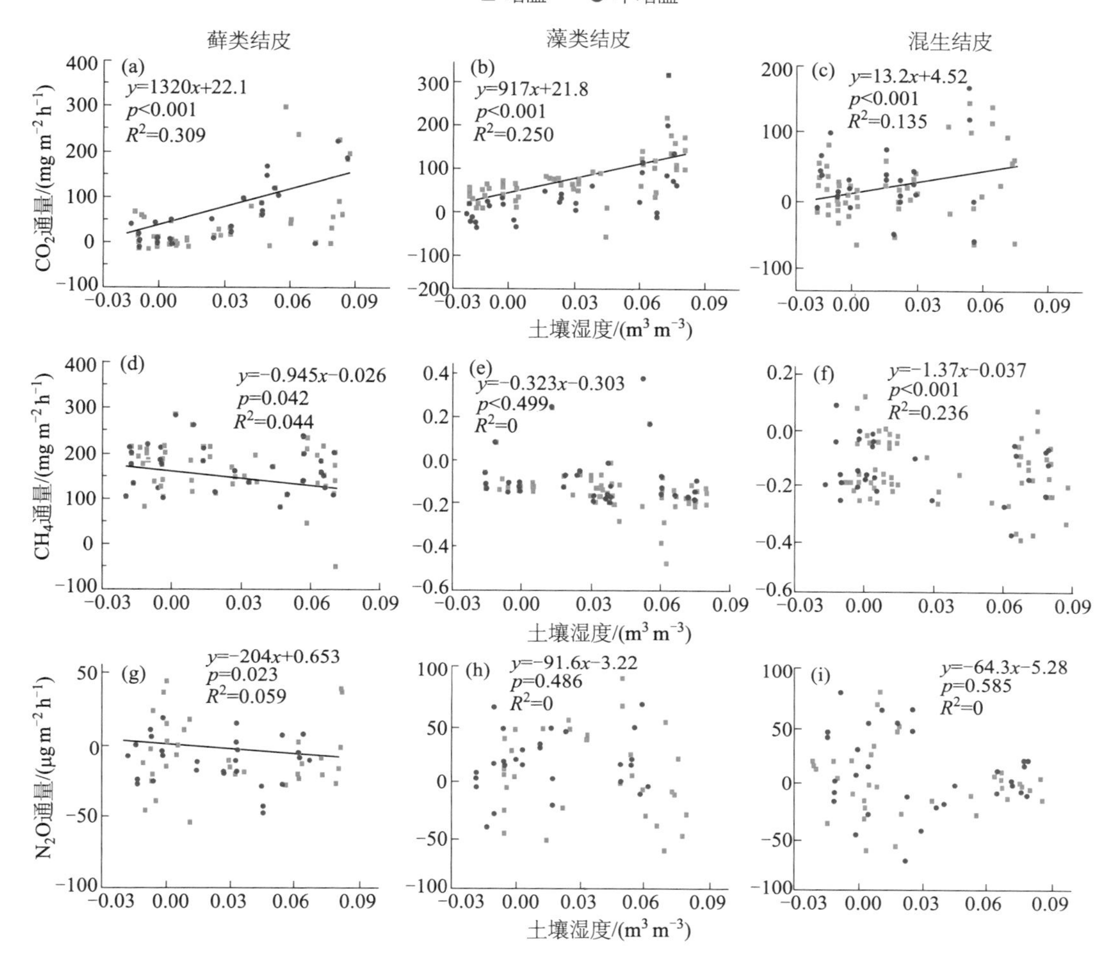

图 3-79　CO_2、CH_4 和 N_2O 通量与 10 cm 深处的土壤湿度的关系：(a)、(d)、(g) 藓类结皮；(b)、(e)、(h) 藻类结皮；(c)、(f)、(i) 混生结皮

Figure 3-79　Relationships of CO_2, CH_4 and N_2O fluxes with soil moisture at 10 cm depth：(a)、(d)、(g) moss crusts；(b)、(e)、(h) algae crusts；(c)、(f)、(i) mixed crusts

类结皮N_2O通量的差异呈近似正相关关系（p=0.051），与其他两种BSC类型的N_2O通量差异不相关。说明温差越大，藻类结皮的N_2O通量变化越大。

综上所述，增温对荒漠BSC–土壤系统CO_2和CH_4的影响因BSC类型的不同而异。然而，增温对各BSC年均CO_2、CH_4、N_2O通量和GWP值无显著影响。表明全球变暖背景下，荒漠BSC–土壤系统CO_2、CH_4和N_2O通量不会有显著的改变。土壤温度与荒漠BSC–土壤系统生态系统呼吸（Re）通量的显著正相关主要表现在季节变化上；增温幅度对Re和CH_4通量

表 3-11　增温和不增温处理下 5 cm 深度土壤温度差异与 CO_2、CH_4 和 N_2O 通量差异的回归分析

Table 3-11　Regressions of the differences in soil temperature at 5 cm depth with the differences in fluxes of CO_2, CH_4 and N_2O between warming and non-warming treatments

BSC类型	CO_2			CH_4			N_2O		
	线性方程	p	R^2	线性方程	p	R^2	线性方程	p	R^2
藓类结皮	y=−19.5x+25.3	0.036	0.097	y=−0.043x+0.059	0.018	0.128	y=−6.76x+13.6	0.141	0.035
藻类结皮	y=10.9x−26.1	0.216	0.016	y=0.027x−0.099	0.264	0.008	y=13.1x−27.5	0.051	0.081
混生结皮	y=−3.35x+12.8	0.487	0.000	y=0.022x−0.043	0.125	0.040	y=2.97x−11.4	0.602	0.000

的变化贡献不大；土壤温度与荒漠BSC-土壤系统的N_2O通量无显著相关关系。土壤湿度与CO_2通量显著正相关，与CH_4和N_2O通量之间的关系随BSC类型的不同而呈弱负相关或不相关。土壤湿度可能是决定CO_2通量“源－汇”关系的主要环境因子，而土壤温度并非是决定荒漠BSC-土壤系统三种温室气体“源－汇”关系的主要环境因子。

3.3.3　温度和湿度对干旱过程中荒漠BSC-土壤系统的硝化作用的影响

BSC对荒漠氮循环中的氮矿化过程有重要的调节作用，其中的藻类结皮具有较强的固氮作用（张鹏等，2011b）。然而，不同类型结皮覆盖的土壤对氮硝化作用的影响及其在干旱过程中对不同温度和水分的响应尚不明确。因此，我们以腾格里沙漠东南缘天然植被区发育良好的两种典型BSC（藓类结皮和藻类结皮）下的土壤和流沙为研究对象，研究不同温度（15 ℃和25 ℃）和不同初始湿度（10%和25%质量含水量）条件下的干旱过程中硝化作用的变化特征，以揭示不同类型BSC覆盖的荒漠土壤在干旱过程中的硝化作用及其对温度和湿度的响应。

2013年8月中旬，用5 cm直径的PVC管，采集典型的藓类结皮、藻类结皮覆盖土壤和流沙共三种土壤类型0～10 cm的原状土。样品带回实验室后在室内放置15天，以保证土壤初始含水量的一致性。供试土壤通过加蒸馏水调到10%和25%的质量含水量后，在15 ℃和25 ℃的恒温培养箱（SPX-250B-Z，上海）中进行敞口连续培养，每种处理4个重复。在培

养后的第1、2、5、8、12和20天，测定土壤中的硝态氮（NO_3^-−N）含量，共计288个样品（2个温度处理 ×2个湿度处理 ×3种土壤类型 ×4个重复 ×6次采样）。培养期间每隔24 h称取原状土的质量，每处理3个重复，共计36个样品（2个温度处理 ×2个湿度处理 ×3种土壤类型 ×3个重复）。

（1）土壤含水量与水分散失速率的变化

三种土壤类型的含水量随着培养时间的增加而逐渐下降（图3-80）。除了25 ℃和25%

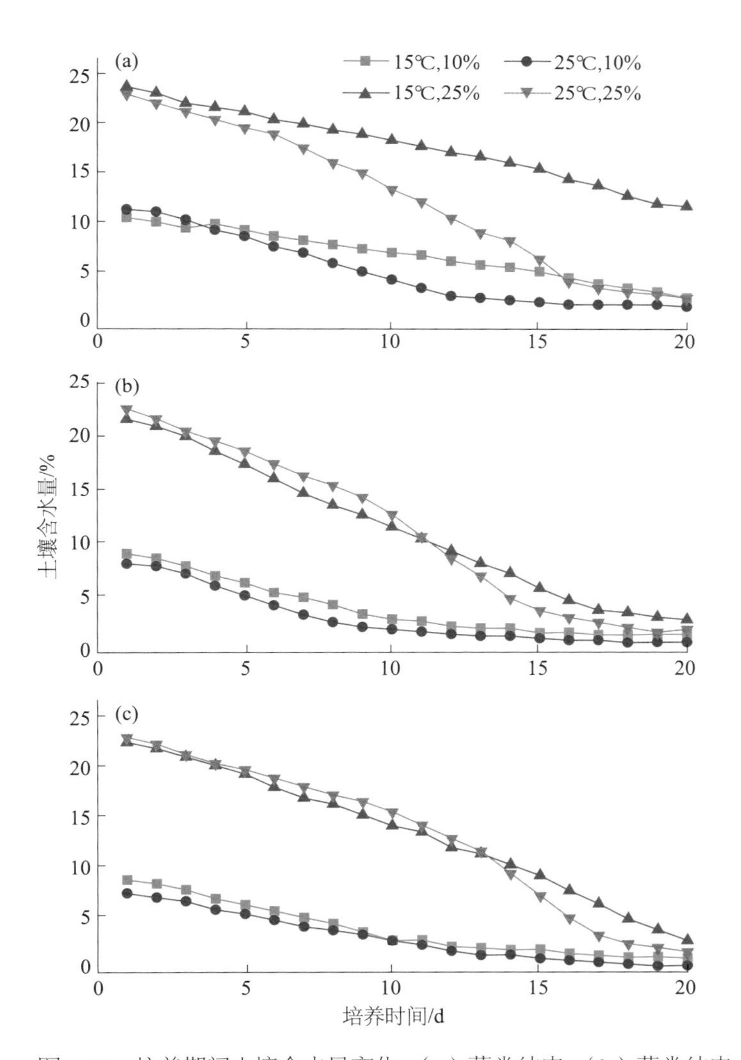

图3-80　培养期间土壤含水量变化：（a）藻类结皮；（b）藓类结皮；（c）流沙

Figure 3-80　Variation of soil water content during incubation：（a）soil covered by algae；（b）soil covered by moss；（c）sand

湿度处理外，其他3种处理下水分散失速率按藻类结皮、藓类结皮和流沙的顺序逐渐增加。15 ℃ 10%和25%湿度培养条件下，藻类结皮、藓类结皮和流沙的水分散失速率分别为0.4% d^{-1}、0.6% d^{-1}、0.7% d^{-1}和0.8% d^{-1}、1.1% d^{-1}、1.5 % d^{-1}。25 ℃ 10%和25%湿度培养条件下，藻类、藓类结皮的水分散失速率分别比15 ℃增加了28.1%、24.2%和99.8%、0.3%，25 ℃下10%湿度的流沙水分散失速率比15 ℃增加15.8%，而25%湿度的流沙减少4.1%。说明低土壤湿度（10%）条件下，培养温度的增加明显加速了三种土壤类型的水分散失速率；而在高湿度下（25%），培养温度的增加对土壤水分散失速率的影响因结皮类型而异，温度增加显著加快了藻类结皮的水分散失速率，而藓类结皮和流沙的变化不显著（图3-81）。

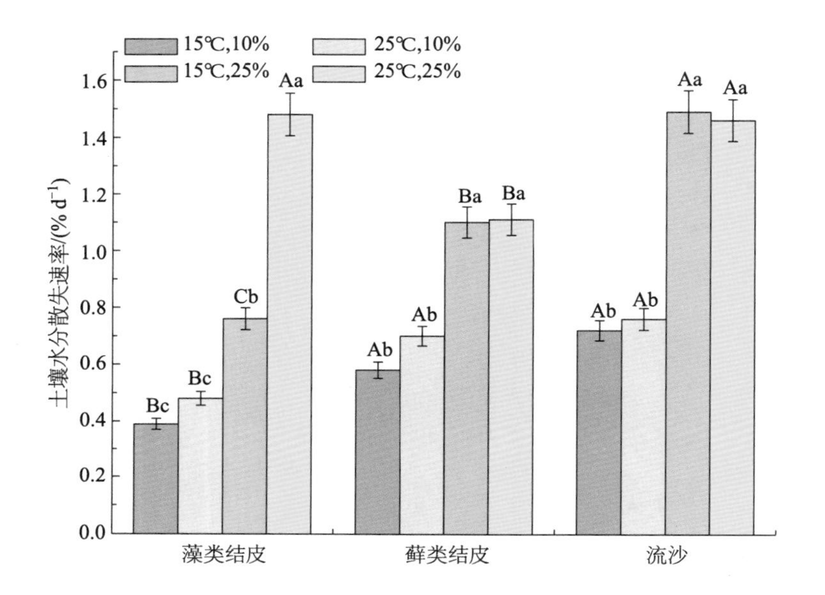

图3-81　土壤水分散失速率。不同小写字母表示同种土壤类型不同处理差异显著，$p<0.05$；不同大写字母表示相同处理不同土壤类型差异显著，$p<0.05$

Figure 3-81　Loss rate of soil water.Different lowercase letters mean significant difference among different treatments for the same soil type at 0.05 level, and different capital letters mean significant difference among three soil types under the same treatment at 0.05 level

（2）硝态氮含量变化

如表3-12所示，培养时间和土壤类型显著影响NO_3^-–N含量，培养温度和土壤湿度对NO_3^-–N含量没有显著影响。培养温度和湿度两者之间的互作未达到显著水平；而培养时间与土壤类型之间及两者分别与土壤湿度、温度之间两者或三者之间的互作部分达到显著水平。

表 3-12　硝态氮含量和净硝化速率多因素方差分析

Table 3-12　Multivariate analysis of NO_3^-−N content and net nitrification rate

因素	硝态氮含量		净硝化速率	
	F	*p*	*F*	*p*
培养时间(D)	9.87	<0.001	6.160	<0.001
培养温度(T)	0.20	0.656	0.028	0.866
土壤湿度(M)	2.64	0.106	0.029	0.865
土壤类型(t)	11.30	<0.001	1.006	0.367
D×T	0.92	0.470	1.330	0.252
D×M	6.36	<0.001	2.150	0.061
D×t	15.90	<0.001	69.200	<0.001
T×M	0.02	0.8788	0.001	0.996
T×t	13.70	<0.001	0.345	0.709
M×t	2.64	0.074	0.521	0.595
D×T×M	1.68	0.023	1.250	0.288
D×T×t	3.41	<0.001	4.510	0.010
D×M×t	2.39	0.011	1.500	1.140
T×M×t	5.87	0.003	0.199	0.820
D×T×M×t	3.97	<0.001	1.380	0.191

培养时间对NO_3^-−N含量的影响随BSC类型和培养条件的变化而变化（图3-82）。例如，随着培养时间的增加，15 ℃和25%湿度条件下的藻类结皮NO_3^-−N含量呈逐渐增加的趋势，培养的第12天NO_3^-−N含量（3.74 mg kg^{-1}）显著高于第1天和第2天；而15 ℃和25%湿度条件下的藓类结皮NO_3^-−N含量在培养后的第12天达到峰值（7.58 mg kg^{-1}），显著高于培养后的第1天、第2天和第20天（图3-82和表3-13）。25 ℃和25 %湿度条件下流沙的NO_3^-−N含量呈先下降后逐渐升高的趋势，培养后的第2天（0.277 mg kg^{-1}）显著低于其他培养时间，其他培养时间之间差异不显著。温度和湿度对NO_3^-−N含量的影响因结皮类型而异。15 ℃ 10%和25%湿度培养条件下，藻类、藓类和流沙的平均NO_3^-−N含量分别为2.38 mg kg^{-1}、2.18 mg kg^{-1}、0.98 mg kg^{-1}和2.13 mg kg^{-1}、2.71 mg kg^{-1}、1.21 mg kg^{-1}。与15 ℃相比，25 ℃下10%湿度的藻类结皮平均NO_3^-−N含量减少47.5%，藓类结皮和流沙分别增加14.1%和71.3%，而

25%湿度的藻类结皮和藓类结皮平均NO_3^--N含量分别减少24.9%和33.8%，流沙增加107%。与低温低湿条件（15 ℃和10%）相比，增温增湿条件下（25 ℃和25%）藻类结皮和藓类结皮平均NO_3^--N含量分别减少32.9%和18.0%，流沙显著增加156%（$p<0.05$）。

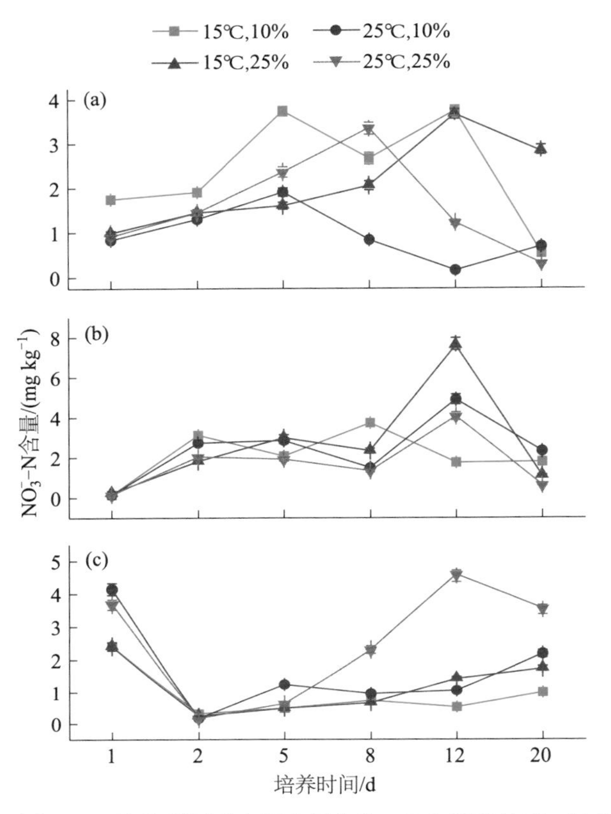

图3-82 培养期间硝态氮含量变化：（a）藻类结皮；（b）藓类结皮；（c）流沙

Figure 3-82 Variation of NO_3^--N content during incubation :（a）algae crust ;（b）moss crusts ;（c）sand

表3-13 硝态氮含量的多重比较结果

Table 3-13 Results of multi-comparison of NO_3^--N content

土壤类型	温度/℃	湿度/%	培养时间/d					
			1	2	5	8	12	20
藻类结皮	15	10	Aab	Aab	Aa	Aab	Aa	Ab
	25	10	Aab	Aa	Ba	Abc	Bc	Abc
	15	25	Ab	Ab	Bab	Aab	Aa	Aab
	25	25	Aab	Aab	ABab	Aa	Bab	Ab

续表

土壤类型	温度 /℃	湿度 /%	培养时间 /d					
			1	2	5	8	12	20
藓类结皮	15	10	Cb	Aab	Aab	Aa	Aab	Aab
	25	10	Aa	Aa	Aa	Ba	Aa	Aa
	15	25	BCb	Ab	Aab	Bab	Aa	Ab
	25	25	ABb	Aab	Aab	Bb	Aa	Aab
流沙	15	10	Aa	Ab	Bb	Bb	Ab	Ab
	25	10	Aa	Ab	Ab	Ba	Ab	Ab
	15	25	Aa	Aa	Ba	Ba	Aa	Aa
	25	25	Aa	Ab	Bab	Aab	Aab	Aa

注：不同小写字母表示同种土壤类型、相同处理下不同培养时间差异显著（$p<0.05$），不同大写字母表示同种土壤类型、同一培养时间不同处理间差异显著（$p<0.05$）。表 3-14 同。

NO_3^--N含量因结皮类型而异（图3-83）。藻类结皮、藓类结皮和流沙在培养期间所有处理的平均NO_3^--N含量分别为1.84 mg kg^{-1}、2.29 mg kg^{-1}和1.59 mg kg^{-1}。藓类结皮与藻类结皮之间差异不显著，但与流沙之间差异显著（$p<0.05$）。

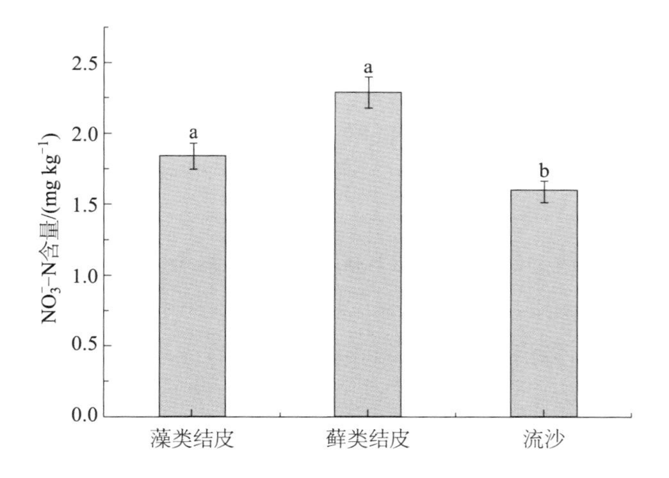

图 3-83　不同类型土壤的所有处理的平均硝态氮含量
Figure 3-83　Average NO_3^--N content of different soil types

（3）净硝化速率的变化

如表3-14所示，培养时间显著影响土壤净硝化速率，培养温度、土壤湿度和土壤类型对净硝化速率的影响不显著。培养时间与土壤类型两者之间以及培养时间、培养温度和土壤类型三者之间的互作对净硝化速率的影响均达到显著水平。

表3-14 净硝化速率多重比较结果

Table 3-14 Results of multi-comparison of net nitrification rate

土壤类型	温度/℃	湿度/%	培养时间/d					
			1	2	5	8	12	20
藻类结皮	15	10	Aabc	Aabc	Aa	Abc	Aab	Bc
	25	10	Ab	Aa	Bb	Ab	Ab	Ab
	15	25	Ab	Aa	Aab	Aab	Aa	Bab
	25	25	Ab	Aab	Aa	Aa	Bab	Ab
藓类结皮	15	10	Cc	Aa	Bb	Ab	Ab	Ab
	25	10	Ac	Aa	Ab	Bb	Aab	Ab
	15	25	BCc	Aa	Aab	Bb	Aab	Ab
	25	25	Ad	Aa	Ac	Bc	Ab	Ac
流沙	15	10	Aa	Ab	Ba	Ba	Aa	Aa
	25	10	Aa	Ab	Aa	Ca	Aa	Aa
	15	25	Aa	Ab	Ba	Ba	Aa	Aa
	25	25	Aa	Ac	Bb	Aab	Ab	Ab

从图3-84可以看出，培养时间对净硝化速率的影响随着BSC类型和培养条件的不同而变化。例如，25 ℃和10%湿度条件下藻类结皮的净硝化速率于培养第2天达到峰值（0.77 mg kg^{-1} d^{-1}），显著高于其他培养时间；而藓类结皮在25 ℃和10%湿度条件下的净硝化速率呈先升高后下降的趋势，培养第2天的净硝化速率（2.43 mg kg^{-1} d^{-1}）显著高于第1、5、8和20天（图3-84和表3-14）。在15 ℃和25%湿度条件下，流沙的净硝化速率变化规律为先下降再升高后无明显变化，培养第2天的净硝化速率（−2.01 mg kg^{-1} d^{-1}）显著低于其他培养时间。温度和湿度对净硝化速率的影响随着BSC类型的变化而不同。15 ℃下，10%和25%湿度培养条件下，藻类结皮、藓类结皮和流沙的平均净硝化速率分别为0.07 mg kg^{-1} d^{-1}、

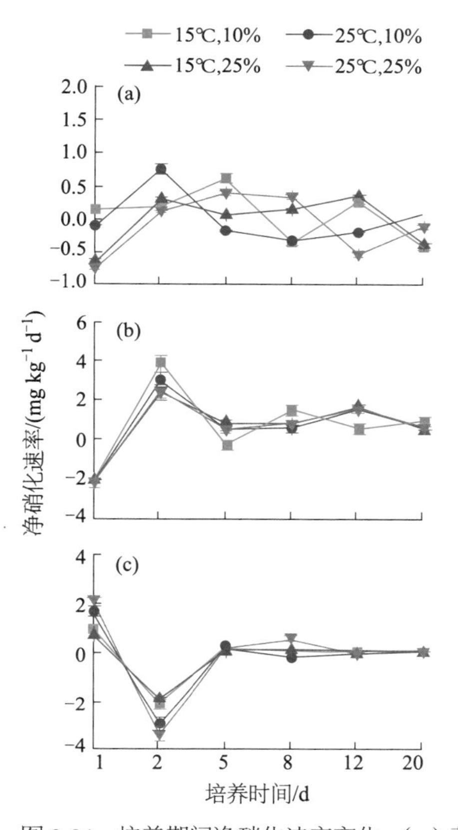

图 3-84　培养期间净硝化速率变化：（a）藻类结皮；（b）藓类结皮；（c）流沙
Figure 3-84　Variation of net nitrification rate during incubation：（a）algae crusts；（b）moss crusts；（c）sand

0.02 mg kg^{-1} d^{-1}、−0.18 mg kg^{-1} d^{-1}和−6.97 × 10^{-4} mg kg^{-1} d^{-1}、−0.09 mg kg^{-1} d^{-1}、−0.15 mg kg^{-1} d^{-1}；25 ℃下，10%湿度的藻类结皮和藓类结皮平均净硝化速率分别比15 ℃减少75.1%和0.7%，流沙增加5.0%；25%湿度的藻类结皮和藓类结皮平均净硝化速率分别减少99.1%和21.3%，流沙增加42.3%。与增温增湿培养条件相比，低温低湿培养条件下藻类结皮和藓类结皮平均净硝化速率分别增加193.4%和107.3%，流沙减少109.6%。

培养期间所有处理的藻类结皮和藓类结皮覆盖土壤以及流沙的平均净硝化速率无显著差异（p>0.05，图3-85），藻类结皮覆盖土壤的平均净硝化速率为0. 003 mg kg^{-1} d^{-1}，分别比藓类结皮覆盖土壤和流沙高100.2%和102.0%。

综上所述，干旱过程中，荒漠BSC−土壤系统的NO_3^-−N含量受培养时间和土壤类型的显著影响，而净硝化速率仅受培养时间的显著影响。培养温度和湿度对NO_3^-−N含量和净硝化速率均没有显著影响。藓类结皮的NO_3^-−N含量高于藻类结皮和流沙，而净硝化速率低于藻类结皮。在增温以及增温增湿条件下，藓类结皮和藻类结皮的净硝化速率均降低。这表明在全球变暖背景下，无论降雨增加或减少，荒漠BSC－土壤系统干旱过程中的净硝化作用将会在一定程度上受到抑制，但不会有显著的变化。

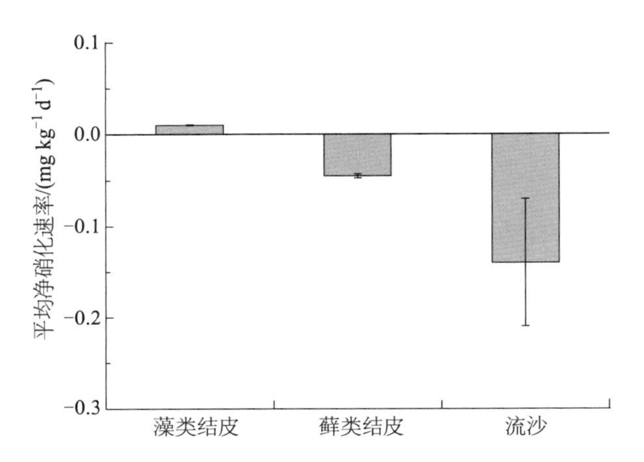

图 3-85　不同类型土壤所有处理的平均的净硝化速率
Figure 3-85　Average net nitrification rate of different soil types

3.3.4　冬季低温及模拟升温对BSC固氮活性的影响

在湿润和光照条件下，温度是控制BSC固氮活性的关键因子（Belnap and Lange，2003；张鹏等，2011a）。但有关极端低温及模拟升温对BSC固氮活性的影响目前尚不明确，尤其缺乏来自野外的实验数据。我们以腾格里沙漠东南缘沙坡头地区人工固沙植被区和相邻天然植被区广泛发育的两类BSC为研究对象，利用OTC增温，采用乙炔还原法（ARA）研究了冬季低温及短期模拟升温对BSC固氮活性的影响，分析并阐述了其影响机理。本研究可为深入认识和准确评价全球变化背景下BSC对区域生态系统的氮贡献提供理论依据和基础数据。

采用样线法在沙坡头人工固沙植被区（1964年建植）和相邻红卫天然植被区选择BSC发育良好的区域采集典型BSC样品（藻类结皮和藓类结皮），采样时间为2010年10月中旬。为避免地貌和灌木及草本植物对样品的影响，采样时选择各种类型BSC均有发育的迎风坡和垄间低地，距离灌丛2 m外的区域采集BSC样品。为保证样品的完整性，采样前先湿润地表，之后用PVC管（直径10 cm，高10 cm）采集原状BSC土壤样品。将野外采集的原状样品带回沙坡头沙漠试验研究站气候观测场摆放在不同规格的OTCs（大OTC：边长1.3 m、高2 m八棱柱；小OTC：边长1 m、高1.5 m长方体）内中心位置及OTC外自然环境下（对照，

CK）。两种结皮类型（藻类结皮和藓类结皮），3个温度处理（大、小OTCs和对照），各3个重复。

2010年12月25日、2011年1月16日、2月27日有降雪（降水量分别为1.0 mm、1.2 mm和0.4 mm），等样品表面积雪融化后，将有机玻璃罩（直径12 cm，高15 cm，下端开口，上端密封，中间留一圆孔用橡皮塞密封）罩在样品上，下端插入沙子中，创造一个密闭环境。之后向培养器中注入乙炔气体，使容器乙炔体积百分比（*V*/*V*）为10%。在自然状态下培养4 h，同时每隔1 h记录大、小OTCs内及室外的空气、地表、地下5 cm和10 cm的温度。培养结束后用注射器收集培养器顶部气体5 mL。收气结束后即刻去除收气罩。将收集的气体带回实验室后即刻用气相色谱仪（Agilent GC6820，美国）测定乙烯生成量。BSC固氮活性结果以乙烯生成速率（nmol C_2H_4 m^{-2} h^{-1}）表示。

（1）不同规格OTCs冬季模拟升温效果

试验期间（2010年12月—2011年2月）不同处理下全天平均气温在−12.6～0.7 ℃变化，平均为−4.7 ℃；大、小OTCs和CK处理下的全天平均气温分别为−4.4 ℃、−4.6 ℃和−5.2 ℃，大、小OTCs分别比CK升温0.8 ℃和0.6 ℃（图3-86）。白天（9：00—19：00）平均气温在−6.6～4.0 ℃变化，平均为0.4 ℃，大、小OTCs和CK下的白天平均气温分别为0.4 ℃、1.1 ℃和0.1 ℃，大、

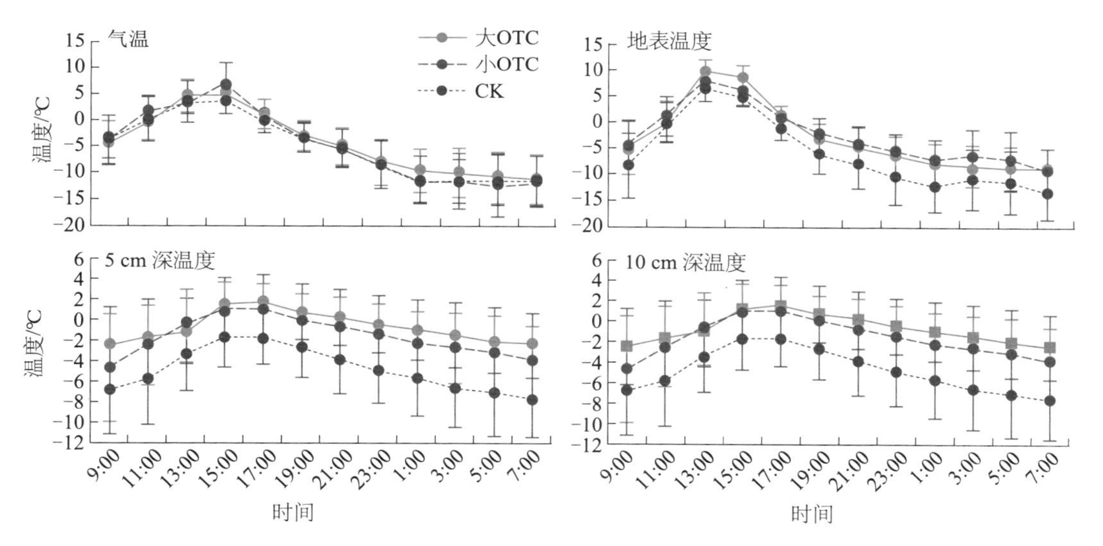

图3-86 不同规格OTCs及CK处理下冬季空气温度与土壤温度的全天变化（平均值 ± 标准误）
Figure 3-86 Variation of air temperature and soil temperature (mean ± SE) of different depths of different OTCs and CK in winter

小OTCs分别比CK升温0.3 ℃和1.0 ℃。夜间（21:00—7:00）平均气温在−19.7～−4.4 ℃变化，平均为−9.8 ℃，大、小OTCs和CK下的夜间平均气温分别为−9.1 ℃、−9.8 ℃和−9.9 ℃，大、小OTCs分别比CK升温0.7 ℃和0.1 ℃。大OTC白天升温幅度小于夜间，小OTC则正好相反。整个试验期大、小OTCs处理下地表、地下5 cm和10 cm温度分别升高了3.1 ℃、3.4 ℃、3.0 ℃和3.2 ℃、3.0 ℃、2.5 ℃（图3-87）。

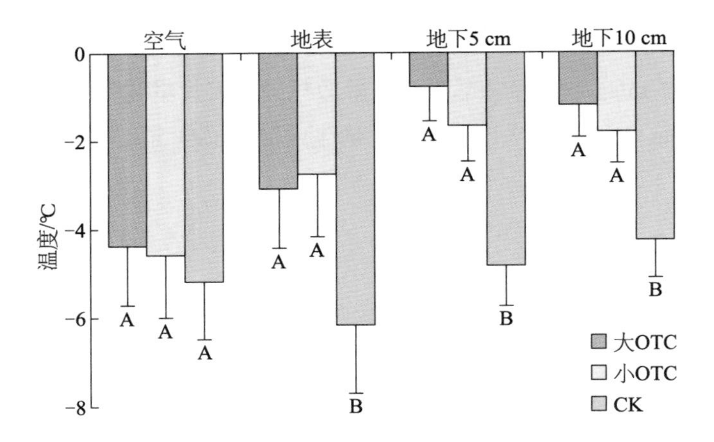

图3-87　试验期OTCs内外空气温度与不同深度土壤温度的变化。数据为平均值 ± 标准误，不同大写字母表示空气、地表、地下5 cm和10 cm温度在不同处理下差异显著，$p<0.05$

Figure 3-87　Variation of air temperature, soil surface temperature and soil temperature (mean ± SE) at 5 and 10 cm depth of different OTCs and CK in winter.Different capital letters indicate significant difference in temperature under different treatments, $p<0.05$

（2）冬季模拟升温对BSC固氮活性的影响

整个试验期，培养期（11:00—15:00）平均气温在−1.6～5.3 ℃变化，大、小OTCs分别比CK平均升温1.0 ℃和1.9 ℃（图3-88）。试验期藻类结皮和藓类结皮的平均固氮活性分别为1.2×10^4 nmol C_2H_4 m^{-2} h^{-1}和0.4×10^4 nmol C_2H_4 m^{-2} h^{-1}，藻类结皮的固氮活性显著高于藓类结皮（$p<0.01$）。自然植被区藻类结皮和藓类结皮平均固氮活性均略高于人工植被区，但差异不显著（$p>0.05$）。模拟升温对藻类结皮固氮活性的影响不显著，但升温处理下其固氮活性均有所提高，大、小OTCs处理下藻类结皮的平均固氮活性是CK的1.1倍和1.3倍。大、小OTCs处理下藓类结皮的固氮活性显著高于CK（$p<0.05$），是CK的1.4倍和1.5倍（图3-88）。

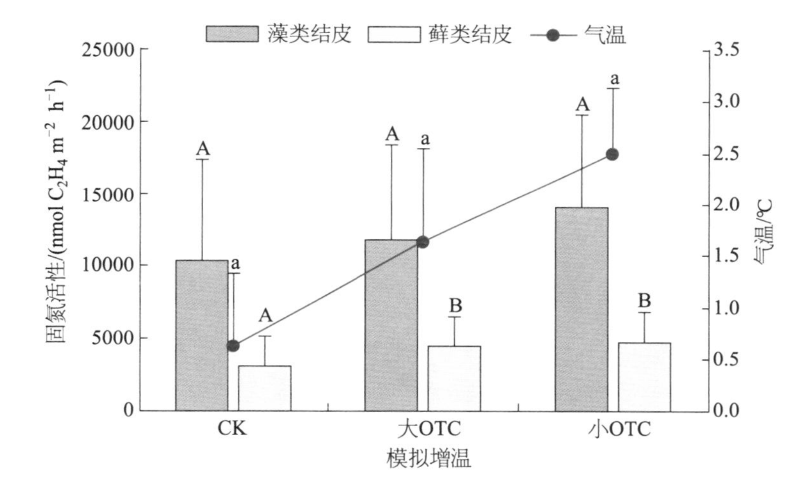

图3-88　模拟升温对BSC固氮活性的影响及培养期气温的变化

Figure 3-88　Effects of simulated warming on nitrogen-fixating activity on BSC and air temperature changes during incubation in winter

（3）BSC固氮活性与温度和降水的关系

对试验期实验前3天的降水量及培养期气温与BSC固氮活性的相关分析发现，藻类结皮和藓类结皮固氮活性均与培养期气温显著正相关（$r_{藻}$=0.93，p<0.001；$r_{藓}$=0.84，p<0.001）（图3-89a）；与试验前3天降水量也均呈显著正相关关系（$r_{藻}$=0.69，p<0.001；$r_{藓}$=0.67，p<0.001）（图3-89b）。

腾格里沙漠东南缘广泛分布的藻类结皮和藓类结皮在冬季降雪发生后湿润条件下均具有固氮能力，藻类结皮的固氮活性显著高于藓类结皮。不同规格OTCs装置的冬季模拟升温效果显著，试验期两类BSC固氮活性与培养期气温和试验前3天降水量均呈显著正相关。在冬季湿润条件下，升温能促进BSC固氮活性的提高。冬季低温冷冻环境下BSC生物体胞内冰晶形成而导致的固氮酶体系受损可能是造成BSC固氮活性较低的主要原因。以上结果说明，在未来全球变暖和降水格局变化背景下，冬季变暖能促进BSC对区域生态系统的氮贡献。

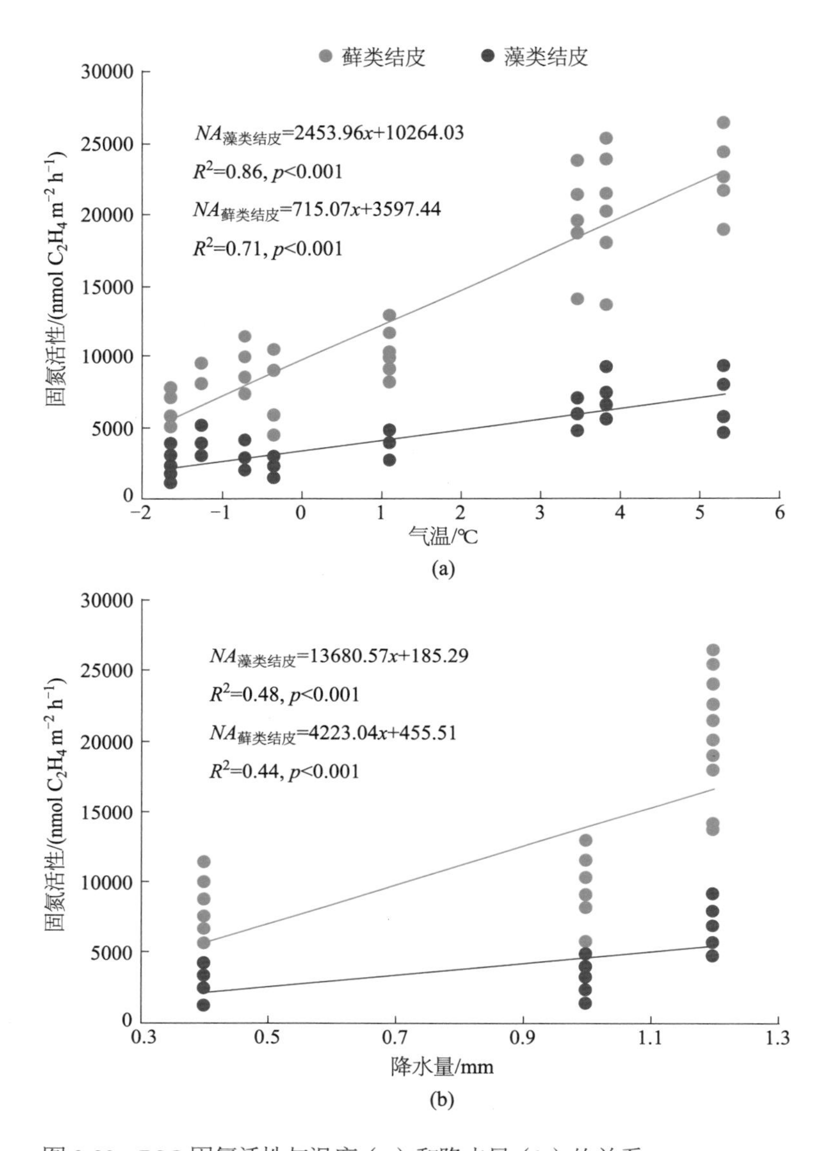

图 3-89　BSC 固氮活性与温度（a）和降水量（b）的关系

Figure 3-89　Relationship between air temperature (a), precipitation (b) and nitrogenase activity of BSC

3.4 BSC对UV-B辐射的生态生理响应

近年来， 由于人类活动而产生大量氯氟烃（chlorofluorocarbons，CFCs） 和氮氧化物，这些化合物长期存留在大气中导致臭氧层衰减，使到达地面的紫外辐射，尤其是波长280～315 nm的UV-B辐射增强（Laube *et al.*，2014）。有研究表明，平流层臭氧浓度每减少1 %，到达地表的UV-B辐射将增加2%（McKenzie *et al.*，2007）。尽管到达地面的UV-B辐射仅占太阳短波辐射的1.5%，但易被一些重要的生物大分子（核酸、蛋白质等）有效吸收（Sicora *et al.*，2006），引起植物的生长发育和生理生化过程发生一系列的变化。目前UV-B辐射增强对植物生态生理特性的影响已成为全球变化研究领域的前沿和热点问题之一（李元等，2006；Pradhan *et al.*，2008；Jansen *et al.*，2010）。

在荒漠地区，高强度的UV-B辐射是影响隐花植物生存及其结皮形成的重要环境因子（饶本强等，2011）。但由于构成结皮的隐花植物形体微小、种类鉴定困难而引起的研究技术复杂等因素的限制，导致UV-B辐射增强对结皮层隐花植物影响的研究仅见少量报道。仅有的研究也多集中在极地、亚极地、北美热带荒漠等地区（Belnap *et al.*，2007）。研究表明，结皮层不同种类隐花植物对UV-B辐射的响应不同。如Gehrke（1999）通过室内模拟亚极地地区UV-B辐射增强30%，探讨对大金发藓（*Polytrichum commune* Hedw.）和塔藓（*Hylocomium splendens* Hedw.）生物量的影响，发现塔藓生物量持续降低，对UV-B辐射反应敏感，而大金发藓抵抗UV-B辐射的能力较强，生物量基本没有受到影响。Niemi等（2002）比较了UV-B辐射增强条件下波罗的海疣泥炭藓（*Sphagnum papillosum* Lindb.）叶绿素a、叶绿素b和类胡萝卜素含量的变化，发现色素含量相比对照都有所增加。此外，从不同生境BSC分离得到的隐花植物对UV-B辐射的响应也不尽相同。Kitzing等（2014）发现从阿尔卑斯山脉不同海拔结皮层中分离的5株绿藻（*Klebsormidium fluitans*（Streptophyta））对UV-B辐射胁迫的耐受性不同，其中2株绿藻的生长受到UV-B辐射影响，而其余3株绿藻的生长未受影响。Csintalan等（2001）测量了8种藓类植物在UV-B辐射胁迫下叶绿素荧光特性的变化，发现大多数藓类植物对UV-B辐射不敏感，且处于干旱环境中的藓类植物具有较强的UV-B辐射耐受性。

对于我国温性荒漠区发育良好、结构和种类组成复杂多样的隐花植物对UV-B辐射响应和适应机制的研究较少。截至目前，能够检索到的来自国内相关研究论文仅有十余篇，且研究对象主要涉及荒漠藻类和藓类，主要研究内容有：① UV-B辐射增强对隐花植物形态学特征的影响。饶本强等（2011）在室内条件下发现UV-B辐射导致具鞘微鞘藻和爪哇伪枝藻［*Scytonema javanicum*（Kuetz.）Bornet］的细胞超微结构遭到明显破坏，其藻丝体数量减少，缢缩成条索状。② UV-B辐射增强对隐花植物生理生化及分子生物学特性的影响。Wang等（2008）和邓松强等（2012）研究发现UV-B辐射降低了念珠藻（*Nostoc* sp.）的光合活性，而外源抗坏血酸和半胱氨酸可以显著提高UV-B辐射条件下藻体的光合活性，有效保护藻类。Chen等（2009）研究了UV-B辐射对荒漠结皮中具鞘微鞘藻的影响，结果表明随着UV-B辐射增强，具鞘微鞘藻光合活性降低、DNA链损伤程度显著增加，而胞外聚合物具有降低DNA链损伤程度的能力。Xie等（2009）也指出UV-B辐射影响具鞘微鞘藻生长，并导致细胞氧化损伤。苏延桂等（2011）的研究表明，UV-B辐射通过降低荒漠藻的光合色素含量，减少了结皮的净光合速率，从而对荒漠区藻类结皮的生产力产生影响。

3.4.1 BSC中真藓和土生对齿藓对UV-B辐射增强的生态生理响应

考虑到目前对我国干旱、半干旱生态脆弱区BSC对UV-B辐射变化的生态生理响应方面缺乏系统研究，选择腾格里沙漠东南缘人工固沙植被区常见藓类结皮中的真藓和土生对齿藓为材料，通过室内模拟短期UV-B辐射增强实验，系统研究UV-B辐射增强对隐花植物光合特性、渗透调节物质、UV-B吸收物质、抗氧化酶系统及超微结构的影响，从形态学特征、生理生化及分子生物学特性、防御体系等方面揭示隐花植物对UV-B辐射增强的响应和适应机制，拓展国内外对隐花植物及其结皮的研究内涵。

沙坡头地区固沙植被区BSC层藓类植物共2科、7属、16种（表3-15），优势种为真藓和土生对齿藓（图3-90；田桂泉等，2005）。

（1）增强UV-B辐射对真藓和土生对齿藓光合特性的影响

在中国科学院寒区旱区环境与工程研究所植物逆境生理生态与生物技术实验室内进行短期UV-B辐射增强模拟试验（2011年8月21—30日）。将40 W UV-B 313型紫外灯管（上海晨辰照明电器有限公司）悬挂于真藓和土生对齿藓结皮正上方，通过调节结皮顶端距灯管的高度设定4个UV-B辐射强度，分别为2.75 W m^{-2}（对照）、3.08 W m^{-2}、3.25 W m^{-2}和3.41 W m^{-2}（相当于沙坡头地区臭氧损耗0%、6%、9%、12%时所达到的UV-B辐射强度），UV-B辐射强度采用742型紫外辐照计（北京师范大学光电仪器厂）测量。紫外灯管外包裹0.13 mm乙酸纤维素膜（Courtaulds Chemicals，英国），以过滤280 nm以下的波段（Wang *et al.*，2010），为防止乙酸纤维素膜因光解而导致UV-B辐射过滤不均，每隔5天更换一次。在

表 3-15　固沙植被区 BSC 层藓类物种组成及群落分布

Table 3-15　Species composition and community distribution of moss crust in fixed dunes vegetation.

物种		分布和群落组成			
		人工固沙植被区			自然固定沙丘
		1956 年	1964 年	1981 年	
真藓科	Bryaceae				
真藓	*Bryum argenteum*	+++	+++	+++	+++
丛藓科	Pottiaceae				
黑对齿藓	*Didymodon nigrescens*	++	++		++
细叶对齿藓	*D. perobtusus*				+
硬叶对齿藓长尖变种	*D. rigidulus* Hedw. var. *ditrichoides*（Broth.）Zand.				+
芽胞对齿藓	*D. reedii*				+
土生对齿藓	*D. vinealis*（Brid.）Zand.	+++	+++	+++	+++
盐土藓	*Pterygoneurum subsessile*				++
短丝流苏藓	*Crossidium aberrans*				+
绿色流苏藓	*C. chloronotos*				+
钝叶芦荟藓	*Aloina rigida*				+
短喙芦荟藓	*A. brevirostris*				+
刺叶芦荟藓	*A. cornifolia*				+
斜叶芦荟藓	*A. obliquifolia*				+
卷叶墙藓	*Tortula atrovirens*				+
刺叶赤藓	*Syntrichia caninervis*	++			++
双齿赤藓	*S. bidentata*	+++	+++		+++

注：+++ 优势种；++ 伴生种；+ 偶见种。

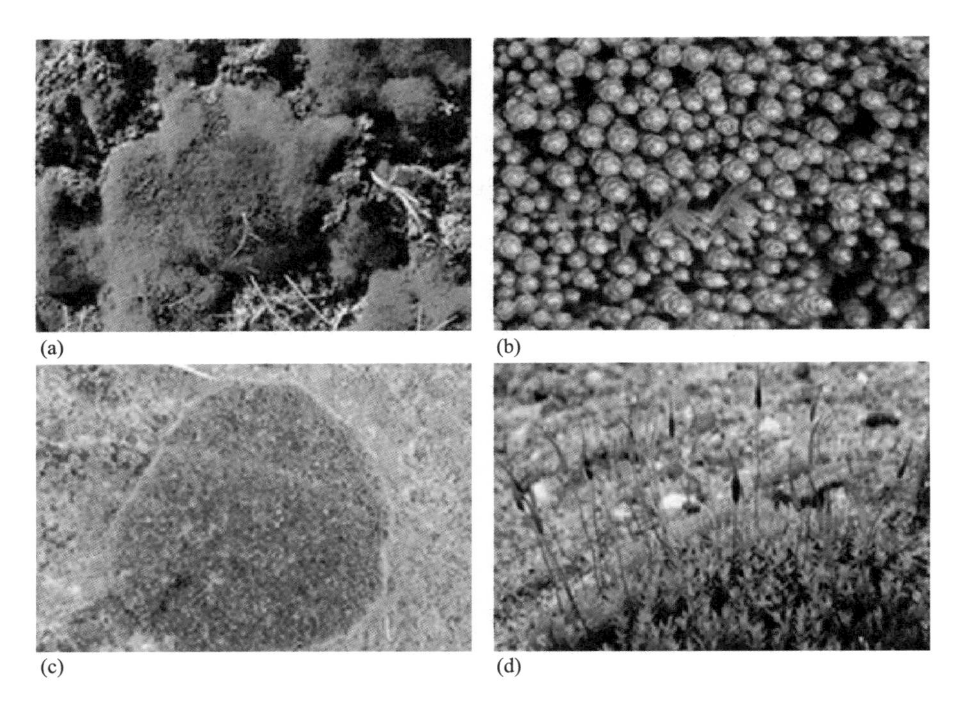

图 3-90　本研究的两种 BSC 类型 ：(a)、(b) 真藓 ；(c)、(d) 土生对齿藓
Figure 3-90　Two types of BSC in this research ：(a)、(b) *B. argenteum* ；(c)、(d) *D. vinealis*

紫外灯管两侧安装40 W日光灯（上海晨辰照明电器有限公司），以提供一定的光合有效辐射（400 ~ 700 nm），强度由光量子传感器（LI-COR，Inc.，美国）测量。在整个实验过程中，培养室内的光合有效辐射为150 μmol $m^{-2}s^{-1}$，温度保持在25 ℃，每天给结皮补充10 mL水分，以提供充足的水分。每个辐射处理重复3次，处理时间8 h（9：00—17：00），连续照射10天，并在第5天和第10天测定真藓和土生对齿藓的叶绿素（Chl-a）荧光诱导动力学参数、光合色素、类黄酮含量；提取UV-B辐射处理10天的真藓和土生对齿藓叶绿体类囊体膜蛋白，通过蓝绿温和聚丙烯酰胺凝胶电泳（BN-PAGE）和SDS-聚丙烯酰胺凝胶电泳（SDS-PAGE）分析UV-B辐射增强过程中类囊体膜蛋白组分的响应特点。

增强UV-B辐射对真藓和土生对齿藓Chl-a荧光诱导动力学参数的影响：光合系统是UV-B辐射最初和最主要的作用靶（Zu *et al.*，2010）。Chl-a荧光是光合作用的良好指标和内在探针，通过对各种荧光参数的分析，可以了解光合机构内部一系列重要的调节过程（Guo *et al.*，2005a）。F_v/F_m、Y、ETR和qP均是反映PS Ⅱ反应中心状态的Chl-a荧光诱导动力学参数。F_v/F_m反映的是PS Ⅱ反应中心均处于开放状态时的量子产量（Guidi *et al.*，2007）。当植物体受到外界环境胁迫时，F_v/F_m一般会降低，表明植物受到了光抑制（Li *et al.*，2010b），光抑制的产生可能与胁迫降低了Cytb6/f的合成速率和D1蛋白被破坏后的修复速率有关（Murata *et al.*，2007）；*Y*反映了用于光化学途径的激发能占进入PS Ⅱ总激发能的比例，是PS Ⅱ反应中心部分关闭时的光化学效率；*ETR*反映了实际光强条件下的表观电子传递速率，其值的下降表明Q_A—Q_B的电子传递受到抑制（梁英等，2009）；*qP*反映的是PS Ⅱ原初电子受体Q_A的氧化还原状态和PS Ⅱ开放中心的数目，*qP*的降低表明PS Ⅱ反应中心受体Q_A的还原程度更高，开放部分比例下降，关闭部分比例提高，这些关闭状态的PS Ⅱ反应中心不能进行稳定的电荷分离，因而不能参与光合电子的传递（李燕宏等，2006）。增强UV-B辐射对真藓Chl-a荧光诱导动力学参数的影响如图3-91所示。F_v/F_m、*Y*、*ETR*和*qP*均表现出随着UV-B辐射强度的增加而显著降低的趋势（$p<0.05$）。当UV-B辐射处理5天后，F_v/F_m、*Y*、*ETR*和*qP*持续减小，且在UV-B辐射强度为3.41 W m^{-2}时分别出现最小值0.329、0.160、14.7 μmol m^{-2} s^{-1}和0.456，与对照相比，分别降低了52.6%（图3-91a）、49.0%（图3-91b）、51.0%（图3-91c）和38.3%（图3-91d）。当UV-B辐射处理10天后，Chl-a荧光诱导动力学参数表现出与处理5天后相似的结果。F_v/F_m、*Y*、*ETR*和*qP*在不同处理间差异显著（$p<0.05$），并随着辐射强度的增大而降幅增大。在UV-B辐射强度达到3.41 W m^{-2}时，各参数受到的抑制最大，与对照相比，F_v/F_m、*Y*、*ETR*和*qP*分别下降了58.7%（图3-91a）、43.6%（图3-91b）、47.6%（图3-91c）和40.2%（图3-91d）。双因素方差分析表明，在对真藓Chl-a荧光诱导动力学参数的影响上UV-B辐射时间和辐射强度的交互作用不显著（$p>0.05$，表3-16），说明辐射时间对真藓的UV-B辐射敏感性影响不大。

对土生对齿藓Chl-a荧光诱导动力学参数变化的研究结果如图3-92所示。增强UV-B辐射显著降低了土生对齿藓的Chl-a荧光诱导动力学参数，且随着UV-B辐射强度的增加，降低幅度越大。在UV-B辐射处理5天后，F_v/F_m分别下降13.82%、33.88%和38.63%（图3-92a）；*Y*下降幅度分别为10.88%、26.53%和35.37%（图3-92b）；*ETR*分别下降20.91%、34.15%和46.34%（图3-92c）；*qP*则分别下降了5.87%、18.18%和24.93%（图3-92d），与对照相比，除UV-B辐射强度在3.08 W m^{-2}时，*Y*和*qP*降低幅度不大，其余均呈显著水平（$p<0.05$）。当UV-B辐射处理土生对齿藓10天后，Chl-a荧光诱导动力学参数也表现出不同程度的降低，各处理组与对照组相比均有显著差异（$p<0.05$）。在辐射强度为3.41 W m^{-2}处理时，各参数受到的抑制最大，F_v/F_m、*Y*、*ETR*和*qP*分别下降了52.02%（图3-92a）、45.95%（图3-92b）、43.17%（图3-92c）和31.99%（图3-92d），表明UV-B辐射导致叶绿体PS Ⅱ活性中心损伤，电子传递受到抑制，原初光能转换效率下降，光合效率降低。双因素方差分析表明，在对土生对齿藓Chl-a荧光诱导动力学参数的影响上UV-B辐射时间和辐射强度不存在交互作用（$p>0.05$，表3-17），5天和10天UV-B辐射处理下，Chl-a荧光诱导动力学参数均极显著地高于对照，说明无论辐射时间长短，UV-B辐射强度对土生对齿藓Chl-a荧光诱导动力学参数均具有显著性影响。

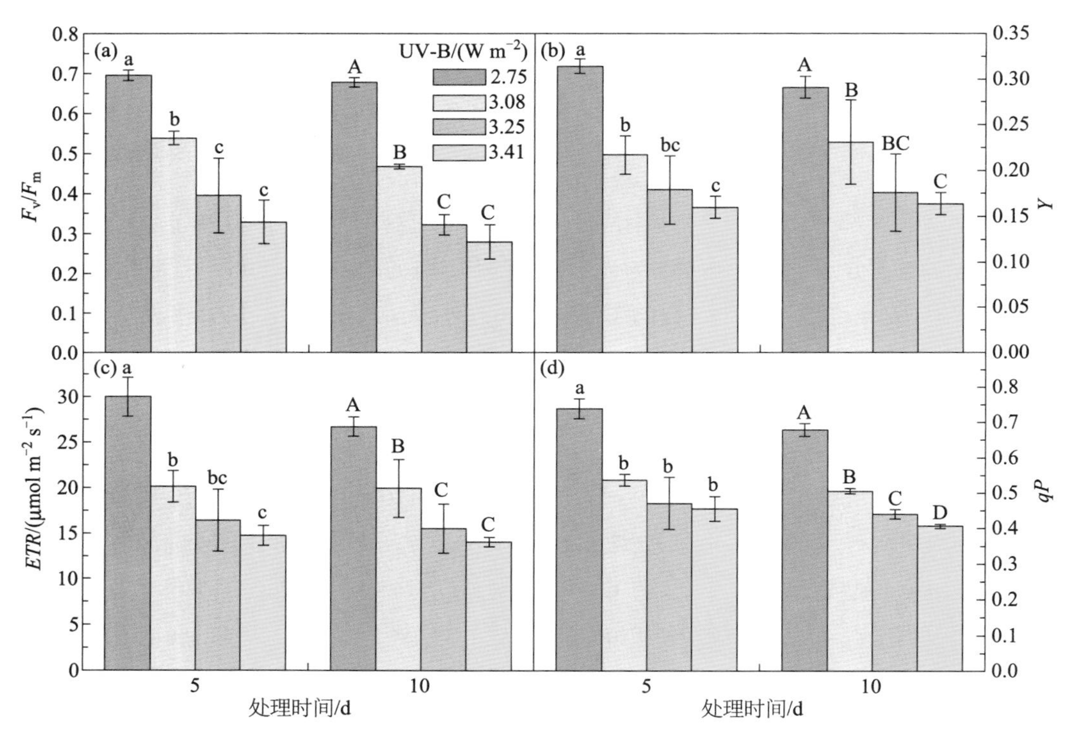

图 3-91　增强 UV-B 辐射对真藓 Chl-a 荧光诱导动力学参数的影响。数据为平均值 ± 标准差，$n=3$；不同字母表示相同处理时间下各参数差异显著，$p<0.05$。图 3-92 ~ 图 3-96 同

Figure 3-91　Effects of enhanced UV-B radiation on chlorophyll a fluorescence induction kinetics parameters of *B. argenteum*. Values are means ± SD，$n=3$；columns with different letters are significantly different，$p<0.05$. It's the same as Figure 3-92 ~ Figure 3-96

两种藓类Chl-a荧光诱导动力学参数随UV-B辐射强度变化可以看出，UV-B辐射使F_v/F_m、Y、ETR和qP呈降低趋势，且随着UV-B辐射强度的增加，降低幅度越大。这与Fabón等（2010）在室内条件下处理水生地钱（*Jungermannia exsertifolia* subsp. Cordifolia）的结果一致，增强UV-B辐射31天导致地钱F_v/F_m显著降低。Surabhi等（2009）研究发现增强UV-B辐射降低了豇豆［*Vigna unguiculata*（L.）Walp.］的ETR。左圆圆等（2005）研究了短期增强UV-B辐射对青榨槭幼苗Chl-a荧光诱导动力学参数的影响，得到相同的结果。说明植物在UV-B辐射胁迫下PS Ⅱ反应中心出现了明显的损伤，发生了光抑制现象，对真藓和土生对齿藓的光合活性带来了直接而严重的损伤。本研究中，UV-B辐射造成Chl-a荧光降低的原因可能是由于核酮糖-1,5-二磷酸羧化酶/加氧酶（Rubisco）含量减少，其CO_2同化能力降低，造成对叶绿体合成的腺苷三磷酸（ATP）、还原型辅酶Ⅱ（NADPH）同化力的需求减少，引起过剩光能的积累，导致PS Ⅱ受到损伤。然而也有一些研究表明增强UV-B辐射对真藓Chl-a荧光没有影响（Green *et al.*，2000）。造成这种差异的原因可能是植物体的生

表 3-16 UV-B 辐射时间和辐射强度对真藓叶绿素光合特性的影响

Table 3-16 Effects of exposure time and UV-B radiation on photosynthetic performances of *B. argenteum* based on two-way ANOVA

	辐射时间 /d		辐射强度 / (W m^{-2})				*p*		
	5	10	2.75	3.08	3.25	3.41	辐射时间	辐射强度	辐射时间 × 辐射强度
F_v/F_m	0.489 ± 0.154	0.437 ± 0.164	0.686 ± 0.014a	0.502 ± 0.040b	0.359 ± 0.072c	0.305 ± 0.051d	0.009	<0.001	0.653
Y	0.218 ± 0.065	0.215 ± 0.060	0.303 ± 0.015a	0.224 ± 0.032b	0.177 ± 0.036c	0.162 ± 0.011c	0.834	<0.001	0.715
ETR/ (μmol m^{-2}s^{-1})	20.3 ± 4.48	19.0 ± 5.47	28.3 ± 2.35a	20.0 ± 2.30b	15.9 ± 2.80c	14.4 ± 0.85c	0.176	<0.001	0.640
qP	0.551 ± 0.124	0.508 ± 0.111	0.710 ± 0.039a	0.522 ± 0.020b	0.456 ± 0.050c	0.431 ± 0.035c	0.005	<0.001	0.830
Chl-a/ (mg g^{-1}FW)	0.318 ± 0.076	0.224 ± 0.126	0.399 ± 0.013a	0.329 ± 0.044b	0.194 ± 0.081c	0.162 ± 0.071d	<0.001	<0.001	<0.001
Chl-b/ (mg g^{-1}FW)	0.099 ± 0.021	0.076 ± 0.031	0.116 ± 0.007a	0.108 ± 0.009b	0.066 ± 0.019c	0.059 ± 0.018d	<0.001	<0.001	<0.001
Chl (a+b) / (mg g^{-1}FW)	0.417 ± 0.097	0.300 ± 0.157	0.514 ± 0.020a	0.437 ± 0.053b	0.260 ± 0.099c	0.221 ± 0.089d	<0.001	<0.001	<0.001
Chl-a/Chl-b	3.185 ± 0.145	2.798 ± 0.516	3.452 ± 0.120a	3.023 ± 0.177b	2.838 ± 0.429c	2.651 ± 0.397d	<0.001	<0.001	<0.001
Car/ (mg g^{-1}FW)	1.806 ± 0.418	1.772 ± 0.856	2.608 ± 0.556a	2.017 ± 0.296b	1.355 ± 0.168c	1.176 ± 0.167d	0.008	<0.001	<0.001
类黄酮 / (μg g^{-1}FW)	1824.9 ± 638.01	1547.1 ± 646.28	2336.0 ± 96.52a	2184.9 ± 417.71b	1296.0 ± 141.86c	927.11 ± 154.27d	<0.001	<0.001	<0.001

注：数据为平均值 ± 标准差，$n \geq 3$；Duncan 法分析不同强度之间的差异显著性，不同字母表示差异显著，$p<0.05$。表 3-17 同。

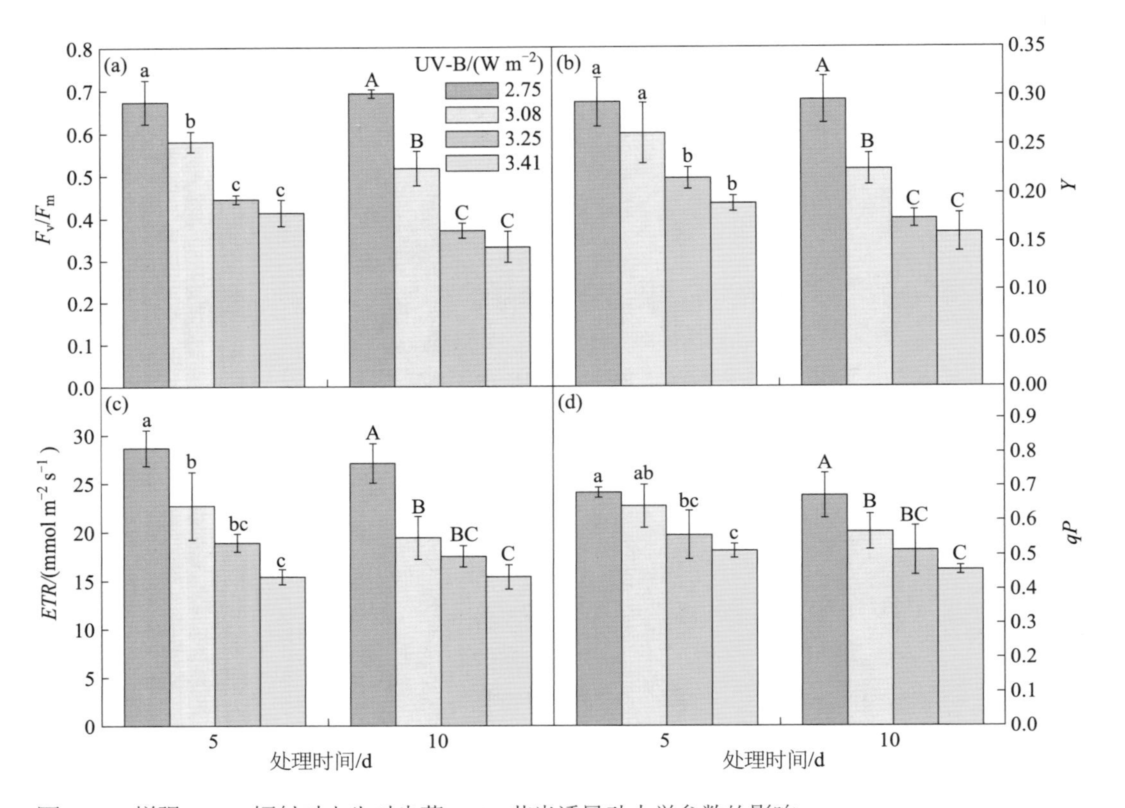

图 3-92　增强 UV-B 辐射对土生对齿藓 Chl-a 荧光诱导动力学参数的影响
Figure 3-92　Effects of enhanced UV-B radiation on chlorophyll a fluorescence induction kinetics parameters of *D. vinealis*

长阶段、试验环境、UV-B辐射强度和处理时间等不同。我们的研究结果表明增强UV-B辐射对真藓Chl-a荧光诱导动力学参数的抑制作用较土生对齿藓强，真藓F_v/F_m、Y、ETR和qP的减少较土生对齿藓更迅速。

增强UV-B辐射对真藓和土生对齿藓光合色素的影响：光合色素是自然界中对光能吸收、传递和转换的一类色素（Lusk，2002）。叶绿素（Chl）作为植物光合作用的主要色素，直接反映绿色植物光合作用的能力强弱。大量研究表明（Niemi *et al.*，2002；左圆圆等，2005），UV-B辐射破坏了植物叶绿体结构，促进Chl分解或阻碍Chl合成，进而使植物体Chl和Car含量降低，导致光合系统反应中心失活，最终导致植物光合作用下降。但也有一些研究表明（刘敏等，2007），UV-B辐射对植物的光合色素没有影响。不同强度UV-B辐射处理的真藓Chl和Car含量变化如图3-93所示。结果表明真藓Chl-a、Chl-b、Chl（a+b）和Chl-a/Chl-b都随着UV-B辐射强度增加而降低，与对照相比，差异显著（$p<0.05$）。在UV-B辐射处理5天后，随着辐射强度的增大，Chl-a的降幅分别为9.34%、34.15%和44.23%

表 3-17　UV-B 辐射时间和辐射强度对土生对齿藓叶绿素光合特性的影响

Table 3-17　Effects of exposure time and UV-B radiation on photosynthetic performances of *D. vinealis* based on two-way ANOVA

	辐射时间 /d		辐射强度 /（$W\,m^{-2}$）				p		
	5	10	2.75	3.08	3.25	3.41	辐射时间	辐射强度	辐射时间 × 辐射强度
F_v/F_m	0.528±0.113	0.478±0.150	0.683±0.034a	0.549±0.045b	0.408±0.043c	0.373±0.054c	0.001	<0.001	0.042
Y	0.241±0.046	0.213±0.058	0.295±0.022a	0.244±0.030b	0.195±0.025c	0.175±0.022c	0.004	<0.001	0.241
ETR/（$\mu mol\,m^{-2}s^{-1}$）	21.4±5.42	19.9±4.85	27.9±1.93a	21.1±3.16b	18.2±1.18c	15.4±0.93d	0.061	<0.001	0.548
qP	0.599±0.081	0.552±0.095	0.677±0.043a	0.605±0.065b	0.536±0.068c	0.485±0.034c	0.044	<0.001	0.753
Chl-a/（$mg\,g^{-1}$ FW）	0.210±0.063	0.206±0.067	0.288±0.008a	0.247±0.010b	0.165±0.006c	0.133±0.004d	0.070	<0.001	<0.001
Chl-b/（$mg\,g^{-1}$ FW）	0.089±0.018	0.064±0.015	0.095±0.013a	0.089±0.018b	0.064±0.014c	0.057±0.010d	<0.001	<0.001	<0.001
Chl（a+b）/（$mg\,g^{-1}$ FW）	0.299±0.082	0.270±0.083	0.383±0.007a	0.336±0.028b	0.229±0.018c	0.190±0.013d	<0.001	<0.001	<0.001
Chl-a/Chl-b	2.317±0.248	3.157±0.313	3.086±0.487a	2.853±0.469b	2.644±0.502c	2.368±0.390d	<0.001	<0.001	0.008
Car/（$mg\,g^{-1}$ FW）	1.711±0.260	1.837±0.466	2.161±0.337a	2.030±0.017b	1.538±0.053c	1.366±0.060d	<0.001	<0.001	<0.001
类黄酮 /（$\mu g\,g^{-1}$FW）	1093.8±170.36	822.67±206.91	1136.0±40.44a	1100.4±242.31a	847.11±175.81b	749.33±160.00c	<0.001	<0.001	<0.001

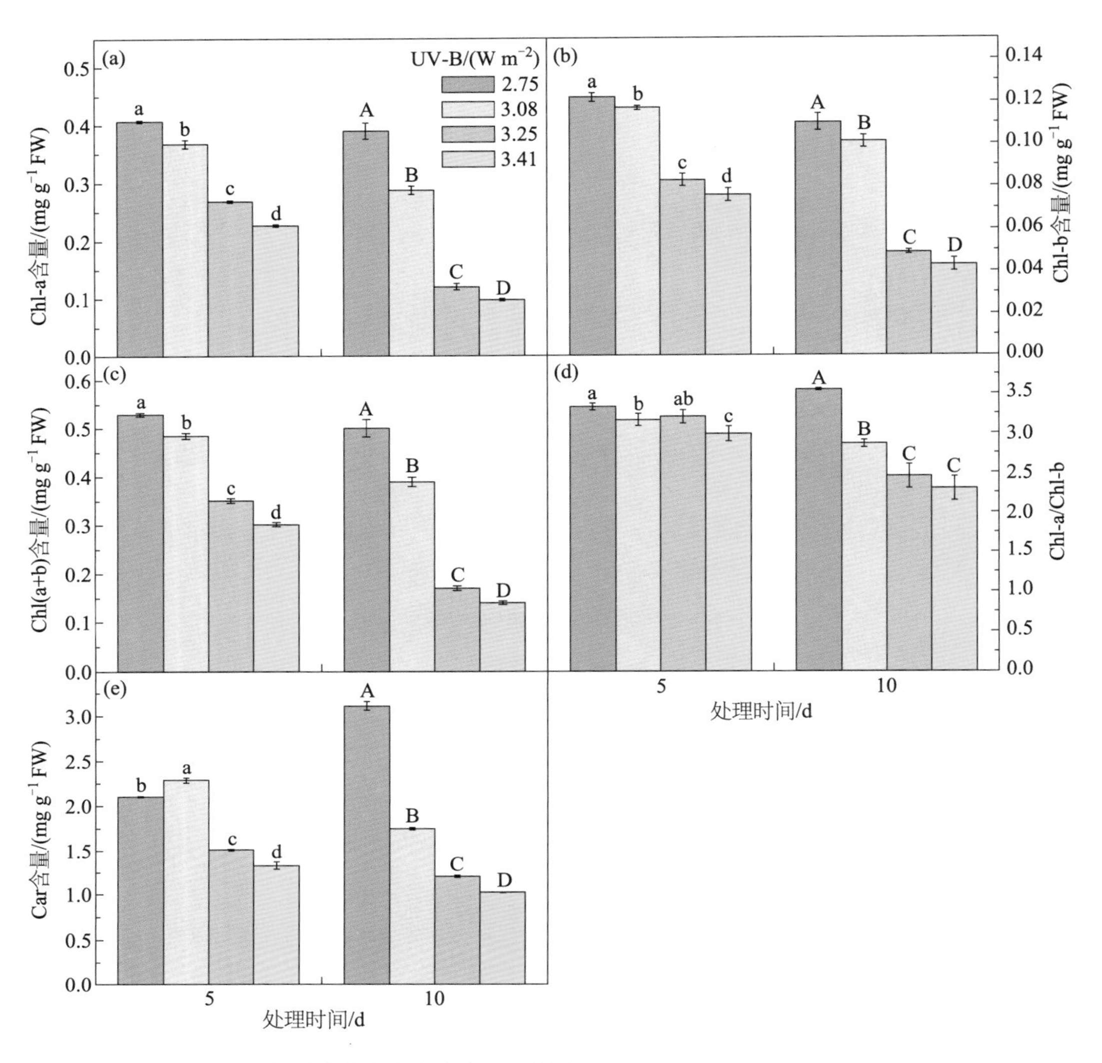

图 3-93 增强 UV-B 辐射对真藓光合色素含量的影响

Figure 3-93 Effects of enhanced UV-B radiation on photosynthetic pigments content of *B. argenteum*

（图3-93a）；而Chl-b在同一辐射水平上降幅比Chl-a小，分别为4.92%、31.97%和37.70%（图3-93b）。因此，Chl-a/Chl-b随辐射强度增加而降低，说明UV-B辐射对Chl-a的破坏作用大于对Chl-b的破坏作用。Chl（a+b）随着UV-B强度的增大，分别降低了8.32%、33.65%和42.91%（图3-93c）。另外，Car含量在强度为3.08 W m^{-2}时升高，与对照相比显著升高了8.80%（图3-93e，p<0.05），随着辐射强度的增大Car含量逐渐降低，在3.25 W m^{-2}和3.41 W m^{-2}处理时，降低幅度分别为28.26%和36.92%（图3-93e）。随着UV-B辐射时间的延长，真藓光合色素的降低幅度逐渐增大，且在辐射强度为3.41 W m^{-2}时达到最小值，Chl-a、

Chl-b、Chl（a+b）、Chl-a/Chl-b和Car含量分别为0.098 mg g^{-1} FW（图3-93a）、0.043 mg g^{-1} FW（图3-93b）、0.140 mg g^{-1} FW（图3-93c）、2.303 mg g^{-1} FW（图3-93d）和1.026 mg g^{-1} FW（图3-93e），与对照相比分别显著下降了74.87%、60.90%、72.00%、35.29%和67.05%（$p<0.05$）。双因素方差分析表明在对真藓光合色素的影响上UV-B辐射时间和辐射强度间存在显著的交互作用（$p<0.05$，表3-16），上述结果说明延长辐射时间增强了UV-B辐射对真藓光合色素含量的影响。

不同强度UV-B辐射处理下，土生对齿藓Chl和Car含量变化的结果如图3-94所示。土生对齿藓经UV-B辐射处理5天后，Chl-a、Chl-b和Chl（a+b）含量随着UV-B辐射强度的增加而显著下降（$p<0.05$），Chl-a较对照减少9.22%、40.43%和52.48%（图3-94a）；Chl-b减少1.87%、28.04%和37.38%（图3-94b）；Chl（a+b）减少6.96%、37.11%和48.20%（图3-94c）。由于Chl-a的降低幅度大于Chl-b的降低幅度，所以Chl-a/Chl-b随着UV-B辐射强度增加逐渐减少，与对照相比分别减少8.18%、17.15%和23.70%（图3-94d）。Car含量在UV-B辐射强度为3.08 W m^{-2}时达到峰值，较对照显著增加9.17%，当UV-B辐射强度继续增加到3.25 W m^{-2}和3.41 W m^{-2}时，Car含量显著降低，与对照相比，分别下降了17.43%和23.42%（$p<0.05$，图3-94e）。随着UV-B辐射时间的延长，土生对齿藓Chl和Car含量下降幅度增大，且随着辐射强度的增加，含量越低。当辐射强度为3.41 W m^{-2}时，Chl-a、Chl-b、Chl（a+b）、Chl-a/Chl-b和Car含量达到最小值，分别为0.131 mg g^{-1} FW（图3-94a）、0.048 mg g^{-1} FW（图3-94b）、0.179 mg g^{-1} FW（图3-94c）、2.720（图3-94d）和1.313 mg g^{-1} FW（图3-94e），与对照相比分别显著下降了55.59%、42.17%、52.65%、22.95%和46.82%（$p<0.05$）。经双因素方差分析可见，UV-B辐射时间和辐射强度间的互作在对土生对齿藓光合色素含量的影响上表现显著（$p<0.05$，表3-17）。该结果说明相同剂量的UV-B辐射在不同时间处理下，对土生对齿藓光合色素含量的抑制是不同的。

不同强度UV-B辐射下真藓和土生对齿藓Chl含量的变化趋势基本一致，随着辐射强度的增加和辐射时间的延长，Chl-a、Chl-b、Chl（a+b）和Chl-a/Chl-b逐渐降低，除UV-B辐射强度为3.08 W m^{-2}处理5天时，Car含量与对照相比略微升高。Chl含量的降低必然导致光合作用的减弱（鲍思伟等，2001），这与Gehrke（1999）的结果一致，研究发现UV-B辐射导致泥炭藓和大金发藓Chl含量降低，且Chl-b的含量明显低于Chl-a。两种藓类Chl含量的降低可能与UV-B诱导的活性氧（ROS）增加有关（Yan and Dai，1996），而ROS能增加Chl的降解速率，且Chl-a对ROS更加敏感，导致Chl-b在同一辐射水平上降幅比Chl-a小（伍泽堂，1991）。Car是UV-B辐射吸收物质之一，可作为UV-B辐射的淬灭剂，清除UV-B辐射诱导产生的ROS，防止膜质发生过氧化，保护Chl，增强植物抵御UV-B辐射胁迫的能力（谢作明，

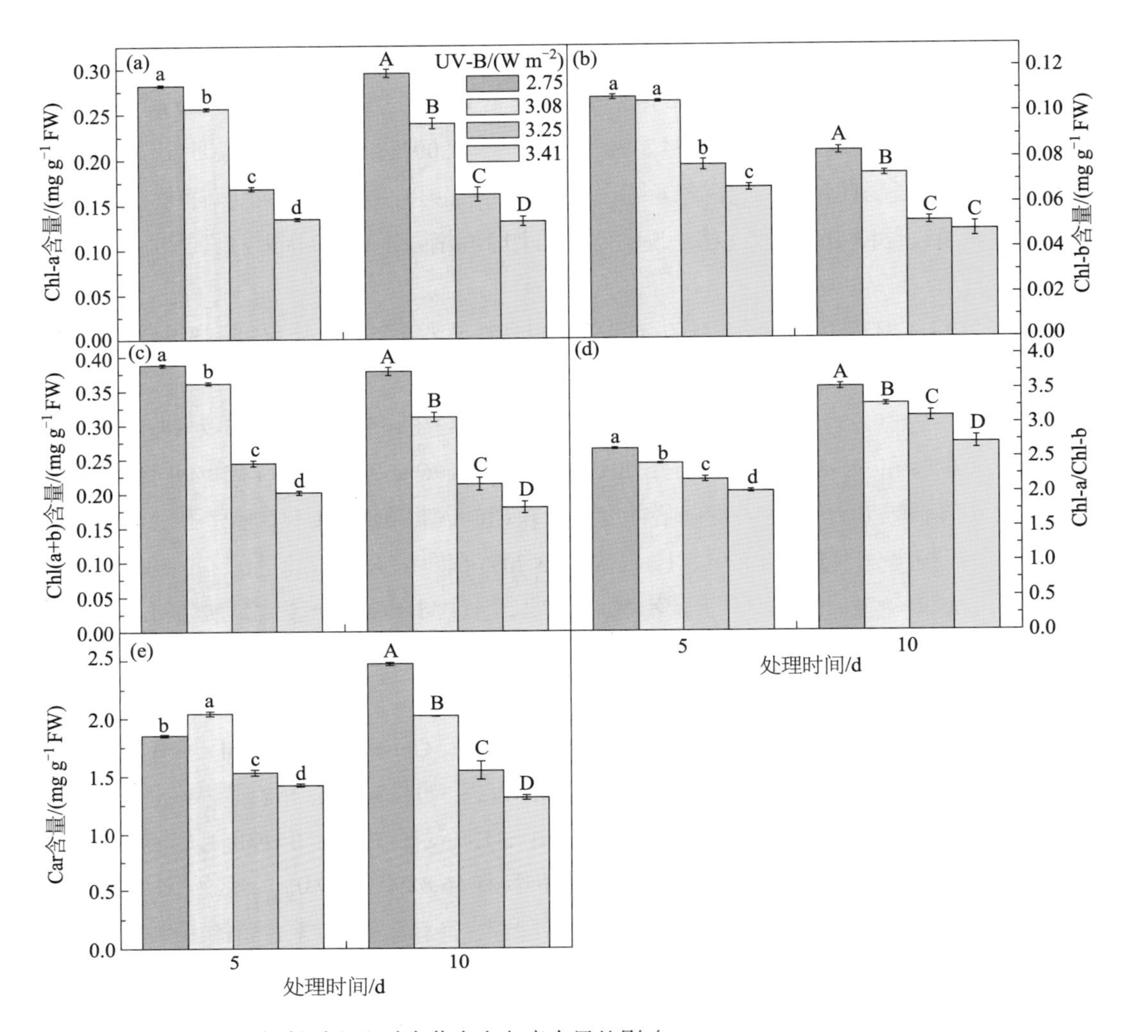

图 3-94　增强 UV-B 辐射对土生对齿藓光合色素含量的影响

Figure 3-94　Effects of enhanced UV-B radiation on photosynthetic pigments content of *D. vinealis*

2006）。本试验结果表明，短时期内低强度UV-B辐射处理下，两种藓类Car含量较大，随着辐射强度增大和辐射时间延长，Car含量显著降低（图3-93，图3-94）。在低辐射强度和短时期内Car含量的增加可能是两种藓类在自身诱导产生的一种保护机制，避免或减轻UV-B辐射的伤害（Newsham，2003）。肖媛等（2010）的研究结果表明，在UV-B辐射胁迫下，雨生红球藻Car含量出现显著增加。而Car含量随UV-B辐射强度增加和时间延长而降低，这可能与诱导Car累积途径达到饱和有关。这与Zhao等（2003）的研究结果一致，在高强度UV-B辐射胁迫下，棉花（*Gossypium hirsutum* L.）的Car含量与对照相比降低16%。从总体趋势上看，UV-B辐射对真藓光合色素合成的抑制效应较土生对齿藓显著，真藓光合色素随UV-B辐射强度的增加而降低的速率较土生对齿藓更显著。

增强 UV-B 辐射对真藓和土生对齿藓类黄酮含量的影响： 类黄酮对紫外光波段有较强的吸收， 是植物在生理生化层面抵御 UV-B 辐射伤害的有效防御机制（Frohnmeyer and Staiger，2003）。目前，研究认为植物类黄酮抵抗UV-B辐射的机理有两个方面。一方面，UV-B辐射引起的增加的类黄酮主要作为自由基的清除剂，从而起到对UV-B辐射的抵抗作用（冯虎元等，2001）。另一方面认为，类黄酮在UV-B辐射波长范围内具有光吸收作用，从而减少UV-B辐射对生物大分子的破坏作用。增强UV-B辐射对真藓类黄酮含量的影响如图3-95所示。UV-B辐射处理5天后，在3.08 W m^{-2} UV-B辐射处理时，类黄酮含量增大至2562.7 μg g^{-1} FW，与对照相比，显著增大了13.4%；随着UV-B辐射强度的增加，类黄酮含量逐渐降低， 当UV-B 辐射强度为3.25 W m^{-2}时， 类黄酮含量较对照显著减少 37.36%；随着 UV-B 强度增大至 3.41 W m^{-2} 时， 类黄酮含量较对照显著减少 53.09%（p<0.05， 图 3-95）。 随着 UV-B 辐射时间延长至 10 天时， 类黄酮含量持续减少为 1807.1 μg g^{-1} FW、1176.0 μg g^{-1} FW 和 793.8 μg g^{-1} FW， 较对照分别显著减少了 25.07%、51.24% 和 67.08%（p<0.05，图3-95）。双因素方差分析进一步显示UV-B辐射时间和辐射强度间的交互作用使真藓类黄酮含量差异显著（p<0.05，表3-16），类黄酮含量随着辐射强度增加和辐射时间延长而逐渐降低。

UV-B辐射处理下土生对齿藓类黄酮含量的变化如图3-96所示。3.08 W m^{-2} UV-B辐射处理5天后，土生对齿藓类黄酮含量最高，为1318.2 μg g^{-1} FW，与对照差异显著（p<0.05）；

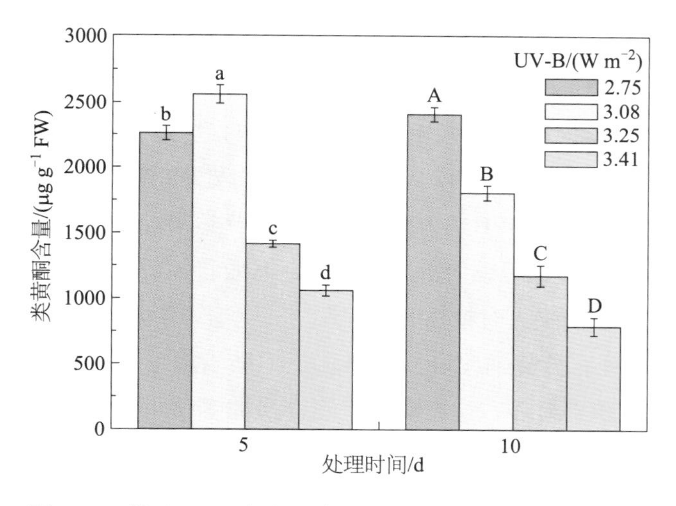

图 3-95 增强 UV-B 辐射对真藓类黄酮含量的影响

Figure 3-95 Effects of enhanced UV-B radiation on total flavonoid content of *B. argenteum*

随着辐射强度的增大，类黄酮含量显著减小，分别为1007.1 μg g^{-1} FW和891.6 μg g^{-1} FW，与对照相比，分别减少了13.05%和23.02%（图3-96）。当UV-B辐射处理10天后，在所有处理强度下，土生对齿藓类黄酮含量均表现出持续减小趋势。在UV-B辐射强度为3.08 W m^{-2}时，类黄酮含量为882.7 μg g^{-1} FW，与对照相比减少了20.75%；辐射强度为3.25 W m^{-2}时，类黄酮含量为687.1 μg g^{-1} FW，与对照相比减少了38.31%；当UV-B辐射强度为3.41 W m^{-2}时，类黄酮含量出现最低值607.1 μg g^{-1} FW，与对照相比减少了45.49%（图3-96）。双因素方差分析表明，辐射时间和辐射强度间交互作用对土生对齿藓类黄酮含量的影响显著（$p<0.05$，表3-17），说明延长辐射时间对土生对齿藓类黄酮合成的抑制有叠加作用。

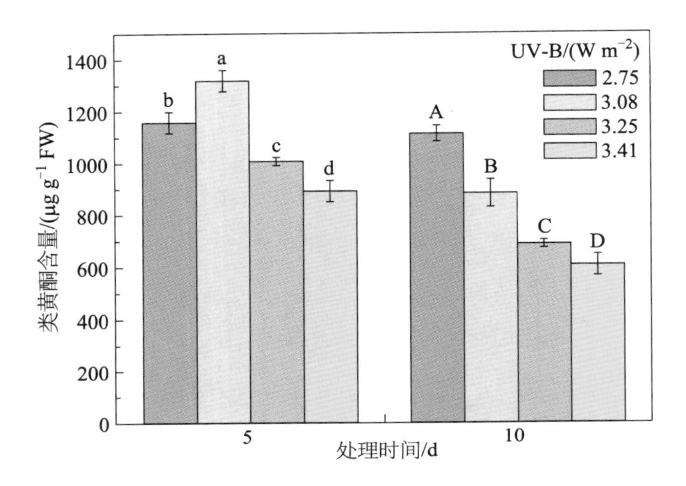

图 3-96　增强 UV-B 辐射对土生对齿藓类黄酮含量的影响

Figure 3-96　Effects of enhanced UV-B radiation on total flavonoid content of *D. vinealis*

增强UV-B辐射处理使两种藓类类黄酮含量均表现出先增大后减小的趋势。随着辐射强度和辐射时间的延长，类黄酮含量逐渐降低，在3.41 W m^{-2} UV-B辐射处理10天时，类黄酮含量达到最小值。这表明短时间内低强度UV-B辐射处理后，两种藓类都可以通过类黄酮含量的升高来吸收部分UV-B辐射，从而减轻辐射带来的危害，而随着辐射强度增强和辐射时间延长，两种藓类类黄酮合成困难，失去对植物的保护作用。这与周新明（2007）的结果相似，UV-B辐射增强下，葡萄叶片类黄酮含量显著增加，但随着辐射时间的延长，类黄酮含量持续降低，与对照相比差异显著，这表明类黄酮含量的变化可能会受到辐射强度和辐射时间的影响。董新纯等（2006）对UV-B辐射增强条件下苦荞体内类黄酮含量的研究表明，增强UV-B辐射诱导类黄酮含量提高。但是，如果辐射强度较大或时间较长，类黄酮的含量反而降低，这可能是UV-B辐射对植物造成的不可逆伤害引起的，这种伤害可以导致植物细胞类黄酮合成困难，降解加速。增强UV-B辐射处理下，类黄酮含量的增加，主要是苯丙氨酸裂解酶（PAL）和有关合成酶活性增强造成，后者能促进查耳酮合成酶活力提高并使合成类黄酮的能力增强（Caldwell *et al.*，2007）。因此，我们假设3.08 W m^{-2} UV-B辐射处理两种藓类5天后导致类黄酮的生物合成增强，而高强度的UV-B辐射很可能抑制了苯基丙酸类合成路径。Eichholz等（2012）的试验支持了该假设，他发现低强度的UV-B辐射导致白芦笋（*Asparagus officinalis* L.）中PAL活性增强。另外，谢灵玲等（2000）也发现紫光、红光和蓝光均能促进PAL的信使RNA（mRNA）的合成，进而提高PAL活性，加速类黄酮合成。真藓类黄酮含量的降低趋势较土生对齿藓更大，说明UV-B辐射处理对真藓类黄酮含量的损伤更加显著。

增强UV-B辐射对真藓和土生对齿藓叶绿体类囊体膜蛋白的影响：类囊体是植物进行光合作用的关键部位，光反应及大部分与光合作用有关的蛋白质和蛋白复合物均定位于类囊体膜及类囊体腔中（Albertsson，2001），所以研究植物类囊体膜蛋白质组有助于进一步了解UV-B辐射胁迫影响植物光合生理变化的内在机制。BN-PAGE作为一种分离膜蛋白质复合物的新技术，受到广泛应用。BN-PAGE系统不采用离子去垢剂，在电泳之前，将蛋白质与考马斯亮蓝染液共存，使蛋白质带上负电荷，屏蔽了各复合物自身电荷，使蛋白质复合物根据自身分子质量的不同在蓝绿温和胶电泳系统得到分离，它可以分离10 kD到100 MD的蛋白质复合物。在凝胶过程中，不需要染色即可显示出蓝色和绿色两种颜色的条带，含有Chl的蛋白复合物在蓝绿温和胶中呈现绿色条带，不含Chl的呈现蓝色条带（李贝贝等，2003）。进一步在变性条件下进行SDS-PAGE凝胶电泳，将各个蛋白质亚基根据其分子质量而分开。UV-B辐射对真藓叶绿体类囊体膜蛋白的影响如图3-97所示。图3-97a显示，类囊体膜蛋白经过蓝绿温和胶分离成4条主要条带，分别为PS Ⅱ超级复合物、PS Ⅰ与捕光色素复合物Ⅰ（LHC Ⅰ）二聚体、PS Ⅱ单体和捕光色素复合物Ⅱ（LHC Ⅱ）聚合体（Cline and Mori，2001；Guo *et al.*，2005a），且随着UV-B辐射强度的增加，其表达量明显减少。将蓝绿温和胶电泳后获得的条带进行SDS-PAGE，结果见图3-97b。UV-B辐射处理的真藓类囊体膜蛋白组分分子质量从不足100 kD到600 kD不等。应用PDQuest 7.3软件对电泳图谱进一步分析，结果检测到高分子量区域的多肽（属于PS Ⅱ反应中心的CP47、CP43和D1蛋白）含量均明显低于对照组，同时PS Ⅱ超级复合物的外周蛋白LHC Ⅱ含量也低于对照组；PS Ⅰ反应中心的PsaA/B蛋白含量也有所下降。随着UV-B辐射强度的增加，其中CP47蛋白相对表达量依次降低为对照组的75.6%、60.8%和47.2%；CP43蛋白相对表达量依次降低为对照组的81.2%、64.1%和50.5%；D1蛋白相对表达量依次降低为对照组的56.8%、35.2%和22.3%；PsaA/B蛋白相对表达量依次降低为对照组的82.2%、71.5%和62.8%。

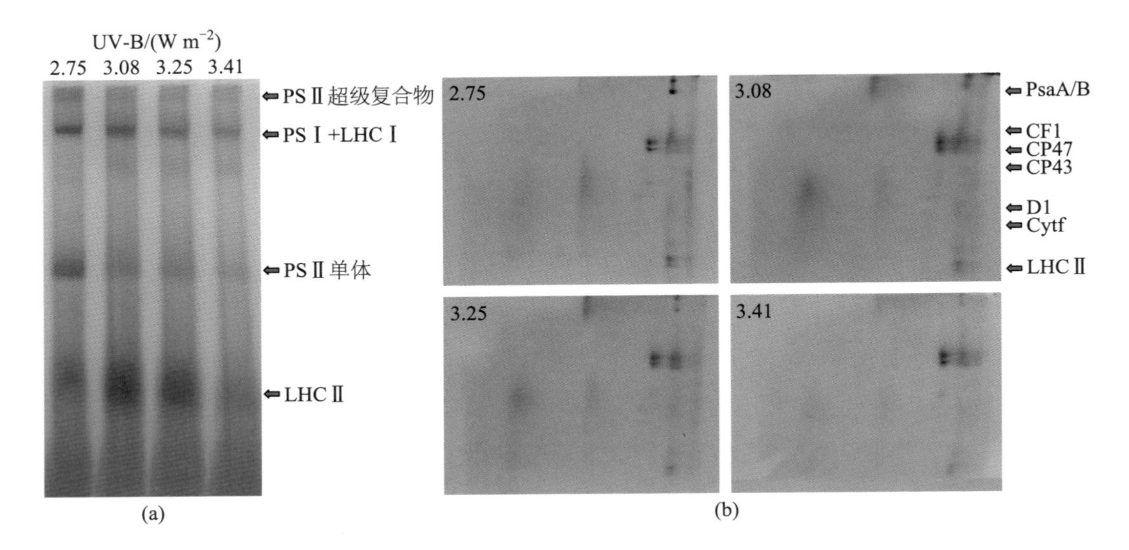

图 3-97　增强 UV-B 辐射对真藓类囊体膜蛋白的影响

Figure 3-97　Effects of enhanced UV-B radiation on thylakoid membrane proteins of *B. argenteum*

土生对齿藓叶绿体类囊体膜蛋白对UV-B辐射强度的响应如图3-98所示。图3-98a显示，土生对齿藓叶绿体类囊体膜蛋白经BN-PAGE分离出4个主要的蛋白复合体条带，分别为PS Ⅱ超级复合物、PS Ⅰ与LHC Ⅰ二聚体、PS Ⅱ单体和LHC Ⅱ聚合体。在辐射强度达到3.25 W m^{-2}和3.41 W m^{-2}时，4个条带蛋白表达量都有不同程度的降低。二向SDS-PAGE结果如图3-98b所示，通过PDQuest 7.3软件分析得到CP47、CP43、LHC Ⅱ和D1蛋白表达量均明显低于对照组，而PS Ⅰ反应中心的PsaA/B蛋白也有少量下降，且随着辐射强度的增加，蛋白表达量越低。当UV-B辐射强度达到3.41 W m^{-2}时，土生对齿藓的CP47、CP43、D1蛋白和PsaA/B相对蛋白表达量进一步降低，分别降低为对照组的52.8%、53.6%、28.1%和70.8%。

通过分离真藓和土生对齿藓的类囊体膜的主要蛋白复合物，UV-B 辐射条件下两种藓类的 BN-PAGE 分析表明，类囊体膜蛋白被分离成 4 条主要条带，分子质量从不足100 kD到600 kD不等。通过第二向SDS-PAGE，可以清晰地观察到CP47、CP43、D1蛋白和LHC Ⅱ多肽含量随着UV-B辐射强度的增加而降低，显示出PS Ⅱ反应中心受到UV-B辐射的破坏。使用PDQuest 7.3软件分析处理组和对照组的类囊体膜蛋白的相对表达量，发现两种藓类的PS Ⅱ相关蛋白的表达量降低幅度大于PS Ⅰ相关蛋白，尤其是PS Ⅱ反应中心D1蛋白的相对表达量减少最显著，且真藓类囊体膜蛋白相对表达量的降低率略高于土生对齿藓，

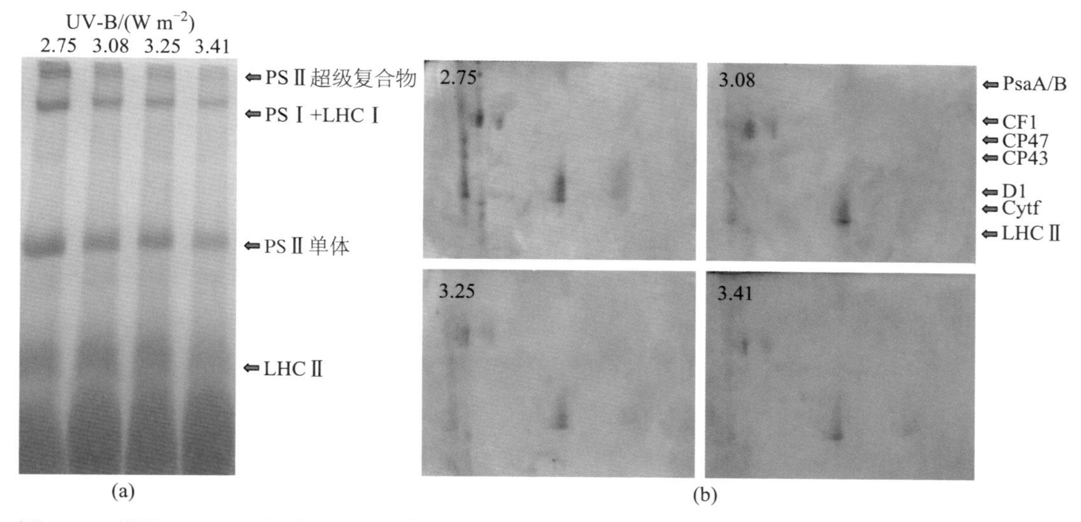

图 3-98　增强 UV-B 辐射对土生对齿藓类囊体膜蛋白的影响
Figure 3-98　Effects of enhanced UV-B radiation on thylakoid membrane proteins of *D. vinealis*

说明真藓类囊体膜蛋白相对表达量对UV-B辐射更敏感（图3-97和图3-98）。这与Aráoz等（1998）得到的结果一致，暴露于UV-B辐射中的念珠藻（*Nostoc* sp.）PS Ⅱ蛋白合成受到严重影响，尤其是D1蛋白表达量。D1蛋白表达量的降低可能会加剧光抑制。PS Ⅰ的多肽PsaA/B含量虽然也有降低，但降低幅度相比前者较小。说明UV-B辐射对PS Ⅱ反应中心的损伤大于PS Ⅰ，PS Ⅱ是UV-B辐射伤害植物的敏感位点之一。Renger等（1989）研究认为UV-B辐射处理下菠菜PS Ⅱ相关蛋白活性的降低程度要大于PS Ⅰ，这与我们的研究结果一致。UV-B辐射通过降低光系统相关蛋白的含量来抑制光合作用，这与我们之前的Chl-a荧光诱导动力学参数、光合色素等的变化一致。

（2）增强UV-B辐射对真藓和土生对齿藓渗透调节物质的影响

将40 W UV-B 313型紫外灯管（上海晨辰照明电器有限公司）悬挂于真藓和土生对齿藓结皮正上方，通过调节结皮顶端距灯管的高度设定4个UV-B辐射强度，分别为2.75 W m^{-2}（对照）、3.08 W m^{-2}、3.25 W m^{-2}和3.41 W m^{-2}（相当于沙坡头地区臭氧损耗0%、6%、9%、12%时所达到的UV-B辐射强度）。连续照射10天，并在第5天和第10天测定真藓和土生对齿藓的脯氨酸含量和可溶性糖含量。

增强UV-B辐射对真藓和土生对齿藓脯氨酸含量的影响：脯氨酸是分布最为广泛的渗透调节物质，以游离状态广泛存在于植物体中。在逆境条件下，大多数植物体都会增加脯氨酸合成和减少其降解导致体内累积大量脯氨酸。植物体内游离脯氨酸的积累能提高原生质的亲水性，防止水分散失，具有解氨毒和保护膜完整性的作用。Smirnoff和Cumbes（1989）认为，植物体内脯氨酸可能具有清除ROS的作用，且脯氨酸对ROS的清除具有一定的专一性，只能清除OH^-和O_2^-。然而，也有一些研究指出（汪天等，2005），逆境下脯氨酸含量提高是植物的一种受害症状。真藓脯氨酸含量随UV-B辐射的动态变化如图3-99所示。当UV-B辐射处理5天后，真藓脯氨酸含量随UV-B辐射强度增加而升高。当UV-B辐射强度为3.08 W m^{-2}时，脯氨酸含量为28.18 μg g^{-1} FW，较对照略有升高但不显著；当UV-B辐射强度为3.25 W m^{-2}时，脯氨酸含量升高为34.14 μg g^{-1} FW，较对照显著上升25.88%；当UV-B辐射强度为3.41 W m^{-2}时，脯氨酸含量进一步升高为37.12 μg g^{-1} FW，较对照显著上升36.87%（$p<0.05$，图3-99）。随着UV-B辐射时间延长至10天后，各UV-B辐射处理条件下脯氨酸含量均有显著积累。在UV-B辐射强度为3.08 W m^{-2}和3.25 W m^{-2}时，脯氨酸含量升高到28.18 μg g^{-1} FW和35.85 μg g^{-1} FW，较对照分别显著增加了9.95%和39.88%；在UV-B辐射强度为3.41 W m^{-2}时，脯氨酸含量达到峰值38.83 μg g^{-1} FW，较对照显著增加了51.50%（$p<0.05$，图3-99）。另外，双因素方差分析表明在对真藓脯氨酸的累积效应上辐射时间和辐射强度间交互作用不显著（表3-18，$p>0.05$），说明辐射时间对真藓脯氨酸的UV-B辐射敏感性影响不大。

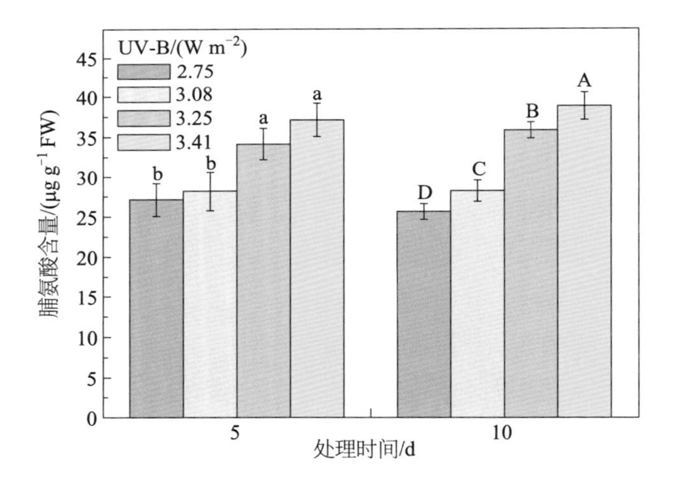

图3-99　增强UV-B辐射对真藓脯氨酸含量的影响

Figure 3-99　Effects of enhanced UV-B radiation on proline content of *B. argenteum*

表 3-18　UV-B 辐射时间和辐射强度对真藓渗透调节物质的影响

Table 3-18　Effects of exposure time and UV-B radiation on osmotic adjustment substances of *B. argenteum* based on two-way ANOVA

	辐射时间 /d		辐射强度 /（W m^{-2}）				p		
	5	10	2.75	3.08	3.25	3.41	辐射时间	辐射强度	辐射时间 × 辐射强度
脯氨酸含量 /（μg g^{-1} FW）	31.6 ± 4.69	32.1 ± 5.74	26.4 ± 1.65^{c}	28.2 ± 1.74^{c}	35.0 ± 1.67^{b}	38.0 ± 1.92^{a}	0.513	<0.001	0.358
可溶性糖含量 /（mg g^{-1} FW）	299.0 ± 76.36	268.8 ± 78.12	378.3 ± 19.76^{a}	322.7 ± 24.93^{b}	248.7 ± 20.87^{c}	186.0 ± 20.78^{d}	<0.001	<0.001	0.783

增强UV-B辐射对土生对齿藓脯氨酸含量的影响如图3-100所示。从图中可以看出，不同强度UV-B辐射使土生对齿藓的脯氨酸含量表现出不同程度的积累，且各处理与对照相比均有显著差异（p<0.05）。UV-B辐射处理5天后，随着辐射强度增大，脯氨酸含量依次升高为34.14 μg g^{-1} FW、37.34 μg g^{-1} FW和39.25 μg g^{-1} FW，与对照相比，分别显著上升10.31%、20.65%和26.82%（p<0.05，图3-100）。随着辐射时间的延长，当UV-B辐射处理10天后，土生对齿藓脯氨酸含量持续升高，当UV-B辐射强度为3.08 W m^{-2}时，脯氨酸含量为36.27 μg g^{-1} FW，较对照显著上升14.81%；当UV-B辐射强度为3.25 W m^{-2}时，脯氨酸含量为39.46 μg g^{-1} FW，较对照显著上升24.91%；当UV-B辐射强度为3.41 W m^{-2}时，脯氨酸含量达到最高值41.59 μg g^{-1} FW，较对照显著上升31.66%（p<0.05，图3-100）。双因素方差分析表明在对土生对齿藓脯氨酸含量的影响上，辐射时间和辐射强度间不存在交互作用（表3-19，p>0.05），说明相同剂量的UV-B辐射下，不同时间处理的土生对齿藓脯氨酸累积量没有显著差异。

两种藓类脯氨酸含量随UV-B辐射强度变化可以看出，UV-B辐射使脯氨酸含量呈升高趋势，且随着UV-B辐射强度的增加，升高幅度越大。这与前人对烟草、银杏、螺旋藻等植物研究的结果一致（刘清华和钟章成，2007；薛林贵等，2011）。增强UV-B辐射对真藓脯氨酸含量的累积速率较土生对齿藓快，真藓脯氨酸含量上升幅度更大。

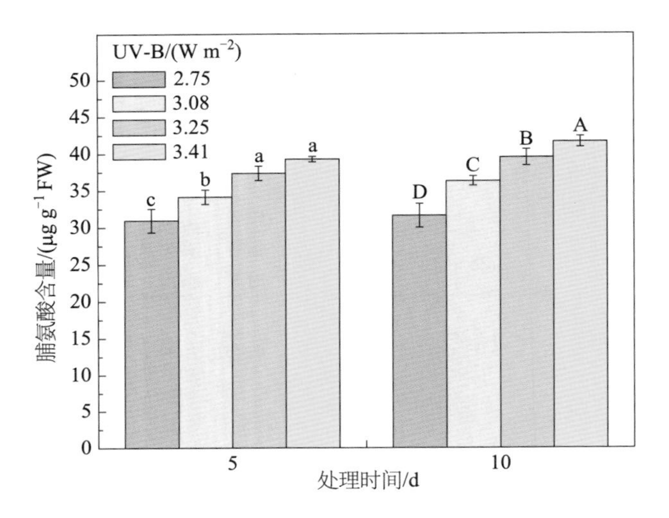

图 3-100　增强 UV-B 辐射对土生对齿藓脯氨酸含量的影响
Figure 3-100　Effects of enhanced UV-B radiation on proline content of *D. vinealis*

表 3-19　UV-B 辐射时间和辐射强度对土生对齿藓渗透调节物质的影响
Table 3-19　Effects of exposure time and UV-B radiation on osmotic adjustment substances of *D. vinealis* based on two-way ANOVA

	辐射时间 /d		辐射强度 /（W m^{-2}）				*p*		
	5	10	2.75	3.08	3.25	3.41	辐射时间	辐射强度	辐射时间 × 辐射强度
脯氨酸含量 /（μg g^{-1} FW）	35.4 ± 3.43	37.2 ± 4.04	31.3 ± 1.48d	35.2 ± 1.38c	38.4 ± 1.49b	40.4 ± 1.38a	0.001	<0.001	0.516
可溶性糖含量 /（mg g^{-1} FW）	413.0 ± 52.56	357.2 ± 77.22	460.3 ± 14.45a	423.6 ± 31.84b	351.6 ± 48.42c	304.9 ± 41.97d	<0.001	<0.001	0.001

增强UV-B辐射对真藓和土生对齿藓可溶性糖含量的影响：可溶性糖是一种重要的渗透调节物质，主要包括葡萄糖、海藻糖、蔗糖等小分子溶质，在植物体内除对细胞膜和原生质体有一定的保护作用，还对抗氧化酶类起重要的保护作用（方志红和董宽虎，2010）。关于增强UV-B辐射对植物可溶性糖含量的影响，不同学者的研究结果不尽相同。吴能表和马红群（2012）的研究表明，UV-B辐射胁迫处理后胡椒和薄荷体内可溶性糖质量浓度与对照相比均明显升高。而李晓阳等（2013）和黎峥等（2003）的研究结果与之相反，他们发现不

同剂量的UV-B辐射处理对拟南芥叶片可溶性糖含量有显著的影响，并且认为适量UV-B辐射有利于促进拟南芥叶片可溶性糖的合成，而过高剂量的UV-B辐射则抑制可溶性糖的合成。UV-B辐射处理下，真藓可溶性糖含量的变化如图3-101所示。从图中可以看出，随着UV-B辐射强度的增加，真藓可溶性糖含量显著降低（$p<0.05$），说明UV-B辐射对真藓的可溶性糖有破坏作用。UV-B辐射处理5天后，可溶性糖依次降低为341.8 mg g^{-1} FW、264.9 mg g^{-1} FW和200.9 mg g^{-1} FW，与对照相比显著降低了12.01%、31.81%和48.28%（$p<0.05$，图3-101）。随着UV-B辐射时间延长至10天，可溶性糖含量持续下降，在UV-B辐射强度为3.08 W m^{-2}时，可溶性糖含量为303.6 mg g^{-1} FW，较对照显著降低了17.51%；在UV-B辐射强度为3.25 W m^{-2}时，可溶性糖含量为232.5 mg g^{-1} FW，较对照显著降低了36.83%；当UV-B辐射强度增大至3.41 W m^{-2}时，可溶性糖含量降低至最小值171.2 mg g^{-1} FW，较对照显著降低了53.49%（$p<0.05$，图3-101）。双因素方差分析表明，UV-B辐射时间和辐射强度对真藓可溶性糖含量不存在交互作用（表3-18，$p>0.05$），说明辐射时间对真藓可溶性糖合成的UV-B辐射敏感性影响不大。

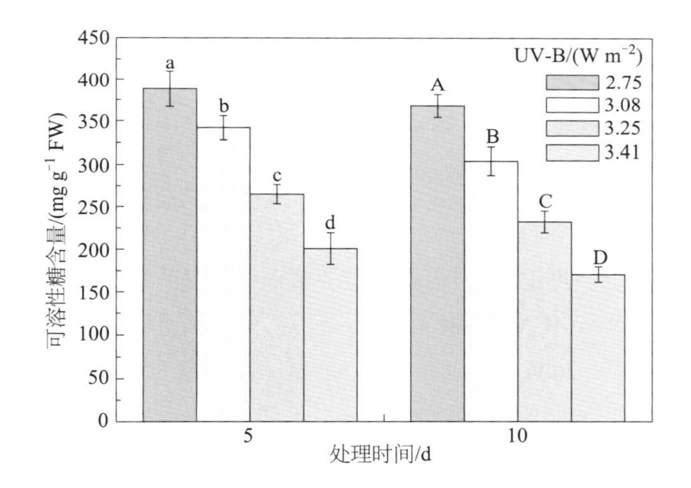

图3-101　增强UV-B辐射对真藓可溶性糖含量的影响

Figure 3-101　Effects of enhanced UV-B radiation on soluble sugar content of *B. argenteum*

增强UV-B辐射处理下，土生对齿藓可溶性糖含量的动态变化如图3-102所示。增强UV-B辐射处理土生对齿藓5天后，可溶性糖含量逐渐降低，当辐射强度为3.08 W m^{-2}时，可溶性糖含量为450.3 mg g^{-1} FW；随着辐射强度的增大，可溶性糖含量持续降低，在辐射强度为3.25 W m^{-2}和3.41 W m^{-2}时，可溶性糖含量分别为394.7 mg g^{-1} FW和342.3 mg g^{-1} FW，与对照相比，分别显著降低了15.10%和26.38%（$p<0.05$，图3-102）。当UV-B辐射处理土生

对齿藓10天后，可溶性糖含量下降幅度更大。随着辐射强度的增加，可溶性糖含量依次降低为396.9 mg g^{-1} FW、308.5 mg g^{-1} FW和267.6 mg g^{-1} FW，分别比对照显著降低了12.88%、32.29%和41.26%（$p<0.05$，图3-102）。双因素方差分析表明在对土生对齿藓可溶性糖含量的影响上，辐射时间和辐射强度间存在显著的交互作用（表3-19，$p<0.05$），随着辐射时间的延长，越强的UV-B辐射对可溶性糖合成的抑制作用越显著。

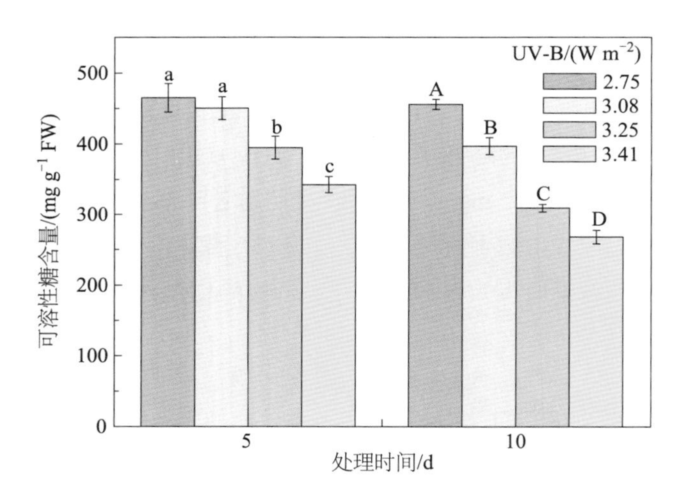

图 3-102　增强 UV-B 辐射对土生对齿藓可溶性糖含量的影响
Figure 3-102　Effects of enhanced UV-B radiation on soluble sugar content of *D. vinealis*

从两种藓类可溶性糖含量随UV-B辐射处理的变化可以看出，随着辐射强度的增加，两种藓类可溶性糖含量都显著降低，在3.41 W m^{-2} UV-B辐射处理10天后，真藓和土生对齿藓可溶性糖含量均达到最低值。这与田继远等（2006）、牛传坡等（2007）和Liu等（2005）对小新月菱形藻、冬小麦和桉树的研究结果一致。可溶性糖含量的降低可能是Hill反应活力降低、Rubisco活性受抑制的结果，这与我们对两种藓类光合作用的研究结果一致。两种藓类相比较，真藓可溶性糖含量减少较土生对齿藓迅速，即增强UV-B辐射处理对真藓可溶性糖含量的抑制效应较土生对齿藓显著。

（3）增强UV-B辐射对真藓和土生对齿藓抗氧化酶系统的影响

将40 W UV-B 313型紫外灯管（上海晨辰照明电器有限公司）悬挂于真藓和土生对齿藓结皮正上方，通过调节结皮顶端距灯管的高度设定4个UV-B辐射强度，分别为2.75 W m^{-2}（对照）、3.08 W m^{-2}、3.25 W m^{-2}和3.41 W m^{-2}（相当于沙坡头地区臭氧损耗0%、6%、9%和12%时所达到的UV-B辐射强度）。连续照射10天，并在第5天和第10天测定真藓和土生对齿藓的MDA含量、可溶性蛋白含量和抗氧化酶活性。

增强UV-B辐射对真藓和土生对齿藓MDA含量的影响：细胞膜是植物接受外界环境因素信息的受体，其完整性是细胞进行正常生理功能的基础（蔡妙珍等，2005）。在受到外界胁迫时，细胞膜的结构和功能首先受到伤害。Murphy（1990）认为，细胞生物膜系统是紫外辐射胁迫伤害的靶位。UV-B辐射处理导致植物细胞膜结构受损，主要由以下两个原因造成：① UV-B辐射能够诱导植物产生O_2^-、OH^-、H_2O_2等自由基，导致细胞内ROS水平升高，使膜脂发生过氧化，导致膜结构变化，从而改变膜的透性（Bowler *et al.*，1992）。② 增强UV-B辐射使植物膜系统多不饱和脂肪酸（PUFA）被ROS氧化，从而降低质膜的SFA/PUFA比值（SFA：饱和脂肪酸），降低膜的流动性，发生膜脂过氧化作用（Kramer *et al.*，1991）。MDA是膜质过氧化的产物，其浓度的高低表示细胞膜系统受伤害的程度（刘丽丽等，2010）。增强UV-B辐射对真藓MDA含量的影响如图3-103所示。真藓MDA含量随UV-B辐射强度增加呈显著升高趋势（$p<0.05$）。UV-B辐射处理5天后，MDA含量依次为0.044 μmol g^{-1} FW、0.056 μmol g^{-1} FW和0.057 μmol g^{-1} FW，与对照相比，分别显著升高10.00%、40.00%和42.50%（$p<0.05$，图3-103）。随着UV-B辐射时间的延长，当UV-B辐射处理10天后，真藓MDA含量进一步升高，分别达到0.047 μmol g^{-1} FW、0.103 μmol g^{-1} FW和0.122 μmol g^{-1} FW，与对照相比，显著增加17.50%、157.50%和205.00%（$p<0.05$，图3-103）。双因素方差分析表明，UV-B辐射时间和辐射强度对真藓MDA含量累积有显著的交互作用（表3-20，$p<0.05$），说明延长辐射时间增强了UV-B辐射对真藓MDA含量的影响。

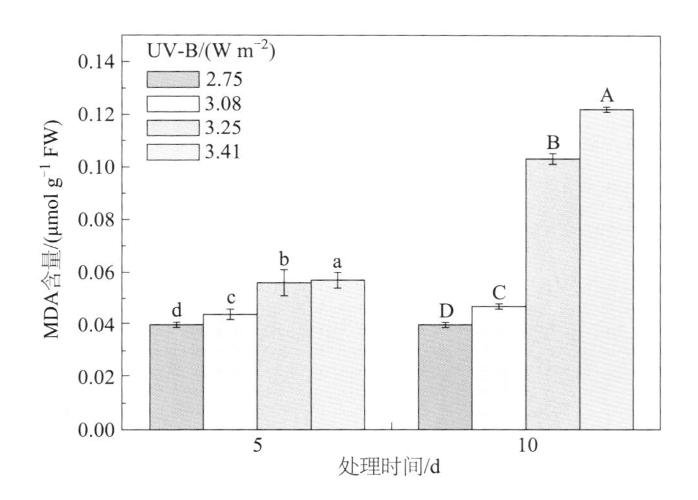

图3-103　增强UV-B辐射对真藓MDA含量的影响

Figure 3-103　Effects of enhanced UV-B radiation on MDA content of *B. argenteum*

增强UV-B辐射对土生对齿藓MDA含量的影响如图3-104所示。在所有的处理中，土生对齿藓的MDA含量明显上升，与对照相比，均达到显著水平（$p<0.05$），表明随着辐射强度的增强，膜脂过氧化现象加剧。UV-B辐射处理5天后，土生对齿藓MDA含量分别比对照显著提高了31.43%、57.14%和65.71%（$p<0.05$，图3-104）；随着辐射达到10天后，土生对齿藓MDA含量持续升高，依次为0.054 μmol g^{-1} FW、0.076 μmol g^{-1} FW和0.086 μmol g^{-1} FW，分别比对照显著提高了38.46%、94.87%和120.51%（$p<0.05$，图3-104）。双因素方差分析也表明，在对土生对齿藓MDA含量的影响上，UV-B辐射时间和辐射强度间的交互作用显著（表3-21，$p<0.05$），说明延长辐射时间明显提高了土生对齿藓MDA含量对UV-B辐射的敏感性。

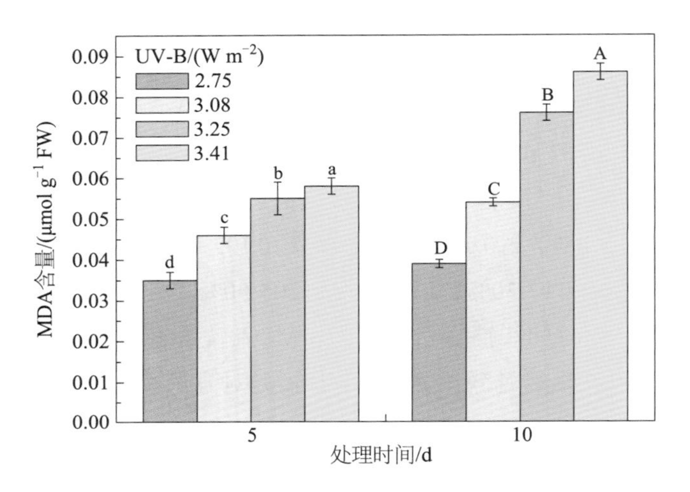

图 3-104　增强 UV-B 辐射对土生对齿藓 MDA 含量的影响
Figure 3-104　Effects of enhanced UV-B radiation on MDA content of *D. vinealis*

随着UV-B辐射强度的增加和辐射时间的延长，两种藓类MDA含量均显著升高（$p<0.05$），说明增强UV-B辐射均加剧了两种藓类的膜脂过氧化作用。这与冯虎元等（2002）对增强UV-B辐射胁迫下小麦幼苗MDA含量变化的研究结果一致。Wang等（2010）的研究结果表明，增强UV-B辐射导致玉米（*Zea mays* L.）花粉膜脂发生过氧化。MDA含量的增加可能是由于UV-B辐射使两种藓类植物细胞内产生大量自由基，而UV-B辐射又降低了活性氧自由基的清除能力，导致质膜发生过氧化（晏斌和戴秋杰，1996）。Allen（1968）发现，细胞膜的稳定性关系着植物的色素含量，细胞膜的破坏很可能导致色素含量的降低，这与我们光合色素的试验结果一致。真藓MDA含量积累的速率较土生对齿藓快，说明增强UV-B辐射对真藓膜脂过氧化程度更为显著，细胞膜生理功能受到了更大的损伤。

表 3-20　UV-B 辐射时间和辐射强度对真藓抗氧化酶系统的影响

Table 3-20　Effects of exposure time and UV-B radiation on antioxidative enzymes systems of *B. argenteum* based on two-way ANOVA

	辐射时间 /d		辐射强度 / ($W\ m^{-2}$)				p		
	5	10	2.75	3.08	3.25	3.41	辐射时间	辐射强度	辐射时间 × 辐射强度
MDA/ ($\mu mol\ g^{-1}$ FW)	0.049 ± 0.008	0.078 ± 0.037	0.040 ± 0.001a	0.046 ± 0.002b	0.080 ± 0.026c	0.090 ± 0.036d	<0.001	<0.001	<0.001
可溶性蛋白 / ($mg\ g^{-1}$ FW)	475.4 ± 98.1	419.8 ± 126.3	596.9 ± 10.6a	497.6 ± 42.3b	368.4 ± 35.6c	327.5 ± 47.0d	<0.001	<0.001	<0.001
SOD 活性 / ($U\ mg^{-1}$ pro)	0.100 ± 0.012	0.098 ± 0.018	0.121 ± 0.007a	0.098 ± 0.006b	0.091 ± 0.006c	0.085 ± 0.005c	0.346	<0.001	0.169
CAT 活性 / ($U\ mg^{-1}$ pro)	1.609 ± 0.228	1.582 ± 0.302	1.955 ± 0.113a	1.647 ± 0.115b	1.460 ± 0.112c	1.321 ± 0.079d	0.541	<0.001	0.344

表 3-21　UV-B 辐射时间和辐射强度对土生对齿藓抗氧化酶系统的影响

Table 3-21　Effects of exposure time and UV-B radiation on antioxidative enzymes systems of *D. vinealis* based on two-way ANOVA

	辐射时间 /d		辐射强度 / ($W\ m^{-2}$)				p		
	5	10	2.75	3.08	3.25	3.41	辐射时间	辐射强度	辐射时间 × 辐射强度
MDA/ ($\mu mol\ g^{-1}$ FW)	0.049 ± 0.009	0.064 ± 0.020	0.037 ± 0.002a	0.050 ± 0.005b	0.066 ± 0.012c	0.072 ± 0.016d	<0.001	<0.001	<0.001
可溶性蛋白 / ($mg\ g^{-1}$ FW)	451.4 ± 25.2	410.7 ± 66.7	489.7 ± 12.5a	452.7 ± 15.4b	404.6 ± 44.1c	377.1 ± 42.7d	<0.001	<0.001	<0.001
SOD 活性 / ($U\ mg^{-1}$ pro)	0.105 ± 0.013	0.100 ± 0.015	0.120 ± 0.008a	0.106 ± 0.008b	0.095 ± 0.004c	0.088 ± 0.006c	0.082	<0.001	0.477
CAT 活性 / ($U\ mg^{-1}$ pro)	1.776 ± 0.228	1.626 ± 0.256	1.999 ± 0.131d	1.808 ± 0.097c	1.571 ± 0.123b	1.426 ± 0.114a	<0.001	<0.001	0.589

增强UV-B辐射对真藓和土生对齿藓可溶性蛋白的影响：蛋白质是植物体内重要的结构和功能物质，叶片中50%以上可溶性蛋白是Rubisco。由于蛋白质的最大吸收波长为280 nm，属于UV-B辐射波长范围（朱鹏锦等，2011），因此，UV-B辐射对植物蛋白质将产生较大影响。UV-B辐射通过影响组成蛋白质的色氨酸的光降解、—SH基的修饰、提高膜蛋白在水中的溶解度、加速多肽链的断裂等修饰，导致酶的失活和蛋白质结构的改变（Quaite *et al.*，1992）。由于试验条件、辐射强度、受试物种等的不同，关于UV-B辐射对植物体可溶性蛋白含量影响的结论也不尽相同。研究表明（唐莉娜等，2004），UV-B辐射增强抑制蛋白质的合成，使蛋白质含量降低。而也有一些报道认为UV-B辐射会增强蛋白质的合成，这可能是合成类黄酮前体的芳香族氨基酸合成加强的结果，而黄酮类物质可以保护生物体结构和功能免受UV-B辐射的损伤（Mackerness *et al.*，2001）。增强UV-B辐射条件下，真藓可溶性蛋白含量的动态变化如图3-105所示。从图中可以看出，随着UV-B辐射强度的增加和辐射时间的延长，可溶性蛋白含量显著降低（$p<0.05$），说明UV-B辐射对真藓的可溶性蛋白有破坏作用。UV-B辐射处理5天后，可溶性蛋白含量与对照相比，分别显著降低了10.21%、32.80%和38.09%（$p<0.05$，图3-105）；随着UV-B辐射时间延长至10天，可溶性蛋白含量持续下降为459.76 mg g^{-1} FW、336.18 mg g^{-1} FW和285.81 mg g^{-1} FW，仅为对照的76.94%、56.26%和47.83%（$p<0.05$，图3-105）。双因素方差分析进一步显示，辐射时间和辐射强度间的交互作用在真藓可溶性蛋白含量上表现显著（表3-20，$p<0.05$），随着辐射时间的延长，UV-B辐射越强，对真藓可溶性蛋白合成的抑制作用越强。

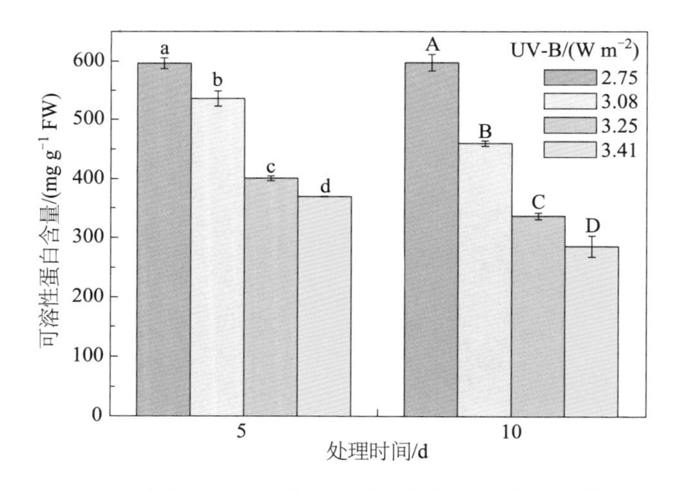

图 3-105　增强 UV-B 辐射对真藓可溶性蛋白含量的影响

Figure 3-105　Effects of enhanced UV-B radiation on soluble protein content of *B. argenteum*

增强UV-B辐射对土生对齿藓可溶性蛋白含量的影响如图3-106所示。不同强度UV-B辐射处理使土生对齿藓可溶性蛋白含量表现出不同程度的降低，各处理组与对照组相比均有显著差异（$p<0.05$）。UV-B辐射处理5天后，随着辐射强度的增加，可溶性蛋白含量依次降低为466.04 mg g^{-1} FW、444.64 mg g^{-1} FW和416.05 mg g^{-1} FW，与对照相比，分别显著降低了2.67%、7.14%和13.11%（$p<0.05$，图3-106）。随着辐射时间的延长，当UV-B辐射处理10天后，土生对齿藓可溶性蛋白含量持续下降，分别为439.34 mg g^{-1} FW、364.62 mg g^{-1} FW和338.22 mg g^{-1} FW，仅为对照的87.76%、72.84%和67.56%（$p<0.05$，图3-106）。双因素方差分析表明，辐射时间、辐射强度及两者对土生对齿藓可溶性蛋白的交互作用的显著性水平均小于0.001，即存在显著差异（表3-21，$p<0.05$）。

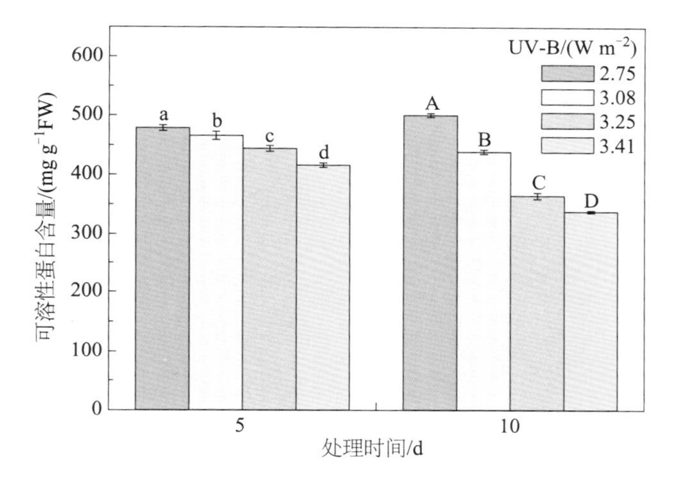

图3-106 增强UV-B辐射对土生对齿藓可溶性蛋白含量的影响
Figure 3-106 Effects of enhanced UV-B radiation on soluble protein content of *D. vinealis*

两种藓类可溶性蛋白含量随UV-B辐射处理的变化可以看出，随着辐射强度的增加和辐射时间的延长，两种藓类可溶性蛋白含量都显著减少。在3.41 W m^{-2} UV-B辐射处理10天后，真藓和土生对齿藓可溶性蛋白含量均达到最低值，说明UV-B辐射抑制了两种藓类植物可溶性蛋白的合成。可溶性蛋白含量的降低可能是由于UV-B辐射使组成蛋白质的一些氨基酸发生开环或与其他物质结合，使蛋白质分子的空间结构改变，导致蛋白质失活（张跃群等，2009）。这与Yu等（2004）对增强UV-B辐射胁迫下亚心形扁藻（*Platymonas subcordiformis*）可溶性蛋白含量变化的研究结果一致。Rubisco作为光合作用的关键酶，约占可溶性蛋白的1/2。可溶性蛋白含量的下降会导致Rubisco活性和光合能力的降低（Strid，1993），抑制光合作用，这与我们对真藓的Chl-a荧光诱导动力学参数的测定结果一致。两种藓类相比较，增强UV-B辐射处理对真藓可溶性蛋白含量的抑制效应较土生对齿藓显著，真藓可溶性蛋白含量减少较土生对齿藓迅速。

增强UV-B辐射对真藓和土生对齿藓SOD活性的影响：植物体酶促抗氧化系统主要包括SOD、CAT、POD、APX、GR等（刘宛等，2003）。其中，SOD作为植物体内以自由基为底物的酶，它能催化O_2^-歧化成为O_2和H_2O_2。增强UV-B辐射对真藓SOD活性的影响如图3-107所示。在UV-B辐射处理5天后，真藓SOD活性呈下降趋势。当UV-B辐射强度为3.08 W m^{-2}时，SOD活性为0.100 U mg^{-1} pro，与对照相比减少了15.05%；随着辐射强度的增大，SOD活性显著降低，在3.25 W m^{-2}和3.41 W m^{-2}时，SOD分别为0.093 U mg^{-1} pro和0.089 U mg^{-1} pro，分别与对照相比降低了20.92%和24.40%（$p<0.05$，图3-107）。随着UV-B辐射时间的延长，当UV-B辐射处理真藓10天后，SOD活性持续降低，依次为0.096 U mg^{-1} pro、0.088 U mg^{-1} pro和0.082 U mg^{-1} pro，分别降低了22.57%、29.24%和34.38%（$p<0.05$，图3-107）。双因素方差分析表明，在对真藓SOD活性的影响上，辐射时间和辐射强度的交互作用不显著（表3-20，$p>0.05$），说明辐射时间对真藓SOD活性的UV-B辐射敏感性影响不大。

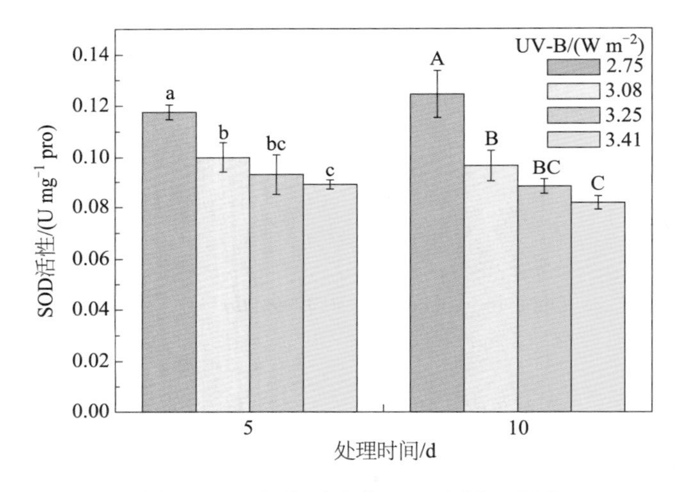

图3-107 增强UV-B辐射对真藓SOD活性的影响

Figure 3-107 Effects of enhanced UV-B radiation on SOD activity of *B. argenteum*

土生对齿藓SOD活性对UV-B辐射增强的响应如图3-108所示。与真藓SOD活性的变化规律一致，土生对齿藓SOD活性随着UV-B辐射增强而逐渐降低。UV-B辐射处理5天后，随着UV-B辐射强度的增加，SOD活性较对照分别降低6.95%、19.26%和23.28%（图3-108）。UV-B辐射时间达到10天后，SOD活性持续降低，在辐射强度为3.08 W m^{-2}时，SOD活性降低到0.101 U mg^{-1} pro，较对照降低16.02%；UV-B辐射强度为3.25 W m^{-2}时，SOD活性降低到0.093 U mg^{-1} pro，较对照降低22.49%；当UV-B辐射强度达到3.41 W m^{-2}时，SOD活性最低，仅为0.084 U mg^{-1} pro，与对照相比显著降低30.12%（$p<0.05$，图3-108）。双因素方差分析表明，UV-B辐射时间和辐射强度对SOD活性的交互作用不显著（$p>0.05$，表3-21），说明UV-B辐射强度对土生对齿藓SOD活性起主要作用。

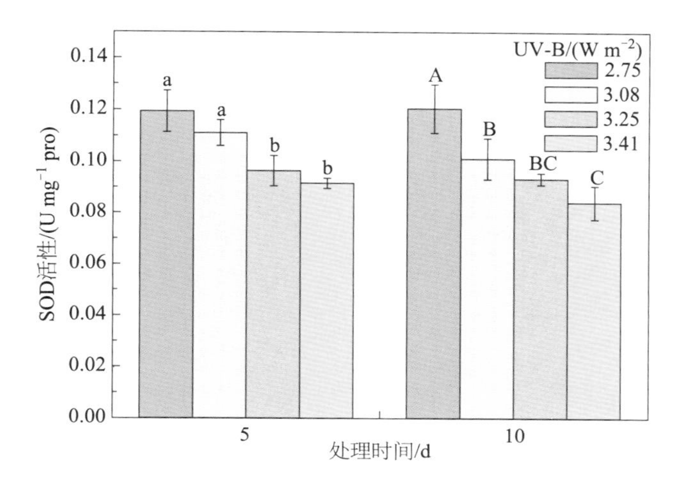

图3-108　增强UV-B辐射对土生对齿藓SOD活性的影响
Figure 3-108　Effects of enhanced UV-B radiation on SOD activity of *D. vinealis*

比较真藓和土生对齿藓SOD活性对UV-B辐射增强的响应发现，两种藓类SOD活性均随着UV-B辐射强度增强而逐渐降低。在3.41 W m^{-2} UV-B辐射处理10天时，SOD活性达到最低值。Xu等（2008）对大豆SOD活性的研究也证实了我们的结论。真藓SOD活性降低速率高于土生对齿藓，说明增强UV-B辐射对真藓SOD活性的影响大于土生对齿藓。

增强UV-B辐射对真藓和土生对齿藓CAT活性的影响：CAT与H_2O_2专一结合，使H_2O_2转化为H_2O和O_2，避免H_2O_2在植物体内的积累，它与SOD共同作用，使植物体内的ROS维持在一个较低的水平，防止ROS产生的毒害。增强UV-B辐射处理对真藓CAT活性的影响如图3-109所示。随着UV-B辐射强度的增加，CAT活性显著降低（$p<0.05$）。当UV-B辐射处理5天后，真藓CAT活性依次为1.681 U mg^{-1} pro、1.504 U mg^{-1} pro和1.354 U mg^{-1} pro，与对照相比，分别下降了11.42%、20.76%和28.62%（图3-109）。随着UV-B辐射处理时间延长至10天，CAT活性持续下降，在3.08 W m^{-2} UV-B辐射处理时，CAT活性为1.614 U mg^{-1} pro，较对照显著降低了19.77%；3.25 W m^{-2} UV-B辐射处理时，CAT活性降低到1.416 U mg^{-1} pro，较对照显著降低了29.61%；当UV-B辐射强度达到3.41 W m^{-2}时，CAT

活性最低，仅为1.288 U mg^{-1} pro，与对照相比显著降低了35.99%（$p<0.05$，图3-109）。双因素方差分析结果表明，UV-B 辐射时间和辐射强度对真藓 CAT 活性的交互作用不显著（$p>0.05$，表3-20），说明辐射时间对真藓CAT活性的UV-B辐射敏感性影响不大。

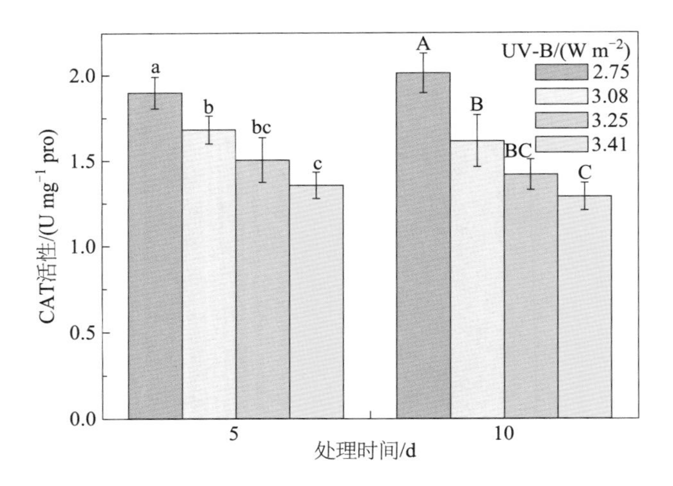

图 3-109　增强 UV-B 辐射对真藓 CAT 活性的影响

Figure 3-109　Effects of enhanced UV-B radiation on CAT activity of *B. argenteum*

增强 UV-B 辐射处理下， 土生对齿藓 CAT 活性的动态变化如图 3-110 所示。UV-B 辐射处理土生对齿藓5天后，CAT活性逐渐降低。当辐射强度为3.08 W m^{-2}时，CAT活性为1.880 U mg^{-1} pro；随着辐射强度的增大，CAT活性持续降低，在辐射强度为3.25 W m^{-2}和3.41 W m^{-2}时，CAT活性分别为1.676 U mg^{-1} pro和1.514 U mg^{-1} pro，与对照相比，分别降低了17.71%和25.64%（图3-110）。当UV-B辐射处理土生对齿藓10天后，CAT活性变化与处理5天后相似。随辐射强度的增大，CAT活性依次为1.736 U mg^{-1} pro、1.467 U mg^{-1} pro和1.339 U mg^{-1} pro，分别比对照显著降低了11.57%、25.24%和31.84%（$p<0.05$，图3-110）。双因素方差分析表明在对土生对齿藓CAT活性的影响上，UV-B辐射时间和辐射强度间的交互作用不显著（$p>0.05$，表3-21），说明辐射时间对土生对齿藓CAT活性的UV-B辐射敏感性影响不大。

两种藓类CAT活性随UV-B辐射处理的动态变化可以看出，随着辐射强度的增加，两种藓类CAT活性都显著降低（$p<0.05$）。在UV-B辐射处理10天后，UV-B辐射强度为3.41 W m^{-2}处理的真藓和土生对齿藓CAT活性均降到最低值。两种藓类的抗氧化酶在UV-B辐射处理后已经失去在抗氧化酶防御系统中的保护作用，不能对自由基和过氧化物进行清除，保护酶活性降低导致了膜脂过氧化的不断加剧，使它们受到更大的损伤。这与Wang等（2010）对UV-B辐射增强条件下玉米（*Zea mays* L.）花粉CAT活性的研究结果一致。两种藓类相比较，

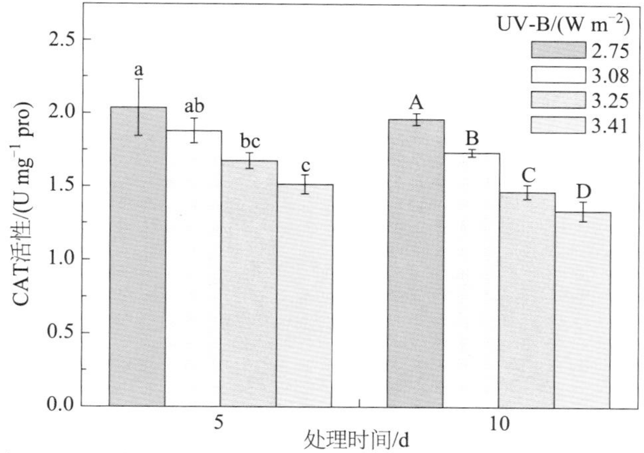

图3-110　增强UV-B辐射对土生对齿藓CAT活性的影响

Figure 3-110　Effects of enhanced UV-B radiation on CAT activity of *D. vinealis*

真藓CAT活性降低较土生对齿藓迅速，即增强UV-B辐射处理对真藓CAT活性的抑制效应较土生对齿藓显著。

（4）增强UV-B辐射对真藓和土生对齿藓超微结构的影响

将40 W UV-B 313型紫外灯管（上海晨辰照明电器有限公司）悬挂于真藓和土生对齿藓结皮正上方，通过调节结皮顶端距灯管的高度设定4个UV-B辐射强度，分别为2.75 W m^{-2}（对照）、3.08 W m^{-2}、3.25 W m^{-2}和3.41 W m^{-2}（相当于沙坡头地区臭氧损耗0%、6%、9%、12%时所达到的UV-B辐射强度）。连续照射10天。取UV-B辐射处理10天的真藓和土生对齿藓茎叶体，通过透射电镜（TEM）分析细胞和叶绿体超微结构变化。

增强UV-B辐射对真藓和土生对齿藓细胞超微结构的影响：完整的细胞结构是植物体完成各项生命活动的基础。增强UV-B辐射引起的真藓细胞超微结构的变化如图3-111所示。在对照处理下，真藓叶绿体呈椭圆形，形态饱满；类囊体片层结构垛叠整齐，排列紧密；细胞核膜结构清晰，核质均匀，有一个明显的核仁区；细胞壁厚度较为均匀，细胞膜与细胞壁紧紧黏在一起，无间隙（图3-111a）。随UV-B辐射强度的增加，真藓细胞超微结构发生不同的变化。在增强UV-B辐射处理下，真藓叶绿体肿胀变圆，呈球形；叶绿体膜系统受到不同程度的破坏，基质类囊体片层变得松散，类囊体膜也受到不同程度的破坏（图3-111b，c，d）。在各处理中，真藓细胞中均没有发现细胞核结构，可能其细胞核在UV-B辐射胁迫下发生降解。在不同UV-B辐射处理下，细胞膜与细胞壁交界的地方变得模糊不清，细胞壁内侧变得凹凸不平；细胞器溶解，细胞内出现大量淀粉粒。尤其在3.41 W m^{-2}UV-B辐射强度下，真藓细胞受损严重，细胞结构仅剩残留细胞壁、淀粉粒和部分细胞质基质（图3-111d）。

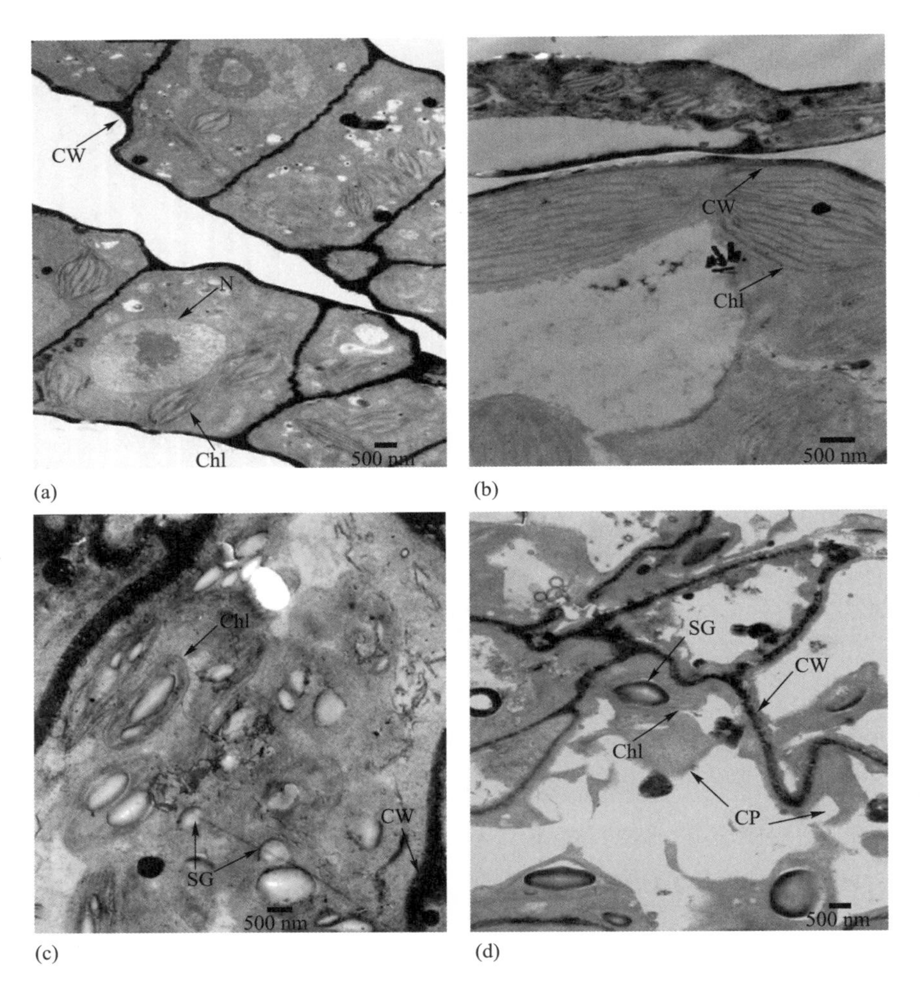

图 3-111　增强 UV-B 辐射对真藓细胞超微结构的影响：2.75 W m^{-2}（a）、3.08 W m^{-2}（b）、3.25 W m^{-2}（c）、3.41 W m^{-2}（d）UV-B 辐射处理下细胞超微结构。Chl：叶绿体；CP：细胞质；CW：细胞壁；N：细胞核；SG：淀粉粒

Figure 3-111　Effects of enhanced UV−B radiation on cell ultrastructure of *B. argenteum*：Cell ultrastructure under 2.75 W m^{-2}（a）、3.08 W m^{-2}（b）、3.25 W m^{-2}（c）、3.41 W m^{-2}（d）UV-B treatment. Chl：chloroplast；CP：cytoplasm；CW：cell wall；N：nucleus；SG：starch grain

增强UV-B辐射对土生对齿藓细胞超微结构的影响如图3-112所示。在正常UV-B辐射条件下生长的土生对齿藓，细胞中的叶绿体为椭圆形；类囊体排列整齐紧密且清晰；细胞核膜完整，平滑清晰；核质与染色质分布均匀；细胞壁厚度均匀，细胞膜与细胞壁紧紧黏在一起，无间隙（图3-112a）。UV-B辐射增强处理后，土生对齿藓叶绿体不同程度上肿胀变形，叶绿体类囊体片层松散变形，电子密度降低；膜系统降解，双层膜结构模糊；嗜锇颗粒增多；叶绿体内膜系统受到不同程度的破坏（图3-112b,c,d）。在3.08 W m^{-2} UV-B辐射处理下，

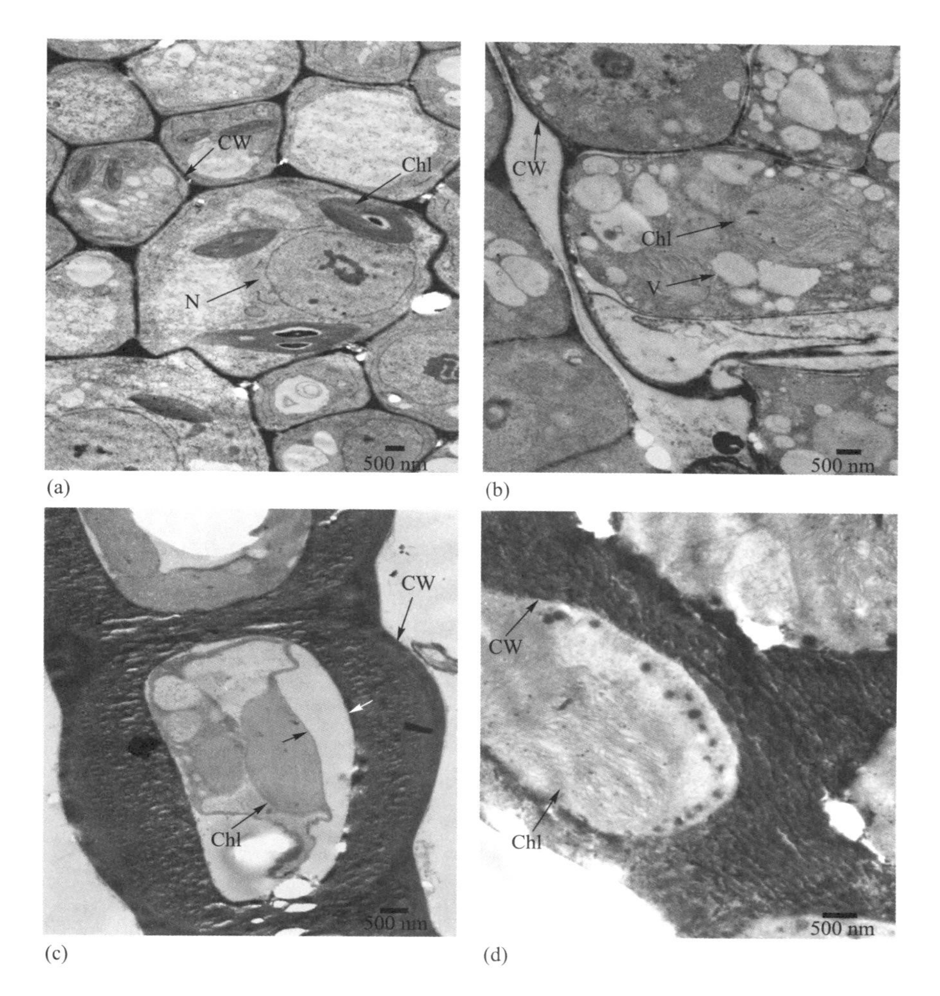

图 3-112　增强 UV-B 辐射对土生对齿藓细胞超微结构的影响：2.75 W m^{-2}（a）、3.08 W m^{-2}（b）、3.25 W m^{-2}（c）、3.41 W m^{-2}（d）UV-B 辐射处理下细胞超微结构。Chl：叶绿体；CW：细胞壁；N：细胞核；V：囊泡

Figure 3-112　Effects of enhanced UV-B radiation on cell ultrastructure of *D. vinealis*: Cell ultrastructure under 2.75 W m^{-2} (a)、3.08 W m^{-2} (b)、3.25 W m^{-2} (c)、3.41 W m^{-2} (d) UV-B treatment. Chl: chloroplast; CW: cell wall; N: nucleus; V: vesicle

土生对齿藓细胞发生质壁分离；细胞内囊泡数目增多（图3-112b）。在3.25 W m^{-2} UV-B辐射胁迫下，土生对齿藓细胞的细胞壁变厚极为明显，厚度不均一；细胞发生质壁分离；多数细胞器溶解（图3-112c）。尤其在3.41 W m^{-2} UV-B 辐射强度处理下，土生对齿藓细胞受损严重，细胞器溶解，细胞壁内侧变得凹凸不平，部分细胞壁破碎，细胞膜大部分地方成了碎片，散落在细胞质中，细胞质基质外渗（图3-112d）。

比较真藓和土生对齿藓细胞超微结构在UV-B辐射增强处理下的变化发现，两种藓类细胞超微结构均随着UV-B辐射强度增强而受到损伤。随着UV-B辐射强度增加，真藓和土生对齿藓叶绿体肿胀变圆，细胞器溶解消失，膜系统降解消失，且真藓细胞随着强度的增强，细胞器结构消失，产生大量淀粉粒，细胞破裂死亡。以上结果说明在UV-B辐射增强处理下，真藓细胞超微结构受损程度明显高于土生对齿藓。蒋霞敏等（2003）对UV-B辐射条件下雨生红球藻（*Haematococcus pluvialis*）细胞超微结构变化的研究表明，UV-B辐射增强对藻类细胞结构有较大的损伤，细胞壁、叶绿体、线粒体等均出现了明显的伤害症状，尤其是叶绿体被严重破坏，类囊体断裂、弯曲，叶绿体内含物流出。另外，吴杏春等（2007）和Kakani等（2003）对水稻叶片超微结构研究显示，UV-B辐射处理下叶绿体结构变形，基质片层排列稀疏紊乱，嗜锇颗粒体积增大。类囊体膜的破坏导致叶绿体蛋白稳定性降低，并加速了Chl的降解。这与我们对两种藓类的Chl-a荧光诱导动力学参数、光合色素、叶绿体类囊体膜蛋白等的测定结果一致。本试验TEM观察到的膜结构损伤与MDA含量累积的结果一致。细胞壁是植物细胞应对UV-B辐射的第一道屏障。庞海河（2010）以自然光为对照，分别增补UV-B辐射强度3.25 μW cm^{-2} nm^{-1}和9.76 μW $cm^{-2}nm^{-1}$为处理组。结果表明，UV-B辐射增强处理后，南方红豆杉叶片表皮细胞壁的厚度分别较对照增加了17.6%和29.5%。这与我们的研究结果一致，随着辐射强度的增加，土生对齿藓细胞发生质壁分离，细胞壁增厚极为明显（图3-112）。然而，真藓中并没有上述现象发生。这可能与两种藓类适应UV-B辐射的能力和自身不同的应对机制有关。细胞壁的增厚可能增强了土生对齿藓对UV-B辐射的抵御能力，这与我们的其他结果一致。

增强UV-B辐射对真藓和土生对齿藓叶绿体超微结构的影响：大量的研究表明，叶绿体是重要的光合器官，也是最易受到UV-B辐射损伤的亚细胞结构（Allen *et al.*，2011）。增强UV-B辐射对真藓叶绿体超微结构的影响如图3-113所示。不同强度UV-B辐射处理下，真藓叶绿体超微结构发生了较为明显的变化。在对照处理下，细胞内叶绿体的形状表现为形态饱满、结构完整的长椭球形；基粒片层整齐有序呈平行排列，同时类囊体片层垛叠整齐；嗜锇颗粒较少；整个叶绿体膜系统结构完整，双层膜结构清晰（图3-113a）。在3.08 W m^{-2} UV-B辐射处理下，叶绿体明显肿胀变大，呈近球形；叶绿体类囊体片层结构松散，部分类囊体结构模糊，类囊体膜结构受到一定程度的破坏；嗜锇颗粒增多（图3-113b）。在3.25 W m^{-2} UV-B辐射处理下，叶绿体双层膜系统只有部分结构完整，其余模糊消失；叶绿体中大量类囊体结构模糊；嗜锇颗粒聚集变大；淀粉粒边缘出现糊化（图3-113c）。在3.41 W m^{-2} UV-B辐射处理下，叶绿体膜系统结构降解；类囊体膜系统降解，基粒片层变得模糊不清，只剩下边缘极度糊化的淀粉粒（图3-113d）。

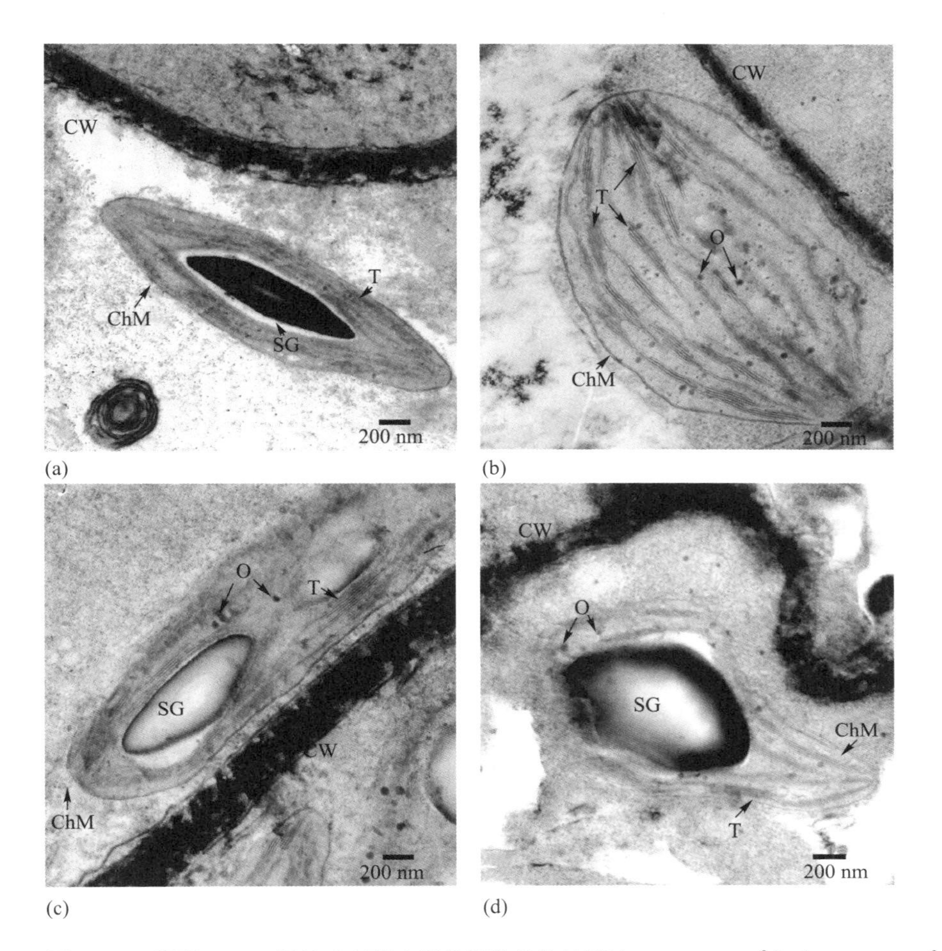

图 3-113　增强 UV-B 辐射对真藓叶绿体超微结构的影响：2.75 W m^{-2}（a）、3.08 W m^{-2}（b）、3.25 W m^{-2}（c）、3.41 W m^{-2}（d）UV-B 辐射处理下叶绿体超微结构。ChM：叶绿体膜；CW：细胞壁；O：嗜锇颗粒；SG：淀粉粒；T：类囊体

Figure 3-113　Effects of enhanced UV-B radiation on chloroplast ultrastructure of *B. argenteum*: Chloroplast ultrastructure under 2.75 W m^{-2}（a）、3.08 W m^{-2}（b）、3.25 W m^{-2}（c）、3.41 W m^{-2}（d）UV-B treatment. ChM: chloroplast membrane; CW: cell wall; O: osmiophilic granule; SG: starch grain; T: thylakoid

不同强度UV-B辐射处理对土生对齿藓叶绿体超微结构的影响如图3-114所示。对照处理时，土生对齿藓细胞内叶绿体呈长椭球形；基粒片层整齐有序呈平行排列，同时类囊体片层垛叠整齐；叶绿体膜系统结构完整，双层膜结构清晰（图3-114a）。在3.08 W m^{-2} UV-B辐射处理下，叶绿体肿胀明显，呈球形；叶绿体类囊体片层结构松散，部分类囊体结构模糊，类囊体膜受到一定程度破坏；嗜锇颗粒聚集变大（图3-114b）。在3.25 W m^{-2} UV-B辐射处理下，叶绿体肿胀变形，叶绿体膜系统开始降解；叶绿体类囊体片层结构松散变形、相互交织；

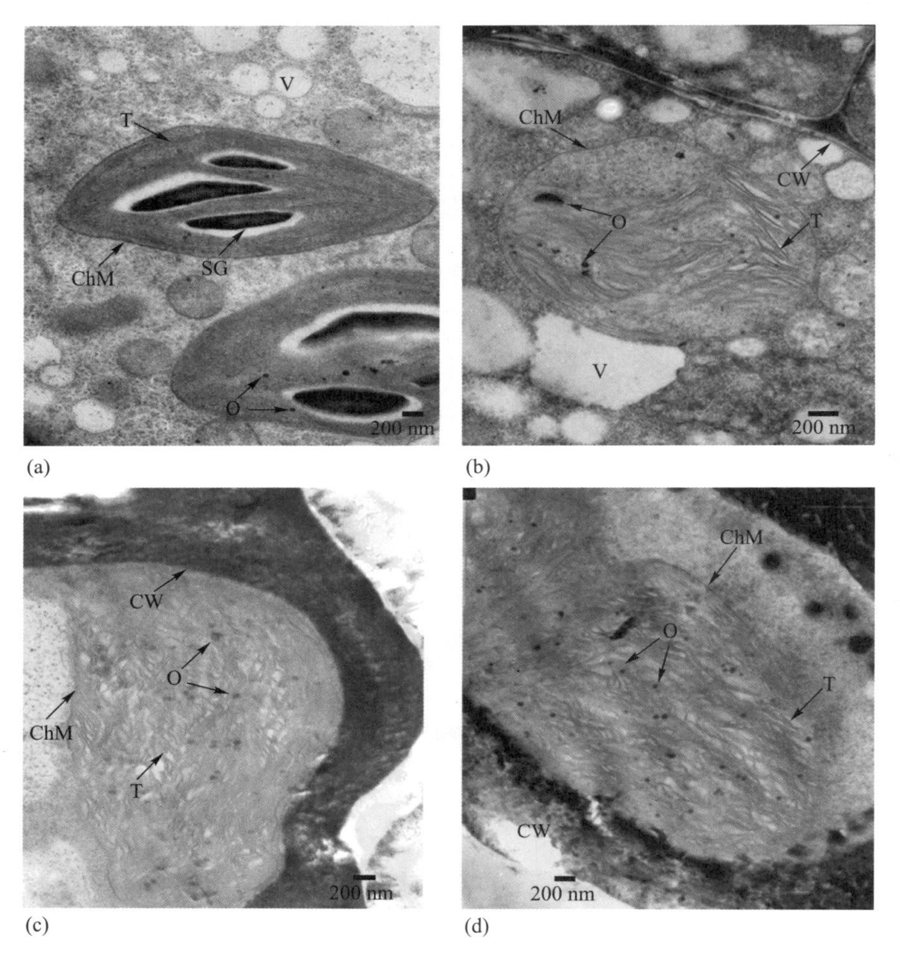

图 3-114　增强 UV-B 辐射对土生对齿藓叶绿体超微结构的影响：2.75 W m^{-2}（a）、3.08 W m^{-2}（b）、3.25 W m^{-2}（c）、3.41 W m^{-2}（d）UV-B 辐射处理下叶绿体超微结构。ChM：叶绿体膜；CW：细胞壁；O：嗜锇颗粒；SG：淀粉粒；T：类囊体；V：囊泡

Figure 3-114　Effects of enhanced UV-B radiation on chloroplast ultrastructure of *D. vinealis*：Chloroplast ultrastructure under 2.75 W m^{-2}（a）、3.08 W m^{-2}（b）、3.25 W m^{-2}（c）、3.41 W m^{-2}（d）UV-B treatment. ChM：chloroplast membrane；CW：cell wall；O：osmiophilic granule；SG：starch grain；T：thylakoid；V：vesicle

嗜锇颗粒增多（图3-114c）。在3.41 W m^{-2} UV-B辐射处理下，叶绿体肿胀变形，叶绿体类囊体片层松散变形，电子密度降低；部分膜系统降解；嗜锇颗粒增多（图3-114d）。

在增强UV-B辐射处理下，真藓和土生对齿藓叶绿体超微结构损伤程度均随辐射强度增大而加剧。随着辐射强度的增加，最终真藓叶绿体膜系统结构彻底降解，双层膜结构模糊，类囊体膜系统降解，淀粉粒边缘极度糊化。而土生对齿藓叶绿体肿胀变形，叶绿体膜系统开始降解，叶绿体类囊体片层结构松散变形，相互交织，电子密度降低，部分膜系统降解。土生对齿藓叶绿体中并没有发生叶绿体结构彻底降解的现象。唐莉娜（2003）以水稻为研究材料发现UV-B辐射胁迫破坏了其叶绿体的形态结构，叶绿体发生扭曲变形，叶绿体膜模糊或消失，叶绿体中的基粒垛叠的片层减少，基粒片层和基质片层膨胀形成空泡。Buchanan等（2000）的研究表明，在增强UV-B辐射处理下，叶绿体膜结构的损伤导致叶绿体蛋白的稳定性降低，并且加速Chl的降解，从而引起叶绿体结构的损伤。

3.4.2 UV-B 辐射增强对不同发育阶段荒漠藻类结皮光合作用的影响

采集沙坡头沙漠试验研究站包兰铁路以北1956年（51龄）、1981年（26龄）人工植被区和自然植被区丘顶发育良好、完整连片分布的藻类结皮（采样时间为2007年9月）。采集结皮之前先用蒸馏水充分湿润以防止在采集过程中破碎，然后利用铝制土壤盒（直径为4.5 cm，样品的表面积约为15.9 cm^2）扣取完整的藻类结皮带回实验室并用蒸馏水冲洗，尽量去除结皮上的土壤盒沙粒。沙坡头地区正午最大UV-B辐射强度为0.25 mW cm^{-2}，UV-B辐射增加处理设置为增加最大辐射强度的2%，即0.005 mW cm^{-2}。UV-B辐射处理采用40 W的UV-B 313型紫外灯管，通过调节样品与灯管的距离获得0.005 mW cm^{-2}的辐射强度，每天将三个植被区半数的藻类结皮置于UV-B灯下12 h（08：00—20：00），持续3天。UV-B辐射处理后对藻类结皮Chl-a含量及光合速率进行测量。

（1）UV-B辐射对三个植被区藻类结皮Chl-a的影响

自然环境条件下，自然植被区、51龄和26龄植被区藻类结皮的Chl-a含量分别为2.10 μg cm^{-2}、2.86 μg cm^{-2}和1.96 μg cm^{-2}，UV-B辐射处理后，以上三个植被区藻类结皮的Chl-a含量分别为1.48 μg cm^{-2}、1.60 μg cm^{-2}和1.35 μg cm^{-2}。UV-B辐射的增强显著降低了51龄和26龄人工植被区藻类结皮的Chl-a含量（图3-115，51龄：p=0.03；26龄：p=0.02），但对自然植被区藻类结皮的Chl-a影响不显著（图3-115，p=0.08）。自然环境中，51龄人工植被区藻类结皮的Chl-a含量显著高于26龄人工植被区和自然植被区（图3-115，p<0.05），而UV-B辐射处理降低了三个植被区藻类结皮的Chl-a含量（图3-115，p>0.05）。

（2）UV-B辐射处理前后三个植被区藻类结皮净光合速率

当藻类结皮含水量为100%时，UV-B辐射增强显著抑制了三个植被区藻类结皮的净光合速率（P_n）（图3-116a，d；p<0.05），在环境温度为25 ℃时，自然植被区、51龄和26龄

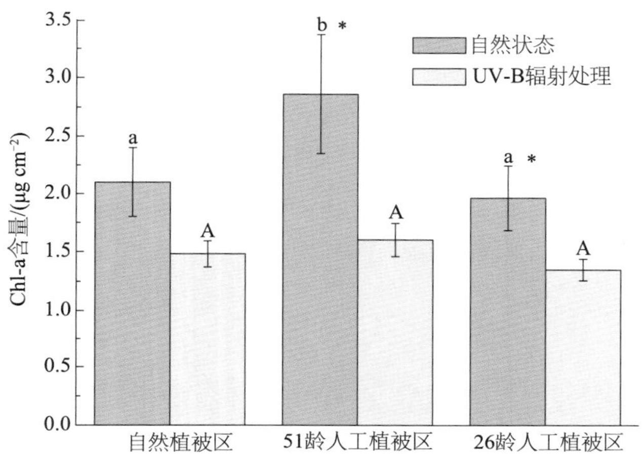

图 3-115 UV-B 辐射处理前后三个植被区（自然植被区、51 龄和 26 龄人工植被区）藻类结皮的 Chl-a 含量。不同小写字母表示自然状态下三个植被区藻类结皮 Chl-a 含量差异显著，不同大写字母表示 UV-B 辐射处理后 3 个植被区藻类结皮 Chl-a 含量差异显著，*UV-B 辐射处理前后 Chl-a 含量差异显著，所有检验都是在 p=0.05 水平上进行

Figure 3-115 The Chl-a content of algal crusts at three successions（natural vegetation，51 years and 26 years of artificial vegetation areas）with or without enhanced UV-B treatment. Different lowercase letters indicates significant difference under natural condition；different capital letters indicates significant difference after enhanced UV-B treatment；*indicates significant difference with or without UV-B treatment，p<0.05

人工植被区藻类结皮的P_n分别下降45%、55%和48%；在环境温度为15 ℃时，三个植被区的P_n分别下降60%、62%和75%。当藻类结皮含水量为60%时，UV-B辐射的增强显著抑制了26龄人工植被区藻类结皮的P_n（图3-116b，e；p<0.05），当环境温度为25 ℃和15 ℃时，26龄人工植被区藻类结皮的P_n分别下降了24%和59%，此外，自然植被区藻类结皮的P_n在环境温度为25 ℃时受到显著抑制，P_n下降48%（图3-116b；p<0.05）。当藻类结皮含水量为20%时，UV-B辐射的增强显著抑制了51龄人工植被区藻类结皮的P_n，当环境温度为25 ℃和15 ℃时，51龄人工植被区藻类结皮的P_n分别下降了29%和59%（图3-116c，f），并且在15 ℃的环境温度时，26龄人工植被区藻类结皮的P_n显著下降46%（图3-116f，p<0.05）。

（3）UV-B辐射增强对三个植被区藻类结皮P_n的影响

25 ℃时，UV-B辐射增强削弱了不同藻类结皮P_n的差异（图3-117a），并且60%含水量时藻类结皮P_n高于其余两个含水量。100%含水量时，自然植被区、51龄和26龄人工植被区藻类结皮的P_n为0.77～0.8 μmol m^{-2} s^{-1}；60%含水量时，三种藻类结皮的P_n分别为1.8 μmol m^{-2} s^{-1}、2.2 μmol m^{-2} s^{-1}和1.9 μmol m^{-2} s^{-1}；20%含水量时，分别为0.85 μmol m^{-2} s^{-1}、1 μmol m^{-2} s^{-1}和0.91 μmol m^{-2} s^{-1}。

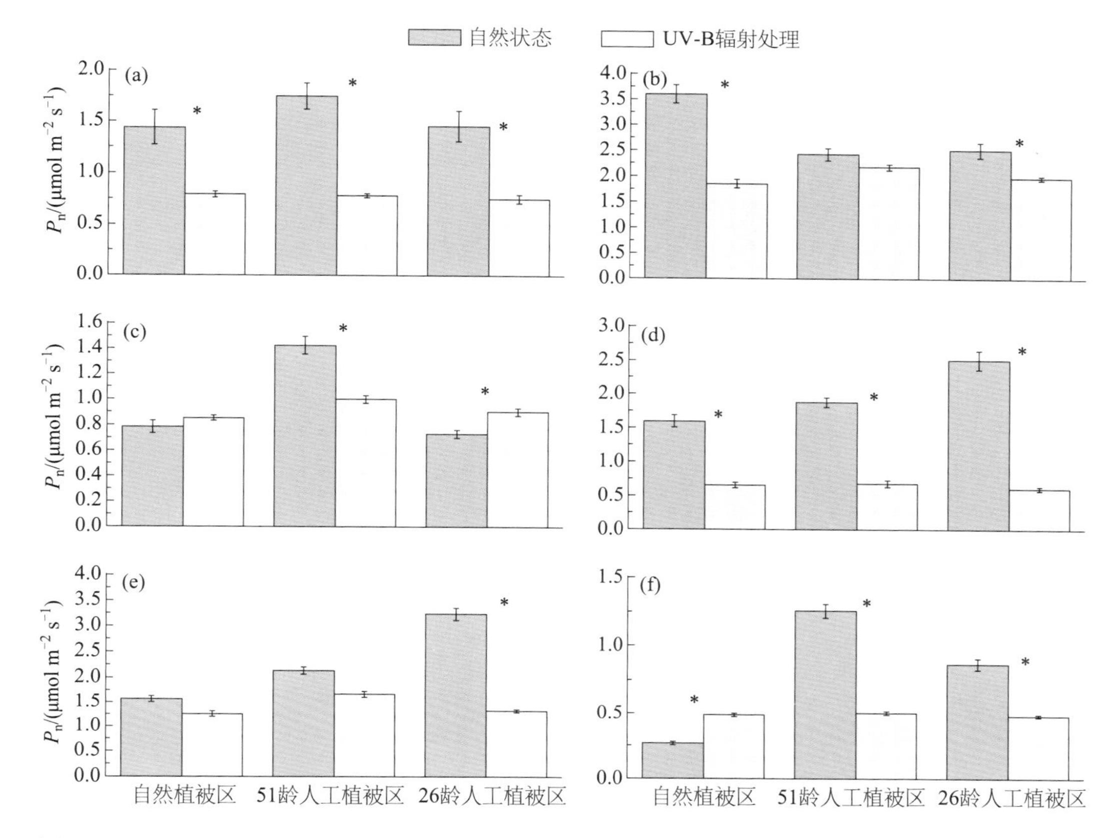

图 3-116　UV-B 辐射处理前后三个植被区（自然植被区、51 龄和 26 龄人工植被区）藻类结皮 P_n。（a）、（b）、（c）表示 25 ℃时三个植被区藻类结皮的 P_n，（d）、（e）、（f）表示 15 ℃时三个植被区藻类结皮的 P_n，（a）和（d）表示藻类结皮含水量为 100%，（b）和（e）表示藻类结皮含水量为 60%，（c）和（f）表示藻类结皮含水量为 20%。* 藻类结皮 P_n 在 UV-B 辐射处理前后差异显著，p=0.05

Figure 3-116　The photosynthesis of algal crusts at three successional stages (natural vegetation area, 51 years and 26 years of artificial vegetation areas) with or without enhanced UV-B treatment. Net photosynthetic rate of algal crust under 100% moisture and 25 ℃ (a) and 15 ℃ (d); Net photosynthetic rate of algal crust under 60% moisture and 25 ℃ (b) and 15 ℃ (e); Net photosynthetic rate of algal crust under 20% moisture and 25 ℃ (c) and 15 ℃ (f); *indicates significant difference between before and after UV-B process of algal crust net photosynthetic rate

与25 ℃时不同，当环境温度为15 ℃，藻类结皮含水量为60%时，51龄人工植被区的P_n显著高于26龄人工植被区和自然植被区（图3-117b，p<0.05），并且此时自然植被区、51龄和26龄人工植被区藻类结皮的P_n分别为1.28 μmol m^{-2} s^{-1}、1.68 μmol m^{-2} s^{-1}和1.33 μmol m^{-2} s^{-1}。而在20%和100%含水量时，不同植被区藻类结皮的P_n无显著差异（图3-117b，p>0.05）。高含水量时，P_n为0.63 ~ 0.71 μmol m^{-2} s^{-1}，低含水量时为0.47 ~ 0.5 μmol m^{-2} s^{-1}。

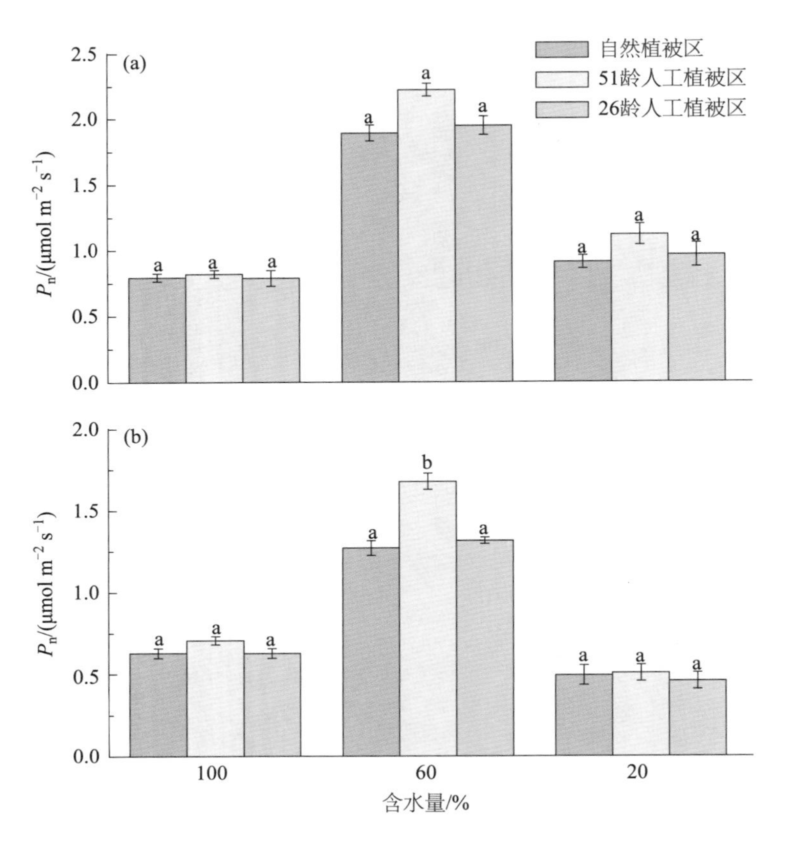

图 3-117　25 ℃（a）和 15 ℃（b）经 UV-B 辐射增强处理的三个植被区（自然植被区、51 龄和 26 龄人工植被区）藻类结皮的 P_n。不同的小写字母表示三个植被区藻类结皮的 P_n 差异显著，$p<0.05$

Figure 3-117　The photosynthesis of algal crusts at three successional stages (natural vegetation area, 51 years and 26 years of artificial vegetation areas) treated by enhanced UV-B. Different lowerbase letters indicates significant difference among three vegetation areas, $p<0.05$

UV-B辐射增强显著降低了藻类结皮的P_n，这一结果同极地藻类结皮和海洋中藻类的结果相一致（Quesada *et al.*，1998；Prasad and Zeeshan，2005）。此外，UV-B辐射增强处理下，藻类结皮光合速率的变化可以从其Chl-a含量的变化上得到反映，UV-B辐射增强显著降低了藻类结皮Chl-a的含量，使得三个植被区藻类结皮Chl-a含量均下降。这说明，UV-B辐射对藻类结皮P_n的抑制是通过减少叶绿素含量的间接途径和对光合系统影响的直接途径实现的（Wu *et al.*，2005；Xue *et al.*，2005）。Rech等（2005）研究了紫外辐射强度和持续时间对5种硅藻光合作用的影响，结果表明，紫外辐射强度和持续时间的增加降低了5种硅藻的P_n，紫外辐射对最大P_n的减少大于40%。紫外辐射还可以通过改变结皮有机体的物种组成对藻类结皮有机体P_n产生影响。以往的研究指出，紫外辐射的增强会使易产生光保护色素的物种数量增加，在湿润条件下，这些物种会扩展到结皮有机体的表面并产生光保护色素，而在表面进行光合作用的有机体则会向下运动，使光合器官得以受到保护（Nadeaul *et al.*，1999；Hernandoa and Ferreyra，2005）。我们的研究中，结皮优势种具鞘微鞘藻是一种移动性很强的藻，一些研究指出，在有强紫外辐射的情况下，它们将向下层土壤移动，由于下层土壤光照条件的显著下降，有机体的光合速率急剧降低（Hernandoa and Ferreyra，2005）。模拟试验的结果表明，在气候变化的影响下，UV-B辐射的持续增强将降低荒漠藻类结皮的光合速率，降低荒漠生态系统的初级生产力，从而影响荒漠植被恢复演替的进程和恢复能力。

在对照处理下，不同发育阶段藻类结皮P_n存在显著差异。然而，UV-B辐射增强处理后，三个植被区藻类结皮的P_n没有显著差异。这一结果说明，UV-B辐射增强对不同发育阶段藻类结皮P_n的影响也不相同，特别是显著降低了自然植被区和51龄人工植被区藻类结皮的P_n，UV-B辐射增强对发育初期藻类结皮光合的影响要大于发育后期。造成这一结果的原因可能是发育初期藻类结皮的抗紫外线辐射保护色素含量较低和结皮种类较为单一，藻类结皮对UV-B辐射增强缺乏相应的补偿机制；发育后期的藻类结皮垂直结构完整，结皮中不同功能藻类之间具有共生和互利关系，在UV-B辐射增强后，不同藻类间生理活性和位置的变化会部分降低UV-B辐射的影响。

3.5 BSC对CO_2浓度的生态生理响应

3.5.1 BSC-土壤系统温室气体通量特征

CO_2、甲烷（CH_4）和氧化亚氮（N_2O）是三种最重要的温室气体。近一个世纪以来，大气中CO_2、CH_4和N_2O的浓度明显升高，导致过去30年全球年均气温按每10年2 ℃的幅度增加（IPCC，2007）。大气中的CO_2、CH_4和N_2O在生物圈和大气圈之间发生着复杂的交换，其“源-汇”关系和交换量因生态系统类型的不同有很大差异（Chapuis-Lardy *et al.*，2006；Dalal and Allen，2008），并受土壤理化性质和植被类型（Grogan and Jonasson，2005；Dalal and Allen，2008）、环境因子（Flanagan and Johnson，2005；Jones *et al.*，2005）、土地利用方式（Jones *et al.*，2005；Hu *et al.*，2010）等多重因素的影响。目前有关CO_2、CH_4和N_2O通量的观测主要集中在农田、森林、草地、湿地和冻原生态系统（Grogan and Jonasson，2005；Dalal and Allen，2008），而针对荒漠生态系统的监测数据相当匮乏。其中，绝大部分研究只集中在生长季，对非生长季的报道十分有限（Jones *et al.*，2005；Mastepanov *et al.*，2008），这给区域尺度上CO_2、CH_4和N_2O通量的估算带来很大的不确定性。

在腾格里沙漠东南缘的包兰铁路两侧，利用草方格固定沙丘表面并种植旱生灌木形成人工固沙植被后，BSC开始发育并不断拓殖和演替，50年后形成了高等植物和BSC隐花植物镶嵌分布的稳定格局，成为沙漠化逆转和生态恢复的典型案例。研究发现，在这一生态恢复过程中，表层土壤的营养状况得以改善（Li *et al.*，2007），BSC及蚂蚁等昆虫的多样性逐渐增加（Li *et al.*，2010d；Liu *et al.*，2011），土壤酶活性逐渐增强（Zhang *et al.*，2012）。在该恢复过程中，BSC-土壤系统中温室气体（CO_2、CH_4和N_2O）通量发生了怎样的变化？目前还不清楚。本研究在沙坡头地区不同年代建植的人工植被固沙区，选择不同类型和不同演替阶段的BSC-土壤系统，通过在全年尺度上对其CO_2、CH_4和N_2O通量的监测，剖析不同类型BSC覆盖的土壤以及沙漠化逆转过程中BSC-土壤系统与大气之间的温室气体交换特征，以及BSC-土壤系统中CO_2、CH_4和N_2O通量与土壤温度和水分之间的关系，旨在为客观地判别荒漠生态系统碳、氮循环以及“源-汇”关系提供基础数据。

选择1964年建植的人工植被固沙区盖度均一的藻类结皮和藓类结皮、1981年建植的人工植被固沙区的藓类结皮和1981年无BSC覆盖的活化沙丘为研究对象，共有4个类型，其BSC盖度分别为91%、98%、94%和0。以1981年无BSC覆盖的活化沙丘作为对照（CK），每个类型3次重复。采样点选择在沙丘顶部地势相对平坦、微地形相对一致的位置。在2011年11月至2012年11月持续1年内，采用静态箱-气相色谱法测定CO_2、CH_4和N_2O的通量（Hu *et al.*，2010）。每个月中旬左右采集气体样品，采样前先人工去除BSC外的其他草本和枯枝落叶，采样时间为9：00—10：00，4个样点同时进行采样。具体方法是将顶箱（长×宽×高=40 cm×40 cm×40 cm）合扣在预先安装在土壤中的不锈钢底座（长×宽×深=40 cm×40 cm×15 cm）上，顶箱与底箱用密封条密封。顶箱内装有风扇，用于箱内气体混合均匀。在密封后的0 min、10 min、20 min和30 min，用50 mL的注射器通过三通阀采集静态箱中的气体30 mL。用手持式温度记录仪（JM624，今明仪器有限公司，中国）测定采样始末静态箱内的温度和5 cm的土壤温度（图3-118）。同时，用土壤水分测定仪（TDR300，美国）测定5 cm深处的土壤湿度。所采集的气体样品带回室内后，在气相色谱仪（Agilent GC6820，美国）上分析气体中CO_2、CH_4和N_2O的浓度。所有气体样品在采集后的24 h内完成分析。

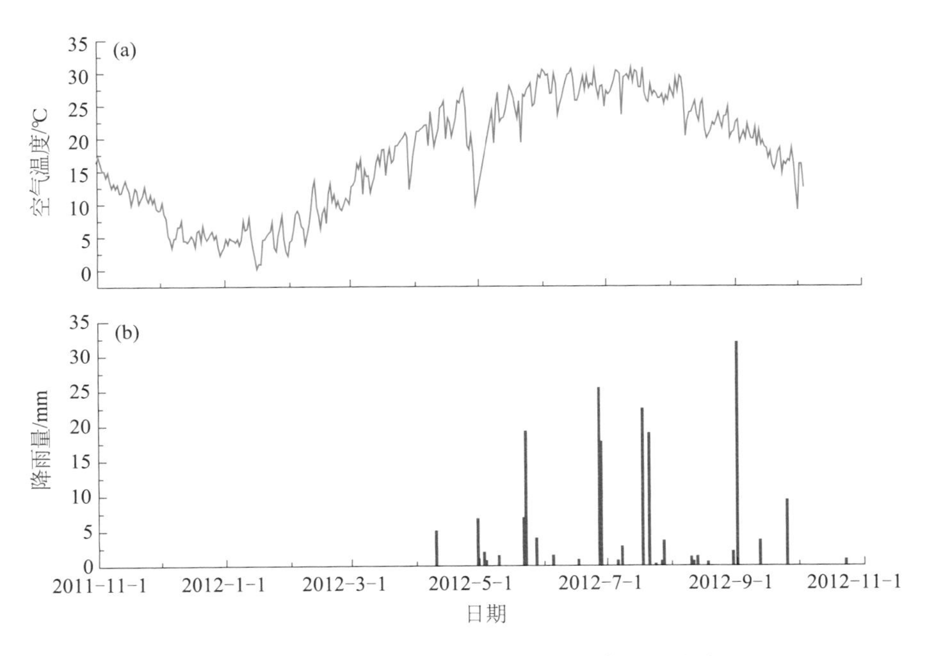

图3-118 2011年和2012年研究区的空气温度（a）和降雨量（b）
Figure 3-118 Air temperature (a) and precipitation (b) in 2011 and 2012

（1）人工植被固沙区BSC-土壤系统的CO_2通量特征

由表3-22可以看出，BSC类型、恢复时间、采样日期以及BSC类型与采样日期、恢复时间与采样日期的相互作用显著影响人工植被固沙区BSC-土壤系统的CO_2通量。BSC-土壤系统CO_2通量的变化因采样日期的不同差异很大，在–23.3 ~ 698.2 mg m^{-2} h^{-1}之间变化。藓类结皮年均通量（105.1 mg m^{-2} h^{-1}）显著高于藻类结皮（37.7 mg m^{-2} h^{-1}），特别是生长季降雨后（7月26日和9月25日），藻类结皮和藓类结皮之间的差异非常明显；非生长季（11月至次年3月）藻类结皮和藓类结皮平均CO_2通量很低，分别为0.7 mg m^{-2} h^{-1}和8.8 mg m^{-2} h^{-1}，显著低于生长季（4—10月）平均通量（74.4 mg m^{-2} h^{-1}和209.0 mg m^{-2} h^{-1}）（图3-119a）。不同建植年代的藓类结皮和活化沙丘年均CO_2通量按1964年藓类结皮、1981年藓类结皮和1981年活化沙丘的顺序逐渐递减，1981年活化沙丘的年均CO_2通量最低（34.8 mg m^{-2} h^{-1}），显著低于1964年藓类结皮（111.6 mg m^{-2} h^{-1}），表明在人工植被固沙区，随着BSC的发育和演替的深入，BSC-土壤系统呼吸逐渐增加。但1981年藓类结皮与活化沙丘、1964年藓类结皮和1981年藓类结皮之间的差异不显著（图3-120b）。

表 3-22　BSC- 土壤系统的 CO_2、CH_4 和 N_2O 通量多因素方差分析

Table 3-22　Multivariate analysis of CO_2，CH_4 and N_2O fluxes in crust-soil system

因素	CO_2		CH_4		N_2O	
	F	*p*	*F*	*p*	*F*	*p*
BSC 类型（T）	33.17	<0.01	0.677	0.413	0.008	0.928
恢复时间（Y）	8.77	0.004	7.77	0.006	0.146	0.703
采样日期（D）	29.89	<0.01	6.10	<0.001	3.49	<0.01
T×D	16.38	<0.01	2.51	0.008	0.194	0.998
Y×D	5.41	<0.01	0.895	0.548	0.747	0.691

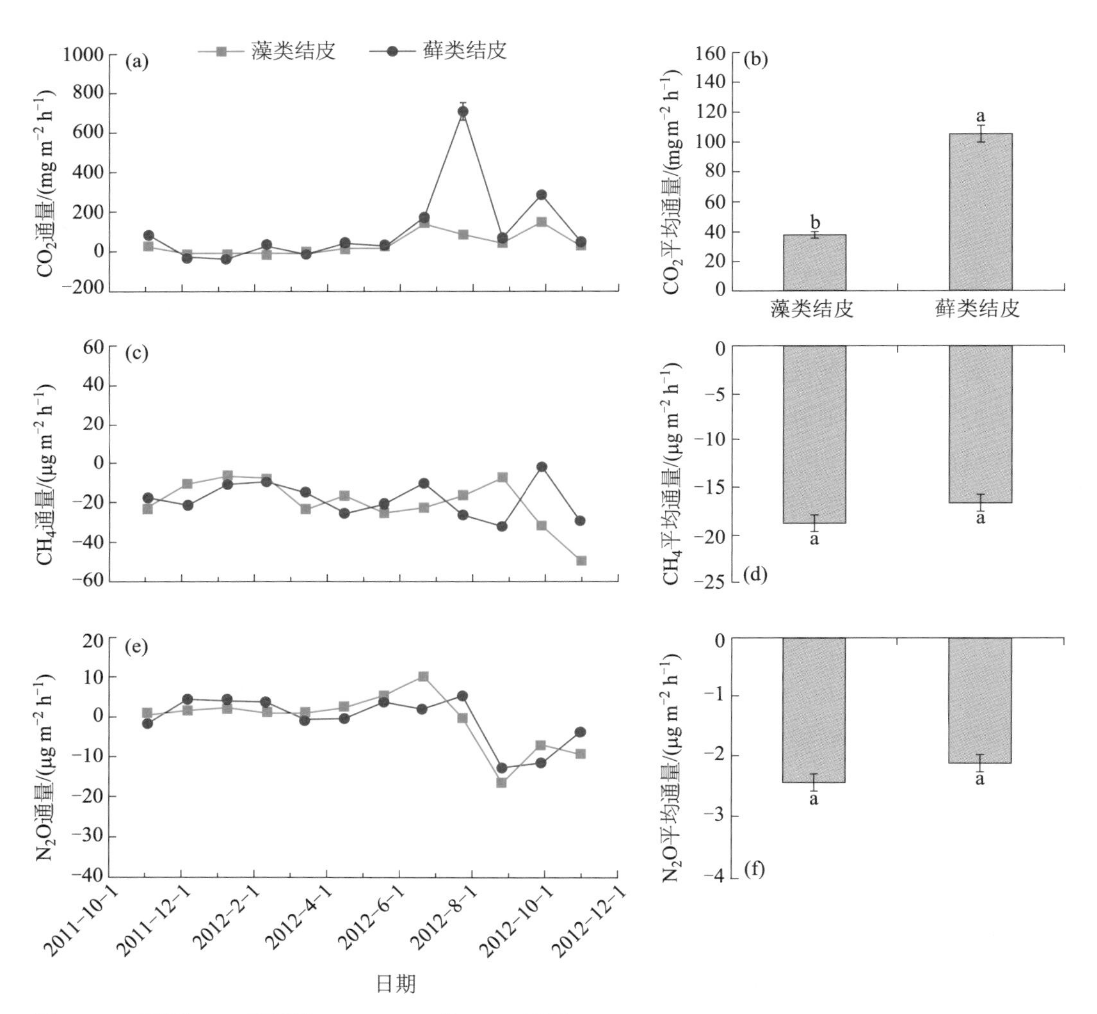

图 3-119　不同 BSC 类型的 CO_2、CH_4 和 N_2O 通量

Figure 3-119　Dynamics of CO_2，CH_4 and N_2O fluxes for various BSC types

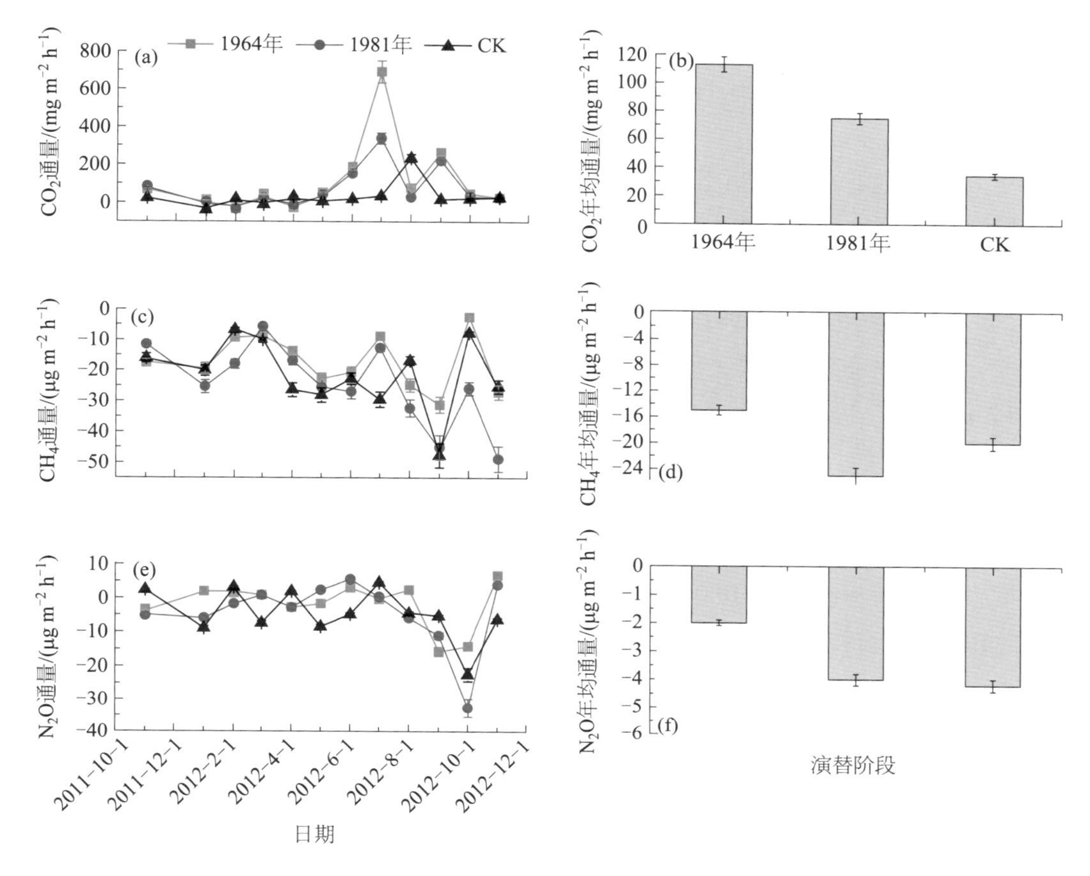

图 3-120　不同演替阶段藓类结皮的 CO_2、CH_4 和 N_2O 通量

Figure 3-120　Fluxes of CO_2, CH_4 and N_2O for moss crust in various succession stages

(2) 人工植被固沙区BSC-土壤系统的CH_4通量

恢复时间、采样日期，以及BSC类型与采样日期的互作显著影响BSC-土壤系统的CH_4通量，BSC类型对CH_4通量的影响不显著，恢复时间与采样日期之间没有互作效应（表3-22）。如图3-119c和图3-120c所示，CH_4通量并没有显示出明显的季节变化，除1964年藓类结皮个别日期（9月25日）CH_4通量为正值外，其他日期均为负值。说明在绝大多数时间里，荒漠人工植被固沙区BSC-土壤系统表现为吸收CH_4，是CH_4的汇。所有BSC类型CH_4年均通量为–19.9（–77.6～41.7）μg m^{-2} h^{-1}。藻类结皮年均CH_4通量（–18.8 μg m^{-2} h^{-1}）略低于藓类结皮（–16.6 μg m^{-2} h^{-1}），但两者差异并不显著（图3-119d）。1981年藓类结皮年均CH_4吸收通量（24.0 μg m^{-2} h^{-1}）显著高于1964年藓类结皮，但两者与1981年活化沙丘CH_4吸收通量差异均不显著（图3-120d）。所有BSC类型（包括活化沙丘）生长季CH_4平均吸收通量

（23.9 μg m^{-2} h^{-1}）高于非生长季（14.3 μg m^{-2} h^{-1}），是非生长季CH_4平均吸收通量的1.7（1.5~ 1.9）倍，说明生长季荒漠人工植被固沙区BSC–土壤系统CH_4的吸收能力比非生长季更强。

（3）人工植被固沙区BSC–土壤系统的N_2O通量

采样日期显著影响N_2O通量，恢复时间和BSC类型对N_2O通量的影响不显著，恢复时间、BSC类型与采样日期之间也不存在互作效应（表3-22）。N_2O通量并没有表现出明显的季节变化规律。所有BSC类型（包括1981年活化沙丘）N_2O年均通量为–3.4（–54.0 ~ 31.5）μg m^{-2} h^{-1}，表明BSC–土壤系统在全年水平上吸收N_2O，是N_2O的汇。生长季（4—10月）N_2O平均吸收通量（5.2 μg m^{-2} h^{-1}）显著高于非生长季（11月至次年3月）（0.9 μg m^{-2} h^{-1}），是非生长季的5.8（2.3 ~ 7.7）倍，说明生长季BSC–土壤系统N_2O的吸收能力更强。藻类结皮年均N_2O吸收通量略高于藓类，但两者之间差异不显著（图3-119f）。从不同年代建植的人工植被固沙区来看，1981年活化沙丘、1981年藓类结皮和1964年藓类结皮年均N_2O通量差异均不显著，并按照1981年活化沙丘、1981年藓类结皮和1964年藓类结皮的顺序降低（图3-120f），说明随着BSC的发育和演替的深入，荒漠人工植被固沙区BSC–土壤系统吸收N_2O的能力逐渐减弱。

如图3-121a和c所示，BSC–土壤系统CO_2通量与5 cm土壤温度和湿度呈显著正相关关系，这表明随着土壤温度和湿度的增加，荒漠BSC–土壤系统呼吸逐渐加强。温度和湿度分别解释21.1%和11.0%的CO_2通量变化，两者结合解释26.2%的CO_2通量变化，其二元回归方程为 Y=5.017T+6.488M+2.301（R^2=0.262，p<0.001）。式中，T为土壤温度，M为土壤湿度。就不同年代和类型的BSC而言，5 cm土壤温度和湿度分别能解释 26.8%~ 37.6%和9.4%~ 51.3%的CO_2通量变化（表3-23）。CH_4通量与土壤温度呈显著负相关，与土壤湿度显著正相关，5 cm土壤温度和湿度分别解释4.7%和8.6%的CH_4通量变化（图3-121b，d），两者结合解释11.0%的CH_4通量变化，其二元回归方程为Y=–0.209T+1.060M–25.109（R^2=0.110，p=0.001）。说明CH_4吸收通量随着温度的增加和湿度的降低而逐渐增强。N_2O通量与5 cm土壤温度和湿度均不相关。

5 cm土壤温度和湿度与CO_2通量之间线性回归方程的斜率可以用来表征生态系统呼吸的温度和湿度敏感性，斜率越大表示敏感性越强。由表3-23可以看出，1964年藓类结皮–土壤系统CO_2通量对土壤温度和湿度的斜率分别为8.8和15.8，明显大于藻类结皮（2.7和7.9），表明藓类结皮呼吸对温度和湿度变化更敏感。就不同年代的BSC类型来看，其CO_2通量与温度的线性回归方程的斜率按照1981年活化沙丘、1981年藓类结皮和1964年藓类结皮的顺序增加。1981年活化沙丘–土壤系统呼吸与湿度不相关，表明活化沙丘呼吸对湿度不敏感。1964年藓类结皮呼吸与湿度线性回归方程的斜率（15.8）明显高于1981年的藓类结皮（10.0）。

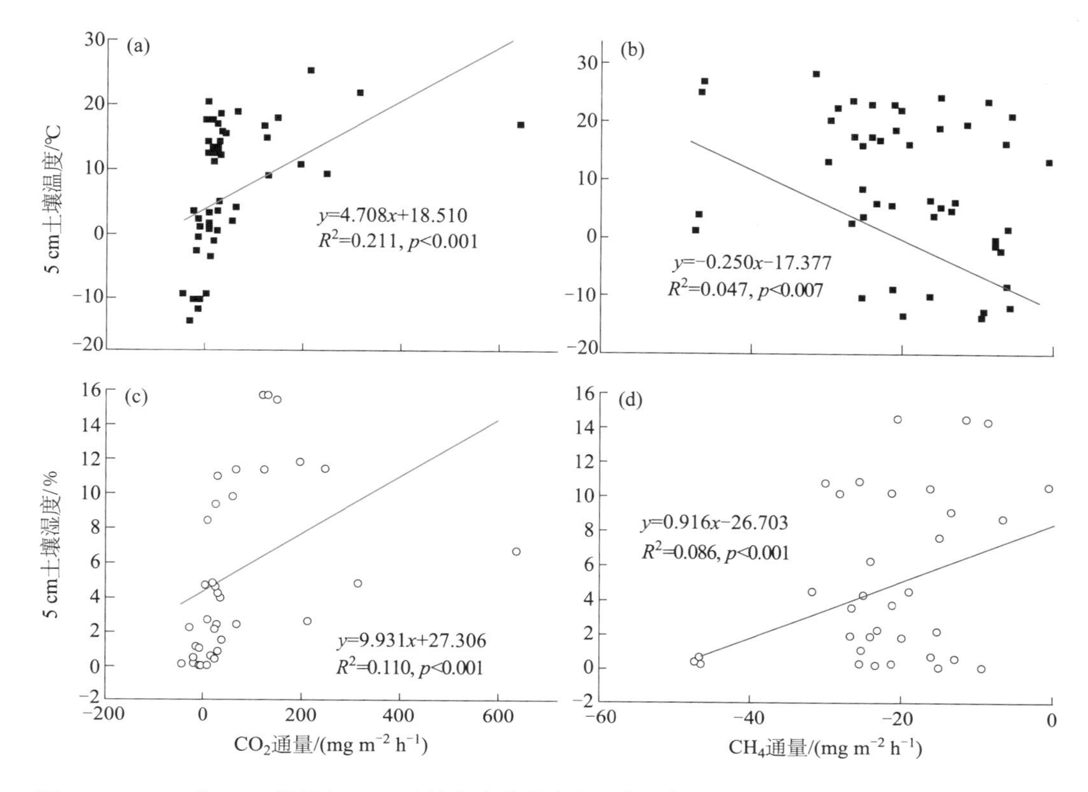

图 3-121　CO_2 和 CH_4 通量与 5 cm 土壤温度和湿度的回归分析

Figure 3-121　Regression between fluxes of CO_2 and CH_4 with soil temperature and moisture at 5 cm soil depth

表 3-23　CO_2 通量与 5 cm 土壤温度和湿度的线性回归分析

Table 3-23　Regression between CO_2 flux and soil temperature or moisture at 5 cm soil depth

年份	类型	5 cm 土壤温度 / ℃			5 cm 土壤湿度 /%		
		线性方程	p	R^2	线性方程	p	R^2
1964	藻类结皮	Y=2.707T+15.874	<0.001	0.376	Y=7.874M+8.220	<0.001	0.513
1964	藓类结皮	Y=8.839T+34.260	0.001	0.268	Y=15.821M+34.188	0.097	0.094
1981	藓类结皮	Y=5.223T+22.711	<0.001	0.374	Y=9.964M+41.444	0.012	0.203
1981	活化沙丘	Y=3.155T−2.199	<0.001	0.364	—	—	

3.5.2 BSC类型下土壤CO_2浓度变化特征及其驱动因子研究

在陆地生态系统中，伴随着土壤的形成、发育过程，土壤CO_2成为土壤空气的主要气体组成成分之一，每年排放到大气中的CO_2约占全球总释放量的5%～20%，是全球碳循环的重要形式和环节，其轻微的变化也会导致大气中CO_2浓度明显改变（张丽华等，2007）。因此，土壤CO_2的研究对估算未来大气CO_2浓度及全球变化具有举足轻重的意义（Liang *et al.*，2004；Trumbore，2006）。BSC的形成使得土层逐渐增厚，表层土壤的理化性质逐渐改善（贾晓红等，2003；杨丽雯等，2009）。而土壤气体中CO_2主要来源于植物根系呼吸和微生物呼吸（吴楠和张元明，2010；刘冉等，2011），其浓度主要受控于生物因素（植物根系、土壤微生物活性等）（吴楠和张元明，2010；刘冉等，2011）和环境因素（土壤理化性质、温度和含水量等）（赵拥华等，2006），因此，BSC的形成必将影响着下层土壤中CO_2的浓度。

荒漠生态系统作为陆地生态系统的重要组成部分，与其他类型的生态系统一样在维持全球生态健康和安全中起着重要的作用，它不仅深刻地影响着整个陆地生态系统的碳循环，而且也是重要的碳贮存地。因此，了解BSC对土壤CO_2释放的影响是全面解析荒漠地区碳循环不可或缺的重要环节，同时也是进一步研究荒漠生态系统如何适应和响应全球变化机制、提出区域科学对策的重要前提。本研究区位于腾格里沙漠东南缘沙坡头地区，该区在经过半个多世纪的人工植被固沙措施的实施后，区域内已有处于不同演替阶段的典型BSC分布。在刁一伟等（2002）研究的基础上自制气体采样器从野外取样，在实验室注入气相色谱（gas chromatography），测得不同深度土壤CO_2浓度变化规律及土壤温度和土壤水分对其的影响。

气体采样器如图3-122所示，主要由注射器、三通阀门、不锈钢钢管、聚四氟导气管和集气管组成，其中集气管为圆柱形的聚四氟乙烯管（外径1.2 cm，内径0.9 cm，长15 cm），管壁两侧针状通气孔均匀分布，以便于管内外土壤气体和水的交换。集气管水平置于土壤内一定深度，垂直向上连接不锈钢钢管，阻止管内气体与地面大气的交换，钢管内装有聚四氟导气管，一端伸到集气管中央，另一端垂直伸出地面约10 cm，露出部分接三通阀门与10 mL注射器相连接。采样器所用均是化学惰性材料，并且实际观测中每次采样间隔6～7天，管内外气体有足够的平衡时间，可以忽略集气管及管内原有气体对观测结果的影响（刁一伟等，2002）。在沙坡头1956年人工植被区内选择发育较好且没有被破坏的藓类结皮和藻类结皮样地，对照选择在流沙样地，并且地形一致。如图3-123所示，首先用土钻在各个采样器埋设处取剖面，然后将气体采样器的集气管垂直插入在藓类结皮、藻类结皮和流沙下5 cm、10 cm、15 cm、20 cm、30 cm、40 cm的土壤中，安装完毕后将土壤原位回填。每个采样器

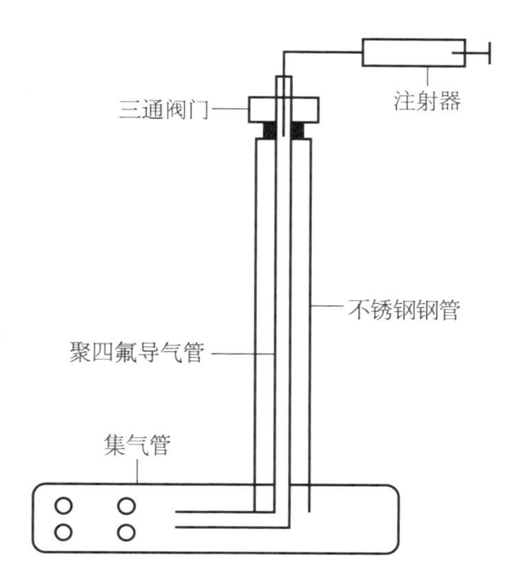

图 3-122　气体采样器
Figure 3-122　Schematic diagram of gas sampler

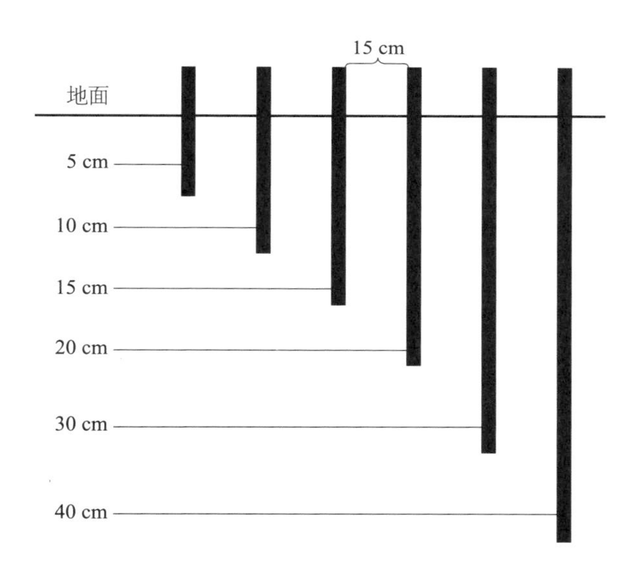

图 3-123　气体采样器安装示意图
Figure 3-123　Schematic diagram of gas sampler fixing in the soil profile

之间的水平间距为15 cm，并设置3个重复，尽量避免采样过程中各土层之间的气体发生相互扰动。本试验中我们埋设的最大深度为40 cm，原因在于结皮层下40 cm以下土壤都为流沙分布，和对照差异不大。

采样时间为2010年6月—2011年6月，固定间隔6天，采样时间固定在每个采样日的上午09：00—11：00。采样时用10 mL标准注射器抽取地表气体5 mL作为0 cm的样品，然后用注射器接入三通阀门，依次从各深度的采样器中各抽取5 mL气体，迅速送到实验室，用气相色谱仪（Agilent GC6820，美国）分析CO_2浓度，气体样品中的CO_2组分在柱箱内经色谱分离后，首先通过镍触媒转化器与H_2反应转化为CH_4，再由氢火焰离子化检测器（FID）检测CH_4信号，载气为高纯N_2，流速15 mL min^{-1}，燃气和反应气H_2流速30 mL min^{-1}，空气流速300 mL min^{-1}，检测器、分离柱温度分别为275 ℃和375 ℃。每个样品的分析时间为2 min。测定前用纯CO_2气体作标准曲线，比照标准曲线计算土壤空气CO_2浓度。

（1）土壤空气CO_2浓度的剖面分布

观测期间藻类结皮、藓类结皮和流沙下的土壤空气CO_2浓度的剖面分布特征如图3-124所示，藻类结皮和藓类结皮在0 cm、5 cm、10 cm、15 cm、20 cm、30 cm、40 cm处的土壤空气CO_2浓度平均值分别为617.94 μmol mol^{-1}、676.69 μmol mol^{-1}、717.54 μmol mol^{-1}、785.86 μmol mol^{-1}、859.96 μmol mol^{-1}、948.01 μmol mol^{-1}和1004.12 μmol mol^{-1}，随着土层深度的增加而增加，主要是因为土壤上层的CO_2气体易和大气进行气体的自由交换，能够快速扩散、逸出土壤，而深层土壤容重大、孔隙度小的特征及地表BSC层的存在等限制了CO_2的扩散，并且随着深度的增加，土壤气体交换受阻，部分微生物活动不断消耗土壤中的O_2并释放出CO_2，造成土壤CO_2含量在较深土层积累较多，从而形成了土壤空气CO_2浓度上低下高的剖面分布特征。两种BSC下的CO_2浓度高于同一深度流沙下土壤空气CO_2浓度值（分别为587.41 μmol mol^{-1}、636.15 μmol mol^{-1}、696.22 μmol mol^{-1}、745.26 μmol mol^{-1}、780.41 μmol mol^{-1}和919.05 μmol mol^{-1}），主要原因在于BSC的形成改善了其下层土壤的理化性质，土壤微生物呼吸增加，因此，BSC层下的土壤空气CO_2浓度要明显高于流沙，但三者之间的差异并不显著（F=0.581，p=0.570）。

（2）土壤温度的影响

土壤温度是土壤空气CO_2浓度变化的重要因素，它主要通过影响微生物量和酶活性而强烈控制微生物活性，如随着温度升高，微生物活动增强，加速了土壤中含碳物质的分解和CO_2产生，同时土壤中生物的呼吸作用也会增强，这些都促使土壤空气CO_2浓度的增加。以观测5 cm土壤温度和土壤空气CO_2浓度数据为例，如图3-125所示两者变化趋势基本一致，但在9月22日土壤空气CO_2浓度急剧增加，主要原因在于当日有降雨发生，影响其浓度值。表3-24相关分析表明，土壤温度对土壤空气CO_2浓度影响程度一般为流沙>藓类结皮>藻类

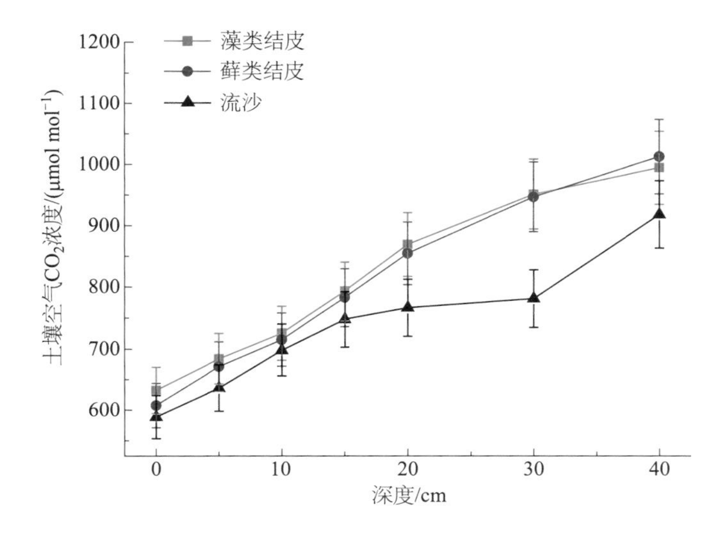

图 3-124　不同深度土壤空气 CO_2 浓度
Figure 3-124　Soil CO_2 concentration at different soil layers

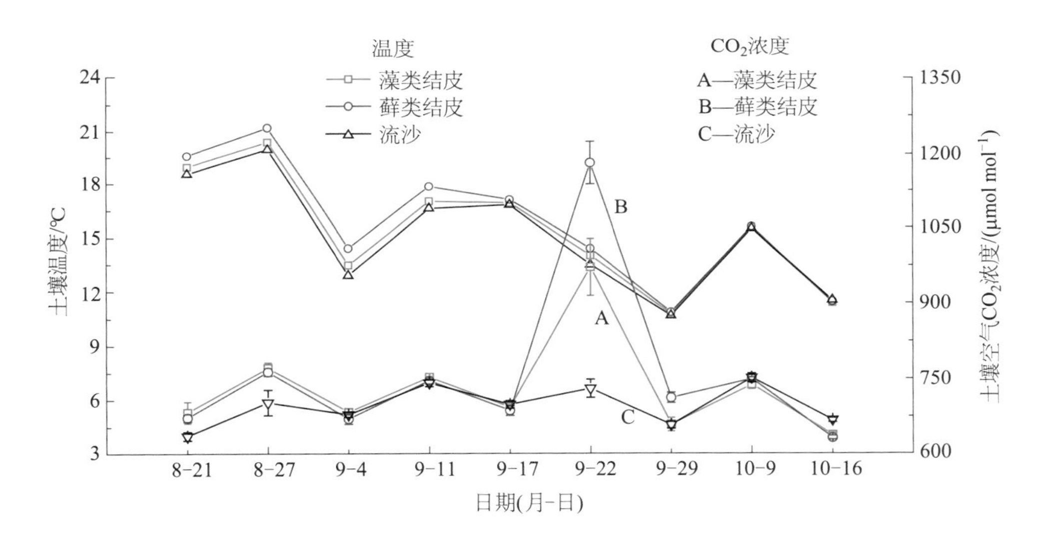

图 3-125　5 cm 处土壤温度与土壤空气 CO_2 浓度值变化
Figure 3-125　Changes of soil temperature and soil CO_2 concentration at 5 cm soil layer

结皮。特别是流沙，在表层相关系数达到了0.868，呈极显著的水平。同时，土壤温度对3个样地土壤空气 CO_2 浓度的影响在0～5 cm范围内较大，然后在5～10 cm降低，但在15～20 cm处又有增加的趋势。

表 3-24 土壤空气 CO_2 浓度与土壤温度的相关性

Table 3-24 Relativity between soil air CO_2 concentration and soil temperature

样地类型	土壤温度				
	0 cm	5 cm	10 cm	15 cm	20 cm
藻类结皮	0.301	0.065	0.059	0.103	0.201
藓类结皮	0.362*	0.164	0.223	0.288	0.262
流沙	0.868**	0.531	0.230	0.360	0.652

注：*p=0.05 水平显著；**p=0.01 水平极显著。

（3）土壤水分的影响

土壤水分是影响土壤空气CO_2浓度的关键环境因子，但影响的方向和程度较为复杂。当土壤相对较干的时候，土壤水分强烈影响着土壤微生物活性及其所参与的生物过程，其随土壤有效水分的增加而增强，而当土壤水分含量充足接近饱和时，氧气的缺乏造成微生物活性降低，抑制了好氧微生物的呼吸。在整个试验期间，土壤水分含量基本上保持在2%～3%，以5 cm土壤水分含量和土壤空气CO_2浓度为例，从图3-126可以看出，两者变化趋势也基本一致。表3-25相关分析表明，土壤水分对土壤空气CO_2浓度的影响程度在表层0～5 cm为流沙>藻类结皮>藓类结皮，但在下层10～40 cm处为藻类结皮>藓类结皮>流沙。特别是藻类结皮，在30 cm和40 cm处的相关系数达到了0.842和0.819，为极显著的水平，成为影响土壤CO_2浓度的主要环境因子。

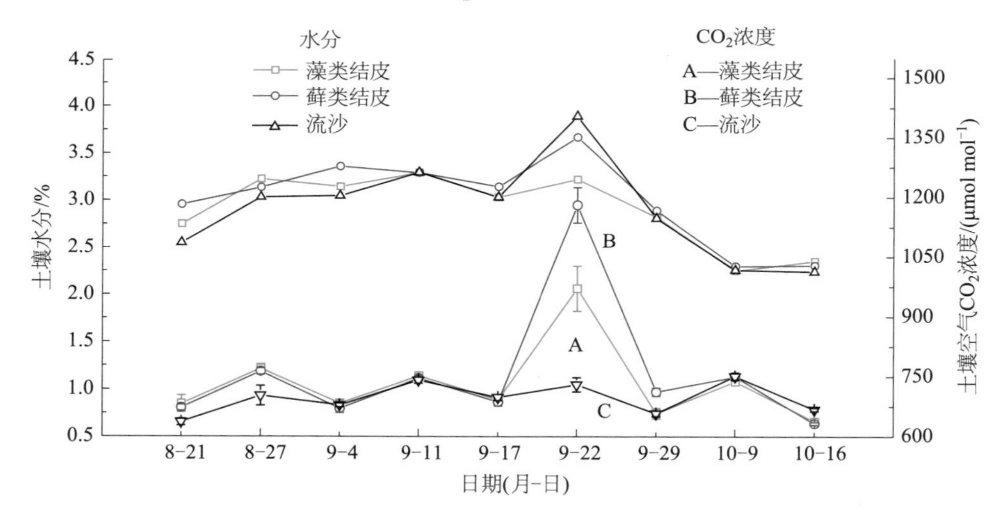

图 3-126 5 cm 处土壤水分与土壤空气 CO_2 浓度值变化

Figure 3-126 Changes of soil moisture and soil CO_2 concentration at 5 cm soil layer

表 3-25　土壤空气 CO_2 浓度与土壤水分的相关性分析

Table 3-25　Relativity between soil air CO_2 concentration and soil moisture

样地类型	土壤水分						
	0 cm	5 cm	10 cm	15 cm	20 cm	30 cm	40 cm
藻类结皮	0.065	0.507	0.550	0.513	0.595	0.842**	0.819**
藓类结皮	0.045	0.428	0.515	0.482	0.479	0.531	0.697**
流沙	0.515	0.579	0.430	0.420	0.263	0.224	0.053

注：*p=0.05 水平显著；**p=0.01 水平极显著。

（4）降雨的影响

降雨作为干旱区水分重要的补给源，对干旱区所有的生态过程都有影响（李新荣等，2009b）。降雨过后一方面土壤水分增加，促进了土壤呼吸，另一方面也阻断了土壤向大气排放 CO_2 的通道，出现了土壤空气 CO_2 浓度的高峰值。试验期间在9月11日和9月22日的降雨量分别是1.8 mm和3.4 mm。其中在9月11日，各层土壤空气 CO_2 浓度均比无降雨条件下增大，但土壤 CO_2 在不同深度的浓度值存在差异性。从图3-127可以看出，在降雨的条件下，土壤空气 CO_2 浓度值为藓类结皮>藻类结皮>流沙。主要原因在于两类BSC在理化性质方面存在差异，如藓类结皮的厚度一般为8～20 mm，而藻类结皮为0.02～2.5 mm，并且藓类结皮下的土壤有机质含量要明显大于藻类结皮和流沙，其下的微生物活性也更高，并由此导致其土壤空气 CO_2 浓度值较高。在1.8 mm的降雨条件下，由于土壤水分再分配使得三者的土壤空气 CO_2 浓度随着土层深度的增加而增加，差异不显著（F=0.882，p=0.431）；但在3.4 mm的降雨条件下，藓类结皮和藻类结皮由于较低的容重和较高的持水性能而在5 cm处达到 CO_2 浓度最大值，其后变化较为平缓，三者之间的差异显著（F=11.094，p<0.0001）。

固沙植被区演替过程中藻类结皮、藓类结皮和流沙影响下的土壤 CO_2 浓度平均值保持在600～1100 μmol mol^{-1}，低于高寒干草原（1052～3050 mL m^{-3}）、高寒草原（3425～39144 mL m^{-3}）、高寒草甸（984～12250 mL m^{-3}）（赵拥华等，2006）和太湖地区典型水稻土（3500～5500 μmol mol^{-1}）（刁一伟等，2002）、塿土（762～13294 μL L^{-1}）（戴万宏等，2004）等土壤剖面的土壤气体 CO_2 浓度变化范围，也比国外报道的农田（560～20900 μL L^{-1}）（Bajracharya *et al.*，2000）和草地（500～22600 μL L^{-1}）（Kammann *et al.*，2001）低。主要原因在于固沙植被区土壤表层结构疏松、孔隙度大，有利于土壤空气和大气进行自由交换，减少了 CO_2 气体在土壤中的累积量，并且荒漠地表植被稀疏，气候干燥，使得微生物活性降

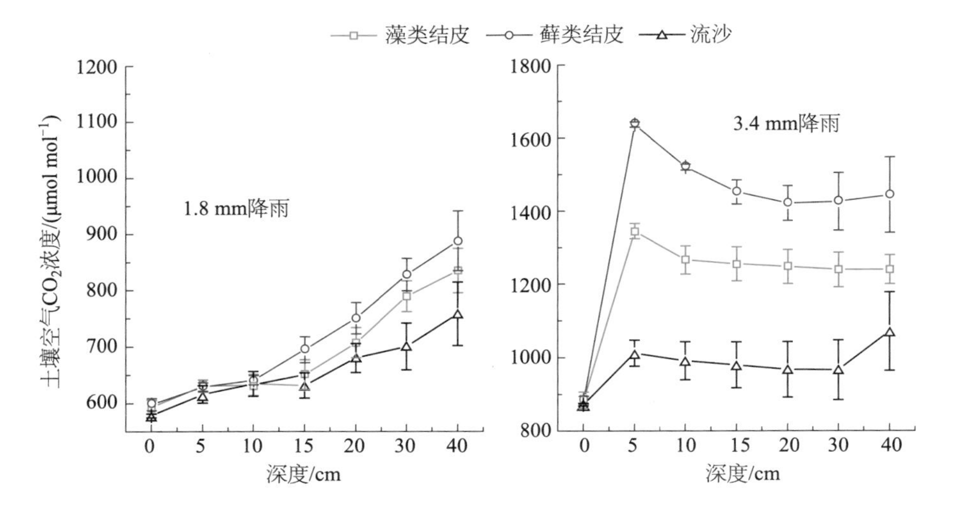

图 3-127　不同降雨量影响下的土壤空气 CO_2 浓度值
Figure 3-127　Response of soil CO_2 concentration to rainfall amount

低，根系呼吸也少，因此，其地下土壤CO_2浓度变化范围均比农田、草原、森林、高寒草甸较小。在沙坡头BSC发育较早的固沙区，BSC的形成使沙面形成了一层紧实的"保护膜"，阻止了土壤中CO_2气体向大气的扩散，藻类结皮和藓类结皮的土壤空气CO_2浓度值都大于同一深度流沙，但三者之间并无明显差异，主要是由于BSC厚度一般在5 mm左右，对土壤CO_2浓度的影响有限，而且在沙坡头地区BSC的形成过程中，藻类结皮和藓类结皮一直是混生的，在所选择的样地中只能判断出优势种，而没有绝对的单纯种，因此差异不显著。

土壤温度和水分是影响土壤空气CO_2浓度的最重要的环境因素，其与土壤温度具有明显的一致性，总体呈现出随土壤温度升高而增加的趋势；而与土壤水分的关系比较复杂，主要是受降雨的影响。土壤温度对不同深度藻类结皮、藓类结皮和流沙的土壤空气CO_2浓度的影响表现为先增大、后减小、再增大的趋势，比较0 ~ 20 cm的土壤水分情况，说明表层土壤温度是主要的影响因子。而在20 cm以下土壤水分对土壤空气CO_2浓度的影响逐渐增大，特别是对于藻类结皮和藓类结皮，原因在于BSC层的存在使得其下层土壤水分的蒸发减小，充足的水分保证地下微生物的活动，使得土壤空气CO_2浓度升高。而对于流沙，则情况相反，土壤水分易于丧失使得其下层土壤微生物活性降低，因此，主要影响其表层土壤空气CO_2浓度。从两次降雨事件中可以看出，小降雨可以使土壤下层空气CO_2浓度急剧升高，当降雨达到一定的阈值之后便不再变化，如图3-127所示，当降雨在3.4 mm时，从10 cm开始土壤空气CO_2浓度基本呈直线。但大降雨事件也使得藻类结皮、藓类结皮和流沙下土壤空气CO_2浓度的差异明显显现出来，一方面体现了两类BSC在土壤修复过程中下层土壤微生物分布情况，因为土壤下的CO_2主要是由微生物呼吸所产生的，刘艳梅等（2012）研究发现藓类结皮下土壤微生物量碳含量明显高于藻类结皮，说明藓类结皮更有利于微生物的生长与繁殖，即其下土壤微生物数量更多；另一方面也验证了BSC对土壤水分的影响，BSC层的存在阻止降雨入渗，但同时也抑制了下层土壤水分的蒸发（李守中等，2005），这种相互作用的关系使得其下层土壤水分含量变化较为复杂。同时土壤温度和水分对土壤空气CO_2浓度的影响作用也并非独立，两者之间存在着相互制约的关系，因此，有待于开展进一步的控制试验，确定水、热因子对土壤空气CO_2浓度的调控机理。

3.6 BSC对风沙活动的生态生理响应

我国是世界上沙化危害最为严重的国家之一，土地沙化已经成为我国特别是西北干旱地区最为严重的生态环境问题之一。强烈的风沙活动使得西北地区每年都有大量乡村和城镇，以及交通、通讯、水利、能源等工程设施受到流沙侵袭和掩埋（卢琦和杨有林，2001），在造成当地巨大经济损失、生态环境恶化的同时，也严重影响到我国东部经济-社会核心区的生态环境安全（国家林业局，2011）。有关风沙对BSC影响的研究几乎全部集中在BSC可以大幅度提高土壤起动风速、减小地表风蚀（王雪芹等，2004；李晓丽和申向东，2006）及BSC破损所造成的土壤细粒和有机质损失（West，1990）等方面，而缺乏风沙对BSC生态生理学特征，尤其是光合作用影响的研究。风的存在一方面加快了空气的流通，提高土壤和植物水分蒸散速率（Chintakovid *et al.*，2002；于云江等，2003），使干旱区植物较早进入干燥状态，而在无风环境下也不利于植物进行光合作用（Kitaya *et al.*，2003）；另一方面，强风还可作为一个机械胁迫，影响植物叶片的形态和解剖结构，进而影响生理功能（Jaffe and Forbes，1993）。不难想象，上述两个方面都会对BSC生物产生影响，因为BSC光合作用产量大小不仅与受水分含量影响的净光合速率有关，还与BSC自身的水分保持能力有关（Lange，2003），而风沙作用的存在会通过改变BSC保水能力影响BSC生物的光合作用产量。

沙埋是沙区，尤其是风沙活动频繁的干旱荒漠生态系统最常见的干扰因素之一（Brown，1997；赵哈林，2013）。沙埋通过改变植物生境的光照、温度和土壤理化性质影响植物个体的生长和存活，进而影响沙漠生态系统植物分布和组成（Maun，2004；马洋等，2014）。有关沙埋对植物影响的研究表明，沙埋可以改变植物茎的形态（Maun，2004），促进植物生长、生物量的积累和新分株的产生（Brown，1997），并能提高植物净光合速率（Perumal and

Maun，2006）。但也有沙埋降低植物光合面积和生产能力的报道（赵哈林等，2013）。虽然在过去的几十年里，中央与当地各级政府投入了大量人力、物力和财力相继开展了一系列的生态修复工程，而且近年来土地沙化出现了整体上逆转的趋势，但是，依然存在工程效果难以持久的问题，尤其是处在风沙活跃前沿的人工植被区因沙埋危害严重而出现退化趋势，而大面积分布的天然植被也一直面临沙埋的威胁，导致出现局部沙化恶化的局面（赵哈林等，2004）。造成这种局面的原因除了生态环境脆弱、人口压力较大、投入有限等因素外，认识不足、理论滞后和缺乏有效的应对措施也是重要原因（王涛，2007）。因此，全面、科学地认知沙埋对生态系统的影响，进而探索正确合理的治理对策和措施是保障西部生态安全和社会经济全面发展的重要前提。然而，在以往的生态工程建设和生态系统管理过程中，人们往往只注意到了高大的固沙灌木与草本植物，而忽略了形体微小的隐花植物及其BSC的利用和保护。

由于隐花植物形体矮小、地表生境等特点，使其相比高大的维管植物更易受到沙埋的影响。因而，沙埋成为干旱沙区隐花植物及其BSC遭受的最为频繁的干扰因素之一（Brown，1997），除了最为常见的风沙活动可引起隐花植物及其BSC被沙掩埋外，动物活动以及人类工程建设等也会造成不同规模的沙埋景观（图3-128）。沙埋除了会给隐花植物及其BSC产生一个由重力引起的机械压迫外，还可以显著改变其生境的光照、温度和土壤水分、土壤机械组成成分、氧气及营养元素含量等条件（Maun，2004；Jia *et al.*，2008），而这些环境因素的变化势必会导致敏感性较强的隐花植物及其BSC发生前面所述的变化，并最终给整个生态系统的稳定性带来极大的不确定性。由此可见，正确认知和处理沙埋对隐花植物多样性及其BSC生态功能的影响，是生态环境建设者与生态系统管理者必须考虑的一个关键问题，也成为摆在科研人员面前亟须解决的一个重要课题。

虽然沙埋对隐花植物及其BSC的影响早已引起人们的注意，如Birse等（1957）早在1957年就发现藓类植物可以忍耐1～4 cm深度不等的沙埋胁迫，但是由于隐花植物形体微小，种类鉴定困难及相关研究方法、手段和理论知识的不足，致使相关研究进展非常缓慢，已发表的相关文献也屈指可数。在已发表的文章中，主要涉及① 沙埋对BSC生理状况的影响。Wang等（2007）在控制条件下发现沙埋干扰显著降低了人工藻类结皮叶绿素a含量、PS Ⅱ光化学效率和胞外多糖含量；Jia等（2008，2012）通过室内控制实验发现沙埋抑制了腾格里沙漠人工固沙植被区的三种藓类结皮和一种地衣结皮的呼吸作用，而对光合作用影响不明显。② 沙埋后隐花植物的恢复机制，包括湿润状态时藻类的向上运动（Campbell，1979）、藓类植物的茎向上伸长（Jia *et al.*，2008）和无性繁殖生长加速（白学良等，2003；聂华丽等，2006）。③ 恢复速率和耐沙埋能力。Birse等（1957）发现藓类植物可以忍耐1～4 cm深度不等的沙埋胁迫；Potts（1999）发现蓝藻标本在干燥状态下被掩埋达一个世纪仍保持活性；

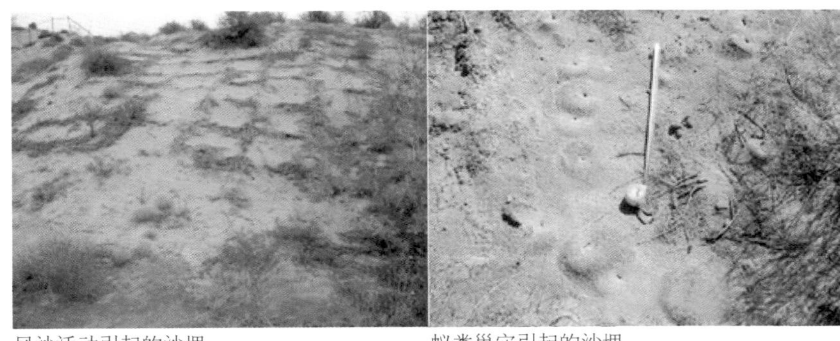

风沙活动引起的沙埋　　蚁类巢穴引起的沙埋

高速公路建设工程引起的沙埋

图 3-128　干旱沙区生态系统中隐花植物及其 BSC 遭受沙埋现象
Figure 3-128　Sand burial on cryptogam and BSC in arid sand area ecosystem

Martínez和Maun（1999）利用盖度变化计算了休伦湖沿岸11种藓类植物遭受沙埋后的恢复速率和耐沙埋指数；Jia等（2008）研究发现腾格里沙漠人工固沙植被区四种BSC的恢复速率和抗沙埋能力与BSC演替顺序一致，证明了沙埋是影响该区BSC演替顺序的重要因素之一。通过分析这些研究发现，尽管这些研究为我们了解沙埋对隐花植物及其BSC的影响提供了宝贵的方法和经验，特别是我国学者在其中做出了最大的贡献，但是不难看出，这些研究还存在较大局限性：一是研究对象不全，集中于藻类和苔藓，而对地衣研究较少，且所涉及的隐花植物种类很少，无法进行研究结果间的比较；二是缺乏室内控制实验和符合自然实际的长期观测研究的相互印证，降低了研究结果可信度；三是研究数量太少，很多重要问题尚不清楚，尤其是有关沙埋对隐花植物多样性的和功能的研究仍然是空白。

鉴于以上研究不足，我们以沙坡头沙漠试验研究站封育多年、具有不同发育和演替阶段、种类丰富且具有代表性（代表了温带荒漠隐花植物及其BSC基本特性）的人工固沙植被区的隐花植物及其BSC为研究对象，利用野外实验和室内实验相结合的方法，探讨沙埋对该区隐花植物多样性及其BSC生态功能的影响，为全球范围内全面、正确认识沙埋在生态系统中的影响，为我国干旱沙区生态环境建设与管理提供理论依据。同时，本项目拓展了国内外的BSC研究内容，为BSC生态学研究向前发展提供动力。

3.6.1 沙埋干扰解除后BSC光合生理恢复机制

通过对沙埋干扰后BSC光合速率和呼吸速率变化的分析，证明了沙埋后呼吸作用的降低是BSC减少碳损失、抵御沙埋的重要光合生理机制之一（Jia *et al.*，2008）。那么BSC是否也存在另外一个机制——沙埋干扰解除后光合作用对碳快速补充的能力呢？带着这样的问题，我们试图通过对沙埋干扰解除后BSC光合作用、呼吸作用及叶绿素荧光动力学参数的跟踪测定，揭示沙埋干扰解除后BSC光合生理恢复机制。

我们选择了真藓结皮、土生对齿藓结皮、刺叶赤藓结皮和坚韧胶衣结皮作为研究对象，设置了0 mm（对照）、1 mm、2 mm、3 mm和4 mm共5个水平的沙埋深度处理及1 mm、3 mm和5 mm共3个水平的水分处理（每隔2天喷水一次）。每个沙埋深度 × 水分处理设置了6个重复。在沙埋处理8周后，进行沙埋去除处理。由于BSC含有胞外分泌物，部分沙粒会黏在藓类和地衣表面，借助吸耳球轻轻将其吹掉，并保证不对BSC产生破坏，影响实验结果。

沙埋解除后，BSC样品继续放置在植物生长箱（Thermoline培养箱，Thermoline Scientific Equipment Pty. Ltd.，澳大利亚）内，光照强度、空气温度、空气相对湿度和CO_2浓度的控制条件保持不变，即在白天（08：00—19：00）分别设定在1000 $\mu mol\ m^{-2}\ s^{-1}$、25 ℃、70%和390 $\mu mol\ m^{-2}\ s^{-1}$，晚上（19：00至第二天08：00）分别为0 $\mu mol\ m^{-2}\ s^{-1}$、15 ℃、65%和400 $\mu mol\ m^{-2}\ s^{-1}$。为便于比较，全部样品喷水量统一增加至饱和水平，每6 h操作一次。

（1）沙埋干扰解除后BSC光合作用变化特征

由图3-129可以看出，沙埋干扰解除后4种BSC光合作用表现出相似的变化规律，随着时间的增加，P_n逐渐增加，其中第1周增加最为快速，第2周增加明显放慢。不同深度沙埋处理间P_n值差异显著，随着沙埋深度的增加逐渐降低（4种BSC，$p<0.05$）。同时，从图3-129可以明显看到沙埋处理时的水分条件也显著影响了沙埋解除后P_n值的大小，随着施水量的减少逐渐降低（4种BSC，$p<0.05$）。

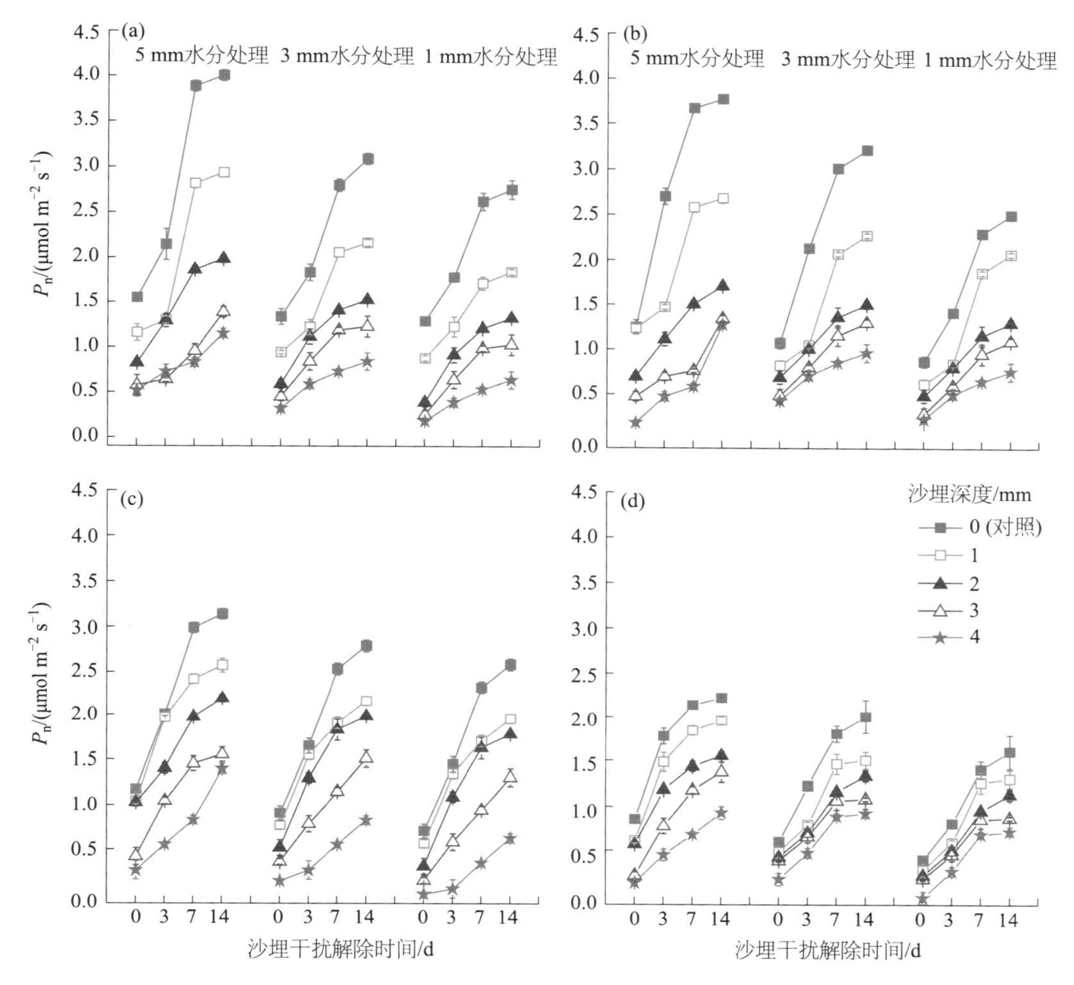

图 3-129 沙埋干扰解除后四种 BSC 的 P_n 值变化特征：(a) 真藓结皮；(b) 土生对齿藓结皮；(c) 刺叶赤藓结皮；(d) 坚韧胶衣结皮

Figure 3-129 Variations of net photosynthetic rates of four BSC with time lapsing after removal of sand burial : (a) *Bryum argenteum* ; (b) *Didymodon vinealis* ; (c) *Syntrichia caninervis* ; (d) *Collema tenax*

(2) 沙埋干扰解除后BSC暗呼吸作用变化特征

沙埋干扰解除后四种BSC暗呼吸作用表现出相似的变化趋势，随着时间的增加，R_d 突然降低后逐渐增加，而对照一直在增加（图3-130）。与 P_n 变化规律相似，随着沙埋深度的增加，R_d 值也逐渐降低（4 种BSC，$p<0.05$）。同样，沙埋处理时的水分条件也显著影响了沙埋解除后 R_d 值的大小，随着施水量的减少，R_d 值也逐渐降低（4 种BSC，$p<0.05$）。

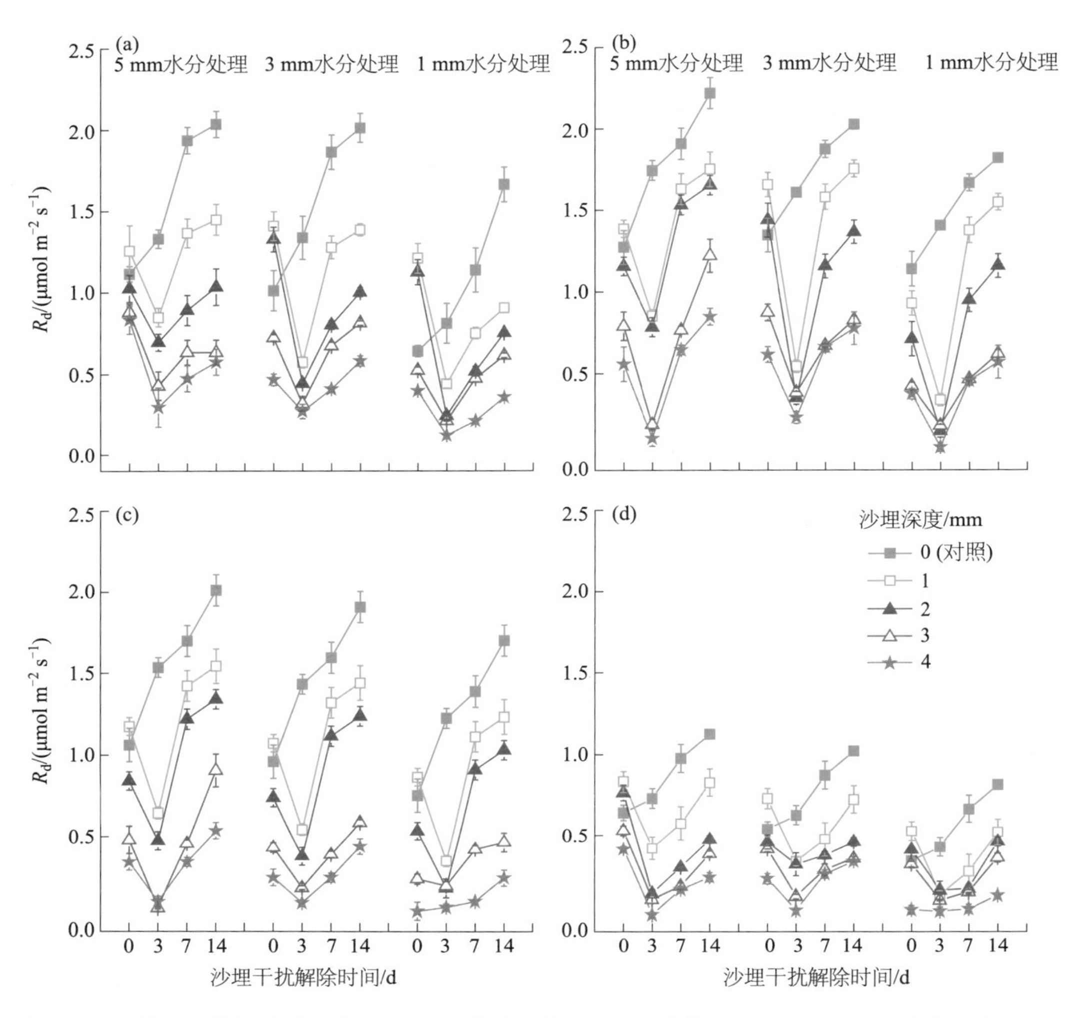

图 3-130　沙埋干扰解除后 4 种 BSC 的 R_d 值变化特征：（a）真藓结皮；（b）土生对齿藓结皮；（c）刺叶赤藓结皮；（d）坚韧胶衣结皮

Figure 3-130　Variations of dark respiratory rates of four BSC with time lapsing after removal of sand burial：（a）*Bryum argenteum*；（b）*Didymodon vinealis*；（c）*Syntrichia caninervis*；（d）*Collema tenax*

（3）沙埋干扰解除后BSC叶绿素荧光动力学参数变化

与P_n变化规律相似，沙埋干扰解除后4种BSC PS Ⅱ光化学效率表现出相似的变化规律，随着时间的增加，F_v/F_m值逐渐增加，其中第1周增加较快，第2周增加速度明显放慢（图3-131）。而且沙埋深度和沙埋处理时的水分条件也显著影响了F_v/F_m的变化，随着沙埋深度的增加或施水量的减少，F_v/F_m值逐渐降低（4种BSC，$p<0.05$）。

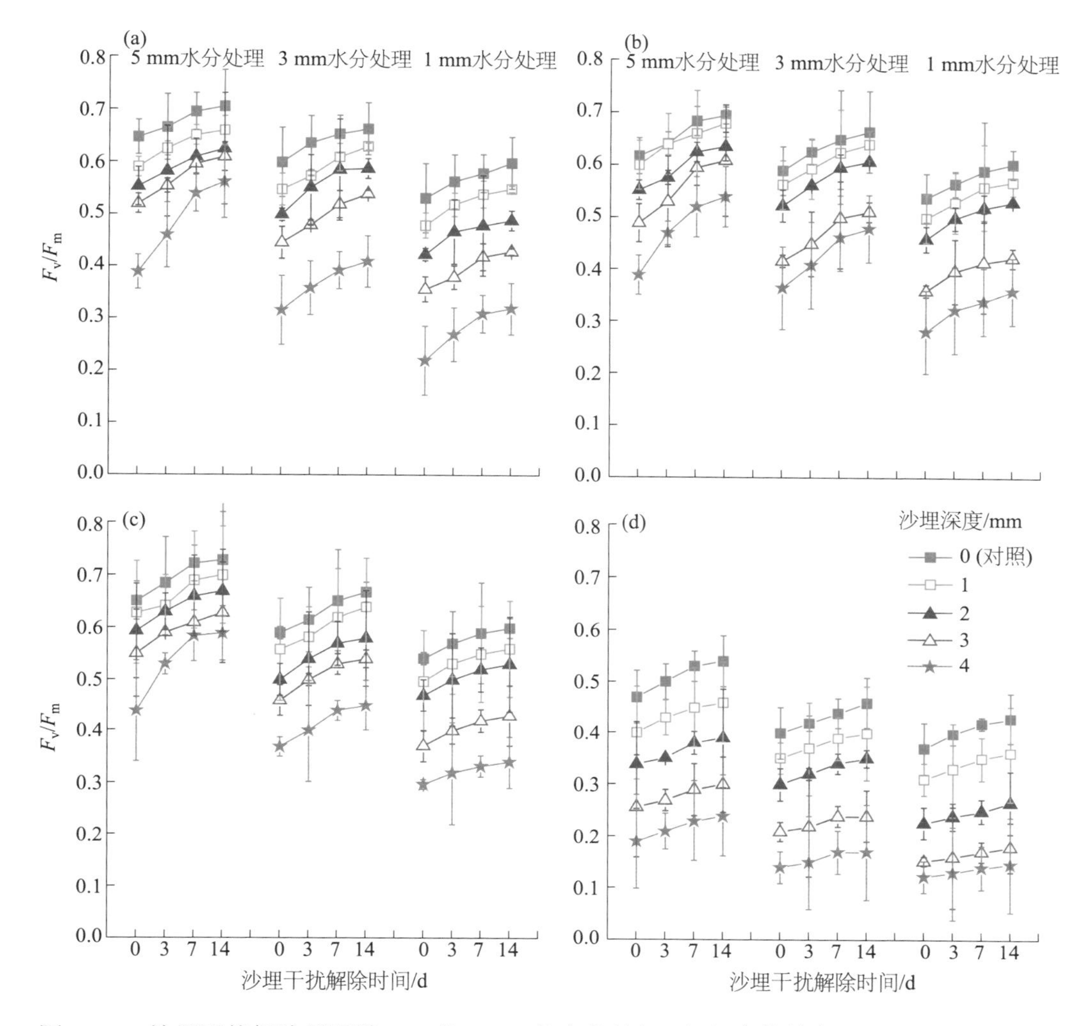

图 3-131　沙埋干扰解除后四种 BSC 的 F_v/F_m 值变化特征：(a) 真藓结皮；(b) 土生对齿藓结皮；(c) 刺叶赤藓结皮；(d) 坚韧胶衣结皮

Figure 3-131　Variations of F_v/F_m values of four BSC with time lapsing after removal of sand burial : (a) *Bryum argenteum* ; (b) *Didymodon vinealis* ; (c) *Syntrichia caninervis* ; (d) *Collema tenax*

沙埋干扰解除后BSC光合作用的恢复受到时间、沙埋深度和沙埋时的水分条件的影响（图3-129）。首先，从时间变化上看，沙埋干扰解除后的第1周是4种BSC光合作用恢复时期，之后增加速度放缓，但2周后光合速率值仍然低于对照，这表明沙埋干扰对BSC光合作用的破坏需要更长的时间才能修复。其次，沙埋干扰显著抑制了其解除后BSC光合作用，这与Kent等（2005）对维管植物的研究结果相似，而且这种抑制作用随着沙埋深度的增加而增加。沙埋解除后光合作用与沙埋深度间的反比关系与Larcher（1995）研究结果——沙埋解除后光合作用修复所需时间与沙埋干扰时间成正比关系——有相通之处，即沙埋干扰越大对植物光合作用器官的破坏也越严重。再者，沙埋干扰时施水量大小也显著影响了沙埋干扰解除后BSC光合作用功能。这不难理解，因为BSC的大部分生物组分为变水植物，只有在施水湿润后才开始生理代谢活动，所以水分条件的好坏直接影响了沙埋干扰时BSC生物的生理活性和生长状况，其中最明显的例子就是施水量的增加促进了藓类植物茎的伸长（Jia *et al.*，2008），增加了光合面积，进而提高了沙埋干扰解除后藓类结皮的光合作用能力。

呼吸作用的降低是植物遭受严重遮光干扰的一个重要生理适应对策（Larcher，1995）。这种适应对策已被Kent等（2005）对沙埋解除后维管植物群落与Jia等（2008）对沙埋时BSC的研究结果所证实。那么沙埋干扰解除后BSC呼吸作用会降低吗？而我们的实验结果并没有实验前想的那么简单。沙埋干扰解除后四种BSC的暗呼吸速率表现出一致的变化规律：随时间的增加，暗呼吸速率突然降低后逐渐增加，而对照一直在增加（图3-130）。首先，造成这种变化规律差别的原因主要与沙埋干扰解除后的第三个效果有关。空气流通限制的突然解除，势必会使原来较为“平静”的BSC土壤–空气交界层气体交换活动产生强烈震荡，而这种“物理性”气体交换速率的增加，远大于“生物性”暗呼吸速率。所以我们认为在沙埋解除后立即测得的“暗呼吸速率”并非真正的BSC生物暗呼吸速率，如果将这部分测定值去掉，遭受沙埋处理与对照的暗呼吸速率变化规律应该是一致的，但这一点还需要进一步验证。其次，沙埋干扰解除后四种BSC暗呼吸作用也并没有按照我们实验前假设变化，相反，从沙埋解除后的第3天开始，暗呼吸速率一直在增加。这可能与沙埋解除后导致BSC生物微环境的改变，特别是O_2利用量的增加（Maun，1994）以及BSC生物生理活性增强有关，而后面这一点可以由沙埋干扰解除后净光合速率和PS Ⅱ光化学效率升高来证实。

有关沙埋干扰对BSC生物叶绿素荧光动力学参数影响的研究极少，所见报道只有Wang等（2007）通过控制实验发现沙埋干扰显著降低了人造藻类结皮PS Ⅱ光化学效率，而对沙埋干扰解除后BSC生物PS Ⅱ光化学效率的变化情况还未见报道。我们的研究结果表明，沙埋干扰解除后4种BSC的PS Ⅱ光化学效率随着时间的增加而增加，这反映了沙埋干扰解除后环境的改善和BSC生物光合器官积极自行修复的内在机制。

3.6.2 沙埋对栖息于BSC中的可培养小型真菌群落的影响

众所周知，沙埋可以减慢种子萌发和出苗（Ren *et al.*，2002）。这种干扰可能是藻类结皮遭受严重胁迫的来源之一，可造成藻类结皮的叶绿素含量和生物量降低，包括总碳水化合物（尤其是多糖）储备的减少（Wang *et al.*，2007；Rao *et al.*，2012）。沙埋还通过改变需氧和厌氧微生物之间的比例以及减少菌根真菌的数量影响土壤微生物群落（Maun，2004）。

非寄生的小型真菌是BSC的重要组成部分。与异养细菌、蓝藻、绿藻、地衣和藓类共同在BSC的组成和功能中起重要作用（States *et al.*，2001）。我们在腾格里沙漠人工植被区进行的真菌学研究显示，134个物种组成了各种BSC的真菌区系（Grishkan *et al.*，2015）。在本研究中，我们检测了该地区沙埋对栖息于不同种类BSC中的可培养小型真菌群落的影响。基于以往的研究结果，我们假设沙埋将影响这些BSC真菌群落组成和结构（体现在不同生活史对策类群的丰度）、物种多样性水平和小型真菌的数量。为了验证假设，本研究分析了以下群落特征：物种组成、主要分类群和生态种群对群落结构贡献、优势种群、菌株密度和物种多样性水平（物种丰富度、异质性和均匀度）。

分别从2个试验地采集BSC和BSC下层土壤样品（2014年9月），2个试验地的BSC分别以藓类结皮（真藓）和混生结皮（比例一致的地衣、蓝藻和藓类）为主。试验地位于1964年人工植被区的迎风坡。将PVC管（直径为10.4 cm，深度为20 cm）于2013年3月随机放置在土壤中（图3-132），每个试验地放置9根管子。其中3个管子分别用6.5 g风干的流沙均匀覆盖，使其形成0.5 mm的流沙覆盖层（轻度沙埋）。另外3个管子分别用130 g流沙覆盖，使其形成10 mm的流沙覆盖层（重度沙埋），其余3个管子不做沙埋处理（对照）。为避免流沙被吹出管外，PVC管的上沿比BSC和流沙处理平面高0.5 cm。每个管子中，采集0.5 ~ 1 cm（混生结皮）和1 ~ 1.5 cm（藓类结皮）深的BSC下沙土样品。为避免边缘效应，所有样品均用直径为5 cm的无菌PVC管在管子中间部分采样。此外，从麦草方格固定的无BSC覆盖沙丘采集3个样品作为对照（表层土）。样品从麦草方格的中间和边缘采集，深度为0 ~ 0.5 cm。共采集42个样品。采样完成后4 ~ 5周内，这些样品在以色列海法大学进化研究所的实验室进行处理和分析。

（1）小型真菌群落的组成和物种多样性

共分离得到来自Mucoromycotina（4个物种）、teleomorphic（形态学意义的有性）Ascomycota（19个物种）和anamorphic（无性）Ascomycota（42个物种）的65个菌种。这些菌种属于43个属，大多数物种属毛壳菌属（9种）、曲霉菌属（4种）和青霉菌属（3种）。8个类型的小型真菌菌株在培养过程中无孢子产生。

对混生结皮而言，分离得到小型真菌物种数最低和最高的处理分别为无沙埋处理和重度沙埋处理。然而在藓类结皮中则恰好相反。此外，相比轻度和无沙埋处理BSC，重度沙埋处理的混生结皮中小型真菌群落异质性和均匀度更高。然而，轻度和无沙埋处理藓类结皮的群落均匀度相似。即使与其他BSC群落相比，表层土的小型真菌群落多样性最高（表3-26）。

图 3-132　藓类结皮（由实线环绕）和混生结皮（由虚线环绕）试验地，1 号管为对照，2 号管为模拟轻度沙埋，3 号管为模拟重度沙埋

Figure 3-132　Experimental plots covered by moss-dominated crusts (surrounded by solid line) and mixed crusts (surrounded by dashed line), with tubes for control (1) and for simulation of shallow (2) and deep (3) sand burial

表 3-26　沙坡头沙漠试验研究站无沙埋 BSC、沙埋 BSC、表层土和 BSC 下沙土层小型真菌群落多样性特征

Table 3-26　Diversity characteristics of micro-fungal communities

位置	BSC/ 表层土			BSC 下沙土层		
	S	*H*	*J*	*S*	*H*	*J*
重度沙埋下混生结皮	28ab	2.20ab	0.66abc	17cde	1.68bcd	0.59bcd
轻度沙埋下混生结皮	25abc	1.29cde	0.40de	9e	0.84f	0.38ef
无沙埋混生结皮	18cde	1.28cde	0.44cde	8e	1.39cde	0.67abc
重度沙埋藓类结皮	20bcd	1.79bcd	0.60bcd	14de	2.03abc	0.77a
轻度沙埋藓类结皮	23abc	1.90abc	0.61bcd	8e	0.73f	0.35f
无沙埋藓类结皮	30a	2.07ab	0.61bcd	16de	0.98dfe	0.35f
麦草方格中央表层土	26ab	2.36a	0.72abc	—	—	—
麦草方格边缘表层土	18bcd	2.14ab	0.74ab	—	—	—

注：*S*—物种数，包括不育菌株；*H*—Shannon 指数；*J*—均匀度；比较所有生境的每一个特性，采用单因素方差分析，the Tukey（HSD）检验；比较 5% 水平上的差异显著性，相同字母表示无显著差异。

与沙埋和无沙埋BSC之间的小型真菌群落多样性特征差异相比，BSC和BSC下沙土层中小型真菌群落多样性特征差异在大部分两两比较中更为显著。不管在什么样的情况下，BSC所在区域的物种丰富度显著高于其他区域。此外，无沙埋混生结皮和重度沙埋藓类结皮下的沙土真菌群落甚至比单独的BSC群落具有更高的均匀度和异质性（表3-26）。

含黑色素的小型真菌在所有表层土真菌群落中占据主导地位，而且在重度沙埋BSC下沙土层中占据优势地位（表3-27）。含黑色素菌种在BSC表层和BSC下沙土层群落中分别占42.4% ~ 95.3%和28% ~ 80.5% 的贡献率（图3-133）。在所有BSC群落中（除了重度沙埋的混生结皮），含黑色素真菌的核心是具有大型（>20 μm）多细胞孢子的物种，诸如埃里格孢菌、链格孢和毛球腔菌（表3-28）（占贡献率的31.4% ~ 82.3%，图3-133）。在沙丘表层土中，这些物种的贡献显著降低（图3-133）。与无沙埋BSC相比，沙埋BSC真菌群落中含黑色素真菌（尤其是具有大型多细胞孢子的真菌）的丰富度更低（图3-133）。

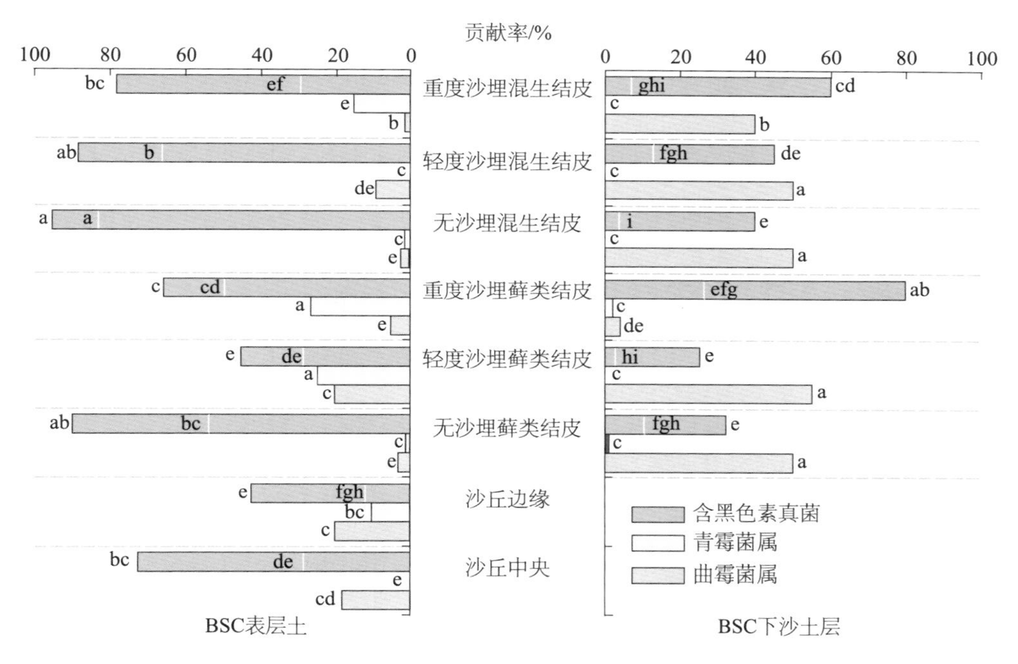

图3-133　沙坡头沙漠试验研究站无沙埋BSC、沙埋BSC、沙丘表面和BSC下沙土层中主要小型真菌群落贡献率差异。灰色：青霉菌属；绿色：曲霉菌属；紫色：含黑色素真菌。含黑色素真菌中的白线代表含大型多细胞孢子的贡献率。相同字母代表差异不显著

Figure 3-133　Differences in contribution of main micro-fungal groupings in unburied crusts, buried crusts, sandy surfaces, and sandy below-crust layers at SRS. Grey : *Penicillium* spp. ; Green : *Aspergillus* spp. ; Purple : melanin-containing spp. The area below the white line on the bars of melanin-containing micro-fungi indicates contributions of species with large multicellular spores. Means with the same letters are not significantly different

在BSC和BSC下沙土层生境中，曲霉菌属的丰富度差异性最显著（主要是烟曲霉菌），曲霉菌属的丰富度在BSC下生境中是BSC层中的30倍之多。与无沙埋BSC群落相比，沙丘表层土和轻度沙埋BSC中真菌群落以曲霉菌属的贡献较高为特征。但是，这些特征在深度方面没有太大差异（图3-133）。青霉菌属（主要是*Penicillium aurantiogriseum*）对轻度和重度沙埋藓类结皮真菌群落组成的贡献率最大（24.4%和26.4%），是重度沙埋混生结皮和沙丘边缘生境的2倍。然而，在其他小型真菌群落中，青霉菌属的丰富度很低，仅有0%～1.6%（图3-133）。

基于物种相对丰富度的聚类分析把小型真菌群落分成几个具有不同相似度的群组（图3-134）。其中一个群组包含来自BSC生境的群落（除了重度沙埋的混生结皮），这些群落之所以聚在一起是由上面所述的具有大型多细胞孢子的含黑色素真菌在群落中占据优势所致。另一个群组由来自无沙埋和轻度沙埋BSC下沙土层的沙地真菌群落组成，这些群落之所以彼此高度相似是由于烟曲霉菌在群落中占据绝对优势（表3-27）。与重度沙埋混生结皮和BSC下生境（由于高的毛壳菌丰富度）一样，具有极高相似度的沙丘表层土真菌群落也聚在一起。重度沙埋藓类结皮下沙土层的真菌群落与其他群落分离是由于其群落中存在高丰富度的*Pyrenochaeta cava*（表3-27）。

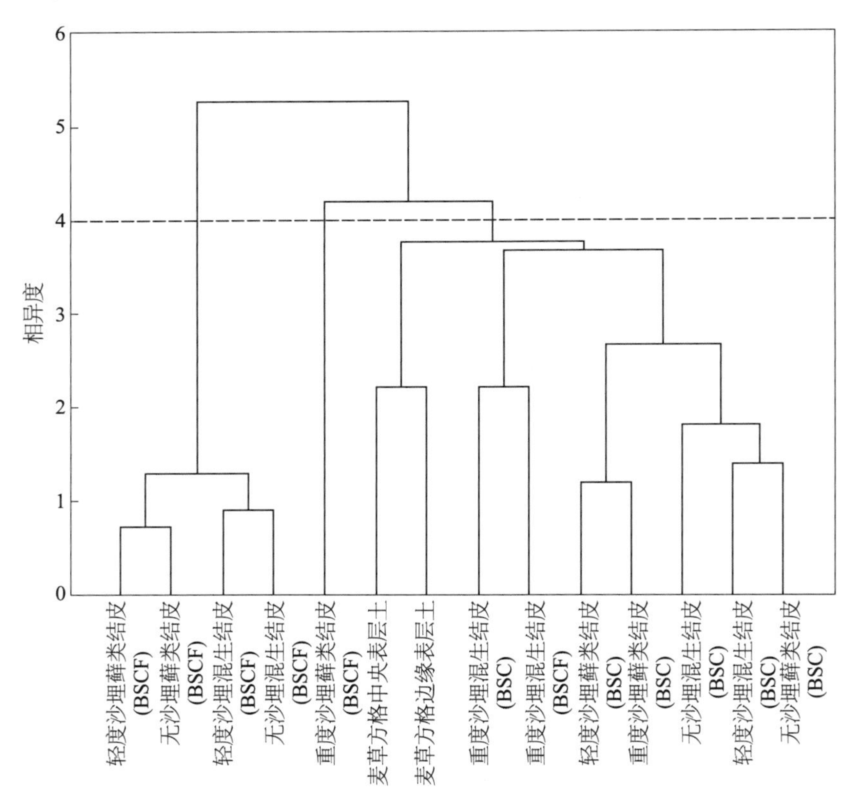

图3-134　沙坡头沙漠试验研究站无沙埋BSC、沙埋BSC、沙丘表面和BSC下沙土层中小型真菌群落聚类分析

Figure 3-134　Clustering the micro-fungal communities from unburied crusts, buried crusts, sandy surfaces, and sandy below-crust layers at SRS

表 3-27　沙坡头沙漠试验研究站无沙埋 BSC、沙埋 BSC、BSC 下沙土层和沙丘表面的小型真菌总名录，包括其平均相对丰度。粗体字标注物种为含黑色素真菌

Table 3-27　Common micro-fungal species from unburied crusts, buried crusts, sandy below-crust layers, and sandy surfaces at SRS, with their average relative abundance (%). Species in bold are melanin-containing

物种	轻度沙埋藓类结皮		重度沙埋藓类结皮		轻度沙埋混生结皮		重度沙埋混生结皮		无沙埋藓类结皮		无沙埋混生结皮		沙丘表面	
	BSC	BSC 下	BSC	BSC 下	BSC	BSC 下	BSC	BSC 下	BSC	BSC 下	BSC	BSC 下	BSC	BSC 下
Mucoromycotina														
Absidia corymbifera	2.4	—	0.9	0	0.4	—	—	—	—	—	—	—	—	—
Mortierella humilis	—	1.1	—	—	0.7	0.4	—	5.4	—	2.9	—	—	—	—
Rhizopus oryzae	2.2	2.2	0.3	16.4	0.7	16.2	1.4	3.3	—	0.4	0.9	15	0.5	—
teleomorphic Ascomycota														
Aspergillus nidulans	—	—	—	—	—	—	0.8	0.3	1.3	—	—	1.9	—	—
Chaetomium strumarium	—	—	2	1.4	0.7	0.9	1.1	—	2.1	0.8	0.2	—	0.4	—
Ch. succineum	—	—	—	—	0.4	0.4	25.9	38.3	1	—	—	—	—	—
***Chaetomium* sp.**	0.2	—	—	5.5	—	—	—	—	—	—	—	—	0.7	—
Pleospora tarda	—	—	—	—	—	—	—	—	0.3	0.4	0.2	—	0.2	—
Sordaria fimicola	—	—	0.9	—	1.1	—	0.3	—	0.5	—	—	—	—	—
Sporormiella minima	—	—	0.9	4.1	—	—	—	—	—	—	—	—	—	—
anamorphic Ascomycota														
Alternaria alternata	4.8	—	3.2	1.4	4	—	3.8	0.9	9.4	—	14.9	—	24.4	5.4
Aspergillus fumigatus	23.5	93.9	2.3	5.5	3.6	75	1.1	33.3	0.5	78	0.7	64.3	16.4	21.8
As. ustus	—	—	—	—	—	—	—	8.6	—	—	—	—	—	—
Boeremia exigua	—	—	—	1.4	0.7	—	0.5	0.3	7.3	0.4	0.9	—	10.6	5.1
Cladosporium cladosporioides	0.3	—	—	—	—	—	—	—	—	0.4	—	0.5	9.2	1.1
Coleophoma empetri	0.5	—	0.9	—	2.8	—	7.2	0.3	3.4	—	—	—	0.4	0.2
Coniothyrium olivaceum	—	—	—	—	—	—	4	—	—	—	—	—	—	—
Curvularia inaequalis	—	—	—	—	0.4	—	—	—	—	—	6.5	—	—	—
Cu. spicifera	0.2	—	0.3	—	—	—	0.3	—	0.8	—	0.2	—	—	0.2
Embellisia phragmospora	20.4	0.3	44.8	12.3	71.1	5.5	26.3	3.9	50.8	6.6	47.5	—	6	—
Fusarium oxysporum	0.2	—	0.6	—	—	—	0.8	—	0.25	1.6	—	—	0.2	14.3
Gibberella acuminata	1	0.5	2.6	5.5	1.4	—	2.7	2.4	3	2	0.9	5.6	3.1	—
Myrothecium verrucaria	—	—	—	—	—	—	1.1	0.3	1.6	—	—	—	7.8	11.4
Papulaspora pannosa	1	—	—	2.7	0.4	0.85	—	0.9	1	0.4	—	—	—	—
Penicillium aurantiogriseum	33.4	—	28.8	—	—	0.4	10	0.3	1	0.4	0.4	—	0.5	8.4
Pyrenochaeta cava	5.8	—	7.7	38.3	—	—	0.8	—	2.6	3.7	19.4	12.7	0.5	—
Setosphaeriar rostrata	0.7	0.2	0.6	—	0.7	—	—	—	1.3	0.4	1.7	—	0.2	0.3
Stachybotrys chartarum	0.2	—	0.3	—	—	—	—	—	1.6	—	—	—	10.8	19.9
Trichoderma koningii	—	—	1.1	—	—	—	0.3	—	—	0.4	—	—	—	0.1
Ulocladium atrum	0.3	0.2	0.9	—	3.2	—	0.8	—	5.2	—	5.4	—	3.1	8
Westerdykella capitulum	1.2	—	—	—	2.2	—	2.2	—	0.3	—	—	—	1.2	—
Mycelia sterilia														
Dark	0.2	—	—	—	—	—	—	—	1.3	—	—	—	1.5	—
Dark 1	—	—	—	—	1.8	—	—	0.3	—	0.8	—	—	—	—
Dark 2	—	—	—	—	0.7	—	1.1	1.2	—	—	—	—	—	—
Light	—	0.6	—	—	0.4	0.4	0.8	—	—	—	—	—	—	—

（2）小型真菌菌株密度

在混生结皮中，无沙埋BSC的菌落形成单位（CFU）值显著（1.7～2.3倍）高于沙埋BSC。然而，在藓类结皮中，最高和最低的菌株密度分别出现在轻度和重度沙埋BSC（图3-135）。在BSC下层生境中，CFU值均显著低于BSC层（图3-135），重度沙埋藓类结皮下的CFU值最低。

（3）样地类型、沙埋深度和土壤深度对真菌特性的影响

在研究的真菌参数中，上述环境因子对物种多样性特征（尤其是物种丰富度）的影响较小。但是，不同小型真菌群组和菌株密度均受到每个因子的单独和交互作用的影响（表3-28）。在环境因子中，土壤深度对小型真菌群落的影响最为显著，其次为沙埋深度（表3-28）。

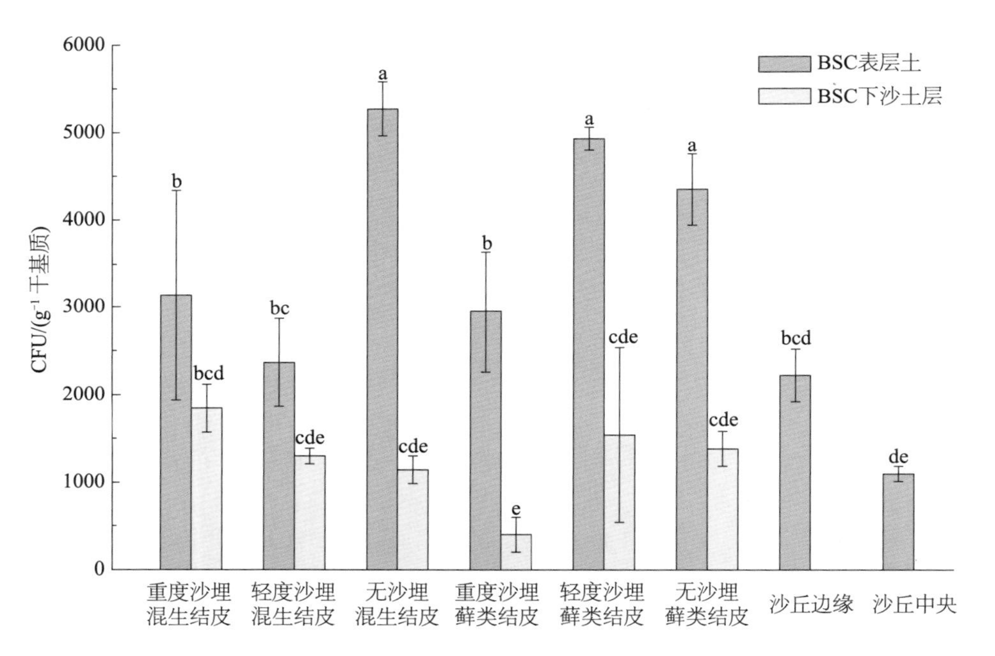

图3-135 沙坡头沙漠试验研究站无沙埋BSC、沙埋BSC、沙丘表面和BSC下沙土层中的菌株密度

Figure 3-135 Isolate density in unburied crusts, buried crusts, sandy surfaces, and sandy below-crust layers at SRS

表 3-28　沙坡头沙漠试验研究站样地类型、沙埋深度、土壤深度及其交互作用对不同小型真菌群落参数影响的双因素非均衡方差分析结果

Table 3-28 Data of two-way unbalanced ANOVA analysis for the effect of locality type, burial depth, soil depth, and the interactions among them on different parameters of micro-fungal communities at SRS

参数	样地	沙埋深度	土壤深度	样地 × 沙埋深度	样地 × 土壤深度	沙埋深度 × 土壤深度	样地 × 沙埋深度 × 土壤深度
物种丰富度	8.5**	NS	117.5****	17.8****	NS	NS	NS
H	4.1@	19.3****	54.3****	NS	4.4@	7**	5.7*
J	NS	28.2****	NS	7**	6@	8.9***	15.3****
含黑色素菌种	42.4****	38****	217.6****	31.3****	37.9****	81.6****	9.9**
含大型多细胞孢子的黑色素菌种	6.8**	14.6****	447.8****	32.9****	27.4****	47.6****	8.5***
青霉菌属物种	25.6****	31.5****	101****	11.1**	29.9****	23.6****	16.7****
曲霉菌属物种	NS	84.3****	769.5****	27.9****	39.3****	37.1****	43.7****
菌株密度	6**	16.2****	302.1****	22.5***	8.7**	13.4****	13.7****

注：@$p \leqslant 0.05$；*$p \leqslant 0.01$；**$p \leqslant 0.005$；***$p \leqslant 0.001$；****$p \leqslant 0.0001$；NS：无显著差异。

沙埋是一个值得关注的环境因子，其改变了干旱和半干旱区土壤的各种理化性质。长时间的沙埋可能造成藻类结皮中氮生物利用度的增加（Williams and Eldridge，2011），这种增加很可能是由BSC生物的死亡、自我分解和细胞外氮的释放引起。由于碱性沙子渗入BSC层和沙层对BSC水分蒸发的屏蔽效应，积沙能轻微地增加藻类结皮pH和土壤含水量（Rao *et al.*，2012）。假定土壤真菌群落的组成和活性与土壤的物理、化学和结构性质密切相关是既定的事实（Ritz and Young，2004），可以认为沙埋会影响腾格里沙漠里BSC中小型真菌的物种多样性和分布。

沙埋的影响体现在对小型真菌群落的物种组成上。含黑色素真菌占优势是所有BSC群落的特征，也是几乎所有干旱区土壤——尤其是BSC（Bates and Garcia-Pichel，2009；Bates *et al.*，2010；Porras-Alfaro *et al.*，2011；Bates *et al.*，2012；Grishkan and Kidron，2013）——真菌学研究的典型特征（例如Ranzoni，1968；Christensen，1981；Halwagy *et al.*，1982；Skujins，1984；Abdullah *et al.*，1986；Hashem，1991；Ciccarone and Rambelli，1998；Zak，2005；Grishkan and Nevo，2010）。但是，无沙埋处理的BSC小型真菌群落间无显著性差异（图3-135），而沙埋BSC群落在含黑色素真菌总丰富度和有大型多细胞孢子黑色素菌种丰富度方面差异显著。由于孢子具有能起到保护作用的形态特征，所以后者是

最耐UV辐射和干旱胁迫的群组。积沙还增加了中温青霉菌属的贡献（轻度沙埋混生结皮除外）。沙层在一定程度上能够避免BSC层水分蒸发和受到UV辐射，可能创造出更适合青霉菌生存和发育的条件。另一方面，耐热曲霉菌属真菌丰富度在轻度沙埋处理的藓类结皮中更高，与沙丘表层土生境中的结果相似。

在两种BSC的研究中，小型真菌群落多样性特征随着沙埋深度而变化。在混生结皮中，沙埋增加了群落物种丰富度和均匀度。然而，在藓类结皮中沙埋降低了物种丰富度，群落均匀度对沙埋不敏感。同样，菌株密度也存在不同的变化趋势。在混生结皮中，无沙埋处理的BSC真菌群落以具有最大菌株密度和具有大型多细胞孢子的含黑色素菌种占80%为特征。同时，在藓类结皮中，轻度沙埋处理下不含黑色素的青霉菌属具有相对较高的密度值，而单细胞孢子的曲霉菌属的CFU值最高。重度沙埋处理下，小型真菌群落菌株密度在两种BSC中类似。与位于北坡和南坡中间位置的混生结皮相比，北坡的藓类结皮具有更低的正午地表温度和更长的湿润持续时间，因而具有更高的叶绿素含量（Grishkan *et al.*，2015）。不同BSC间小气候的差异不仅可以解释无沙埋BSC间小型真菌群落的差异，而且还可以解释真菌群落对相同水平沙埋的不同响应。有趣的是，这种差异在重度沙埋处理下并不明显。

所有分析结果表明，群落参数中与土壤深度相关的变化比与沙埋相关的变化以及BSC和沙丘表面群落的变化更显著。上述事实有一个合理的解释：鉴于BSC层的积沙一定程度上改变了土壤小型真菌生存的基质质量和小气候条件，从BSC到沙丘的转变（尤其是BSC下层）大幅度改变了环境条件。营养状况从相对富含有机质的BSC层到矿物质贫乏的沙土层的这种改变显著降低了菌株密度（真菌生物量的间接特征）。这种改变还伴随着物种丰富度随深度而显著降低，以及主要与优势种差异相关的小型真菌群落组成和结构的明显改变。在BSC下沙土层，烟曲霉菌（具有小的单细胞孢子的耐热浅色真菌）取代了具有大型多细胞孢子的深色真菌（其是BSC群落的优势种）。烟曲霉菌是一种分布范围极广的宽温域腐生真菌（Domsch *et al.*，2007），其最适生长温度为35～38 ℃（Magan *et al.*，2007）。经常从不同干旱区土壤中分离到数量较高的烟曲霉菌（Ranzoni，1968；Halwagy *et al.*，1982；Moubasher *et al.*，1985；Abdullah *et al.*，1986；Abdel-Hafez *et al.*，1989；Bokhary，1998；Mandeel，2002；Sharma *et al.*，2010；Baeshen *et al.*，2014）。

在内盖夫沙漠（以色列）和腾格里沙漠的BSC与其他内盖夫地区的土壤中，在37 ℃下分离得到的耐热真菌群落中烟曲霉菌占支配或共支配地位（Grishkan and Nevo，2010；Grishkan and Kidron，2013；Grishkanet *et al.*，2015）。最近的分类学研究结果显示，烟曲霉菌是由几个物种组成的一个复合体，这些物种形态上几乎一致，但一些基因序列、生长温度范围（所有的物种都是耐热的真菌）和胞外分泌物表达谱有所不同（Hong *et al.*，2005）。很显然，烟曲霉菌在沙基质（尤其是BSC下层和亚表土层）中获得最有利的发育条件，因此避免了与具多细胞孢子的含黑色素真菌（在BSC小型真菌群落中占绝对优势地位）的激烈竞争。

本研究阐明了沙埋对中国腾格里沙漠BSC中小型真菌群落定性和定量特征的影响。其可能与沙层的屏蔽效应有关联（即使是较浅的沙层），沙层可以避免BSC层水分蒸发和受到UV辐射。沙埋的影响表现在含黑色素真菌丰富度的降低（尤其是具大型多细胞孢子的真菌），同时产小型单细胞孢子的浅色真菌（主要是中温的青霉菌属真菌）丰富度增加。沙埋还造成菌株密度降低，其在混生结皮中尤为显著。比较沙埋处理，从BSC层到BSC下沙土层的转变中小型真菌群落变化更显著。土壤条件的改变（主要是营养状况，从相对富含有机质的BSC层到矿物质贫乏的沙土层的改变）导致菌株密度和物种丰富度显著降低和优势种的替代——在BSC下沙土层中耐热广义烟曲霉菌取代了BSC层中具有大型多细胞孢子的含黑色素真菌。因此，该地区BSC持续和深度沙埋可能破坏BSC小型真菌群落的稳定性和功能，从而可能减少小型真菌的数量、降低其物种多样性和改变物种组成（尤其是优势种和常见种种群）。

3.7 BSC对氮沉降的生态生理响应

氮（N）是控制许多陆地生态系统的物种组成、多样性和生产力的关键营养元素之一（Zechmeister-Boltenstern *et al.*，2011），据IPCC（2007）报道，在过去的一个多世纪，大气活性氮（主要是氮氧化物和氨）沉降增加了3～5倍，Lamarque等（2005）预测下个世纪陆地生态系统的大气氮沉降将会进一步增长250%。学术界普遍认为氮添加会对环境造成广泛的影响，包括引起气候变化，促进温室气体的排放，导致物种消失，甚至会威胁到人类健康（Nemergut *et al.*，2008；Ramirez *et al.*，2010）。氮沉降对生态系统初级生产力的影响已得到广泛研究，例如氮沉降增加会以增加植物生物量的形式导致碳储量升高，影响植物种类组成和导致物种多样性的丧失等（Clark *et al.*，2007）。

3.7.1 氮添加对BSC中微生物的影响

氮添加可以改变微生物群落结构及其多样性和随后的生态功能（Waldrop *et al.*，2004；

Allison *et al.*，2008；Campbell *et al.*，2010）。其中，可以预期的是微生物群落会转变为以细菌为主（Tietema，1998）。近年来，氮添加对各生态系统土壤微生物多样性和群落组成的影响受到了越来越多的重视，包括森林（Entwistle *et al.*，2013）、草原（Zhang *et al.*，2008）、草地和农田系统（Ramirez *et al.*，2010）和高山苔原（Nemergut *et al.*，2008）等。近期研究的共识表明，过量的土壤氮素减少了微生物量和生物活性，并降低了土壤微生物呼吸水平（Ramirez *et al.*，2010，2012）。据我们所知，在整个生态系统中，土壤微生物对施氮的响应罕有一致的结论。例如，Eisenlord和Zak（2010）的研究发现模拟大气氮沉降对森林地表放线菌的丰度没有影响，而Ramirez等（2010）的研究结果表明，在草地和农田系统中，放线菌的丰度随施氮强度增加而增加。此外，不同的氮添加实验不断报道真菌群落的物种多样性和群落结构会升高、降低或没有净变化（Johnson，1993；Frey *et al.*，2004；Jumpponen and Johnson，2005；Allison *et al.*，2007）。这些变化可能是由于不同的施氮量和不同的处理持续时间引起的，Meta分析表明，微生物（如细菌和真菌）丰度的下降和微生物群落结构的改变随着施氮量的增加和处理时间的延长表现得更为明显（Treseder，2008）。这也有可能归因于土壤的独特的最初特性，例如，不同的土壤特性和微生物群落结构会造成土壤对氮添加的不同的、无法量化的响应（Zeglin *et al.*，2007）。

对于BSC的形成和维持的生物过程而言，关键是要保持表土层微生物多样性、群落结构和微生物丰度之间的平衡。如果土壤微生物受到干扰和破坏，将会对BSC的生态作用和功能产生影响。因此，从氮管理和BSC的保护角度而言，我们必须基于对微生物响应氮限制机理的彻底理解的基础上进行解释和预测。由于细菌和真菌是BSC中最重要的两种组成部分（Gundlapally and Garcia-Pichel，2006；Green *et al.*，2008）和重要的有机质分解者，而且大多数干旱和半干旱生态系统是氮贫瘠的（McCalley and Sparks，2009），我们有充分的理由认为，氮添加具有改变微生物群落的物种多样性和群落结构（例如，细菌和真菌的主要种系型的转变）和（或）改变微生物丰度的潜力和趋势，特别是改变细菌和真菌的比例，其中任意一种变化都很可能影响到BSC的微结构和随后的生态功能，例如对荒漠生态系统中的碳循环和储存具有重要的影响。

作为评估氮添加对植物物种多样性、生产力和动态的长期影响的实验的一部分（Su *et al.*，2013a），我们同时在腾格里沙漠沙坡头–翠柳沟地区对石果衣结皮开展了7年的野外施氮实验，施氮量分别为对照：大气氮沉降；低氮：3.5 g N m^{-2} a^{-1}；中氮：7.0 g N m^{-2} a^{-1}；高氮：14.0 g N m^{-2} a^{-1}。对采集的样品开展了微生物多样性和定量分析。

（1）氮添加对BSC中微生物 α 多样性的影响

454高通量测序实验结果表明，将每个样品测序序列标准化为5500条后，对照样品中细菌丰度最高，可操作分类单元（OTU）数最多，达到2403，而高氮施加后，样品中细菌OTU数目最少，仅1915。相应地OTU丰度指数ACE和Chao1也发生了明显一致的下降趋势；OTU多样性指数，如Shannon指数也相应地呈现下降趋势。总的来说，细菌多样性随着氮添加的增加而线性下降。对BSC中真菌而言，对氮添加的响应与细菌相比，并不完全一致。将每个样品测序序列标准化为3800条后我们发现，高氮施加后，样品中真菌丰度最高，OTU数目最多，达到498，而中氮施加后OTU数目最少，仅310。但是OTU丰度指数ACE和Chao1显示低氮处理后的样品比其他样品具有更高的物种多样性（表3-29）。

表3-29 氮添加后BSC中微生物多样性指数的变化特征

Table 3-29 Diversity indices for the 16S rDNA sequences/ITS rDNA sequences after seven years of N additions

	测序样品	优化	标准化	OTU丰度			OTU多样性指数		盖度/%
				观测丰度	ACE	Chao1	Shannon	Simpson	
细菌	对照	8119	5500	2403	8118	5902	7.26	0.0018	0.7508
	低氮	8153	5500	2234	7575	5464	7.02	0.0032	0.7692
	中氮	7122	5500	2150	6320	4456	6.85	0.0044	0.7650
	高氮	7090	5500	1915	5405	4075	6.84	0.0043	0.8126
真菌	对照	5464	3800	461	776	788	4.28	0.0593	0.9537
	低氮	6233	3800	427	1126	869	3.75	0.1354	0.9574
	中氮	4324	3800	310	715	572	3.81	0.0879	0.9636
	高氮	6693	3800	498	860	847	4.26	0.0473	0.9584

（2）氮添加对BSC中微生物丰度的影响

在本研究中，共测定出了细菌域下的19个门和真菌域下的8个门。在细菌群落中，高丰度细菌门（OTU数>总OTU数的10%）包括放线菌门（Actinobacteria）、绿弯菌门（Chloroflexi）、蓝藻门（Cyanobacteria）和变形菌门（Proteobacteria），而低丰度细菌门（总OTU数的1%<OTU数<总OTU数的10%）包括酸杆菌门（Acidobacteria）、装甲菌门（Armatimonadetes）、拟杆菌门（Bacteroidetes）、芽单胞菌门（Gemmatimonadetes）和浮霉菌门（Planctomycetes）（图3-136a）。在真菌群落中，高丰度真菌门仅包括子囊菌门

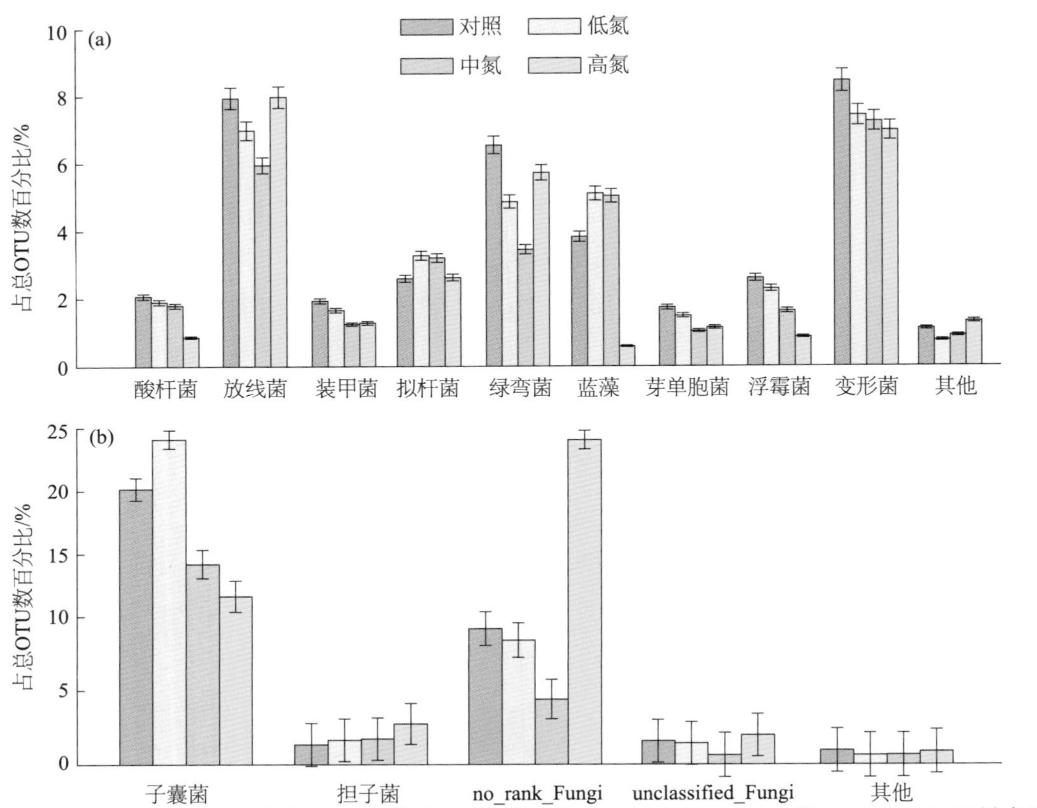

图 3-136 氮添加后细菌（a）和真菌（b）群落中高丰度门（占总 OTU 数 10% 以上）和低丰度门（占总 OTU 数 1%～10%）的分布特征（根据相似度 97% 划分 OTU）

Figure 3-136 Abundant phyla（>10% of total OTUs）and low-abundance phyla（1%～10% of total OTUs）distributed in N-added samples for bacteria（a）and fungi（b），respectively（Data are defined at a 3% OTU genetic distance）

（Ascomycota）和no_rank_Fungi。低丰度真菌门为担子菌门（Basidiomycota）和unclassified_Fungi（图3-136b）。

与以往的藻类结皮中蓝藻是最主要的优势类群不同（Nagy *et al.*，2005；Gundlapally and Garcia-Pichel，2006；Abed *et al.*，2010），在本研究的对照地衣结皮中，变形菌门是最主要的优势菌门，占细菌群落的25.0%左右，这与其他研究区域中地衣结皮中的细菌分类结果相似（Grube *et al.*，2009；Maier *et al.*，2014）。变形菌门以 α－变形菌为主，而 β－变形菌和 δ－变形菌相对丰度只有1.0%～3.1%，γ－变形菌只占0.2%。蓝藻是第二大优势菌门，占细菌群落的19.5%，放线菌门占19.4%，绿弯菌门占12.2%，拟杆菌门占6.6%，酸杆菌门占5.0%，浮霉菌门占4.6%，装甲菌门占3.1%，芽单胞菌门占3.0%，其他门相对丰度均小于1%（图3-137a）。

在属水平上，施氮处理导致细菌类群减少：在对照样品、低氮、中氮和高氮处理样品中分别鉴定出275、265、259和239个分类群，在四个处理样品中最优势的属分别是no_rank_*Chloroflexi*（9.8%）、no_rank_*Cyanobacteria*（12.2%）、*Microcoleus*（12.1%）和no_rank_*Chloroflexi*（12.4%）。值得注意的是，蓝藻的微鞘藻属（*Microcoleus*）和绿弯菌门的no_rank_*Chloroflexi*属相对丰度的转变显示样品间明显不同的反应模式：从对照到中氮处理，*Microcoleus*属的相对丰度从6.1%升高到12.1%，但在高氮处理下显著下降到0.3%。然而，从对照到中氮处理，no_rank_*Chloroflexi*属的相对丰度从9.8%降到5.7%，但在高氮处理后却上升至12.4%左右（图3-137b）。

土壤pH通常会强烈影响微生物群落结构，这可能是控制酸杆菌门和浮霉菌门相对丰度的主要因子，因为它们的相对丰度随着土壤酸化而下降。细菌相对丰度的变化符合富营养假设理论（copiotrophic hypothesis）（Fierer *et al.*，2007；Ramirez *et al.*，2010），酸杆菌门和拟杆菌门通常认为是寡营养微生物，即在低碳环境下生长更好（Fierer *et al.*，2007；Davis *et al.*，2011），例如本研究中发现酸杆菌门相对丰度随样品中总碳含量的升高而降低（图3-137a和表3-30），但是前人预测的寡营养微生物拟杆菌门在各样品中相对丰度差别并不大。

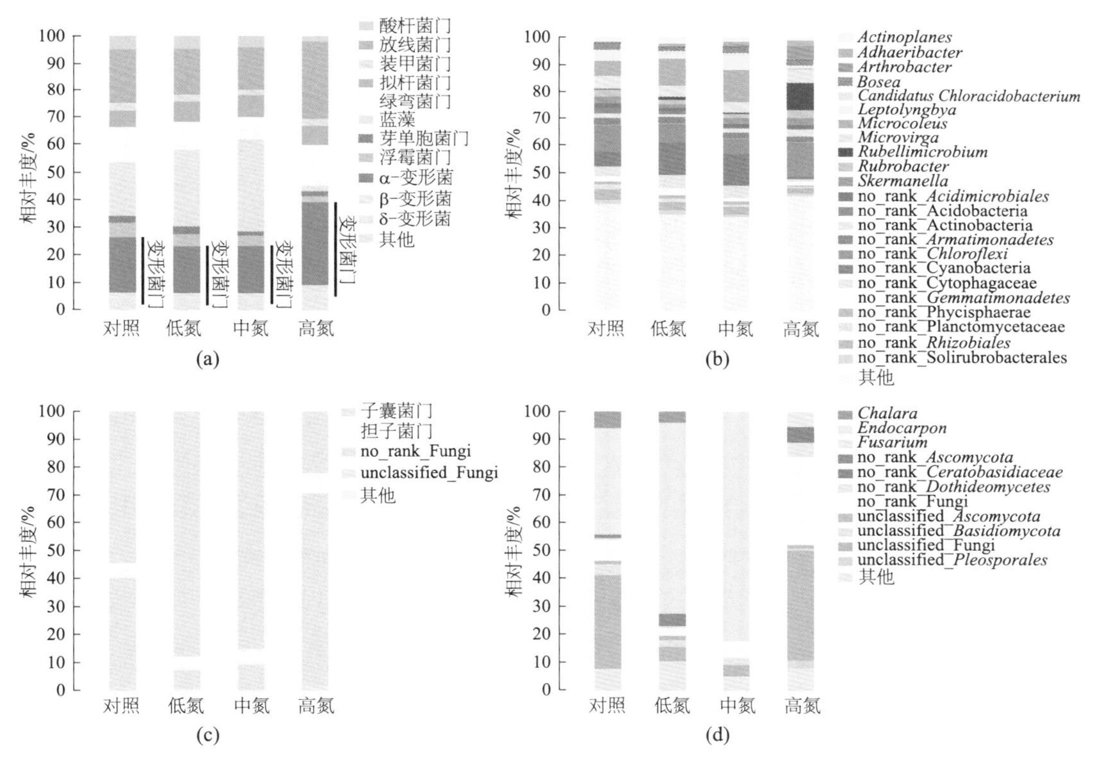

图3-137 氮添加后分别在门和属水平上优势细菌（a，b）和真菌（c，d）群落结构变化特征

Figure 3-137 Relative abundance of dominant bacteria（>1% of total reads）（a）at the phylum level with classes of Proteobacteria，（b）at the genus level，and dominant fungi（>1% of total reads）（c）at the phylum level，（d）at the genus level

表 3-30　氮添加后土壤理化性质的变化（平均值 ± 标准差，n=5）

Table 3-30　Physico-chemical properties of seven years of increasing N-added soils （mean ± SD，n=5）

处理	总氮 /（g kg^{-1}）	总磷 /（g kg^{-1}）	速效氮 /（mg kg^{-1}）	硝态氮 /（mg kg^{-1}）	铵态氮 /（mg kg^{-1}）	有机碳 /（g kg^{-1}）	碳氮比	pH
对照	0.34 ± 0.03c	0.34 ± 0.02a	14.98 ± 0.64c	2.47 ± 0.21c	0.58 ± 0.02c	2.97 ± 0.6c	8.62 ± 0.4a	7.77 ± 0.04a
低氮	0.42 ± 0.01b	0.39 ± 0.01a	18.44 ± 1.84c	2.66 ± 0.22c	1.07 ± 0.11bc	3.42 ± 0.4b	8.14 ± 0.5b	7.48 ± 0.03b
中氮	0.54 ± 0.02a	0.41 ± 0.02a	22.28 ± 2.06b	3.86 ± 0.23b	2.15 ± 0.19b	3.76 ± 0.9b	7.01 ± 0.8c	7.41 ± 0.05b
高氮	0.68 ± 0.04a	0.43 ± 0.01a	32.88 ± 1.13a	7.14 ± 0.66a	4.66 ± 0.17a	4.41 ± 0.8a	6.49 ± 0.6c	7.31 ± 0.02c

研究中发现β－变形菌、放线菌和绿弯菌在高氮添加后有明显的增加趋势，一般认为β－变形菌和放线菌是富营养细菌，随着土壤中的总有机碳含量的升高而增加（Fierer *et al.*，2007）。绿弯菌在本研究中有富营养细菌化趋势，因为它们是重要的有机质分解者，高碳环境通常会刺激它们的拓殖（Freeman *et al.*，2009；Eisenlord and Zak，2010）。需要注意的是放线菌和绿弯菌相对丰度在低氮和中氮处理后呈轻微下降趋势，而在高氮处理后上升比较明显，这说明土壤pH和土壤有机碳（或者说碳矿化）同时影响着这两大类群，在低氮和中氮处理下土壤酸化效应控制为主，而在高氮处理后，碳矿化作用占据优势。氮饱和效应也是需要考虑的一个因素，微生物对过多的氮添加响应通常不是线性的（Ramirez *et al.*，2012），因此很难孤立地解释微生物的氮添加响应。研究中还发现蓝藻在低氮和中氮添加后，相对数量大量增加，而在高氮添加后急剧下降，说明中等程度的氮添加导致的土壤总氮和速效氮的增加有助于蓝藻的拓殖，但是过多的氮明显对蓝藻的生长不利（Schulz *et al.*，2013）。

细菌 *Rubellimicrobium* 属主要分解淀粉和纤维素多糖（Dastager *et al.*，2008），高氮添加后该类群的大幅增加说明大量的氮添加能缓解氮贫瘠并有效刺激纤维素分解细菌的繁殖。蓝藻的 *Leptolyngbya* 和 *Microcoleus* 属能大量分泌胞外多糖，在BSC的形成和稳定过程中发挥重要作用，低氮和中氮添加能显著增加这些类群的相对丰度，但是受到高氮的明显抑制。另外，*Rubrobacter* 和 *Microvirga* 属具有抗高温和干旱的特性（Ferreira *et al.*，1999；Laiz *et al.*，2009），*Rubrobacter* 属能抵抗强辐射（Yoshinaka *et al.*，1973），*Actinoplanes* 属能形成菌丝（Vobis，2006），尽管上述这些类群在不同氮添加后相对丰度没有发生明显变化，但是在干旱区BSC的形成和维持过程中均具有重要的作用。

历时7年的氮添加也造成了地衣为主的BSC中真菌群落明显的转变（图3-137c）。在对照BSC中，子囊菌门（Ascomycota）占据绝对数量优势，达到总数的54.8%左右，其次是“非分配序列”（no assigned）（占总测序条数的40.6%：其中no_rank_Fungi占39.4%；

unclassified_Fungi占1.2%）和担子菌门（Basidiomycota）(4.6%)。最突出的变化是子囊菌门和担子菌门：在低氮和中氮添加后子囊菌门占据了绝对主导地位（分别占87.3%和84.8%）。在4个处理所有样品序列中，（非分配序列OTU数占总OTU数的25.6%；其中no_rank_Fungi占23.7%，unclassified_Fungi占1.9%。）高氮处理后，非分配序列升高至70.6%，这可能是由于高氮处理下非分配序列分担了真菌样品的多样性。然而担子菌门的相对丰度却随着氮添加的增加而升高：从4.6%上升到7.3%。

我们的研究结果表明，在对照、低氮和中氮处理样品中，子囊菌门的石果衣属（*Endocarpon*）是最优势真菌属，但在高氮处理样品中的最优势属是镰刀菌属（*Fusarium*）（子囊菌门）（非分配序列未计算在内）。在鉴定出的三个确定属（全部是子囊菌门）中，横节霉菌属（*Chalara*）的相对丰度在氮添加后呈现一直下降的趋势，而镰刀霉属在高氮添加后急剧增加。从对照到中氮处理，石果衣属的相对丰度呈上升趋势（从38.0%升高到82.2%），但在高氮处理7年后下降到接近0（图3-137d）。

真菌群落中鉴定出的高丰度的非分配序列在以往的研究也有发现（Green *et al.*，2008），说明地衣结皮中特有的真菌仍处于人类未认知的状态。我们发现非分配序列在对照、低氮和中氮处理后差别并不明显，但是在高氮处理后有高达70%的序列被划分为非分配序列，说明高氮处理后多样性（OTU）的增加可能是由非分配序列驱动所致。

我们的研究认为子囊菌门是地衣结皮中的最优势真菌门，这与以往众多的研究结果一致（Green *et al.*，2008；Bates and Garcia-Pichel，2009；Abed *et al.*，2013），但是在本研究中发现子囊菌门的Verrucariales目（主要是石果衣属*Endocarpon*）是主要的优势真菌目，这与以往的研究认为BSC中以能分泌抗辐射色素的格孢腔菌目（Pleosporales）占优势（Porras-Alfaro *et al.*，2011；Bates *et al.*，2012；Abed *et al.*，2013）不同，这与本地区的地衣结皮的构成有关，因为本地区是以石果衣地衣结皮占优势（Zhang and Wei，2011）。低氮和中氮均能显著增加子囊菌门的相对丰度，但是明显受高氮抑制。

氮添加会导致速效氮增加，通常会抑制木质素降解（Craine *et al.*，2007），担子菌门能合成一系列的木质素降解有关的酶，该类群通常被认为是木质素主要的降解者。来自森林地表真菌的研究一般都认为氮添加会显著降低其相对丰度（Blackwood *et al.*，2007），但是我们的结果截然不同，发现担子菌门相对丰度的变化与木质素降解并不相关，因为荒漠中植物稀疏分布，BSC层中来自植物的聚合物并不多见（Bates and Garcia-Pichel，2009）。与森林地面相比，BSC层中木质素含量少得多，而且森林地面木质素降解的主要类群是担子菌门的Agaricales目（Osono and Takeda，2006；Snajdr *et al.*，2010），但是这个类群在本实验中相对丰度都很低（均低于1%）。因此，本实验中担子菌门的丰度变化可能与氮添加导致的低木质素但高纤维素有机质有关，因为氮添加会刺激低木质素含量的凋落物的分解（Blackwood *et al.*，2007）。

（3）氮添加对BSC中微生物群落β多样性的影响

我们对来自4个不同氮处理的微生物群落用非加权配对算术平均法聚类分析发现，与中氮和高氮相比，对照中的细菌群落与低氮的关系更紧密，而高氮的细菌群落明显与其他处理中的细菌群落不同，在真菌群落中也有类似的发现。Libshuff成对分析结果揭示，所有细菌或真菌群落中的两两比对结果均为显著差异。我们对优势类群（相对丰度大于1%）进行的组间显著性差异分析发现，在细菌群落中，除了在对照和中氮处理之外，红杆菌目（Rhodobacterales）相对丰度在所有的处理中两两比较均差异显著。而no_rank_Actinobacteria只对高氮敏感，在比较组如对照和低氮（p=0.764），对照和中氮（p=0.176）以及低氮和中氮（p=0.300）中差异均不显著。然而蓝藻的三个优势目颤藻目（Oscillatoriales）、原绿菌目（Prochlorales）和Cyanobacteria_uncultured在所有的氮处理样品的两两比对结果均为差异显著，表明这些类群可能对施氮处理极其敏感，即使低氮处理亦能显著改变其相对丰度。在真菌群落中，鸡油菌目（Cantharellales）和肉座菌目（Hypocreales）对氮添加有相似的响应，即在对照和中氮组中差异不显著（p=1.000和p=0.903），其他组两两比对均显著。格孢腔菌目（Pleosporales）只对高氮添加敏感，尽管柔膜菌目（Helotiales）和Lichinales对氮添加敏感，但是在中氮和高氮处理下差异不显著（p=0.166和p=1.000），另外，Verrucariales、no_rank_Fungi和unclassified_Fungi这3个优势目对氮添加有一致的响应趋势，即在不同组中均差异显著。

（4）氮添加对地衣结皮中微生物丰度的影响

荧光定量PCR实验表明，该地区对照地衣结皮中细菌数量达到每克土壤10^{10}个拷贝左右（以16S rRNA基因来衡量）。低氮和中氮添加对细菌数量并没有显著的影响（与对照BSC相比，p=0.205和p=0.356），而高氮添加明显降低了BSC中的细菌数量（与对照、低氮和中氮相比，p=0.020、p=0.003和p=0.005）；对照地衣结皮中真菌数量达到每克土壤10^{8}个拷贝左右（以ITS rRNA基因来衡量）。低氮和中氮处理显著增加了BSC中真菌数量（与对照BSC相比，分别为p<0.001和p=0.003），而高氮处理使得BSC中真菌数量显著下降了一个数量级（与对照、低氮和中氮相比，分别为p=0.038、p<0.001和p<0.001）。我们发现在所有的BSC样品中，细菌数量均高于真菌数量，高氮处理显著增加了样品中细菌/真菌比（与对照、低氮和中氮相比，分别为p=0.022，p<0.004和p=0.005；表3-31）。

表 3-31　氮添加后地衣结皮中细菌和真菌丰度的变化以及相应的比值变化

Table 3-31　Absolute abundances of bacteria and fungi (copies of ribosomal genes per gram of soil) in lichen-dominated soil crusts quantified by qPCR and their paired ratios

氮添加	细菌丰度		真菌丰度		细菌 / 真菌比（16S/ITS1f-5.8s）	
	范围	平均值 ± 标准差	范围	平均值 ± 标准差	范围	平均值 ± 标准差
对照	1.43×10^{10} ~ 2.67×10^{10}	2.03×10^{10} ± 6.20×10^{9}a	1.85×10^{8} ~ 3.36×10^{8}	2.53×10^{8} ± 7.68×10^{7}b	4.27×10^{1} ~ 1.12×10^{2}	8.73×10^{1} ± 3.88×10^{1}b
低氮	1.88×10^{10} ~ 4.02×10^{10}	2.83×10^{10} ± 1.09×10^{10}a	6.52×10^{8} ~ 9.86×10^{8}	8.01×10^{8} ± 1.70×10^{8}a	1.91×10^{1} ~ 5.25×10^{1}	3.71×10^{1} ± 1.69×10^{1}b
中氮	2.12×10^{10} ~ 3.36×10^{10}	2.60×10^{10} ± 6.64×10^{9}a	5.57×10^{8} ~ 8.24×10^{8}	6.64×10^{8} ± 1.41×10^{8}a	2.57×10^{1} ~ 5.50×10^{1}	4.08×10^{1} ± 1.46×10^{1}b
高氮	2.23×10^{9} ~ 4.99×10^{9}	3.36×10^{9} ± 1.44×10^{9}b	1.28×10^{7} ~ 2.03×10^{7}	1.64×10^{7} ± 3.76×10^{6}c	1.10×10^{2} ~ 3.10×10^{2}	2.15×10^{2} ± 1.00×10^{2}a

注：以每克土壤含有的核糖体基因拷贝数来计算，平均值 ± 标准差，n=4。

3.8　BSC 对火烧干扰的生态生理响应

火烧是荒漠生态系统重要的干扰因子，尤其是在非洲Savanna地区，对荒漠植物种子更新、种群动态、群落组成和结构、生物多样性等具有显著的影响。火烧因子常被认为是自然生态系统维持稳定性与生态健康的重要干扰因子，并日益受到重视。探讨荒漠植被对火烧干扰的响应和采取的生态适应对策，对实现荒漠植被生态稳定和生态系统管理有重要的科学意义。尽管国际上针对草原草本植物生长对火烧干扰的响应已有大量的研究，但这些研究主要集中在非洲荒漠草原、热带稀树草原区、地中海气候区和澳洲等地，在我国主要集中于火烧因子对以羊草和大针茅为建群种的典型草原生态系统的影响，但很少有研究涉及荒漠草本植物对火烧因子的响应。我国荒漠植被中BSC作为群落的重要组成成分，得到了很好的发育，较高的盖度在防风固沙中起着十分重要的作用。荒漠火烧的概率相对较小，长期以来，火烧成为人为控制和预防的特殊因子，因而火烧作为荒漠生态系统干扰因子，对荒漠植被，尤其是BSC如何影响的研究十分匮乏。

于2008年4月底（晚春）选择4块不同群落的样地（10 m×10 m），另于2009年2月底（早春）选择3块不同群落（表3-32）的样地（10 m×10 m）进行相同的火烧处理（将地上干枯草本层植物原地充分燃烧；图3-138），并在每块样地旁边留出植被、土壤条件基本相同的未烧对照样地供比较研究，共设样地14块。在火烧处理后，分别在每块样方中设置6个1 m×1 m的小样方，共84个样方。在2008年、2009年的生长季节，对样方中草本植物进行逐月调查，分别记录每个样方中当月的物种数、每种的个体数，并测定每种草本植物株高（5个重复）。在生长季节结束时采用收割法测定了不同处理样方中草本植物地上部生物量。以后均以字母代表不同群落的样地。我们分别在2009年和2014年对样地A～D，2010年和2015年对样地E～G，即火烧1年和5年后进行调查，与火烧处理前样地BSC基本特征（种类组成和盖度）进行了分析对比（图3-139和图3-140）。

图3-138　春季火烧处理对BSC影响的实验设置：（a）处理前的样地；（b）火烧处理；（c）处理后的观测样地之一；（d）处理后地表BSC差异。图左侧未受火烧，图右侧为烧后情况

Figure 3-138　The effects of filed fire on BSC，conducted in spring in the southern of the Tengger Desert.（a）plot before fire treatment；（b）a fire treatment has been conducting；（c）plot after the treatment；（d）soil surface of fired plot. The left showed no-fired surface，right was surface after fire

表 3-32 火烧处理前所选样地植被和 BSC 概况

Table 3-32 The status of sample plot of different vegetation composition

火烧时间	样地号	优势种	盖度 /%		丰富度 /m^{-2}		BSC 总盖度 /%
			草本	灌木	草本	灌木	
2008 年	样地 A	华北驼绒藜（*Ceratoides arborescens*）、红砂（*Reaumuria songarica*）、碱韭（*Allium polyrhizum*）、阿拉善单刺蓬（*Cornulaca alaschanica*）、无芒隐子草（*Cleistogenes songorica*）	5	15	4	3	55
	样地 B	碱韭（*Allium polyrhizum*）、青蒿（*Artemisia carvifolia*）、阿拉善单刺蓬（*Cornulaca alaschanica*）、无芒隐子草（*Cleistogenes songorica*）	10	0	6	0	45
	样地 C	喀什霸王（*Sarcozygium kaschgaricum*）、华北驼绒藜（*Ceratoides arborescens*）、狭叶锦鸡儿（*Caragana stenophylla*）、猫头刺（*Oxytropis aciphylla*）、青蒿（*Artemisia carvifolia*）	3	7	3	4	50
	样地 D	红砂（*Reaumuria songarica*）、珍珠猪毛菜（*Salsola passerina*）、喀什霸王（*Sarcozygium kaschgaricum*）、华北驼绒藜（*Ceratoides arborescens*）、碱韭（*Allium polyrhizum*）、无芒隐子草（*Cleistogenes songorica*）	5	20	3	5	55
2009 年	样地 E	狭叶锦鸡儿（*Caragana stenophylla*）、华北驼绒藜（*Ceratoides arborescens*）、碱韭（*Allium polyrhizum*）、青蒿（*Artemisia carvifolia*）、阿拉善单刺蓬（*Cornulaca alaschanica*）	10	8	5	3	50
	样地 F	华北驼绒藜（*Ceratoides arborescens*）、碱韭（*Allium polyrhizum*）	2	5	3	1	45
	样地 G	红砂（*Reaumuria songarica*）、华北驼绒藜（*Ceratoides arborescens*）、紫茎泽兰（*Eupatorium adenophora*）、青蒿（*Artemisia carvifolia*）	10	25	5	4	53

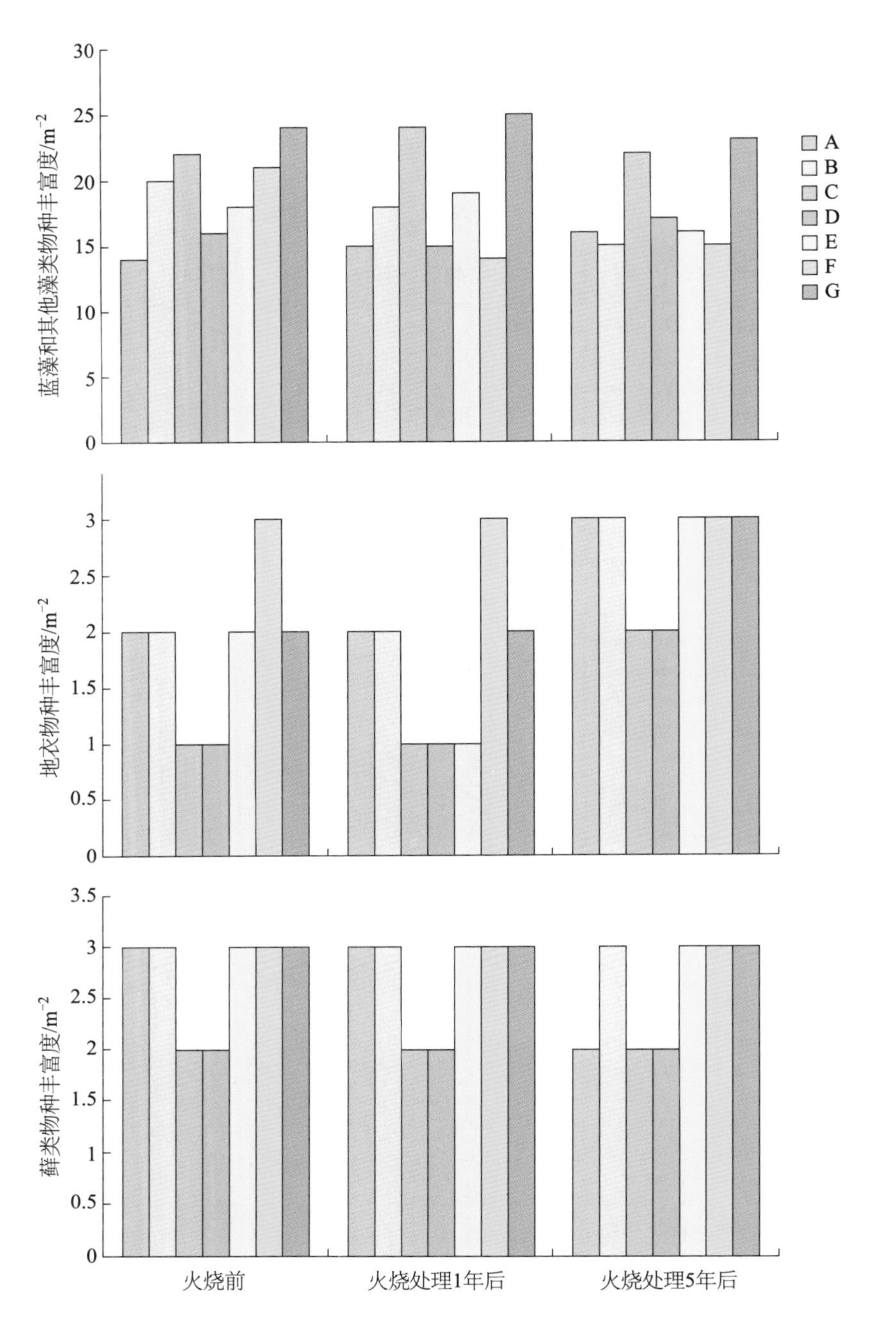

图 3-139　火烧处理 1 年和 5 年后 BSC 群落物种丰富度变化。图中 A~ G 为样地，见表 3-32

Figure 3-139　Changes in species richness of crustal composition (cyanobacteria and algae, lichens, mosses) after 1 and 5 years since fire treatment in the southern of the Tengger Desert. A-G present different plot, see Table 3-32

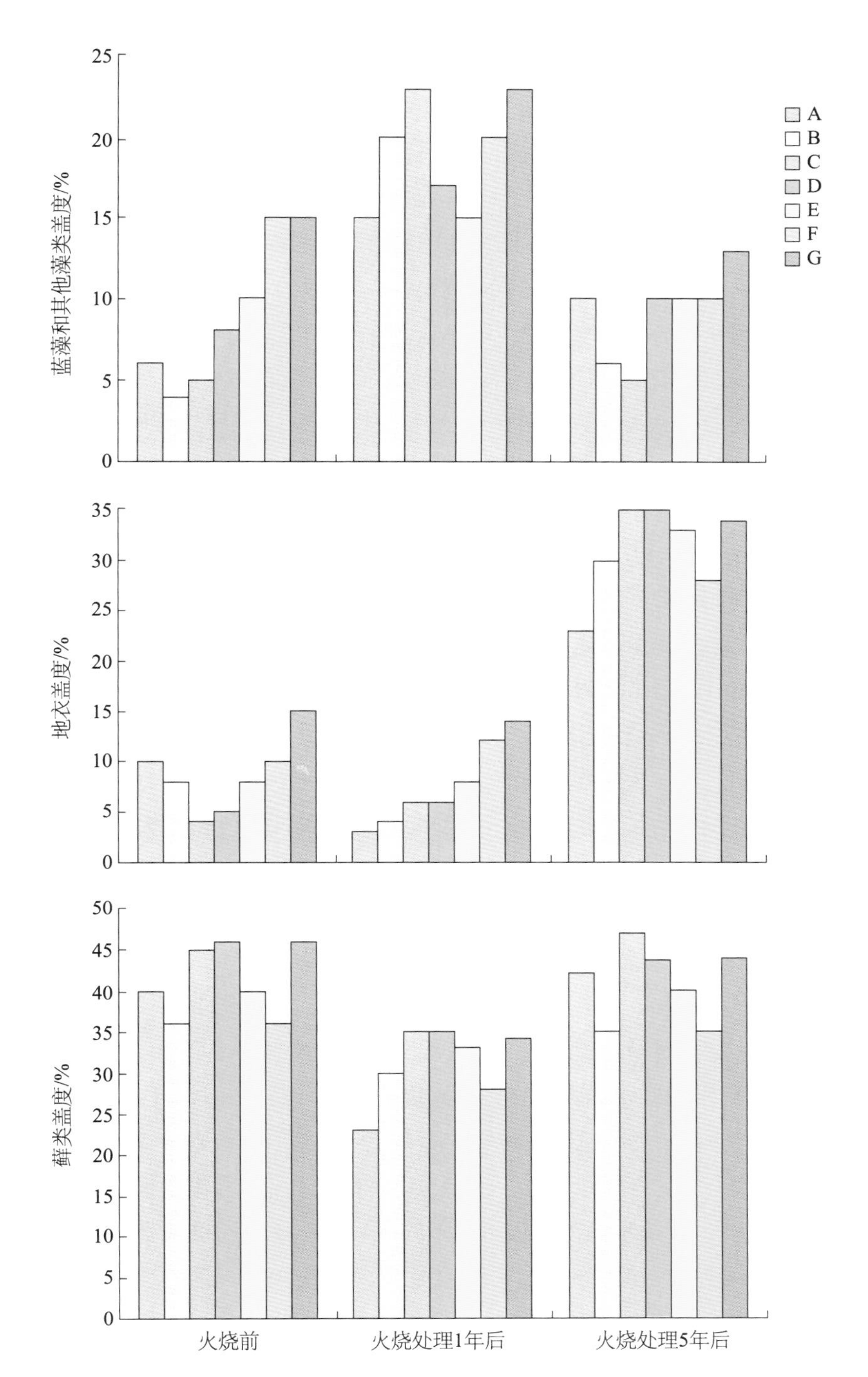

图 3-140　火烧处理 1 年和 5 年后 BSC 群落优势类群的盖度变化。图中 A~ G 为样地，见表 3-32
Figure 3-140　Changes in the cover of crustal cyanobacteria and algae，lichens，mosses after 1 and 5 years since fire treatment in the southern of the Tengger Desert. A-G present different plot，see Table 3-32

通过以上实验观测，我们认为，火烧总体上对BSC有一定的影响。对BSC群落中的蓝藻来说，其一部分生活在土层中，比如具鞘微鞘藻，只有降水后其叶鞘、菌丝体和部分器官才露出土壤表面，当干燥时再返回土中，留在土表的菌丝体则和土壤胶结在一起，加固了BSC层。进行火烧处理的春季一般少有降水，地表干燥，这可能是对BSC群落中蓝藻和一些荒漠藻影响较小的重要原因。当然，另外一些常贴伏在地表的蓝藻可能会被烧毁或部分烧毁，如发菜和葛仙米等。对地衣和藓类的影响相对于藻类较大，主要是因为两者均有地上部分，尤其是藓类，火烧时地上部分容易被烧掉，特别是春季这些变水植物组织均处于干燥状态。从图3-139可以看出，火烧对BSC群落中蓝藻和隐花植物种的丰富度影响不显著，尽管我们所选样地之间的物种丰富度有较大差异，但对某一个样地如样地A，火烧处理后1年和5年的观测表明，火烧无论对蓝藻和其他藻类还是地衣和藓类都很少有影响。图中数据的微小差异可能是实验室供试样品或鉴定的误差所致。由于BSC中地衣和藓类在野外条件下容易鉴定，除了地衣在火烧处理前为2种，而5年后调查发现有3种外，藓类的丰富度几乎没有变化。

相对于BSC群落中蓝藻和其他藻类、地衣和藓类种的丰富度变化，火烧处理对它们的相对盖度影响较大（图3-140）。除了蓝藻和其他藻类结皮火烧1年后其盖度明显增加外，地衣和藓类结皮的盖度均在火烧1年后降低，但降低幅度不大；火烧处理5年后，蓝藻和其他藻类结皮的盖度又降低，基本维持在火烧之前的水平，但是地衣和藓类的盖度增加明显，藓类增加到火烧前的盖度水平，而地衣的增加确实显著，远远高于火烧处理前的盖度，这说明在腾格里沙漠南缘草原化荒漠地区，火烧干扰有利于地衣为主的BSC的繁衍、拓殖。

以上我们的火烧处理实验结果，即火烧对BSC群落物种丰富度影响较小，对BSC群落种蓝藻和隐花植物的相对盖度影响较大，火烧后显著增加了蓝藻和其他藻类的盖度，减少了地衣和藓类的盖度，但后两者又能在短期内较快地恢复其盖度，且火烧有利于增加地衣的覆盖。我们对此结果的解释是：① 研究区属于腾格里沙漠南缘的沙漠向黄土的过渡区，其植被类型为草原化荒漠，土壤质地中黏粒和粉粒物质含量远远高于固定沙区和流动沙丘区，且火烧前样地中地表干扰较小，表土层没有风蚀现象，这样的环境有利于地衣在维管植物板块之间的空间拓殖、繁衍（见第1章），灌木冠幅下相对荫蔽的区域有利于藓类的覆盖，在此稳定的景观中作为BSC演替初期阶段的优势种，蓝藻和其他荒漠藻在BSC群落的盖度就相对较小。② 火烧处理后，由于地上植被被烧，一年生和多年生草本植物被烧尽，部分多年生植物被烧死，而木本灌木多被烧伤（火烧后部分枝茎被烧死，冠幅明显减小），这样地表植被盖度显著下降，开阔空间增加，有利于蓝藻和其他荒漠藻的繁衍，而5年后在没有其他干扰再发生的情况下，随着地上植被的恢复，蓝藻结皮生存的生态位又回到了火烧之前，使之仍然保持原来的盖度。③ 由于火烧处理仅仅烧掉了可燃和易燃的植物地上部分，对地表和土壤结构没有产生大的影响，特别是草原化荒漠地区，多风环境很难使灰烬直接进入土壤参与养分循环，也就是说，一次性的火烧对土壤的理化性状和生物学性状的影响较小，土壤表层亦稳定，使地衣的盖度减少较小，反而有了较大的增加。④ 不同于典型草原区，草原化荒漠植被稀疏、生物量小，火烧持续时间短，火烧干扰的强度相对较低，虽然降低了植被的盖度，对高出地表的藓类破坏较大，但灌丛下和小土丘周围的藓类得以保存，在持续不再干扰的情况下，藓类恢复很快。此外，BSC群落物种丰富度以蓝藻和其他藻类较高，而地衣和藓类仅为3种，较短的火烧干扰和稳定的地表，以及此后无重复的干扰可能是对它们丰富度造成影响较小的重要原因之一。较高强度的火烧干扰，如1年多次或连续多年几次的干扰可能会对BSC产生重要的影响，是未来荒漠生态系统恢复实践、调控和管理中需考虑的科学问题。

3.9 BSC对增温与降水减少的生态生理响应

全球变暖是一个众所周知的事实，是人类迄今面临的最重大环境问题之一，它涉及国家发展空间问题。我国《气候变化国家评估报告》预测到2020年，中国年平均气温可能增加1.1～2.1 ℃，年平均降水量可能增加2%～3%，降水日数在北方显著增加，降水区域差异更为明显。由于平均气温增加，蒸发增强，总体上北方荒漠地区水资源短缺状况将进一步加剧，未来极端天气气候事件呈增加趋势。因此，基于全球变暖背景下的BSC群落中隐花植物多样性及其BSC功能演变的研究是认知我国乃至全球干旱荒漠生态系统如何响应全球变化，以及探讨应采取的适应和减缓对策等科学问题的重要前提。

众所周知，气候变化深刻地改变着陆地生态系统的组成、结构和功能（Maestre *et al.*，2012；Lu *et al.*，2013）。这种改变能够直接通过生理胁迫（Hui *et al.*，2014，2015）和间接地通过改变生物种之间的关系（Harley，2011）以及增加生物入侵的风险来影响组成生态系统的生物体（organisms）（Bellard *et al.*，2013；Lu *et al.*，2013）。未来气候变化模型预测荒漠地区降水的改变将会发生在持续时间和强度上（Belnap *et al.*，2004）。也有其他研究预测在干旱荒漠这一BSC广泛分布区会提高干旱的频率和持续增温（Zelikova *et al.*，2012）。我国广袤的荒漠地区，特别是在腾格里沙漠东南缘、毛乌素沙地周边、阿拉善荒漠、柴达木盆地和古尔班通古特沙漠，BSC都得到了良好的发育。随干扰后BSC的恢复明显地出现了蓝藻结皮、荒漠藻结皮、地衣结皮和藓类结皮等演替阶段（Li *et al.*，2004），这与全球其他荒漠地区结皮演变过程相似（Eldridge and Greene，1994；Li *et al.*，2002），因而在全球范围内有很好的代表性（代表了温带荒漠隐花植物与BSC的基本生态特征）（Li *et al.*，2004）。值得注意的是，荒漠生态系统是一个非生物因素强烈胁迫的系统，BSC在没有干扰的情况下趋于一个稳定阶段系统中的永久组成成分，其主要构建者隐花植物的丧失或增加可以引发系统两个稳定阶段之间的转换，从而增加了荒漠生态系统的不确定性或不稳定性。因此，BSC群落中隐花植物多样性及其本身属性的改变或丧失不仅影响到BSC的组成、结构和功能的变化，而且会直接影响到荒漠系统的生态功能与稳定性（Bowker，2007）。

来自国内的预测报道，至2050年我国西北荒漠区的气温将上升1.9 ~ 2.3 ℃，降水将有一个较大的不确定性（Qin，2002）。中国科学院沙坡头沙漠试验研究站的长期监测也部分地支持了这一预测。腾格里沙漠沙坡头地区年均气温从1994—2003年的10.4 ℃上升至2004—2013年的10.9 ℃，增温0.5 ℃，而降水则从1994—2003年的平均年降水量180.6 mm下降至2004—2013年平均年降水量176.7 mm，下降了3.9 mm（图3-141）。

可以肯定的是，全球变暖和降水的变化会改变BSC的组成、结构和多元功能的发挥。但到底是如何影响的，特别是在增温和降水减少的情况下，其作用机理如何，很少有专门的研究涉及。本章的相关内容就全球变化，如增温、水分变化、CO_2浓度升高、UV-B增强和氮沉降升高对BSC的非生物胁迫做了详细研究，本节基于增温和降水减少对BSC影响及其响应的机理进行了研究。

基于长期监测和前期模拟实验，我们在腾格里沙漠东南缘沙坡头地区建立了2组OTCs，每组为3个OTCs（图3-142）。为了使增温与实际监测温度上升的结果有较好的一致性，我们设计OTC时通过多种选择和2年的预实验调整了OTC的大小，包括底面积和桶体的高度，通过调节桶体高度来实现对降水的减少，而不是利用传统的遮雨棚方法。一组OTC为体积较大的OTC（简称大OTC），其为八棱柱，上、下底边和高度各为100 cm × 100 cm × 200 cm，材料为铝合金与玻璃；另一组小OTC为长方体，底边及高为100 cm × 100 cm × 150 cm。小型OTC相比于大型OTC有较高的增温和较大减少降水的效果，因为玻璃墙体能够防止部分降水进入OTC内，OTC越小排除的降水就越多。与对照相比，大、小OTC均有增温和减少降水的效果，且对温度的增加和对降水的减少均在中国科学院沙坡头沙漠试验研究站长期观测与模型预测的范围之内（Qin，2002）。两组OTC对空气温度、湿度、降水和光合有效辐射的影响采用全自动气象系统（HOBO U30，Onset Computer Corporation，美国）（图3-142）进行监测，OTC具体的增温和减少降水的效果见表3-33。

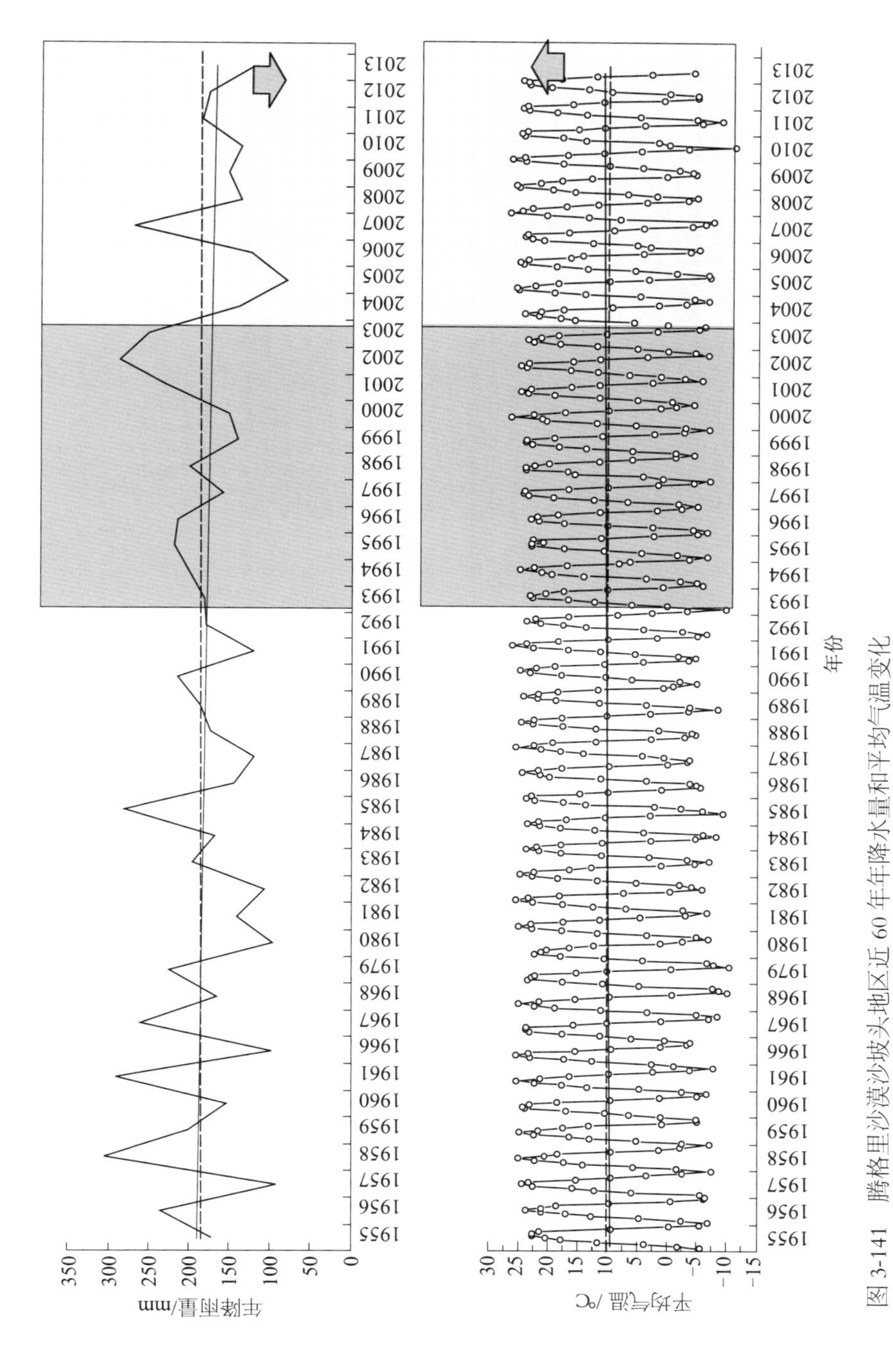

图 3-141 腾格里沙漠沙坡头地区近 60 年年降水量和平均气温变化

Figure 3-141 Sixty-year changes in mean air temperature and mean annual precipitation in Shapotou region of the Tengger Desert

图 3-142　用于模拟增温的不同大小的开顶式增温设置及对照实验（2007 年摄）
Figure 3-142　The different sizes of Open-Top Chambers（OTCs）were employed to simulate warming and a control experiment was also designed（photo in 2007）

表 3-33　实验期间（2008—2013 年）OTC 对年均气温、年降水量、空气湿度和光合有效辐射的影响以及 1990—2000 年和 2001—2013 年沙坡头沙漠试验研究站气象站记录的年均气温、降水量、空气湿度和光合有效辐射
Table 3-33　Effects of OTCs for annual warming, rainfall reduction, humidity and PAR （mean ± SE） during 2008-2013, and records from 1990-2000 and 2001-2013 based on meteorological observations by SDES, CAS

	年均气温 / ℃	降水量 /mm	空气湿度 /%	光合有效辐射 /（μmol m^{-2} s^{-1}）
大 OTC	12.3 ± 2.7	147.0 ± 19.8	43.2 ± 7.7	264.1 ± 32.5
小 OTC	12.8 ± 2.7	139.7 ± 19.9	43.6 ± 6.6	229.0 ± 59.3
对照	11.3 ± 2.8	157.7 ± 22.1	45.8 ± 6.8	312.0 ± 23.0
1990—2000 年	10.4 ± 0.6	180.6 ± 33.5	42.6 ± 1.6	316.5 ± 31.4
2001—2013 年	10.9 ± 0.5	176.7 ± 63.9	42.5 ± 3.1	318.3 ± 28.6

在野外进行BSC取样时（2007年春季），根据BSC中优势类群的盖度，将BSC划分为三类，即蓝藻、地衣和藓类为优势的BSC。在采样点使用有机玻璃盒采集完整BSC样品，包括下层沙土，选择的采样点BSC盖度>90%，且地表相对平整均一。每种BSC取样重复90个，并等同地放入OTC和对照实验地，即每个OTC中和对照地各放入10个相同类型BSC的样品，共三类BSC样品，每个OTC有30个样品，同样对照地也有30个BSC样品。

对BSC的变化测定分为3个时段进行：放置到OTC和对照实验地时，增温处理6年后（2013年8月）和增温处理7年后（2014年8月）。我们通过测定的系列参数来评价BSC功能在6～7年持续增温和减少降水处理期间的变化。我们鉴定了三个测定阶段BSC群落中蓝藻、地衣和藓类的组成及其物种多样性，并测定了生物量与盖度的相对变化。具体测定方法见《荒漠生物土壤结皮生态与水文学研究》（李新荣，2012）一书。

由表3-33结果表明，2组增温处理分别使年均气温上升了1.0 ℃和1.5 ℃，降水量减少了10.7 mm和18 mm，空气湿度和光合有效辐射也有所降低。BSC之间对吸湿凝结水的捕获有着明显的差异，以藓类为优势的BSC有较强的凝结水捕获能力，蓝藻为优势的BSC相对较弱（图3-143和图3-144）。无论是哪一类BSC，其对吸湿凝结水的捕获的高峰值均发生在3：00–11：00之间，它们对增温的响应各异。总的来说，增温和降水减少均导致了BSC对凝结水的捕获量的改变，其中显著降低了藓类结皮和地衣结皮对凝结水的捕获量，而对蓝藻为优势的BSC凝结水捕获的影响不显著（$p>0.05$，图3-144）。增温幅度在1.0 ℃和1.5 ℃之间，伴随着降水的减少，藓类结皮和地衣结皮的凝结水捕获量显著下降（$p<0.05$），而在2组OTC之间蓝藻结皮的凝结水捕获量没有明显的差异（$p>0.05$，图3-144）。

对于增温和降水减少后BSC及表土层最大持水能力的变化，图3-145表明，三类BSC中，藓类结皮拥有最大的土壤持水能力，蓝藻结皮的持水能力较差。不同BSC最大持水能力对增温和降水减少的响应不同。当增温1.0 ℃及降水减少10.7 mm时，藓类结皮的最大持水能力明显下降（$p<0.05$），但增温1.5 ℃和降水减少18 mm时，其最大持水能力不再有显著的改变（$p>0.05$，图3-145）。对于蓝藻结皮而言，无论增温1.0 ℃还是1.5 ℃，其最大持水能力均没有发生显著的改变（$p>0.05$），而地衣结皮在增温1.5 ℃及降水减少18 mm时，其最大持水能力明显下降（$p<0.05$），当增温幅度较小时，这种改变是不明显的（$p>0.05$，图3-145）。

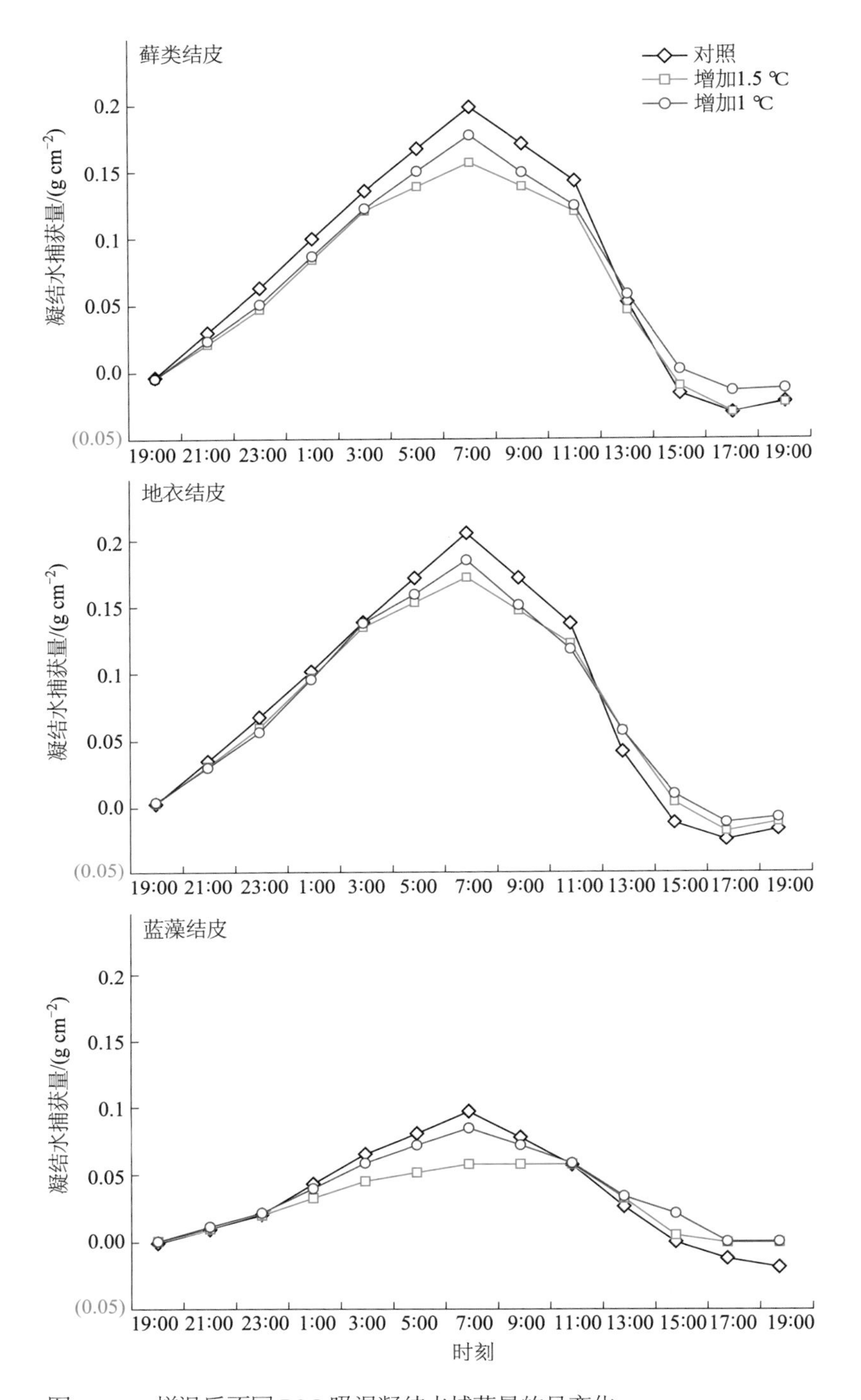

图 3-143　增温后不同 BSC 吸湿凝结水捕获量的日变化

Figure 3-143　The daily changes in dew entrapment of different BSCs after warming

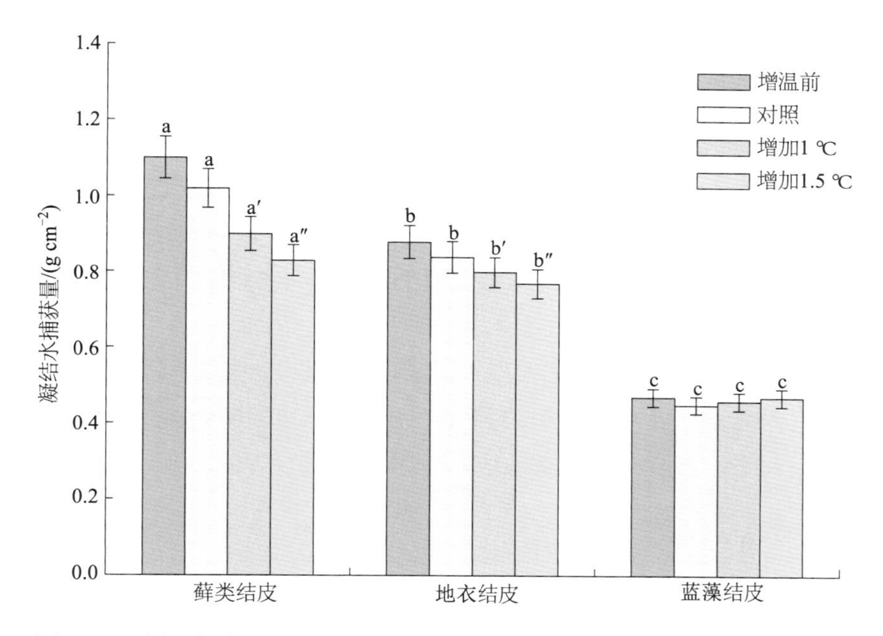

图 3-144　增温条件下不同 BSC 的凝结水捕获量（平均值 ± 标准误）的变化。不同字母表示观测值之间存在显著差异，$p<0.05$

Figure 3-144　Changes in dew entrapment (mean ± SE) of different BSC under warming. Within a given measured parameter，different superscripts indicate a significanct difference among treatments at $p<0.05$

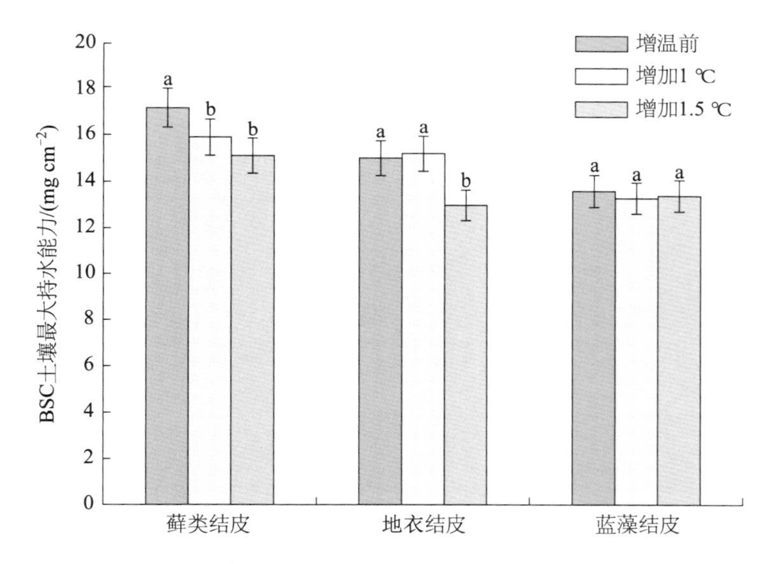

图 3-145　增温条件下不同 BSC 土壤最大持水能力（平均值 ± 标准误）的变化。对于每一个给定的测量参数，不同字母表示观测值之间存在显著差异，$p<0.05$

Figure 3-145　Changes in WHC of different BSC under warming. Within a given measured parameter，Different superscripts indicate a significanct difference among treatments at $p<0.05$.

如表3-34所示，BSC群落物种丰富度对两个增温和降水减少处理均没有明显的响应。对以藓类为优势的BSC群落，藓类种随增温和降水减少有减少的趋势，而蓝藻为优势的BSC群落中蓝藻种则有增加的趋势，尽管这些变化是不显著的，而对地衣种的多样性也看不出有任何影响（$p>0.05$）。随着增温和降水减少，不论是藓类还是地衣为优势的BSC，其盖度均呈下降趋势，地衣盖度下降不显著，但藓类在BSC中的盖度下降明显，从对照的92.74%下降至增温1.5 ℃后的74.07%。增温与降水减少似乎对蓝藻结皮盖度不会引发显著的变化或影响很小。

BSC生物量对增温和降水减少的响应因BSC类型和增温与降水减少的强度不同而存在较大差异。藓类结皮的生物量随增温和降水减少而显著下降，并与增温与降水减少的幅度呈负相关关系。相反，增温与降水减少对地衣结皮和蓝藻结皮的生物量没有明显的影响（表3-34）。

表3-34　增温前后BSC群落物种丰富度、盖度和生物量的变化

Table 3-34　Changes in crustal species richness，crustal cover and biomass under warming treatments

测量参数	BSC	对照	增温1 ℃	增温1.5 ℃
物种丰富度 /cm^{-2}	藓类结皮	苔藓（4.2±0.4）a	苔藓（3.9±0.2）a	苔藓（3.9±0.6）a
		地衣（1.2±0.3）a	地衣（1.2±0.1）a	地衣（1.0±0.3）a
		蓝藻（10.3±1.2）a	蓝藻（10.6±1.1）a	蓝藻（11.3±0.8）a
	地衣结皮	苔藓（1.3±0.3）a	苔藓（1.2±0.3）a	苔藓（1.2±0.4）a
		地衣（2.5±0.4）a	地衣（2.4±0.2）a	地衣（3.0±0.4）a
		蓝藻（10.5±1.1）a	蓝藻（11.6±1.3）a	蓝藻（11.4±1.3）a
	蓝藻结皮	苔藓（0）	苔藓（0）	苔藓（0）
		地衣（0）	地衣（0）	地衣（0）
		蓝藻（16.5±2.3）a	蓝藻（17.3±2.1）a	蓝藻（17.9±1.6）a
BSC盖度 /%	藓类结皮	92.74±2.23a	80.43±3.71b	74.07±4.95c
	地衣结皮	90.59±3.03a	89.13±3.33a	88.56±2.69a
	蓝藻结皮	90.12±2.83a	89.36±3.05a	90.10±3.35a
BSC生物量 /（mg cm^{-2}）	藓类结皮	8.21±0.58a	6.03±0.55b	4.76±0.44c
	地衣结皮	5.57±0.45a	5.52±0.33a	5.14±0.53a
	蓝藻结皮	3.48±0.41a	3.52±0.39a	3.46±0.29a

注：不同的字母表示测定参数在不同处理间有显著差异，$p<0.05$。

此外，我们对增温和降水减少处理后不同BSC的光合特性进行了对比研究。不同OTC中经过增温和降水减少的BSC被放置在对照BSC旁边，2013年夏季一场约10 mm的降雨事件后，我们马上连续2天进行了不同处理BSC的光合速率日变化的测定。除了2组不同BSC经过大、小OTC增温和降水减少处理6年外，其他环境因子与对照BSC的环境因子一致。

图3-146表明，当发生10 mm的降水事件时，首先不同的BSC被湿润，相对于藓类，地衣和蓝藻很快进入光合作用，其净光合速率迅速增加。藓类吸湿水合后呼吸过程剧烈，呼吸作用大于光合作用，使其净光合速率长时间为负值。但是，从测定结果可以看出，经受不同增温和降水减少处理后，藓类的这种先强呼吸作用后较大光合作用的特性有所改变，主要体现在处理后藓类的呼吸强度似乎减弱，较小增温和降水减少处理过的BSC（大OTC中），在13：00已出现光合速率大于呼吸速率而净光合速率为正值，在较大增温和降水减少处理（小OTC）后的BSC则在15：00出现净光合速率正值。这说明增温和降水减少对藓类的光合固碳影响较大。尽管如此，从测定结果可以看出，在随后的测定过程中来自大、小OTC中的BSC净光合速率均下降。与对照相比，藓类BSC的光合固碳能力下降。对于地衣BSC而言，增温和降水减少对其光合固碳能力有降低的趋势，但没有显著的差异。增温和降水减少对蓝藻结皮的影响是提高了其光合固碳能力，至少湿润后其净光合速率得到显著提高。

以上通过增温和降水减少模拟BSC对未来气候变化响应的结果可以解释为BSC对吸湿凝结水的捕获能力大小差异主要归因于BSC的组成和结构（Li *et al.*，2010c；Pan *et al.*，2010）。藓类结皮中藓类常生长在地表，其高度达到0.5～1 cm，与其他类型BSC相比较，更容易形成并捕获凝结水，易获得较多凝结水（Liu *et al.*，2006；Pan *et al.*，2010）。蓝藻为优势的BSC群落中，由于大多数藻类生活在表土层中，对地表粗糙度的改变较小，形成的凝结水也较少，因此，对整个BSC而言其群落中藓类所占的比例和盖度对其凝结水的捕获起着至关重要的作用。然而增温和降水减少实验结果表明，在增温1 ℃和1.5 ℃，及降水分别减少10.7 cm和18 cm时，BSC群落中蓝藻和藻类、地衣种的丰富度变化不显著，藓类种的丰富度有降低趋势，但亦不显著，而BSC群落中藓类的盖度和生物量明显降低。这样的改变直接导致藓类在BSC群落组成和功能中的优势地位丧失，在持续增温和降水减少的情境下将发生藓类在群落中被地衣和蓝藻类替代的现象。这一发现也预示着在持续增温的背景下，BSC的功能将发生显著改变，如凝结水的捕获量减少、土壤持水能力下降、浅层土壤水分有效性降低、降水入渗增加、BSC光合固碳能力下降，这些均归因于BSC群落组成和结构的改变，即演替后期的类型被初期的类型所替代。

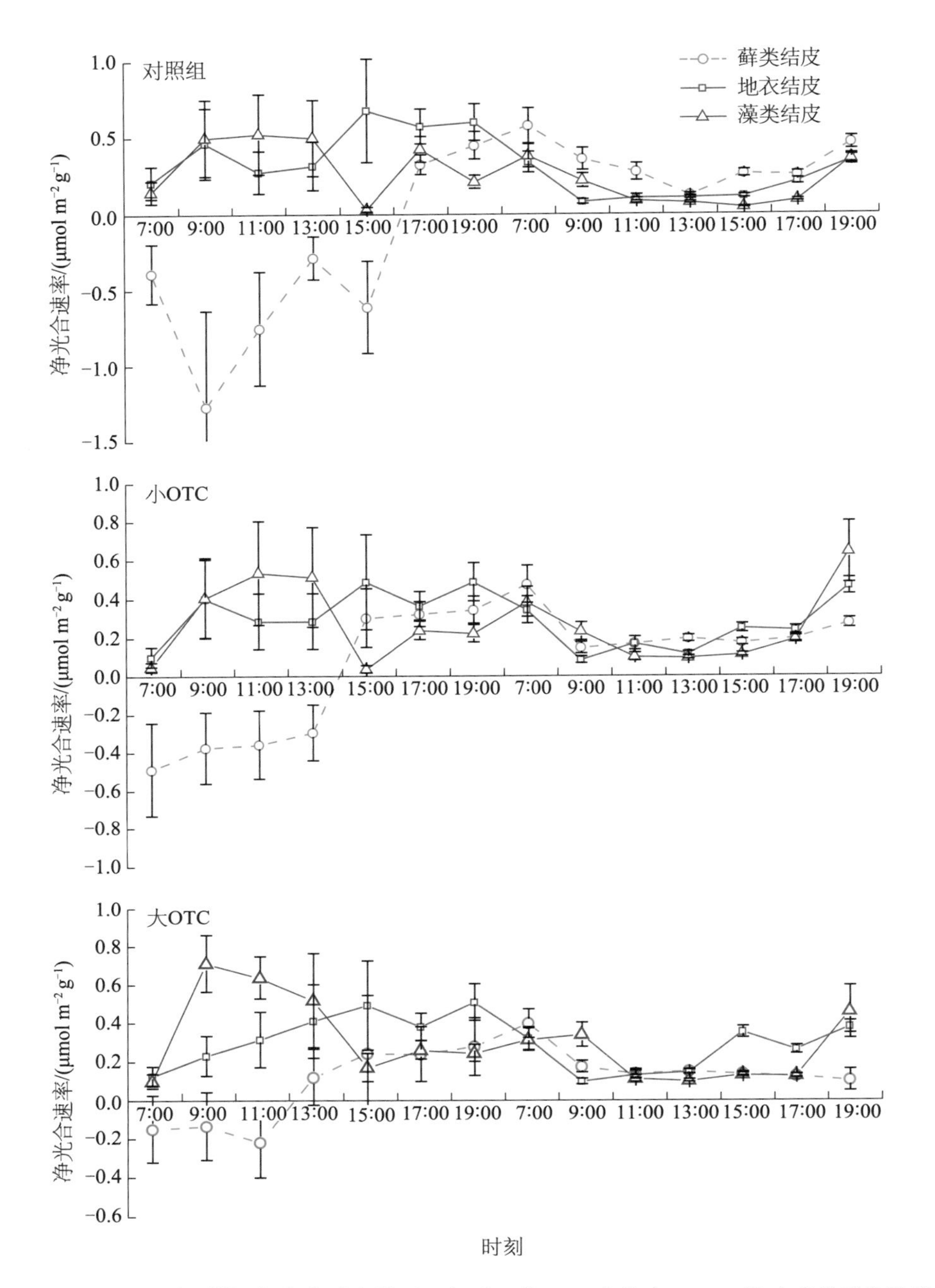

图 3-146 不同增温与降水减少处理 6 年后三类 BSC 在发生 10 mm 降水事件后的净光合速率的变化
Figure 3-146 Changes in P_n of moss-, algae (cyanobacteria) - and lichen-dominated BSC after six-year treatment for warming and reducing rainfall using two size of OTCs after 10 mm precipitation event

地衣在BSC群落中对增温和降水减少的响应不明显，说明在未来持续气候变化的趋势下，地衣占优势的BSC是相对稳定的，该BSC的功能也能得到较好的维持。同样，蓝藻和其他荒漠藻对增温和降水减少并不十分敏感，我们的增温和降水减少模拟处理对其没有显著影响，这说明持续的增温和降水减少对蓝藻为优势的BSC影响也较小。BSC的重要水文功能体现在对入渗的影响，而直接决定入渗的是BSC覆盖的表土层和结皮层的厚度，及BSC的盖度和生物量（Belnap and Lange，2003；Li *et al.*，2010c）。而在我国沙区藓类结皮覆盖的表土层厚度和生物量最大，蓝藻结皮表土层的厚度和生物量最小。这也就不难理解BSC组成和结构改变直接导致其功能变化的结论了。

另外，BSC对增温和降水减少的光合生理响应的重要性在于它直接关联着BSC对整个荒漠生态系统的碳输送。BSC中隐花植物独特的生活史特点增强了其在严酷环境中的拓殖能力，例如微小繁殖体的风播特点、对干燥的耐性、特殊的光保护色素，以及固氮和固碳的能力。往往被人们忽略的是隐花植物及其BSC的碳固定和呼吸是干旱、半干旱荒漠系统碳循环必不可少的重要环节（Beymer and Klopatek，1991；Belnap，2002）。有研究认为BSC中地衣的光能利用率为0.5%～2%，和维管植物处于同一量级（Palmqvist，2000）；而在条件适宜时它们的光合效率在两个量级之间变化，即0.1～11.5 μmol CO_2 m^{-2} s^{-1}，后者与同一地区的旱生灌木相当（Lange，2003）。尽管隐花植物的瞬时光合速率较维管植物低，但如果考虑到在干旱、半干旱地区BSC所占的盖度，那么BSC的光合固碳是这些地区主要碳源的结论是毋庸置疑的（West，1990；Harper and Marble，1998）；另外如果考虑到干旱区严酷的生境条件不利于维管植物的发育，那么隐花植物的固碳量将更加不容忽视（Belnap，1996）。以*Microcoleus*属为主的早期演替阶段的BSC年固碳量为0.4～2.3 g C m^{-2} a^{-1}，而以地衣为主的后期演替阶段的BSC年固碳量为12～37 g C m^{-2} a^{-1}。然而，最新的研究表明干旱区降水格局的变化与升温将引起BSC结构和组成的改变，从而导致其碳贮量和通量的变化（Belnap *et al.*，2004）。也有研究断言，任何干扰包括气候的波动都有可能影响隐花植物固有的习性（Lange *et al.*，1992；Zaady *et al.*，2000；Belnap and Lange，2003）。BSC对环境因子极度敏感的特性常被用作生态系统健康和环境因子变化的指示剂（Patrick，2002；Li *et al.*，2004），持续的增温与降水格局变化，以及综合其他全球变化如CO_2浓度升高、氮沉降增加和UV-B增强等对BSC固碳和其他功能的影响，以及BSC群落生物从分子、器官、个体和种群多尺度上对此的响应的研究，是未来了解荒漠/沙地生态系统恢复与重建，以及系统可持续发展生态系统管理的重要理论基础和前沿方向。

参考文献

白学良，王瑶，徐杰，李新荣，张景光．沙坡头地区固定沙丘结皮层藓类植物的繁殖和生长特性研究．中国沙漠，2003，23(2)：171–173.

包维楷，冷俐．相同环境下 3 种藓类植物光合色素含量的比较．植物资源与环境学报，2005，14(3)：53–54.

鲍思伟，谈锋，廖志华．蚕豆 (*Vicia faba L.*) 对不同水分胁迫的光合适应性研究．西南师范大学学报，2001，26(4)：448–451.

卜楠，俞丽蓉，马万里，马铁娟，周鸿升，毕建瓴，孙克．不同水分条件下沙漠豆生理指标的变化．中国水土保持科学，2012，10(6)：77–81.

蔡妙珍，刘鹏，徐根娣，沈丽萍．蓼科、禾本科植物细胞膜对铝胁迫反应的比较研究．水土保持学报，2005，19(6)：122–125.

陈少裕．膜脂过氧化与植物逆境胁迫．植物学通报，1989，6(4)：211–217.

陈文佳，张楠，杭璐璐，王媛，季梦成．干旱胁迫与复水过程中遮光对细叶小羽藓的生理生化影响．应用生态学报，2013，24(1)：57–62.

戴万宏，王益权，黄耀，刘军，赵磊．塿土剖面 CO_2 浓度的动态变化及其受环境因素的影响．土壤学报，2004，41(5)：170–175.

邓松强，杨翠琨，安晓亮，胡超珍，陈兰洲．荒漠念珠藻对 UV-B 辐射的反应及外源抗氧化剂的保护作用．干旱区研究，2012，29(5)：816–819.

刁一伟，郑循华，王跃思，徐仲均，韩圣慧，朱建国．开放式空气 CO_2 浓度增高条件下旱地土壤气体 CO_2 浓度廓线测定．应用生态学报，2002，13(10)：1249–1252.

董新纯，赵世杰，郭珊珊，孟庆伟．增强 UV-B 条件下类黄酮与苦荞逆境伤害和抗氧化酶的关系．山东农业大学学报，2006，37(2)：157–162.

方志红，董宽虎．NaCl 胁迫对碱蒿可溶性糖和可溶性蛋白含量的影响．中国农学通报，2010，26(16)：147–149.

冯虎元，安黎哲，陈书燕，王勋陵，程国栋．增强 UV-B 辐射与干旱复合处理对小麦幼苗生理特性的影响．生态学报，2002，22(9)：1546–1568.

冯虎元，安黎哲，徐世健，强维亚，陈拓，王勋陵．紫外线 –B 辐射增强对大豆生长、发育、色素和产量的影响．作物学报，2001，27(3)：319–323.

伏毅，戴媛，谭晓荣，万谦．干旱对小麦幼苗脂类和蛋白质氧化损伤的影响．作物杂志，2010，26(3)：45–50.

国春晖，沙伟，李孝凯．干旱胁迫对三种藓类植物生理特性的影响．北方园艺，2014，38(9)：78–82.

国家林业局．中国荒漠化与沙化状况公报．2011.

何明珠，李新荣，张景光，周海燕，金文杰，张志山．土壤生物结皮蒸散特征研究．中国沙漠，2006，26(2)：159–165.

黄磊，张志山，胡宜刚，张鹏，赵洋．固沙植被区典型生物土壤结皮类型下土壤 CO_2 浓度变化特征及其驱动因子研究．中国沙漠，2012，32(6)：1583–1589.

回嵘，李新荣，贾荣亮，赵昕，刘艳梅，陈翠云．增强 UV-B 辐射对真藓结皮生理特性的影响．生态学杂志，2012，31(1)：38–43.

贾子毅，吴波，卢琦．半干旱区生物土壤结皮光合固碳研究进展．安徽农业科学，2011，39(2)：12768–12770.

蒋霞敏，翟兴文，王丽，陆开形．雨生红球藻对紫外辐射的生理适应及超微结构变化．水产科学，2003，27(2)：105–112.

黎峥，段舜山，武宝玕．UV-B 对两种藻光合色素和多糖含量的影响．生态科学，2003，22(1)：42–44.

李贝贝，郭进魁，周云，张珠珠，张立新．一种分析叶绿体类囊体膜色素蛋白复合物的蓝绿温和胶电泳系统．生物化学与生物物理进展，2003，30(4)：639–643.

李刚，刘立超，高艳红，赵杰才，杨昊天．降雪对生物土壤结皮光合及呼吸作用的影响．中国沙漠，2014，34(4)：998–1006.

李守中，肖洪浪，罗芳，宋耀选，刘立超，李守丽．沙坡头植被固沙区生物结皮对土壤水文过程的调控作用．中国沙漠，2005，25(2)：228–233.

李晓丽，申向东．结皮土壤的抗风蚀性分析．干旱区资源与环境，2006，20(2)：203–207.

李晓阳，陈慧泽，韩榕．UV-B 辐射对拟南芥种子萌发和幼苗生长的影响．植物学报，2013，48(1)：52–58.

李新荣．荒漠生物土壤结皮生态与水文学研究．北京：高等教育出版社，2012.

李亚敏，肖红利 .2 种藓类植物对 Cd–Cu 复合胁迫的生理响应．江苏农业科学，2011，39(4)：441–443.

李燕宏，洪健，谢礼，杨勇，周雪平，蒋德安．蚕豆萎蔫病毒 2 号分离物侵染对蚕豆叶片光合活性和叶绿体超微结构的影响．植物生理与分子生物学学报，2006，32(4)：490–496.

李元，何永美，祖艳群．增强 UV-B 辐射对作物生理代谢、DNA 和蛋白质的影响研究进展．应用生态学报，2006，17(1)：123–126.

梁英，冯力霞，田传远．高温胁迫对球等鞭金藻 3011 和 8701 叶绿素荧光特性的影响．水产学报，2009，33(1)：37–44.
刘丽丽，张文会，范颖伦，吕艳伟，苗秀莲．不同剂量 UV-B 辐射对冬小麦幼苗形态及生理指标的影响．生态学杂志，2010，29(2)：314–318.
刘敏，李荣贵，范海，杜桂彩．UV-B 辐射对烟草光合色素和几种酶的影响．西北植物学报，2007，27(2)：291–296.
刘清华，钟章成．紫外线 –B 辐射对银杏活性氧代谢及膜系统的影响．西华师范大学学报，2007，28(2)：148–152.
刘冉，李彦，王勤学，许皓，郑新军．盐生荒漠生态系统二氧化碳通量的年内、年际变异特征．中国沙漠，2011，31(1)：108–114.
刘宛，李培军，周启星，孙铁珩，台培东，许华夏，张海荣．短期菲胁迫对大豆幼苗超氧化物歧化酶活性及丙二醛含量的影响．应用生态学报，2003，14(4)：581–584.
刘艳梅，李新荣，何明珠，贾荣亮，李小军，张志山．生物土壤结皮对土壤微生物量碳的影响．中国沙漠，2012，32(3)：669–673.
卢琦，杨有林．全球沙尘暴警世录．北京：中国环境科学出版社，2001.
马洋，王雪芹，张波，刘进辉，韩章勇，唐钢梁．风蚀和沙埋对塔克拉玛干沙漠南缘骆驼刺水分和光合作用的影响．植物生态学报，2014，38(5)：491–498.
聂华丽，吴楠，梁少民，王红玲，张元明．不同沙埋深度对刺叶墙藓植株碎片生长的影响．干旱区研究，2006，23(1)：66–70.
牛传坡，蒋静艳，黄耀．UV-B 辐射强度变化对冬小麦碳氮代谢的影响．农业环境科学学报，2007，26(4)：1327–1332.
庞海河．增强 UV-B 辐射下南方红豆杉形态、结构及代谢的研究．东北林业大学博士学位论文，2010.
饶本强，吴沛沛，Dauta A，李敦海，刘永定．温室条件下 UV-B 辐射对蓝藻结皮生长和超微结构的影响．环境科学学报，2011，31(3)：649–657.
石莎，马风云，刘立超，周易君，冯金朝．沙坡头地区不同地表和植物表面温度的日变化．干旱区研究，2004，21(3)：275–279.
苏延桂，李新荣，赵昕，王正宁．紫外辐射增强对不同阶段荒漠藻结皮光合作用的影响．中国沙漠，2011，31(4)：889–893.
唐莉娜．水稻对 UV-B 辐射增强的生理响应及其抗性动态遗传的研究．福建农林大学博士学位论文，2003.
唐莉娜，林文雄，梁义元，陈芳育．UV-B 辐射增强对水稻蛋白质及核酸的影响研究．中国生态农业学报，2004，12(1)：40–42.
田桂泉，白学良，徐杰，张建升．固定沙丘生物结皮层藓类植物形态结构及其适应性研究．中国沙漠，2005，25(2)：107–113.
田继远，唐学玺，于娟，肖惠，冯蕾．海洋微藻对 UV-B 辐射的生理生化响应．海洋科学，2006，30(4)：54–58.
汪天，王素平，郭世荣．低氧胁迫下黄瓜根系细胞壁结合态多胺氧化酶活性变化及测定方法的确立．西北植物学报，2005，25(3)：503–506.
王凤玉，周广胜，贾丙瑞，王玉辉．水热因子对退化草原羊草恢复演替群落土壤呼吸的影响．植物生态学报，2003，2(5)：644–649.
王涛．干旱区主要陆表过程与人类活动和气候变化研究进展．中国沙漠，2007，27(5)：711–718.
王兴梅，张勃，戴声佩，张凯，马中华，王亚敏．甘肃省黄土高原区夏季极端降水的时空特征．中国沙漠，2011，31(1)：223–229.
王雪芹，张元明，张伟民，韩致文．古尔班通古特沙漠生物结皮对地表风蚀作用影响的风洞实验．冰川冻土，2004，26(5)：632–638.
卫伟，陈利顶，傅伯杰，巩杰，黄志霖．黄土丘陵沟壑区极端降雨事件及其对径流泥沙的影响．干旱区地理，2007，30(6)：896–901.
魏美丽，张元明．脱水对生物结皮中齿肋赤藓光合色素含量和叶绿体结构的影响．中国沙漠，2010，30(6)：1311–1318.
吴楠，张元明．古尔班通古特沙漠生物土壤结皮影响下的土壤酶分布特征．中国沙漠，2010，30(5)：1128–1136.
吴能表，马红群．增强 UV-B 辐射对胡椒薄荷逆境生理指标的影响．西南大学学报，2012，34(10)：1–6.
吴杏春，林文雄，黄忠良．UV-B 辐射增强对两种不同抗性水稻叶片光合生理及超显微结构的影响．生态学报，2007，27(2)：554–564.
伍泽堂．超氧自由基与叶片衰老时叶绿素破坏的关系 (简报)．植物生理学通讯，1991，27(4)：277–279.
肖媛，王高鸿，刘永定．UV-B 辐射对雨生红球藻光合特性和虾青素含量的影响及其响应．水生生物学报，2010，34(6)：1077–1082.
谢灵玲，赵武灵，沈黎明．光照对大豆叶片苯丙氨酸裂解酶 (PAL) 基因表达及异黄酮合成的调节．植物学通报，2000，17(5)：443–449.
谢作明．荒漠藻类对紫外辐射的响应及其结皮形成的研究．中国科学院水生生物研究所博士学位论文，2006.
许长城，邹琦．大豆叶片旱促衰老及其与膜脂过氧化的关系．作物学报，1993，19(4)：359–364.

薛林贵，石小霞，褚可成，陈志梅，李师翁．螺旋藻对短期增强 UV-B 辐射的生理生化响应．中国水产科学，2011，18(5)：1108–1114.
晏斌，戴秋杰．紫外线 B 对水稻叶组织中活性氧代谢及膜系统的影响．植物生理学报，1996，22(4)：373–378.
杨丽雯，周海燕，樊恒文，贾晓红，刘立超，李爱霞．沙坡头人工固沙植被生态系统土壤恢复研究进展．中国沙漠，2009，29(6)：1116–1123.
杨武．苔藓植物适应环境的形态结构及生理学机制．浙江师范大学硕士学位论文，2008.
于云江，史培军，鲁春霞，刘家琼．不同风沙条件对几种植物生态生理特征的影响．植物生态学报，2003，27(1)：53–58.
张丽华，陈亚宁，李卫红，赵锐锋．准噶尔盆地梭梭群落下土壤 CO_2 释放规律及其影响因子的研究．中国沙漠，2007，27(2)：266–272.
张鹏，李新荣，何明珠，李小军，高艳红．冬季低温及模拟升温对生物土壤结皮固氮活性的影响．生态学杂志，2012，31(7)：1653–1658.
张鹏，李新荣，胡宜刚，黄磊，冯丽，赵洋．湿润持续时间对生物土壤结皮固氮活性的影响．生态学报，2011b，31(20)：6116–6124.
张鹏，李新荣，贾荣亮，胡宜刚，黄磊．沙坡头地区生物土壤结皮的固氮活性及其对水热因子的响应．植物生态学报，2011a，35(9)：906–913.
张显强，罗在柒，唐金刚，卢文芸，乙引．高温和干旱胁迫对鳞叶藓游离脯氨酸和可溶性糖含量的影响．广西植物，2004，24(6)：570 –573.
张元明，曹同．干旱与半干旱地区苔藓植物生态学研究综述．生态学报，2002，22(7)：1129–1134.
张跃群，陆德祥，王勇军．紫外线辐照对 3 种海洋微藻蛋白质含量的效应．安徽农业科学，2009，37(20)：9350–9351.
张志山，何明珠，谭会娟，陈应武，潘颜霞．沙漠人工植被区生物结皮类土壤的蒸发特性．土壤学报，2007，44(3)：404–410.
赵哈林，曲浩，周瑞莲，王进，云建英，李瑾．小叶锦鸡儿幼苗对沙埋的生态适应和生理响应．西北植物学报，2013，33(7)：1388–1394.
赵哈林，赵学勇，张铜会，周海燕．沙漠化过程中植物的适应对策及植被的稳定性．北京：中国海洋出版社，2004.
赵拥华，赵林，武天云，唐素然．冬春季青藏高原北麓河多年冻土活动层中气体 CO_2 浓度分布特征．冰川冻土，2006，28(2)：183–190.
赵允格，许明祥，Belnap J. 生物结皮光合作用对光温水的响应及其对结皮空间分布格局的解译——以黄土丘陵区为例．生态学报，2010，30 (17)：4668–4675.
周新明．UV-B 辐射增强对葡萄叶片光合生理及总酚含量的影响．西北农林科技大学硕士学位论文，2007.
朱鹏锦，尚艳霞，师生波，韩发．植物对 UV-B 辐射胁迫响应的研究进展．热带生物学报，2011，2(1)：89–96.
左圆圆，刘庆，林波，何海．短期增强 UV-B 辐射对青榨槭幼苗生理特性的影响．应用生态学报，2005，16(9)：1682–1686.
Abdel-Hafez SII，Mohawed SM，El-Said AHM. Seasonal fluctuations of soil fungi of Wadi Qena at eastern desert of Egypt. *Acta Mycologica*，1989，25：113–125.
Abdullah SK，Al-Khesraji TO，Al-Edany TY. Soil mycoflora of the southern desert of Iraq. *Sydowia*，1986，39：8–16.
Abed RMM，Al-Sadi AM，Al-Shehi M，Al-Hinai S，Robinson MD. Diversity of free-living and lichenized fungal communities in biological soil crusts of the Sultanate of Oman and their role in improving soil properties. *Soil Biology and Biochemistry*，2013，57：695–705.
Abed RMM，Kharusi SA，Schramm A，Robinson MD. Bacterial diversity，pigments and nitrogen fixation of biological desert crusts from the Sultanate of Oman. *FEMS Microbiology Ecology*，2010，72：418–428.
Albertsson P. A quantitative model of the domain structure of the photosynthetic membrane. *Trends in Plant Science*，2001，6：349–354.
Allen JF，Paula de WBM，Puthiyaveetil S，Nield J. A structural phylogenetic map for chloroplast photosynthesis. *Trends in Plant Science*，2011，16：645–655.
Allen MM. Photosynthetic membrane system in *Anacystis nidulans*. *Journal of Bacteriology*，1968，96：836–841.
Alley RB，Berntsen T，Bindoff NL. *Climate Change*：*The Physical Science Basis*，*Summary for Policy Makers*. Geneva：IPCC Secretariat，2007.
Allison SD，Czimczik CA，Treseder KK. Microbial activity and soil respiration under nitrogen addition in Alaskan boreal forest. *Global Change Biology*，2008，14：1156–1168.
Allison SD，Hanson CA，Treseder KK. Nitrogen fertilization reduces diversity and alters community structure of active fungi in boreal ecosystems. *Soil Biology and Biochemistry*，2007，39：1878–1887.
Apel K，Hurt H. Reactive oxygen species：Metabolism，oxidative stress，and signal transduction. *Annual Review of Plant Biology*，2004，55：373–599.
Aranibar JN，Andersonw IC，Ringrosez S，Macko SA. Importance of nitrogen fixation in soil crusts of southern African arid

ecosystems: Acetylene reduction and stable isotope studies. *Journal of Arid Environments*, 2003, 54: 345–358.
Aráoz R, Lebert M, Häder DP. Translation activity under ultraviolet radiation and temperature stress in the cyanobacterium *Nostoc* sp. *Journal of Photochemistry and Photobiology B*: *Biology*, 1998, 47: 115–120.
Aronson EL, McNulty SG. Appropriate experimental ecosystem warming methods by ecosystem, objective, and practicality. *Agricultural and Forest Meteorology*, 2009, 149: 1791–1799.
Baeshen NA, Sabir JS, Zainy MM, Baeshen MN, Abo-Aba SEM, Moussa TAA, Ramadan HAI. Biodiversity and DNA barcoding of soil fungal flora associated with *Rhazya stricta* in Saudi Arabia. *Bothalia*, 2014, 44: 301–314.
Bajracharya RM, Lai R, Kimble JM. Erosion effects on carbon dioxide concentration and carbon flux from an Ohio Alfisol. *Soil Science Society American Journal*, 2000, 64: 694–700.
Bartels D, Sunkar R. Drought and salt tolerance in plants. *Critical Reviews in Plant Sciences*, 2005, 24: 23–58.
Bates ST, Garcia-Pichel F. A culture-independent study of free-living fungi in biological soil crusts of the Colorado Plateau: Their diversity and relative contribution to microbial biomass. *Environmental Microbiology*, 2009, 11: 56–67.
Bates ST, Nash Ⅲ TH, Garcial-Pichel F. Patterns of diversity for fungal assemblages of biological soil crusts from the southwestern United States. *Mycologia*, 2012, 104: 353–361.
Bates ST, Nash Ⅲ TH, Sweat KG, Garcial-Pichel F. Fungal communities of lichen-dominated biological soil crusts: Diversity, relative microbial biomass, and their relationship to disturbance and crust cover. *Journal of Arid Environments*, 2010, 74: 1192–1199.
Beare MH, Gregorich EG, St-Georges P. Compaction effects on CO_2 and N_2O production during drying and rewetting of soil. *Soil Biology and Biochemistry*, 2009, 41: 611–621.
Bellard C, Thuiller W, Leroy B, Genovesi P, Bakkenes M, Courchamp F. Will climate change promote future invasions? *Global Change Biology*, 2013, 19: 3740–3748.
Belnap J, Lange OL. Biological soil crusts: Structure, function and management. Berlin: Springer-Verlag, 2003.
Belnap J. Soil surface disturbances in cold deserts: Effects on nitrogenase activity in cyanobacterial-lichen soil crusts. *Biology and Fertility of Soils*, 1996, 23: 362–367.
Belnap J, Phillips SL, Miller ME. Response of desert biological soil crusts to alterations in precipitation frequency. *Oecologia*, 2004, 141: 306–316.
Belnap J, Phillips SL, Smith SD. Dynamics of cover, UV-protective pigments, and quantum yield in biological soil crust communities of an undisturbed Mojave Desert shrubland. *Flora*, 2007, 202: 674–686.
Belnap J. Nitrogen fixation in biological soil crusts from southeast Utah, USA. *Biology and Fertility of Soils*, 2002, 35: 128–135.
Bennett EM, Peterson GD, Levitt EA. Looking to the future of ecosystem services. *Ecosystems*, 2005, 8: 125–132.
Bewley JD. Physiological aspects of desiccation tolerance. *Annual Review of Plant Physiology*, 1979, 30: 195–238.
Beymer RJ, Klopatek JM. Potential contribution of carbon by microphytic crusts in Pinyon-juniper woodlands. *Arid Soil Research and Rehabilitation*, 1991, 5: 187–198.
Birse EM, Landsberg SY, Gimingham CH. Effects of burial by sand on dune mosses. *Transactions of the British Bryological Society*, 1957, 3: 285–301.
Blackwood CB, Waldrop MP, Zak DR, Sinsabaugh RL. Molecular analysis of fungal communities and laccase genes in decomposing litter reveals differences among forest types but no impact of nitrogen deposition. *Environmental Microbiology*, 2007, 9: 1306–1316.
Bohnert HJ, Sheveleva E. Plant stress adaptations—Making metabolism move. *Current Opinion in Plant Biology*, 1998, 1: 267–274.
Bokhary HA. Mycoflora of desert sand dunes of Riyadh region, Saudi Arabia. *Journal of King Saud University*, 1998, 10: 15–29.
Bowker MA. Biological soil crust rehabilitation in theory and practice: An underexploited opportunity. *Restoration Ecology*, 2007, 15: 13–23.
Bowler CW, Montagu V, Ince D. Superoxide dismutase and stress tolerance. *Annual Review of Plant Physiology and Plant Molecular Biology*, 1992, 43: 83–116.
Brown JF. Effects of experimental burial on survival, growth and resource allocation of three species of dune plants. *Journal of Ecology*, 1997, 85: 151–158.
Buchanan B, Gruissem W, Jones R. *Biochemistry and Molecular Biology of Plants*. New York: Wiley, 2000: 1158–1203.
Buitink J, Hemmings MA, Hoekstra FA. Is there a role for oligosaccharides in seed longevity? An assessment of intracellular glass stability. *Plant Physiology*, 2000, 122: 1217–1224.
Butterly CR, Marschner P, McNeill AM, Baldock JA. Rewetting CO_2 pulses in Australian agricultural soils and the influence of

soil properties. *Biology and Fertility of Soils*, 2010, 46: 739–753.

Caldwell MM, Bornman JF, Ballaré CL, Flint SD, Kulandaivelu G. Terrestrial ecosystems, increased solar ultraviolet radiation, and interactions with other climate change factors. *Photochemistry and Photobiology*, 2007, 6: 252–266.

Campbell BJ, Polson SW, Hanson TE, Mack MC, Schuur EAG. The effect of nutrient deposition on bacterial communities in Arctic tundra soil. *Environmental Microbiology*, 2010, 12: 1842–1854.

Campbell SE. Soil stabilization by a prokaryotic desert crust: Implications for Precambrian land biota. *Origins of Life*, 1979, 9: 335–348.

Castillo-Monroy AP, Maestre FT, Rey A, Soliveres S, Garcia-Palacios P. Biological soil crust micro-sites are the main contributor to soil respiration in a semiarid ecosystem. *Ecosystems*, 2011, 14: 835–847.

Chapman SB. Some interrelationships between soil and root respiration in lowland *Calluna* heathland in Southern England. *Journal of Ecology*, 1979, 67: 1–20.

Chapuis-Lardy L, Wrage-Mönnig N, Metay A, Chotte JL, Bernoux M. Soils, a sink for N_2O? A review. *Global Change Biology*, 2006, 13: 1–17.

Chen G, Hu WY, Xie FT, Zhang LJ. Solvent for extracting malondialdehyde in plant as an index of senescence. *Plant Physiology Communications*, 1991, 27: 44–46.

Chen LZ, Wang GH, Hong S, Liu A, Li C, Liu YD. UV-B-induced oxidative damage and protective role of exopolysaccharides in desert cyanobacterium *Microcoleus vaginatus*. *Journal of Integrative Plant Biology*, 2009, 51: 194–200.

Chimner RA, Welker JM. Ecosystem respiration responses to experimental manipulations of winter and summer precipitation in a Mixedgrass Prairie, WY, USA. *Biogeochemistry*, 2005, 73: 257–270.

Chintakovid W, Kubota C, Bostick WM, Kozai T. Effect of air current speed on evapotranspiration rate of transplant canopy under artificial light. *Journal of Society of High Technology in Agriculture*, 2002, 14: 25–31.

Christensen JH, Hewitson B. Regional climate projections. In: Solomon S, Qin D, Manning M, Chen Z, Marquis M, Averyt KB, Tignor M, Miller HL (eds.). *Climate Change* 2007: *The Physical Science Basis. Contribution of Working Group I to the Fourth Assessment Report of the Intergovernmental Panel on Climate Change.* Cambridge. New York: Cambridge University Press, 2007, 847–940.

Christensen M. Species diversity and dominance in fungal community. In: Carroll GW, Wicklow DT (eds.). *The Fungal Community, Its Organization and Role in the Ecosystem.* New York: Marcell Dekker, 1981, 201–232.

Christensen TR, Prentice IC, Kaplan J, Haxeltine A, Sitch S. Methane flux from northern wetlands and tundra. *Tellus B*, 1996, 48: 652–661.

Ciccarone C, Rambelli A. A study on micro-fungi in arid areas. Notes on stress-tolerant fungi. *Plant Biosystems*, 1998, 132: 17–20.

Clark RM, Schweikert G, Toomajian C, Ossowski S, Zeller G, Shinn P, Warthmann N, Hu TT, Fu G, Hinds DA, Chen H, Frazer KA, Huson DH, Schölkopf B, Nordborg M, Rätsch G, Ecker JR, Weigel D. Common sequence polymorphisms shaping genetic diversity in *Arabidopsis thaliana*. *Science*, 2007, 317: 338–342.

Cline K, Mori H. Thylakoid ΔpH-dependent precursor proteins bind to a cpTatC-Hcf106 complex before Tha4-dependent transport. *Journal of Cell Biology*, 2001, 154: 719–730.

Cooper K, Farrant JM. Recovery of the resurrection plant *Craterostigma wilmsii* from desiccation: Protection versus repair. *Journal of Experimental Botany*, 2002, 53: 1805–1813.

Cooper RL, Martin RJ, St. Martin SK, Calip-DuBois A, Fioritto RJ, Schmitthenne AF. Registration of 'Stressland' soybean. *Crop Science*, 1999, 39: 590–591.

Crum H. *Structural Diversity of Bryophytes*. Herbarium: University of Michigan, 2001.

Csintalan Z, Tuba Z, Takács Z, Laitat E. Responses of nine bryophyte and one lichen species from different microhabitats to elevated UV-B radiation. *Photosynthetica*, 2001, 39: 317–320.

Cuming AC. LEA proteins. In: Shewry PR, Casey R. (eds.). *Seed Proteins.* Dordrecht: Kluwer Academic Publishers, 1999, 753–780.

Dalal RC, Allen DE. Turner review No. 18. Greenhouse gas fluxes from natural ecosystems. *Australian Journal of Botany*, 2008, 56: 369–407.

Dastager SG, Lee JC, Ju YJ, Park DJ, Kim CJ. *Rubellimicrobium mesophilum* sp. nov., a mesophilic, pigmented bacterium isolated from soil. *International Journal of Systematic and Evolutionary Microbiology*, 2008, 58: 1797–1800.

Davey MC. Effects of continuous and repeated dehydration on carbon fixation by bryophytes from the maritime Antarctic. *Oecologia*, 1997, 110: 25–31.

Davis KER, Sangwan P, Janssen PH. *Acidobacteria*, *Rubrobacteridae* and *Chloroflexi* are abundant among very slow-growing and mini-colony-forming soil bacteria. *Environmental Microbiology*, 2011, 13: 798–805.

de Oliveira TS, de Costa LM, Schaefer CE. Water-dispersible clay after wetting and drying cycles in four Brazilian oxisols. *Soil and Tillage Research*, 2005, 83: 260–269.

Domsch KH, Gams W, Anderson TH. *Compendium of Soil Fungi*. New York: Academic Press, 2007.

Du R, Lu D, Wang GC. Diurnal, seasonal, and inter-annual variations of N_2O fluxes from native semi-arid grassland soils of Inner Mongolia. *Soil Biology and Biochemistry*, 2006, 38: 3474–3482.

Eichholz I, Rohn S, Gamm A, Beesk N, Herppich WB, Kroh LW. UV-B-mediated flavonoid synthesis in white asparagus (*Asparagus officinalis* L.). *Food Research International*, 2012, 48: 196–201.

Eisenlord SD, Zak DR. Simulated atmospheric nitrogen deposition alters Actinobacterial community composition in forest soils. *Soil Science Society of America Journal*, 2010, 74: 1157–1166.

Eldridge DJ, Greene RSB. Microbiotic soil crusts: A view of their roles in soil and ecological processes in the rangelands of Australia. *Australian Journal of Soil Research*, 1994, 32: 389–415.

Entwistle EM, Zak DR, Edwards IP. Long-term experimental nitrogen deposition alters the composition of the active fungal community in the forest floor. *Soil Science Society of America Journal*, 2013, 77: 1648–1658.

Fabón G, Martínez-Abaigar J, Tomás R, Núñez-Olivera E. Effects of enhanced UV-B radiation on hydroxycinnamic acid derivatives extracted from different cell compartments in the aquatic liverwort *Jungermannia exsertifolia* subsp. *cordifolia*. *Physiologia Plantarum*, 2010, 140: 269–279.

Farrant JM. A comparison of mechanisms of desiccation tolerance among three angiosperm resurrection plant species. *Plant Ecology*, 2000, 151: 29–39.

Fay PA, Kaufman DM, Nippert JB, Carlisle JD, Harper CW. Changes in grassland ecosystem function due to extreme rainfall events: Implications for responses to climate change. *Global Change Biology*, 2008, 14: 1600–1608.

Ferreira AC, Nobre MF, Moore E, Rainey FA, Battista JR, da Costa MS. Characterization and radiation resistance of new isolates of *Rubrobacter radiotolerans* and *Rubrobacter xylanophilus*. *Extremophiles*, 1999, 3: 235–238.

Fierer N, Schimel JP. Effects of drying-rewetting frequency on soil carbon and nitrogen transformations. *Soil Biology and Biochemistry*, 2002, 34: 777–787.

Fierer N, Bradford MA, Jackson RB. Toward an ecological classification of soil bacteria. *Ecology*, 2007, 88: 1354–1364.

Flanagan LB, Johnson BG. Interacting effects of temperature, soil moisture and plant biomass production on ecosystem respiration in a northern temperate grassland. *Agricultural and Forest Meteorology*, 2005, 130: 237–253.

Franzluebbers AJ, Haney RL, Honeycutt CW, Schomberg HH, Hons FM. Flush of carbon dioxide following rewetting of dried soil relates to active organic pools. *Soil Science Society of America Journal*, 2000, 64: 613–623.

Franzluebbers K, Weaver RW, Juo ASR, Franzluebbers AJ. Carbon and nitrogen mineralization from cowpea plants part decomposing in moist and in repeatedly dried and wetted soil. *Soil Biology and Biochemistry*, 1994, 26: 1379–1387.

Freeman KR, Pescador MY, Reed SC, Costello EK, Robeson MS, Schmidt SK. Soil CO_2 flux and photoautotrophic community composition in high-elevation, 'barren' soil. *Environmental Microbiology*, 2009, 11: 674–686.

Frey SD, Knorr M, Parrent JL, Simpson RT. Chronic nitrogen enrichment affects the structure and function of the soil microbial community in temperate hardwood and pine forests. *Forest Ecology and Management*, 2004, 196: 159–171.

Frohnmeyer H, Staiger D. Ultraviolet-B radiation-mediated responses in plants balancing damage and protection. *Plant Physiology*, 2003, 133: 1420–1428.

Gehrke C. Impacts of enhanced ultraviolet-b radiation on mosses in a subarctic heath ecosystem. *Ecology*, 1999, 80: 1844–1851.

Goffinet B, Show AJ. *Bryophyte Biology: Second Edition*. New York: Cambridge University Press, 2008.

Gradstein S, Churchill S, Salazar-Allen N. *Guide to the Bryophytes of Tropical America*. New York: New York Botanical Garden Press, 2001.

Green LE, Porras-Alfaro A, Sinsabaugh RL. Translocation of nitrogen and carbon integrates biotic crust and grass production in desert grassland. *Journal of Ecology*, 2008, 96: 1076–1085.

Green TGA, Schroeter B, Seppelt R. Effect of temperature, light and ambient UV on the photosynthesis of the moss *Bryum argenteum* Hedw. in continental Antarctica. In: Davison W, Howard-Williams C, Broady P (eds.). *Antarctic Ecosystems: Models for Wider Understanding*. Christchurch: Caxton Press, 2000: 165–170.

Grishkan I, Kidron GJ. Biocrust-inhabiting cultured micro-fungi along a dune catena in the western Negev Desert, Israel. *European Journal of Soil Biology*, 2013, 56: 107–114.

Grishkan I, Nevo E. Soil micro-fungi of the Negev Desert, Israel—Adaptive strategies to climatic stress. In: Veress B, Szigethy J (eds.). *Horizons in Earth Science Research*, Vol. 1. New York: Nova Science Publishers Inc., 2010: 313–333.

Grishkan I, Jia RL, Kidron GJ, Li XR. Cultivable micro-fungal communities inhabiting biological soil crusts in the Tengger Desert, China. *Pedosphere*, 2015, 25: 351–363.

Grogan P, Jonasson S. Temperature and substrate controls on intra: Annual variation in ecosystem respiration in two subarctic

vegetation types. *Global Change Biology*, 2005, 11: 465–475.
Groisman PY, Knight RW, Easterling DR, Karl TR, Hegerl GC, Razuvaev VN. Trends in intense precipitation in the climate record. *Journal of Climate*, 2005, 18: 1326–1350.
Grube M, Cardinale M, de Castro Jr JV, Müller H, Berg G. Species-specific structural and functional diversity of bacterial communities in lichen symbioses. *The ISME Journal*, 2009, 3: 1105–1115.
Guidi L, Mori S, Degl' Innocenti E, Pecchia S. Effects of ozone exposure or fungal pathogen on white lupin leaves as determined by imaging of chlorophyll a fluorescence. *Plant Physiology and Biochemistry*, 2007, 45: 851–857.
Gundlapally SR, Garcia-Pichel F. The community and phylogenetic diversity of biological soil crusts in the Colorado Plateau studied by molecular fingerprinting and intensive cultivation. *Microbial Ecology*, 2006, 52: 345–357.
Guo DP, Guo YP, Zhao JP, Liu H, Peng Y, Wang QM, Chen JS, Rao GZ. Photosynthetic rate and chlorophyll fluorescence in leaves of stem mustard (*Brassica juncea* var. *tsatsai*)after turnip mosaic virus infection. *Plant Science*, 2005a, 168: 57–63.
Guo JK, Zhang ZZ, Bi YR, Yang W, Xu YN, Zhang LX. Decreased stability of photosystem I in dgd1 mutant of *Arabidopsis thaliana*. *FEBS Letters*, 2005b, 579: 3619–3624.
Guschina IA, Harwood JL, Smith M, Beckett RP. Abscisic acid modifies the changes in lipids brought about by water stress in the moss *Atrichum androgynum*. *New Phytologist*, 2002, 156: 255–264.
Gwózdz EA, Bewley JD, Tucker EB. Studies on protein synthesis in *Tortula ruralis*: Polyribosome reformation following desiccation. *Journal of Experimental Botany*, 1974, 25: 599–608.
Halwagy R, Moustafa AF, Kamel S. Ecology of the soil mycoflora in the desert of Kuwait. *Journal of Arid Environments*, 1982, 5: 109–125.
Harley CDG. Climate change, keystone predation, and biodiversity loss. *Science*, 2011, 334: 1124–1127.
Harley PC, Tenhunen JD, Murray KJ, Beyers J. Irradiance and temperature effects on photosynthesis of tussock tundra Sphagnum mosses from the foothills of the Philip Smith Mountains, Alaska. *Oecologia*, 1989, 79: 251–259.
Harper KT, Marble JR. A role for non-vascular plants in management of arid and semi-arid rangelands. In: Tueller PT (ed.). *Application of Plant Sciences to Rangeland Management*. Amsterdam: Martinus Nijhoff/W. Junk, 1998, 135–169.
Hartley AE, Schlesinger WH. Potential environmental controls on nitrogenase activity in biological crusts of the northern Chihuahuan Desert. *Journal of Arid Environments*, 2002, 52: 293–304.
Hasegawa PM, Bressan RA, Zhu JK, Bohnert HJ. Plant cellular and molecular response to high salinity. *Annual Review of Plant Physiology and Plant Molecular Biology*, 2000, 51: 463–499.
Hashem AR. Studies on the fungal flora of Saudi Arabian soil. *Cryptogamic Botany*, 1991, 2/3: 179–182.
Hernandoa MP, Ferreyra GA. The effects of UV radiation on photosynthesis in an Antarctic diatom (*Thalassiosira* sp.): Dose vertical mixing matter? *Journal of Experimental Marine Biology and Ecology*, 2005, 325: 35–45.
Hong SB, Go SJ, Shin HD, Frisvad JC, Samson RA. Polyphasic taxonomy of *Aspergillus fumigatus* and related species. *Mycologia*, 2005, 97: 1316–1329.
Hu YG, Chang XF, Lin XW, Wang YF, Wang SP, Duan JC. Effects of warming and grazing on N_2O fluxes in an alpine meadow ecosystem on the Tibetan Plateau. *Soil Biology and Biochemistry*, 2010, 42: 944–952.
Hui R, Li XR, Jia RL, Liu LC, Zhao RM, Zhao X, Wei YP. Photosynthesis, fluorescence, soluble protein content and chloroplast ultrastructure of two moss crusts from the Tengger Desert with contrasting sensitivity to supplementary UV-B radiation. *Photosynthetica*, 2014, 52: 36–49.
Hui R, Li XR, Zhao RM, Liu LC, Gao YH, Zhao X, Wei YP. UV-B radiation suppresses chlorophyll fluorescence, photosynthetic pigment and antioxidant systems of two key species in soil crusts from the Tengger Desert, China. *Journal of Arid Environments*, 2015, 113: 6–15.
Ingram J, Bartels D. The molecular basis of dehydration tolerance in plants. *Annual Review of Plant Physiology and Plant Molecular Biology*, 1996, 47: 377–403.
IPCC(Intergovernmental Panel on Climate Change). *Climate Change* 2007: *the Physical Science Basis*. Contribution of Working Group I to the Fourth Assessment Report of the Intergovernmental Panel on Climate Change. Cambridge: Cambridge University Press, 2007.
Jaffe MJ, Forbes S. Thigmomorphogenesis: The effect of mechanical perturbation on plants. *Plant Growth Regulation*, 1993, 12: 313–324.
Jansen MAK, Lemartret B, Koornneef M. Variations in constitutive and inducible UV-B tolerance; dissecting photosystem Ⅱ protection in *Arabidopsis thaliana* accessions. *Physiologia Plantarum*, 2010, 138: 22–34.
Jia RL, Li XR, Liu LC, Gao YH, Li XJ. Responses of biological soil crusts to sand burial in a revegetated area of the Tengger Desert, Northern China. *Soil Biology and Biochemistry*, 2008, 40: 2827–2834.

Jia RL, Li XR, Liu LC, Gao YH, Zhang XT. Differential wind tolerance of soil crust mosses explains their micro-distribution in nature. *Soil Biology and Biochemistry*, 2012, 45: 31–39.

John K. Acetylene reduction in the dark by mats of blue-green algae mats in sub-tropical grassland. *Annals of Botany*, 1977, 41: 807–812.

Johnson NC. Can fertilization of soil select less mutualistic mycorrhizae? *Ecological Applications*, 1993, 3: 749–757.

Jones SK, Rees RM, Skiba U, Ball BC. Greenhouse gas emissions from a managed grassland. *Global and Planetary Change*, 2005, 47: 201–211.

Jumpponen A, Johnson LC. Can rDNA analyses of diverse fungal communities in soil and roots detect effects of environmental manipulations—A case study from tallgrass prairie. *Mycologia*, 2005, 97: 1177–1194.

Kakani VG, Reddy KR, Zhao D, Sailaja K. Field crop responses to ultraviolet-B radiation: A review. *Agricultural and Forest Meteorology*, 2003, 120: 191–218.

Kammann C, Grunhage L, Jager HJ. A new sampling technique to monitor concentration of CO_2, CH_4 and N_2O in air at well-defined depths in soils with varied water potential. *European Journal of Soil Science*, 2001, 52: 297–303.

Kappen L, Lange OL. Die Kälteresistenz einiger Makrolichenen. *Flora*, 1972, 161: 1–29.

Keever C. Establishment of *Grimmia laevigata* on bare granite. *Ecology*, 1957, 38: 422–429.

Keller M, Kaplan WA, Wofsy SC. Emissions of N_2O, CH_4 and CO_2 from tropical forest soils. *Journal of Geophysical Research: Atmospheres* (1984–2012), 1986, 91: 11791–11802.

Kent M, Owen NW, Dale MP. Photosynthetic responses of plant communities to sand burial on the machair dune system of the Outer Hebrides, Scotland. *Annual of Botany*, 2005, 95: 869–877.

Kermode AR, Finch-Savage WE. Desiccation sensitivity in orthodox and recalcitrant seeds in relation to development. In: Black M, Pritchard HW (eds.). *Desiccation and Survival in Plants: Drying Without Dying*. Wallingford: CABI Publishing, 2002, 149–184.

Kitaya Y, Tsuruyama J, Shibuya T, Yoshida M, Kiyota M. Effects of air current speed on gas exchange in plant leaves and plant canopies. *Advances in Space Research*, 2003, 31: 177–182.

Kitzing C, Pröschold T, Karsten U. UV-induced effects on growth, photosynthetic performance and sunscreen contents in different populations of the green alga *Klebsormidium fluitans* (Streptophyta)from Alpine soil crusts. *Microbial Ecology*, 2014, 67: 327–340.

Knight H, Knight MR. Abiotic stress signalling pathways: Specificity and cross-talk. *Trends in Plant Science*, 2001, 6: 262–267.

Kramer GF, Norman HA, Krizek DT, Mirecki RM. Influence of UV-B radiation on polyamines, lipid peroxidation and membrane lipids in cucumber. *Phytochemistry*, 1991, 30: 2101–2108.

Krochko JE, Bewley JD, Pacey J. The effects of rapid and very slow speeds of drying on the ultrastructure and metabolism of the desiccation-sensitive moss *Cratoneuron filicinum*. *Journal of Experimental Botany*, 1978, 29: 905–917.

Kursar TA. Elevation of soil respiration and soil CO_2 concentration in a low land moist forest in Panama. *Plant and Soil*, 1989, 113: 21–29.

Laiz L, Miller AZ, Jurado V, Akatova E, Sanchez-Moral S, Gonzalez JM, Dionísio A, Macedo MF, Saiz-Jimenez C. Isolation of five *Rubrobacter* strains from biodeteriorated monuments. *Naturwissenschaften*, 2009, 96: 71–79.

Lamarque JF, Kiehl JT, Brasseur GP, Butler T, Cameron-Smith P, Collins WD, Collins WJ, Granier C, Hauglustaine D, Hess PG, Holland EA, Horowitz L, Lawrence MG, McKenna D, Merilees P, Prather MJ, Rasch PJ, Rotman D, Shindell D, Thornton P. Assessing future nitrogen deposition and carbon cycle feedback using a multimodel approach: Analysis of nitrogen deposition. *Journal of Geophysical Research*, 2005, 110: D19303.

Lange OL, Belnap J, Reichenberger J, Meyer A. Photosynthesis of green algal soil crust lichens from arid lands in southern Utah, USA: Role of water content on light and temperature responses of CO_2 exchange. *Flora*, 1997, 192: 115–129.

Lange OL, Green TGA, Reichenberger H. The response of lichen photosynthesis to external CO_2 concentration and its interaction with thallus water-status. *Journal of Plant Physiology*, 2003, 154: 157–166.

Lange OL, Kidron GJ, Büdel B, Meyer A, Kilian E, Abeliovich A. Taxonomic composition and photosynthetic characteristics of the biological crusts covering sand dunes in the western Negev. *Functional Ecology*, 1992, 6: 519–527.

Lange OL. Photosynthesis of soil-crust biota as dependent on environmental factors. In: Belnap J, Lange OL. *Biological Soil Crusts: Structure, Function, and Management*. Berlin: Springer-Verlag, 2003, 217–240.

Larcher W. *Physiological Plant Ecology* (3rd edn). Berlin: Springer-Verlag, 1995.

Laube JC, Newland MJ, Hogan C, Brenninkmeijer CAM, Fraser PJ, Martinerie P, Oram DE, Reeves CE, Röckmann T, Schwander J, Witrant E, Sturges WT. Newly detected ozone-depleting substances in the atmosphere. *Nature Geoscience*, 2014, 7: 266–269.

Levitt J. *Responses of Plants to Environmental Stresses*. New York, London: Academic Press, 1972.

Li G, Wan SW, Zhou J, Yang ZY, Qin P. Leaf chlorophyll fluorescence, hyperspectral reflectance, pigments content, malondialdehyde and proline accumulation responses of castor bean (*Ricinus communis* L.)seedlings to salt stress levels. *Industrial Crops and Products*, 2010b, 31: 13–19.

Li XR, Chen YW, Yang LW. Cryptogam diversity and formation of soil crusts in temperate desert. *Annals of Arid Zone*, 2004, 43: 335–353.

Li XR, He MZ, Duan ZH, Xiao HL, Jia XH. Recovery of topsoil physicochemical properties in revegetated sites in the sand-burial ecosystems of the Tengger Desert, Northern China. *Geomorphology*, 2007, 88: 254–265.

Li XR, Tian F, Jia RL, Zhang ZS, Liu LC. Do biological soil crusts determine vegetation changes in sandy deserts? Implications for managing artificial vegetation. *Hydrological Processes*, 2010a, 24: 3621–3630.

Li XR, He MZ, Stefan Z, Li XJ, Liu LC. Micro-geomorphology determines community structure of BSCs at small scale. *Earth Surface Processes and Landforms*, 2010c, 35: 932–940.

Li XR, Jia RL, Chen YW, Huang L, Zhang P. Association of ant nests with successional stages of biological soil crusts in the Tengger Desert, Northern China. *Applied Soil Ecology*, 2010d, 47: 59–66.

Li XR, Wang XP, Li T, Zhang JG. Microbiotic soil crust and its effect on vegetation and habitat on artificially stabilized desert dunes in Tengger Desert, North China. *Biology and Fertility of Soils*, 2002, 35: 147–154.

Li XR, Zhou HY, Wang XP, Zhu YG, O'Conner PJ. The effects of re-vegetation on cryptogam species diversity in Tengger Desert, Northern China. *Plant and Soil*, 2003, 251: 237–245.

Liang NS, Nakadai T, Hirano T, Qu LY, Koike T, Fujinuma Y, Inoue G. In situ comparison of four approaches to estimating soil CO_2 efflux in a northern larch (*Larix kaempferi* Sarg.)forest. *Agricultural and Forest Meteorology*, 2004, 123: 97–117.

Lin XW, Zhang ZH, Wang SP, Hu YG, Xu GP, Luo CY, Chang XF, Duan JC, Lin QY, Xu BX, Wang YF, Zhao XQ, Xie ZB. Response of ecosystem respiration to warming and grazing during the growing seasons in the alpine meadow on the Tibetan Plateau. *Agricultural and Forest Meteorology*, 2011, 151: 792–802.

Liu LC, Li SZ, Duan ZH, Wang T, Li XR. Effects of microbiotic crust on dew deposition in the restored vegetated areas at Shapotou, Northern China. *Journal of Hydrology*, 2006, 238: 331–337.

Liu LC, Song YX, Gao YH, Wang T, Li XR. Effects of microbiotic crusts on evaporation from the revegetated area in a Chinese desert. *Soil Research*, 2007, 45: 422–427.

Liu LX, Xu SM, Woo KC. Solar UV-B radiation on growth, photosynthesis and the xanthophyll cycle in tropical acacias and eucalyptus. *Environmental and Experimental Botany*, 2005, 54: 121–130.

Liu YM, Li XR, Jia RL, Huang L, Zhou YY, Gao YH. Effects of biological soil crusts on soil nematode communities following dune stabilization in the Tengger Desert, Northern China. *Applied Soil Ecology*, 2011, 49: 118–124.

Lu XM, Siemann E, Shao X, Wei H, Ding JQ. Climate warming affects biological invasions by shifting interactions of plants and herbivores. *Global Change Biology*, 2013, 19: 2339–2347.

Lusk CH. Leaf area accumulation helps juvenile evergreen trees tolerate shade in a temperate rainforest. *Oecologia*, 2002, 132: 188–196.

Mackerness SAH, John CF, Jordan B, Thomas B. Early signaling components in ultraviolet-B responses: Distinct roles for different reactive oxygen species and nitric oxide. *FEBS Letters*, 2001, 489: 237–242.

Maestre FT, Salguero-Gomez R, Quero JL. It is getting hotter in here: Determining and projecting the impacts of global environmental change on drylands. *Philosophical Transactions of the Royal Society of London Series B: Biological Sciences*, 2012, 367: 3062–3075.

Maier S, Schmidt TSB, Zheng L, Peer T, Wagner V, Grube M. Analyses of dryland biological soil crusts highlight lichens as an important regulator of microbial communities. *Biodiversity Conservation*, 2014, 23: 1735–1755.

Mandeel QA. Micro-fungal community associated with rhizosphere soil of *Zygophyllum qatarense* in arid habitats of Bahrain. *Journal of Arid Environments*, 2002, 50: 665–681.

Martínez ML, Maun MA. Responses of dune mosses to experimental burial by sand under natural and greenhouse conditions. *Plant Ecology*, 1999, 145: 209–219.

Mastepanov M, Sigsgaard C, Dlugokencky EJ, Houweling S, Ström L, Tamstorf MP, Christensen TR. Large tundra methane burst during onset of freezing. *Nature*, 2008, 456: 628–630.

Maun MA. Adaptations enhancing survival and establishment of seedlings on coastal dune systems. *Vegetatio*, 1994, 111: 59–70.

Maun MA. Burial of plants as a selective force in sand dunes. In: Martunez ML, Psuty NP (eds.). *Coastal Dunes: Ecology and Management Ecological Studies*. Berlin, New York: Springer-Verlag, 2004, 119–136.

McCalley CK, Sparks JP. Abiotic gas formation drives nitrogen loss from a desert ecosystem. *Science*, 2009, 326: 837–840.

McKenzie RL, Aucamp PJ, Bais AF, Björn LO, Ilyas M. Changes in biologically-active ultraviolet radiation reaching the

earth's surface. *Photochemical and Photobiological Sciences*, 2007, 6: 218–231.
Mehdy MC. Active oxygen species in plant defense against pathogens. *Plant Physiology*, 1994, 105: 467–472.
Meurs C, Basra AS, Karssen CM, van Loon LC. Role of abscisic acid in the induction of desiccation tolerance in developing seeds of *Arabidopsis thaliana*. *Plant Physiology*, 1992, 98: 1484–1493.
Mikha MM, Rice CW, Milliken GA. Carbon and nitrogen mineralization as affected by drying and wetting cycles. *Soil Biology and Biochemistry*, 2005, 37: 339–347.
Moubasher AH, Abdel-Hafez SII, El-Maghraby OMO. Studies on soil mycoflora of Wadi Bir-El-Ain, eastern desert, Egypt. *Cryptogamie Mycologie*, 1985, 6: 129–143.
Mudd JB. Biochemical basis for the toxicity of ozone. In: Yunus M, Iqbal M (eds.). *Plant Response to Air Pollution*. New York: John Wiley and Sons, 1996, 267–284.
Muhr J, Franke J, Borken W. Drying-rewetting events reduce C and N losses from a Norway spruce forest floor. *Soil Biology and Biochemistry*, 2010, 42: 1303–1312.
Murata N, Takahashi S, Nishiyama Y, Allakhverdiev SI. Photoinhibition of photosystem Ⅱ under environmental stress. *Biochimica et Biophysica Acta*, 2007, 1767: 414–421.
Murphy TF. Effects of broad-band ultraviolet and visible radiation on hydrogen peroxide formation by cultured rose cells. *Physiologia Plantarum*, 1990, 80: 63–68.
Nadeaul TL, Clive H, Richard W. Effects of solar UV and visible irradiance on photosynthesis and vertical migration of *Oscillatoria* sp. (Cyanobacteria)in an Antarctic microbial mat. *Aquatic Microbial Ecology*, 1999, 20: 231–243.
Nagy ML, Pérez A, Garcia-Pichel F. The prokaryotic diversity of biological soil crusts in the Sonoran Desert (Organ Pipe Cactus National Monument, AZ). *FEMS Microbiology Ecology*, 2005, 54: 233–245.
Nash TH Ⅲ. *Lichen Biology*. Cambridge: Cambridge University Press, 1996.
Neil SJ, Desikan R, Clarke A, Hurst RD, Hancock JT. Hydrogen peroxide and nitric oxide as signaling molecules in plants. *Journal of Experimental Botany*, 2002, 53: 1237–1247.
Nemergut DR, Townsend AR, Sattin SR, Freeman KR, Fierer N, Neff JC, Bowman WD, Schadt CW, Weintraub MN, Schmidt SK. The effects of chronic nitrogen fertilization on alpine tundra soil microbial communities: Implications for carbon and nitrogen cycling. *Environmental Microbiology*, 2008, 10: 3093–3105.
New M, Todd M, Hulme M, Jones P. Precipitation measurements and trends in the twentieth century. *International Journal of Climatology*, 2001, 21: 1899–1922.
Newsham KK. UV-B radiation arising from stratospheric ozone depletion influences the pigmentation of the Antarctic moss *Andreaea regularis*. Oecologia, 2003, 135: 327–331.
Niemi R, Martikainen PJ, Silvola J, Sonninen E, Wulff A, Holopainen T. Responses of two *Sphagnum* moss species and *Eriophorum vaginatum* to enhanced UV-B in a summer of low UV intensity. *New Phytologist*, 2002, 156: 509–515.
Noctor G, Foyer CH. Ascorbate and glutathione: Keeping active oxygen under control. *Annual Review of Plant Physiology and Plant Molecular Biology*, 1998, 49: 249–279.
Oberbauer SF, Tweedie CE, Welker JM, Fahnestock JT, Henry GH, Webber PJ, Hollister RD, Walker MD, Kuchy A, Elmore E, Starr G. Tundra CO_2 fluxes in response to experimental warming across latitudinal and moisture gradients. *Ecological Monographs*, 2007, 77: 221–238.
Oliver MJ. Influence of protoplasmic water loss on the control of protein synthesis in the desiccation-tolerant moss *Tortula ruralis*: Ramifications for a repair-based mechanism of desiccation tolerance. *Plant Physiology*, 1991, 97: 1501–1511.
Oliver MJ, Bewley JD. Desiccation-tolerance in plant tissues. A mechanistic overview. *Horticultural Reviews*, 1997, 18: 171–214.
Oliver MJ, Mishler BD, Quisenberry JE. Comparative measures of desiccation-tolerance in the *Tortula ruralis* complex. I. Variation in damage control and repair. *American Journal of Botany*, 1993, 80: 127–136.
Oliver MJ, Wood AJ, Mahony PO. "To dryness and beyond"—Preparation for the dried state and rehydration in vegetative desiccation-tolerant plants. *Plant Growth Regular*, 1998, 24: 193–201.
Osborne DJ, Boubriak I, Leprince O. Rehydration of dried systems: Membranes and the nuclear genome. In: Black M, Pritchard HW (eds.). *Desiccation and Survival in Plants: Drying Without Dying*. Wallingford: CABI Publishing, 2002, 343–364.
Osono T, Takeda H. Fungal decomposition of *Abies* needle and *Betula* leaf litter. *Mycologia*, 2006, 98: 172–179.
Palmqvist K. Carbon economy in lichens. *New Phytologist*, 2000, 148: 11–36.
Pan YX, Wang XP, Zhang YF. Dew formation characteristics in a revegetation-stabilized desert ecosystem in Shapotou area, Northern China. *Journal of Hydrology*, 2010, 387: 265–272.
Patrick E. Researching crusting soils: Themes, trends, recent developments and implications for managing soil and water

resources in dry areas. *Progress in Physical Geography*, 2002, 26: 442–461.
Perumal VJ, Maun MA. Ecophysiological response of dune species to experimental burial under field and controlled conditions. *Plant Ecology*, 2006, 184: 89–104.
Porras-Alfaro A, Herrera J, Natvig DO, Lipinski K, Sinsabaugh RL. Diversity and distribution of soil fungal communities in a semiarid grassland. *Mycologia*, 2011, 103: 10–21.
Potts M. *Nostoc*. In: Whitton BA, Potts M (eds.). *The Ecology of Cyanobacteria:Their Diversity in Time and Space*. The Netherland: Kluwer Academic Publishers, 1999, 465–504.
Pradhan MK, Nayak L, Joshi PN, Mohapatra PK, Patro L, Biswal B, Biswal UC. Developmental phase-dependent photosynthetic responses to ultraviolet-B radiation: Damage, defence, and adaptation of primary leaves of wheat seedlings. *Photosynthetica*, 2008, 46: 370–377.
Prasad SM, Zeeshan M. UV-B radiation and cadmium induced changes in growth, photosynthesis, and antioxidant enzymes of cyanobacterium *Plectonema boryanum*. *Biologia Plantarum*, 2005, 49: 229–236.
Proctor MCF. Patterns of desiccation tolerance and recovery in bryophytes. *Plant Growth Regulation*, 2001, 35: 147–156.
Proctor MCF, Smirnoff N. Rapid recovery of photosystems on rewetting desiccation-tolerant mosses: Chlorophyll fluorescence and inhibitor experiments. *Journal of Experimental Botany*, 2000, 51: 1695–1704.
Proctor MCF, Ligrone R, Duckett JG. Desiccation tolerance in the moss *Polytrichum formosum*: Physiological and fine-structural changes during desiccation and recovery. *Annals of Botany*, 2007a, 99: 75–93.
Proctor MCF, Oliver MJ, Wood AJ, Alpert P, Stark LR, Cleavitt NL, Mishler BD. Desiccation-tolerance in bryophytes: A review. *Bryologist*, 2007b, 110: 595–621.
Qin DH. *An Integrated Report on the Evaluation of Environmental Changes of Western China*. Beijing: Science Press, 2002, 56–60.
Quaite FE, Sutherland BM, Sutherland JC. Action spectrum for DNA damage in alfalfa lower predicted impact of ozone depletion. *Nature*, 1992, 358: 576–578.
Quesada SM, Goff L, Karentz D. Effects of natural UV radiation on Antarctic cyanobacterial mats. *Proceeding of the NIPR Symposium on Polar Biology*, 1998, 11: 98–111.
Ramirez KS, Craine JM, Fierer N. Consistent effects of nitrogen amendments on soil microbial communities and processes across biomes. *Global Change Biology*, 2012, 18: 1918–1927.
Ramirez KS, Lauber CL, Knight R, Bradford MA, Fierer N. Consistent effects of N fertilization on soil bacterial communities in contrasting systems. *Ecology*, 2010, 91: 3463–3470.
Ranzoni FV. Fungi isolated in culture from soils of the Sonoran Desert. *Mycologia*, 1968, 60: 356–371.
Rao BQ, Liu YD, Lan SB, Wu PP, Wang WB, Li DH. Effects of sand burial stress on the early developments of cyanobacterial crusts in the field. *European Journal of Soil Biology*, 2012, 48: 48–55.
Rech M, Mouget JL, Morant-Manceau A, Philippe R, Tremblin G. Long-term acclimation to UV radiation: Effects on growth, photosynthesis and carbonic anhydrase activity in marine diatoms. *Botanica Marine*, 2005, 48: 407–420.
Reiners WA. Carbon dioxide evolution from the floor of three Minnesota forests. *Ecology*, 1968, 49: 471–483.
Ren J, Tao L, Liu XM. Effect of sand burial depth on seed germination and seedling emergence of *Calligonum* L. species. *Journal of Arid Environments*, 2002, 51: 603–611.
Renger G, Völker M, Eckert HJ, Fromme R, Hohm-veit S, Gräber P. On the mechanism of photosystem Ⅱ deterioration by UV-B irradiation. *Photochemistry and Photobiology*, 1989, 49: 97–105.
Ritz K, Young IM. Interaction between soil structure and fungi. *Mycologist*, 2004, 18: 52–59.
Rustad LE, Fernandez IJ. Experimental soil warming effects on CO_2 and CH_4 flux from a low elevation spruce-fir forest soil in Maine, USA. *Global Change Biology*, 1998, 4: 597–605.
Sakaki T, Kondo N, Sugahara K. Breakdown of photosynthetic pigments and lipids in spinach leaves with ozone fumigation: Role of active oxygens. *Physiologia Plantarum*, 1983, 59: 28–34.
Schimel DS, House JI, Hibbard KA, Bousquet P, Ciais P, Peylin P, Braswell BH, Apps MJ, Baker D, Bondeau A, Canadell J, Churkina G, Cramer W, Denning AS, Field CB, Friedlingstein P, Goodale C, Heimann M, Houghton RA, Melillo JM, Moore Ⅲ B, Murdiyarso D, Noble I, Pacala SW, Prentice IC, Raupach MR, Rayner PJ, Scholes RJ, Steffen WL, Wirth C. Recent patterns and mechanisms of carbon exchange by terrestrial ecosystems. *Nature*, 2001, 414: 169–172.
Schulz S, Brankatschk R, Dümig A, Kögel-Knabner I, Schloter M, Zeyer J. The role of microorganisms at different stages of ecosystem development for soil formation. *Biogeosciences*, 2013, 10: 3983–3996.
Schuster RM. New manual of bryology. *The Bryologist*, 1983, 87: 288–290.
Sharma K, Swaranjeet K, Gehlot P. Diversity of thermophilic fungi in the soil of Indian Thar Desert. *Indian Phytopathology*, 2010, 63: 452–453.

Sicora C, Scilárd A, Sass L, Turcsányi E, Máté Z, Vass I. UV-B and UV-A radiation effects on photosynthesis at the molecular level. In: Ghetti F, Checcucci G, Bornmann JF (eds.). *Environmental UV Radiation: Impact on Ecosystems and Human Health and Protective Models*. Dordrecht: Springer-Verlag, 2006: 121–135.

Skujins J. Microbial ecology of desert soils. *Advances in Microbial Ecology*, 1984, 7: 49–91.

Smirnoff N. *Antioxidants and Reactive Oxygen Species in Plants*. Oxford: Blackwell Publishing, 2005.

Smirnoff N, Cumbes QJ. Hydroxyl radical scavenging activity of compatible solutes. *Phytochemistry*, 1989, 28: 1057–1060.

Šnajdr J, Steffen KT, Hofrichter M, Baldrian P. Transformation of 14C-labelled lignin and humic substances in forest soil by saprobic basidiomycetes *Gymnopus erythropus* and *Hypholoma fasciculare*. *Soil Biology and Biochemistry*, 2010, 42: 1541–1548.

States JS, Christensen M, Kinter CK. Soil fungi as components of biological soil crusts. In: Belnap J, Lange O (eds.). *Biological Soil Crusts: Structure, Function, and Management*. Berlin, New York: Springer-Verlag, 2001: 155–166.

Strid A. Alteration in expression of defense genes in *Pisum sativum* after exposure to supplementary ultraviolet-B radiation. *Plant and Cell Physiology*, 1993, 34: 949–953.

Surabhi GK, Reddy KR, Singh SK. Photosynthesis, fluorescence, shoot biomass and seed weigh responses of three cowpea (*Vigna unguiculata* (L.)Walp.)cultivars with contrasting sensitivity to UV-B radiation. *Environmental and Experimental Botany*, 2009, 66: 160–171.

Surewicz WK, Mantsch HH, Capman D. Determination of protein secondary structure by Fourier transform infrared spectroscopy: A critical assessment. *Biochemistry*, 1993, 32: 389–439.

Takezawa D, Komatsu K, Sakata Y. ABA in bryophytes: How a universal growth regulator in life became a plant hormone? *Journal of Plant Research*, 2011, 124: 437–453.

Thomas AD, Hoon SR. Carbon dioxide fluxes from biologically-crusted Kalahari Sands after simulated wetting. *Journal of Arid Environments*, 2010, 74: 131–139.

Thomas AD, Hoon SR, Linton P. Carbon dioxide fluxes from cyanobacteria crusted soils in the Kalahari. *Applied Soil Ecology*, 2008, 39: 254–263.

Tietema A. Microbial carbon and nitrogen dynamics in coniferous forest floor material collected along a European nitrogen deposition gradient. *Forest Ecology and Management*, 1998, 101: 29–36.

Treseder KK. Nitrogen additions and microbial biomass: A meta-analysis of ecosystem studies. *Ecology Letters*, 2008, 11: 1111–1120.

Trumbore S. Carbon respired by terrestrial ecosystems-recent progress and challenges. *Global Change Biology*, 2006, 12: 141–153.

Tuba Z, Lichtenthaler H, Csintalan Z, Nagy Z, Szente K. Loss of chlorophylls, cessation of photosynthetic CO_2 assimilation and respiration in the poikilochlorophyllous plant *Xerophyta scabrida* during desiccation. *Physiologia Plantarum*, 1996, 96: 383–388.

Tuba Z, Proctor MC, Csintalan Z. Ecophysiological responses of homoiochlorophyllous and poikilochlorophyllous desiccation-tolerant plants: A comparison and ecological perspective. *Plant Growth Regulation*, 1998, 24: 211–217.

Velten J, Oliver MJ. Tr288: A rehydrin with a dehydrin twist. *Plant Molecular Biology*, 2001, 45: 713–722.

Vobis G. The genus *Actinoplanes* and related genera. In: Dworkin M, Falkow S, Rosenberg E, Schleifer KH, Stackebrandt E (eds.). *The Prokaryotes*, 3rd ed., vol. 3. New York: Springer Science and Business Media, 2006, 623–653.

Waldrop MP, Zak DR, Sinsabaugh RL. Microbial community response to nitrogen deposition in northern forest ecosystems. *Soil Biology and Biochemistry*, 2004, 36: 1443–1451.

Walters MJ, Wayman GA, Notis JC, Goodman RH, Soderling TR, Christian JL. Calmodulin-dependent protein kinase Ⅳ mediated antagonism of BMP signaling regulates lineage and survival of hematopoietic progenitors. *Development*, 2002, 129: 1455–1466.

Wan SQ, Hui DF, Wallace L, Luo YQ. Direct and indirect effects of experimental warming on ecosystem carbon processes in a tallgrass prairie. *Global Biogeochemical Cycles*, 2005, 19: GB2014.

Wang GH, Chen K, Chen LZ, Hu CX, Zhang DL, Liu YD. The involvement of the antioxidant system in protection of desert cyanobacterium *Nostoc* sp. against UV-B radiation and the effects of exogenous antioxidants. *Ecotoxicology and Environmental* Safety, 2008, 69: 150–157.

Wang SW, Xie BT, Yin LN, Duan LS, Li ZH, Eneji AE, Tsuji W, Tsunekawa A. Increased UV-B radiation affects the viability, reactive oxygen species accumulation and antioxidant enzyme activities in maize (*Zea mays* L.)pollen. *Photochemistry and Photobiology*, 2010, 86: 110–116.

Wang WB, Yang CY, Tang DS, Li DH, Liu YD, Hu CX. Effects of sand burial on biomass, chlorophyll fluorescence and extracellular polysaccharides of man-made cyanobacterial crusts under experimental conditions. *Science in China (Life*

Sciences), 2007, 50: 530–534.

Welker JM, Fahnestock JT, Henry GH, O'Dea KW, Chimner RA. CO_2 exchange in three Canadian High Arctic ecosystems: response to long-term experimental warming. *Global Change Biology*, 2004, 10: 1981–1995.

West NE. Structure and function of microphytic soil crusts in wild land ecosystem of arid and semi-arid regions. *Advance of Ecological Research*, 1990, 20: 179–223.

Willekens H, Langebartels C, Tiré C, van Montagu M, Inzé D, van Camp W. Differential expression of catalase genes in *Nicotiana plumbaginifolia* (L.). *Proceedings of the National Academy of Sciences of the United States of America*, 1994, 91: 10450–10454.

Williams WJ, Eldridge DJ. Deposition of sand over a cyanobacterial soil crust increases nitrogen bioavailability in a semi-arid woodland. *Applied Soil Ecology*, 2011, 49: 26–31.

Wise MJ, Tunnacliffe A. POPP the question: What do LEA proteins do? *Trends in Plant Science*, 2004, 9: 13–17.

Wood AJ, Oliver MJ. Translational control in plant stress: The formation of messenger ribonucleoprotein particles (mRNPs)in response to desiccation of *Tortula ruralis* gametophytes. *The Plant Journal*, 1999, 18: 359–370.

Wood AJ, Oliver MJ. Molecular biology and genomics of the desiccation-tolerant moss *Tortula ruralis*. In: Wood AJ, Oliver MJ (eds.). *New Frontiers in Bryology: Physiology, Molecular biology and Functional Genomics*. Dodrecht: Kluwer Academic Publisher, 2004: 71–90.

Wood AJ. The nature and distribution of Vegetative desiccation-tolerance in hornworts, liverworts and mosses. *Bryologist*, 2007, 110: 163–177.

Wu HY, Gao KS, Villafane VE, Watanabe T, Helbling EW. Effects of solar UV radiation on morphology and photosynthesis of filamentous cyanobacterium *Arthrospira platensis*. *Applied and Environmental Microbiology*, 2005, 71: 5004–5013.

Wu J, Brookes PC. The proportional mineralization of microbial biomass and organic matter caused by air-drying and rewetting of a grassland soil. *Soil Biology and Biochemistry*, 2005, 37: 507–515.

Xia JY, Niu SL, Wan SQ. Response of ecosystem carbon exchange to warming and nitrogen addition during two hydrologically contrasting growing seasons in a temperate steppe. *Global Change Biology*, 2009, 15: 1544–1556.

Xie ZM, Wang YX, Liu YD, Liu YM. Ultraviolet-B exposure induces photo-oxidative damage and subsequent repair strategies in a desert cyanobacterium *Microcoleus vaginatus* Gom. *European Journal of Soil Biology*, 2009, 45: 377–382.

Xu CP, Natarajan S, Sullivan JH. Impact of solar ultraviolet-B radiation on the antioxidant defense system in soybean lines differing in flavonoid contents. *Environmental and Experimental Botany*, 2008, 63: 39–48.

Xu D, Duan X, Wang B, Hong B, Ho THD, Wu R. Expression of a late embryogenesis abundant protein gene, HVA1, from barley confers tolerance to water deficit and salt stress in transgenic rice. *Plant Physiology*, 1996, 110: 249–257.

Xue LG, Zhang Y, Zhang TG, An LZ, Wang XL. Effects of enhanced ultraviolet-B radiation on algae and cyanobacteria. *Critical Reviews in Microbiology*, 2005, 31: 79–89.

Yan B, Dai QJ. Effects of ulrtaviolet-B radiation on active oxygen metabolism and membrane system of rice leaves. *Acta Phytophysiologica Sinica*, 1996, 22: 373–378.

Yang J, Yen H. Early salt stress effects on the changes in chemical composition in leaves of ice plant and *Arabidopsis*. A fourier transform infrared spectroscopy study. *Plant Physiology*, 2002, 130: 1032–1042.

Yoshinaka T, Yano K, Yamaguchi H. Isolation of a highly radioresistant bacterium, *Arthrobacter radiotolerans* nov. sp. *Agricultural and Biological Chemistry*, 1973, 37: 2269–2275.

Yu J, Tang XX, Zhang PY, Tian JY, Cai HJ. Effects of CO_2 enrichment on photosynthesis, lipid peroxidation and activities of antioxidative enzymes of *Platymonas subcordiformis* subjected to UV-B radiation stress. *Acta Botanica Sinica*, 2004, 46: 682–690.

Zaady E, Groffman P, Shachak M. Nitrogen fixation in macro- and microphytic patches in the Negev Desert. *Soil Biology and Biochemistry*, 1998, 30: 449–454.

Zaady E, Kuhn U, Wilske B, Sandoval-Soto L, Kesselmeier J. Patterns of CO_2 exchange in biological soil crusts of successional age. *Soil Biology and Biochemistry*, 2000, 32: 959–966.

Zak J. Fungal communities of desert ecosystems: Links to climate change. In: Dighton J, White Jr JF, White J, Oudemans P (eds.). *The Fungal Community: Its Organization and Role in the Ecosystem*. Boca Raton: CRC Press, 2005, 659–681.

Zechmeister-Boltenstern S, Michel K, Pfeffer M. Soil microbial community structure in European forests in relation to forest type and atmospheric nitrogen deposition. *Plant and Soil*, 2011, 343: 37–50.

Zeglin LH, Stursova M, Sinsabaugh RL, Collins SL. Microbial responses to nitrogen addition in three contrasting grassland ecosystems. *Oecologia*, 2007, 154: 349–359.

Zelikova TJ, Housman DC, Grote EdE, Neher DA, BelnapJ. Warming and increased precipitation frequency on the Colorado Plateau: Implications for biological soil crusts and soil processes. *Plant and Soil*, 2012, 335: 265–282.

Zhang NL, Wan SQ, Li LH, Bi J, Zhao MM, Ma KP. Impacts of urea N addition on soil microbial community in a semi-arid temperate steppe in Northern China. *Plant and Soil*, 2008, 311: 19–28.

Zhang T, Wei JC. Survival analyses of symbionts isolated from *Endocarpon pusillum* Hedwig to desiccation and starvation stress. *Science China Life Sciences*, 2011, 54: 480–489.

Zhang YF, Wang XP, Hu R, Pan YX, Zhang H. Variation of albedo to soil moisture for sand dunes and biological soil crusts in arid desert ecosystems. *Environmental Earth Sciences*, 2014, 71: 1281–1288.

Zhang YF, Wang XP, Pan YX, Hu R. Diurnal relationship between the surface albedo and surface temperature in revegetated desert ecosystems, Northwestern China. *Arid Land Research and Management*, 2012, 26: 32–43.

Zhao DB, Rdedy KR, Kakani VG, Read JJ, Sullivan J. Growth and physiological responses of cotton (*Gossypium hirsutum* L.)to elevated carbon dioxide and ultraviolet-B radiation under controlled environmental conditions. *Plant, Cell and Environment*, 2003, 25: 771–782.

Zhao X, Shi Y, Liu Y, Jia RL, Li XR. Osmotic adjustment of soil biocrust mosses in response to desiccation stress. *Pedosphere*, 2015, 25: 459–467.

Zhao Y, Li XR, Zhang ZS, Hu YG, Chen YL. Biological soil crusts influence carbon release responses following rainfall in a temperate desert, Northern China. *Ecological Research*, 2014, 29: 889–896.

Zhou CP, Ouyang H. Effect of temperature on nitrogen mineralization at optimum and saturated soil water content in two types of forest in Changbai Mountain. *Acta Ecologica Sinica*, 2001, 21: 1469–1473.

Zotz CG, Schweikert A, Jetz W, Westerman H. Water relations and gain are closely related to cushion size in the moss *Grimmia pulvinata*. *New Phytologist*, 2000, 148: 59–67.

Zu YG, Pang HH, Yu JH, Li DW, Wei XX, Gao YX, Tong L. Responses in the morphology, physiology and biochemistry of *Taxus chinensis* var. *mairei* grown under supplementary UV-B radiation. *Journal of Photochemistry and Photobiology B: Biology*, 2010, 98: 152–158.

第4章　荒漠BSC对生物因子的生态生理响应

4.1　维管植物与BSC的关系

BSC与维管植物之间的关系很复杂，其中与维管植物群落的盖度和组成结构（一年生植物与多年生植物盖度）之间的关系及互馈机理在第1章中已充分讨论。BSC对维管植物个体的影响主要表现在维管植物种子散布、萌发、定居以及维管植物的繁衍等方面。许多干旱、半干旱地区维管植物的垂直和水平结构有利于BSC的形成，维管植物结构多样性较高会导致BSC的组分多样性较高。维管植物能抵御风沙，为BSC提供荫蔽，并影响到达BSC表面的水分和光照。

4.1.1　BSC对维管植物土壤种子库的影响

当种子从母株上散落后，许多环境因子影响着种子的运动。在不考虑种子大小和形状的情况下，在干旱、平滑的土壤表面，许多种子极易被风和水带走。然而，由于一些种子有黏性的外壳和冠毛，所以它们极易黏结在湿润的土壤表面，即使当土壤表面光滑时也会发生这种情况。BSC的存在改变了地表的光滑度，对维管植物种子在地表的分布产生了重要的影响。在热带地区，由于没有冻融作用，BSC的存在使得地表趋于光滑，维管植物种子在BSC表面不易被捕获（Prasse，1999）；而在寒带地区，由于结霜和冻胀作用的存在，结皮的存在增加了地表粗糙度，为维管植物种子在结皮表面定居创造了条件（Harper and Clair，1985a）。土壤种子库是指一定面积土体中有生活力或发芽能力的种子总和。本研究对腾格里沙漠东南缘沙坡头人工植被区（24龄、41龄和50龄）及相邻天然植被区BSC上的土壤种子库进行了研究，旨在探讨荒漠化逆转过程中BSC对土壤种子库的组成、数量和特征的影响。

在中国科学院沙坡头沙漠试验研究站铁路以北采集人工植被区（24龄、41龄和50龄）

的完整藓类结皮，在天然植被区采集完整的藓类结皮和藻类结皮（2005年4月18—21日）。将采集样品放在装有流沙的花盆中，置于温室。每天向花盆表面均匀地喷洒300 mL自来水，使BSC表面保持湿润。用大头针作标记，每隔1天记录BSC表面的出苗数以及出苗种类，将已鉴定的幼苗移出花盆并取出大头针，直到花盆中连续7天不出幼苗为止结束试验，每个处理12个重复，共60个花盆供试。对天然植被区BSC厚度进行测量，得出藻类结皮为0.55 ± 0.12 cm，土层厚度为1.55 ± 0.18 cm；藓类结皮高度为2.3 ± 0.28 cm，土层厚度为1.5 ± 0.2 cm。

（1）BSC演替对荒漠土壤种子库的影响

BSC种子库分布：对4个时间序列BSC种子库的方差分析结果表明（表4-1），不同演替阶段的BSC种子库差异极显著（F=11.878，p=0）。BSC种子库总密度从小到大排序为24龄<41龄<50龄<天然植被区，从人工植被区24龄BSC到天然植被区BSC，土壤种子库总密度从1263.36 ± 309.94有效种子数 m^{-2}（平均值 ± 标准误）增加到了4263.83 ± 564.17有效种子数 m^{-2}。人工植被区50龄和41龄土壤种子库总密度分别是24龄植被区土壤种子库总密度的1.9倍和1.6倍，天然植被区种子库总密度增加最快，分别是人工植被区种子库总密度的3.4倍、2.1倍和1.8倍。以上表明，土壤种子库总密度与BSC的演替正相关，但在不同的演替阶段，BSC种子库总密度增加的速度不同。

从表4-1中我们也可以看出，不同植物种的种子库密度变化特征与BSC的演替阶段相关。在人工植被区，禾本科植物小画眉草、狗尾草和虎尾草种子库密度从24龄人工植被区的649.23 ± 223.19有效种子数 m^{-2}、122.83 ± 17.55有效种子数 m^{-2}、35.09 ± 22.19有效种子数 m^{-2} 增加到了50龄人工植被区的1070.34 ± 191.09有效种子数 m^{-2}、210.56 ± 47.08有效种子数 m^{-2}、105.28 ± 38.44有效种子数 m^{-2}，分别为原来的1.6倍、1.7倍和3倍；但在天然植被区，三者的种子库密度分别为210.56 ± 81.55有效种子数 m^{-2}、105.28 ± 47.08有效种子数 m^{-2}、70.18 ± 35.09有效种子数 m^{-2}，与50龄人工植被区种子库密度相比，土壤种子库密度分别下降了80%、50%、33%。小画眉草种子库密度在BSC演替阶段之间差异显著（F=5.599，p<0.05），而狗尾草和虎尾草种子库密度在BSC演替阶段之间差异不显著（F=1.333，p>0.05；F=0.720，p>0.05）（表4-1）。半灌木油蒿和一年生草本雾冰藜种子库密度在24龄人工植被区BSC上分别为140.37 ± 52.05有效种子数 m^{-2}、70.19 ± 22.19有效种子数 m^{-2}，在50龄人工植被区BSC上为193.01 ± 17.55有效种子数 m^{-2}、122.83 ± 32.35有效种子数 m^{-2}，分别为原来的1.4倍和1.7倍；而在天然植被区这两种植物的种子库密度为0。油蒿和雾冰藜种子库密度在不同植被区BSC上差异显著（F=3.6，p<0.05；F=3.921，p<0.05）。以上表明，不同植物种子库密度的变化趋势与BSC的发育程度并不一致。

表 4-1　不同演替阶段 BSC 土壤种子库总密度（平均值 ± 标准误；有效种子数 m^{-2}）的变化特征
Table 4-1　Density（mean ± SE；viable seeds m^{-2}）of seed bank in BSC at different successional stages

植物名	科	生活型	人工植被区（有效种子数 m^{-2}）			天然植被区（有效种子数 m^{-2}）	F	p
			24 龄（1981 年）	41 龄（1964 年）	50 龄（1956 年）			
总量[a]			1263.36 ± 309.94	2017 ± 179.11	2421.44 ± 318.17	4263.83 ± 564.17	11.878	0
油蒿	菊科	S	140.37 ± 52.05	193.01 ± 78.86	193.01 ± 17.55	0	3.6	0
小画眉草	禾本科	A	649.23 ± 223.19	1035.25 ± 149.92	1070.34 ± 191.09	210.56 ± 81.55	5.599	0.006
狗尾草	禾本科	A	122.83 ± 17.55	193.01 ± 57.13	210.56 ± 47.08	105.28 ± 47.08	1.333	0.292
雾冰藜	藜科	A	70.19 ± 22.19	105.28 ± 38.44	122.83 ± 32.35	0	3.912	0.024
虎尾草	禾本科	A	35.09 ± 22.19	87.83 ± 42.26	105.28 ± 38.44	70.18 ± 35.09	0.720	0.552
糙隐子草	禾本科	P	0	0	0	87.73 ± 68.86	1.632	0.216
茵陈蒿	菊科	P	0	0	0	2333.70 ± 341.05	46.821	0
蒺藜	蒺藜科	A	0	0	0	193.01 ± 50.25	14.756	0
锋芒草	禾本科	A	0	0	0	87.73 ± 42.26	4.3	0.017
地锦	大戟科	A	0	0	0	140.37 ± 35.09	16	0
苣荬菜	菊科	A	0	0	0	35.09 ± 22.19	2.5	0.089
达乌里胡枝子	豆科	S	0	0	0	17.55 ± 17.55	1	0.413

注：a. R^2=0.493（adjusted R^2=0.398），p=0.05；S：半灌木（semi-shrub）；P：多年生草本（perennial herbage）；A：一年生草本（annual herbage）。

BSC种子库植物种类及生活型：在人工植被区，BSC种子库的物种数均为5种，从50龄人工植被区到天然植被区，土壤种子库物种数从5增加到10，增加了2倍（表4-1）。这表明，在人工植被区，土壤种子库物种数增加很少，但从人工植被区到天然植被区，土壤种子库物种数急剧增加。人工植被区土壤种子库的物种组成基本上相似，主要组成是油蒿、小画眉草、狗尾草、雾冰藜以及虎尾草，而天然植被区土壤种子库的物种除了与人工植被区中相同的物种小画眉草、狗尾草、虎尾草以外，还有独有的植物种：糙隐子草、茵陈蒿、蒺藜、锋芒草、地锦、苣荬菜、达乌里胡枝子（表4-1）。

在四个不同演替阶段的BSC上，土壤种子库中各生活型所占的百分比如图4-1所示。不同演替阶段土壤种子库种类组成均以一年生草本植物为主，在人工植被区一年生草本所占的比例均为80%左右，天然植被区一年生草本所占的比例为70%。从表4-1中可以看出，人工植被区和天然植被区土壤种子库的优势植物种不同，人工植被区的优势植物种是油蒿和小画眉草，约占24龄、41龄、50龄人工植被区土壤种子库密度的63%、61%、52%，天然植被区的优势植物种是茵陈蒿，占到了55%。小画眉草，狗尾草和虎尾草在不同演替阶段的BSC上均出现，说明这三种植物的生态位广，生态适应能力较强。油蒿仅出现在人工植被区，这是由于油蒿是在固沙时期作为先锋物种引入而存在的，并且油蒿在人工植被区得到了很好的更新，也是其存在的主要原因。一些具有草原性质的植物种如达乌里胡枝子、茵陈蒿、苣荬菜、糙隐子草等是固定沙地的优势种，它们仅出现在荒漠化草原性质显著的天然植被区。

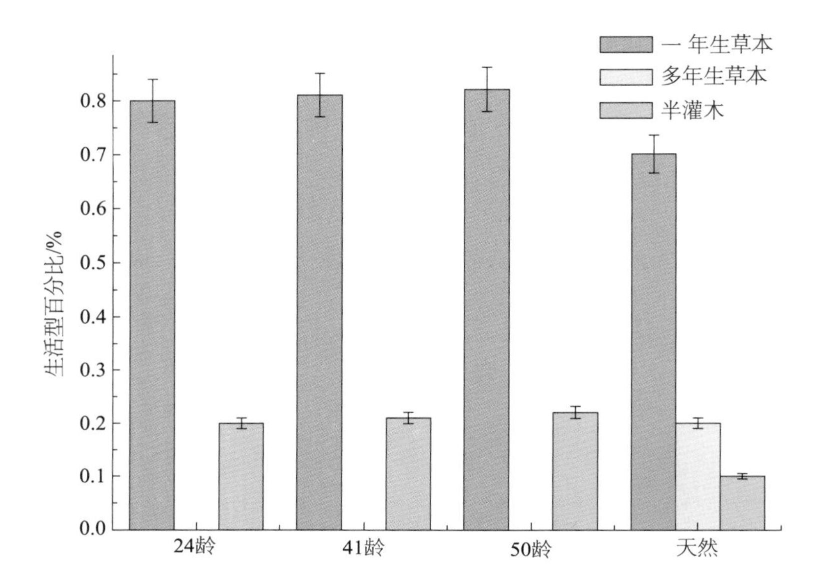

图4-1　不同演替阶段BSC种子库各植物生活型所占的百分比

Figure 4-1　Percentage of life forms of species in crusts seed bank at different successional stages

计算Simpson多样性指数和Shannon-Weaver多样性指数表明，土壤种子库物种多样性指数从大到小排序为天然植被区>24龄>41龄>50龄。在人工植被区，物种多样性与BSC发育负相关（表4-2）。与BSC种子库密度作比较可知，天然植物区种子库总密度和物种多样性均为最高，而人工植被区种子库总密度和物种多样性呈现相反的变化趋势。

BSC种子库相似性指数：BSC的四个演替阶段土壤种子库物种组成相似性系数变化范围是0.4 ~ 1（表4-3），人工植被区之间的土壤种子库物种组成相似性系数最大（均为1），天然植被区和人工植被区物种组成的相似性系数较小，均为0.4（表4-3）。

表 4-2　不同演替阶段 BSC 土壤种子库物种多样性、丰富度、均匀度指数

Table 4-2　Species diversity, richness and evenness of seed bank in BSC at different successional stages

BSC 年龄	24 龄	41 龄	50 龄	天然植被区
物种数	5	5	5	10
Margalet 丰富度指数 *R*	0.56	0.53	0.51	1.08
Simpson 多样性指数 *D*	0.615	0.556	0.421	0.693
Shannon-Weaver 多样性指数 *H*	0.837	0.666	0.435	1.112
Peilow 均匀度指数 *E*	0.167	0.133	0.087	0.111

表 4-3　不同演替阶段 BSC 物种相似性指数

Table 4-3　Similarities of seed bank in BSC at different successional stages

	24 龄	41 龄	50 龄	天然植被区
24 龄	1			
41 龄	1	1		
50 龄	1	1	1	
天然植被区	0.4	0.4	0.4	1

（2）BSC类型对荒漠土壤种子库的影响

BSC含水量与种子库储量的相关性分析：将BSC种子库的储量与BSC含水量作相关性分析得出，经过干燥处理的藓类结皮与藻类结皮上两因素之间的相关性很高（R^2=0.907和R^2=0.924），经过湿润处理的BSC上的相关性稍低（R^2=0.857和R^2=0.912）。这说明BSC要通过与水分的相互作用来影响种子库的数量特征。分别对湿润处理的藓类结皮和藻类结皮的含水量作独立样本T检验得出，两种BSC的含水量没有达到显著差异的水平（p=0.368），用同样的方法对干燥处理的两种BSC上的含水量做检验，得出两种BSC之间的差异亦没有达到显著的程度（p=0.154），在相同处理下，两种BSC种子库数量的显著差异与种子萌发的基质有关。李新荣等（2000）对沙坡头植被多样性的研究认为，BSC及下层土壤发育层

中微植物根系密集分布，BSC表面被有密集绒状的藻类、藓类等形成紧实的板状结构，即使一年生植物种子因遇有限降水而萌发，其根系也很难在BSC表层侵入，很快因失水而枯死。对干燥、湿润处理的两种BSC含水量作配对样本T检验得出两者之间存在着显著差异（$p<0.001$），两种BSC经过不同的处理后，种子的萌发量存在着显著差异（表4-4），说明在干燥条件下不能使植物种子的幼苗全部萌发，也说明在干旱、半干旱地区，水分是阻止植物幼苗萌发的主要限制因子。

对湿润处理的藓类结皮和藻类结皮的表面温度进行差异显著性分析，得出两者之间没有达到显著差异（$p=0.068$）；对干燥处理的两种BSC的表面温度进行显著性检验，得出两者之间也没有达到显著差异（$p=0.135$）；但对同一类型的BSC进行干、湿处理后，BSC的表面温度达到了显著差异，其中，藓类结皮为$p=0.025$，藻类结皮为$p=0.035$，这说明由于含水量的不同导致的BSC表面温度的不同也影响着种子的萌发，但基质的不同并没有使表面温度的差异达到显著程度。

BSC种子库的植物组成：经分析统计，两种类型的BSC上种子库的植物种类组成以及有效的植物种子数见表4-4。从表中可以看出，在整个试验过程中，共有9种幼苗出现在BSC上，它们分别是独行菜、油蒿、灰绿藜、多根葱、小画眉草、雾冰藜、马齿苋、狗尾草和苋菜。可以明显地看出，该地区的植被结构相对简单，物种多样性相对较低。李新荣等（2000）对我国干旱区植被演替中植物种的多样性做了研究，结果表明，沙坡头地区植被演替进行至今，植物多样性指数较低，植被的结构相对简单，群落的稳定性较低，对干扰的反应敏感。

在出现的植物种幼苗中，独行菜、灰绿藜、小画眉草、雾冰藜、马齿苋、苋菜和狗尾草等7种植物属于一年生草本植物，占到了总植物种数的77.8%，油蒿为多年生半灌木，多根葱为多年生草本，分别占到了总植物种数的11.1%。对沙坡头植被演替动态的研究表明，在整个演替过程中，多年生半灌木油蒿一直存在，并且随着流沙成土过程的开始，沙层表面BSC层逐渐增厚，同时一年生植物也大量侵入，逐渐占据了优势地位（王刚和梁学功，1995）。李新荣等（2000）对沙坡头流沙和不同发育年代的BSC及其上定植的植物的多样性进行了研究，结果表明，BSC的厚度为0.3～2 cm时与植物种的多样性呈正相关关系。BSC的存在使沙漠区域环境得到了改善，为一年生草本植物多样性的提高创造了条件。随着演替的进行，人工栽植的灌木除油蒿在BSC上的更新较好外，其他灌木种逐渐退出，而一年生草本植物所占的比例增大，这是由于BSC越厚，对降水的阻碍作用越大，使得一些深根系的灌木得不到降水的补给而死亡；另一方面，BSC有较好的持水性，当它被较少的降水湿润后，很容易为许多一年生草本植物的种子萌发创造条件。

在不同类型的BSC上，出现的植物种不同。独行菜、油蒿、灰绿藜、多根葱和小画眉草

表 4-4　不同 BSC 土壤种子库密度（平均值 ± 标准误）的变化特征

Table 4-4　Changes in seedling densities （mean ± SE） of different BSC soil

植物种	生活型	科	萌发幼苗总密度				百分比 / %				总量
			A	B	C	D	A	B	C	D	
植物种数			6	7	7	8					9
萌发幼苗总密度			1168 ± 52a	794 ± 48b	820 ± 41b	631 ± 36d					3478
独行菜	一年生草本	十字花科	836 ± 42a	583 ± 29b	507 ± 25c	415 ± 21d	71.6	73.4	61.3	65.8	2341
油蒿	半灌木	菊科	86 ± 4b	34 ± 2a	134 ± 7c	58 ± 3d	7.36	4.3	16	9.2	312
灰绿藜	一年生草本	藜科	133 ± 6.7b	64 ± 3a	48 ± 2.9c	76 ± 3.8d	11.4	8.1	5.8	12	321
多根葱	多年生草本	百合科	54 ± 2.7a	58 ± 2.9a	29 ± 1.9c	5 ± 0.25d	0.47	7	3.5	7.9	146
小画眉草	一年生草本	禾本科	54 ± 2.7a	66 ± 3.3b	71 ± 3.5c	34 ± 1.7d	4.6	8.3	8.6	5.4	225
雾冰藜	一年生草本	藜科	5 ± 0.25a	5 ± 0.25b	—	—	0.43	0.63	—	—	10
马齿苋	一年生草本	马齿苋科	—	—	7 ± 0.35c	19 ± 0.95c	—	0.89	—	2.3	26
狗尾草	一年生草本	禾本科	—	—	34 ± 1.7c	5 ± 0.25d	—	4.3	—	0.6	39
苋菜	一年生草本	苋科	—	9 ± 0.45b	—	19 ± 0.45b	—	1.1	—	3	28

注：A、B、C、D 分别表示湿润藓类结皮、干燥藓类结皮、湿润藻类结皮、干燥藻类结皮，不同字母表示差异显著（$p<0.05$）。

在两种类型的BSC中均存在，说明藻类结皮和藓类结皮对这些植物的种子均能捕获，并且植物种子在两种BSC上均能萌发。雾冰藜的幼苗仅出现在藓类结皮上，可能是幼苗出土的基质不同所造成的。龙利群和李新荣（2002）曾对沙坡头地区雾冰藜在藓类结皮和藻类结皮两种基质上种子的出苗情况做了定量研究，结果表明，由于完整致密的藻类结皮对雾冰藜种子萌发后根的扎入设置了物理障碍，大量种子萌发后因无法穿透BSC层扎根于土壤而很快失水干死；雾冰藜在完整藻类结皮上的出苗率最低，仅为13.5%。而藓类结皮因其植物体疏松，为种子的出苗提供了较为疏松的发芽床，所以在藓类结皮上雾冰藜种子的萌发率高达85%以上。这说明种子萌发基质的不同会影响种子的发芽率进而直接影响到种子库的储量。马齿苋和狗尾草在藻类结皮上出现，这可能是由于BSC外在结构的不同造成了BSC对种子捕获能力的差异。从直观上来看，BSC的存在增加了土表的粗糙度，提供了一个相对有利的生境而增加了维管植物的繁殖和存活能力。在一些蓝藻占优势的地区，BSC上呈现了一层光滑的藻类层外貌，增加了表面粗糙度和微地貌的变异性，保护了其上生长的植物免受风沙掩盖，为种子的捕获与存活提供了有利条件。李新荣等（2000）的研究则认为，在沙坡头建立较早的固沙区（1956年固定的沙丘），BSC的形成使沙面形成了一层紧实的“保护膜”（特别是在丘间低地），加之沙漠地区频繁而强大的风力作用（年均风速2.9 m s^{-1}），植物的种子很难在其上保存。种子雨多汇集在沙面粗糙度大、有枯死的植株和半灌木油蒿等植物分布的根际范围，由于这些区域粗糙度大，风速相对较小，植物的种子容易得到保存而不被风吹走。所以BSC的外部形态以及种子的外部结构特征直接影响着种子被捕获的概率。同时，BSC对种子萌发的影响也因种子的类型不同而存在着差异（李新荣等，2001）。苋菜虽然在藓类结皮和藻类结皮上均有出现，只是它仅出现在经过干燥处理的BSC上，这说明BSC在一定水平上改变着水分的入渗与再分配状况（李守中等，2002），直接影响着其上植物种子的萌发情况。

BSC种子库的数量特征：经统计分析，两种BSC种子库的萌发幼苗总密度和每种植物在种子库中所占的比例见表4-4。从表中可以看出，在两种处理的BSC上，种子库的组成、种子库的总密度以及每种植物的密度存在着不同程度的差异。在湿润处理的两种BSC上，萌发幼苗总密度的差异达到了极显著水平，在干燥处理的两种BSC上，萌发幼苗的总密度达到了显著差异。在两种湿润BSC上，独行菜、油蒿以及灰绿藜的幼苗密度存在着极显著的差异，小画眉草幼苗密度以及幼苗总密度存在着显著差异；在两种干燥处理的BSC，独行菜、灰绿藜、多根葱、小画眉草的幼苗密度和幼苗总密度存在着极显著的差异，油蒿的幼苗密度存在着显著的差异（p=0.007），苋菜在经过干燥处理的两种BSC上的幼苗密度没有显著差异。这说明不同的BSC对种子的捕获能力是不同的，从直观的角度来看，不同的BSC由于表面粗糙度的不同以及种子自身外部形状的差异而直接影响着种子是否进入BSC层。在本试验中，两种BSC对独行菜、灰绿藜、多根葱以及油蒿的捕获能力差异很大，而对苋菜的捕获能力没有显著差异。同时也表明萌发基质的不同也会影响进入BSC的种子的萌发量，进而对种子库的数量特征产生影响。关于BSC与种子萌发关系的研究存在着三种不同的观点，即它们之间是互利关系、中性关系还是竞争关系。

藓类结皮和藻类结皮在经过干、湿两种处理后，萌发幼苗的总密度均存在着极显著的差异。对于藓类结皮的两种处理进行方差分析得出，除了多根葱的幼苗密度以及幼苗的总密度没有达到显著的差异外，其余各种植物的幼苗密度均达到了极显著的程度。对于藻类结皮的两种处理的方差分析得出，独行菜、油蒿、灰绿藜、小画眉草、狗尾草的幼苗密度以及幼苗总密度存在着极显著的差异，多根葱的幼苗密度存在着显著的差异，马齿苋的幼苗密度没有差异。比较明显的是，无论是在藓类结皮还是在藻类结皮上，湿润处理的萌发量都大于干燥处理的，在湿润处理的BSC上，油蒿、小画眉草在藓类结皮上的种子库密度小于藻类结皮的，而灰绿藜和多根葱在藓类结皮上的种子库密度大于藻类结皮的，与湿润处理的BSC不同，在干燥处理的BSC上，所有植物种在藓类结皮上的种子库密度均大于藻类结皮的。这说明两种BSC通过改变水分在BSC层的入渗、分配状况而极大地影响着种子库的数量特征。对于不同的植物种，它们之间的差异程度随着BSC类型以及BSC含水量的不同而表现出很大的差异。

4.1.2 BSC对两种一年生植物种子萌发和出苗的影响

一年生植物是沙坡头人工植被固沙区建立后随着生境的改善而逐渐自然入侵和繁衍的。它们分布较广、盖度较大，在人工植被区起着重要的作用。一年生植物在干旱区的一个显著特征是植株遇干旱年份时生长矮小，却仍能开花、结实完成生活史。一年生植物的盖度随着年降雨量及其分配状况的变化呈现出很大的波动。湿润年里，一年生植物在表土层较厚而BSC明显的地段上能形成浓密的绿色景观，其盖度一般都在40%以上；干旱年因植被稀少而又发育孱弱，盖度一般只有20%左右。

选择沙坡头沙漠试验研究站人工植被区草本盖度最大的两个种（小画眉草和雾冰藜），在温室可控制条件下研究不同类型的BSC、不同破坏程度的BSC在不同水分处理下对两种一年生草本植物萌发和出苗的影响。实验所用BSC分别采自1981年固沙区的藻类结皮，

BSC厚度为0.45 ± 0.12 cm，土层厚度为1.85 ± 0.2 cm；采自1956年的长尖扭口藓（*Barbula ditrichoides*）结皮，藓类植株高度为2.5 ± 0.25 cm，土层厚度为3.2 ± 0.14 cm。将BSC铺在花盆中的沙子上，在萌发实验前进行材料预处理：每天对BSC浇水以消耗掉其中的土壤种子库。处理从2001年4月26日开始到BSC中连续7天不再有幼苗出现而停止。该实验为研究自然条件中不同降雨量及其分配状况下，人工植被固沙区中两种草本植物的盖度变化及不同的空间分配格局提供生态生理学参考，也为进一步探讨人工植被固沙区的演变规律及其稳定性维持的机制提供科学依据。

实验用来验证两种草本植物的种子萌发及出苗状况如何受BSC类型及BSC的不同破坏程度的影响。两种类型的BSC分别设完整BSC组和人为破坏BSC组，流沙对照组分别设种子适度掩埋组和非掩埋组。流沙适度掩埋的深度为0 ~ 5 cm。两种植物的种子每100粒一份，各6个重复，均匀撒入花盆中。水分处理分干、湿两种处理。湿润处理为每天浇2 ~ 3次水，每次对每个花盆浇500 mL水，以保证花盆表面湿润，避免干旱胁迫对种子萌发的抑制。干燥处理为每星期浇一次水，每次对每个花盆浇500 mL水，两次浇水期间花盆内保持干燥状态。所有花盆均放置于温室内，室内温度18 ~ 35 ℃，相对湿度63% ~ 91%。实验于2001年6月28日开始，每天下午17时开始统计各花盆中的出苗数，用镊子去除已出苗的种子。重复以上实验过程直到连续7天没有新苗出现而停止实验，实验共持续65天。

藓类结皮、藻类结皮、流沙另分别设空白组：即不播种于其上，在干燥处理中，两种植物的种子出苗一般集中在浇水后24 h左右，在浇水后24 h分别测定各空白组BSC表层、流沙表层、各BSC 0 ~ 5 cm深度和流沙0 ~ 5 cm深度的土壤水分。

图4-2表明在湿润处理下，小画眉草种子在两种完整的BSC上和流沙非掩埋状态下的出苗率无显著差异。在完整的藓类结皮上出苗率为41.33%，完整的藻类结皮上为37.17%，流沙非掩埋状态下为38.83%。而在人为破坏的两种BSC中和流沙掩埋状态下其出苗率分别下降至28%、26.33%和24%。

小画眉草在流沙上的出苗进程表明（图4-3），小画眉草种子在浇水24 h后才有种子出芽，较高的日出苗率出现在第2、3、4天，分别为8.33%、11.17%和6%。其后种子出苗呈间断分布状态，出苗率为0% ~ 2.67%，持续时间可达17天。

在干燥处理下（图4-4），小画眉草出苗率状况与在湿润处理下类似，最高值出现在完整的藓类结皮上为40%，最低出现在流沙掩埋组中为21.17%，人为破坏的藻类结皮上为25.5%，低于完整的藻类结皮上的38.67%，人为破坏的藓类结皮上为29.17%，流沙非掩埋状态下为34%。

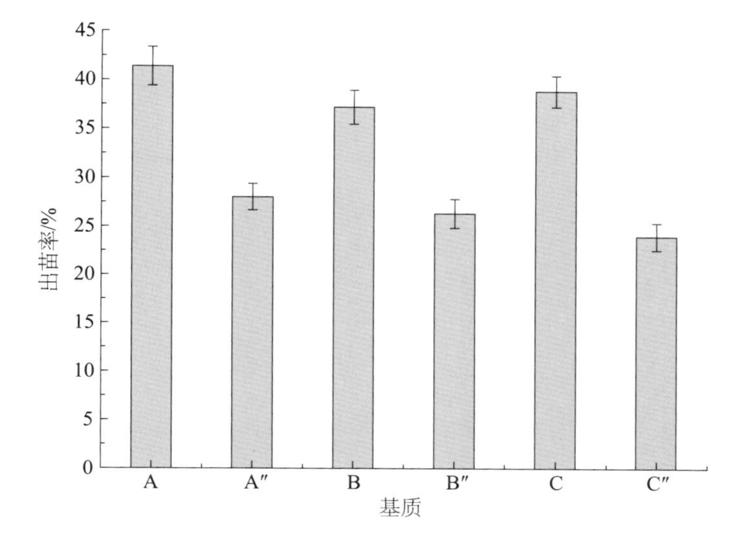

图 4-2 湿润处理下小画眉草在不同基质上的出苗率（平均值 ± 标准误）。A：完整的藓类结皮；A″：人为破坏的藓类结皮；B：完整的藻类结皮；B″：人为破坏的藻类结皮；C：流沙非掩埋处理；C″：流沙掩埋处理，以下同

Figure 4-2 Mean（±SE）seedling emergence of *Eragrostis poaeoides* on different substrates in the moist treatment. A. intact moss crust；A″ . destructive moss crust；B. intact algae crust；B″ . destructive algae crust；C. no moving sand bury treatment；C″ . moving sand bury treatment.It is the same as figures below

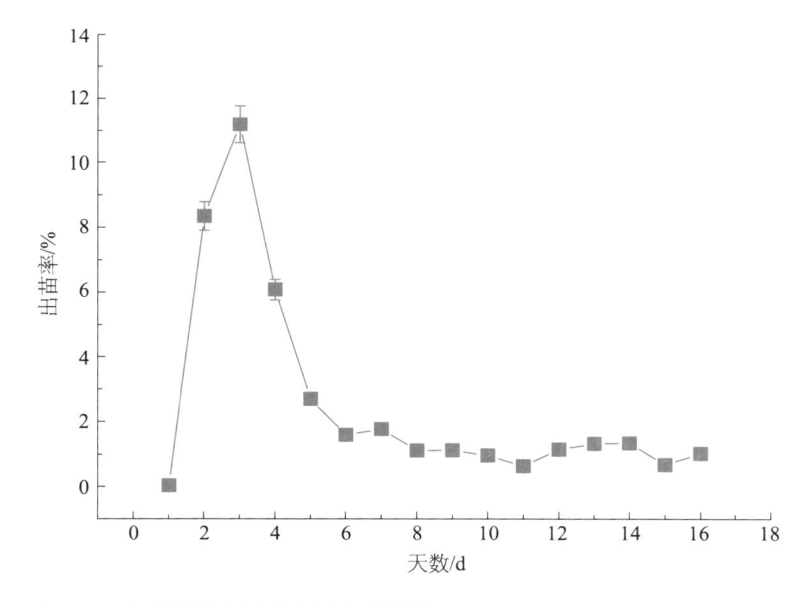

图 4-3 小画眉草在流沙上的出苗进程

Figure 4-3 Progress of seedling emergence of *Eragrostis poaeoides* on moving sand

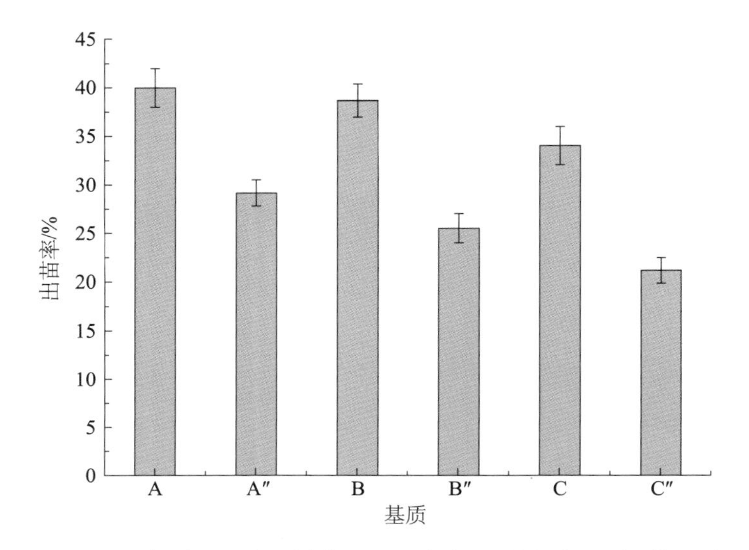

图 4-4 干燥处理下小画眉草在不同基质上的出苗率（平均值 ± 标准误）
Figure 4-4 Mean（±SE）seedling emergence of *Eragrostis poaeoides* on different substrates in the dry treatment

在湿润处理下（图4-5），雾冰藜出苗率在完整的藻类结皮上最低，仅为13.5%。在其他基质上的出苗率较为一致，集中在86.33%～91.83%。雾冰藜具有快速的萌发能力，浇水处理后3 h即有种子萌发，且萌发非常集中（图4-6）。在水分充足的条件下，雾冰藜种子24 h内出苗率高达69%以上，5天内有活力的种子全部萌发。湿润处理下流沙掩埋与否对其种子出苗率无明显影响。在实验观测中，雾冰藜种子在完整的藻类结皮表面上萌发率高达85%以上，但完整致密的藻类结皮对雾冰藜种子萌发后根的扎入设置了物理屏障，大量种子萌发后因无法穿透BSC层扎根于土壤而很快失水干死。干扰破坏藻类结皮能很大程度削弱致密BSC层对种子扎根的物理屏障作用，出苗率从13.5%提高至89.67%。长尖扭口藓结皮因其植物体疏丛生、植株高2～3 cm，为种子出苗提供较疏松的发芽床。长尖扭口藓结皮不同于致密的藻类结皮和一些密集丛生成毡垫状的银叶真藓结皮，不会对植物种子出苗产生物理屏障作用。

干燥处理下雾冰藜在各种基质处理下的出苗率都大大降低（图4-7），其中在完整的藻类结皮上最低，仅为2%；流沙非掩埋组次之为26.5%；最高的出苗率出现在流沙掩埋组为61%；藓类结皮完整组和人为破坏组分别为53.67%和52.83%，略高于藻类结皮人为破坏组的40.17%。

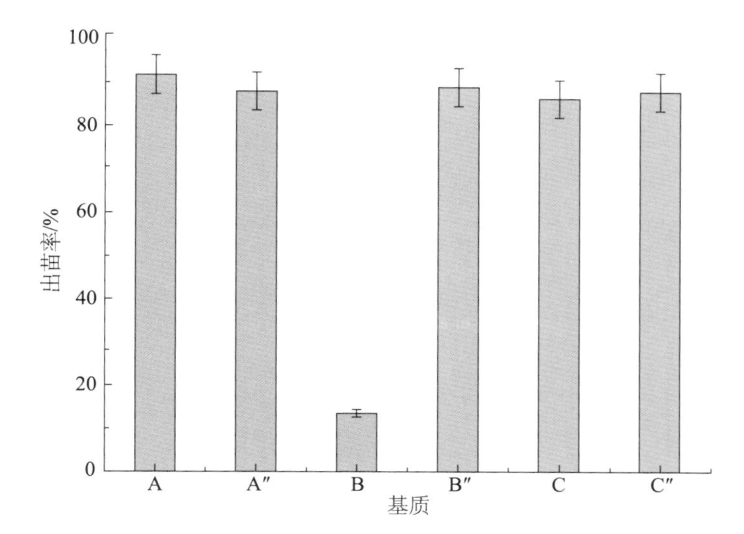

图 4-5　湿润处理下雾冰藜在不同基质上的出苗率（平均值 ± 标准误）
Figure 4-5　Mean（±SE）seedling emergence of *Bassia dasyphylla* on different substrates in the moist treatment

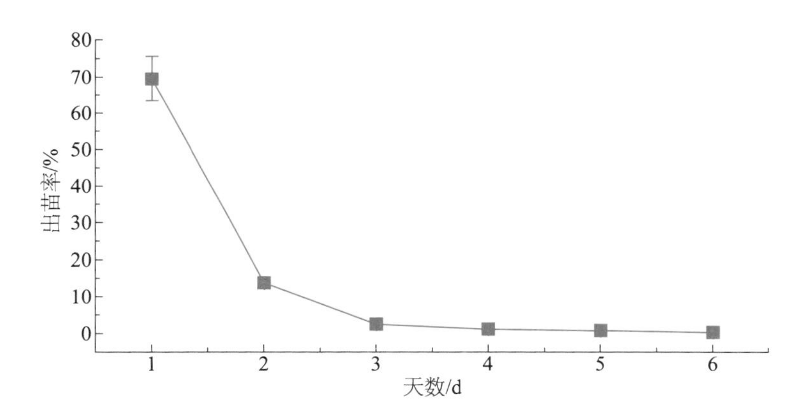

图 4-6　雾冰藜在流沙上的出苗进程
Figure 4-6　Progress of seedling emergence of *Bassia dasyphylla* on moving sand

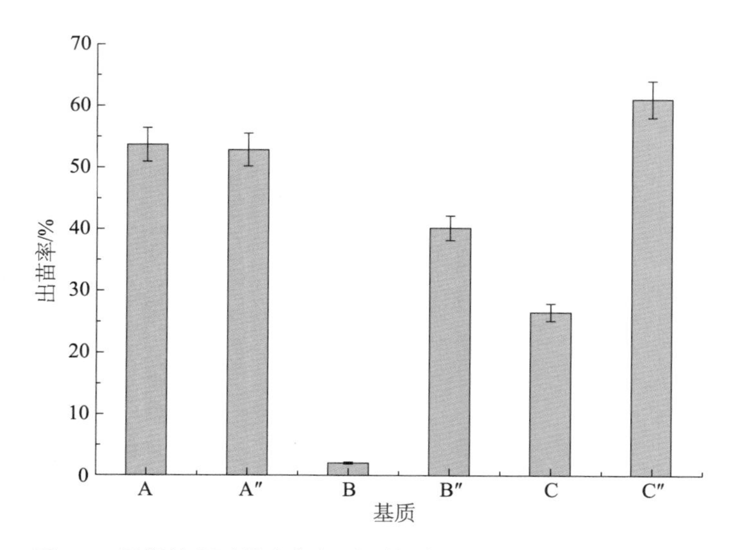

图 4-7　干燥处理下雾冰藜在不同基质上的出苗率（平均值 ± 标准误）
Figure 4-7　Mean（±SE）seedling emergence of *Bassia dasyphylla* on different substrates in the dry treatment

在干燥处理空白组中，浇水24 h后分别测定藓类表层平均水分含量为30.37%，藓类结皮0～5 cm层为18.64%；藻类结皮表层为16.05%，藻类结皮0～5 cm层为10.32%；流沙表层为3.51%，流沙0～5 cm层为8.79%。藓类结皮的持水能力最大，藻类结皮次之，流沙最低。

就机理而言，认为BSC通过改变以下生境来影响维管植物的萌发、定居和存活：①改变土壤表面的微地形。在热带荒漠，由于缺少土壤冻融，土壤表面粗糙度的改变较小，其中藻类能使土壤表面光滑化，而地衣和藓类在土表的生长增加了土表的粗糙度，但藓类的发育在温带荒漠特殊的地貌条件下如在固定沙丘丘间低地也使土表光滑化（Li *et al.*，2005；Su *et al.*，2007）。相反，在寒冷荒漠地区由于地表和各种BSC土壤的侵蚀，使土表的粗糙度明显增加，以至于土表高度能隆升15 cm（Belnap and Gardner，1993）。这些相对大的地形变化特征可捕获风和水带来的种子、有机质、土壤微粉粒和水分（Li *et al.*，2003；Su *et al.*，2007）。②通过影响种子的捕获间接地影响参与萌发与定居的种子数量。当土壤表面光滑时对种子的捕获（seed entrapment）量很低，而在那些因BSC存在而显著增加地表粗糙度的生境（发育已有的BSC和冻胀丘）种子的捕获量则很高。光滑BSC表面对种子散播的影响在以色列Nizzana沙丘地区得到了很好的描述和解释。Prasse和Bornkamm（2000）指出，种子在这些蓝藻形成的光滑BSC表面停留的能力很低。进行干扰试验使这些BSC粗糙化，种子在土表的停留和捕获量显著增加。③影响维管植物养分。BSC对土壤表面化学性质的改变与相关植物种子组织体中主要生命元素的含量变化密切相关（Harper and Pendleton，1993；

Harper and Belnap，2001）。总的来说，BSC的存在提高了植物对镁、钾、铜和锌的获得，而同时减少了植物对铁的吸收。④ 种子本身的生物学特性。种子萌发对水分的需求存在差异。在空气干燥的荒漠，许多种子要求一定的植被覆盖和土壤来保持充分的湿度从而进行萌发。土表很小的裂缝和断裂微地形对小颗粒种子植物萌发已足够（Gutterman，1994），但对大颗粒种子则需要额外土壤或凋落物的覆盖（Evans and Young，1984）。一些缺少特殊穿透结构的植物种子（Boeken and Sbachak，1994），通常是一年生的物种，分布于土壤－土表干扰相对较大的地区，它们的萌发在那些凋落物较少、BSC完整稳定以及土表干扰较低的地区就会受到抑制。这一现象已被许多研究证实，如在美国和以色列的一年生植物*Bromus*（Gutterman，1994；Prasse and Bornkamm，2000），澳大利亚和以色列的植物*Schismus*，以色列的植物*Plantago*和*Reboudia*（Zaady *et al.*，1997）以及澳大利亚和美国的植物*Salsola*（West，1990）。

我们通过野外对照观测和控制实验发现，BSC对植物种子萌发、定居和存活的影响主要取决于BSC的类型、研究区自然条件（风蚀与降水节律）以及种子的生物学特性（种子大小、结构和种子休眠特性等）。对供试植物本身生物学特性的忽视也是学术界对这一科学问题产生争论的主要原因之一（Li *et al.*，2005）。实验证明，在多风的荒漠环境中表面完整的BSC和演替后期的BSC降低了对种子雨的捕获，直接减少了参与萌发的种子数量；演替后期的BSC因表面颜色深（增温效应）、持水能力强，有利于种子的萌发，且后期的BSC如藓类和地衣为优势的混生结皮具有很高的碳氮固定能力，为已定居的植物提供了养分，有利于植物的存活（李新荣，2012）。

4.1.3 BSC对外来植物种子萌发及生长的影响

生物入侵（biological invasion）是全球最严重的环境问题之一，其危害主要体现在对生态系统结构和功能的破坏（Chisholm，2010；van Driesche *et al.*，2010）；压迫、排斥本地物种，使得当地特有的物种生存受到威胁，甚至使一些本土优势种趋于绝灭，改变了生态系统服务功能（Raven and Johnson，1992；Wan *et al.*，2002）。在全球已确定的1000多万物种中有近150万物种及其生境遭到了外来物种不同程度的破坏，造成严重的经济损失（Pimentel，2001；Theoharides and Dukes，2007）。我国是受到生物入侵影响最大的国家之一，外来物种的入侵造成1/3的物种濒危。入侵生物每年引起的生物灾害给农业带来的损失占粮食产量的10%～15%，棉花产量的5%～20%和水果蔬菜产量的20%～30%，农林业损失每年可达

57.4亿美元，总体经济损失高达150亿美元左右（Xu *et al.*，2006；刘婷婷等，2010；丁晖等，2011）。生物入侵已引起了公众和科学界的高度关注（Ding *et al.*，2008；Weber *et al.*，2008）。

针对植物的入侵途径，研究表明传播是其先决条件，而适宜的气候、良好的水分和养分等环境因素是成功入侵的必要条件（Drexler and Bedford，2002；Olde Venterink *et al.*，2002；Wilson *et al.*，2009）。因此，农田、森林、海滨和湿地生态系统往往更有利于外来植物的入侵（Zedler and Kercher，2004；van Driesche *et al.*，2010）。而干旱、半干旱荒漠和沙区生态系统，因严酷的自然条件、不便利的交通和相对较低的人类干扰及相对落后的经济水平，如贸易往来较少，大部分地方长期处于封闭或半封闭的状态。此外，水流、动物等其他种子传播的媒介也相对缺乏，使入侵植物进入该区的机会减少，水分限制和可用养分的短缺，也使得植物的萌发和定居困难，因而遭受外来植物入侵的风险相对较小，但这并不意味着该区没有外来植物入侵的风险（van Driesche *et al.*，2010）。

随着西部大开发，沙区资源开发利用和经济、交通及旅游业的发展，为外来物种的传播提供了机遇，特别是全球变化引发的降水的不确定性和空间异质性导致阶段性局地水分和养分等生境因子的改善，为来自其他生态系统的植物入侵创造了条件。此外，沙地生态脆弱，生态系统自我修复或抵御能力相对较弱（Whitford，2002），一旦遭受外来植物的侵害将很难恢复，而干旱沙区大约3亿人口的生计与该区特有的植物资源密切相关。因此，维持我国沙地生态系统的稳定和健康具有重要意义（李新荣和王涛，2004）。

除了传播、区域气候和严酷的生境条件限制了外来物种的入侵外，荒漠或沙地生态系统自身特殊的生物组成和格局是否在抵御植物入侵、维持系统稳定性中同样发挥着重要的作用？BSC作为系统的主要构建者（West，1990；Belnap and Lange，2001），其在植物入侵中扮演什么角色？BSC是通过哪些途径来影响外来植物入侵的？

BSC作为联系非生物与生物因素之间的"生态系统工程师"，是荒漠和沙地生态系统重要的组成和地表的生物覆盖体之一（李新荣，2012），其盖度达到生物活体覆盖的40%以上（Bowker，2007）。尽管目前已有大量文献报道了BSC与土壤理化性质、碳氮循环、植被格局和景观过程间的关系（West，1990；Eldridge and Greene，1994；Belnap and Lange，2003；Bowker，2007；李新荣等，2009），但是对BSC影响外来物种入侵及其机理的报道却很少。

为此，本研究中我们假设在沙区BSC的存在通过以下三种途径来影响外来植物的入侵。① BSC的存在通过减少进入土壤种子库的植物种子来降低入侵植物参与萌发和定居的概率。拓殖于已固定沙丘表面的BSC在沙地水文过程中发挥着重要的作用，影响着表层土壤水分的含量、土壤的持水能力以及降水的入渗（Li *et al.*，2010）。Li等（2005）的研究表明，在水分充足的条件下无论是完整的还是受到破坏的BSC对种子的萌发率影响均不显著，但在较低的水分条件下，受到破坏的BSC上种子的萌发率显著高于完整的BSC。在空气干燥的荒漠，许多植物种子要求一定的植被覆盖和土壤来保持充分的湿度才能萌发。种子通常有其适合的埋藏深度，增深和变浅都会有碍于萌发。由于BSC促进了土壤的稳定性，它们的存在能够不同程度地影响埋藏和随后的种子萌发（Baskin and Baskin，2000）。参与萌发的种子数量及其在地表停留时间，是影响种子萌发和定居的又一重要因素。BSC的存在使土壤表面形成"保护膜"（Eldridge *et al.*，2000），进而影响了种子进入土壤种子库的数量及其在土壤表面的停留时间。这是因为沙漠地区频繁而强大的风力作用（年均风速3.5 m s^{-1}），种子很难在BSC表面停留。种子雨多汇集在粗糙度大、有枯死的植株和半灌木油蒿等植物存在的地表，一年生植物种子进入土壤种子库的数量显著降低，减少了参与萌发的种子数量（Li *et al.*，2005）。此外，BSC及下层土壤层中植物根系密集分布，表面有密集绒状的藻类、苔藓等，形成了紧实的板状结构，即使一年生植物种子能够在有限降水的条件下萌发，但其根系却很难扎入BSC表层，很快会因失水而枯死（Larsen，1995）。种子的萌发和定居是外来植物成功入侵最重要的环节，大多数入侵植物为*r*–策略，具有较快的生长速率，因此需要大量水分作为生长和发育的保证。BSC是否减少了有效水分的量，延长了种子萌发的时间，提高了萌发所需的土壤含水量阈值进而阻碍了外来植物的入侵？BSC是否通过改变地表特征，影响参与萌发的种子数量和定居进而阻碍了外来植物入侵？这是本项目研究的科学问题之一。

② 与维管植物相比，BSC更有利于对贫瘠资源的竞争，从而降低了外来植物的定居机会。土壤养分水平对决定外来植物入侵性的强弱起着重要作用，土壤肥沃（如碳、氮、磷含量高的地方）的区域往往容易遭受外来物种的入侵，相对贫瘠的土壤不易被入侵（Davis *et al.*，2000）。土壤中可利用性养分含量的高低，直接影响着外来植物的生长和发育，进而影响了外来植物是否能够成功入侵（Hoopes and hall，2002）。沙地生态系统是一个土壤养分十分贫瘠的系统（Whitford，2002），BSC的存在对荒漠系统的养分循环产生了重要影响和贡献。BSC的存在改变了土壤表层的化学性质，增加了大多数生命重要元素包括氮、磷、钾、钙、镁、铁等元素在表土层中的含量（Harper and Pendleton，1993；Fearnehough *et al.*，1998）。BSC中的蓝藻和地衣具有固氮作用，研究证实BSC的存在增加了土壤环境氮量的200%，这些氮能够被生物体所直接利用。但是，BSC也通过反硝化作用（denitrification）和挥发（volatilization）过程造成氮的损伤，从而也限制了其他生物体利用氮的量。藓类结皮中的养分很难被其他生物体所利用，除非当藓类结皮死

亡、分解；另一方面由于藓类结皮自身需要大量的养分来维持生长，增加了与维管植物间的养分竞争（Harper and Belnap，2001）。对许多荒漠/沙地系统而言，沙质石灰土壤中通常缺乏铁，因此，BSC可能与维管植物的根系竞争这些元素。竞争可能对分布在地表的那些微小的、短命的植物不利（Black，1968）。Belnap等（2006）对一年生植物旱雀麦（*Bromus tectorum*）的研究表明在藓类和地衣结皮盖度较低的地区有利于其定居。这种现象的发生是否与BSC中的养分可利用性有关？BSC的存在是否阻碍了外来植物直接利用养分，进而降低了入侵的成功率，需试验验证。较高的可利用性氮含量往往与外来植物入侵相联系。较高的可利用性氮含量更有利于快速生长的外来入侵植物，而不利于本地植物生长。使可利用性氮的含量维持在较低的水平是解决这一问题的核心。大多数学者认可增加碳的含量来降低可利用性氮含量这一方法，即通过添加碳以增加碳氮比，同时微生物的量也会增加，微生物氮矿化增强，使得氮的可利用性降低，进而减少可利用性氮的含量（Blumenthal，2008；Perry *et al.*，2010）。但是，这一方法的最大弊端是碳添加一旦停止，可利用性氮的含量将不会继续维持在较低的水平；另外，依靠人工进行碳添加需要很高的成本（Prober and Lunt，2009）。因此，维持较低水平可利用氮含量的最好办法是通过植物–土壤间的反馈作用来控制（Perry *et al.*，2010）。BSC生物体中的蓝藻、绿藻、地衣和藓类等具有光合固碳能力，以蓝藻和绿藻为优势种的BSC年固碳量可达到1.1 ~ 1.7 g C m^{-2}，以藓类为优势种的BSC年固碳量可达到1.5 ~ 2.7 g C m^{-2}，以地衣为优势种的BSC的年固碳量可达到1.2 ~ 3.7 g C m^{-2}（Li *et al.*，2010）。在适宜的条件下，BSC均可进行光合固定作用，BSC的存在有效解决了碳持续添加的难点，也实现了与土壤间的反馈作用。但是，BSC能否通过自身的固碳作用减少可利用性氮的含量而限制外来植物入侵，是我们需要验证的机理之一。③ BSC通过维持本地植物的多样性来抑制外来植物的入侵。本地植物物种多样性与外来植物入侵间的关系一直存在争议（Dark，2004；Crutsinger *et al.*，2008）。物种多样性阻抗假说认为，多样性较高的群落增强竞争从而有效抵抗了外来物种入侵（Elton，1958；Crawley，1987；Levine and D' Antonio，1999），这个假说得到了理论（Case，1990）和试验（Levine，2000；Naeem *et al.*，2000；Symstad，2000）两方面的证实。但是，一部分学者认为，本地物种的多样性与外来植物的入侵呈正相关关系（Stohlgren *et al.*，2001；Wardle，2001；Sax，2002；Stohlgren *et al.*，2002）。本地种多样性和入侵程度间的关系受到空间尺度大小的影响，Brown和Peet（2003）研究发现，小尺度上，本地种多样性和入侵程度间呈负相关关系；而在大尺度上（100 km^2），本地种多样性和入侵程度间呈正相关关系（Liu *et al.*，2005）。在试验研究中，资源的可用性被用于解释物种多样性阻碍生物入侵的机理。较高的物种多样性导致了外来植物只有很少的资源可以使用，进而阻碍了外来植物入侵（Ruijven *et al.*，2003）。另外一些研究认为，物种多样性较高的群落中的某些物种起到了阻碍外来植物入侵的重要作用（Wardle，2001）。BSC覆盖区域的本地植物多样性如何影响外来植物入侵尚缺少研究报道。

我们以不同沙地发育的BSC为研究对象，基于以上BSC三种影响途径的假说，通过野外调查和控制试验模拟和验证BSC对具有地带性分布的草原植物种（3～5种）的定居和存活的影响，实验验证和理论解析BSC对外来植物入侵的影响及其机理，从防止生物入侵的视角阐明BSC的存在对维护沙地生态系统健康的重要意义，全面认知BSC在沙地生态系统中的功能和地位。

正如前面所述，相对于其他生物气候带的陆地生态系统，干旱、半干旱沙地生态系统具有资源有效性低、限制新陈代谢以及限制资源获取（极端温度）等特点，从而受生物入侵的风险较小。除了区域气候和生境条件限制了外来物种的入侵外，是否与整个沙地生态系统的生物组成和结构有关？是哪些生物组成在限制或抑制入侵植物定居和生存中发挥着重要的作用？本研究通过研究沙坡头人工固沙植被区不同处理BSC上外来物种对本地物种萌发以及生长的影响，探讨BSC对外来植物入侵是否具有抑制作用，从而为干旱区生态系统管理提供科学依据。

沙生针茅（*Stipa glareosa*）根系发达，具有耐寒、耐旱、耐高温和分蘖能力，种子数量巨大且一年两次开花结实（李发明等，2013；刘克彪等，2015），这些特性基本符合典型入侵杂草的Baker特征（倪丽萍和郭水良，2005）。因此，本研究假设广泛分布于包兰铁路人工植被区以西20 km的红卫草原化荒漠的沙生针茅为入侵植物种，人工固沙植被区的优势物种小画眉草及其伴生种茵陈蒿为本地植物种。本地植物种子和入侵植物种子分别采集于腾格里沙漠东南缘人工固沙植被区（2013年8、9月）和红卫草原化荒漠区（2014年7月）。将采集的植物种子进行晾干，收集，存放于种子袋中备用。

藻类结皮采自腾格里沙漠东南缘沙坡头人工固沙植被区（2014年7月），采用PVC自制取样器以不破坏BSC结构的方式采集含BSC表层0～10 cm的土壤，采集前先用水湿润土壤以保证土壤结构的稳定；流沙样品采自沙丘顶部的流动带。采集的BSC分别均匀平摊在花盆中，盆中铺垫约5 cm厚的无种子沙土。将花盆置于温室中培养，定期向花盆中喷洒适量水分，促使土壤种子库种子萌发。定期观察种子萌发情况，持续6周无新幼苗出现则结束观测，除去萌发的幼苗，以去除BSC自身土壤种子库对试验的干扰。植物种子采回后经过晾晒、除杂、收集后分别存放于种子袋中备用，其中沙生针茅种子需在实验开始前对种子进行4周的低温（5 ℃）处理以打破内生休眠（黄振英，2003）。在实验开始前对藻类结皮进行部分破碎（干扰）处理，使每种BSC形成100%覆盖、50%覆盖和无BSC覆盖（流沙）三种处理。其中，50%BSC覆盖是将采集的完整BSC样品用人工破碎的方法，使藻类结皮盖度降低至50%左右。

试验在中国科学院沙坡头沙漠试验研究站进行（2014年8月28日至10月26日）。本试验是在温室条件下，以不同处理的BSC为土壤基质，设计乡土种及其与外来种4种组合：小画眉草（X）、小画眉草+针茅（XZ）、茵陈蒿（Y）、茵陈蒿+针茅（YZ）。每种组合共20粒，每种处理5个重复。用记号笔对每个花盆进行编号，每天19：00—21：00给花盆喷洒充足水分，以保证土壤湿润。每天观察并记录种子萌发数，植株个体数量、株高和盖度等每5天记录一次。持续8周无新幼苗出现则结束试验。待到植物停止生长后，将不同种类的植物分物种装在纸袋中带回实验室，测定植物地上部分单株生物量。

（1）藻类结皮对本地植物种子萌发率和生长的影响

随着藻类结皮盖度的降低，小画眉草和茵陈蒿的种子萌发率升高（图4-8）。小画眉草在100%BSC覆盖、50%BSC覆盖和无BSC覆盖下的萌发率分别为1%、21%和45%，方差分析结果表明，小画眉草在不同藻类结皮盖度下种子萌发率差异达到极显著水平（F=51.82，p=0）。茵陈蒿在三种BSC盖度下的萌发率分别为9%、11%和59%，多重比较结果显示，茵陈蒿种子萌发率在无BSC覆盖与100%BSC覆盖和50%BSC覆盖间的差异达到极显著水平，而100%BSC覆盖与50%BSC覆盖间的差异不显著。小画眉草在50%BSC覆盖的地上部分的单株生物量大于其他两种BSC盖度处理，差异显著性均达到极显著水平，100%BSC覆盖与50%BSC覆盖间的差异不显著。茵陈蒿的地上部分的单株生物量在三种藻类结皮盖度下没有显著差异。

（2）藻类结皮对外来植物种子萌发率和生长的影响

对于外来植物沙生针茅，种子萌发率在不同藻类结皮盖度下具有极显著差异（表4-5，F=16.708，p=0），萌发率从高到低依次为无BSC覆盖>50% BSC覆盖>100% BSC覆盖（图4-9），说明藻类结皮对沙生针茅种子的萌发具有抑制作用。而不同结皮覆盖和物种组合（分别与本地植物小画眉草和茵陈蒿组合）及其二维交互效应对沙生针茅单株生物量均没有显著影响。说明在水分充足的情况下，不同藻类结皮覆盖对外来植物沙生针茅的生长影响较小，本地植物与沙生针茅也没有在水分利用上形成竞争作用。

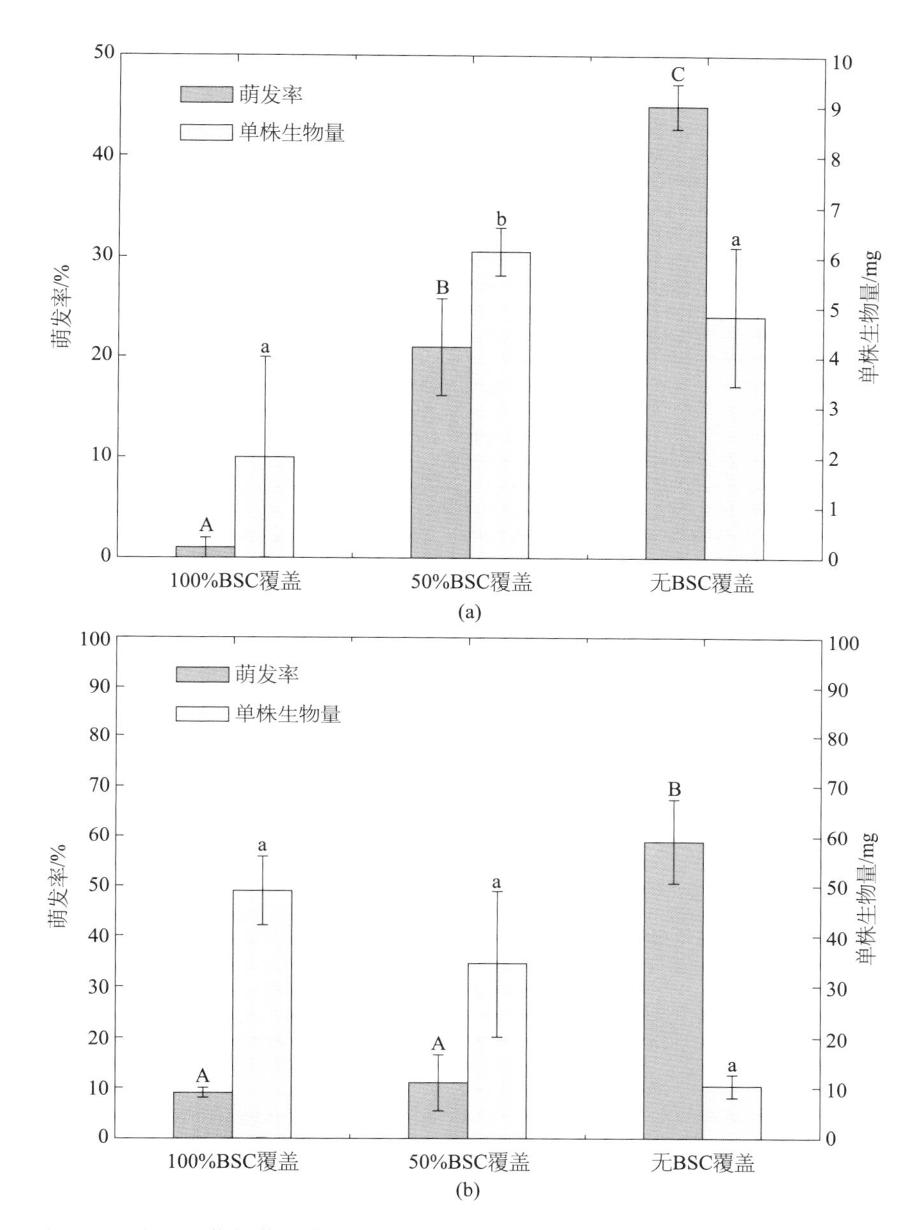

图 4-8　小画眉草与茵陈蒿在不同 BSC 盖度处理下的萌发率与单株生物量：(a) 小画眉草；(b) 茵陈蒿。不同大写字母表示在不同 BSC 盖度处理间萌发率差异显著（$p<0.05$），不同小写字母代表在不同 BSC 盖度处理间单株生物量差异显著（$p<0.05$）

Figure 4-8　The seed germination and individual biomass of *E. poaeoides* and *A. capillaries* in different coverage of algal crust: (a) *E. poaeoides*; (b) *A. capillaries*. Values with different capital letters show significant differences of seed germination among different algal crust coverage at 0.05 level, values with different lowercase letters show significant differences of individual biomass among different algal crust coverage at 0.05 level

表 4-5　藻类结皮盖度与物种组合对沙生针茅种子萌发率与单株生物量影响的双因素方差分析

Table 4-5　Two-way ANOVA analysis on germination rate（%）and individual biomass of *S. glareosa* in different coverage of algal crusts and species combinations

	变异来源	Ⅲ型平方和	自由度	均方	F	p
萌发率	校正模型	14077.500[a]	5	2815.500	7.827	0.000
	截距	50676.300	1	50676.300	140.878	0.000
	BSC 盖度	12020.600	2	6010.300	16.708	0.000
	物种组合	381.633	1	381.633	1.061	0.313
	BSC 盖度 × 物种组合	1675.267	2	837.633	2.329	0.119
	误差	8633.200	24	359.717		
	总计	73387.000	30			
	校正的总计	22710.700	29			
单株生物量	校正模型	8.015[b]	5	1.603	0.769	0.581
	截距	254.468	1	254.468	122.128	0.000
	BSC 盖度	4.093	2	2.047	0.982	0.389
	物种组合	0.031	1	0.031	0.015	0.904
	BSC 盖度 × 物种组合	3.890	2	1.945	0.934	0.407
	误差	50.007	24	2.084		
	总计	312.489	30			
	校正的总计	58.022	29			

注：a：R^2=0.541，b：R^2=0.138。

（3）外来植物对本地植物生长的影响

双因素方差分析结果显示（表4-6），小画眉草在不同BSC盖度间单株生物量差异显著（F=5.267，p=0.013），在不同物种组合间差异不显著（F=0.372，p=0.548），但BSC盖度与物种组合的二维交互效应对小画眉草单株生物量有显著影响（F=4.937，p=0.016）。对于茵陈蒿，BSC盖度与物种组合对其单株生物量均无显著影响（BSC盖度：F=2.417，p=0.111；物种组合：F=1.437，p=0.242），而两者的二维交互效应对茵陈蒿单株生物量影响极显著

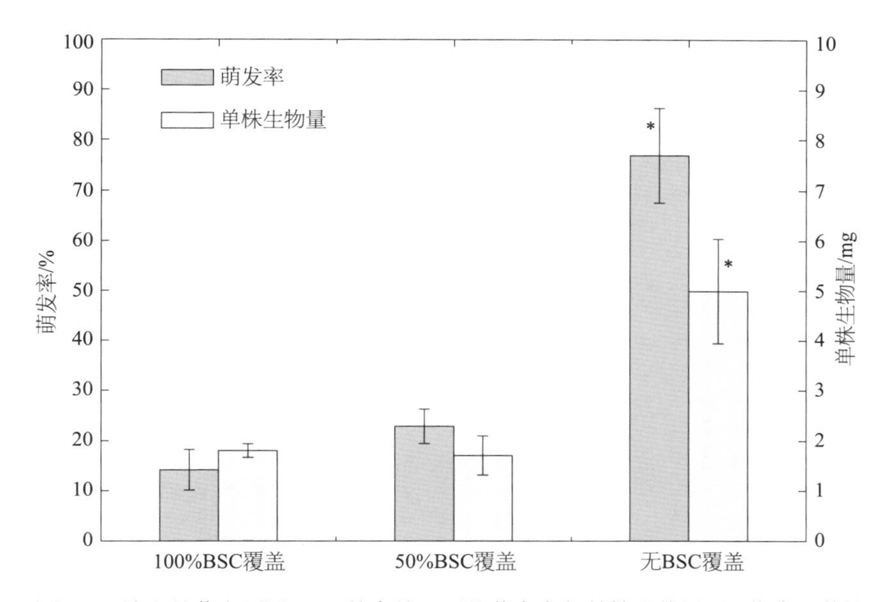

图 4-9　沙生针茅在不同 BSC 盖度处理下的萌发率与单株生物量（* 代表显著性差异）
Figure 4-9　The seed germination rate and individual biomass of *S. glareosa* for different BSC coverage treatments（* represent significant difference）

（*F*=10.060，*p*=0.001）。

本研究结果表明，小画眉草和茵陈蒿在三种藻类结皮盖度处理下的萌发率从高到低为无BSC覆盖>50%BSC覆盖>100%BSC覆盖。BSC对植物种子萌发的影响存在促进作用（苏延桂等，2007；Harper and Clair，1985b；Rivera-Aguilar *et al.*，2005）、抑制作用（Shem-Tov *et al.*，1999；Prasse and Bornkamm，2000）和因种而异（Zaady *et al.*，1997；Hawkes，2004）3种不同的观点。李新荣等（2009）认为产生争议的原因是这些研究分别在不同气候条件、不同土壤质地、不同BSC种的组成以及不同维管植物生活条件下完成的。聂华丽等（2009）研究认为这与种子和土壤间接触面积有关：BSC增加了地表粗糙度，减小了种子与土壤的接触面积，从而抑制植物种子萌发。外来植物沙生针茅在三种藻类结皮盖度下的萌发率均大于本地植物（图4-9和表4-7），说明沙生针茅具有极强的生态适应性和扩展能力，从而在新的栖息地保持较高的种群数量（王伯荪等，2005），而入侵植物在入侵区域的种群扩散是入侵植物对入侵生态系统造成危害的根本原因（张润志等，2004）。

表 4-6　藻类结皮盖度与物种组合对小画眉草和茵陈蒿单株生物量影响的双因素方差分析

Table 4-6　Two-way ANOVA analysis on individual biomass of *E. poaeoides* and *A. capillaris* in different coverage of algal crusts and species combinations

	变异来源	Ⅲ型平方和	自由度	均方	*F*	*p*
小画眉草	校正模型	83.369[a]	5	16.674	4.156	0.007
	截距	385.776	1	385.776	96.157	0.000
	BSC 盖度	42.262	2	21.131	5.267	0.013
	物种组合	1.493	1	1.493	0.372	0.548
	BSC 盖度 × 物种组合	39.615	2	19.807	4.937	0.016
	误差	96.287	24	4.012		
	总计	565.433	30			
	校正的总计	179.656	29			
茵陈蒿	校正模型	106.000[b]	5	21.200	5.278	0.002
	截距	248.131	1	248.131	61.779	0.000
	BSC 盖度	19.419	2	9.710	2.417	0.111
	物种组合	5.772	1	5.772	1.437	0.242
	BSC 盖度 × 物种组合	80.809	2	40.405	10.060	0.001
	误差	96.395	24	4.016		
	总计	450.526	30			
	校正的总计	202.395	29			

注：a：R^2=0.352，b：R^2=0.425。

外来植物种子萌发率在不同藻类结皮盖度处理下与本地植物的趋势一致，说明藻类结皮的存在能够抑制外来植物种子萌发。Larsen（1995）研究发现一年生外来草本植物早雀

表 4-7　沙生针茅在不同处理藻类结皮上的萌发率与单株生物量

Table 4-7　The seed germination rate and individual biomass of *S. glareosa* under different processing conditions

物种	土壤处理	萌发率 / %	单株生物量 /mg
小画眉草 + 针茅（XZ）	100%BSC 覆盖	14 ± 3.99	1.78 ± 0.14
	50%BSC 覆盖	23 ± 3.41	1.68 ± 0.39
	无 BSC 覆盖	76 ± 9.26	4.93 ± 1.03
茵陈蒿 + 针茅（YZ）	100%BSC 覆盖	32 ± 8.59	2.70 ± 0.66
	50%BSC 覆盖	40 ± 15.14	2.10 ± 0.80
	无 BSC 覆盖	62 ± 3.74	3.67 ± 0.77

麦的萌发和定居受到BSC的抑制，Crisp（1975）在澳大利亚的研究也得到相同的结果，即BSC对一年生入侵种*Schismus*属草本种子的萌发和定居具有抑制作用。这是因为在水分充足条件下，藻类结皮有机体会分泌一种抑制种子萌发的化学物质（deRivera *et al.*，2005），在沙坡头地区野外调查发现，藻类结皮表面光滑，植物种子在风的作用下很难在藻类结皮表面停留。因此，BSC成为外来植物入侵BSC覆盖地区的一道生物屏障。

外来植物的自身特性对其入侵成功具有极其重要的作用，例如对资源的竞争力强、对环境的适应性强以及具有较高的抗干扰性等（Walker and Smith，1997）。 沙生针茅在长期适应环境过程中形成耐干旱、抗沙埋等较强的适应能力，从而扩大其适生范围（黄振英等，1995，1997）。研究结果表明，沙生针茅在无BSC覆盖处理上的单株生物量最大（表4-7），说明藻类结皮对入侵植物的生长具有抑制作用。已有研究认为蓝藻在生长季节会分泌大量胞外多聚糖，该物质能够降低水分在土壤中的运动速率（Hu *et al.*，2000；Johansen，1993）。藻类结皮对水分的这种保持作用可能会降低水分对沙生针茅的有效性，使沙生针茅的单株生物量表现为无BSC覆盖高于有BSC覆盖。

藻类结皮对土壤水分的保持作用使沙生针茅与小画眉草和茵陈蒿间对水分的竞争加强。沙生针茅通过其较大的地下生物量配置（李发明等，2013），从土壤中获取较多水分，使小画眉草的可利用水分减少而导致生物量降低，致使小画眉草虽然在100%BSC覆盖的基质上萌发率低而生物量反而大。生物阻抗假说（Hu *et al.*，2000）认为，入侵生态系统中的多种生物因子和（或）生物过程都能够抵御外来植物的入侵，作为干旱区“生态系统工程师”，BSC可能会成为干旱区生态系统中的生物入侵阻力，在抵御外来植物入侵中起到积极作用。

入侵植物对本地植物的种子萌发以及生长过程都会产生一定影响（张天瑞等，2011），以达到其入侵到本地生态系统的目的。在干旱区，BSC的存在会对入侵植物种子萌发和生长产生抑制作用，对外来植物的入侵形成了一道天然的生物屏障，对人工固沙植被系统的稳定性起到保护作用。但入侵植物沙生针茅对本地植物以及BSC对入侵植物种子萌发和生长的作用机制尚不清楚，还需进行深入研究，具体技术线路如图4-10。

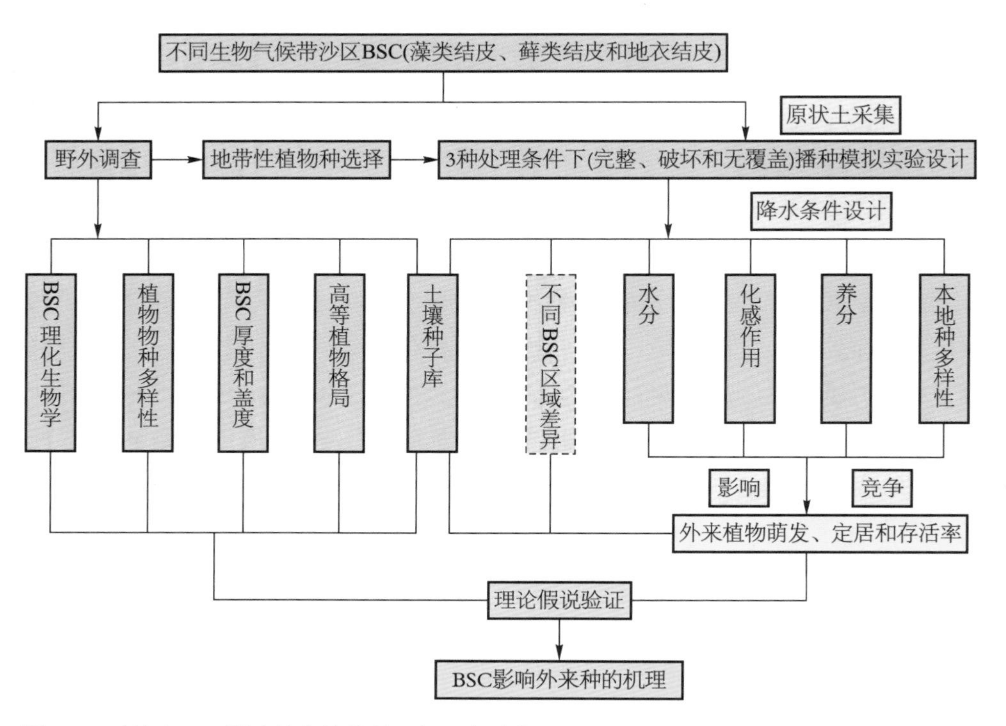

图 4-10　沙区 BSC 影响外来植物种入侵研究总体思路

Figure 4-10　A study frame on the mechanism of BSC preventing from biological invasion in sandy ecosystems

4.2 土壤微生物与BSC的相互作用

4.2.1 固沙植被区BSC中蓝藻群落的物种多样性

蓝藻是BSC中的先锋藻类，也是荒漠藻类中的主要类群，严格地讲它是细菌，与绿藻不同，非隐花植物，它在藻类结皮的形成及BSC行使生态功能中发挥重要作用。在固沙植被区蓝藻的多样性特征是如何变化的等相关信息目前知之甚少。我们采集了沙坡头天然固沙植被区（自然恢复N年，$N>100$）和人工固沙植被区（1956年始建）的BSC（采样时间2012年9月），对BSC中蓝藻进行了分析，这有助于了解长期封育对BSC中蓝藻多样性的影响过程，以及从微生物多样性恢复的角度阐述构建人工生态系统的可行性及其生态学意义。

在沙坡头铁路北1956年始建的固沙植被区（sptA）和翠柳沟天然植被区（sptN）取样，并以流沙区沙土作为对照（采样时间为2012年9月）。沙坡头地区BSC主要有蓝藻占优势的结皮（藻类结皮）、地衣占优势的结皮（地衣结皮）和藓占优势的结皮（藓类结皮）等三种类型。本研究重点关注的是天然植被区和人工植被区BSC中蓝藻群落的差异变化，故采样时在每个样地中根据BSC的三种主要类型，每种类型设6个样点，每个样点采用正方形对角线五点采样法采集BSC（0～5 cm）。每种类型单独混合后，再根据不同植被区的BSC盖度将同一植被区的BSC样品混合均匀后，取适量置于冰盒保存，尽快带回实验室，−20 ℃下保存，DNA提取在1个月内完成。

（1）天然植被区和人工植被区蓝藻16S rDNA文库多样性

由于从流沙区蓝藻16S rDNA文库中只获得了一个阳性克隆，因此未进行后续的分析。从沙坡头－翠柳沟天然植被区（sptN）和沙坡头铁路北1956年始建人工植被区（sptA）蓝藻16S rDNA文库中，分别随机筛选了117个和103个阳性克隆进行分析和测序。基因克隆文库的库容值C，其计算公式为$C=1-n1/N$，其中$n1$代表16S rRNA基因克隆文库中仅出现1次的OTU数，N代表克隆文库的库容。在sptA和sptN文库仅出现1次的OTU数分别为20个和27个，克隆文库的库容为103，因此sptA文库$C=0.81$，而sptN文库$C=0.74$，说明两个克隆文库已经涵盖了本地区的大多数蓝藻样本，符合采样要求，基本能反映本地区的蓝藻多样性特征。与天然植被区相比，人工植被区蓝藻文库除了Shannon-Weaver多样性指数略小，Simpson优势度指数接近外，Chao指数和ACE丰富度指数均明显小于天然植被区（表4-8），通过Libshuff软件分析了两区蓝藻群落结构的差异，设定$p<0.05$水平为差异显著，即临界阈值=0.05/2=0.025，分析结果见表4-9，差异结果均低于0.025，说明天然植被区和人工植被区蓝藻群落结构差异显著。

表 4-8　腾格里沙漠沙坡头 1956 年始建人工植被区（sptA）和沙坡头 – 翠柳沟天然植被区（sptN）蓝藻群落多样性指数

Table 4-8　Diversity index of cyanobacterial community in Shapotou artificial vegetation region （restored since 1956，sptA） and Shapotou-Cuiliugou natural vegetation region （sptN） of Tengger Desert

分析水平	群落	Chao 指数	ACE 丰富度指数	Shannon-Weaver 多样性指数	Simpson 优势度指数
0.03	sptA	46.00	58.43	3.10	0.06
0.03	sptN	71.11	97.97	3.22	0.07

表 4-9　腾格里沙漠沙坡头 1956 年始建人工植被区（sptA）和沙坡头 – 翠柳沟天然植被区（sptN）蓝藻群落结构 Libshuff 分析（设定显著性水平为 $p<0.05$）

Table 4-9　Libshuff analysis of cyanobacterial community in Shapotou artificial vegetation region（restored since 1956，sptA）and Shapotou-Cuiliugou natural vegetation region （sptN） of Tengger Desert（with a significant level of $p<0.05$）

比较	dCXY Score	p
sptA — sptN	0.00080106	0.0120
sptN — sptA	0.00176189	0.0005

（2）BSC蓝藻16S rDNA克隆文库序列

把同源性不低于97%的序列定义为同一OTU，两个文库共获得69个OTU（17个OTU是两区共有的），人工植被区获得38个OTU，天然植被区获得48个OTU，代表性序列经过NCBI的BLAST程序比对并进行了注释。定义OTU所占比例大于3%的为优势种（武传东等，2011），分类群OTU3、OTU18、OTU22和OTU13在腾格里沙漠沙坡头地区分别占20.45%、10.91%、7.73%和3.18%，为该地区优势蓝藻群。在人工植被区，分类群OTU3、OTU18、OTU22、OTU50和OTU39分别占18.45%、14.56%、9.71%、4.85%和3.88%，为该区优势蓝藻群，而天然植被区优势种群却有比较明显的不同，除了占据前三位的分类群OTU3、OTU18和OTU22分别占22.22%、7.69%和5.98%外，其余优势分类群OTU14、OTU13和

OTU15分别达到了4.27%、3.42%和3.42%。经过NCBI比对分析，发现其中等位基因百分比含量（某物种分属各个OTU的序列数之和与总的克隆文库序列数目的比值）最高的为暗绿颤藻（*Oscillatoria nigro-viridis*），占20.45%，其次为斯氏微鞘藻（*Microcoleus steenstrupii*），占12.27%。

（3）基于蓝藻16S rDNA序列的系统发育

根据腾格里沙漠沙坡头固沙植被区所获得的蓝藻16S rDNA序列构建蓝藻系统发育树（图4-11），表明该地区蓝藻可分为五大类：颤藻目（Oscillatoriales）64.55%（sptA和sptN分别占35%和29.55%），色球藻目（Chroococcales）1.82%（sptA和sptN分别占0和1.82%），念珠藻目（Nostocales）16.36%（sptA和sptN分别占6.81%和9.55%），宽球藻目（Pleurocapsales）6.82%（sptA和sptN分别占1.82%和5%）和未分类蓝藻（unclassified）10.45%（sptA和sptN分别占3.18%和7.27%）。颤藻目中按照所占总文库的比例大小分为颤藻属（*Oscillatoria*）22.73%、瘦鞘丝藻属（*Leptolyngbya*）18.64%、微鞘藻属（*Microcoleus*）14.10%、席藻属（*Phormidium*）4.09%、伪鱼腥藻科（Pseudanabaenaceae）2.27%、颤藻目未知属种1.81%以及常丝藻属（*Tychonema*）0.91%。宽球藻目和色球藻目分别仅有一属，即拟色球藻属（*Chroococcidiopsis*）6.82%和隐球藻属（*Aphanocapsa*）1.82%。念珠藻目按照所占总文库的比例大小分为伪枝藻属（*Scytonema*）7.74%、未知属种5.91%、念珠藻属（*Nostoc*）1.81%和单岐藻属（*Tolypothrix*）0.90%。未分类蓝藻（unclassified）共23个克隆，占总文库的10.45%。从图4-11系统发育树结果可以发现，微鞘藻属OTU占总OTU比例最高，说明该属在所有的属中贡献了最高的蓝藻群落多样性。对人工植被区和天然植被区蓝藻相对数量的比较分析发现，颤藻目总体呈下降趋势，而宽球藻目、色球藻目、念珠藻目和未分类蓝藻都呈上升趋势（表4-10）。

OTU14经NCBI上BLAST比对与颤藻目发毛针藻属蓝藻（*Crinalium epipsammum*，序列号NR_102461）相似度达到99%，但是Oren（2004）认为该命名属于无效命名，从系统发育树上看，该类群与颤藻目其他属亲缘关系较远，而是与未分类蓝藻归类到一个分支：unclassified *Cyanobacterium* Ⅰ genera（图4-12），也直接支持了上文所述，因此将它划分到未分类蓝藻。

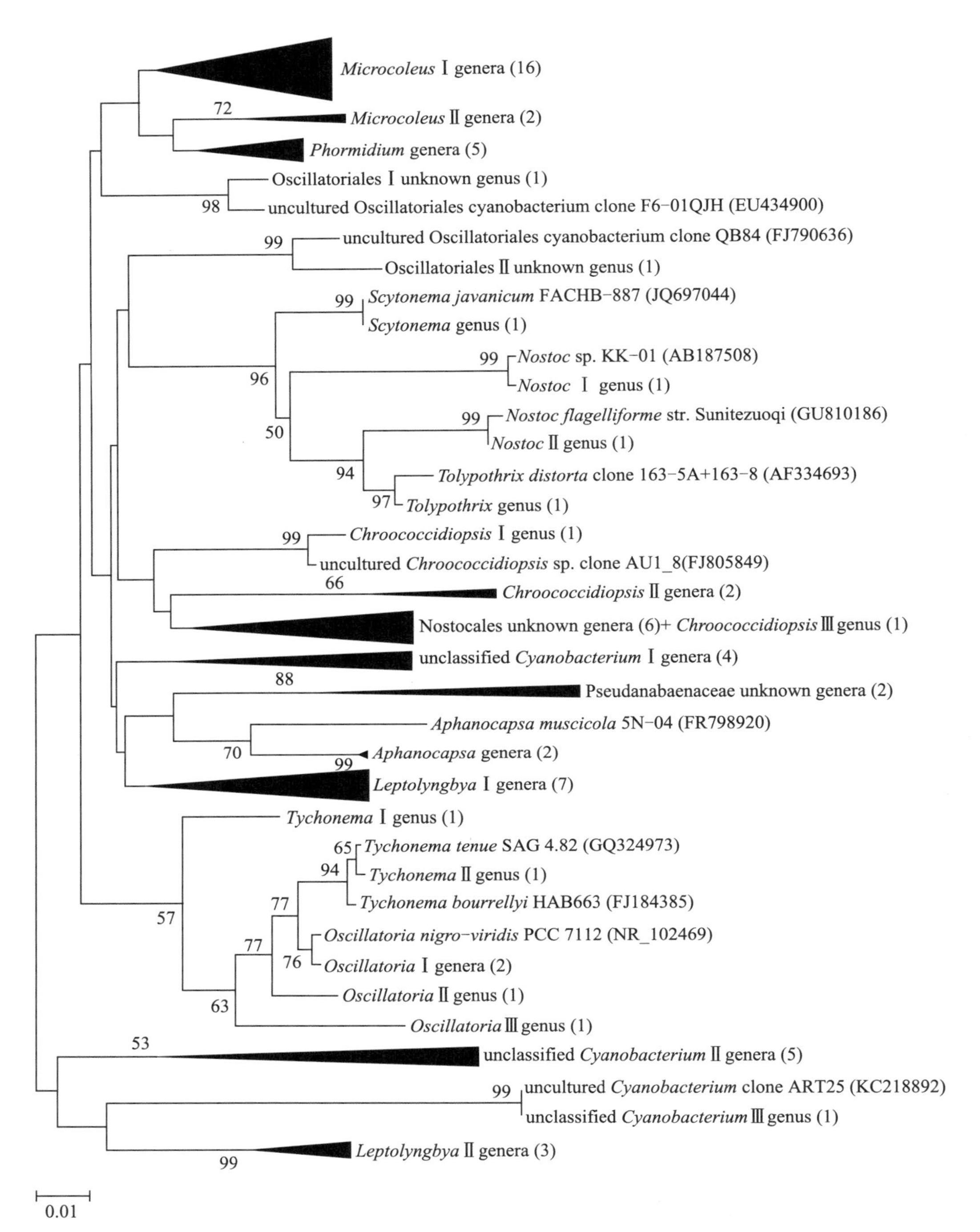

图 4-11 腾格里沙漠沙坡头地区蓝藻系统发育树。按照 Neighbor-joining 算法，自展次数设定为 1000 次，自展检验结果低于 50 的未显示

Figure 4-11 Phylogenetic tree of cyanobacteria in Shapotou region of Tengger Desert. Constructed according to the Neighbor-joining algorithm. The times of bootstrap is set up to 1000 and values lower than 50 are not shown

表 4-10　腾格里沙漠沙坡头 1956 年始建人工植被区（sptA）和沙波头 – 翠柳沟天然植被区（sptN）蓝藻相对数量变化

Table 4-10　Shifts of the relative amount of cyanobacteria in Shapotou artificial vegetation region （restored since 1956，sptA） and Shapotou-Cuiliugou natural vegetation region （sptN）

蓝藻	克隆所占总文库百分比 / %			变化趋势（sptN 相对于 sptA）
	sptA	sptN	合计	
颤藻目	35	29.55	64.55	–
颤藻属	9.55	13.18	22.73	+
瘦鞘丝藻属	11.82	6.82	18.64	–
微鞘藻属	9.55	4.55	14.10	–
席藻属	2.27	1.82	4.09	–
伪鱼腥藻科	1.36	0.91	2.27	–
颤藻目未知属	0.45	1.36	1.81	+
常丝藻属	0	0.91	0.91	+
色球藻目	0	1.82	1.82	+
隐球藻属	0	1.82	1.82	+
宽球藻目	1.82	5	6.82	+
拟色球藻属	1.82	5	6.82	+
念珠藻目	6.81	9.55	16.36	+
伪枝藻属	4.55	3.19	7.74	–
念珠藻目未知属	1.36	4.55	5.91	+
念珠藻属	0.45	1.36	1.81	+
单岐藻属	0.45	0.45	0.90	0
未分类蓝藻	3.18	7.27	10.45	+

注：+：增加；–：降低；0：不变。

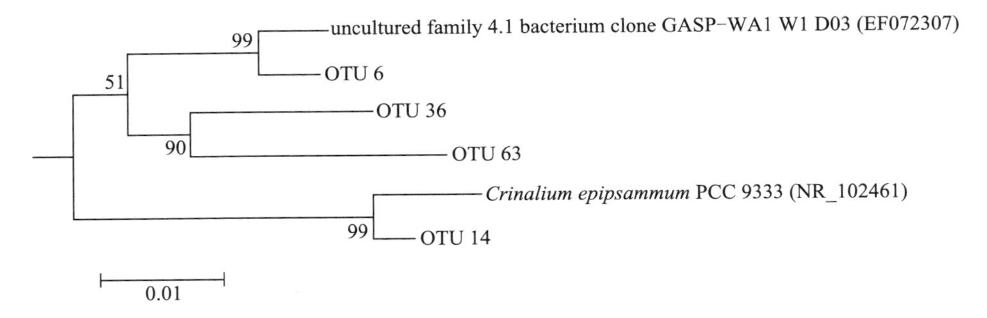

图 4-12　未分类蓝藻属 I 分支
Figure 4-12　Subtree of unclassified Cyanobacteria I genus

构建系统发育树时发现，宽球藻目拟色球藻属（Caudales *et al.*，2000）的一个克隆OTU1与拟色球藻属其他种（NCBI序列号FJ805849和AJ344552）亲缘关系相对较远，而与念珠藻目未知属种归类到一个分支（图4-13），这可能是拟色球藻属一些种和念珠藻属一样具有固氮功能，长期共同进化过程中发生基因横向转移所致，需要进一步验证。

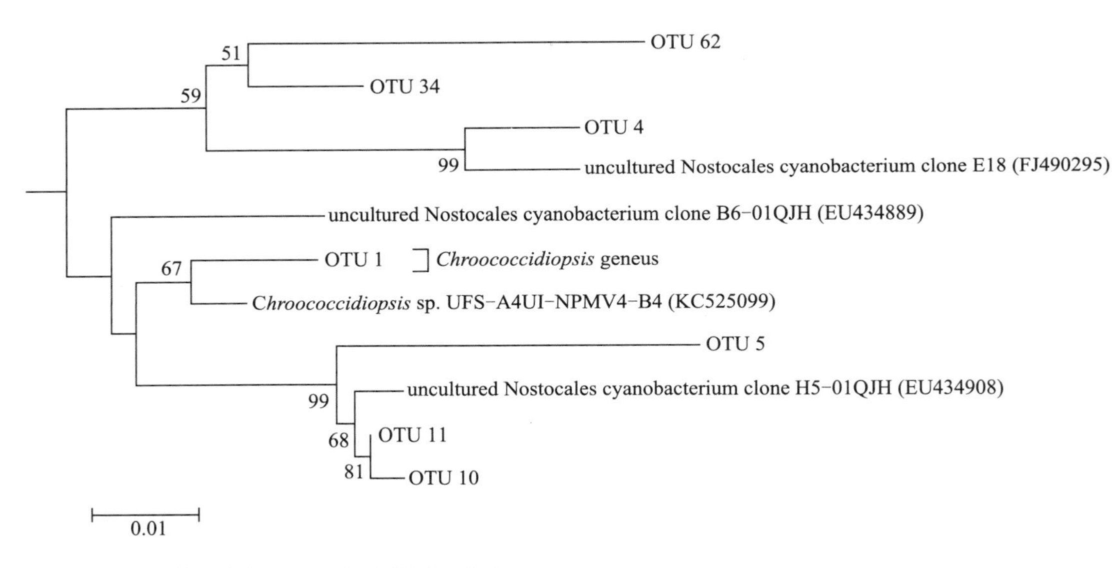

图 4-13　念珠藻目未知属和拟色球藻属Ⅲ分支
Figure 4-13　Subtree of Nostocales unknown genus and *Chroococcidiopsis* Ⅲ genus

在干旱半干旱地区，BSC大多起于蓝藻，而且往往是以蓝藻为优势种（Belnap *et al.*，2001）。BSC中的蓝藻可以增强土壤稳定性和土壤持水性，保护土壤免受侵蚀，而且其中的固氮蓝藻增加了土壤氮库量（Brotherson and Rushford，1983；Harper and Marple，1988；Belnap and Gillette，1998）。研究蓝藻的多样性通常有两种方法：① 表型多样性分析，即通过培养法进行形态鉴定。② 基因型多样性分析，通过PCR扩增、克隆和DNA测序分析完成。国内对于BSC中蓝藻的鉴定主要基于其表型的分析（胡春香等，1999；张丙昌等，2005），这种传统的鉴定常常依赖于已知物种的形态特征，或者直接进行显微镜下观察或经过分离培养后显微镜下观察，前者受制于微生物的生存环境，导致形态特征区别不明显，尤其难以准确鉴定近似物种，给分类和系统发育的研究带来不确定性。而分离培养方法会改变微生物群落原有的结构，快速生长的菌株会抑制其他菌株的生长，而且培养条件的非通用性会造成有些蓝藻丢失，从而形成并不完全相同的新的群落结构。而且实验室可培养的微生物有限，通常低于1%（Whitman *et al.*，1998），因此难以准确重现原有的物种多样性特征。尽管采用分子生物学方法也存在一些缺陷，例如一些具厚鞘的丝状蓝藻或具有异形胞的蓝藻难以破壁，对其DNA提取效率低，与直接镜检结果差别比较大，分子生物学方法很可能低估了具鞘蓝藻的丰度（Gundlapally and Garcia-Pichel，2006）；但总的来说，分子生物学方法已经成为蓝藻研究最重要的手段之一（庄丽和陈月琴，1999；Olsson-Francis *et al.*，2010；刘梅等，2011）。

本试验发现丝状蓝藻暗绿颤藻（*Oscillatoria nigro-viridis*）（序列百分比为20.45%）和斯氏微鞘藻（*Microcoleus steenstrupii*）（序列百分比为12.27%）为沙坡头地区的优势蓝藻，这与早期显微镜下野外样品未培养直接鉴定的结果差别较大（胡春香等，1999）（其多以具鞘微鞘藻为优势种，小席藻为主要种）。最近的研究结果认为蓝藻在温性荒漠与寒性荒漠以及热带荒漠中存在不同的分布格局，具鞘微鞘藻主要是寒性荒漠中的蓝藻优势种，而斯氏微鞘藻主要是温性荒漠中的蓝藻优势种（Garcia-Pichel *et al.*，2013），腾格里沙漠是温性沙漠，本研究结果与之基本符合。另外，不同地区BSC中蓝藻群落的构成差别很大，可能与当地自然环境及土质密切相关，如古尔班通古特沙漠BSC中通常以具鞘微鞘藻为优势种（Zhang *et al.*，2011）；美国科罗拉多高原BSC中优势蓝藻为具鞘微鞘藻和斯氏微鞘藻（Gundlapally and Garcia-Pichel，2006；Garcia-Pichel *et al.*，2001）；斯氏微鞘藻则是索诺拉沙漠BSC中的优势蓝藻（Nagy *et al.*，2005）；印度的塔尔沙漠BSC中以席藻属（*Phormidium*）、颤藻属（*Oscillatoria*）和鱼腥藻属（*Anabaena*）为优势种（Bhatnagar *et al.*，2008）；以色列西部内盖夫沙漠BSC以伪枝藻属（*Scytonema*）为主，依次分别为微鞘藻属（*Microcoleus*）、裂须藻属（*Schizothrix*）和念珠藻属（*Nostoc*）（Danin，1991）。

以蓝藻16S rDNA序列分析的非培养技术和统计学分析相结合为基础的研究表明沙坡头地区存在着较为丰富的蓝藻类群，但是，尽管人工固沙植被区（1956年始建）恢复了56年，其蓝藻Chao指数和ACE丰富度指数均明显小于天然植被区（恢复了*N*年）。与恢复了56年的植被区BSC中蓝藻相比，天然植被区颤藻目蓝藻数目减少，但其中颤藻属和未知属却呈增加趋势，天然区还发现常丝藻属和色球藻目蓝藻，宽球藻目和念珠藻目蓝藻也相对较多，在总的克隆文库中，未确定分类的蓝藻在天然植被区中的比例是恢复了56年的人工植被区的2倍多（表4-10），而且Libshuff分析发现人工植被区蓝藻群落结构与天然植被区仍然存在着统计学的显著差异。这与Hu和Liu（2003）的研究结果并不完全相同，他们采用显微镜直接观察的方法对BSC演替过程中蓝藻群落进行了研究，认为随着演替时间的增加，群落多样性逐渐增加，念珠藻目的爪哇伪枝藻（*Scytonema javanicum*）比例逐渐减少，而纤细席藻（*Phormidium tenue*）呈增加趋势。但是较为一致的结论是蓝藻多样性随着恢复年限的延长而增加，而且荒漠区生态系统微生物多样性的恢复是一个长期的过程。因此，本研究可以为我们今后科学评价人工治理荒漠措施的合理性提供理论参考依据，对完善荒漠区生态恢复理论和指导荒漠区生态修复实践具有重要意义。

4.2.2 腾格里沙漠BSC中可培养小型真菌群落的研究

与异养细菌、蓝藻细菌、绿藻、地衣和藓类结合在一起的非寄生小型真菌，在BSC的组成和功能中起着重要作用。然而，对于BSC中真菌多样性尚缺乏研究。

在美国荒漠和半干旱草原，通过培养法和非培养分子法做了大量研究（States and Christensen，2001；States *et al.*，2001），揭示了BSC中的真菌主要属于子囊菌门。从阿拉伯沙漠的BSC中获得了不同的真菌区系（Abed *et al.*，2013）。目前的研究主要集中在真菌群落的优势种和常见种以及不同BSC种类和地理区域中真菌群落特性相关内容。

在以色列的内盖夫沙漠西侧Nizzana研究点（NRS），研究了随着降水梯度（Grishkan *et al.*，2006） 不同 BSC 种类覆盖的沙丘链（Grishkan and Kidron，2013） 中的真菌区系。研究发现，该地区BSC小型真菌群落与内盖夫沙漠其他地区相似，都是以具有大型多细胞孢子的产黑色素真菌为优势种。由北向南沿着降水梯度，沙丘链中存在更多的耐旱BSC，耐旱种群的丰度增加。分离出的小型真菌菌株密度与叶绿素含量呈正相关关系，表明干旱程度可能显著影响真菌生物量的有机物含量和湿润持续时间（Grishkan and Kidron，2013）。以藓类为优势的BSC在物种相对丰度、物种多样性水平、菌株密度等水平上与蓝藻为优势的BSC显著不同。BSC小型真菌群落的变化在Nizzana研究点的沙丘链和沿着降水梯度变化的内盖夫沙漠相似，这说明小气候差异和区域气候变化对小型真菌的影响很可能具有同等效力。

我们的研究通过分析物种的组成、主要分类种群和生态种群对群落结构贡献、优势种群、菌株密度和物种多样性水平（物种丰富度、异质性和均匀度）的影响，探明腾格里沙漠不同种类BSC中小型真菌的组成和结构，同时根据早期对反映BSC生物量的参数（叶绿素含量和表面湿润持续时间）的分析，评估了以上真菌参数和叶绿素含量的关系（Kidron *et al.*, 2009）。

在沙坡头沙漠试验研究站附近采集BSC样品（2011年7月和2013年8月）。按照植被建立年代，分别对1956年、1964年、1981年和1987年建立的4个人工植被区的BSC进行采样。采集4个不同种类BSC，分别为藻类结皮、混生结皮（地衣、绿藻、蓝藻和藓类混生）和以藓类为主的两种BSC（真藓和土生对齿藓，以藓类结皮-1和藓类结皮-2表示）。在1987年建立的人工植被区，仅两种BSC被采集，表4-11为主要的BSC属性。天然植被区（位于红卫附近，2011年采集样品）BSC和一个未被固定且无BSC覆盖的沙丘（裸地）阳坡（SFS）和阴坡（NFS）样品作为对照。

表4-11 腾格里沙漠1956年人工植被区BSC的选择特性

Table 4-11 Selected properties of the BSC from artificial vegetation region with revegetation in 1956 in the Tengger Desert, China

结皮类型	厚度/mm	pH	有机碳/($g\ kg^{-1}$)	Chl-a、Chl-b/($mg\ m^{-2}$)	光合自养物种		
					蓝藻	地衣	苔藓
藻类结皮	2～4	7.42±0.26	0.6±0.2	40.4±5.4	MIC，PHO，NOS，OSC，SCH		
混生结皮	5～10	7.41±0.39	1.0±0.5	74.4±14.0	MIC，PHO，NOS，OSC，SCH	COL，END	BRY
藓类结皮-1	10～15	7.93±0.05	13.2±0.5	108.5±21.5	MIC，PHO，NOS，OSC，SCH		BRY，TORb，RORd
藓类结皮-2	15～25	7.99±0.10	13.4±0.8	153.6±44.2	MIC，PHO，NOS，OSC，SCH		DID，BAR，BRY

注：MIC：具鞘微鞘藻；PHO：席藻；NOS：念珠藻属；OSC：颤藻；SCH：*Schizothrix rupicola*；COL：坚韧胶衣；END：石果衣；BRY：真藓；TORb：双齿墙藓；RORd：刺叶墙藓；DID：土生对齿藓；BAR：*Barula ditrichoides*。

根据BSC厚度（表4-11）采集BSC（2011年）。按照BSC表层（0.0～0.4 cm）和亚表土层（0.4～0.8 cm）收集样品（2013年），藻类结皮仅采集0.0～0.5 cm表层样品。每个位点的每种BSC采集3～5个样品，采样后的2～6周，在以色列海法大学进化研究所实验室内进行样品分析。

（1）小型真菌群落的组成和物种多样性

从2011年和2013年采集的BSC样品中分别分离得到123个和67个菌种。134个菌种被分离，共6个亚门、22个来自有性型的子囊菌门和106个来自无性型的子囊菌门。这些菌种属于66个属，大多数物种属曲霉菌属（12种）、青霉菌属（8种）、毛壳菌属（8种）、茎点霉属（7种）。8个类型的小型真菌菌株在培养过程中无孢子产生。一些常见菌种的照片详见图4-14。

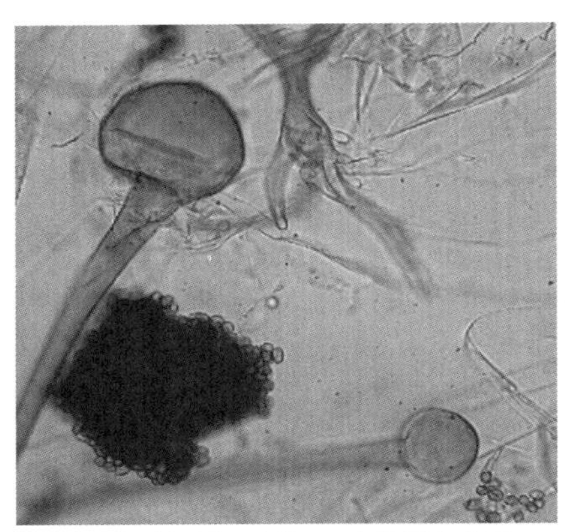

匍枝根霉(黑根霉)，假根、孢囊梗和孢子

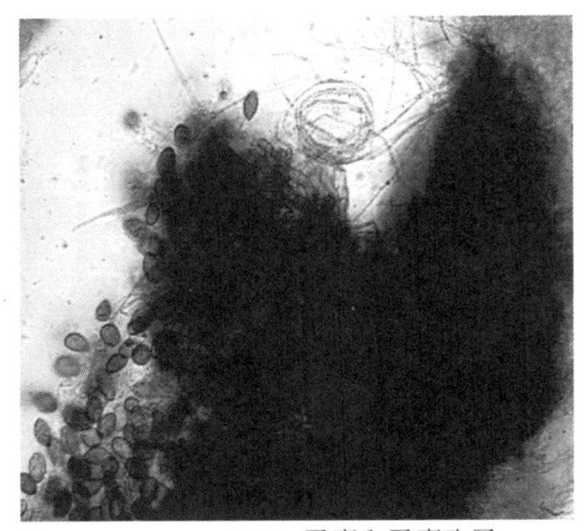

Canaryomyces notabilis 子囊和子囊孢子

球毛壳菌，子囊壳和子囊孢子

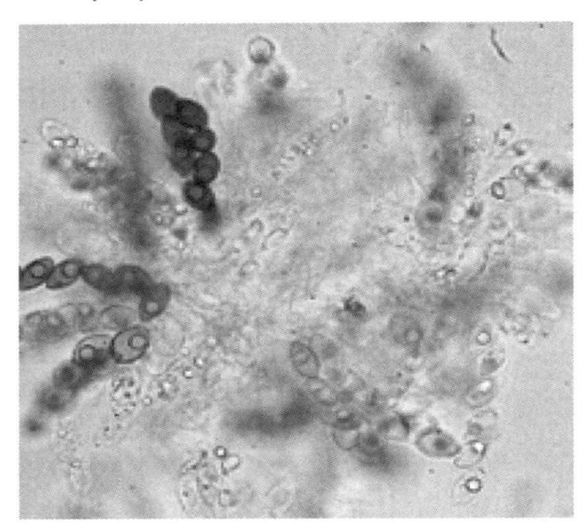

瘤突毛壳，子囊壳和子囊孢子

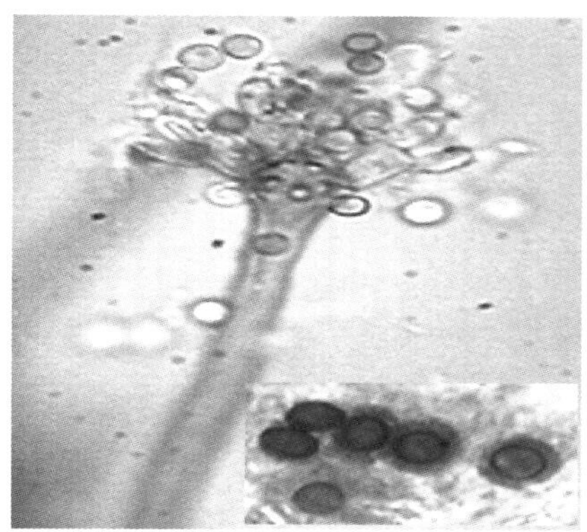

构巢裸孢壳，分生孢子和子囊孢子

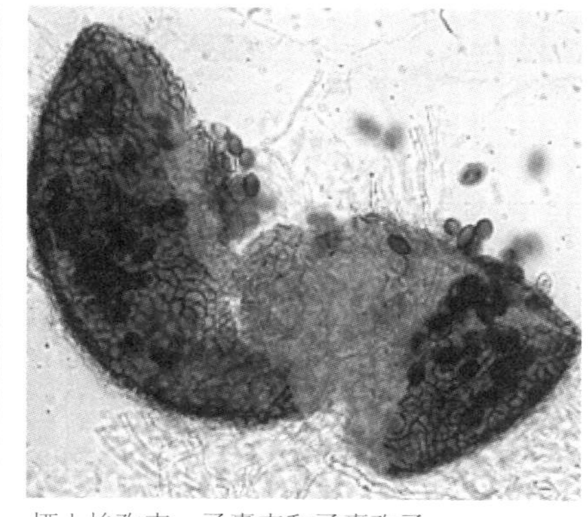

栖土梭孢壳，子囊壳和子囊孢子

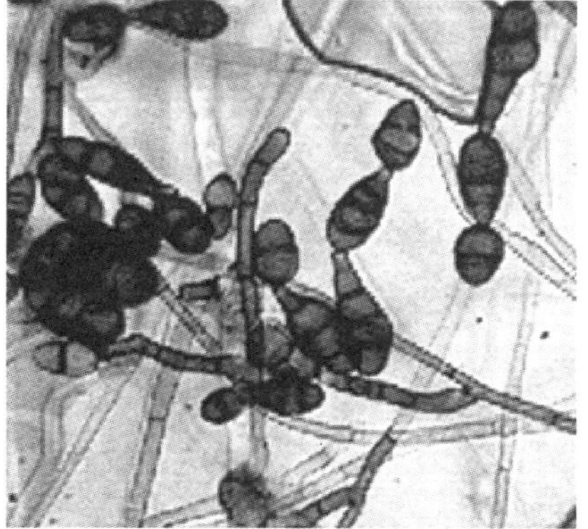

烟草赤星病菌，分生孢子梗和分生孢子

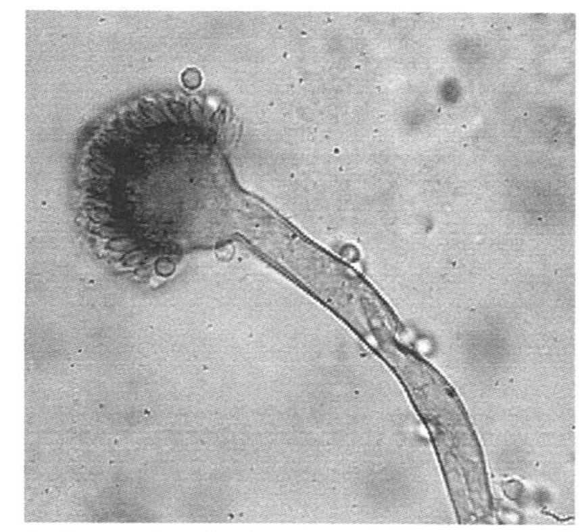

烟曲霉，分生孢子

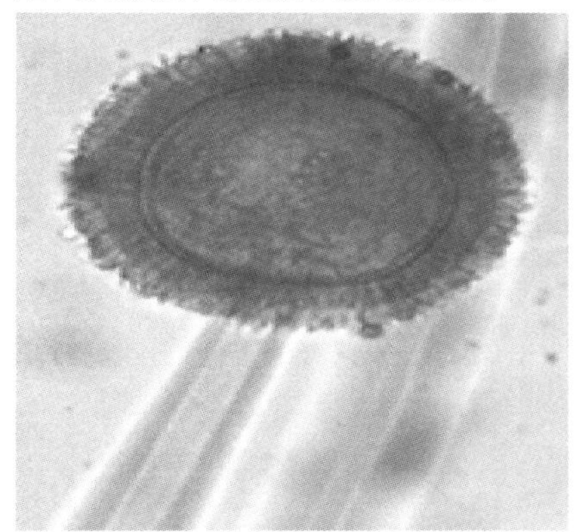

黑曲霉，分生孢子

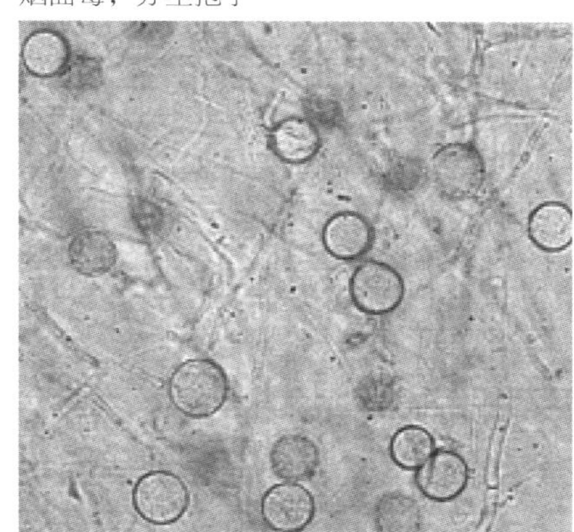

被毛枝葡萄孢，分生孢子

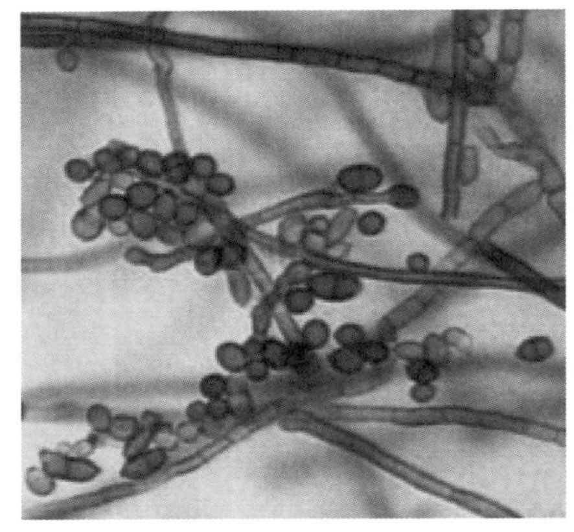

枝孢样枝孢霉，分生孢子梗和分生孢子

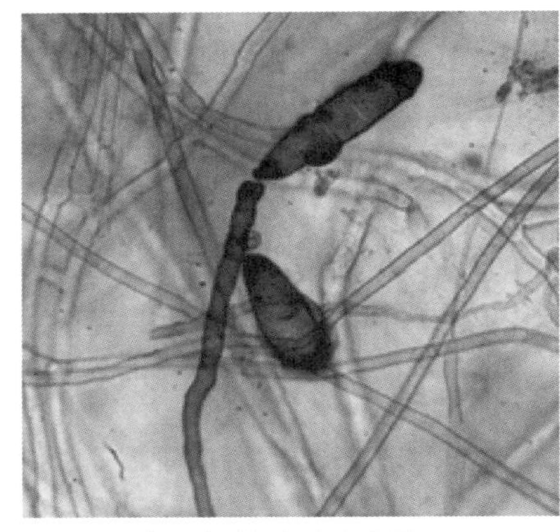

德斯霉，分生孢子梗和分生孢子

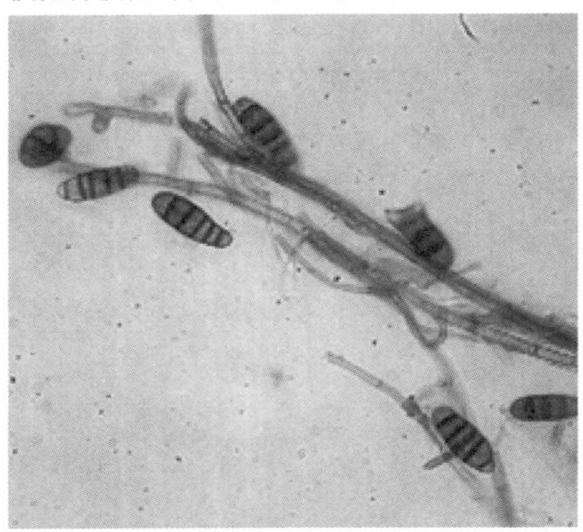

多孢埃里格孢菌，分生孢子梗和分生孢子

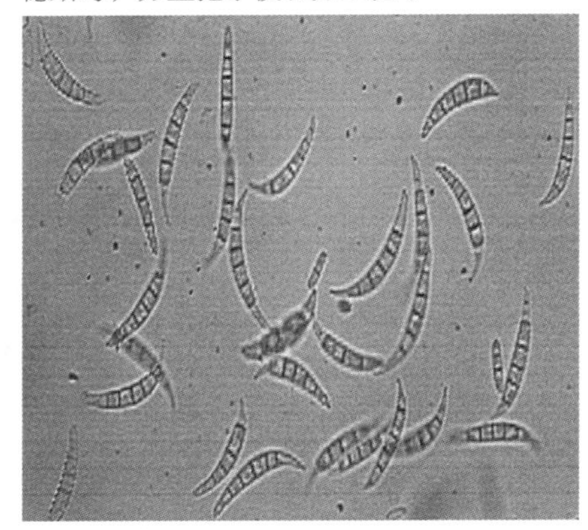

木贼镰刀菌，分生孢子

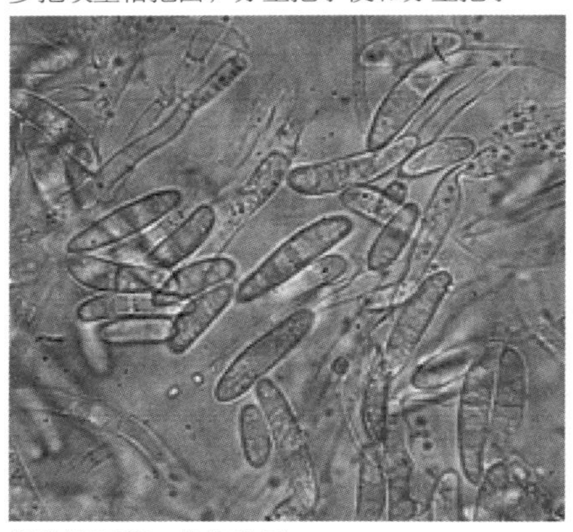

尖孢镰刀菌，分生孢子

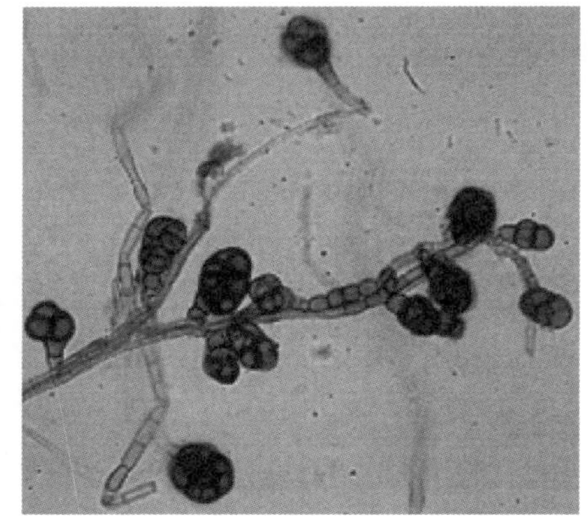

Monodyctis fluctuata 分生孢子梗和分生孢子

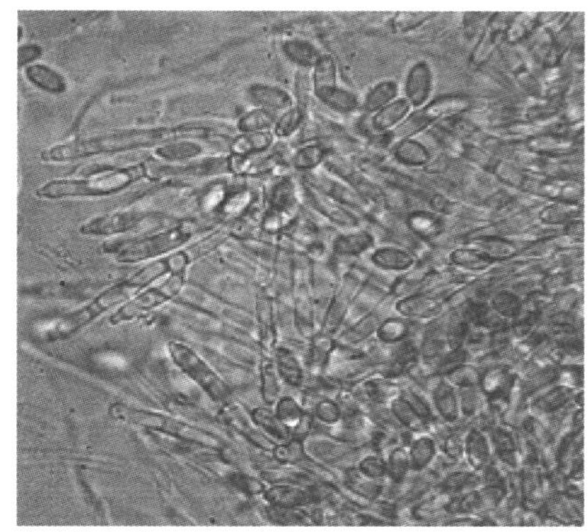
疣孢漆斑菌，分生孢子

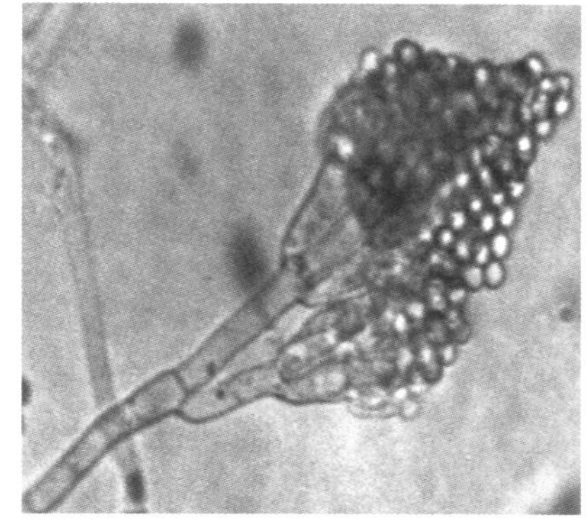
黄灰青霉，分生孢子

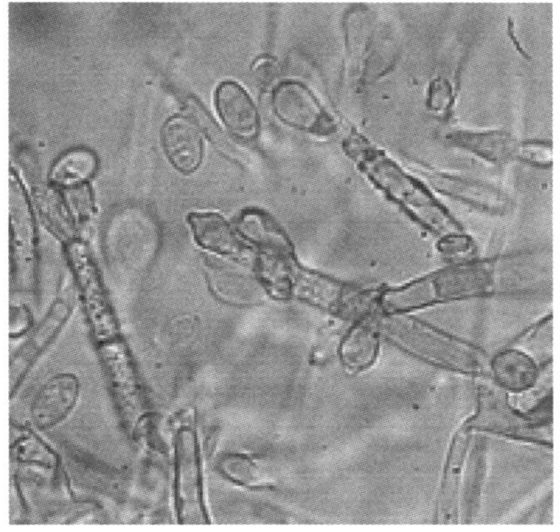
纸葡萄穗霉，分生孢子

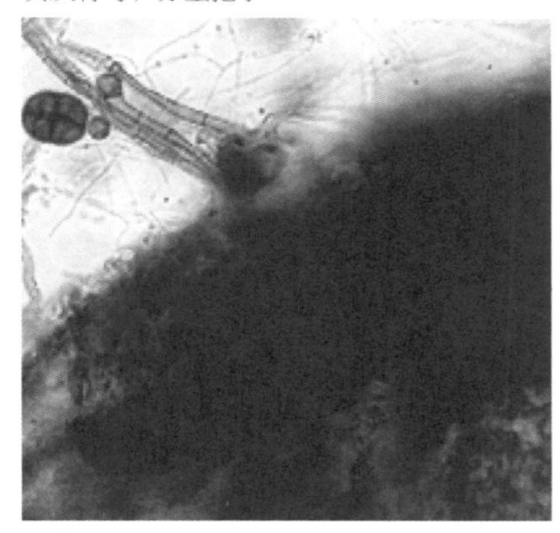
迟熟格孢腔菌，分生孢子梗和分生孢子

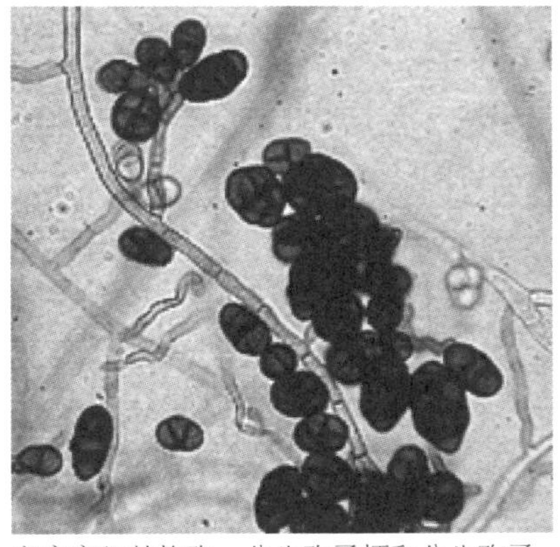
密疣突细基格孢，分生孢子梗和分生孢子

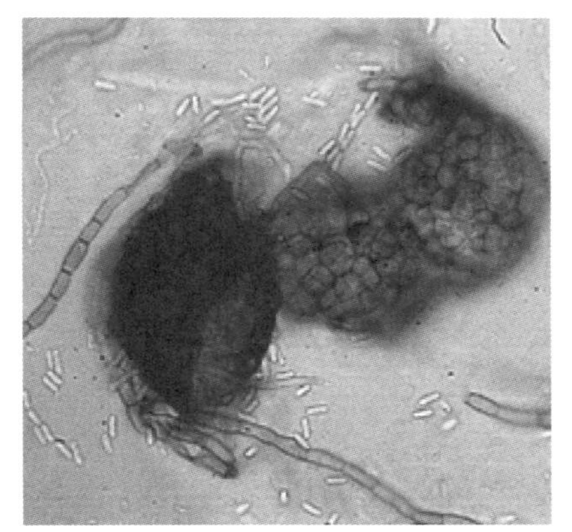
鞘茎点霉，分生孢子器和分生孢子晕圈

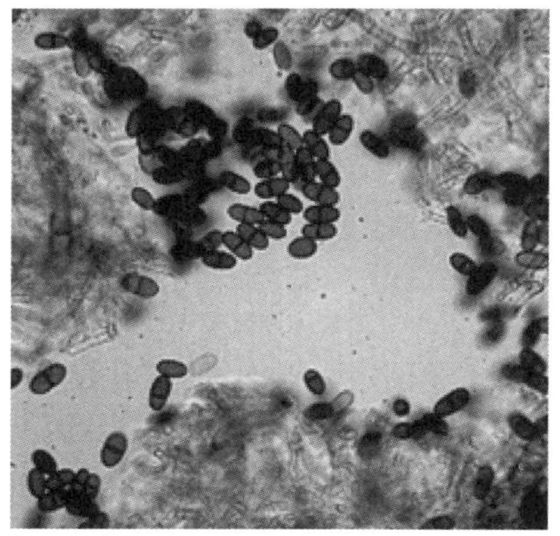
色二孢菌，分生孢子

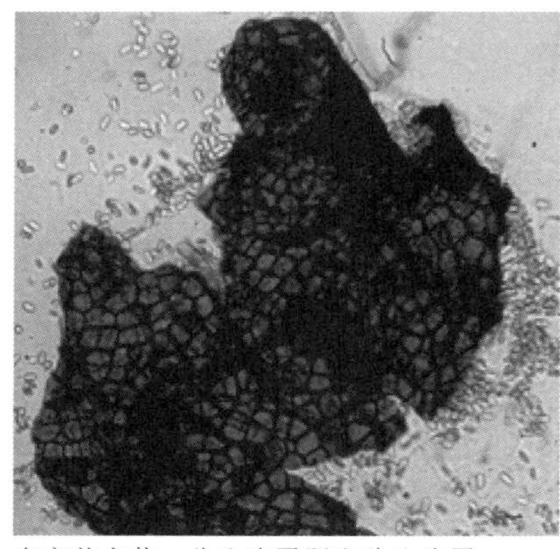
多变茎点菌，分生孢子器和分生孢子

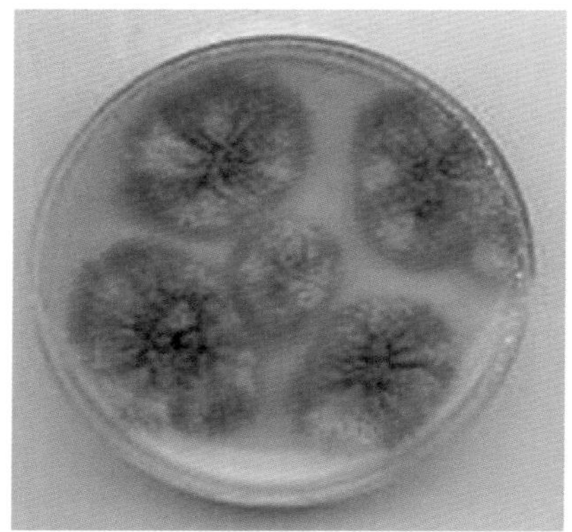
构巢裸孢壳

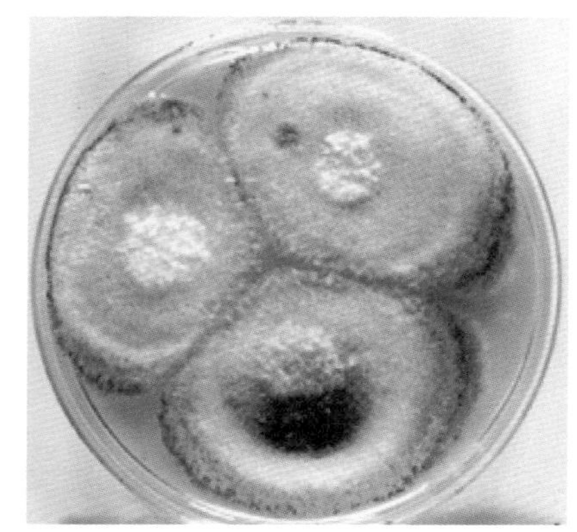
烟曲霉

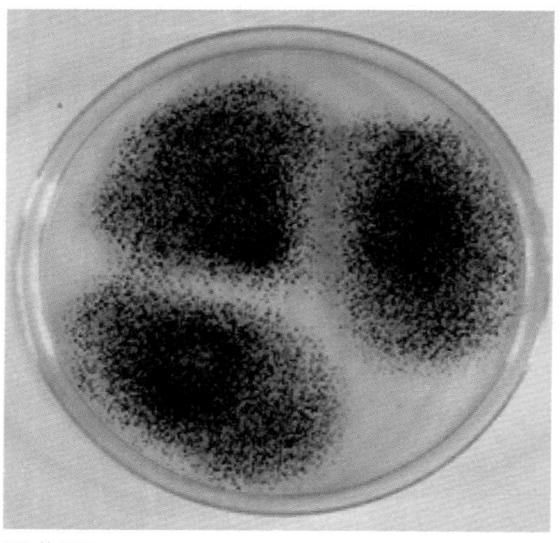
黑曲霉

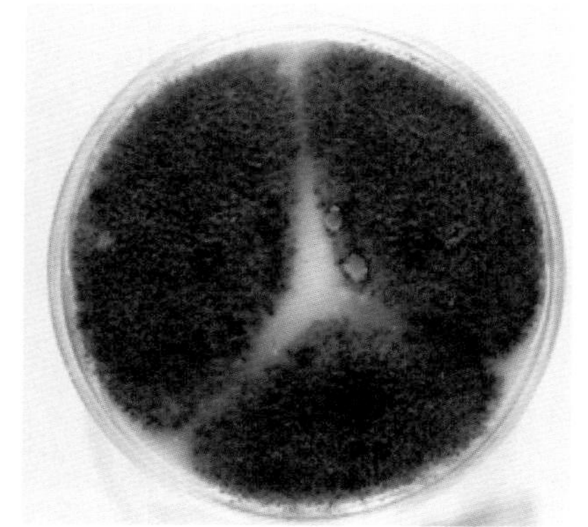
垣孢埃里砖格孢

黄灰青霉

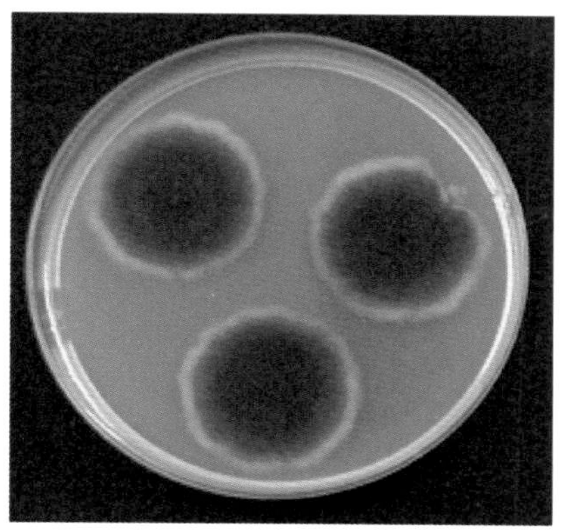

密疣突细基格孢

图 4-14　腾格里沙漠 BSC 群落中常见的真菌

Figure 4-14　The common fungus species in BSC communities in the Tengger Desert, China

我们对2011年采集的1956年、1964年、1981年、1987年人工植被区和天然植被区的BSC样品进行分离，分别得到72、69、80、58和76个菌种。人工植被区的藻类结皮群落具有低的Shannon指数，而土生对齿藓结皮含有大多数小型真菌群落，但两者差异不显著（表4-12）。相反，在天然植被区以藓类为优势的BSC群落的Shannon指数和均匀度指数均为最低。在所有研究群落中，裸地中小型真菌群落多样性最高（表4-12）。随着1956年到1987年不同年代序列的变化，藻类结皮和混生结皮物种多样性指数呈增加趋势。对所有类型BSC而言，1956年的BSC中小型真菌群落具有最低的异质性和稳定性。

表 4-12　2011 年采集的腾格里沙漠 1956 年、1964 年、1981 年和 1987 年人工植被区 BSC、天然植被区 BSC 及裸地阳坡（SFS）和阴坡（NFS）沙丘的小型真菌群落多样性

Table 4-12　Diversity characteristics of microfungal communities in the BSC sampled in 2011 from artificial vegetation region with revegetation in 1956, 1964, 1981 and 1987, in natural crust localities, and on the south-facing slope（SFS）and north-facing slope（NFS）of bare localities of the Tengger Desert, China

生境	1956 年			1964 年			1981 年		
	S	*H*	*J*	*S*	*H*	*J*	*S*	*H*	*J*
藻类结皮	26c	1.88b	0.58ab	36abc	2.22ab	0.62ab	37abc	2.36ab	0.65ab
混生结皮	38abc	2.18ab	0.60ab	39abc	2.56ab	0.70ab	38abc	2.50ab	0.69ab
苔藓结皮 -1	44ab	2.42ab	0.64ab	39abc	2.45ab	0.67ab	43ab	2.50ab	0.66ab
苔藓结皮 -2	38abc	2.41ab	0.66ab	41ab	2.86a	0.77ab	47a	2.88a	0.75ab
SFS	—	—	—	—	—	—	—	—	—
NFS	—	—	—	—	—	—	—	—	—

续表

生境	1987 年			天然植被区			裸地		
	S	*H*	*J*	*S*	*H*	*J*	*S*	*H*	*J*
藻类结皮	42ab	2.61ab	0.70ab	43ab	2.87a	0.76ab	—	—	—
混生结皮	43ab	2.72ab	0.72ab	38abc	2.62ab	0.72ab	—	—	—
藓类结皮 -1	—	—	—	32bc	2.33ab	0.64ab	—	—	—
藓类结皮 -2	—	—	—	38abc	2.53ab	0.70ab	—	—	—
SFS	—	—	—	—	—	—	43a	2.86a	0.76ab
NFS	—	—	—	—	—	—	37abc	0.93a	0.81a

注：① *S* = 物种数，包括无菌菌株；*H*=Shannon-Weaver 指数；*J* = 均匀度；② 每一 BSC 类型的描述同表 4-11；③ 每一特征的相同字母表示没有显著差异（one-way ANOVA，p=0.05）。

我们对2013年采集的1956年、1964年和1981年人工植被区的BSC样品中分别分离得到50、46和58个菌种。从空间趋势上看，2011年采集的BSC更难表达，1956—1981年，仅混生结皮的Shannon指数和均匀度指数增加，而与来自0.4 ~ 0.8 cm BSC层群落相比，0 ~ 0.4 cm BSC层群落异质性和稳定性更低（表4-13）。

表 4-13 2013 年采集的腾格里沙漠 1956 年、1964 年、1981 年人工固沙植被区 0 ~ 0.4 cm BSC 层和 0.4 ~ 0.8 cm BSC 层的小型真菌群落的物种多样性

Table 4-13 Diversity characteristics of microfungal communities in surface（0–0.4 cm）and subsurface （0.4-0.8 cm） layers of the BSC sampled in 2013 from artificial vegetation region with revegetation in 1956, 1964, and 1981 in the Tengger Desert, China

结皮类型	BSC 层深度 /cm	1956 年			1964 年			1981 年		
		S	*H*	*J*	*S*	*H*	*J*	*S*	*H*	*J*
藻类结皮	0.4 ~ 0.8	21bcd	1.88cde	0.62abc	23abc	1.94cde	0.62efg	23abc	1.75de	0.56gh
混生结皮	0.0 ~ 0.4	18bcd	1.38ef	0.48h	23abc	1.73def	0.55gh	27ab	2.08abc	0.63efg
	0.4 ~ 0.8	14de	2.19abc	0.82ab	10e	1.66def	0.72bcd	17cde	2.17abc	0.77abc
藓类结皮 -1	0.0 ~ 0.4	21bcd	1.99bcd	0.65def	24abc	1.99bcd	0.63efg	24abc	2.05bcd	0.65def
	0.4 ~ 0.8	20bcd	2.51ab	0.84a	20bcd	2.01bcd	0.67def	24abc	2.22abc	0.69cde
藓类结皮 -2	0.0 ~ 0.4	19bcd	1.98bcd	0.67cde	24abc	2.41ab	0.76abc	22bcd	1.95cde	0.63efg
	0.4 ~ 0.8	16abc	2.42abc	0.74abc	22bcd	2.01bcd	0.65def	33a	2.44ab	0.70cde

注：① *S* = 物种数，包括无菌菌株；*H*=Shannon 指数；*J* = 均匀度；② 每一结皮类型的描述同表 4-11；③ 每一特征的相同字母表示没有显著差异（one-way ANOVA，p=0.05）。

产黑色素的小型真菌在所研究的群落中占主导地位（表4-14和表4-15），它们占总物种数的67%和空间动态贡献率的51.4%～94%（图4-15）。产黑色素真菌的核心是具有大型（>20 μm）多细胞孢子的物种，如埃里格孢、链格孢、毛球腔菌、内脐蠕孢、弯孢等。在2011年采集的人工植被区和天然植被区BSC中分别包含上述多细胞孢子物种42.4%～72.3%和43.1%～49.5%（图4-15a～e），在裸地上这些物种的贡献率很低（图4-15f）。对于BSC尤其是混生结皮而言，其表层和亚土层产黑色素真菌，特别是具有大型多细胞孢子真菌丰富度差异显著（图4-15g～i）。产黑色素真菌在不同类型BSC中丰富度的差异性被很少表达（蓝藻结皮（2011年）和混生结皮（2013年）的丰富度较高，而以藓类为优势种的BSC丰富度较低）。

除了以上提及的物种，在研究区出现数量最多、频率最高的物种有支孢菌属、尖孢镰刀菌、赤霉菌属、葡萄穗霉属、棘壳孢属、构巢曲霉、曲霉属真菌和漆斑菌。天然植被区BSC和裸地不同于人工植被区BSC，它们的优势种丰富度较低（表4-15）。在大多数天然植被区BSC中，颜色较浅的真菌，尤其是马拉色菌，有更高的丰富度。在裸地上的小型真菌群落，比如米根菌、抗真菌和葡萄穗霉菌，有大量的深色非孢子分离菌。

除了产黑色素真菌的贡献率较低，0.4～0.8 cm BSC亚土层比0～0.4 cm BSC表层含有更多的烟曲霉菌。与2011年所采的样品相比，2013年采集的人工植被区BSC层的小型真菌群落的差异主要表现在毛壳菌属与曲霉菌出现的频率和数量较低，而棘壳孢属的丰富度较高。

在温度为37 ℃时，小型真菌群落的组成与温度为25 ℃时不同。以具有大型多细胞孢子的产黑色素真菌为优势的群落在37 ℃时不能生长，有性型的物种（毛壳菌属、曲霉属真菌）、变形的曲霉属物种（烟曲霉菌、黑曲霉）以及米根菌是最丰富的真菌。

基于物种相对丰富度的聚类分析将2011年的小型真菌群落分为两组。第一组包含大多数来自人工植被区BSC的群落；第二组包含来自天然植被区BSC和裸地的群落，以及那些来自1981年的藓类结皮和来自1987年的藻类结皮和混生结皮的群落（图4-16a）。然而，这两组又有相当高的相似性（除了1956年的土生对齿藓结皮群落，其余不同点均小于30%）。人工植被区BSC的小型真菌群落，不管是在BSC类型还是聚居地方面，并没有表露出明显的模式。类似地，在2013年采集的BSC样品的小型真菌群落中并没有发现明显的模式（图4-16b）。通常情况下，与0.4～0.8 cm BSC亚土层的小型真菌群落相比，0～0.4 cm BSC表层的小型真菌群落之间更为相似。

表 4-14　中国腾格里沙漠区 1956 年、1964 年、1981 年、1987 年人工植被区、天然植被区 BSC 及未被固定和无 BSC 覆盖沙丘（裸地）阳坡（SFS）和阴坡（NFS）沙丘不同类型 BSC［藻类结皮（CB）、混生结皮（MIX）、真藓结皮（M1）和土生对齿藓结皮（M2）］小型真菌的相对丰度（%）

Table 4-14　Average relative abundance of common microfungal species from different BSC types［cyanobacterial（CB），mixed（MIX），moss-1（M1）and moss-2（M2）crusts］from artificial vegetation region with revegetation in 1956，1964，1981 and 1987，in natural crust localities，and on the south-facing slope（SFS）and north-facing slope（NFS）of non-stabilized and non-crusted sand dunes（bare localities）in the Tengger Desert，China

物种	1956 年				1964 年				1981 年				1987 年		天然植被区				裸地	
	CB	MIX	M1	M2	CB	MIX	M1	M2	CB	MIX	M1	M2	CB	MIX	CB	MIX	M1	M2	SFS	NFS
Mucoromycotina																				
Mortierella humilis	—	—	—	—	0.2	—	0.2	—	—	—	—	—	0.4	—	—	0.1	0.1	1.2	—	—
Rhizopus oryzae	0.7	0.6	0.4	—	0.2	—	0.5	0.5	0.3	0.7	0.3	0.3	1.0	0.9	—	0.1	0.2	—	2.3	3.8
Teleomorphic Ascomycota																				
Canariomyces notabilis	0.6	2.6	0.9	1.6	1.2	2.6	2.8	3.9	1.4	0.2	0.5	0.8	0.3	2.1	4.2	0.9	1.4	3.3	0.1	0.1
Chaetomium atrobrunneum	—	0.3	—	—	—	—	—	—	—	0.3	0.2	0.3	0.4	6.8	—	0.9	—	—	1.2	—
Ch. globosum	—	—	—	—	—	—	0.7	0.3	0.3	0.1	—	—	—	—	—	—	0.5	—	—	—
Ch. strumarium	2.3	2.0	1.5	0.4	2.5	5.5	3.1	4.3	2.3	5.6	3.2	1.7	3.4	2.8	4.1	1.8	0.7	1.6	2.3	0.6
Aspergillus nidulans	—	0.6	1.0	—	0.6	0.8	0.9	1.7	—	—	0.2	—	—	—	1.1	0.1	—	1.1	2.3	0.4
As. rugulosus	2.1	0.1	0.8	4.1	0.2	5.6	1.0	2.9	1.1	1.1	1.4	1.3	0.4	0.3	0.3	—	—	0.8	—	—
Thielavia terricola	—	—	—	—	—	—	—	0.2	0.6	—	—	—	—	—	0.3	—	—	—	—	—
Anamorphic Ascomycota																				
Alternaria alternata	12.9	9.9	5.5	6.5	14.1	19.6	3.5	6.6	10.2	15.6	5.2	7.4	19.2	18.7	17.4	20.0	8.0	21.0	18.4	14.2
As. fumigatus	1.4	0.8	1.1	0.1	0.4	0.5	0.8	1.1	1.0	0.4	0.4	1.2	1.4	0.8	0.4	0.2	0.1	0.1	6.1	3.5

续表

物种	1956 年				1964 年				1981 年				1987 年		天然植被区				裸地	
	CB	MIX	M1	M2	CB	MIX	M1	M2	CB	MIX	M1	M2	CB	MIX	CB	MIX	M1	M2	SFS	NFS
As. niger	—	0.1	0.1	0.1	—	—	—	0.2	—	0.5	—	—	0.2	—	0.1	0.1	—	—	0.4	0.4
Boeremia exigua	10.1	4.2	7.0	8.9	4.4	2.0	2.0	1.5	7.9	2.8	4.4	2.6	9.1	7.9	1.5	4.6	1.7	1.9	2.4	5.3
Cladosporium cladosporioides	2.4	1.2	0.1	2.7	1.7	1.0	0.6	1.3	1.5	3.7	1.8	5.9	1.9	3.4	1.8	1.9	2.0	3.4	3.3	9.3
Coleophoma empetri	0.5	—	—	—	0.5	0.4	—	0.8	1.1	0.6	0.4	0.8	0.5	0.4	0.1	—	0.1	0.1	1.6	0.4
Curvularia inaequalis	0.9	—	2.2	0.3	—	0.2	0.9	0.3	2.0	3.0	2.9	0.5	1.4	0.2	—	—	—	—	—	0.4
Cu. spicifera	0.5	0.4	0.3	0.3	0.7	0.5	0.8	—	0.1	0.5	—	0.4	0.9	0.5	0.5	0.4	—	0.2	0.8	0.4
Drechslera dematioidea	—	0.3	0.1	0.1	6.2	0.8	0.5	4.6	0.1	0.4	0.2	1.1	0.3	—	1.1	2.1	2.0	2.4	0.8	—
Embellisia phragmospora	52.0	44.0	44.0	33.7	51.6	38.5	47.3	34.0	47.8	35.7	43.4	24.2	26.8	29.4	18.7	27.5	26.2	21.0	14.7	13.4
Fusarium oxysporum	0.1	7.7	2.9	0.1	1.7	3.6	9.4	7.9	6.1	9.4	15.0	15.6	6.8	10.6	9.0	11.8	29.0	20.0	15.3	9.0
Gibberella intricans	2.6	—	9.3	23.1	0.2	—	—	—	—	1.8	0.6	3.0	—	—	—	—	—	0.3	—	—
Gilmaniella humicola	0.5	8.0	0.5	0.7	—	—	—	—	—	0.6	—	0.2	0.4	—	—	—	—	0.6	—	—
Monodyctis fluctuata	—	—	—	—	—	0.2	—	—	—	1.6	—	—	0.5	—	9.9	—	1.2	—	—	—
Myrothecium verrucaria	0.2	—	0.2	0.3	0.9	0.4	1.5	1.8	0.4	0.2	1.1	—	1.7	0.6	—	1.1	0.2	0.7	—	—
Papulaspora pannosa	—	—	—	—	—	0.4	0.3	0.3	0.2	—	0.4	0.4	0.8	0.2	0.4	—	0.2	0.6	0.8	—
Penicillium aurantiogriseum	—	—	—	—	—	0.2	4.1	2.1	1.2	0.5	1.2	3.4	0.2	0.8	—	0.1	0.1	0.2	—	—
Phoma medicaginis	—	0.8	0.3	0.1	0.2	—	—	—	—	—	—	—	—	—	3.5	2.4	0.9	3.0	0.4	—
Pyrenochaeta cava	1.6	—	1.4	2.2	0.8	1.0	1.7	—	3.3	0.3	1.9	0.6	—	0.5	1.5	2.6	11.0	1.1	0.4	—
Septoria sp.	—	3.3	—	—	—	—	—	—	—	—	—	—	—	—	3.0	9.9	—	3.8	—	—
Setosphaeriar rostrata	0.7	0.4	1.9	1.6	0.2	2.8	3.3	7.2	1.2	0.7	4.7	10.2	1.6	4.3	0.3	1.0	2.2	1.8	—	—

续表

物种	1956 年				1964 年				1981 年				1987 年		天然植被区				裸地	
	CB	MIX	M1	M2	CB	MIX	M1	M2	CB	MIX	M1	M2	CB	MIX	CB	MIX	M1	M2	SFS	NFS
Stachybotrys chartarum	0.6	0.5	1.5	0.6	0.3	0.9	0.6	1.5	0.4	0.5	0.6	2.0	0.6	1.3	1.3	0.3	0.3	0.9	2.6	4.1
Talaromyces variabilis	—	0.8	0.2	0.2	—	0.2	0.2	0.3	0.4	0.2	0.1	0.1	—	0.8	0.3	—	—	—	—	—
Trichoderma koningii	—	—	0.1	—	—	—	1.2	0.5	—	—	—	0.3	—	0.3	—	—	—	—	—	0.6
Ulocladium atrum	3.5	3.2	4.2	3.0	8.5	5.1	7.5	5.1	4.5	5.0	3.2	5.5	10.3	4.7	6.3	5.1	6.3	2.8	1.9	2.5
Westerdykella capitulum	0.4	—	6.9	0.3	—	—	—	0.5	—	1.2	—	—	—	—	—	0.1	—	—	—	—
Diplodia sp.	—	—	—	—	—	—	—	—	0.8	0.5	—	—	—	—	1.6	1.4	—	0.2	—	—
Mycelia sterilia																				
Dark—colored 1	—	0.7	—	—	—	—	1.6	5.2	2.4	3.2	1.0	0.8	—	0.6	—	—	—	0.3	1.2	—
Dark—colored 2	—	—	—	—	0.2	0.2	1.3	—	—	—	0.2	0.2	0.3	0.9	3.0	0.2	0.3	0.9	11.1	8.9

表 4-15 中国腾格里沙漠区 1956 年、1964 年、1981 年、1987 年人工植被区不同类型 BSC［藻类结皮（CB）、混生结皮（MIX）、真藓结皮（M1）和土生对齿藓（M2）结皮］0 ~ 0.4 cm 或 0 ~ 0.5 cm 表层和 0.4 ~ 0.8 cm 亚土层中小型真菌的相对丰度（%）

Table 4-15 Average relative abundance of common microfungal species from the surface （0 ~ 0.4 cm or 0 ~ 0.5 cm） and subsurface （0.4 ~ 0.8 cm） layers of different BSC types［cyanobacterial （CB）, mixed （MIX）, moss-1 （M1）, and moss-2 （M2） crusts］in the localities after sand stabilization with revegetation in 1956, 1964, 1981, and 1987 in the Tengger Desert, China

物种	1956 年							1964 年							1981 年						
	CB	MIX		M1		M2		CB	MIX		M1		M2		CB	MIX		M1		M2	
	0.0~0.5 cm	0.0~0.4 cm	0.4~0.8 cm	0.0~0.4 cm	0.4~0.8 cm	0.0~0.4 cm	0.4~0.8 cm	0.0~0.5 cm	0.0~0.4 cm	0.4~0.8 cm	0.0~0.4 cm	0.4~0.8 cm	0.0~0.4 cm	0.4~0.8 cm	0.0~0.5 cm	0.0~0.4 cm	0.4~0.8 cm	0.0~0.4 cm	0.4~0.8 cm	0.0~0.4 cm	0.4~0.8 cm
Mucoromycotina																					
Mortierella humilis	—	0.9	15.0	—	1.8	—	—	3.5	—	36.0	—	—	—	—	—	0.3	8.7	—	0.4	—	0.8
Rhizopus oryzae	0.3	0.5	—	—	—	—	—	0.2	0.2	—	—	—	0.4	—	—	0.5	10.8	0.2	—	—	0.4
Teleomorphic Ascomycota																					
Canariomyces notabilis	—	—	—	—	—	0.2	1.4	0.2	—	—	1.4	1.5	0.4	—	1.4	1.0	—	—	1.3	—	3.2
Chaetomium atrobrunneum	1.2	0.4	1.4	—	0.9	—	—	—	—	—	—	0.5	—	—	—	—	20.0	0.2	—	—	2.0
Ch. strumarium	0.9	0.7	—	0.9	—	—	0.9	—	0.9	—	0.5	3.0	1.5	1.9	—	0.3	—	1.8	0.4	—	2.0
Aspergillus nidulans	—	0.5	—	—	—	—	0.5	0.5	0.4	8.5	1.0	—	—	—	—	0.3	—	5.8	—	—	0.4
Sporormiella minima	—	—	—	—	—	—	—	—	—	—	—	—	—	—	—	0.5	0.5	—	0.4	—	0.8
Immature fruit bodies	—	—	—	—	—	—	—	0.3	0.4	1.2	—	1.0	—	—	—	—	—	—	0.9	—	—
Anamorphic Ascomycota																					
Alternaria alternata	8.6	5.8	3.5	4.7	—	5.2	3.8	7.5	6.6	—	6.8	3.5	5.9	2.2	5.8	11.5	2.6	6.8	6.0	9.7	5.6
As. fumigatus	1.2	—	10.7	—	12.7	0.2	4.7	0.9	0.5	1.2	—	2.5	—	0.7	—	0.3	23.1	0.6	5.6	0.2	5.2

续表

物种	1956 年							1964 年							1981 年						
	CB	MIX		M1		M2		CB	MIX		M1		M2		CB	MIX		M1		M2	
	0.0~0.5 cm	0.0~0.4 cm	0.4~0.8 cm	0.0~0.4 cm	0.4~0.8 cm	0.0~0.4 cm	0.4~0.8 cm	0.0~0.5 cm	0.0~0.4 cm	0.4~0.8 cm	0.0~0.4 cm	0.4~0.8 cm	0.0~0.4 cm	0.4~0.8 cm	0.0~0.5 cm	0.0~0.4 cm	0.4~0.8 cm	0.0~0.4 cm	0.4~0.8 cm	0.0~0.4 cm	0.4~0.8 cm
As. niger	—	—	17.7	—	0.9	—	—	—	—	—	—	—	—	—	—	—	—	—	—	—	—
Boeremia exigua	4.8	1.1	1.4	9.1	2.7	8.1	5.6	7.0	0.7	—	2.2	0.5	—	—	2.0	3.4	0.5	2.2	—	4.7	2.0
Cladosporium cladosporioides	1.8	1.4	1.4	—	—	—	1.4	2.0	1.7	—	2.6	1.0	4.8	—	1.6	2.4	0.5	1.4	—	1.9	1.2
Coleophoma empetri	16.0	3.3	—	0.2	—	—	0.9	2.2	3.0	—	4.3	—	1.5	0.4	1.6	—	—	2.2	1.7	1.2	2.0
Curvularia inaequalis	—	0.2	—	—	—	0.4	—	—	0.2	—	—	—	—	0.4	0.8	—	—	—	—	—	—
Cu. spicifera	0.9	—	—	0.2	—	—	—	—	0.4	—	0.3	—	—	0.4	—	—	0.5	0.2	—	2.3	0.4
Embellisia phragmospora	41.8	17.7	11.4	37.4	25.5	37.3	42.3	51.3	54.6	37.3	44.3	48.0	37.7	47.0	66.0	47.8	13.9	31.2	37.0	36.0	34.4
Fusarium oxysporum	—	—	—	—	—	—	2.3	—	—	—	0.2	—	—	—	—	0.8	—	1.2	1.7	2.1	4.8
Gibberella intricans	10.0	0.7	—	9.6	12.7	7.0	1.4	1.7	0.9	3.1	5.0	—	6.6	6.3	9.4	3.2	1.5	0.2	1.3	3.1	2.8
Myrothecium verrucaria	—	—	—	0.7	—	—	0.9	10.0	—	—	—	—	2.9	—	—	—	—	—	—	—	—
Papulaspora pannosa	1.2	—	—	—	0.9	0.4	0.5	—	—	—	—	—	0.4	0.7	0.2	—	—	—	—	0.2	0.4
Penicillium aurantiogriseum	—	0.7	—	0.2	—	—	1.9	—	—	—	0.2	—	1.5	1.1	—	—	—	—	—	—	—
Phoma medicaginis	—	—	—	—	1.8	6.0	—	—	—	—	0.3	—	—	0.4	0.4	—	—	—	—	—	—
Pyrenochaeta cava	2.4	62.5	25.6	27.5	11.0	0.9	—	9.4	21.3	15.7	23.5	15.5	19.3	15.6	3.0	8.6	12.3	1.4	14.7	3.0	0.8
Septoria sp.	—	—	—	—	—	25.2	15.7	—	—	—	—	—	—	—	—	0.3	—	33.8	4.3	28.0	6.0
Setosphaeriar rostrata	1.8	1.1	2.1	1.8	0.9	4.0	1.4	1.4	0.7	—	5.1	4.0	2.9	8.5	0.2	2.2	—	4.8	0.4	1.2	2.0

续表

物种	1956年							1964年							1981年						
	CB	MIX		M1		M2		CB	MIX		M1		M2		CB	MIX		M1		M2	
	0.0～0.5 cm	0.0～0.4 cm	0.4～0.8 cm	0.0～0.4 cm	0.4～0.8 cm	0.0～0.4 cm	0.4～0.8 cm	0.0～0.5 cm	0.0～0.4 cm	0.4～0.8 cm	0.0～0.4 cm	0.4～0.8 cm	0.0～0.4 cm	0.4～0.8 cm	0.0～0.5 cm	0.0～0.4 cm	0.4～0.8 cm	0.0～0.4 cm	0.4～0.8 cm	0.0～0.4 cm	0.4～0.8 cm
Stachybotrys chartarum	—	0.7	—	0.2	—	—	1.9	—	—	—	0.2	—	1.5	1.1	—	—	—	—	—	—	—
Talaromyces variabilis	0.3	0.5	—	0.2	2.7	0.2	—	0.2	0.4	—	0.2	1.0	0.7	0.7	—	—	3.6	0.4	0.4	0.2	2.4
Trichoderma koningii	—	—	—	—	—	—	0.5	—	—	—	—	—	—	—	—	2.9	—	0.2	—	—	—
Ulocladium atrum	—	—	—	—	—	—	0.4	—	—	—	0.2	—	1.8	0.7	0.2	—	—	—	—	0.4	—
Westerdykella capitulum	—	—	—	—	6.4	—	—	0.5	—	—	—	7.5	—	—	1.8	2.9	—	—	3.0	0.5	—
Mycelia sterilia																					
Dark—colored 1	—	—	1.4	1.3	—	0.2	5.2	0.5	—	—	0.3	—	3.3	6.3	—	—	—	0.2	13.3	—	13.2
Dark—colored 2	—	—	1.4	—	0.9	—	—	—	—	—	0.2	—	—	—	—	—	—	—	0.4	—	0.4
Dark—colored 3	—	—	—	—	3.6	—	—	—	—	—	—	—	—	—	—	0.6	0.5	—	—	—	—
Light—colored	1.2	—	7.1	—	1.8	—	—	—	—	—	—	—	—	—	—	0.4	—	—	—	—	0.8

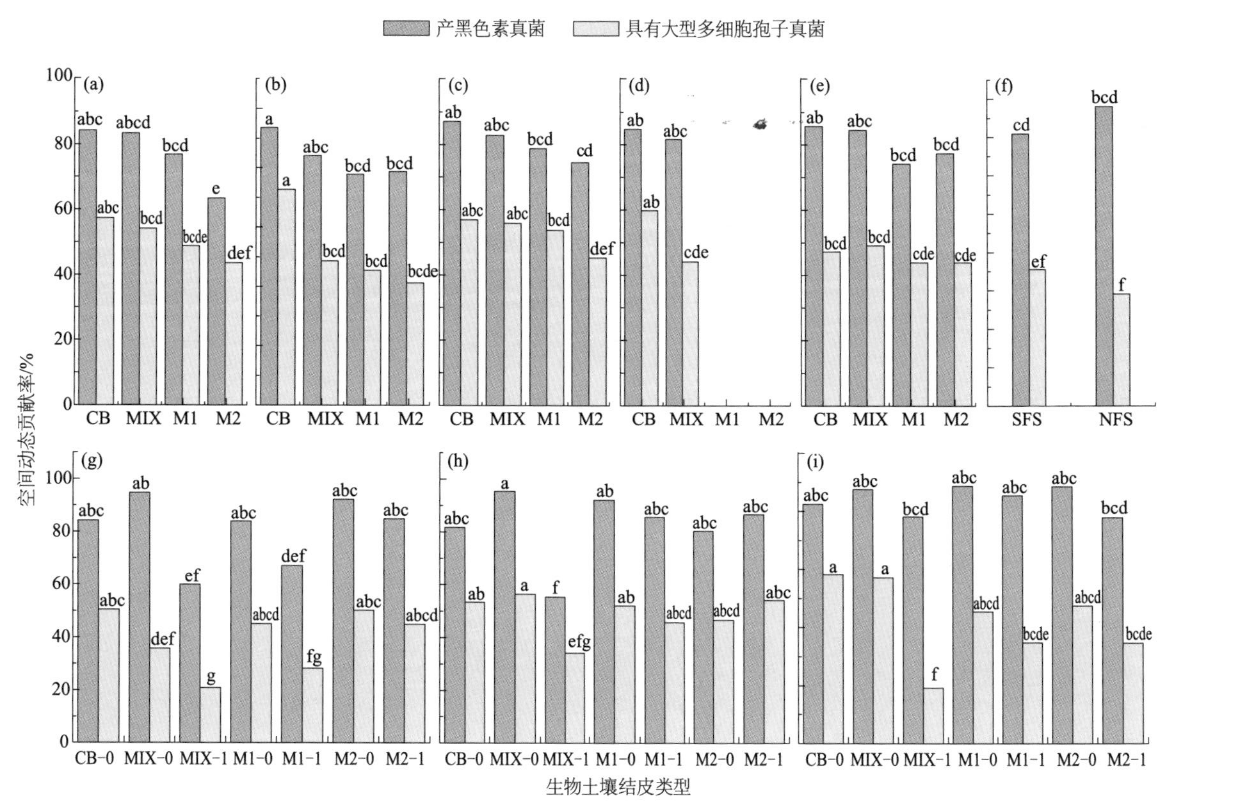

图 4-15　BSC 中产黑色素真菌及具有大型多细胞孢子真菌的空间动态贡献率。所研究的 BSC 为 1956 年（a）、1964 年（b）、1981 年（c）、1987 年（d）年人工植被区 BSC、天然植被区 BSC（e）、未被固定且无 BSC 覆盖沙丘（裸地）阳坡（SFS）和阴坡（NFS）（f）和 1956 年（g）、1964 年（h）、1981 年（i）年人工植被区 BSC 的表层和亚土层。相同的字母表示产黑色素真菌及具有大型多细胞孢子真菌在 5% 水平上没有显著差异。CB：藻类结皮；MIX：混生结皮；M1：真藓结皮；M2：土生对齿藓结皮。-0 和 -1 分别表示 BSC 表层和亚土层

Figure 4-15　Spatial dynamics of contribution of melanin-containing species and melanized fungi with large multicellular spores in BSC from artificial vegetation region with revegetation in 1956 (a), 1964 (b), 1981 (c), and 1987 (d), in natural crust localities (e), on the south-facing slope (SFS) and north-facing slope (NFS) of non-stabilized and non-crusted sand dunes (bare localities) (f), and in surface and subsurface crust layers from artificial vegetation region with revegetation in 1956 (g), 1964 (h), and 1981 (i) in the Tengger Desert, China. Bars with the same letter (s) within the melanin-containing species or the melanized fungi with large multicellular spores are not significantly different at the 5% level of probability. CB, MIX, M1 and M2 are the abbreviations of crust types: cyanobacterial, mixed, moss-1 and moss-2, respectively. Abbreviations with −0 and −1 indicate surface and subsurface crust layers, respectively

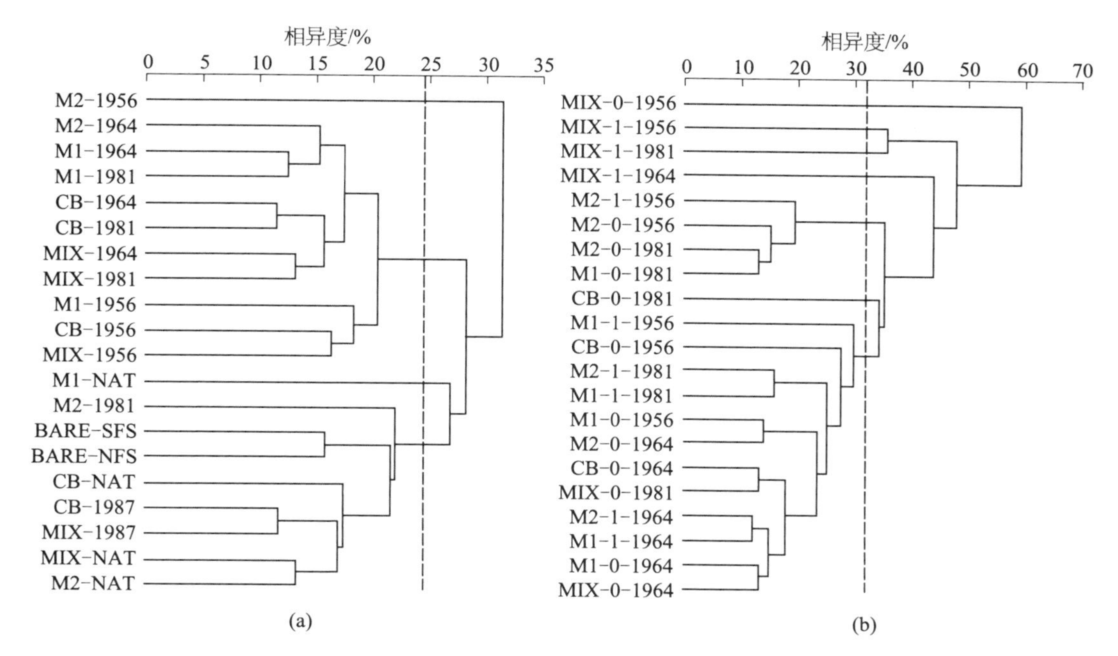

图 4-16 对腾格里沙漠 1956 年、1964 年、1981 年、1987 年人工植被区、天然植被区（NAT）及裸地阳坡（SFS）和阴坡（NFS）(a) 和 1956 年、1964 年、1981 年人工植被区 BSC 表层和亚土层（b）小型真菌群落的聚类分析。CB：藻类结皮；MIX：混生结皮；M1：真藓结皮；M2：土生对齿藓结皮。-0 和 -1 分别表示 BSC 表层和亚土层

Figure 4-16 Clustering of the microfungal communities in the BSC from artificial vegetation region with revegetation in (a) 1956, 1964, 1981 and 1987, in natural crust localities (NAT), and on the south-facing slope (SFS) and north-facing slope (NFS) of non-stabilized and non-crusted sand dunes (bare localities, and in the surface and subsurface layers of the crusts from artificial vegetation region with revegetation in (b) 1956, 1964 and 1981 in the Tengger Desert, China. CB, MIX, M1 and M2 are the abbreviations of crust types: cyanobacterial, mixed, moss-1 and moss-2, respectively. Abbreviations with −0 and −1 indicate surface and subsurface crust layers, respectively

（2）小型真菌菌株密度

在 2011 年所采集的 BSC 样品中，人工植被区 BSC 表层中的菌落数量为 3.6×10^3 ~ 11.6×10^3 CFU，天然植被区BSC表层为4.4×10^3 ~ 9.8×10^3 CFU。在每一个人工植被区区域，土生对齿藓结皮具有更高的菌株密度（图4-17a ~ c）。裸地的菌株密度要比人工植被区BSC层的菌株密度低11 ~ 58倍。

在2013年所采的BSC样品中，0 ~ 0.4 cm BSC表层与0.4 ~ 0.8 cm BSC亚土层的菌株密度有很大差异，前者比后者高2.2 ~ 4.7倍。不同类型BSC的菌株密度差异被较少表达（在以藓类为优势的BSC中较高而在藻类结皮和混生结皮中较低）。

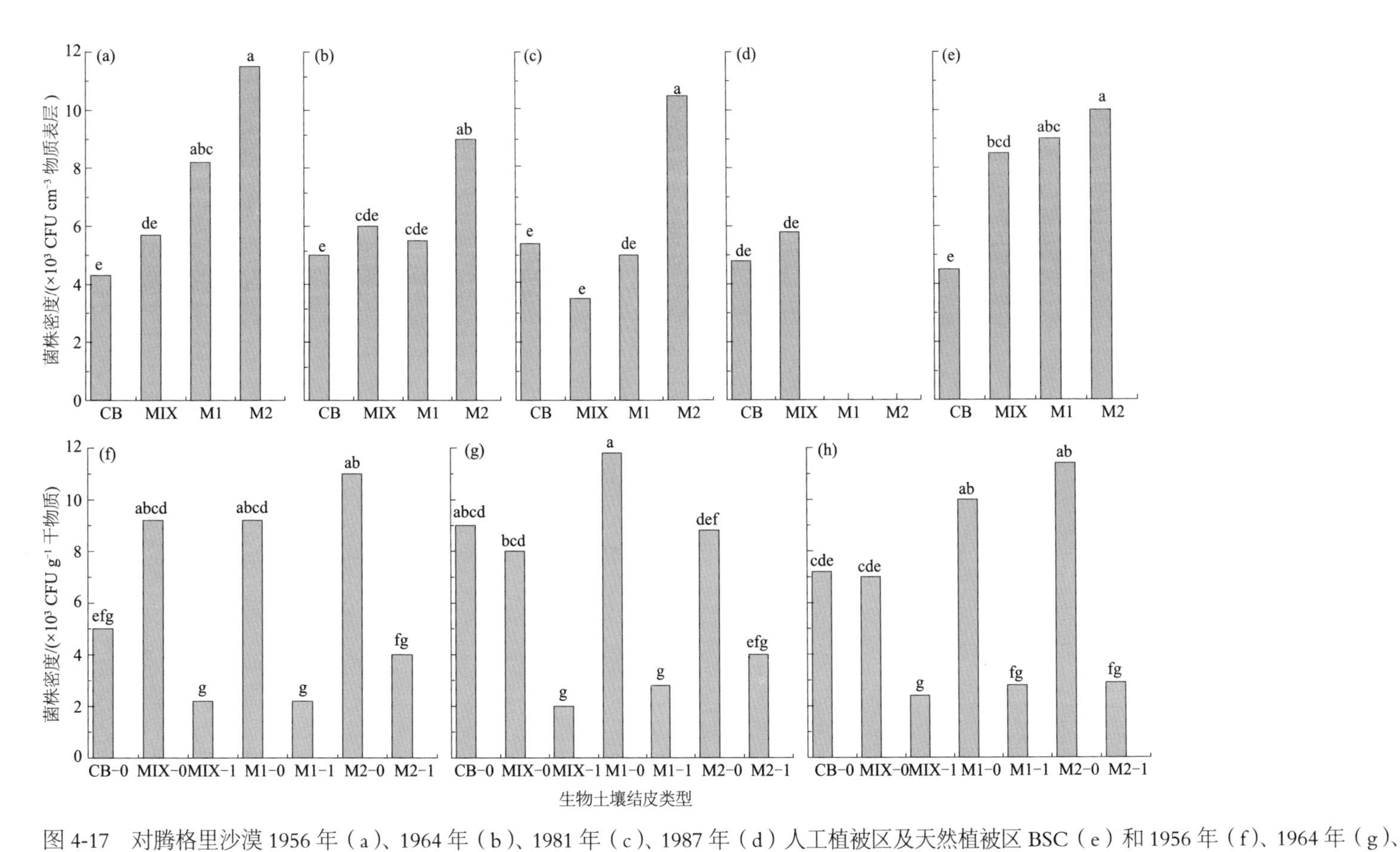

图 4-17 对腾格里沙漠 1956 年（a）、1964 年（b）、1981 年（c）、1987 年（d）人工植被区及天然植被区 BSC（e）和 1956 年（f）、1964 年（g）、1981 年（h）人工植被区 BSC 表层和亚土层 BSC 中真菌菌株密度的空间动态（菌落形成单位表示）。CB：藻类结皮；MIX：混生结皮；M1：真藓结皮；M2：土生对齿藓结皮。-0 和 -1 分别表示 BSC 表层和亚土层

Figure 4-17 Spatial dynamics of fungal isolate density, expressed as colony forming units (CFU), in the BSC from artificial vegetation region with revegetation in 1956 (a), 1964 (b), 1981 (c) and 1987 (d), in natural crust localities (e), and in the surface and subsurface layers of the crusts in localities after sand stabilization with revegetation in 1956 (f), 1964 (g) and 1981 (h) in the Tengger Desert, China. CB, MIX, M1 and M2 are the abbreviations of crust types: cyanobacterial, mixed, moss-1 and moss-2, respectively. Abbreviations with −0 and −1 indicate surface and subsurface crust layers, respectively

（3）真菌生物特性与叶绿素含量的关系

菌株密度与叶绿素含量呈极显著线性正相关（$p<0.001$）（图4-18c）。产黑色素真菌和具有大型多细胞孢子的产黑色素真菌丰富度与叶绿素含量之间呈显著线性负相关（$p<0.05$；图4-18a，b）。物种多样性特征（物种丰富度、异质性、均匀度）与叶绿素含量没有显著的相关性。

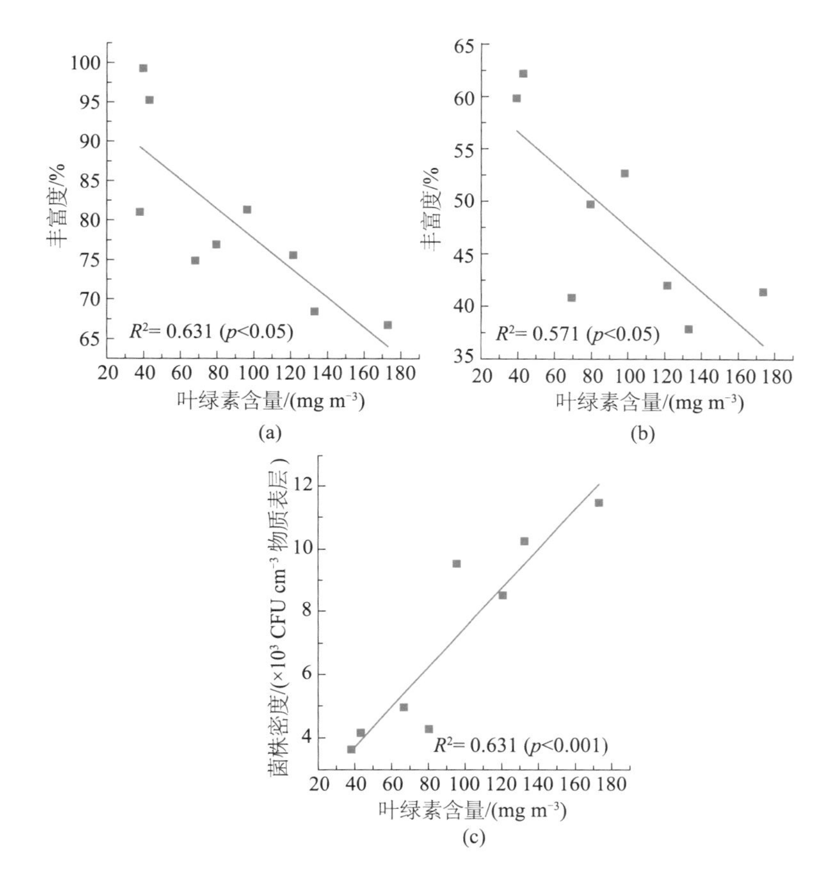

图4-18　腾格里沙漠BSC中产黑色素真菌的丰富度（a）、具有大型多细胞孢子的产黑色素真菌的丰富度（b）、用CFU值表示的真菌菌株密度（c）与BSC中叶绿素含量的关系

Figure 4-18　Relationships of the abundance of melanin-containing species（a）, the abundance of melanized fungi with large multicellular spores（b）, and the density of fungal isolates expressed as colony forming units（CFU）（c）with chlorophyll content in the BSC in the Tengger Desert, China

腾格里沙漠BSC中真菌物种组成的大致状况与其他沙漠是一致的。这主要是由于产黑色素的真菌在数量和丰富度上都占优势，这被认为是干旱土壤中真菌的一个典型特征（Ranzoni，1969；Christensen，1981；Halwagy *et al.*，1982；Skujinš，1984；Abdullah *et al.*，

1986；Hashem，1991；Ciccarone and Rambelli，1998；Zak，2005）。在以色列内盖夫沙漠BSC（Grishkan and Kidron，2013）以及其他内盖夫土壤中（Grishkan *et al.*，2006，2007；Grishkan and Nevo，2010）也发现了这一特征。同样地，在腾格里沙漠发现的带有大型多细胞孢子的产黑色素真菌也是内盖夫沙漠真菌的一个典型特征。在美国新墨西哥州的科罗拉多高原、奇瓦瓦沙漠、索诺拉沙漠以及半干旱草地，那些属于格孢菌目的真菌在BSC的菌落中占优势地位（Bates and Garcia-Pichel，2009；Bates *et al.*，2010，2012；Porras-Alfaro *et al.*，2011）。我们的研究表明，在受风沙胁迫的沙漠地区，黑色素以及大型多细胞孢子形态对沙地有分散和固定的作用。

沙坡头地区不同类型BSC的分布有明显的空间模式，这显著影响BSC中小型真菌的分布。藻类结皮主要分布在沙丘阳坡，藓类结皮则主要分布在阴坡或者低地，BSC的空间分布反映了位点的水分可利用状况。降雨是该区域的主要水分来源（Li *et al.*，2009），但由于正午高于10～15℃的地表温度，阳坡与其他地点相比有更高的蒸发率，这与内盖夫沙漠是一致的（Kidron *et al.*，2000）。结果表明，阳坡的典型特征是湿润持续时间较短（微生物活力也就较短），这与从内盖夫沙漠西侧Nizzana研究点所获得的数据是一致的（Kidron *et al.*，2009）。由于地表湿润持续时间和BSC叶绿素含量密切相关，相对较短时间的微生物活力能够解释藻类结皮的欠发育状态以及较低的叶绿素含量（表4-11）。或者，在阴坡和一些低地，更长的湿润持续时间和随之而来的更长时间的微生物活力能够解释藓类结皮的转变过程（Kidron *et al.*，2010），BSC变厚、叶绿素含量升高，混生结皮受媒介非生命环境影响随后表现出媒介作用特性。对于小型真菌群落而言，菌株密度与叶绿素含量之间表现出极显著线性正相关，这表明BSC表面湿润持续时间不仅影响BSC的自养性要素，而且也影响真菌的数量。在干旱沙漠区菌株密度反映了BSC中孢子的数量，如其潜在的生物量、休眠状况。产黑色素真菌的丰富度，包括具有大型多细胞孢子的产黑色素真菌，与叶绿素含量呈显著负相关关系，这是由于BSC所在的地形位置与湿润持续时间和辐射强度密切相关。

不同类型BSC之间的比较揭示了小型真菌群落物种组成的微小差异。所有BSC都表现出存在大量的含有黑色素真菌，其中包括具有大型多细胞孢子的产黑色素真菌。在所研究的0～0.4 cm BSC表层和0.4～0.8 cm BSC亚土层间，我们发现这些参数有明显的差异。显然，穿透到土壤100 μm深的紫外辐射在小型真菌群落组织方面有重要作用，或许能够解释从地下BSC上层到下层具有大型多细胞孢子的产黑色素真菌丰富度降低的原因。

在不同年代序列间，人工植被区相同BSC类型的微小真菌群株的菌株密度并没有表现出显著差异。这可能是由于所有BSC均表现出半均衡态，随着非生物环境到达它们的演替终期。考虑到内盖夫沙漠西侧Nizzana研究点小于10年和小于25年藻类结皮和藓类结皮的快速恢复速率（Kidron *et al.*，2008），我们认为沙坡头沙漠试验研究站40～60年的BSC已经到达发育的终期阶段，这反过来可以解释，在固沙和植被重建后的不同时期，相同BSC类型间菌株密度表现出相似性。菌株密度随BSC加深而降低可能与有机质含量减少和可利用水分降低有关，这两个与土壤相关的因素影响土壤真菌的生长发育与活性。

一般而言，尽管没有达到显著水平，但群落多样性指数随BSC年代序列而表现出降低趋势（表4-12），这表明随着时间变化，一个或两个普遍种的优势度提高，因此微小真菌群落的异质性和均匀度降低。在裸地所表现出的最高多样性指数与以上趋势一致。同样，在内盖夫沙漠西侧Nizzana研究点沙地群落的多样性指数要比BSC群落高出1.2～1.4倍。与BSC上层相比，BSC下层的微小真菌群落表现出更大的异质性和均匀度（表4-13）。显然，由于缺乏高辐射胁迫的压力，BSC下层没有被具有大型多细胞孢子的产黑色素真菌占据，因此变得更加平衡。

人工和天然植被区微小真菌群落之间具有一定差异性。在物种组成方面，我们在人工植被区发现，优势物种*Embellisia phragmospora*的丰富度更高，而常见种*Fusarium oxysporum*的丰富度较低。在物种多样性水平来看，在人工植被区中，与来自藓类结皮的相比，来自藻类结皮的微小真菌群落的物种丰富度较低，而在天然植被区，来自藓类结皮的群落表现出最小的异质性和均匀度。天然植被区的模式与内盖夫沙漠西侧Nizzana研究点的一致（Grishkan and Kidron，2013）。藓类结皮所具有的菌株密度增加的特征，可归因于一个或两个物种的大量生长发育降低了微小真菌群落的物种丰富度和物种多样性水平。

与内盖夫沙漠相比，除了优势真菌种（分别为*E. phragmospora*和*Ulocladium atrum*）的差异之外，腾格里沙漠中大量存在的耐热真菌要高出10倍之多。这样的差异从根本上来说与中国和以色列沙漠不同的降雨特征相关。内盖夫沙漠的BSC承受5～20 ℃相对低温下的冬季降水，而腾格里沙漠的BSC适应于30～35 ℃高温下的夏季降水。在腾格里沙漠地区，雨季的高温或许能解释耐热藓类比如真藓的优势地位（Convey and Lewis Smith，2006），以及耐热真菌物种更高的丰富度水平。

腾格里沙漠沙坡头试验区附近不同类型的BSC中存在丰富的微小真菌群落，在不同年代人工植被区表现出较小的差异。在腾格里沙漠区，地形所诱发的非生物环境差异造成了BSC中小型真菌群落的物种组成、菌株密度及物种多样性特征的差异。与以色列内盖夫沙漠相似，腾格里沙漠中的真菌区系是以带有大型多细胞孢子的产黑色素真菌为优势种的，尽管在种类上有差异。与内盖夫沙漠相比，腾格里荒漠BSC中的耐热真菌有着更高的丰富度，这主要是因为降水季节特征不同，内盖夫沙漠和腾格里沙漠的降水特征分别是冬季降雨和夏季降雨。

4.2.3 BSC对土壤微生物和线虫的影响

在土壤生态系统中，数量庞大、种类繁多的微生物是土壤生物相中最活跃的组成成分之一，其以丰富的生物多样性成为生态系统中最活跃和最具有决定性作用的组分之一。土壤微生物是土壤有机质和养分转化、循环的动力，它们参与土壤有机质分解、腐殖质形成、土壤养分转化和循环等过程，是土壤肥力发展的原动力，是土壤养分的储存库和植物生长可利用养分的一个重要来源，是土壤肥力的活性指标（Gu *et al.*，2009）。

土壤微生物群落多样性作为土壤微生物指标中最重要的因素，与土壤的生态稳定性密切相关，土壤的细微变化可引起土壤微生物群落多样性的变化（裴雪霞，2010）。土壤微生物量是土壤有机质中最活跃和最易变化的部分。它是土壤中易于利用的养分库及有机物分解和矿化的动力，与土壤中的碳、氮、磷、硫等养分循环密切相关，其变化可反映土壤肥力的变化及土壤的健康程度。土壤酶是土壤中最活跃的有机成分之一，主要来源于土壤微生物细胞，影响土壤微生物活性的因素必然影响到土壤酶的活性。因此，可用土壤酶活性客观地评价土壤微生物总体活性（Bergstorm *et al.*，1998），进而反映土壤质量和土壤肥力的变化（Trasar-Cepeda *et al.*，2008；焦婷等，2011）。因此，土壤微生物群落多样性、微生物量和土壤酶活性等均可作为评价土壤健康程度及生态系统恢复状况的生物学指标，分析这些微生物指标对于如何有效提高土壤质量和促进生态系统的恢复具有重要的指导意义（Bünemann *et al.*，2004；Mckinley *et al.*，2005）。

线虫作为地球上最丰富的后生动物，普遍存在于各种土壤和底泥中，是土壤生物中的优势类群。线虫食性多样，是碎屑食物网的重要组成部分，在物质循环过程中发挥着重要作用（吴纪华等，2007），其多样性的变化可以很好地指示土壤生态系统营养源流向和路径的改变（Ettema，1998；Neher，2001）。线虫与微生物关系非常密切，线虫通过与微生物的相互作用参与有机物分解过程和有机态养分的矿化，进而影响养分在土壤－植物系统中的循环及植物生产力，对土壤生态系统起着重要的调节作用。土壤中栖息的线虫以自由生活线虫占绝对优势，自由生活线虫以食微线虫（食细菌线虫和食真菌线虫）为主要营养功能类群，约占自由生活线虫的80%左右（Griffiths，1989；毛小芳，2006）。这些食微线虫与微生物到底存在着什么样的关系，食微线虫的作用是增加还是降低微生物数量及活性，研究结果不一致，而且当前的研究结果基本上是基于悉生微缩培养（gnotobiotic microcosm）系统，在接种单一微生物和线虫的简单食物链条件下获得的（陈小云等，2004）。从理论上分析，食微线虫对某一种或几种微生物的取食作用会影响到整个微生物群落数量、结构和活性的变化，但迄今为止，在开放土壤条件下，土壤食微线虫与微生物的相互作用目前尚不清楚，国内外也少有涉及。

由于土壤线虫与微生物的研究相对滞后，因此对于BSC与土壤线虫和微生物的关系研究仅见少量报道。边丹丹（2011）的研究表明，黄土丘陵区BSC下的土壤微生物数量和土壤酶活性显著高于无BSC的土壤；阳贵德（2010）的研究发现，铜陵铜尾矿上的BSC能显著增加尾矿中脲酶、脱氢酶、过氧化氢酶、碱性磷酸酶的活性和微生物量碳、氮；Darby等（2007）的研究表明，BSC类型影响土壤线虫的群落，发育晚期的BSC（藓类结皮和地衣结皮）比相对早期的BSC（藻类结皮）下土壤线虫群落更成熟、更复杂。尽管这些研究为我们了解BSC对土壤微生物和线虫的影响提供了宝贵的方法和经验，但是这些研究还存在较大局限性：一是研究不太系统，多用单个指标来说明BSC对生态系统的作用，使试验数据显得很单薄，不能有力地支撑结论；二是研究数据基本是在某一个时间点上得到的，使得试验结果之间没有时间梯度上的相互印证，降低了研究结果的可信度；三是研究数量太少且零散，尤其很少涉及荒漠区中BSC对土壤微生物、线虫及BSC下微生物与线虫关系的影响。

（1）BSC对土壤微生物群落结构、微生物活性及微生物量的影响

本研究以1956年（55龄）、1964年（47龄）、1981年（30龄）和1991年（20龄）建立的腾格里沙漠东南缘的人工植被固沙区、红卫天然植被区BSC覆盖的沙丘土壤为研究对象，以流沙（0龄）为对照，研究BSC对其下土壤微生物群落结构、微生物活性及微生物量的影响。

BSC对土壤微生物量碳的影响：由图4-19a可以看出，人工植被固沙区藻类结皮下0～10 cm土层中微生物量碳的变化规律为：55龄和47龄固沙区藻类结皮下土壤微生物量碳最高，分别为73.92 mg kg^{-1}和70.39 mg kg^{-1}，与流沙对照差异显著（$p<0.05$）；30龄和20龄固沙区次之，分别为52.8 mg kg^{-1}和49.28 mg kg^{-1}，与流沙对照差异显著（$p<0.05$）；流沙对照最低。同样，藻类结皮下10～20 cm土层中微生物量碳的变化规律基本类似，55龄和47龄固沙区最高，分别为42.75 mg kg^{-1}和42.24 mg kg^{-1}，与流沙对照差异显著（$p<0.05$）；30龄和20龄固沙区次之，分别为31.68 mg kg^{-1}和24.64 mg kg^{-1}，与流沙对照差异显著（$p<0.05$）。而20～30 cm土层中55龄、47龄、30龄和20龄固沙区的藻类结皮下土壤微生物量碳与流沙对照相比已无显著差异（$p>0.05$）。因此，在0～10 cm和10～20 cm土层，藻类结皮可显著提高土壤微生物量碳，与流沙对照差异显著（$p<0.05$）。随着土层的加深，其影响逐渐减弱，到20～30 cm土层已无显著差异（$p>0.05$）。

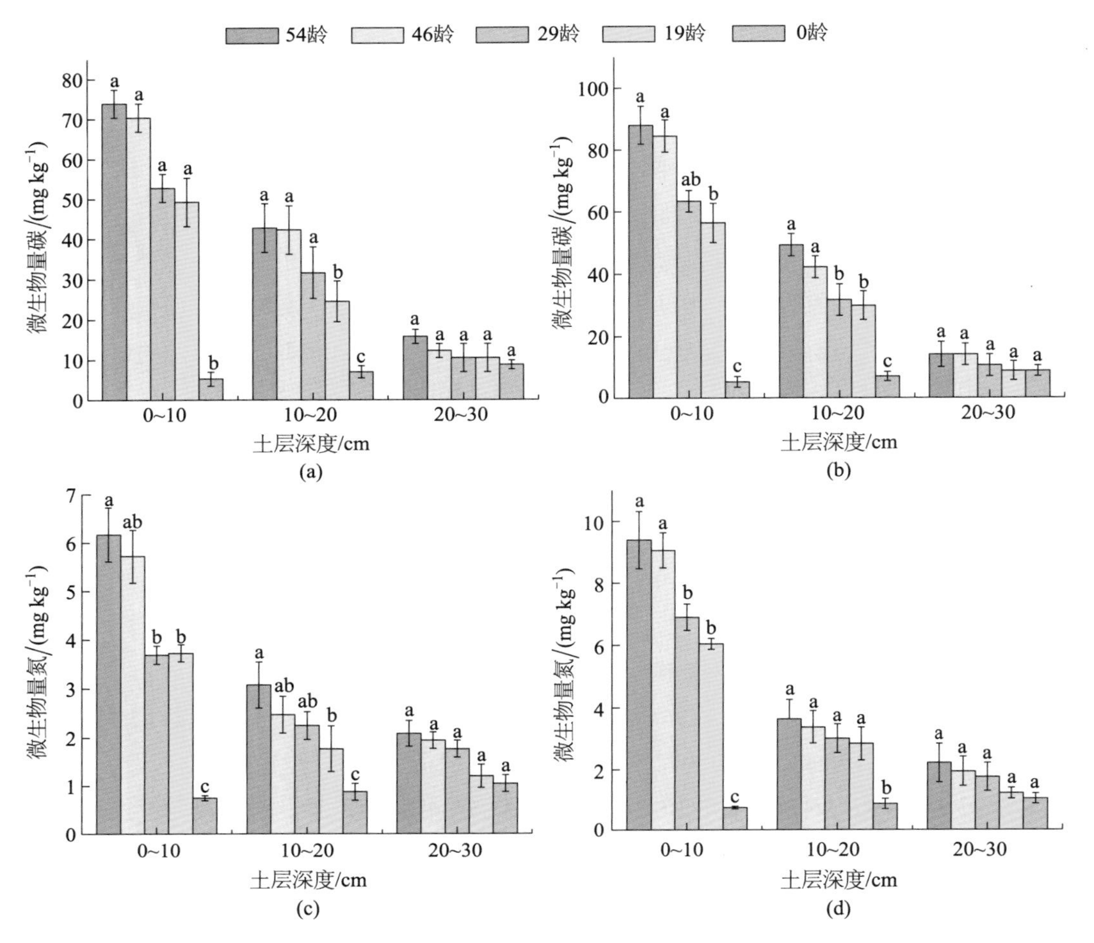

图 4-19　两种 BSC 对土壤微生物量碳、氮（平均值 ± 标准误）的影响：（a）、（c）藻类结皮；（b）、（d）藓类结皮。不同字母表示均值在 $p<0.05$ 水平上差异显著

Figure 4-19　Impact of two crusts on soil microbial biomass carbon and nitrogen（mean ± SE）:（a）、（c）algal crust；（b）、（d）moss crust. Different letters indicate the significant differences among mean values at $p<0.05$

由图4-19b可以看出，人工植被固沙区藓类结皮下0 ~ 10 cm土层中微生物量碳的变化规律为：55龄和47龄固沙区藓类结皮下土壤微生物量碳最高，分别为88 mg kg^{-1}和84.48 mg kg^{-1}，与流沙对照差异显著（$p<0.05$）；30龄和20龄固沙区次之，分别为63.37 mg kg^{-1}和56.32 mg kg^{-1}，与流沙对照差异显著（$p<0.05$）；流沙对照最低。同样，藓类结皮下10 ~ 20 cm土层中微生物量碳的变化规律基本类似，55龄和47龄固沙区最高，分别为49.28 mg kg^{-1}和42.24 mg kg^{-1}，与流沙对照差异显著（$p<0.05$）；30龄和20龄固沙

区次之，分别为31.68 mg kg^{-1}和29.92 mg kg^{-1}，与流沙对照差异显著（$p<0.05$）。但到了20～30 cm土层，55龄、47龄、30龄和20龄固沙区藓类结皮下土壤微生物量碳与流沙对照相比均无显著差异（$p>0.05$）。因此，在0～10 cm和10～20 cm土层，藓类结皮可显著提高土壤微生物量碳，与流沙对照差异显著（$p<0.05$）。随着土层的加深，其影响逐渐减弱，到20～30 cm土层与流沙对照相比已无显著差异（$p>0.05$）。

相关分析结果表明（表4-16），不同固沙年限的藻类结皮和藓类结皮分别与其下土壤微生物量碳存在显著的正相关关系，其相关系数分别为r=0.9869和r=0.9845（$p<0.05$），表明固沙年限越长，BSC定殖时间越久，BSC层越厚，BSC下土壤微生物量碳的含量越高。

表 4-16 不同固沙年限的 BSC 与土壤微生物量和群落的相关分析

Table 4-16 Correction analysis between microbial biomass carbon and nitrogen with the crusts of different sand-fixing time

	项目	线性方程及相关系数	
不同固沙年限的藻类结皮（y）	土壤微生物量碳（x）	y = 7.5928 + 0.7311x	r = 0.9869
	土壤微生物量氮（x）	y = 3.8656 + 0.2183x	r = 0.9826
	磷脂脂肪酸（PLFAs）总含量（x）	y = 17.918 + 0.0996x	r = 0.8837
	细菌 PLFAs（x）	y = 6.009 + 0.0277x	r = 0.9031
	真菌 PLFAs（x）	y = 1.3053 + 0.0423x	r = 0.9722
	细菌 PLFAs/ 真菌 PLFAs（x）	y = 0.2273 + 0.0049x	r = 0.9556
不同固沙年限的藓类结皮（y）	土壤微生物量碳（x）	y = 7.8524 + 0.8587x	r = 0.9845
	土壤微生物量氮（x）	y = 3.9704 + 0.37x	r = 0.9658
	PLFAs 总含量（x）	y = 3.7987 + 4.2223x	r = 0.8257
	细菌 PLFAs（x）	y = 5.593+ 0.0667x	r = 0.9093
	真菌 PLFAs（x）	y = 1.306 + 0.0409x	r = 0.9744
	细菌 PLFAs/ 真菌 PLFAs（x）	y = 0.2509 + 0.0025x	r = 0.9100

BSC对土壤微生物量氮的影响：由图4-19c可以看出，藻类结皮下0～10 cm土层中微生物量氮的变化规律为：55龄和47龄固沙区藻类结皮下土壤微生物量氮最高，分别为6.17 mg kg^{-1}和5.72 mg kg^{-1}，与流沙对照差异显著（$p<0.05$）；30龄和20龄固沙区次之，分别为3.69 mg kg^{-1}和3.73 mg kg^{-1}，与流沙对照差异显著（$p<0.05$）；流沙对照最低。同样，

藻类结皮下10 ~ 20 cm土层中微生物量氮的变化规律基本类似，55龄和47龄固沙区最高，分别为3.08 mg kg^{-1}和2.48 mg kg^{-1}，与流沙对照差异显著（p<0.05）;30龄和20龄固沙区次之，分别为2.25 mg kg^{-1}和1.78 mg kg^{-1}，与流沙对照差异显著（p<0.05）。但到了20 ~ 30 cm土层，55龄、47龄、30龄和20龄固沙区的土壤微生物量氮与流沙对照相比均无显著差异（p>0.05）。因此，在0 ~ 10 cm和10 ~ 20 cm土层，藻类结皮可显著提高土壤微生物量氮，与流沙对照差异显著（p<0.05）。随着土层的加深，其影响逐渐减弱，到20 ~ 30 cm土层已无显著差异（p>0.05）。

由图4-19d可以看出，藓类结皮下0 ~ 10 cm土层中微生物量氮的变化规律为：55龄和47龄固沙区藓类结皮下土壤微生物量氮最高，分别为9.39 mg kg^{-1}和9.07 mg kg^{-1}，与流沙对照差异显著（p<0.05）；30龄和20龄固沙区次之，分别为6.89 mg kg^{-1}和6.06 mg kg^{-1}，与流沙对照差异显著（p<0.05）；流沙对照最低。同样，藓类结皮下10 ~ 20 cm土层中微生物量氮的变化规律基本类似，55龄和47龄固沙区最高，分别为3.65 mg kg^{-1}和3.38 mg kg^{-1}，与流沙对照差异显著（p<0.05）；30龄和20龄固沙区次之，分别为3.01 mg kg^{-1}和2.84 mg kg^{-1}，均与流沙对照差异显著（p<0.05）。 但到了20 ~ 30 cm土层，55龄、47龄、30龄和20龄固沙区的藓类结皮下土壤微生物量氮与流沙对照相比均无显著差异（p>0.05）。因此，在0 ~ 10 cm和10 ~ 20 cm土层，藓类结皮可显著提高土壤微生物量氮，与流沙对照差异显著（p<0.05）。随着土层的加深，其影响逐渐减弱，到20 ~ 30 cm土层与流沙对照相比已无显著差异（p>0.05）。

相关分析结果表明（表4-16），不同固沙年限的藻类结皮和藓类结皮与其下土壤微生物量氮也存在显著的正相关关系，其相关系数分别为r=0.9826和r=0.9658（p<0.05），表明固沙年限越长，BSC定殖时间越久，BSC层越厚，BSC下土壤微生物量氮的含量也越高。可见固沙年限与土壤微生物量碳、氮存在显著的正相关关系（p<0.05），固沙年限越长，BSC层越厚，BSC下土壤微生物量碳、氮的含量越高。

从表4-17可以得出，土壤微生物量碳、氮的含量随着BSC类型、固沙年限和土壤深度的不同而表现出差异。BSC类型影响土壤微生物量碳、氮的含量，藓类结皮下土壤微生物量碳、氮的含量均高于藻类结皮的，差异达到了显著水平（p<0.01）；对相同演替阶段的BSC而言，固沙年限越长，BSC定殖时间越久，BSC层越厚，其下土壤微生物量碳、氮的含量也越高，不同固沙年限之间差异显著（p<0.001）；藻类结皮和藓类结皮下表土层的微生物量碳、氮含量均显著高于下层的（p<0.001），而且这种影响与固沙年限之间存在明显的交互作用。

表 4-17　3 种因素对土壤微生物量碳、氮影响的交互作用

Table 4-17　Tests of among three subjects effects on microbial biomass carbon and nitrogen

变异来源	平方和	自由度	均方	F
微生物量碳				
校正模型	57407.654	29	1979.574	13.773
互作	95061.564	1	95061.564	661.396
BSC 类型	279.046	1	279.046**	1.941
土壤深度	28432.947	2	14216.4733***	98.912
固沙年限	18527.665	4	4631.916***	32.227
BSC 类型 × 土壤深度	373.767	2	186.884	1.300
BSC 类型 × 固沙年限	83.729	4	20.932	0.146
土壤深度 × 固沙年限	9535.755	8	1191.969***	8.293
BSC 类型 × 土壤深度 × 固沙年限	174.745	8	21.843	0.152
误差	8623.718	60	143.729	
微生物量氮				
校正模型	476.367	29	16.426	20.451
互作	856.316	1	856.316	1.006E3
BSC 类型	25.027	1	25.027***	31.160
土壤深度	209.348	2	104.674***	130.321
固沙年限	132.514	4	33.129***	41.245
BSC 类型 × 土壤深度	24.176	2	12.088***	15.050
BSC 类型 × 固沙年限	6.8977	4	1.724	2.147
土壤深度 × 固沙年限	70.165	8	8.771***	10.920
BSC 类型 × 土壤深度 × 固沙年限	8.238	8	1.030	1.282
误差	48.192	60	0.803	

注：**p<0.01，***p<0.001。

BSC对土壤微生物群落的影响：由图4-20a可以看出，藻类结皮和藓类结皮下0 ~ 10 cm土层中磷脂脂肪酸（PLFAs）总含量的变化规律为：55龄、47龄、30龄和20龄固沙区的PLFAs总含量大于流沙对照，且差异显著（$p<0.05$），表明藻类结皮和藓类结皮可显著增加土壤微生物总生物量。由图4-20b可以看出，藻类结皮和藓类结皮下0 ~ 10 cm土层中细菌PLFAs变化规律为55龄、47龄、30龄和20龄固沙区土壤细菌PLFAs含量均大于流沙对照，其中藓类结皮与对照相比差异显著（$p<0.05$），而藻类结皮与对照相比差异不显著，表明藓类结皮可显著增加其下土壤细菌生物量。

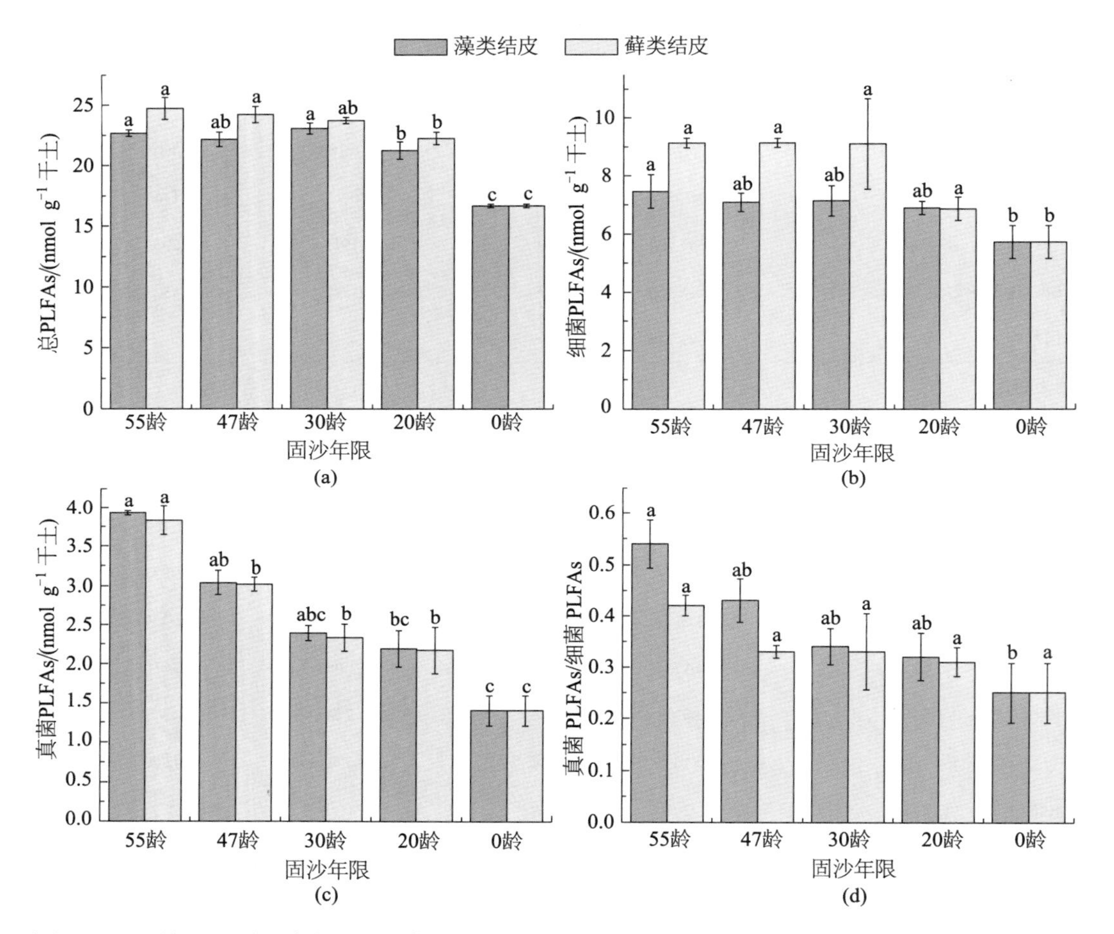

图4-20 两种BSC对土壤微生物群落（平均值 ± 标准误）的影响。不同字母表示均值在$p<0.05$水平上差异显著

Figure 4-20 Impact of two biological soil crusts on soil microbial communities. Mean ± SE ; different letters indicate the significant differences between mean values at $p<0.05$

由图4-20c可以看出，藻类结皮和藓类结皮下0～10 cm土层中真菌PLFAs变化情况为：55龄、47龄、30龄和20龄固沙区的土壤真菌PLFAs含量均大于流沙对照，除30龄和20龄固沙区的藻类结皮下土壤真菌PLFAs含量与对照相比差异不显著外，其他均与对照相比差异显著，表明这两种BSC均可提高其下土壤中的真菌生物量。由图4-20d可以看出，藻类结皮和藓类结皮下0～10 cm土层中真菌PLFAs/细菌PLFAs的变化规律为55龄、47龄、30龄和20龄固沙区的土壤真菌PLFAs/细菌PLFAs的值虽大于流沙对照，但基本上差异不显著（$p>0.05$），表明该固沙区藻类结皮和藓类结皮不能明显改变其下土壤真菌生物量与细菌生物量的比例。

由表4-16相关分析可见，不同固沙年限的藻类结皮和藓类结皮分别与其下土壤PLFAs总含量存在显著的正相关关系，其相关系数分别为r=0.8837和r=0.8257（$p<0.05$），表明固沙年限越长，BSC定殖时间也越久，BSC层越厚，其下土壤微生物总生物量也越高；不同固沙年限的藻类结皮分别与其下土壤细菌PLFAs、真菌PLFAs和真菌PLFAs/细菌PLFAs值存在显著的正相关关系，其相关系数分别为r=0.9031、r=0.9722和r=0.9556（$p<0.05$），表明固沙年限越长，藻类结皮定殖的时间越久，BSC层越厚，其下土壤细菌的生物量和真菌生物量也随之增加，进而引起了真菌生物量与细菌生物量比值的变化；不同固沙年限的藓类结皮分别与其下土壤细菌PLFAs、真菌PLFAs和真菌PLFAs/细菌PLFAs的值存在显著的正相关关系，其相关系数分别为r=0.9093、r=0.9744和r=0.9100（$p<0.05$）。随着固沙年限增加，藓类结皮定殖时间也相应增加，其下土壤细菌生物量和真菌生物量也随之增加，进而引起了真菌生物量与细菌生物量比值的变化。可见，在该人工植被固沙区，土壤微生物总生物量、细菌生物量和真菌生物量随着固沙年限增加而增加，这种增加进而引起了真菌生物量与细菌生物量的比值发生改变。

对两种BSC下土壤PLFAs分析，共检测出13种PLFAs，包括8种细菌PLFAs、1种真菌PLFA和1种放线菌PLFA。其中5种G^+ PLFAs，3种G^- PLFAs。i15:0、i16:0、i17:0、a15:0、a17:0、16:1 ω 7c、18:1 ω 9c和18:1 ω 7c属于细菌，其中i15:0、i16:0、i17:0、a15:0和a17:0属于G^+，16:1 ω 7c、18:1 ω 9c和18:1 ω 7c属于G^-；18:2 ω 6,9c属于真菌；10Me 16:0属于放线菌；16:1 2OH、18:0和16:0也指示微生物群体。由图4-21a和b的13种PLFAs进行主成分分析可以看出，主成分PC1占82.82%、PC2占9.47%。18:1 ω 9c、a15:0、i17:0、16:1 ω 7c和i16:0作为主成分PC1的主要因素，而16:0作为主成分PC2的主要因素。由图4-21a可以看出，藻类结皮和藓类结皮下土壤微生物群落组成在主成分PC1上与流沙对照明显分开，这一结果说明藻类结皮和藓类结皮的存在改变了人工植被固沙区土壤微生物群落组成，这种不同主要是由于G^+和G^-导致的。

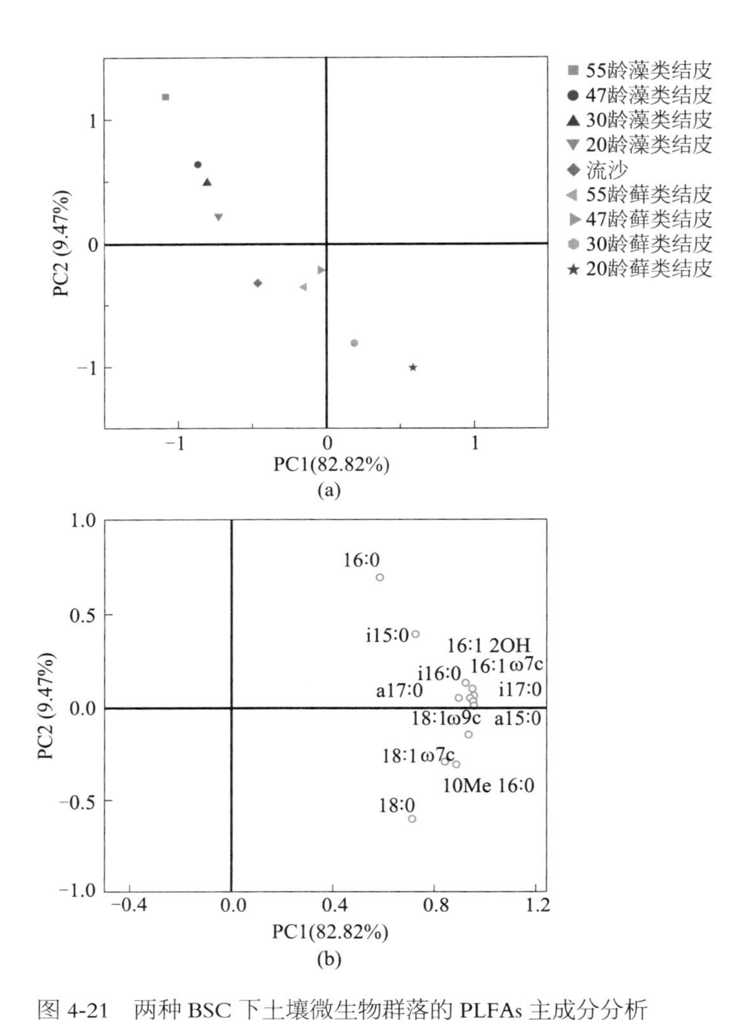

图 4-21　两种 BSC 下土壤微生物群落的 PLFAs 主成分分析

Figure 4-21　Principal component analysis of the variation in the microbial community structure estimated by the concentrations of the PLFAs profiles under two crusts

由表4-18可以得出，BSC类型影响土壤总PLFAs含量，藓类结皮下土壤总PLFAs的含量显著高于藻类结皮（$p<0.01$）。由此可见，相对于演替早期的藻类结皮而言，演替晚期的藓类结皮下土壤微生物总生物量更高，表明藓类结皮更有利于其下土壤微生物的繁衍，但BSC类型没有显著影响土壤细菌PLFAs和真菌PLFAs的含量。固沙年限也是影响土壤总PLFAs、细菌PLFAs和真菌PLFAs含量的一个重要因素，对相同类型的BSC而言，固沙年限越长，BSC定殖时间越久，其下土壤总PLFAs、细菌PLFAs和真菌PLFAs的含量越高，指示微生物总生物量、细菌生物量和真菌生物量越高。相反，固沙年限越短，BSC定殖时间越短，其下土壤总PLFAs、细菌PLFAs和真菌PLFAs的含量越低，指示土壤微生物总生物量、

细菌生物量和真菌生物量越低。然而，固沙年限和BSC类型不能显著影响真菌PLFAs / 细菌PLFAs的值（$p>0.05$）。

表 4-18　两因素对土壤微生物 PLFAs、细菌 PLFAs、真菌 PLFAs 和真菌 PLFAs/ 细菌 PLFAs 的影响及互作

Table 4-18　Tests of between two subjects effects on microbial PLFAs，bacterial PLFAs，fungal PLFAs and the ratio of fungal-to-bacterial PLFAs

变异来源	平方和	自由度	均方	*F*
微生物 PLFAs				
校正模型	222.606	9	24.734	29.372
固沙年限	207.748	4	51.937***	61.676
BSC 类型	9.999	1	9.999**	11.8774
固沙年限 ×BSC 类型	4.858	4	1.214	1.442
误差	16.842	20	0.842	
细菌 PLFAs				
校正模型	45.801	9	5.089	2.072
固沙年限	29.588	4	7.397*	3.012
BSC 类型	9.565	1	9.565	3.895
固沙年限 ×BSC 类型	6.647	4	1.662	0.677
误差	49.117	20	2.456	
真菌 PLFAs				
校正模型	20.950	9	2.328	24.191
固沙年限	20.928	4	5.232***	54.371
BSC 类型	0.011	1	0.011	0.113
固沙年限 ×BSC 类型	0.011	4	0.003	0.030
误差	1.925	20	0.096	
真菌 PLFAs/ 细菌 PLFAs（F/B）				
校正模型	0.205	9	0.023	2.264
固沙年限	0.169	4	0.042	4.184
BSC 类型	0.017	1	0.017	1.668
固沙年限 ×BSC 类型	0.020	4	0.005	0.492
误差	0.202	20	0.010	

注：$*p<0.05$，$**p<0.01$，$***p<0.001$。

BSC下土壤微生物量碳的季节变化：由图4-22a和b可以看出，天然植被区和人工植被固沙区藻类结皮下0～10 cm土层中微生物量碳含量的季节变化规律为：夏季土壤微生物量碳含量最高，分别为105.59 mg kg^{-1}和73.92 mg kg^{-1}，与春季和秋季差异显著（$p<0.05$）；春季次之，分别为68.94 mg kg^{-1}和52.92 mg kg^{-1}，与秋季差异显著（$p<0.05$）；秋季最低。同样，藻类结皮下10～20 cm和20～30 cm土层微生物量碳含量的季节变化规律基本类似，夏季最高，春季次之，秋季最低。因此，天然植被区和人工植被固沙区藻类结皮下0～10 cm、10～20 cm和20～30 cm土层中土壤微生物量碳含量存在明显的季节变化，基本表现为夏季>春季>秋季，秋季的土壤微生物量碳含量低于其他季节。从垂直水平来观察这种季节变化可以发现，藻类结皮下表层土壤微生物量碳含量的季节变化明显，随着土层的加深这种季节变化减弱。

由图4-22c和d可以看出，天然植被区和人工植被固沙区藓类结皮下0～10 cm土层中微生物量碳含量的季节变化规律为：夏季土壤微生物量碳最高，分别为158.39 mg kg^{-1}和88.00 mg kg^{-1}，与秋季差异显著（$p<0.05$）；春季次之，分别为87.07 mg kg^{-1}和61.54 mg kg^{-1}；秋季最低。同样，藓类结皮下10～20 cm和20～30 cm土层微生物量碳含量的季节变化规律基本类似，夏季最高，春季次之，秋季最低。因此，天然植被区和人工植被固沙区藓类结皮下0～10 cm、10～20 cm和20～30 cm土层微生物量碳存在明显的季节变化，基本表现为夏季>春季>秋季，秋季土壤微生物量碳低于其他季节。这种季节变化也受土壤深度的影响，藓类结皮下表层土壤微生物量碳含量的季节变化明显，随着土层的加深这种季节变化有所减弱。

可见，天然植被区和人工植被固沙区藓类结皮和藻类结皮下0～10 cm、10～20 cm和20～30 cm土层中微生物量碳的含量存在明显的季节变化，表现为夏季>春季>秋季，秋季的微生物量碳含量低于其他季节，且这种季节变化也受土壤深度的影响，表层土壤微生物量碳含量季节变化明显，随着土层的加深这种季节变化趋势明显变弱。

BSC下土壤微生物量氮的季节变化：由图4-23a和b可以看出，天然植被区和人工植被固沙区藻类结皮下0～10 cm土层微生物量氮含量的季节变化规律为：夏季土壤微生物量氮的含量最高，分别为7.51 mg kg^{-1}和5.64 mg kg^{-1}，与春季和秋季差异显著（$p<0.05$）；春季次之，分别为5.05 mg kg^{-1}和3.27 mg kg^{-1}；秋季最低。同样，藻类结皮下10～20 cm和20～30 cm土层微生物量氮含量的季节变化规律基本类似，夏季最高，春季次之，秋季最低，但季节之间差异基本不显著。因此，天然植被区和人工植被固沙区藻类结皮下0～10 cm、10～20 cm和20～30 cm土层微生物量氮含量存在明显的季节变化，基本表现为夏季>春

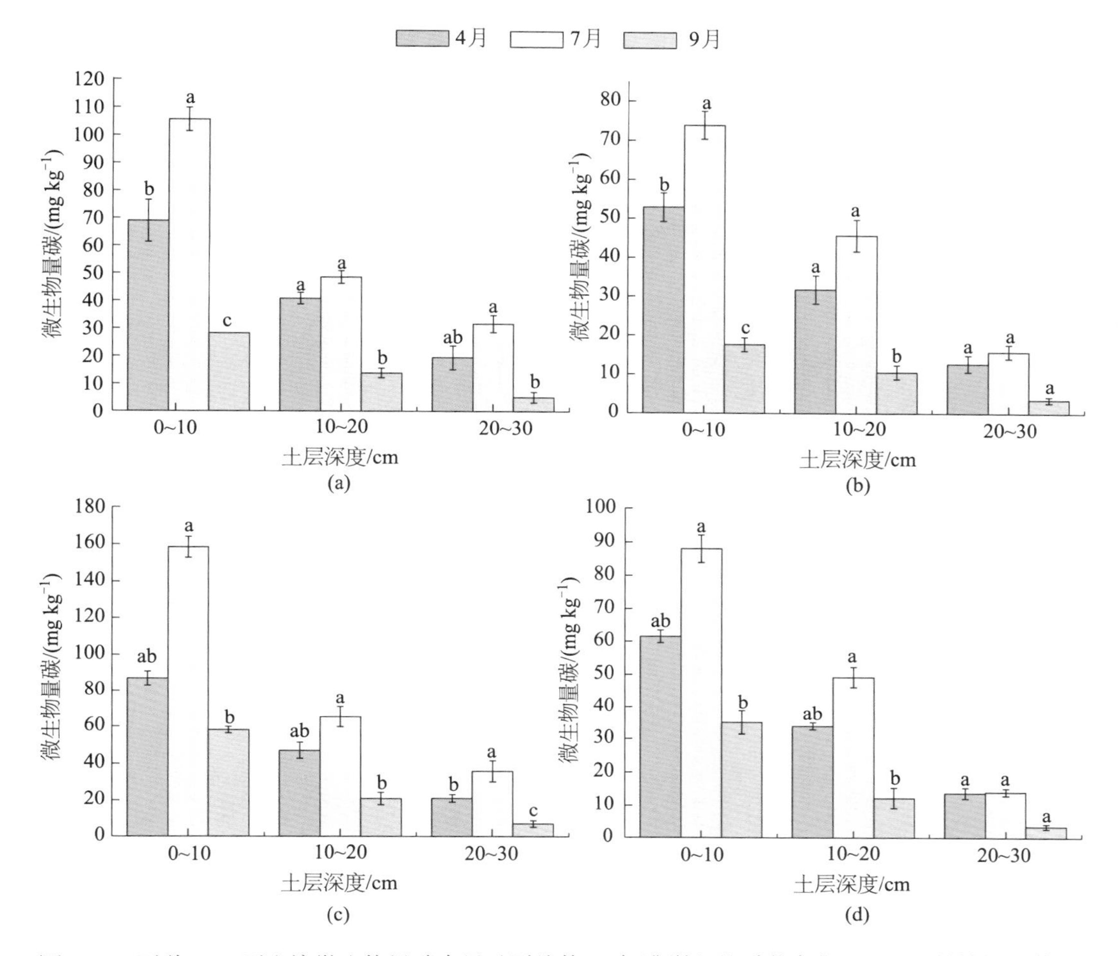

图 4-22　两种 BSC 下土壤微生物量碳含量（平均值 ± 标准误）的季节变化 :（a）天然植被区藻类结皮 ;（b）人工植被固沙区藻类结皮 ;（c）天然植被区藓类结皮 ;（d）人工植被固沙区藓类结皮。不同字母表示均值在 $p<0.05$ 水平上差异显著

Figure 4-22　Seasonal changes of soil microbial biomass carbon (mean ± SE) under two crusts: (a) algal crust in natural vegetation areas ; (b) algal crust in the artificially revegetated desert areas ; (c) moss crust in natural vegetation areas ; (d) moss crust in the artificially revegetated desert areas. Different letters indicate the significant differences between mean values at $p<0.05$

季>秋季，秋季的土壤微生物量氮含量低于其他季节。从垂直水平来观察这种季节变化可以发现，藻类结皮下0 ~ 10 cm土层微生物量氮含量的季节变化明显，但到了10 ~ 20 cm和20 ~ 30 cm土层这种季节变化差异基本不显著。

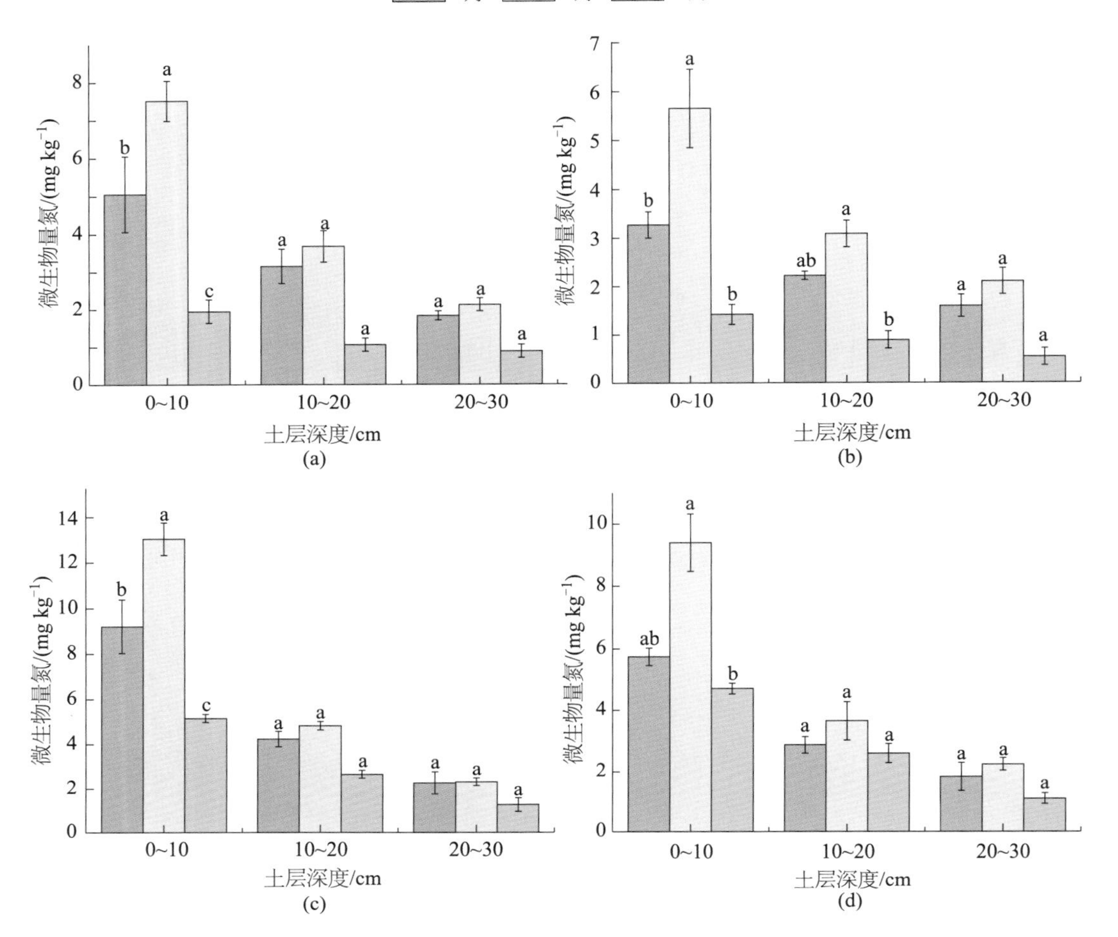

图 4-23　两种 BSC 下土壤微生物量氮含量（平均值 ± 标准误）的季节变化：（a）天然植被区藻类结皮；（b）人工植被固沙区藻类结皮；（c）天然植被区藓类结皮；（d）人工植被固沙区藓类结皮

Figure 4-23　Seasonal changes of soil microbial biomass nitrogen (mean ± SE) under two crusts: (a) algal crust in natural vegetation areas ; (b) algal crust in the artificially revegetated desert areas ; (c) moss crust in natural vegetation areas ; (d) moss crust in the artificially revegetated desert areas

由图4-23c和d可以看出，天然植被区和人工植被固沙区藓类结皮下0 ~ 10 cm土层微生物量氮含量的季节变化规律为：夏季土壤微生物氮最高，分别为13.04 mg kg^{-1}和9.39 mg kg^{-1}，与秋季差异显著（p<0.05）；春季次之，分别为9.20 mg kg^{-1}和5.74 mg kg^{-1}；秋季最低。同样，藓类结皮下10 ~ 20 cm和20 ~ 30 cm土层微生物量氮含量的季节变化规律基本类似，夏季最高，春季次之，秋季最低，但季节之间差异不显著（p>0.05）。因此，天然植被区和人工植被固沙区藓类结皮下土壤微生物量氮含量存在明显的季节变化，基本表现为夏季>春季

>秋季，秋季土壤微生物量氮含量低于其他季节。这种季节变化也受土壤深度的影响，藓类结皮下0 ~ 10 cm土层微生物量氮含量的季节变化差异明显，但到了10 ~ 20 cm和20 ~ 30 cm土层这种季节变化差异不显著。

综上所述，天然植被区和人工植被固沙区藓类结皮和藻类结皮下0 ~ 10 cm、10 ~ 20 cm和20 ~ 30 cm土层微生物量碳和氮含量存在明显的季节变化，均表现为夏季>春季>秋季，秋季的土壤微生物量碳、氮含量低于其他季节，且这种季节变化也受土壤深度的影响，BSC下表层土壤微生物量碳、氮含量季节变化明显，随着土层的加深这种季节变化趋势明显变弱。

BSC对土壤脲酶活性的影响：由图4-24a可以看出，人工植被固沙区藻类结皮下0 ~ 10 cm土层脲酶活性的变化规律为：55龄和47龄固沙区土壤脲酶的活性最高，分别为0.191 mg（g 24 h）$^{-1}$和 0.168 mg（g 24 h）$^{-1}$，与流沙对照差异显著（p<0.05）；30龄和20龄固沙区次之，分别为0.158 mg（g 24 h）$^{-1}$和0.148 mg（g 24 h）$^{-1}$，与流沙对照差异显著（p<0.05）；流沙对照最低。同样，藻类结皮下10 ~ 20 cm土层中脲酶活性的变化规律基本类似，55龄和47龄固沙区最高，30龄和20龄固沙区次之，基本与流沙对照差异显著（p<0.05），但到了20 ~ 30 cm土层，除55龄固沙区外，47龄、30龄和20龄固沙区的土壤脲酶活性与流沙对照相比均无显著差异（p>0.05）。因此，在0 ~ 10 cm和10 ~ 20 cm土层，藻类结皮可显著提高其下土壤脲酶的活性，与流沙对照差异显著，但随着土层的增加，这种影响明显减弱。

由图4-24b可以看出，人工植被固沙区藓类结皮下0 ~ 10 cm土层中脲酶活性的变化规律为55龄和47龄固沙区土壤脲酶的活性最高，分别为0.331 mg（g 24 h）$^{-1}$和0.271 mg（g 24 h）$^{-1}$，与流沙对照差异显著（p<0.05）；30龄和20龄固沙区次之，分别为0.236 mg（g 24 h）$^{-1}$和0.176 mg（g 24 h）$^{-1}$，与流沙对照差异显著（p<0.05）；流沙对照最低。同样，藓类结皮下10 ~ 20 cm和20 ~ 30 cm土层中脲酶活性的变化规律基本类似，55龄和47龄固沙区最高，30龄和20龄固沙区次之，除20龄固沙区藓类结皮下20 ~ 30 cm土层外，它们均与流沙对照差异显著（p<0.05）。因此，在0 ~ 10 cm、10 ~ 20 cm和20 ~ 30 cm土层，藓类结皮均可显著提高土壤脲酶的活性，与流沙对照差异显著（除20龄固沙区的藓类结皮下20 ~ 30 cm土层外），但藓类结皮对其下土壤脲酶活性的影响明显随着土层深度的增加而降低，表层土壤脲酶的活性高。

相关分析结果表明（表4-19），不同固沙年限的藻类结皮和藓类结皮分别与其下土壤脲酶的活性存在显著的正相关关系，其相关系数分别为r=0.9532和r=0.9816（p<0.05）。可见对于同种类型的BSC而言，固沙年限越长，BSC定殖时间也相应越久，BSC层越厚，BSC下土壤脲酶的活性也越高；相反，固沙年限越短，BSC定殖时间也越短，BSC下土壤脲酶的活性也越低。

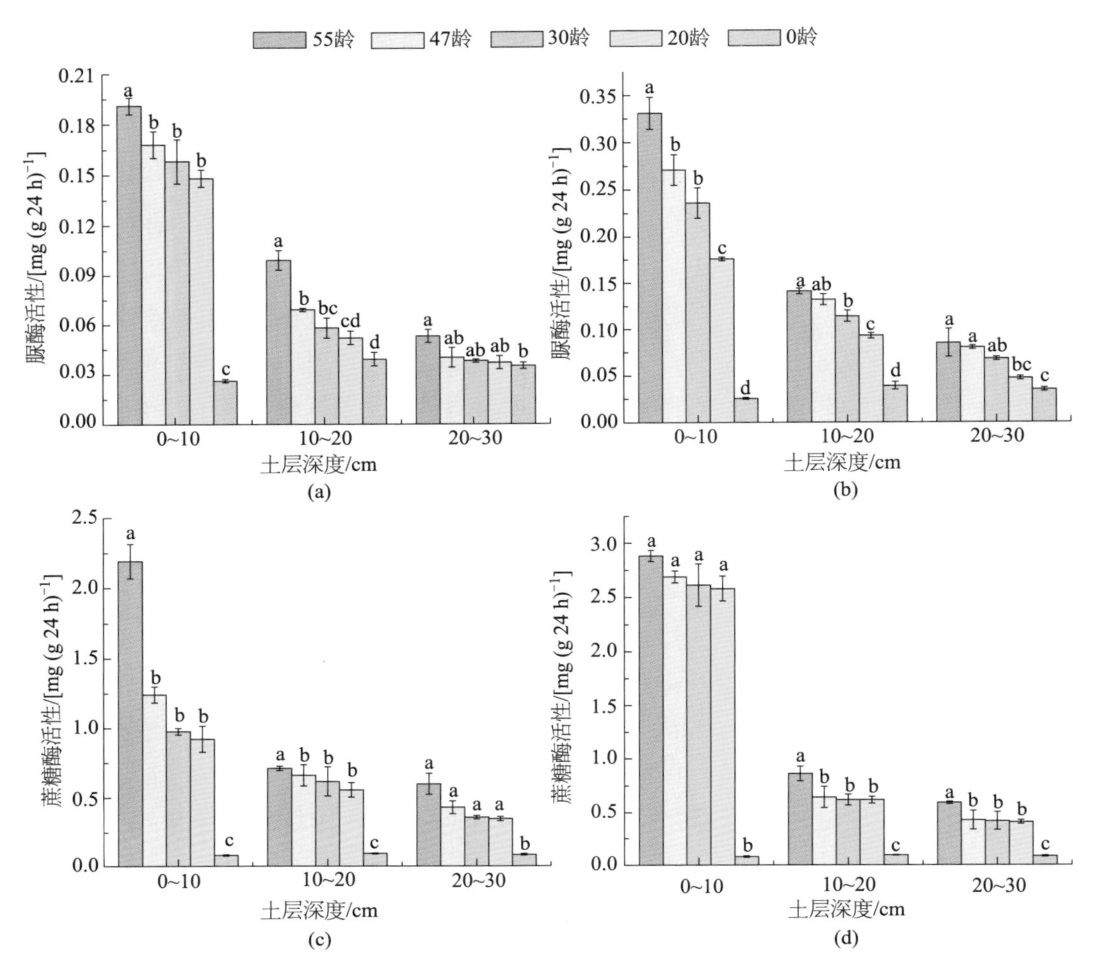

图 4-24　两种 BSC 对土壤脲酶和蔗糖酶活性（平均值 ± 标准误）的影响：(a)、(c) 藻类结皮；(b)、(d) 藓类结皮。不同字母表示均值在 $p<0.05$ 水平上差异显著

Figure 4-24　Impact of two crusts on the activity of soil urease and sucrase (mean ± SE): (a)、(c) algal crust; (b)、(d) moss crust. Different letters indicate the significant differences between mean values at $p<0.05$

BSC 对土壤蔗糖酶活性的影响：由图 4-24c 可以看出，人工植被固沙区藻类结皮下 0～10 cm 土层蔗糖酶活性的变化规律为：55龄和47龄固沙区土壤蔗糖酶的活性最高，分别为2.19 mg(g 24 h)$^{-1}$和1.24 mg(g 24 h)$^{-1}$，与流沙对照差异显著（$p<0.05$）；30龄和20龄固沙区次之，分别为0.977 mg(g 24 h)$^{-1}$和0.921 mg(g 24 h)$^{-1}$，与流沙对照差异显著（$p<0.05$）；流沙对照最低。同样，藻类结皮下 10～20 cm 和 20～30 cm 土层蔗糖酶活性的变化规律基本类似，55龄和47龄固沙区最高，30龄和20龄固沙区次之，它们均与流沙对照差异显著（$p<0.05$）。因此，在 0～10 cm、10～20 cm 和 20～30 cm 土层，藻类结皮可显著提高土壤蔗

表 4-19　不同固沙年限的 BSC 与土壤酶活性的相关分析

Table 4-19　Correction analysis between soil enzyme activity and the crusts of different sand-fixing time

	项目	线性方程及相关系数	
不同固沙年限的藻类结皮（y）	土壤脲酶活性（x）	$y = 0.0013x + 0.0414$	$r = 0.9532$
	土壤蔗糖酶活性（x）	$y = 0.0017x + 0.1414$	$r = 0.9539$
	土壤过氧化氢酶活性（x）	$y = 0.0027x + 0.1722$	$r = 0.8709$
	土壤脱氢酶活性（x）	$y = 0.\ 2704x + 2.4237$	$r = 0.9896$
不同固沙年限的藓类结皮（y）	土壤脲酶活性（x）	$y = 0.0027x + 0.0442$	$r = 0.9816$
	土壤蔗糖酶活性（x）	$y = 0.0213x + 0.3907$	$r = 0.8590$
	土壤过氧化氢酶活性（x）	$y = 0.0057x + 0.1793$	$r = 0.9457$
	土壤脱氢酶活性（x）	$y = 0.4985x + 6.5925$	$r = 0.9221$

糖酶的活性，与流沙对照差异显著。但随着土层深度的增加，这种影响逐渐减弱。

由图4-24d可以看出，人工植被固沙区藓类结皮下0 ~ 10 cm土层中蔗糖酶活性的变化规律为：55龄和47龄固沙区土壤蔗糖酶的活性最高，分别为2.88 mg（g 24 h）$^{-1}$和2.69 mg（g 24 h）$^{-1}$，与流沙对照差异显著（$p<0.05$）；30龄和20龄固沙区次之，分别为2.61 mg（g 24 h）$^{-1}$和2.58 mg（g 24 h）$^{-1}$，与流沙对照差异显著（$p<0.05$）；流沙对照最低。同样，藓类结皮下10 ~ 20 cm和20 ~ 30 cm土层蔗糖酶活性的变化规律基本类似，55龄和47龄固沙区最高，30龄和20龄固沙区次之，它们均与流沙对照差异显著（$p<0.05$）。因此，在0 ~ 10 cm、10 ~ 20 cm和20 ~ 30 cm土层，藓类结皮可显著提高土壤蔗糖酶的活性，与流沙对照差异显著（$p<0.05$），但这种影响随着土层的加深而逐渐减弱。

由表4-19的相关分析结果可见，不同固沙年限的藻类结皮和藓类结皮与其下土壤蔗糖酶的活性存在显著的正相关关系，其相关系数分别为r=0.9539和r=0.8590（$p<0.05$）。可见，对于相同类型的BSC而言，固沙年限越长，BSC定殖时间也越久，BSC层越厚，BSC下土壤蔗糖酶的活性也越高；相反，固沙年限越短，BSC定殖时间也越短，两种BSC下土壤蔗糖酶的活性也越低。

BSC对土壤过氧化氢酶活性的影响：由图4-25a可以看出，人工植被固沙区藻类结皮下0 ~ 10 cm土层过氧化氢酶活性的变化规律为：55龄和47龄固沙区土壤过氧化氢酶的活性最高，分别为0.478 mL（g 20 min）$^{-1}$和0.424 mL（g 20 min）$^{-1}$，与流沙对照差异显著（$p<0.05$）；30龄和20龄固沙区次之，分别为0.418 mL（g 20 min）$^{-1}$和0.411 mL（g 20 min）$^{-1}$，与流沙对照差异显著（$p<0.05$）；流沙对照最低。同样，藻类结皮下10 ~ 20 cm土层中过氧化氢酶活

性的变化规律基本类似，55龄和47龄固沙区最高，30龄和20龄次之，它们均与流沙对照差异显著（$p<0.05$）。但到了20～30 cm土层，55龄、47龄、30龄和20龄固沙区与流沙对照相比均无显著差异（$p>0.05$）。因此，在0～10 cm和10～20 cm，藻类结皮可显著提高其下土壤过氧化氢酶的活性，与流沙对照差异显著（$p<0.05$）。随着土层的加深，这种影响逐渐减弱，到20～30 cm土层与对照相比已无显著差异（$p>0.05$）。

由图4-25b可以看出，人工植被固沙区藓类结皮下0～10 cm土层过氧化氢酶活性的变化规律为：55龄和47龄固沙区土壤过氧化氢酶的活性最高，分别为0.712 mL（g 20 min）$^{-1}$和0.578 mL（g 20 min）$^{-1}$，与流沙对照差异显著（$p<0.05$）；30龄和20龄固沙区次之，分别为0.548 mL（g 20 min）$^{-1}$和0.537 mL（g 20 min）$^{-1}$，与流沙对照差异显著（$p<0.05$）；流沙对照最低。同样，藓类结皮下10～20 cm土层过氧化氢酶活性的变化规律基本类似，55龄和47龄固沙区最高，30龄和20龄固沙区次之，它们均与流沙对照差异显著（$p<0.05$）。但到了20～30 cm土层，除55龄固沙区外，47龄、30龄和20龄固沙区与流沙对照相比均无显著差异（$p>0.05$）。因此，在0～10 cm和10～20 cm土层，藓类结皮可显著提高土壤过氧化氢酶的活性，与流沙对照差异显著（$p<0.05$）。随着土层的加深，这种影响逐渐减弱，到20～30 cm土层已无显著差异（除55龄固沙区的藓类结皮外）（$p>0.05$）。

表4-19的相关分析结果显示，不同固沙年限的藻类结皮和藓类结皮与土壤过氧化氢酶的活性存在显著的正相关关系，其相关系数分别为r=0.8709和r=0.9457（$p<0.05$）。可见对于相同类型的BSC而言，固沙年限越长，BSC定殖时间也越久，BSC层越厚，其下土壤过氧化氢酶的活性也越高。相反，固沙年限越短，BSC定殖时间也越短，BSC层越薄，其下土壤过氧化氢酶的活性也越低。

BSC对土壤脱氢酶活性的影响：由图4-25c可以看出，人工植被固沙区藻类结皮下0～10 cm土层脱氢酶活性的变化规律为：55龄和47龄固沙区土壤脱氢酶的活性最高，分别为40.13 μg（g 24 h）$^{-1}$和38.38 μg（g 24 h）$^{-1}$，与流沙对照差异显著（$p<0.05$）；30龄和20龄固沙区次之，分别为26.78 μg（g 24 h）$^{-1}$和20.38 μg（g 24 h）$^{-1}$，与流沙对照差异显著（$p<0.05$）；流沙对照最低。同样，藻类结皮下10～20 cm土层脱氢酶活性的变化规律基本类似，55龄和47龄固沙区最高，30龄和20龄固沙区次之，它们均与流沙对照差异显著（$p<0.05$）。但到了20～30 cm土层，除55龄固沙区外，47龄、30龄和20龄固沙区与流沙对照相比均无显著差异（$p>0.05$）。因此，在0～10 cm和10～20 cm土层，藻类结皮可显著提高其下土壤脱氢酶的活性，与流沙对照差异显著（$p<0.05$）。随着土层的加深，这种影响逐渐减弱，到20～30 cm土层已无显著影响（除55龄固沙区的藻类结皮外）（$p>0.05$）。

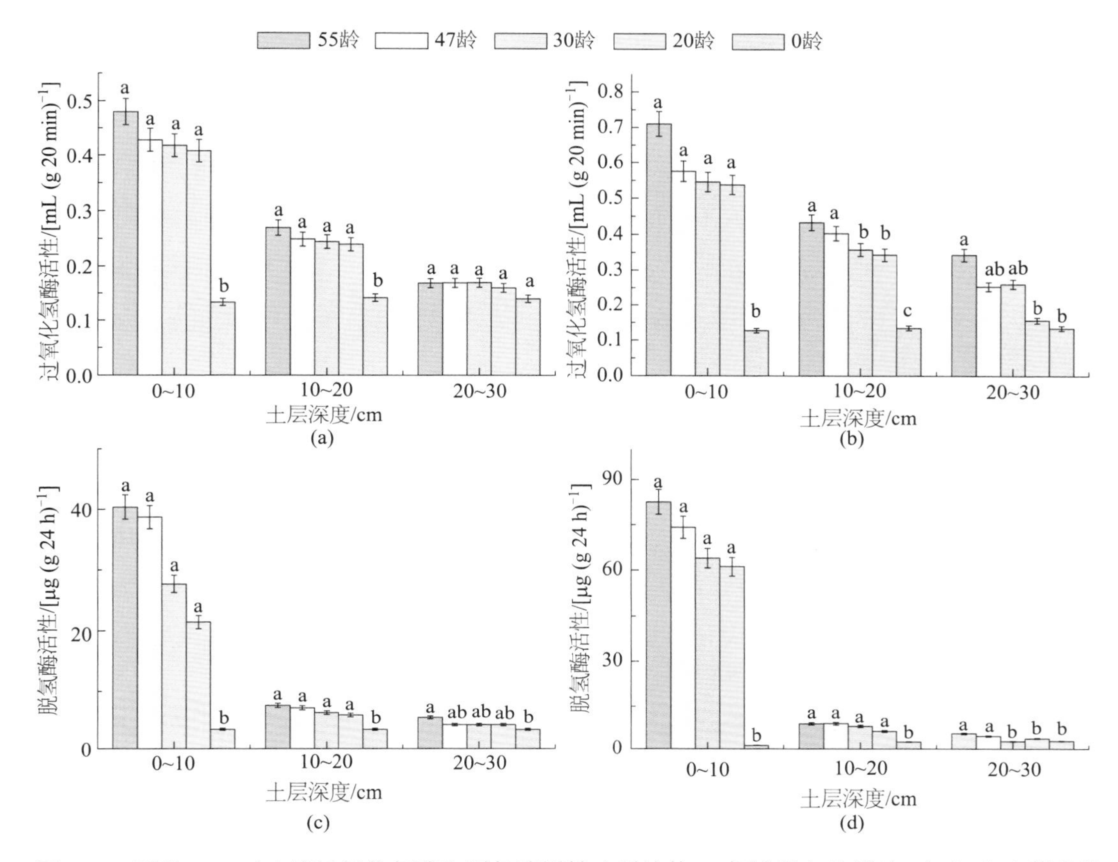

图 4-25　两种 BSC 对土壤过氧化氢酶和脱氢酶活性（平均值 ± 标准误）的影响：(a)、(c) 藻类结皮；(b)、(d) 藓类结皮。不同字母表示均值在 $p<0.05$ 水平上差异显著

Figure 4-25　Impact of two crusts on the activity of soil catalase and dehydrogenase (mean ± SE): (a)、(c) algal crust; (b)、(d) moss crust. Different letters indicate the significant differences between mean values at $p<0.05$

由图4-25d可以看出，人工植被固沙区藓类结皮下0 ~ 10 cm土层脱氢酶活性的变化规律为：55龄和47龄固沙区土壤脱氢酶的活性最高，分别为82.10 μg（g 24 h）$^{-1}$和73.97 μg（g 24 h）$^{-1}$，与流沙对照差异显著（$p<0.05$）；30龄和20龄固沙区次之，分别为63.71 μg（g 24 h）$^{-1}$和61.05 μg（g 24 h）$^{-1}$，与流沙对照差异显著（$p<0.05$）；流沙对照最低。同样，藓类结皮下10 ~ 20 cm和20 ~ 30 cm土层脱氢酶活性的变化规律基本类似，55龄和47龄固沙区最高，30龄和20龄固沙区次之，除30龄和20龄固沙区藓类结皮下20 ~ 30 cm土层外，它们均与流沙对照相比差异显著（$p<0.05$）。因此，在0 ~ 10 cm、10 ~ 20 cm和20 ~ 30 cm土层，藓类结皮可显著提高土壤脱氢酶的活性，与流沙对照差异显著（除30龄和20龄固沙区藓类结皮下20 ~ 30 cm土层外）。但随着土层的加深，这种影响逐渐减弱。

由表4-19的相关分析结果可以看出，不同固沙年限的藻类结皮和藓类结皮与其下土壤脱氢酶的活性存在显著的正相关关系， 其相关系数分别为 r=0.9896 和 r=0.9221（p<0.05）。可见对于相同类型的BSC而言，固沙年限越长，BSC定殖时间也越久，其下土壤脱氢酶的活性也越高。相反，固沙年限越短，BSC定殖时间也越短，其下土壤脱氢酶的活性也越低。综上所述，不同固沙年限的藻类结皮和藓类结皮与其下土壤脲酶、蔗糖酶、过氧化氢酶和脱氢酶的活性均存在正相关关系（p<0.05），表明随着固沙年限增加，藻类结皮和藓类结皮下土壤脲酶、蔗糖酶、过氧化氢酶和脱氢酶的活性增加，指示了土壤微生物活性也增加。

三因素对土壤酶活性影响的交互作用：由表4-20可以得出，BSC覆盖下土壤脲酶的活性随着BSC类型、固沙年限和土壤深度的不同而表现出显著差异（p<0.01）。BSC类型显著影响土壤脲酶的活性，即藓类结皮下土壤脲酶的活性显著高于藻类结皮（p<0.001），而且这种影响分别与固沙年限和土壤深度存在明显互作；固沙年限显著影响土壤脲酶的活性（p<0.001），固沙年限越长，BSC定殖时间也越久，其下土壤脲酶的活性越高，而且这种影响分别与BSC类型和土壤深度存在明显互作；BSC下表层脲酶的活性显著高于下层土壤（p<0.001），而且这种影响分别与固沙年限和BSC类型存在明显互作。

由表4-20也可以得出，BSC覆盖下土壤蔗糖酶的活性随着BSC类型、固沙年限和土壤深度的不同而表现出显著差异（p<0.01）。BSC类型显著影响土壤蔗糖酶的活性，即藓类结皮下土壤蔗糖酶的活性显著高于藻类结皮（p<0.001），且这种影响分别与固沙年限和土壤深度存在明显互作；固沙年限显著影响土壤蔗糖酶活性（p<0.001），固沙年限越长，BSC定殖时间越久，其下土壤蔗糖酶的活性越高，而且这种影响分别与BSC类型和土壤深度存在明显互作；BSC下表层蔗糖酶的活性显著高于下层土壤（p<0.001），而且这种影响分别与固沙年限和BSC类型存在明显互作。可见BSC覆盖下土壤脲酶和蔗糖酶的活性随着BSC类型、固沙年限和土壤深度的不同而表现出显著差异，且这三因素之间交互作用，共同影响土壤脲酶和蔗糖酶的活性。

BSC下土壤酶活性的季节变化——土壤脲酶活性的季节变化：由图4-26a和b可以看出，天然植被区和人工植被固沙区藻类结皮下0～10 cm土层脲酶活性的季节变化规律为：夏季土壤脲酶活性最高，分别为0.311 mg（g 24 h）$^{-1}$和0.191 mg（g 24 h）$^{-1}$，与春季和冬季差异显著（p<0.05）；秋季次之，分别为0.275 mg（g 24 h）$^{-1}$和0.0845 mg（g 24 h）$^{-1}$，与春季差异显著（p<0.05）；春季和冬季最低。同样，藻类结皮下10～20 cm和20～30 cm土层脲酶活性的季节变化规律基本类似，夏季最高，与其他季节差异显著（p<0.05）；秋季次之；春季和冬季最低。因此，天然植被区和人工植被固沙区藻类结皮下0～10 cm、10～20 cm和

表 4-20 三因素对土壤脲酶和蔗糖酶活性影响的交互作用

Table 4-20 Tests of among three subjects effects on the activity of soil urease and sucrase

变异来源	平方和	自由度	均方	*F*
土壤脲酶				
校正模型	0.540	29	0.019	112.086
互作	0.947	1	0.947	5.703E3
BSC 类型	0.045	1	0.045***	2773.854
土壤深度	0.240	2	0.120***	722.831
固沙年限	0.142	4	0.036***	213.935
BSC 类型 × 土壤深度	0.008	2	0.004***	24.304
BSC 类型 × 固沙年限	0.016	4	0.004***	24.525
土壤深度 × 固沙年限	0.080	8	0.010***	60.131
BSC 类型 × 土壤深度 × 固沙年限	0.008	8	0.001***	5.935
误差	0.010	60	0.000	
土壤蔗糖酶				
校正模型	62.477	29	2.154	28.079
互作	64.738	1	64.738	1.144E3
BSC 类型	3.281	1	3.281***	57.985
土壤深度	27.610	2	13.805***	244.010
固沙年限	14.965	4	3.741***	66.1277
BSC 类型 × 土壤深度	5.591	2	2.795***	49.411
BSC 类型 × 固沙年限	1.092	4	0.273***	4.824
土壤深度 × 固沙年限	7.849	8	0.981***	17.343
BSC 类型 × 土壤深度 × 固沙年限	2.089	8	0.261***	4.615
误差	3.395	60	0.057	

注：***$p<0.001$。

20 ~ 30 cm土层的土壤脲酶活性存在明显的季节变化，表现为夏季>秋季>冬季和春季，春季和冬季的脲酶活性低于其他季节。从垂直水平来观察这种季节变化可以发现，藻类结皮下表层土壤脲酶活性的季节变化明显，随着土层的加深这种季节变化明显减弱。

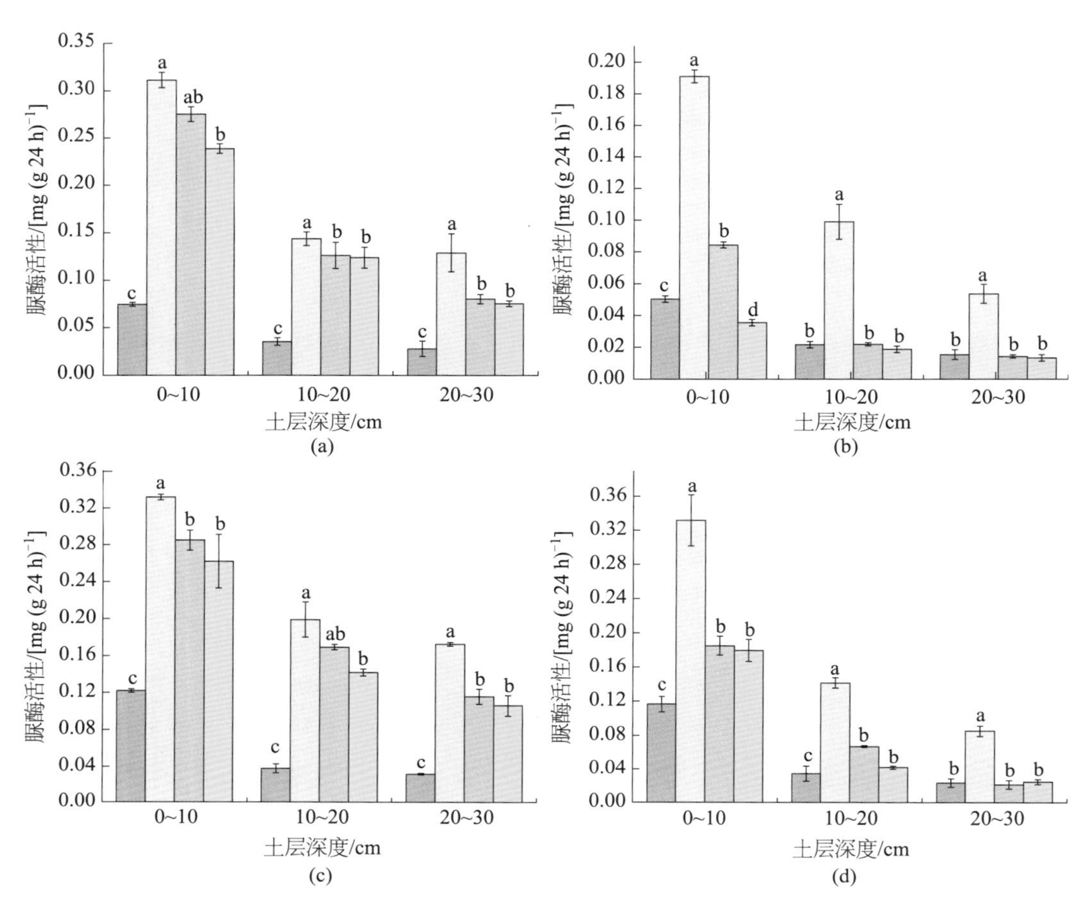

图 4-26　两种 BSC 下土壤脲酶活性（平均值 ± 标准误）的季节变化 ;（a）天然植被区藻类结皮 ;（b）人工植被固沙区藻类结皮 ;（c）天然植被区藓类结皮 ;（d）人工植被固沙区藓类结皮。不同字母表示均值在 $p<0.05$ 水平上差异显著

Figure 4-26　Seasonal changes of soil urease activity (mean ± SE) under two crusts: (a) algal crust in natural vegetation area ; (b) algal crust in the artificially revegetated desert area ; (c) moss crust in natural vegetation area ; (d) moss crust in the artificially revegetated desert area. Different letters indicate the significant differences between mean values at $p<0.05$

由图4-26c和d可以看出，天然植被区和人工植被固沙区藓类结皮下0 ~ 10 cm土层脲酶的季节变化规律为：夏季脲酶的活性最高，分别为0.332 mg（g 24 h）$^{-1}$和0.331 mg（g 24 h）$^{-1}$，与其他季节差异显著（$p<0.05$）；秋季次之，分别为0.2849 mg（g 24 h）$^{-1}$

和0.1851 mg（g 24 h）$^{-1}$，与春季差异显著（p<0.05）；冬季再次之；春季最低。同样，藓类结皮下10～20 cm和20～30 cm土层脲酶的活性季节变化规律基本类似，夏季最高，与春季和冬季差异显著（p<0.05），秋季次之，冬季再次之，春季最低。因此，天然植被区和人工植被固沙区藓类结皮下0～10 cm、10～20 cm和20～30 cm土层脲酶的活性存在明显的季节变化，表现为夏季>秋季>冬季>春季，春季脲酶的活性低于其他季节。从垂直水平来观察这种季节变化可以发现，藓类结皮下表层土壤脲酶活性的季节变化明显，随着土层的加深这种季节变化减弱。

可见，天然植被区和人工植被固沙区藓类结皮和藻类结皮下 0～10 cm、10～20 cm 和 20～30 cm 土层脲酶活性存在明显的季节变化，表现为夏季 > 秋季 > 冬季和春季，春季和冬季脲酶的活性低于其他季节，且随着土层的加深这种季节变化趋势明显变弱。

BSC下土壤酶活性的季节变化——土壤蔗糖酶活性的季节变化：由图4-27a和b可以看出，天然植被区和人工植被固沙区藻类结皮下0～10 cm土层中蔗糖酶活性的季节变化规律为：夏季土壤蔗糖酶活性最高，分别为2.72 mg（g 24 h）$^{-1}$和2.19 mg（g 24 h）$^{-1}$，与其他季节差异显著（p<0.05）；秋季次之，分别为1.57 mg（g 24 h）$^{-1}$和1.52 mg（g 24 h）$^{-1}$，与春季和冬季差异显著（p<0.05）；冬季再次之；春季最低。同样，藻类结皮下10～20 cm和20～30 cm土层蔗糖酶活性季节变化规律基本类似，夏季最高，秋季次之，冬季再次之，春季最低。因此，天然植被区和人工植被固沙区藻类结皮下0～10 cm、10～20 cm和20～30 cm土层土壤蔗糖酶活性存在明显的季节变化，基本表现为夏季>秋季>冬季>春季，春季的蔗糖酶活性低于其他季节。从垂直水平来观察这种季节变化可以发现，藻类结皮下表层土壤蔗糖酶活性的季节变化明显，随着土层的加深这种季节变化减弱。

由图4-27c和d可以看出，天然植被区和人工植被固沙区藓类结皮下0～10 cm土层蔗糖酶活性的季节变化规律为：夏季土壤蔗糖酶活性最高，分别为5.57 mg（g 24 h）$^{-1}$和2.88 mg（g 24 h）$^{-1}$，与春季和冬季差异显著（p<0.05）；秋季次之，分别为3.17 mg（g 24 h）$^{-1}$和2.35 mg（g 24 h）$^{-1}$；冬季再次之；春季最低。同样，藓类结皮下10～20 cm和20～30 cm土层蔗糖酶活性季节变化规律基本类似，夏季最高，与春季和冬季差异显著（p<0.05）；秋季次之；冬季再次之；春季最低。因此，天然植被区和人工植被固沙区藓类结皮下0～10 cm、10～20 cm和20～30 cm土层蔗糖酶活性存在明显的季节变化，基本表现为夏季>秋季>冬季>春季，春季蔗糖酶活性低于其他季节。这种季节变化也受土壤深度的影响，藓类结皮下表层土壤蔗糖酶活性的季节变化明显，随着土层的加深这种季节变化减弱。

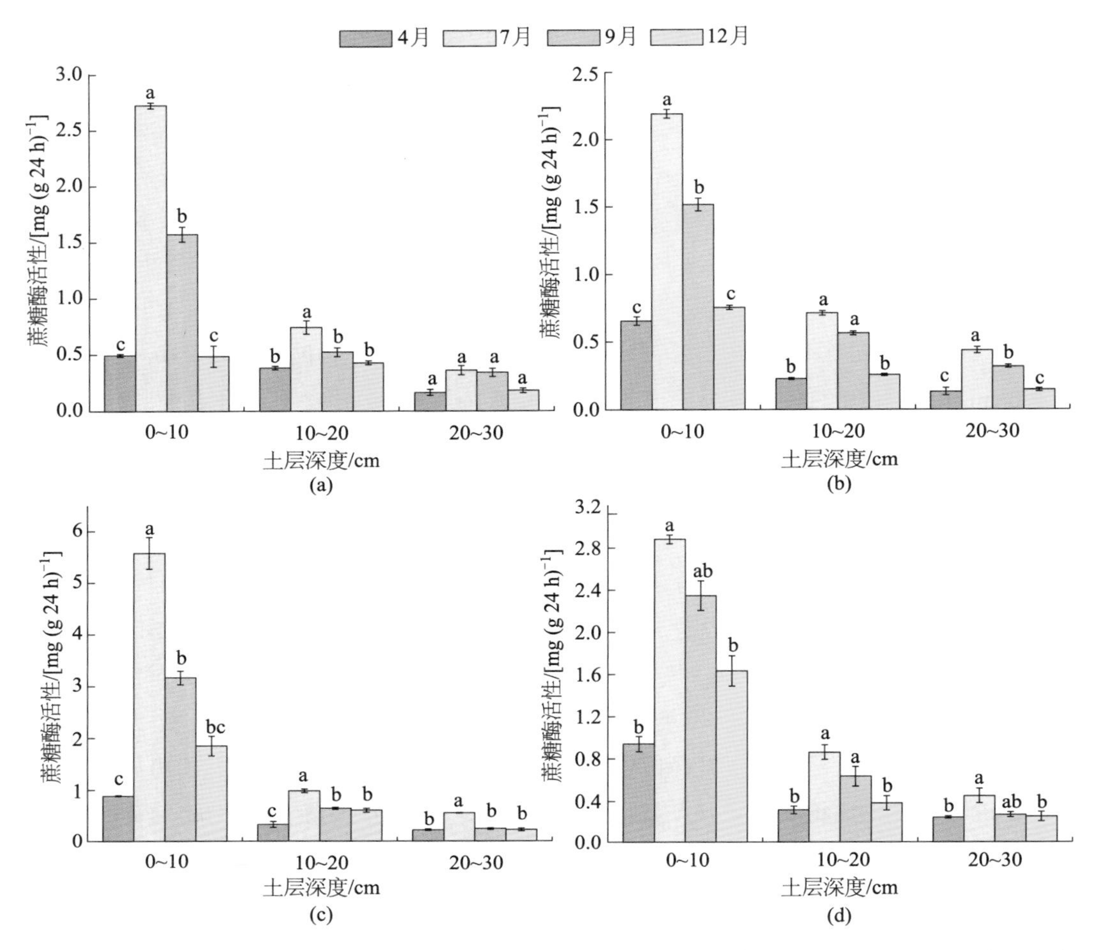

图 4-27　两种 BSC 下土壤蔗糖酶活性（平均值 ± 标准误）的季节变化：(a) 天然植被区藻类结皮；(b) 人工植被固沙区藻类结皮；(c) 天然植被区藓类结皮；(d) 人工植被固沙区藓类结皮。不同字母表示均值在 $p<0.05$ 水平上差异显著

Figure 4-27　Seasonal changes of soil sucrase activity (mean ± SE) under two crusts: (a) algal crust in natural vegetation area; (b) algal crust in the artificially revegetated desert area; (c) moss crust in natural vegetation area; (d) moss crust in the artificially revegetated desert area. Different letters indicate the significant differences between mean values at $p<0.05$

可见，天然植被区和人工植被固沙区藓类结皮和藻类结皮下 0 ~ 10 cm、10 ~ 20 cm 和 20 ~ 30 cm 土层蔗糖酶活性存在明显的季节变化，基本表现为夏季 > 秋季 > 冬季 > 春季，春季的蔗糖酶活性低于其他季节，且土壤蔗糖酶活性季节变化也受土壤深度的影响，表层土壤蔗糖酶季节变化明显，随着土层的加深这种季节变化趋势明显变弱。

BSC下土壤酶活性的季节变化——土壤过氧化氢酶活性的季节变化：由图4-28a和b可以看出，天然植被区和人工植被固沙区藻类结皮下0～10 cm土层过氧化氢酶活性的季节变化规律为：夏季土壤过氧化氢酶活性最高，分别为0.587 mL（g 20 min）$^{-1}$和0.477 mL（g 20 min）$^{-1}$，与春季和冬季差异显著（p<0.05）；秋季次之，分别为0.443 mL（g 20 min）$^{-1}$和0.431 mL（g 20 min）$^{-1}$，与春季和冬季差异显著（p<0.05）；春季再次之，与冬季差异显著（p<0.05）；冬季最低。同样，藻类结皮下10～20 cm和20～30 cm土层过氧化氢酶活性季节变化规律基本类似，夏季最高，秋季次之，春季再次之，冬季最低。因此，天然植被区和人工植被固沙区藻类结皮下0～10 cm、10～20 cm和20～30 cm土层，土壤过氧化氢酶活性存在明显的季节变化，基本表现为夏季>秋季>春季>冬季，冬季的过氧化氢酶活性低于其他季节。从垂直水平来观察这种季节变化可以发现，藻类结皮下表层土壤过氧化氢酶活性的季节变化明显，随着土层的加深这种季节变化减弱。

由图4-28c和d可以看出，天然植被区和人工植被固沙区藓类结皮下0～10 cm土层过氧化氢酶活性的季节变化规律为：夏季土壤过氧化氢酶活性最高，分别为1.020 mL（g 20 min）$^{-1}$和0.699 mL（g 20 min）$^{-1}$，与其他季节差异显著（p<0.05）；秋季次之，分别为0.650 mL（g 20 min）$^{-1}$和0.494 mL（g 20 min）$^{-1}$，与冬季差异显著（p<0.05）；春季再次之，与冬季差异显著（p<0.05）；冬季最低。同样，藓类结皮下10～20 cm和20～30 cm土层过氧化氢酶活性季节变化规律基本类似，夏季最高，秋季次之，春季再次之，冬季最低。因此，天然植被区和人工植被固沙区藓类结皮下0～10 cm、10～20 cm和20～30 cm土层，过氧化氢酶活性存在明显的季节变化，基本表现为夏季>秋季>春季>冬季，冬季的过氧化氢酶活性低于其他季节。从垂直水平来观察这种季节变化可以发现，藓类结皮下表层土壤过氧化氢酶活性的季节变化明显，随着土层的加深这种季节变化减弱。

可见，天然植被区和人工植被固沙区藓类结皮和藻类结皮下0～10 cm、10～20 cm和20～30 cm土层过氧化氢酶的活性存在明显的季节变化，基本表现为夏季 > 秋季 > 春季 > 冬季，冬季的过氧化氢酶活性低于其他季节，且随土层的加深这种季节变化趋势明显变弱。

BSC下土壤酶活性的季节变化——土壤脱氢酶活性的季节变化：由图4-29a和b可以看出，天然植被区和人工植被固沙区藻类结皮下0～10 cm土层脱氢酶活性的季节变化规律为：夏季土壤脱氢酶活性最高，分别为74.46 μg（g 24 h）$^{-1}$和40.13 μg（g 24 h）$^{-1}$，与其他季节差异显著（p<0.05）；秋季次之，分别为41.11 μg（g 24 h）$^{-1}$和26.29 μg（g 24 h）$^{-1}$；春季再次之；冬季最低，但它们之间差异不显著（p>0.05）。同样，藻类结皮下10～20 cm和20～30 cm土层脱氢酶活性的季节变化规律基本类似，夏季最高，秋季次之，春季再次之，冬季最低，但它们之间差异小。因此，天然植被区和人工植被固沙区藻类结皮下0～10 cm、10～20 cm

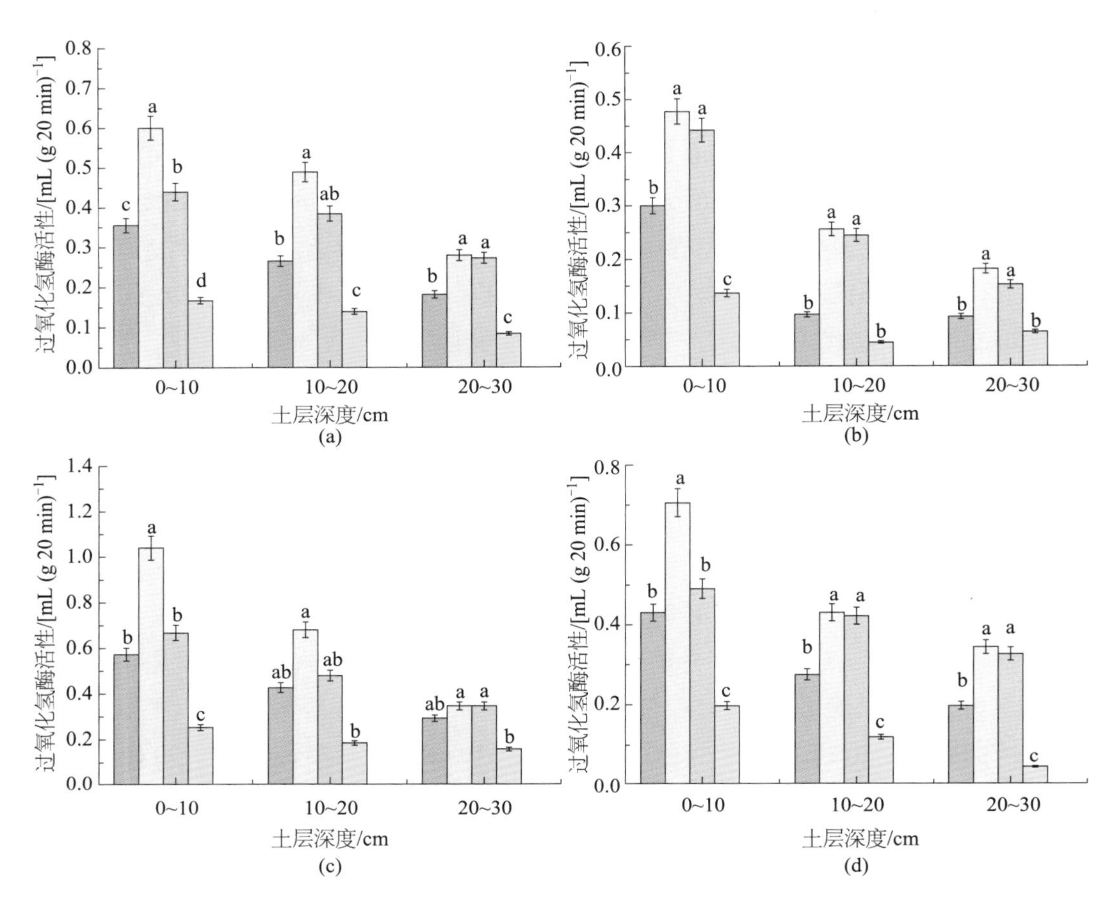

图 4-28　两种 BSC 下土壤过氧化氢酶活性（平均值 ± 标准误）的季节变化：（a）天然植被区藻类结皮；（b）人工植被固沙区藻类结皮；（c）天然植被区藓类结皮；（d）人工植被固沙区藓类结皮。不同字母表示均值在 $p<0.05$ 水平上差异显著

Figure 4-28　Seasonal changes of soil catalase activity（mean ± SE）under two crusts:（a）algal crust in natural vegetation area；（b）algal crust in the artificially revegetated desert area；（c）moss crust in natural vegetation area；（d）moss crust in the artificially revegetated desert area. Different letters indicate the significant differences between mean values at $p<0.05$

和20 ~ 30 cm土层土壤脱氢酶的活性存在明显的季节变化，基本表现为夏季>秋季>春季>冬季，冬季土壤脱氢酶活性低于其他季节。从垂直水平来观察这种季节变化可以发现，藻类结皮下表层土壤脱氢酶活性的季节变化明显，随着土层的加深这种季节变化明显减弱。

由图4-29c和d可以看出，天然植被区和人工植被固沙区藓类结皮下0 ~ 10 cm土层脱氢酶活性的季节变化规律为：夏季土壤脱氢酶活性最高，分别为119.89 μg（g 24 h）$^{-1}$和

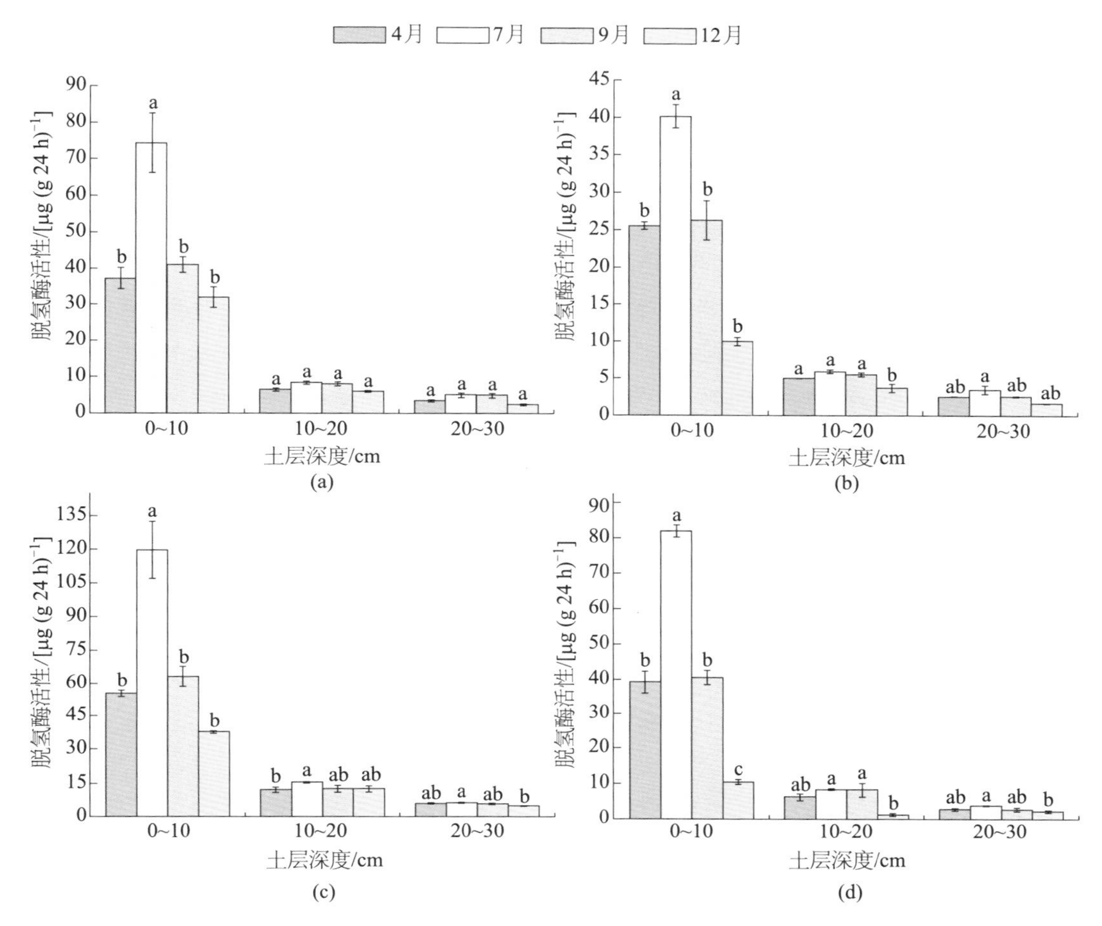

图 4-29　两种 BSC 下土壤脱氢酶活性（平均值 ± 标准误）的季节变化：(a) 天然植被区藻类结皮；(b) 人工植被固沙区藻类结皮；(c) 天然植被区藓类结皮；(d) 人工植被固沙区藓类结皮。不同字母表示均值在 $p<0.05$ 水平上差异显著

Figure 4-29　Seasonal changes of soil dehydrogenase activity (mean ± SE) under two crusts: (a) algal crust in natural vegetation area; (b) algal crust in the artificially revegetated desert area; (c) moss crust in natural vegetation area; (d) moss crust in the artificially revegetated desert area. Different letters indicate the significant differences between mean values at $p<0.05$

82.10 μg(g 24 h)$^{-1}$，与其他季节差异显著（$p<0.05$）；秋季次之，分别为63.45 μg(g 24 h)$^{-1}$和40.69 μg(g 24 h)$^{-1}$；春季再次之；冬季最低。同样，藓类结皮下10～20 cm和20～30 cm土层脱氢酶活性季节变化规律基本类似，夏季最高，秋季次之，春季再次之，冬季最低，但季度之间差异小。因此，天然植被区和人工植被固沙区藓类结皮下0～10 cm、10～20 cm和20～30 cm土层脱氢酶的活性存在明显的季节变化，基本表现为夏季>秋季>春季>冬季，

冬季土壤脱氢酶的活性低于其他季节。从垂直水平来观察这种季节变化可以发现，藓类结皮下表层土壤脱氢酶活性的季节变化明显，随着土层的加深这种季节变化明显减弱。

因此，天然植被区和人工植被固沙区藓类结皮和藻类结皮下0～10 cm、10～20 cm和20～30 cm土层脱氢酶活性存在明显的季节变化，基本表现为夏季>秋季>春季>冬季，冬季土壤脱氢酶的活性低于其他季节，且随着土层的加深这种季节变化趋势明显变弱。

综上所述，在天然植被区和人工植被固沙区藓类结皮和藻类结皮下0～10 cm、10～20 cm和20～30 cm土层脲酶、蔗糖酶、过氧化氢酶和脱氢酶活性存在明显的季节变化，其中脲酶和蔗糖酶的活性基本表现为夏季>秋季>冬季>春季；而过氧化氢酶和脱氢酶活性基本表现为夏季>秋季>春季>冬季。而且，这几种土壤酶活性的季节变化也受土壤深度的影响，表层土壤酶季节变化明显，随着土层的加深这种季节变化趋势明显变弱。

由表4-21可见，BSC覆盖下土壤过氧化氢酶的活性随着BSC类型、固沙年限和土壤深度的不同而表现出显著差异（$p<0.001$）。BSC类型显著影响土壤过氧化氢酶的活性（$p<0.001$），相对于藻类结皮，藓类结皮下土壤过氧化氢酶的活性更高，而且这种影响与固沙年限之间存在明显互作；固沙年限显著影响土壤过氧化氢酶的活性（$p<0.001$），固沙年限越长，BSC定殖时间越久，其下土壤过氧化氢酶的活性越高，而且这种影响分别与BSC类型和土壤深度存在明显互作；BSC下表层过氧化氢酶的活性显著高于下层（$p<0.001$），而且这种影响与固沙年限存在明显互作。

由表4-21可见，BSC覆盖下土壤脱氢酶的活性随着BSC类型、固沙年限和土壤深度的不同而表现出显著差异（$p<0.001$）。BSC类型显著影响土壤过氧化氢酶的活性，藓类结皮下土壤脱氢酶的活性显著高于藻类结皮（$p<0.001$），而且这种影响分别与固沙年限和土壤深度存在明显互作；固沙年限显著影响土壤脱氢酶的活性（$p<0.001$），固沙年限越长，BSC定殖时间越久，BSC下土壤脱氢酶的活性越高，而且这种影响分别与BSC类型和土壤深度存在明显互作；BSC下表层脱氢酶的活性显著高于下层（$p<0.001$），而且这种影响分别与固沙年限和BSC类型存在明显互作。

综上所述，人工植被固沙区BSC覆盖下土壤脲酶、蔗糖酶、过氧化氢酶和脱氢酶的活性随着BSC类型、固沙年限和土壤深度的不同而表现出显著差异，它们之间相互作用共同影响BSC下土壤脲酶、蔗糖酶、过氧化氢酶和脱氢酶的活性。

表 4-21　3 因素对土壤过氧化氢酶和脱氢酶活性影响的交互作用

Table 4-21　Tests of among three subjects effects on the activity of soil catalase and dehydrogenase

变异来源	平方和	自由度	均方	*F*
土壤过氧化氢酶				
校正模型	2.195	29	0.076	26.735
互作	8.274	1	8.274	2.923E3
BSC 类型	0.226	1	0.226***	79.986
土壤深度	0.918	2	0.459***	162.098
固沙年限	0.708	4	0.177***	62.537
BSC 类型 × 土壤深度	0.013	2	0.007	2.381
BSC 类型 × 固沙年限	0.085	4	0.021***	7.469
土壤深度 × 固沙年限	0.234	8	0.029***	10.317
BSC 类型 × 土壤深度 × 固沙年限	0.011	8	0.001	0.477
误差	0.170	60	0.003	
土壤脱氢酶				
校正模型	52391.976	29	1806.620	253.646
互作	24192.510	1	24192.510	3.397E3
BSC 类型	29777.428	1	2977.428***	418.025
土壤深度	28568.652	2	14284.326***	2,005E3
固沙年限	5833.902	4	1458.476***	204.767
BSC 类型 × 土壤深度	4839.502	2	2419.752***	339.7728
BSC 类型 × 固沙年限	759.377	4	189.844***	26.654
土壤深度 × 固沙年限	8190.717	8	1023.840***	143.745
BSC 类型 × 土壤深度 × 固沙年限	1222.399	8	152.800***	21.453
误差	427.356	60	7.123	

注：***p<0.001。

（2）BSC对土壤线虫群落的影响

本研究以1956年、1964年、1981年和1991年建立的腾格里沙漠东南缘的人工植被固沙区BSC覆盖的沙丘土壤为研究对象，研究BSC对土壤线虫群落影响的时空变化。分别采集不同固沙年限藓类结皮和藻类结皮下0～10 cm、10～20 cm、20～30 cm的土壤样品（2010年7月），以流沙为对照，共采集72个BSC下土样和9个对照，分别统计土样中线虫的种类与数量；采集人工植被固沙区藓类结皮和藻类结皮下0～10 cm、10～20 cm、20～30 cm的土壤样品（分别于2011年4月、7月、9月、12月采集样品），以沙坡头地区的红卫天然植被区为参照，4次共采集土壤样品114个，统计分析BSC下土壤线虫群落的季节变化。

BSC下土壤线虫群落组成：本研究对腾格里沙漠东南缘的人工植被固沙区藻类结皮和藓类结皮下土壤线虫进行调查，共鉴定出土壤线虫29属，隶属2纲6目19科，包括食细菌线虫，其占的属数最多，为10属；其次为食真菌线虫共7属；再次为杂食－捕食线虫和植食性线虫分别为6属（表4-22）。藻类结皮下土壤线虫共鉴定出6目14科20属，包括食细菌线虫7属，所占比例最大，占线虫总数的53.57%；其次为食真菌线虫共4属，占线虫总数27.83%；再次为杂食－捕食线虫共6属，占线虫总数18.51%；植食性线虫仅3属，占线虫总数最少，为0.09%。藓类结皮下共鉴定出土壤线虫24属，隶属2纲6目15科，包括食细菌线虫9属，所占比例最大，占线虫总数的61.57%；其次为杂食－捕食线虫共5属，占线虫总数20.08%；再次为食真菌线虫共6属，占线虫总数17.92%；植食性线虫仅4属，占线虫总数最少，为0.43%。藻类结皮和藓类结皮下优势的线虫类群为丽突属（*Acrobeles*）、拟丽突属（*Acrobeloides*）、鹿角唇属（*Cervidellus*）、滑刃属（*Aphelenchoides*）。

由图4-30可以看出，人工植被固沙区藻类结皮和藓类结皮下不同土层线虫营养类群的组成中，食细菌线虫均为数量最丰富的营养类群，占土壤线虫总数的50%左右，食真菌线虫和捕食－杂食线虫数量位居第二，植食性线虫数量最少，占的比例最低；通过相关性分析表明，食细菌线虫、食真菌线虫和捕食－杂食线虫的数量与不同固沙年限的藻类结皮呈显著的正相关关系，相关系数分别为r=0.894、r=0.980和r=0.970（p<0.05）；同样，食细菌线虫、食真菌线虫和捕食－杂食线虫的数量与不同固沙年限的藓类结皮呈显著的正相关关系，相关系数分别为r=0.965、r=0.959和r=0.958（p<0.05）；但植食性线虫与不同固沙年限的BSC均无相关性。

表 4-22　藻类结皮和藓类结皮下土壤线虫属和相对多度

Table 4-22　Soil nematodes genera and mean relative abundance under algal and moss crusts

营养类群	目	科	属	c-p 类群	相对多度 / %	
					藻类结皮	藓类结皮
食细菌线虫					53.57	61.57
	小杆目	头叶科	丽突属	2	34.44	45.15
	小杆目	头叶科	拟丽突属	2	7.83	7.11
	小杆目	短腔科	短腔属	1	2.15	0.58
	小杆目	头叶科	鹿角唇属	2	6.86	7.80
	小杆目	头叶科	板唇属	2	1.52	0.53
	窄咽目	绕线科	唇绕线属	2	0	0.19
	小杆目	头叶科	真头叶属	2	0	0.06
	色矛目	微咽科	微咽属	2	0.71	0.07
	色矛目	色矛科	原色矛属	3	0.06	0
	小杆目	伪双胃总科	伪双胃属	1	0	0.08
食真菌线虫					27.83	17.92
	垫刃目	滑刃科	滑刃属	2	23.23	14.96
	垫刃目	真滑刃科	真滑刃属	2	3.46	2.65
	小杆目	双胃科	双胃属	3	0	0.01
	垫刃目	伪垫刃科	伪垫刃属	2	0.45	0.12
	矛线目	垫咽科	垫咽属	4	0	0.11
	矛线目	膜皮科	巨宫属	3	0.69	0
	矛线目	垫咽科	小剑属	4	0	0.07
植食性线虫					0.09	0.43
	垫刃目	环科	轮属	3	0	0.32
	矛线目	丝尾科	丝尾属	2	0.05	0
	垫刃目	环科	鞘属	3	0	0.07
	垫刃目	环科	半轮属	3	0	0.03
	垫刃目	异皮科	异皮属	3	0.03	0
	垫刃目	垫刃科	垫刃属	2	0.01	0.01
捕食-杂食线虫					18.51	20.08
	单齿目	单齿科	倒齿属	4	0.13	0.09
	单齿目	单齿科	基齿属	4	1.07	0.40
	单齿目	单齿科	等齿属	4	0.05	0
	单齿目	单齿科	单齿属	4	4.07	2.98
	矛线目	小穿咽科	小穿咽属	5	0.49	0.61
	矛线目	穿咽科	穿咽属	5	12.7	16.0

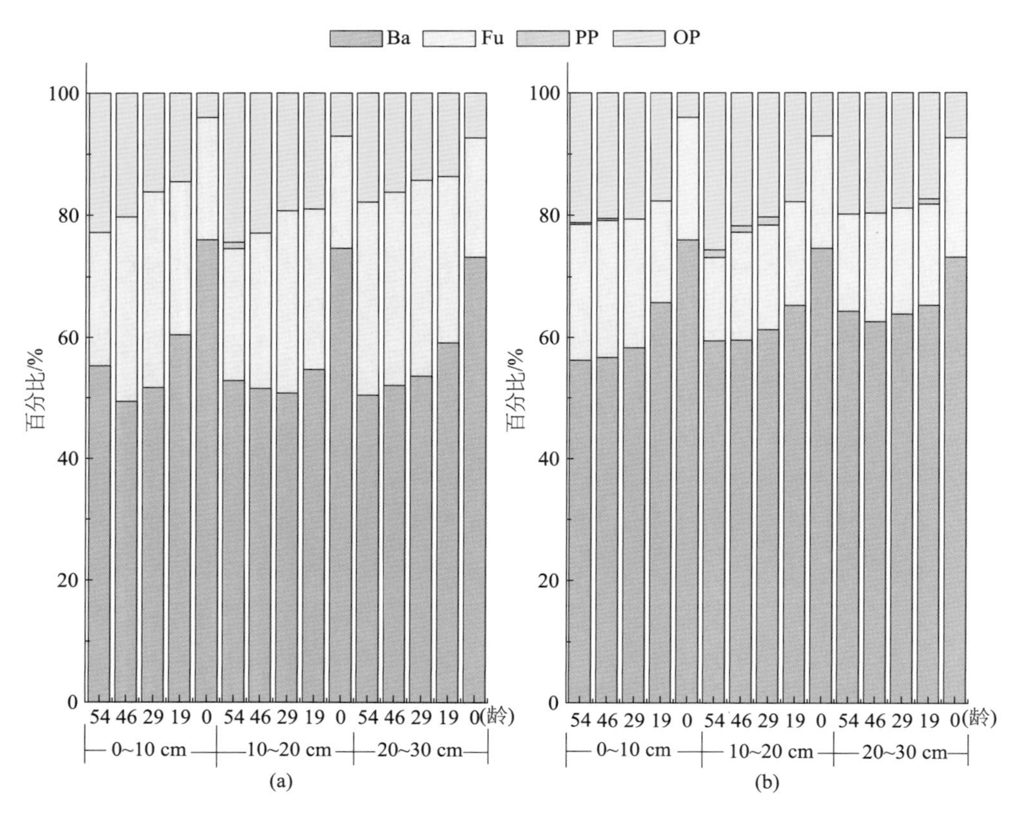

图 4-30　两种 BSC 下不同土层线虫营养类群的组成：(a) 藻类结皮；(b) 藓类结皮。Ba、Fu、PP 和 OP 分别指食细菌线虫、食真菌线虫、植食性线虫和捕食－杂食线虫

Figure 4-30　Soil nematode composition of trophic groups in the different soil layers under two crusts: (a) algal crust ; (b) moss crust. Ba, Fu, PP and OP refer to bacteriovores, fungivores, plant parasites and omnivores-predators, respectively

BSC对土壤线虫多度的影响：由图4-31a可以看出，藻类结皮下0 ~ 10 cm、10 ~ 20 cm和20 ~ 30 cm土层线虫多度的变化规律。54龄和46龄固沙区的土壤线虫多度最高，29龄和19龄固沙区次之，它们均与流沙对照差异显著（$p<0.05$）。同样，藓类结皮下0 ~ 10 cm、10 ~ 20 cm和20 ~ 30 cm土层线虫多度的变化规律基本类似，54龄和46龄固沙区的土壤线虫多度最高，29龄和19龄固沙区次之，均与流沙对照差异显著（图4-31b，$p<0.05$）。可见，无论是藓类结皮还是藻类结皮均对土壤线虫多度有显著影响，即有BSC覆盖的土壤线虫多度明显高于流沙对照，差异达到显著水平（$p<0.05$）。不同固沙年限的藻类结皮和藓类结皮均与线虫多度存在显著的正相关关系，其相关系数分别是r=0.964和r=0.961（表4-23，$p<0.05$），且BSC下表层土壤线虫多度最高，随土层加深而降低。

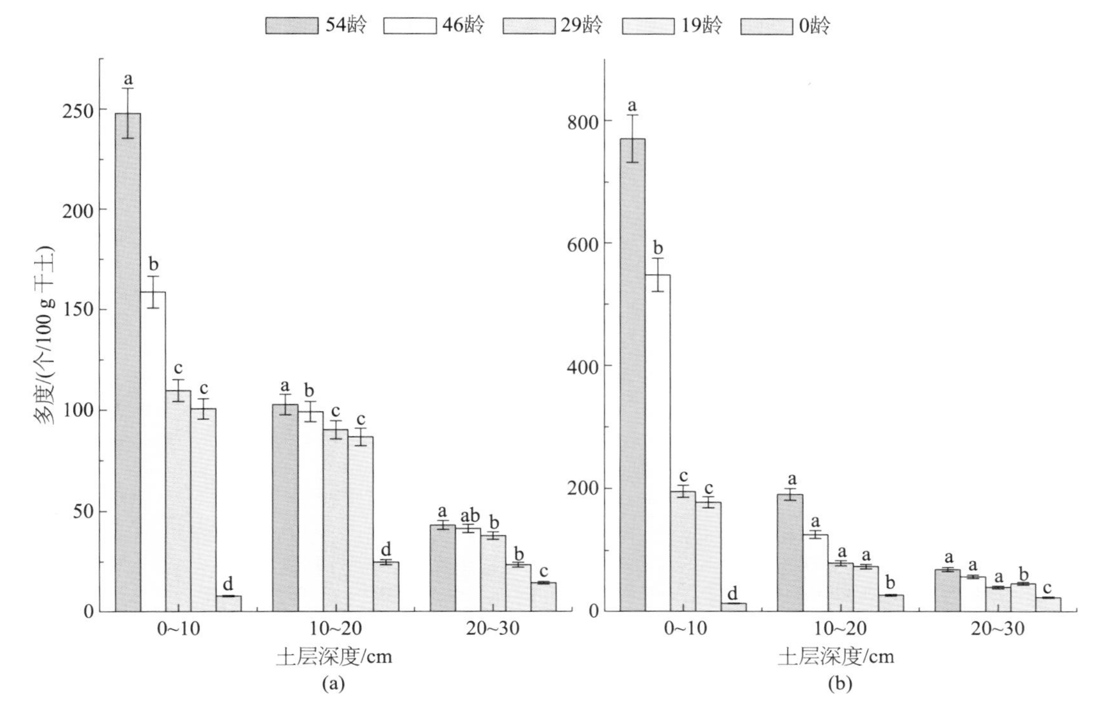

图 4-31　两种 BSC 对土壤线虫多度（平均值 ± 标准误）的影响：（a）藻类结皮；（b）藓类结皮。不同字母表示均值在 $p<0.05$ 水平上差异显著

Figure 4-31　Impact of two crusts on nematode abundance（mean ± SE）:（a）algal crust；（b）moss crust. Different letters indicate the significant differences between mean values at $p<0.05$

BSC 对土壤线虫群落多样性的影响： 由图 4-32a 可以看出， 藻类结皮下 0 ~ 10 cm、10 ~ 20 cm和20 ~ 30 cm土层线虫属丰富度的变化规律为：54 龄和46 龄固沙区土壤线虫属的丰富度最大，29龄和19龄固沙区次之，它们均与流沙对照差异显著（$p<0.05$）。同样，藓类结皮下0 ~ 10 cm、10 ~ 20 cm和20 ~ 30 cm土层线虫属丰富度的变化规律基本类似，54龄和46龄固沙区最高，29龄和19龄固沙区次之，均与流沙对照差异显著（图4-32b，$p<0.05$）。因此，在0 ~ 10 cm、10 ~ 20 cm和20 ~ 30 cm土层，藓类结皮和藻类结皮均可显著提高土壤线虫属丰富度，与流沙对照差异显著（$p<0.05$）。不同固沙年限的藻类结皮和藓类结皮与其下土壤线虫属的丰富度存在显著的正相关关系，其相关系数分别是$r=0.892$和$r=0.980$（表4-23，$p<0.05$），且BSC下土壤线虫属的丰富度在表层最高，随土层加深而有所降低。

表 4-23 不同固沙年限的 BSC 与土壤线虫多度和生态指数的相关性分析

Table 4-23 Correction analysis between abundance and ecological indices of soil nematodes and the crusts of different sand-fixing time

	项目	线性方程及相关系数	
不同固沙年限的藻类结皮（y）	土壤线虫多度（x）	$y = 1.800x + 23.431$	$r = 0.964$
	土壤线虫丰富度（x）	$y = 0.0756x + 4.6944$	$r = 0.892$
	土壤线虫 Shannon-Weaver 多样性指数（x）	$y = 0.0128x + 1.155$	$r = 0.892$
	土壤线虫成熟度指数（x）	$y = 0.0059x + 2.2429$	$r = 0.849$
	土壤线虫富集指数（x）	$y = 0.2607x + 17.55$	$r = 0.959$
	土壤线虫结构指数（x）	$y = 0.0861x + 15.737$	$r = 0.968$
不同固沙年限的藓类结皮（y）	土壤线虫多度（x）	$y = 5.7814x - 10.795$	$r = 0.961$
	土壤线虫丰富度（x）	$y = 0.0988x + 4.4784$	$r = 0.980$
	土壤线虫 Shannon-Weaver 多样性指数（x）	$y = 0.0101x + 1.1592$	$r = 0.835$
	土壤线虫成熟度指数（x）	$y = 0.0079x + 2.2331$	$r = 0.921$
	土壤线虫富集指数（x）	$y = 0.2990x + 8.0318$	$r = 0.944$
	土壤线虫结构指数（x）	$y = 0.290x + 8.7232$	$r = 0.914$

由图4-32c可以看出，藻类结皮下0 ~ 10 cm、10 ~ 20 cm和20 ~ 30 cm土层线虫Shannon-Weaver多样性指数的变化规律为：54龄和46龄固沙区的土壤线虫 Shannon-Weaver多样性指数最大，29龄和19龄固沙区次之，它们均与流沙对照差异显著（$p<0.05$）。同样，藓类结皮下0 ~ 10 cm、10 ~ 20 cm和20 ~ 30 cm土层线虫Shannon-Weaver多样性指数的变化规律基本类似，54龄和46龄固沙区最大，29龄和19龄固沙区次之，它们均与流沙对照差异显著（图4-32d，$p<0.05$）。无论是藓类结皮还是藻类结皮均对土壤线虫Shannon-Weaver多样性有显著影响，即有BSC覆盖的土壤线虫Shannon-Weaver多样性明显高于流沙对照，差异达到显著水平（$p<0.05$）。不同固沙年限的藻类结皮和藓类结皮与其下土壤线虫Shannon-Weaver多样性存在显著的正相关关系，其相关系数分别是r=0.892和r=0.835（表4-23，$p<0.05$）。

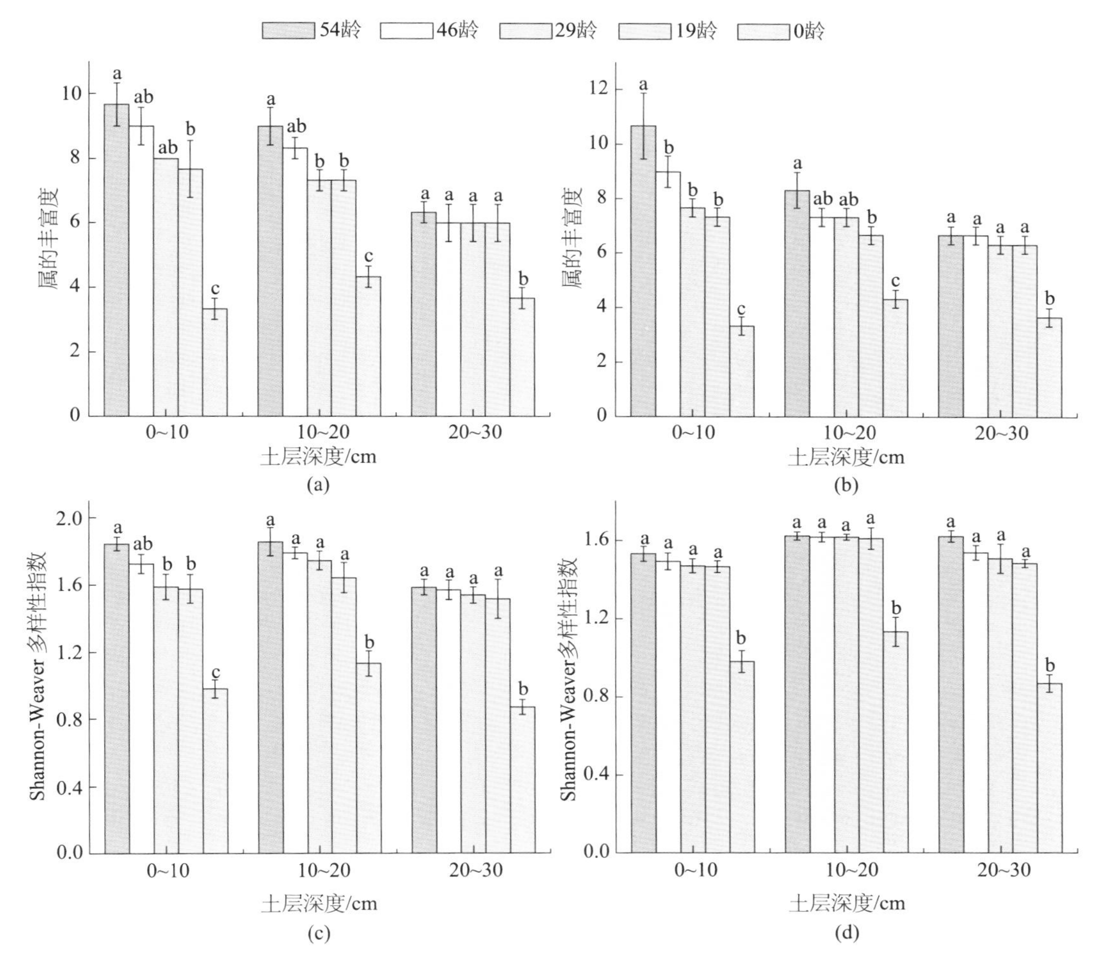

图 4-32 两种 BSC 对土壤线虫属丰富度和 Shannon-Weaver 多样性指数（平均值 ± 标准误）的影响：（a）、（c）藻类结皮；（b）、（d）藓类结皮。不同字母表示均值在 $p<0.05$ 水平上差异显著

Figure 4-32 Impact of two crusts on generic richness and Shannon-Weaver Index of soil nematodes (mean ± SE): (a), (c) algal crust; (b), (d) moss crust. Different letters indicate the significant differences between mean values at $p<0.05$

BSC 对土壤线虫成熟度指数的影响：由图 4-33a 可以看出，藻类结皮下 0 ~ 10 cm 和 10 ~ 20 cm 土层线虫成熟度指数的变化规律为：54龄和46龄固沙区土壤线虫成熟度指数最大，29龄和19龄固沙区次之，它们均与流沙对照差异显著（$p<0.05$）；20 ~ 30 cm 土层，藻类结皮下土壤线虫成熟度指数虽然高于对照，但它们之间差异不显著（$p>0.05$）。同样，藓类结皮下 0 ~ 10 cm 和 10 ~ 20 cm 土层线虫成熟度指数的变化规律基本类似，54龄和46龄固沙区

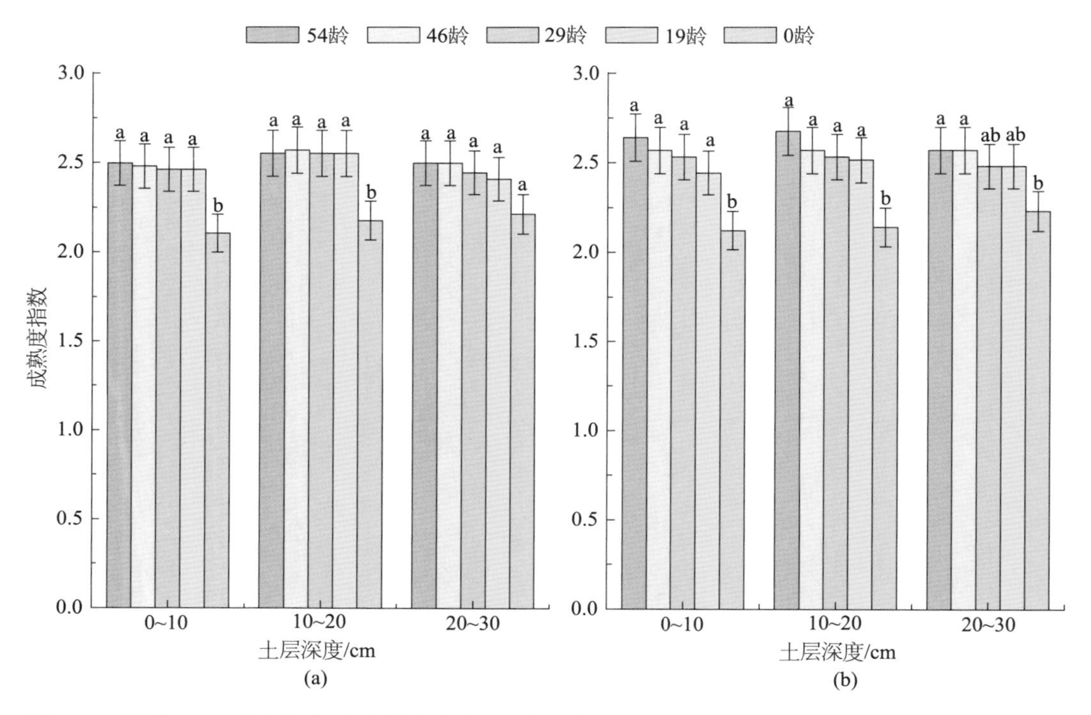

图 4-33　两种 BSC 对土壤线虫成熟度指数（平均值 ± 标准误）的影响：（a）藻类结皮；（b）藓类结皮。不同字母表示均值在 $p<0.05$ 水平上差异显著

Figure 4-33　Impact of two crusts on maturity index of soil nematodes (mean ± SE): (a) algal crust; (b) moss crust. Different letters indicate the significant differences between mean values at $p<0.05$

最大，29龄和19龄固沙区次之，均与流沙对照差异显著（$p<0.05$）；但20～30 cm土层，藓类结皮下土壤线虫成熟度指数虽然高于对照，但除54龄和46龄固沙区外，其他不同年限的固沙区与流沙对照之间差异不显著（图4-33b，$p>0.05$）。因此，对于0～10 cm和10～20 cm土层，藓类结皮和藻类结皮可显著提高其下土壤线虫成熟度指数，与流沙对照差异显著（$p<0.05$）；到20～30 cm土层，土壤线虫成熟度指数虽然高于流沙对照，但部分没有达到差异显著水平。可见藓类结皮和藻类结皮均对表层土壤线虫成熟度指数有显著影响，即有BSC覆盖的土壤线虫成熟度指数明显高于流沙对照，差异达到显著水平（$p<0.05$）。不同固沙年限的藻类结皮和藓类结皮与其下土壤线虫成熟度指数存在显著的正相关关系，其相关系数分别是r=0.849和r=0.921（表4-23，$p<0.05$）。

BSC对土壤线虫富集指数和结构指数的影响：由图4-34a和c可以看出，藻类结皮下0～10 cm、10～20 cm和20～30 cm土层线虫富集指数（EI）和结构指数（SI）的变化规律

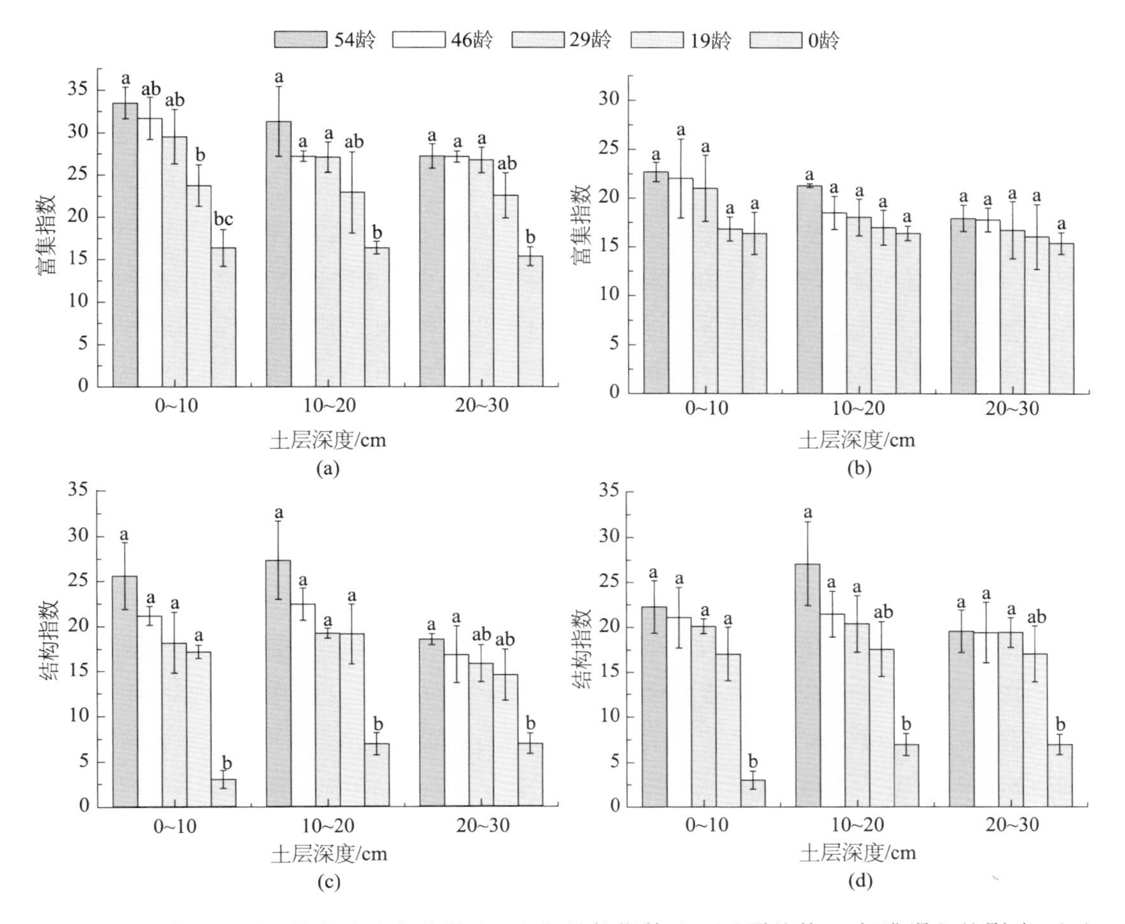

图 4-34 两种BSC对土壤线虫富集指数（EI）与结构指数（SI）（平均值 ± 标准误）的影响：(a)、(c) 藻类结皮；(b)、(d) 藓类结皮。不同字母表示均值在 $p<0.05$ 水平上差异显著

Figure 4-34 Impact of two crusts on enrichment index and structure index of soil nematodes (mean ± SE): (a)、(c) algal crust; (b)、(d) moss crust. Different letters indicate the significant differences between mean values at $p<0.05$

为：54龄和46龄固沙区土壤线虫富集指数和结构指数最大，29龄和19龄固沙区次之，它们均大于流沙对照。同样，藓类结皮下0～10 cm、10～20 cm和20～30 cm土层线虫富集指数和结构指数的变化规律基本类似，54龄和46龄固沙区最大，29龄和19龄固沙区次之，均大于流沙对照（图4-34b，d）。无论是藓类结皮还是藻类结皮均对土壤线虫富集指数和结构指数有影响，即BSC覆盖可提高土壤线虫富集指数和结构指数，这说明人工植被固沙区BSC的生存和演替可改善土壤的健康状态。但富集指数和结构指数的值均低于50，表明目前该

人工植被固沙区土壤的健康状态虽有所提高，但仍处于干扰大、营养水平相对低的状况，土壤的修复还需要更长的时间。藻类结皮不同固沙年限与富集指数和结构指数存在显著的正相关关系，其相关系数分别是r=0.959和r=0.968（p<0.05）；藓类结皮不同固沙年限与土壤线虫富集指数和结构指数存在显著的正相关关系，其相关系数分别是r=0.944和r=0.914（表4-23，p<0.05）。

三因素对土壤线虫多度和属丰富度影响的交互作用：从表4-24可以得出，土壤线虫多度随着BSC类型、固沙年限和土壤深度的变化而有所差异。BSC类型影响土壤线虫多度，即藓类结皮下土壤线虫多度显著高于藻类结皮（p<0.01），且这种影响分别与固沙年限和土壤深度存在明显的交互作用；固沙年限显著影响土壤线虫多度，即不同固沙年限之间土壤线虫多度差异显著（p<0.001），且这种影响分别与BSC类型和土壤深度存在明显的交互作用；BSC下表层线虫多度显著高于下层（p<0.001），且这种影响分别与固沙年限和BSC类型存在明显的交互作用。从表4-24也可以得出，土壤线虫属丰富度随着BSC类型、固沙年限和土壤深度的变化而变化。BSC类型影响土壤线虫属丰富度，藓类结皮下土壤线虫属丰富度显著高于藻类结皮（p<0.01）；固沙年限显著影响土壤线虫属丰富度，即土壤线虫属丰富度在不同固沙年限之间差异显著（p<0.001），且这种影响与土壤深度存在明显交互作用；BSC下表层线虫属的丰富度显著高于下层（p<0.001），且这种影响与固沙年限存在明显的交互作用。可见，土壤线虫多度和属丰富度随着BSC类型、固沙年限和土壤深度的变化而有所不同。

BSC下土壤线虫群落的季节动态：由图4-35a和b可以看出，天然植被区和人工植被固沙区藻类结皮下0～10 cm土层中线虫多度的季节变化规律为：秋季线虫多度最高，分别为238.67个/100 g干土和277.67个/100 g干土，与冬季差异显著（p<0.05）；夏季次之，分别为179.67个/100 g干土和246.00个/100 g干土；春季再次之，但它们之间差异不显著（p>0.05）；冬季最低。同样，藻类结皮下10～20 cm和20～30 cm土层线虫多度的季节变化规律基本类似，秋季最高，夏季次之，春季和冬季最低。因此，天然植被区和人工植被固沙区藻类结皮下0～10 cm、10～20 cm和20～30 cm土层土壤线虫多度存在明显的季节波动，基本表现为秋季>夏季>春季>冬季。

表 4-24　3 因素对土壤线虫多度和属丰富度影响的交互作用

Table 4-24　Tests of among three subjects effects on nematode abundances and generic richness

变异来源	自由度	均方	*F*
线虫多度			
校正模型	29	78907.291	323.641
截距	1	1290964.900	5294.939
BSC 类型	1	152522.500***	625.576
固沙年限	4	129633.067***	531.695
土壤深度	2	310392.633***	1273.086
BSC 类型 × 固沙年限	4	36624.833****	150.218
BSC 类型 × 土壤深度	2	102852.633***	421.854
固沙年限 × 土壤深度	8	58696.300***	240.745
BSC 类型 × 固沙年限 × 土壤深度	8	21837.050***	89.565
误差	60	243.811	
属丰富度			
校正模型	29	14.693	22.040
互作	1	4536.900	6805.350
BSC 类型	1	8.100**	12.150
土壤深度	4	70.956***	106.433
固沙年限	2	40.433***	60.650
BSC 类型 × 土壤深度	4	1.044	1.567
BSC 类型 × 固沙年限	2	0.900	1.350
土壤深度 × 固沙年限	8	5.281***	7.921
BSC 类型 × 土壤深度 × 固沙年限	8	0.636	0.954
误差	60	0.667	

注：**p<0.01，***p<0.001。

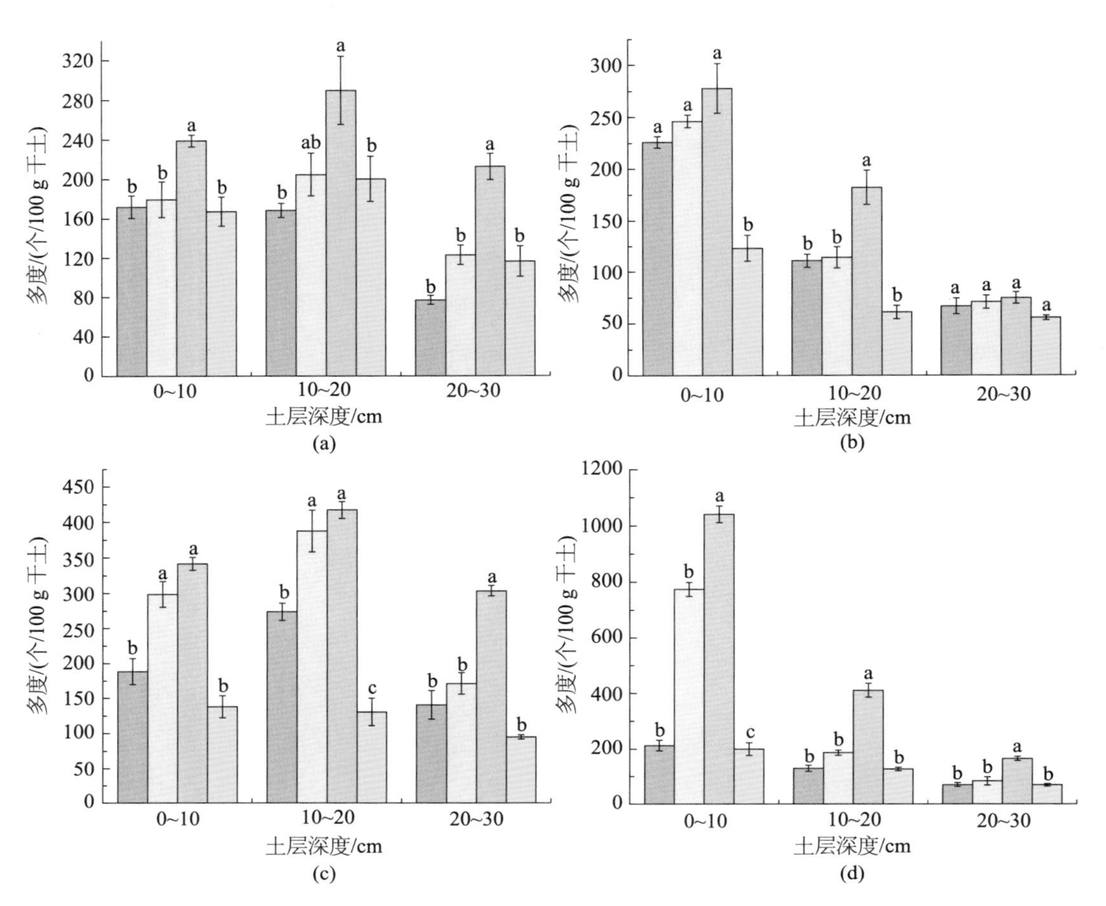

图 4-35 两种BSC下土壤线虫多度（平均值 ± 标准误）的季节变化：(a) 天然植被区藻类结皮；(b) 人工植被固沙区藻类结皮；(c) 天然植被区藓类结皮；(d) 人工植被固沙区藓类结皮。不同字母表示均值在 $p<0.05$ 水平上差异显著

Figure 4-35 Seasonal changes of soil nematode abundances (mean ± SE) under two crusts: (a) algal crust in natural vegetation area; (b) algal crust in the artificially revegetated desert area; (c) moss crust in natural vegetation area; (d) moss crust in the artificially revegetated desert area. Different letters indicate the significant differences between mean values at $p<0.05$

由图4-35c和d可以看出，天然植被区和人工植被固沙区藓类结皮下0 ~ 10 cm土层线虫多度的季节变化规律为：秋季土壤线虫多度最高，分别为341.67个/100 g干土和1040.33个/100 g干土，与春季和冬季差异显著（$p<0.05$）；夏季次之，分别为298.67个/100 g干土和

774.33个/100 g干土，与冬季差异显著（$p<0.05$）；春季再次之；冬季最低。同样，藓类结皮下10 ~ 20 cm和20 ~ 30 cm土层线虫多度变化规律基本类似，秋季最高，与春季和冬季差异显著（$p<0.05$）；夏季次之；春季再次之；冬季最低，但它们之间差异变小。因此，天然植被区和人工植被固沙区藓类结皮下0 ~ 10 cm、10 ~ 20 cm和20 ~ 30 cm土层，土壤线虫多度均存在季节变化，表现为秋季 > 夏季 > 春季 > 冬季。

可见，天然植被区和人工植被固沙区藓类结皮和藻类结皮下0 ~ 10 cm、10 ~ 20 cm和20 ~ 30 cm土层线虫多度均存在明显的季节变化，基本表现为秋季 > 夏季 > 春季 > 冬季，冬季的线虫多度低于其他季节。而且，这种季节变化也受土壤深度的影响，表层土壤线虫多度季节变化明显，随着土层的加深这种季节变化趋势明显变弱。

图4-36为光学显微镜下观察到的腾格里沙漠东南缘沙坡头地区BSC覆盖下固定沙丘中常见的线虫。

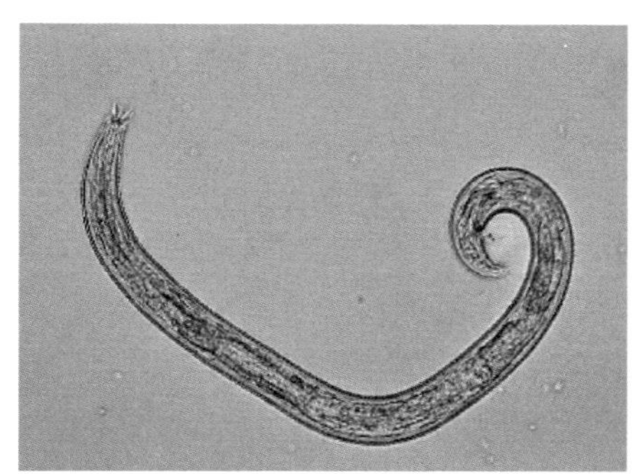

丽突属(*Acrobeles*)

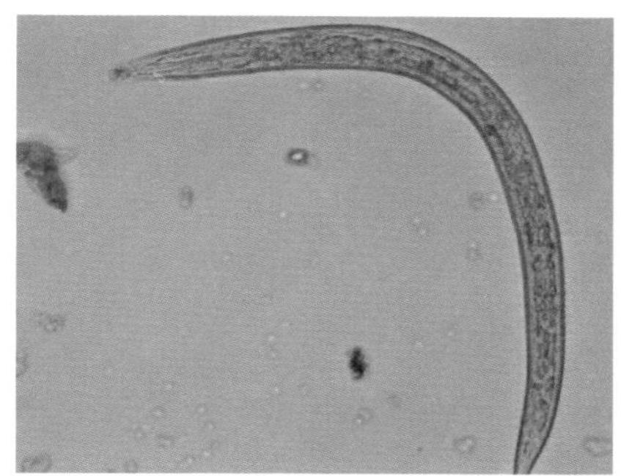

丽突属(*Acrobeles*)

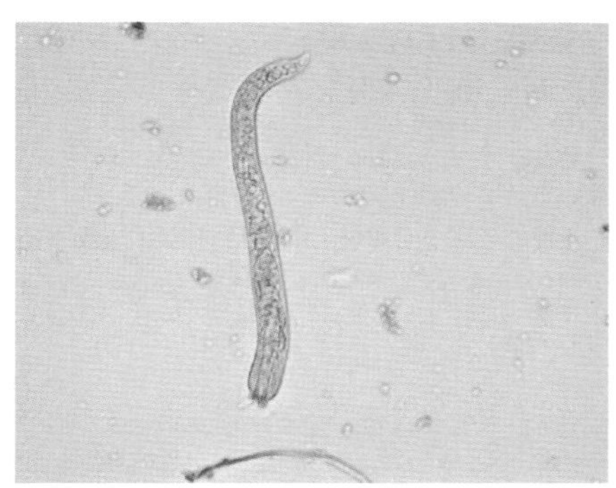

拟丽突属(*Acrobeloides*)

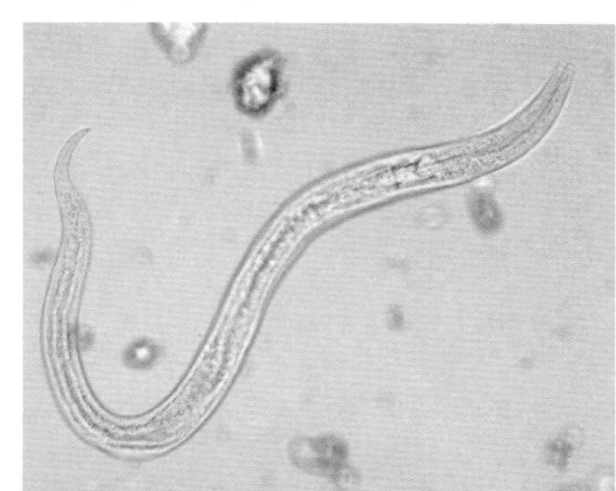

板唇属(*Chiloplacus*)

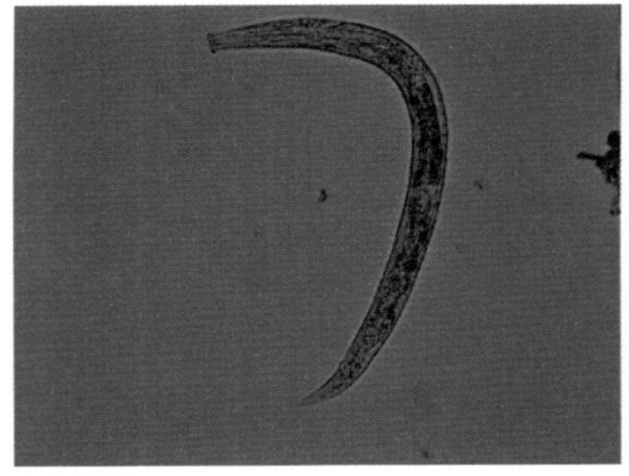

鹿角唇属(*Cervidellus*)

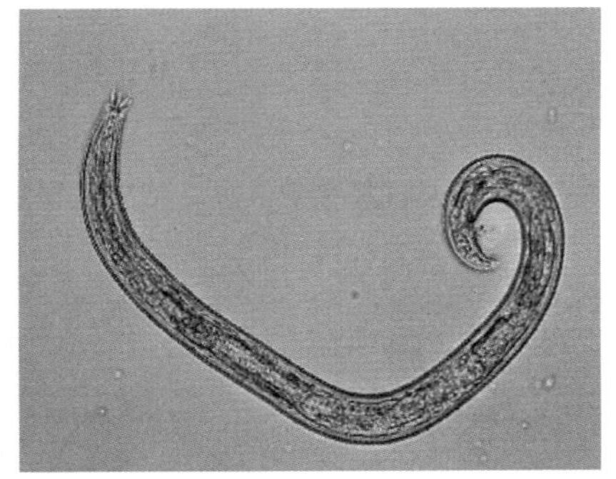

滑刃属(*Aphelenchoides*)

图 4-36　BSC 覆盖下固定沙丘中的常见线虫。采用贝尔曼漏斗法分离线虫，60 ℃温热杀死后，用 4% 甲醛溶液固定后待检。在 Olympus 光学显微镜下鉴定并统计样本中的线虫种类与数量，并进行显微摄影

Figure 4-36　Nematodes in fixed sand dunes covered by BSC. Behrman funnel method was used to separate nematodes, which was killed at 60 ℃, fixed with 4% formaldehyde solution. The kinds and number of nematode samples were identified and figured using Olympus optical microscope and photomicrographed

参考文献

边丹丹 . 黄土丘陵区不同植被状况下土壤生物结皮对土壤生物学性质的影响 . 西北农林科技大学硕士学位论文，2011.

陈小云，李辉信，胡峰，刘满强 . 食细菌线虫对土壤微生物量和微生物群落结构的影响 . 生态学报，2004，24(12): 2825–2830.

丁晖，徐海根，强胜，孟玲，韩正敏，缪锦来，胡白石，孙红英，黄成，雷军成，乐志芳 . 中国生物入侵的现状与趋势 . 生态与农村环境学报，2011，27(3): 35–41.

胡春香，刘永定，宋立荣 . 宁夏沙坡头地区藻类及其分布 . 水生生物学报，1999，23(5): 443–448.

黄振英 . 鄂尔多斯高原固沙禾草沙鞭种子休眠和萌发与环境的关系 . 西北植物学报，2003，23(7): 1128–1133.

黄振英，吴鸿，胡正海 . 新疆 10 种沙生植物旱生结构的解剖学研究 . 西北植物学报，1995，15(6): 56–61.

黄振英，吴鸿，胡正海 . 30 种新疆沙生植物的结构及其对沙漠环境的适应 . 植物生态学报，1997，21(6): 521–530.

焦婷，常根柱，鱼小军，赵桂琴 . 温性荒漠草原土壤酶与肥力的关系 . 中国草地学报，2011，33(5): 88–93.

李发明，刘淑娟，张莹花，朱淑娟，刘克彪，王理德 . 光照和沙埋对沙生针茅种子萌发与幼苗出土的影响 . 中国农学通报，2013，29(31): 47–52.

李守中，肖洪浪，宋耀选，李金贵，刘立超 . 腾格里沙漠人工固沙植被区生物土壤结皮对降水的拦截作用 . 中国沙漠，2002，22(6): 612–616.

李新荣 . 荒漠生物土壤结皮生态与水文学研究 . 北京：高等教育出版社，2012.

李新荣，王涛 . 沙地生态系统研究 . 见李文华，赵景柱 . 生态学回顾与展望 . 北京 : 气象出版社，2004: 625–649.

李新荣，贾玉奎，龙利群，王新平，张景光 . 干旱半干旱地区土壤微生物土壤结皮的生态学意义及若干研究进展 . 中国沙漠，2001，21(1): 4–11.

李新荣，张景光，刘立超，陈怀顺，石庆辉 . 我国干旱沙漠地区人工植被与环境演变过程中植物多样性的研究 . 植物生态学报，2000，24(3): 257–261.

李新荣，张元明，赵允格 . 生物土壤结皮研究：进展、前沿与展望 . 地球科学进展，2009，24(1): 11–24.

刘克彪，李发明，张元恺 . 沙生针茅种子破除休眠的方法 . 草业科学，2015，32(7): 1099–1106.

刘梅，赵秀侠，詹倩，高毅，阳贵德，孙庆业 . 铜陵铜尾矿废弃地生物土壤结皮中的蓝藻多样性 . 生态学报，2011，31(22): 6886–6895.

刘婷婷，张洪军，马忠玉 . 生物入侵造成经济损失评估的研究进展 . 生态经济，2010，2: 173–175.

龙利群，李新荣．微生物土壤结皮对两种一年生植物种子萌发和出苗的影响．中国沙漠，2002，22(6): 581-585.
毛小芳．食细菌线虫与细菌的相互作用及其对土壤氮素矿化和植物根系生长的影响．南京农业大学博士学位论文，2006.
倪丽萍，郭水良．论 DNA C- 值与植物入侵性的关系．生态学报，2005，25(9): 2372-2381.
聂华丽，张元明，吴楠，张静，张丙昌．生物结皮对 5 种不同形态的荒漠植物种子萌发的影响．植物生态学报，2009，33(1): 161-170.
裴雪霞．典型种植制度下长期施肥对土壤微生物群落多样性的影响．中国农业科学院博士学位论文，2010.
任继周．草原研究方法．北京：中国农业出版社，1998: 1-8.
苏延桂，李新荣，黄刚，李小军，郑敬刚．实验室条件下两种生物土壤结皮对荒漠植物种子萌发的影响．生态学报，2007，27(5): 1845-1851.
王伯荪，郝艳茹，王昌伟，彭少麟．生物入侵与入侵生态学．中山大学学报（自然科学版），2005，44(3): 75-77.
王刚，梁学功．沙坡头人工固沙区的种子库动态．植物学报，1995，37(3): 231-237.
吴纪华，宋慈玉，陈家宽．食微线虫对植物生长及土壤养分循环的影响．生物多样性，2007，15(2): 124-133.
武传东，辛亮，李秀颖，王保莉，曲东．长期施肥对黄土旱塬黑垆土氨氧化古菌群落多样性和丰度的影响．中国农业科学，2011，44(20): 4230-4239.
阳贵德．铜陵铜尾矿废弃地生物土壤结皮细菌多样性研究．安徽大学硕士学位论文，2010.
张丙昌，张元明，赵建成，张茹春．准噶尔盆地古尔班通古特沙漠生物结皮蓝藻研究．地理与地理信息科学，2005，21(5): 107-109.
张润志，张大勇，叶万辉，桑卫国，薛大勇，李尉民．农业外来生物入侵种研究现状与发展趋势．植物保护，2004，30(3): 5-9.
张天瑞，皇甫超河，杨殿林，白小明．外来植物黄顶菊的入侵机制及生态调控技术研究进展．草业学报，2011，20(3): 268-278.
庄丽，陈月琴．蓝藻分子系统学研究进展．中山大学学报（自然科学版），1999，38(1): 74-78.
Abdullah SK，Al-Khesraji TO，Al-Edany TY. Soil mycoflora of the Southern Desert of Iraq. *Sydowia*，1986，39: 8-16.
Abed RMM，Al-Sadi AM，Al-Shehi M，Al-Hinai S，Robinson MD. Diversity of free-living and lichenized fungal communities in biological soil crusts of the Sultanate of Oman and their role in improving soil properties. *Soil Biology and Biochemistry*，2013，57: 695-705.
Baskin CC，Baskin JM. *Seeds: Ecology，Biogeography and Evolution of Dormancy and Germination*. New York: Academic Press，2000.
Bates ST，Garcia-Pichel F. A culture independent study of free-living fungi in biological soil crusts of the Colorado Plateau: Their diversity and relative contribution to microbial biomass. *Environmental Microbiology*，2009，11: 56-67.
Bates ST，Nash III TH，Garcia-Pichel F. Patterns of diversity for fungal assemblages of biological soil crusts from the southwestern United States. *Mycologia*，2012，104: 353-361.
Bates ST，Nash III TH，Sweat KG，Garcia-Pichel F. Fungal communities of lichen-dominated biological soil crusts: Diversity，relative microbial biomass，and their relationship to disturbance and crust cover. *Journal of Arid Environments*，2010，74: 1192-1199.
Belnap J，Gardner JS. Soil microstructure in soils of the Colorado Plateau: The role of the cyanobacterium *Microcoleus vaginatus*. *Great Basin Naturist*，1993，53: 40-47.
Belnap J，Gillette DA. Vulnerability of desert biological soil crusts to wind erosion: The influence of crust development，soil texture，and disturbance. *Journal of Arid Environments*，1998，39: 133-142.
Belnap J，Lange OL. Preface. In Belnap J，Lange OL. *Biological Soil Crusts: Structure，Function，and Management*. Berlin: Springer-Verlag，2001，V-IX.
Belnap J，Lange OL. *Biological Soil Crusts: Structure，Function，and Management*. Berlin: Springer-Verlag，2003，281-302.
Belnap J，Büdel B，Lange OL. Biological soil crusts：Characteristics and distribution. In: Belnap J，Lange OL(eds.). *Biological Soil Crusts: Structure，Function and Management*. Berlin: Springer-Verlag，2001.
Belnap J，Phillips SL，Troxler T. Soil lichen and moss cover and species richness can be highly dynamic: The effects of invasion by the annual exotic grass *Bromus tectorum*，precipitation，and temperature on biological soil crusts in SE Utah. *Applied Soil Ecology*，2006，32: 63-76.
Bergstorm DW，Monreal CM，King DJ. Sensitivity of soil enzyme activities to conservation practices. *Soil Science Society of American Journal*，1998，62: 1286-1295.
Bhatnagar A，Basha M，Garg MK，Bhatnagar M. Community structure and diversity of cyanobacteria and green algae in the soils of Thar Desert (India). *Journal of Arid Environments*，2008，72: 73-83.
Black CA. *Soil-Plant Relationships*. New York: John Wiley，1968.
Blumenthal DM. Carbon addition interacts with water availability to reduce invasive forb establishment in a semi-arid grassland. *Biological Invasions*，2008，11(6): 1281-1290.

Boeken B，Sbachak M. Desert plant communities in human-made patches-implication for management. *Ecological Application*，1994，4: 702−716.

Bowker MA. Biological soil crust rehabilitation in theory and practice: An underexploited opportunity. *Restoration Ecology*，2007，15: 13−23.

Brotherson JD，Rushford SR. Influence of cryptogamic crusts on moisture relationships of soils in Navajo National Monument，Arizona. *Great Basin Naturalist*，1983，43: 73−78.

Brown RL，Peet RK. Diversity and invasibility of southern Appalachian plant communities. *Ecology*，2003，84: 32−39.

Bünemann EK，Bossio DA，Smithson PC，Frossard E，Oberson A. Microbial community composition and substrate use in a highly weathered soil as affected by crop rotation and P fertilization. *Soil Biology and Biochemistry*，2004，36: 889−901.

Case TJ. Invasion resistance arises in strongly interacting species-rich model competition communities. *Proceedings of the National Academy of Sciences*，*USA*，1990，87: 9610−9614.

Caudales R，Wells JM，Butterfield JE. Cellular fatty acid composition of cyanobacteria assigned to subsection Ⅱ，order Pleurocapsales. *International Journal of Systematic and Evolutionary Microbiology*，2000，50: 1029−1034.

Chisholm RA. Exponential growth in invasion biology. *BioScience*，2010，60: 314−315.

Christensen M. Species diversity and dominance in fungal community. In: Carroll GW，Wicklow DT (eds). *The Fungal Community*，*Its Organization and Role in the Ecosystem*. New York: Marcell Dekker，1981，201−232.

Ciccarone C，Rambelli A. A study on micro-fungi in arid areas. Notes on stress-tolerant fungi. *Plant Biosystems*，1998，132: 17−20.

Convey P，Lewis Smith RI. Geothermal bryophyte habitats in the South Sandwich Islands，maritime Antarctic. *Journal of Vegetation Science*，2006，17: 529−538.

Crawley MJ. What makes a community invasible? In: Crawley M J，Edwards PJ，Gray，AJ(eds). *Colonization*，*Succession and Stability*. Oxford: Blackwell，1987，429−453.

Crisp MD. Long term changes in arid zone vegetation at Koonamore，South Australia. *The Rangeland Journal*，1975，1: 78−79.

Crutsinger GM，Sauza L，Sanders NJ. Intraspecific diversity and dominant genotypes resist plant invasions. *Ecology Letters*，2008，11(1): 16−23.

Danin. Plant adaptations in desert dunes. *Journal of Arid Environments*，1991，21: 193−212.

Darby BJ，Neher DA，Belnap J. Soil nematode communities are ecologically more mature beneath late-than early-successional stage biological soil crusts. *Applied Soil Ecology*，2007，35: 203−212.

Dark SJ. The biogeography of invasive alien plants in California: An application of GIS and spatial regression analysis. *Diversity and Distributions*，2004，10: 1−9.

Davis MA，Grime JP，Thompson K. Fluctuating resources in plant communities: A general theory of invasibility. *Journal of Ecology*，2000，88: 528−534.

de Rivera CE，Ruiz GM，Hines AH，Jivoff P. Biotic resistance to invasion: Native predator limits abundance and distribution of an introduced crab. *Ecology*，2005，86: 3364−3376.

Ding JQ，Mack RN，Lu P，Ren MX，Huang HW. China’s booming economy is sparking and accelerating biological invasions. *BioScience*，2008，58(4): 317−324.

Drexler JZ，Bedford BL. Pathways of nutrient loading and impacts on plant diversity in a New York peatland. *Wetlands*，2002，22: 263−281.

Eldridge DJ，Greene RSB. Microbiotic soil crusts: A view of their roles in soil and ecological processes in the rangelands of Australia. *Australian Journal of Soil Research*，1994，32: 389−415.

Eldridge DJ，Zaady E，Shachak M. Infiltration through three contrasting biological soil crusts in patterned landscapes in the Negev，Israel. *Catena*，2000，40: 323−336.

Elton CS. *The Ecology of Invasions by Animals and Plants*. London: Methuen，1958.

Ettema CH. Soil nematode diversity: Species coexistence and ecosystem function. *Journal of Nematology*，1998，30: 159−169.

Evans RA，Young JA. Microsite requirements for downy brome (*Bromus tectorum*) infestation and control on sagebrush rangelands. *Weed Science*，1984，13: 13−17.

Fearnehough W，Fullen M，Mitchell DJ，Trueman IC，Zhang J. Aeolian deposition and its effect on soil and vegetation changes on stabilized desert dunes in northern China. *Geomorphology*，1998，23: 171−182.

Garcia-Pichel F，López-Cortés A，Nübel U. Phylogenetic and morphological diversity of cyanobacteria in soil desert crusts from the Colorado Plateau. *Applied Environmental Microbiology*，2001，67: 1902−1910.

Garcia-Pichel F，Loza V，Marusenko Y，Mateo P，Potrafka RM. Temperature drives the continental-scale distribution of key microbes in topsoil communities. *Science*，2013，340: 1574−1577.

Griffiths BS. The role of bacterial feeding nematodes and protozoa in rhizosphere nutrient cycling. *Aspects of Applied Biology*，

1989, 22: 141–145.
Grishkan I, Beharav A, Kirzhner V, Nevo E. Adaptive spatiotemporal distribution of soil microfungi in "Evolution Canyon" Ⅲ, Nahal Shaharut, extreme Southern Negev Desert, Israel. *Biology Journal of the Linnean Society*, 2007, 90: 263–277.
Grishkan I, Kidron GJ. Biocrust-inhabiting cultured microfungi along a dune catena in the western Negev Desert, Israel. *European Journal of Soil Biology*, 2013, 56: 107–114.
Grishkan I, Nevo E. Spatiotemporal distribution of soil microfungi in the Makhtesh Ramon area, central Negev Desert, Israel. *Fungal Ecology*, 2010, 3: 326–337.
Grishkan I, Zaady E, Nevo E. Soil crust microfungi along a southward rainfall aridity gradient in the Negev Desert, Israel. *European Journal of Soil Biology*, 2006, 42: 33–42.
Gu YF, Zhang XP, Tu SH, Lindström K. Soil microbial biomass, crop yields, and bacterial community structure as affected by long-term fertilizer treatments under wheat-rice cropping. *European Journal of Soil Biology*, 2009, 45: 239–246.
Gundlapally SR, Garcia-Pichel F. The Community and phylogenetic diversity of biological soil crusts in the Colorado Plateau studied by molecular fingerprinting and intensive cultivation. *Microbial Ecology*, 2006, 52: 345–357.
Gutterman Y. Strategies of seed dispersal and germination in plants inhibiting desert. *Botanic Review*, 1994, 60: 373–416.
Halwagy R, Moustafa AF, Kamel S. Ecology of the soil mycoflora in the desert of Kuwait. *Journal of Arid Environment*, 1982, 5: 109–125.
Harper KT, Belnap J. The influence of biological soil crusts on mineral uptake by associated vascular plants. *Journal of Arid Environments*, 2001, 47: 347–357.
Harper KT, Clair LLS. Cryptogamic soil crusts on arid and semiarid rangelands in Utah: Effects on seedling establishment and soil stability: Final report on BLM Contract No. BLM AA 851-CT1-48. Salt Lake city: Bureau of Land Management, Utah State Office, 1985a.
Harper KT, Clair LLS. Cryptogamic Soil Crusts on Arid and Semi-arid Rangelands in Utah: Effects on Seedling Establishment and Soil Stability. Provo: Department of Botany Range Science, Brigham Young University, 1985b.
Harper KT, Marple JR. A role for nonvascular plants in management of arid and semiarid rangelands. In: Tueller PT(ed.). *Vegetation Science Applications for Rangeland Analysis and Management*. Dordrecht: Kluwer, 1988.
Harper KT, Pendleton RL. Cyanobacteria and cyanolichens: Can they enhance availability of essential minerals for higher plants? *Great Basin Naturalist*, 1993, 53: 59–72.
Hashem AR. Studies on the fungal flora of Saudi Arabian soil. *Cryptogamic Botany*, 1991, 2: 179–182.
Hawkes CV. Effects of biological soil crusts on seed germination of four endangered herbs in a xeric Florida shrubland during drought. *Plant Ecology*, 2004, 170: 121–134.
Hoopes MF, Hall LM. Edaphic factors and competition affect pattern formation and invasion in a California grassland. *Ecological Applications*, 2002, 12(1): 24–39.
Hu CX, Liu YD. Primary succession of algal community structure in desert soil. *Acta Botanica Sinica*, 2003, 45: 917–924.
Hu CX, Liu YD, Song LR, Huang ZB. Species composition and distribution of algae in semi-desert algal crusts. *Chinese Journal of Applied Ecology*, 2000, 11: 61–65.
Johansen JR. Cryptogamic crusts of semiarid and arid lands of North America. *Journal of Phycology*, 1993, 29: 140–147.
Kidron GJ, Barzilay E, Sachs E. Microclimate control upon sand microbiotic crust, western Negev Desert, Israel. *Geomorphology*, 2000, 36: 1–18.
Kidron GJ, Vonshak A, Abeliovich A. Recovery rates of microbiotic crusts within a dune ecosystem in the Negev Desert. *Geomorphology*, 2008, 100: 444–452.
Kidron GJ, Vonshak A, Abeliovich A. Microbiotic crusts as biomarkers for surface stability and wetness duration in the Negev Desert. *Earth Surface Processes Landforms*, 2009, 34: 1594–1604.
Kidron GJ, Vonshak A, Dor I, Barinova S, Abeliovich A. Properties and spatial distribution of microbiotic crusts in the Negev Desert, Israel. *Catena*, 2010, 82: 92–101.
Larsen KD. Effects of microbiotic crusts on the germination and establishment of three range grasses. Boise: Boise State University, 1995.
Levine J. Plant diversity and biological invasions: Relating local process to community pattern. *Science*, 2000, 288: 852–854.
Levine J, D' Antonio CM. Elton revisited: A review of evidence linking diversity and invasibility. *Oikos*, 1999, 87: 15–26.
Li XR, He MZ, Zerbe S, Li XJ, Liu LC. Micro-geomorphology determines community structure of biological soil crusts at small scales. *Earth Surface Processes Landforms*, 2010, 35: 932–940.
Li XR, Jia XH, Long LQ, Zerbe S. Effects of biological soil crusts on seed bank, germination and establishment of two annual plant species in the Tengger Desert (N China). *Plant and Soil*, 2005, 277: 375–385.

Li XR, Zhang ZS, Huang L, Liu LC, Wang XP. The ecohydrology of the soil-vegetation system restoration in arid zones: A review. *Sciences in Cold and Arid Regions*, 2009, 1: 199–206.

Li XR, Zhou HY, Wang XP, Zhu YG, O' Connor PJ. The effects of sand stabilization and revegetation on cryptogam species diversity and soil fertility in Tengger Desert, Northern China. *Plant and Soil*, 2003, 251: 237–245.

Liu J, Liang SC, Liu FH, Wang RQ, Dong M. Invasive alien plant species in China: Regional distribution patterns. *Diversity and Distributions*, 2005, 11(4): 341–347.

Mckinley VL, Peacock AD, White DC. Microbial community PLFA and PHB responses to ecosystem restoration in tallgrass prairie soils. *Soil Biology and Biochemistry*, 2005, 37: 1946–1958.

Naeem S, Knops JMH, Tilman D, Howe KM, Kennedy T, Gale S. Plant diversity increases resistance to invasion in the absence of covarying extrinsic factors. *Oikos*, 2000, 91: 97–108.

Nagy M, Pérez A, Garcia-Pichel F. The prokaryotic diversity of biological soil crusts in the Sonoran Desert (Organ Pipe Cactus National Monument, AZ). *FEMS Microbiology Ecology*, 2005, 54: 233–245.

Neher DA. Role of nematodes in soil health and their use as indicators. *Journal of Nematology*, 2001, 33: 161–168.

Olde Venterink H, Davidsson TE, Kiehl K, Leonardson L. Impact of drying and re-wetting on N, P and K dynamics in a wetland soil. *Plant and Soil*, 2002, 243: 119–130.

Olsson-Francis K, de la Torre R, Cockell CS. Isolation of novel extreme-tolerant cyanobacteria from a rock-dwelling microbial community by using exposure to low earth orbit. *Applied and Environmental Microbiology*, 2010, 76: 2115–2121.

Oren A. A proposal for further integration of the cyanobacteria under the Bacteriological Code. I*nternational Journal of Systematic and Evolutionary Microbiology*, 2004, 54: 1895–1902.

Perry LG, Blumenthal DM, Monaco TA, Paschke M, Redente EF. Immobilizing nitrogen to control plant invasion. *Oecologia*, 2010,163(1): 13–24.

Pimentel D, McNair S, Janecka J, Wightman J, Simmonds C, O' Connell C, Wong E, Russel L, Zem J, Aquino T, Tsomondo T. Economic and environmental threats of alien plant, animal, and microbe invasions. *Agriculture, Ecosystems and Environment*, 2001, 84: 1–20.

Porras-Alfaro A, Herrera J, Natvig DO, Lipinski K, Sinsabaugh R. Diversity and distribution of soil fungal communities in a semiarid grassland. *Mycologia*, 2011, 103: 10–21.

Prasse R. *Experimentelle Untersuchungen an Gefäßpflanzenpopulationen auf verschiedenen Gefändeoberflächen in einem Sandwüstengebiet*. Osnabriick: Universitatsverlag Rasch, 1999.

Prasse R, Bornkamm R. Effects of microbiotic soil surface crusts on emergence of vascular plants. *Plant Ecology*, 2000, 150: 65–75.

Prober SM, Lunt ID. Restoration of *Themeda australis* swards suppresses soil nitrate and enhances ecological resistance to invasion by exotic annuals. *Biology Invasions*, 2009, 11:171–181.

Ranzoni FV. Fungi isolated in culture from soils of the Sonoran Desert. *Mycologia*, 1969, 60: 356–371.

Raven PH, Johnson GB. *Biology*, 3rd Ed. St. Louis: Mosby Year Book, 1992.

Rivera-Aguilar V, Godinez-Alvarez H, Manuell-Cacheux I, Rodríguez-Zaragoza S. Physical effects of biological soil crusts on seed germination of two desert plants under laboratory conditions. *Journal of Arid Environments*, 2005, 63: 344–352.

Ruijven J, de Deyn GB, Berendse F. Diversity reduces invasibility in experimental plant communities: The role of plant species. *Ecology Letters*, 2003, 6(10): 910–918.

Sax DF. Native and naturalized plant diversity are positively correlated in scrub communities of California and Chile. *Diversity and Distributions*, 2002, 8: 193–210.

Shem-Tov S, Zaady E, Groffman PM, Gutterman Y. Soil carbon content along a rainfall gradient and inhibition of germination: A potential mechanism for regulating distribution of *Plantago coronopus*. *Soil Biology and Biochemistry*, 1999, 31: 1209–1217.

Skujinš J. Microbial ecology of desert soils. *Advances in Microbial Ecology*, 1984, 7: 49–91.

States JS, Christensen M, Kinter CK. Soil fungi as components of biological soil crusts. In: Belnap J, Lange O(eds.). *Biological Soil Crusts: Structure, Function, and Management*. Berlin: Springer-Verlag, 2001, 155–166.

States JS, Christensen M. Fungi associated with biological soil crusts in desert grasslands of Utah and Wyoming. *Mycologia*, 2001, 93: 432–439.

Stohlgren TJ, Chong GW, Schell LD, Villa LCA. Assessing vulnerability to invasion by non-native plant species at multiple spatial scales. *Environmental Management*, 2002, 29: 566–577.

Stohlgren TJ, Otsuki Y, Villa CA, Lee M, Belnap J. Patterns of plant invasions: A case example in native specieshotspots and rare habitats. *Biological Invasions*, 2001, 3: 37–50.

Su YG, Li XR, Cheng YW, Tan HJ, Jia RL. Effects of biological soil crusts on emergence of desert vascular plants in North

China. *Plant Ecology*, 2007, 191: 11–19.

Symstad AJ. A test of the effects of functional group richness and composition on grassland invasibility. *Ecology*, 2000, 81: 99–109.

Theoharides KA, Dukes JS. Plant invasion across space and time: Factors affecting nonindigenous species success during four stages of invasion. *New Phytologist*, 2007, 176: 256–273.

Trasar-Cepeda C, Leirós MC, Gil-Sotres F. Hydrolytic enzyme activities in agricultural and forest soils. Some implications for their use as indicators of soil quality. *Soil Biology and Biochemistry*, 2008, 40: 2146–2155.

van Driesche RG, Carruthers RI, Center T. Classical biological control for the protection of natural ecosystems. *Biological Control*, 2010, 54: S2–S33.

Walker LR, Smith SD. Impacts of invasive plants on community and ecosystem properties. In: Luken JO, Thieret JW(eds.). *Assessment and Management of Plant Invasions*. New York: Springer-Verlag, 1997: 69–86.

Wan FH, Guo JY, Wang DH. Alien invasive species in China: Their damages and management strategies. *Biodiversity Science*, 2002, 10: 119–125.

Wardle DA. Experimental demonstration that plant diversity reduces invasibility—Evidence of a biological mechanism or a consequence of sampling effect? *Oikos*, 2001, 95: 161–170.

Weber E, Sun SG, Li B. Invasive alien plants in China: Diversity and ecological insights. *Biological Invasions*, 2008, 10(8): 1411–1429.

West NE. Structure and function of microphytic soil crusts in wildland ecosystem of arid and semi-arid regions. *Advance of Ecological Research*, 1990, 20: 179–223.

Whitford W. *Ecology of Desert Systems*. San Diego: Academic Press, 2002.

Whitman WB, Coleman DC, Wiebe WJ. Prokaryotes: The unseen majority. *Proceedings of the National Academy of Sciences*, 1998, 95: 6578–6583.

Wilson JRU, Dormontt EE, Prentis PJ, Lowe AJ, Richardson DM. Something in the way you move: Dispersal pathways affect invasion success. *Trends Ecology and Evolution*, 2009, 24: 136–44.

Xu HG, Ding H, Li MY, Qiang S, Guo JY, Han ZM, Huang ZG, Sun HY, He SP, Wu HR, Wan FH. The distribution and economic losses of alien species invasion to China. *Biological Invasions*, 2006, 8(7): 1495–1500.

Zaady E, Gutterman Y, Boeken B. The germination of mucilaginous seeds of *Plantago coronopus*, *Reboudia pinnata*, and *Carrichtera annua* on cynobacterial soil crust from the Negev desert. *Plant and Soil*, 1997, 190: 247–252.

Zak J. Fungal communities of desert ecosystems: Links to climate change. In: Dighton J, White JF, Oudemans P(eds.). *The Fungal Community, Its Organization and Role in the Ecosystems*. 3rd ed. Boca Raton: CRC Press, 2005, 659–681.

Zedler JB, Kercher S. Causes and consequences of invasive plants in wetlands: Opportunities, opportunists and outcomes. *Critical Reviews in Plant sciences*, 2004, 23: 431–452.

Zhang B, Zhang Y, Downing A, Niu YL. Distribution and composition of cyanobacteria and microalgae associated with biological soil crusts in the Gurbantunggut desert, China. *Arid Land Research and Management*, 2011, 25: 275–293.

第5章　BSC人工培养及其在沙区的应用

我国是世界上受沙漠化危害最严重的国家之一，土地沙漠化发展的速率不断加快。沙漠化不仅造成生态系统失衡，而且使耕地面积不断缩小，给我国工农业生产和人民生活带来了严重影响和危害。我国西北干旱区沙漠和沙漠化土地，已成为中国乃至亚太地区沙尘暴主要发源地之一，给国家、社会及经济造成了巨大的损失。因此，沙漠化土地治理是国家在生态建设和环境保护方面的迫切需求。长期以来，植树造林和种草是沙漠治理的主要途径，在实践中也取得一定成效。然而，通过植树造林等传统方式进行沙漠治理有时很难达到治沙目标，例如，降水少于200 mm的地区（魏江春，2005）。因此，沙漠化治理必须要有新的思路。

随着我国沙地人工植被建设历史和经济投入的增加，对人工植被下风沙土的发育机制以及植物的改土作用效果的研究也逐步得到了加强，特别是近些年来，流动沙地固定后，BSC的生态作用研究引起了人们的极大关注。BSC作为荒漠生态系统的重要组成部分，占活体覆盖面积的40%，有的地区甚至达到了70%（Belnap and Lange，2003）。大量研究发现，BSC在防治沙化、维护荒漠生态系统的稳定性、生态平衡和生态系统修复等方面发挥着重要作用（张元明等，2005；Bowker *et al.*，2007；Elbert *et al.*，2012；李新荣，2012）。自然发育形成的BSC具有良好的固沙效果，那么能否将BSC进行人工培育，作为一种防沙治沙的新方法？

人工结皮固沙技术就是利用BSC的固定沙表和抗风蚀的作用，与传统的生物治沙措施相结合，将BSC中的主要生物体（藻类、藓类和地衣）进行人工培育并接种到沙地表面，通过养护成活，地表形成BSC，起到防风固沙的作用，提高了防风固沙效果。大量研究和实践证明，BSC可以通过人工方法进行培育。此外，人工培养的BSC具有形成速度快的特点。在干旱、半干旱荒漠地区天然形成BSC速度缓慢，10年左右才能形成稳定的BSC，人工培育的BSC可以在1年时间内完成自然过程（陈兰周等，2002；胡春香和刘永定，2003）。BSC人工培养可以加速土壤的转化和荒漠生态系统的原生演替。

本章将从BSC的人工培养方法、人工结皮的特征等方面对BSC人工培养及其在沙区的应用进行详细描述。

5.1　藻类结皮人工培养及其应用

藻类是BSC形成的先锋物种。作为先锋拓殖生物，蓝藻能够在恶劣的环境条件下（如干旱、紫外线辐射、营养贫瘠等）生长和繁殖。蓝藻能在沙面表层形成固定流沙的藻类结皮。采用人工培养能够在短期内形成固沙藻类结皮。随着藻类－沙粒结皮的发育及演替，加之细粒物质沉积和大气降尘积累所带来的物质输入，促进了沙面表层营养物质的富集，为微型土壤生物的繁殖和草本植物的拓殖创造了条件，继而推进沙漠生态系统进入良性循环过程（胡春香等，2000；Belnap and Lange，2003；李新荣，2012）。

5.1.1　藻类结皮人工培养的流程和方法

藻类结皮人工培养，就是运用藻类生态、生理学原理和生物结皮理论，分离、选育自然发育形成的藻类结皮中的优良藻种，经大规模人工培养后返接流沙表面，使其在流沙表面快速形成并发育成具有藻类、细菌、真菌在内的人工藻类结皮。该技术主要包括5个方面：① 藻类结皮中优良藻种的分离、纯化与选育；② 种藻的规模培养；③ 工厂化/规模化生产；④ 野外接种；⑤ 管理与维护。在我国腾格里沙漠东南缘、古尔班通古特沙漠和毛乌素沙地等地区，蓝藻是藻类结皮形成初期的先锋种和优势种（李新荣，2012）。因此，BSC中藻类的人工培养以蓝藻的培养为主。藻类结皮人工培养的主要过程和步骤如下（图5-1）。

① 采取自然条件下发育良好的藻类结皮。采样时首先将采样框用力按入土壤中，至采样框完全进入土壤中，用75%酒精消毒过的铲子轻轻取出不锈钢框内约50 cm^2藻类结皮，装入信封中待用。

② 样品过筛。操作前需用75%酒精对手套进行消毒，并佩戴手套；将采集的藻类结皮样品揉碎，过0.2 mm筛子，待用。

③ 清洗。取过筛后的藻类结皮样品20 g，浸泡在50 mL蒸馏水中，静置20 min。将浸泡后的藻类结皮样品放在普通医用纱布上，在蒸馏水中慢慢漂洗，直至样品中的杂质和土壤全部清洗干净，剩下的液体为混合藻类悬浮液。

④ 藻种的分离。将步骤③中的藻类悬浮液在显微镜下检查。如发现需要分离的藻类数量较多时，可立即分离。若数量很少，先进行预培养，待其增多后再分离。

图5-1 人工藻类结皮培养流程：(a)藻种分离和纯化；(b)种藻扩繁；(c)工厂化生产；(d)野外接种
Figure 5-1 Flow diagram of artificial algae crust cultivation：(a) algae species separation and purification；(b) algae propagation；(c) factory production；(d) field inoculation

⑤ 种藻培养。取分离后的单一或混合藻种悬浮液2～5 mL，加入装有300 mL BG11培养液（BG11培养液组分和用量详见表5-1）的三角烧瓶中。将三角烧瓶置于摇床上（转速为140 r min^{-1}）进行初步培养，室内温度控制在25～30 °C，光照强度控制在600 lx。室内培养7～10天后，将进行了初步培养的藻类悬浮液转移至室外进行种藻扩繁；室外培养装置为100 L的塑料收纳箱，每个箱子中装有BG11培养液60 L，并有增氧设施一套（如家用观赏鱼增氧泵）；种藻在室外培养7天左右即可达到藻工厂化生产的用量要求。

⑥ 藻类规模化生产。将步骤⑤中培养好的种藻加入生产池中（每池加入种藻鲜重约200 g，干重约2 g）。培养池中所用培养液为BG11培养液，水深0.5 m左右。蓝藻培养过程中，水体温度在25～30 °C时蓝藻生长情况最佳，水体温度>40 °C时（因为水温超过40 °C时蓝藻将停止生长）需采取降温措施；光强控制在15000 lx。蓝藻培养7～12天后进行收获（夏季7、8月7天左右，春季4—6月和秋季9、10月10～12天）。收获时期的确定：根据蓝藻的生长曲线进行确定（生长曲线的测定详见下节内容），当蓝藻的生长速率或增长量开始下降时（生长量达到最大值后开始下降），进行收获。收获前将培养池中液体静置6～8 h，

然后将培养液排放到空池子中（培养过1次藻的液体仍可以继续培养藻，但是产量会下降30%～40%。因为在培养过程中空气的绿藻会进入培养池中，导致藻种间的竞争加剧和蓝藻种的纯度降低），待池中的水深约为0.5 cm时，将藻液收集至塑料桶中待用。

表 5-1　人工藻类结皮培养 BG11 培养液组分和用量

Table 5-1　Components and dosage of BG11 nutrient solution for artificial algae cultivation

组分	用量 /（$g\,L^{-1}$）	组分	用量 /（$g\,L^{-1}$）
$MgSO_4 \cdot 7H_2O$	0.07	H_3BO_3	2.86
$K_2HPO_4 \cdot 3H_2O$	0.04	$MnCl_2 \cdot 4H_2O$	1.86
$CaCl_2 \cdot 2H_2O$	0.036	$Na_2MoO_4 \cdot 2H_2O$	0.39
Na_2CO_3	1.50	$ZnSO_4 \cdot 7H_2O$	0.22
柠檬酸	0.006	$CuSO_4 \cdot 5H_2O$	0.08
柠檬酸铁铵	0.006	$Co(NO_3)_2 \cdot 6H_2O$	0.05
EDTA	0.001		

⑦ 野外接种：将收集的藻液用喷雾器均匀地接入沙地表面，藻液用量为3 g干重 m^{-2}。接种后前10天，每隔2天用喷雾器浇水（表面湿润即可），下雨期间停止浇水。1个月后检测形成的藻类结皮，检测时采取表土层面积5 cm^2、厚度约为5 mm的样品（由于藻类主要分布在土壤表层5 mm内，所以每次采样尽可能不超过5 mm），测定叶绿素a和叶绿素b的含量。叶绿素a和叶绿素b含量的测定方法：将采集的样品置于研钵中，研磨成粉末状，放入盛有10 mL提取液的试管内，提取液为无水乙醇:丙酮:水=4.5:4.5:1；然后塞上橡皮塞。放入45 ℃的恒温箱中静置24 h，至藻粉变成白色后取出。比色分析用紫外分光光度计测定645 nm、652 nm、663 nm处波长的吸光值，计算叶绿素含量（单位为mg g^{-1}鲜重）。

$$叶绿素a含量=(12.71 \times D663 - 2.59 \times D645) \times V/1000$$

$$叶绿素b含量=(22.88 \times D645 - 4.67 \times D663) \times V/1000$$

$$叶绿素总量=(20.29 \times D645 + 8.04 \times D663) \times V/1000$$

式中，V表示提取液体积。若叶绿素总量不低于接种初期的60%则接种成功；若叶绿素总量低于接种初期的20%，则接种失败，需要重新接种。

5.1.2　工厂化生产中不同藻种的生长特征

我们在沙坡头沙漠试验研究站（以下简称“沙坡头站”）蓝藻规模化生产基地（2015年6月26日—7月10日）进行了14天的蓝藻培养。蓝藻生产池的尺寸为15 m × 1 m × 1 m，分别培养了5个单种藻（念珠藻、席藻、鱼腥藻、单歧藻和伪枝藻）和1个混合藻种（从藻类结皮中直接分离后培养，不纯化）。

图5-2描述了念珠藻、席藻、鱼腥藻、单歧藻、伪枝藻和混合藻的干重日均增长特征。结果表明，念珠藻的日均增长量为0.58 g L^{-1}，单日最大增长量为0.72 g L^{-1}，出现在培养后的第8天；席藻的日均增长量0.59 g L^{-1}，单日最大增长量为0.85 g L^{-1}，出现在培养后的第10天；鱼腥藻的日均增长量0.74 g L^{-1}，单日最大增长量为2.90 g L^{-1}，出现在培养后的第8天；单歧藻的日均增长量0.69 g L^{-1}，单日最大增长量为1.89 g L^{-1}，出现在培养后的第8天；伪枝藻的日均增长量0.61 g L^{-1}，单日最大增长量为1.21 g L^{-1}，出现在培养后的第7天；混合藻的日均增长量0.59 g L^{-1}，单日最大增长量为0.99 g L^{-1}，出现在培养后的第8天。鱼腥藻和单歧藻的日均增长量和单日最大增长量均显著高于其他藻种。培养的6个藻种单日最大增长量出现在培养后的第7～10天，之后增长量呈现出显著的下降趋势。所培养的5个藻种和混合藻的单日最大增长量出现后，尽管每日的产量仍然不断增加，但是从培养成本、管理成本等角度进行综合分析，蓝藻的规模化生产过程中，蓝藻的收获日期应选择在单日最大增长量出现后的第2天。

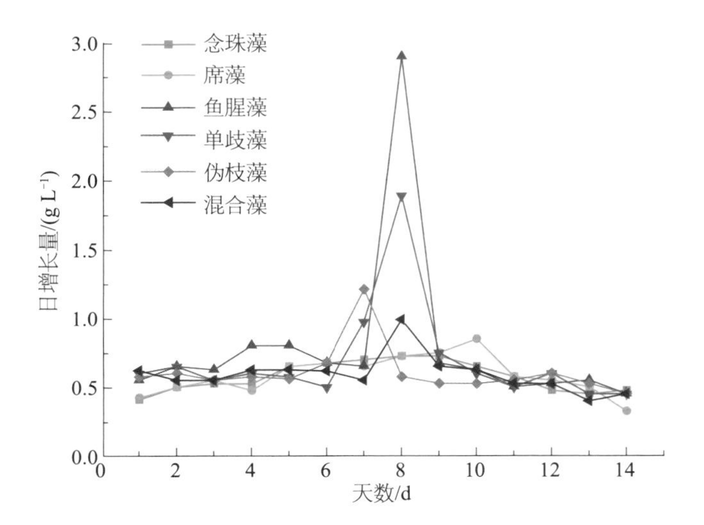

图5-2　蓝藻工厂化生产日增长量

Figure 5-2　Daily average growth amount of cyanobacteria industrialized production

图5-3描述了念珠藻、席藻、鱼腥藻、单歧藻、伪枝藻和混合藻的产量特征。经过14天的培养，念珠藻、席藻、鱼腥藻、单歧藻、伪枝藻和混合藻的鲜重产量分别为44.7 kg、42.3 kg、39.5 kg、35.6 kg、35.8 kg和40.5 kg，念珠藻、席藻和混合藻的鲜重产量较高。念珠藻、席藻、鱼腥藻、单歧藻、伪枝藻和混合藻的干重产量分别为567.5 g、455.6 g、1231.9 g、1045.2 g、658.15 g和607.05 g，鱼腥藻和单歧藻的干重产量显著高于其他藻种。我们发现换算为干重后，鱼腥藻和单歧藻的产量最高，与鲜重产量的结果不一致，主要原因是不同藻种自身的含水量存在差异，上述6个藻种的含水量分别为98.96%、98.92%、95.83%、95.92%、97.45%和98.65%。为更准确地比较蓝藻的产量，应使用干重进行比较。工厂化蓝藻生产可以通过培养单种藻和混合藻两种途径实现。根据以上研究可以发现，单种藻的产量，尤其是鱼腥藻和单歧藻的产量显著高于其他藻种产量。进行单种藻的工厂化生产，首先需要进行藻种的分离和纯化，这个周期为3～4个月，其次才能用于工厂化生产；而混合藻只需将藻类从BSC中分离出来，省略了藻种分离、纯化等步骤，藻种的培养周期约为1个月。从时间成本、人力成本和经济成本来分析，培养混合藻具有节约时间的特点。此外，纯种藻的分离和纯化需要专业的设备和专业人员管理，而混合藻种的分离程序简单，操作方便，经过简单培训，很容易掌握分离和培养技术，任何人员可以随时培养随时繁殖。另外，从藻种保存的技术要求和成本来分析，混合藻种具有节约成本的特点。综上所述，无论从时间成本、技术成本还是经济成本上分析，BSC中蓝藻的工厂化生产，应当优先考虑选择混合藻种。

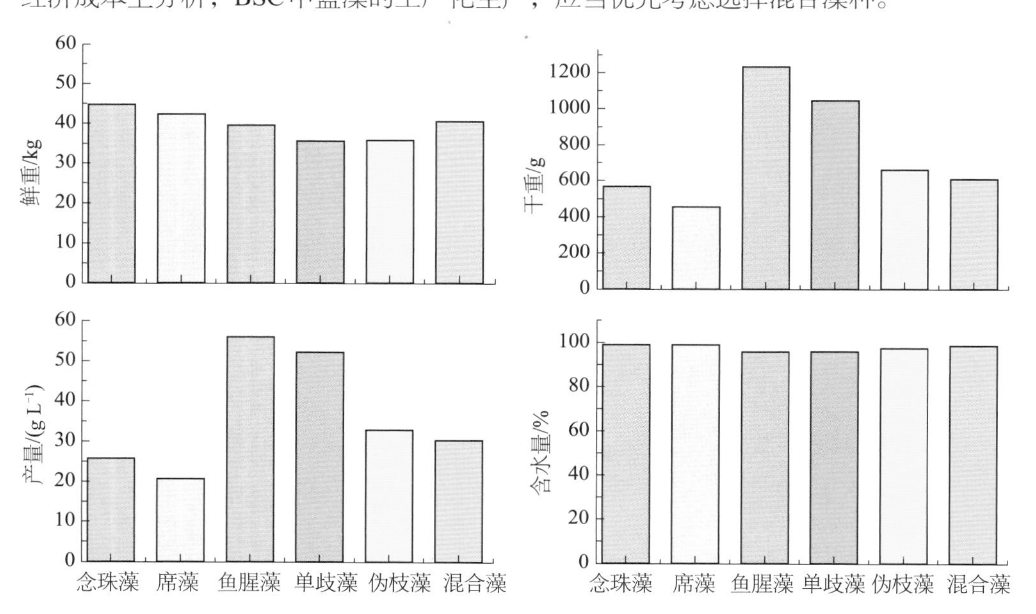

图 5-3　蓝藻工厂化生产产量

Figure 5-3　Yield of cyanobacteria industrialized production

5.1.3 实验室条件下人工藻类结皮的培养——高吸水性聚合物对人工藻类结皮形成的影响

土壤稳定性是决定BSC形成最重要的前提条件之一。受风蚀和水蚀等严重扰动的土壤表面很难形成BSC，尤其是受到沙埋等严重干扰时，BSC生物体将全部死亡，而新的生物体也很难在干扰后的生境中重新定居和发展。在腾格里沙漠东南缘可以发现，尽管在覆沙的土壤表面或流动沙丘也有极少的蓝藻存在，但由于受到流沙频繁运移的干扰，蓝藻和其他细菌及微生物数量少于固定沙丘上的万分之一；再者覆沙地或沙丘表土层极难沉积降尘物质，使得细微的黏粒和粉粒在土壤基质中的含量极低。这些极少的BSC群落生物体无法与体积较大的沙粒胶结形成蓝藻结皮等初级演替阶段的BSC（李新荣，2012）。

一些化学固沙剂，如有机聚合物等，已应用于沙面固定。化学固沙剂能够使沙粒聚合，进而增加沙面的稳定性，同时保护沙面，减少风蚀。在干旱、半干旱地区，为了防止风蚀和水蚀对沙面的侵蚀，化学固沙剂已被广泛使用，如醋酸乙烯聚合体（polymerized by the monomer of vinyl acetate，PVIN）、聚乙烯醇（polyvinyl alcohol，PVA）、聚天冬氨酸（polyaspartic acid，PASP）以及聚丙烯酰胺（polyacrylamide，PAM）。与物理（或机械）固沙材料及植物固沙材料相比，这些化学固沙剂的成本低廉（Liu *et al.*，2012），而且短期内可以有效地固定沙粒，保持沙面稳定。但是从长期的角度来看，化学固沙剂不易降解，也不利于沙面微生物的繁殖和更新（Corti *et al.*，2002），除非受到强烈的紫外线辐射或剧烈的温度波动（Yang *et al.*，2007b）。相反地，BSC的拓殖更适合沙面固定和退化土壤的恢复。化学固沙剂和BSC相结合能否促进BSC的拓殖和发育？高吸水性聚合物（superabsorbent polymer，SAP）短期内能够将细沙粒进行物理固定，同时可以增加土壤的含水量。因此，我们假设SAP+蓝藻处理比单一蓝藻处理能够更快地形成BSC。

本研究以腾格里沙漠东南缘沙坡头天然植被区和人工植被固沙区发育良好的藻类结皮为材料。将从藻类结皮中进行分离、纯化的蓝藻种（念珠藻*Nostoc* sp.），在25 ℃条件下每天光照12 h（光照强度100 μmol m^{-2} s^{-1}），进行3个月恒温的培养。在实验室条件下，分别在0、0.05%（0.025 mg DW cm^{-2}）和0.3%（0.15 mg DW cm^{-2}）3个SAP浓度条件下测定SAP对人工藻类结皮形成和发育的影响。分别测定了固定沙粒尺寸分布特征、沙粒团聚体平均重量直径及其对藻类结皮生物量和胞外多糖（extracellular polysaccharide，EPS）产量的影响，探讨了SAP在人工藻类结皮培养中的作用，阐述了SAP的使用方法。主要研究结果如下。

（1）SAP处理形成的固定沙粒尺寸分布特征

如图5-4所示，与念珠藻（*Nostoc* sp.）处理（AC）相比，SAP并没有显著地影响到沙面沙粒的整体稳定性。在没有使用SAP的处理中（0 mg DW cm^{-2}），尽管未发现>2 mm的沙粒，但是在使用了SAP的处理中（0.025 ~ 0.15 mg DW cm^{-2}），>2 mm的沙粒仅有0.8% ~ 1.9%，而60%的沙粒为0.2 ~ 0.5 mm。但是，AC处理显著地影响土壤颗粒团聚。在快速湿法（fast wetting test）和慢速湿法（slow wetting test）测定条件下，念珠藻+SAP处理（ACSAP）表现出了协同影响土壤颗粒团聚的特征。特别是在快速湿法测定条件下，AC处理和ACSAP处理的大尺寸沙粒所占比例分别为13.7%和22.8%。这些发现证明了ACSAP处理比AC处理能形成更多大尺寸的颗粒，形成较少的小尺寸颗粒。在慢速湿法测定条件下，AC处理和ACSAP处理>2 mm的沙粒所占比例分别为18.9%和22.5%，与在快速湿法测定条件下的结果类似。相反地，在湿法搅拌（wet stirring test）测定条件下，AC处理的大尺寸颗粒所占比例大于ACSAP处理，分别为19.6%和14.2%。从以上测定发现，随着念珠藻细胞浓度的增加，形成的大尺寸颗粒所占比例也随之增加。但是，ACSAP处理中，随着念珠藻细胞浓度的增加，形成的大尺寸颗粒所占比例并未随之增加。这些研究结果表明，ACSAP的处理能够有效增加沙粒的固定。

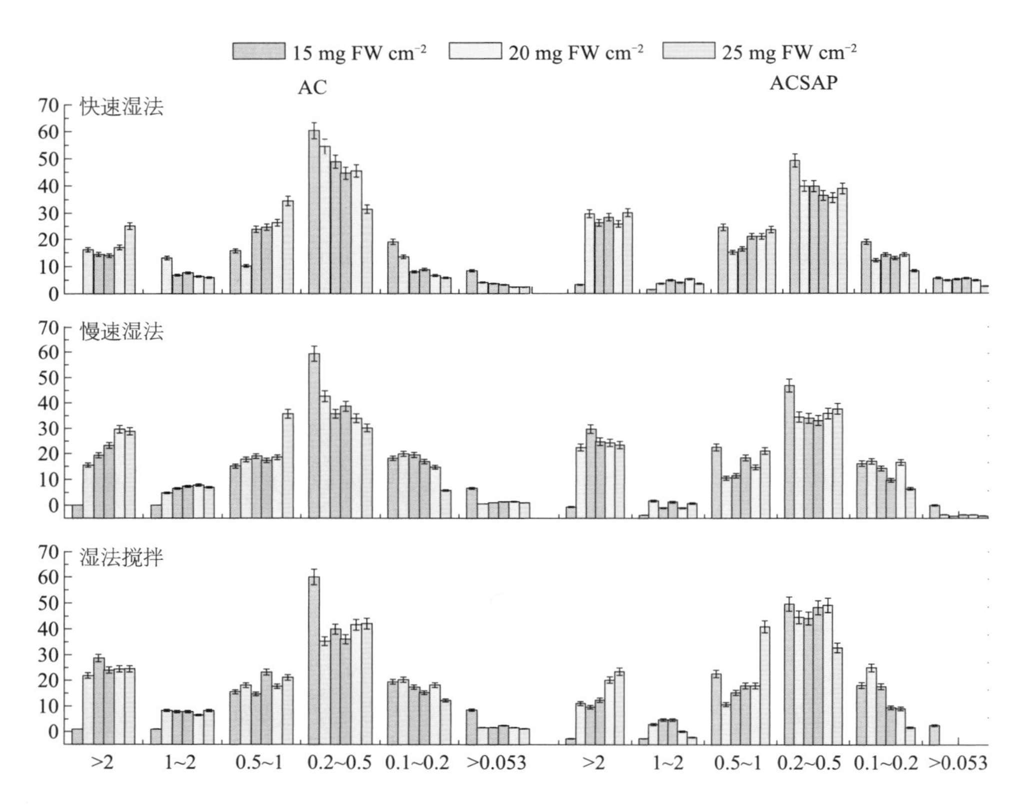

图 5-4　不同处理形成的固定沙粒尺寸分布特征

Figure5-4　Fragment size distribution of soil aggregates treated with different doses of superabsorbent polymer

（2）不同处理形成的沙粒团聚体平均重量直径比较

三种测定方法均证明 SAP 处理增加了沙粒团聚体平均重量直径（mean weight diameter）（图5-5）。在快速湿法测定条件下，无SAP（0 mg DW cm^{-2}）添加的处理，沙粒团聚体平均重量直径为0.34 ~ 0.35 mm，但是添加了SAP的处理，沙粒团聚体平均重量直径为0.37 ~ 0.47 mm，显著高于无SAP添加的处理。其他两种方法测定的结果和快速湿法基本一致。一般而言，沙粒团聚体平均重量直径随着SAP添加量的增加而增加。但是，SAP浓度为0.025 mg DW cm^{-2}、0.05 mg DW cm^{-2}和0.1 mg DW cm^{-2}的处理间差异不显著。基于上述研究我们发现，使沙粒固定所需的最小SAP浓度为0.025 mg DW cm^{-2}。

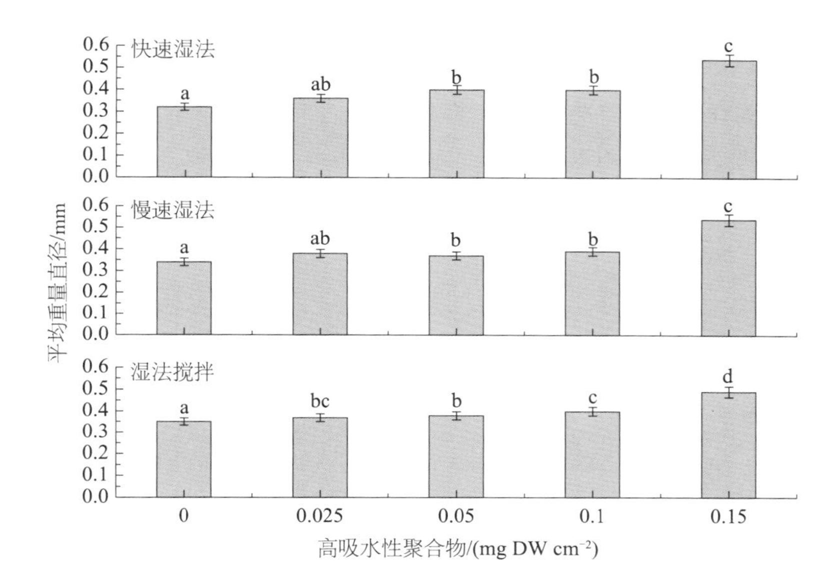

图 5-5　不同处理条件下沙粒团聚体平均重量直径

Figure 5-5　Values of the mean weight diameter（MWD）of sand particles treated with algal（*Nostoc*）cells（AC）and combined application of algal cells and superabsorbent polymer（0.025 mg DW cm^{-2}）（ACSAP）after three testing treatments

三种测定方法发现念珠藻处理（AC）同样能够增加沙粒团聚体平均重量直径，而且极显著高于SAP处理（图5-5）。在快速湿法测定条件下，没有念珠藻细胞（0 mg FW cm^{-2}）时沙粒团聚体平均重量直径为0.35 ~ 0.38 mm，而念珠藻细胞浓度为5 ~ 25 mg FW cm^{-2}时，AC和ACSAP两个处理出现了大量平均重量直径为0.68 ~ 0.93 mm和0.82 ~ 0.93 mm沙粒团聚体。在快速湿法测定条件下发现，ACSAP（0.025 mg DW cm^{-2}）的处理促进了沙粒团聚体平均重量直径的增加。当念珠藻细胞浓度为25 mg FW cm^{-2}时，AC和ACSAP处理的最大的沙粒团聚体平均重量直径均达到了0.93 mm。在慢速湿法测定条件下，AC和ACSAP处理的最大的沙粒团聚体平均重量直径分别为0.69～1.0 mm和0.84～0.89 mm。AC处理中，沙粒团聚体平均重量直径随着念珠藻细胞浓度的增加而增加。采用湿法搅拌测定方法，AC和ACSAP处理的沙粒团聚体平均重量直径分别为0.80～0.87 mm和0.60～0.89 mm。在高浓度念珠藻细胞条件下（20 mg FW cm^{-2}和25 mg FW cm^{-2}），AC处理的沙粒团聚体平均重量直径高于ACSAP处理，但是两者间的差异不显著。同时研究发现AC处理的沙粒团聚体较ACSAP处理更稳定。

（3）不同处理间念珠藻形成结皮的生物量比较

藻类结皮的生物量通常用叶绿素的含量表示。当念珠藻细胞浓度从5 mg FW cm^{-2}增加到25 mg FW cm^{-2}时，AC和ACSAP处理的生物量均显著增加。经过3个月的培养，ACSAP处理的生物量显著高于AC处理。AC和ACSAP处理的生物量分别为8.8～21.7 μg g^{-1}和10.8～23.4 μg g^{-1}，该结果表明SAP的添加有利于念珠藻的生长，促进了藻类结皮的形成（图5-6）。

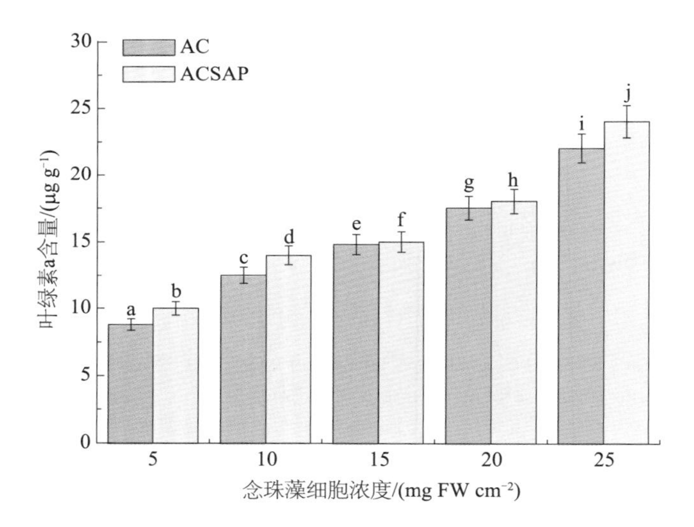

图5-6　不同处理条件下念珠藻叶绿素a含量比较

Figure 5-6　Chlorophyll a contents of soil treated with algal (*Nostoc*) cells (AC) and combined application of algal cells and superabsorbent polymer (0.025 mg DW cm^{-2}) (ACSAP)

（4）不同处理间念珠藻胞外多糖含量比较

图5-7描述了AC和ACSAP处理下念珠藻胞外多糖的含量。念珠藻细胞浓度为5 mg FW cm^{-2}

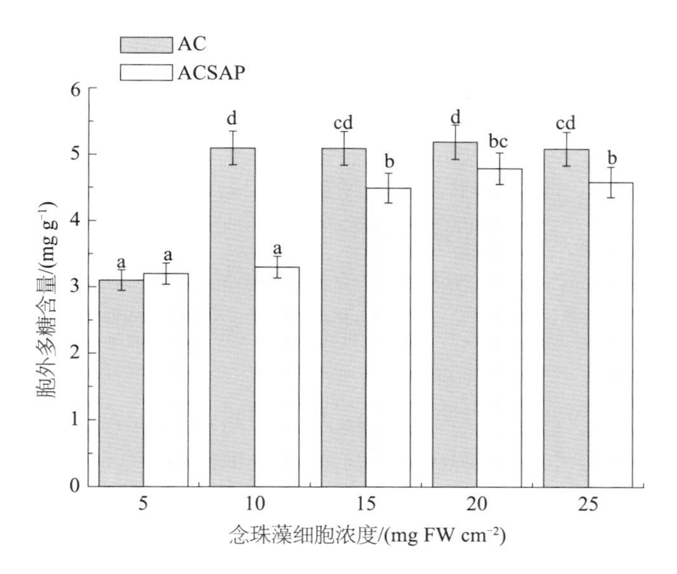

图 5-7　不同处理条件下念珠藻胞外多糖含量比较

Figure 5-7　Extracellular polysaccharide (EPS) contents of the soil treated with algal (*Nostoc*) cells (AC) and combined application of algal cells and superabsorbent polymer (0.025 mg DW cm^{-2}) (ACSAP)

时，AC处理胞外多糖含量为3.13 mg g^{-1}。但是，当念珠藻细胞浓度从10 mg FW cm^{-2}增加到25 mg FW cm^{-2}时，胞外多糖含量值变化不显著，为5.19 ~ 5.23 mg g^{-1}。ACSAP处理的胞外多糖含量为3.20 ~ 4.73 mg g^{-1}，较AC处理胞外多糖含量低

聚乙烯醇（PVA）和聚丙烯酰胺（PAM）等都是水溶性高分子聚合材料，广泛应用于防沙、固沙实践中（Gokcen，2010）。前人的研究表明，用高分子聚合材料固定的沙面能够减少径流和减弱土壤侵蚀程度（Uysal *et al.*，1995）。但是，这些材料是否会影响蓝藻的生长和发育需要进一步的试验。因此，我们试图评估高分子聚合材料对蓝藻生长和发育的影响。SAP是一种高持水性的化学材料，这种属性有利于蓝藻细胞的生长，并能使土壤颗粒胶结在一起。因此，本研究中我们使用了不同浓度的SAP（0.025 ~ 0.15 mg DW cm^{-2}），探索了不同浓度对土壤团聚体形成的影响。SAP的使用显著增加了土壤团聚体的平均重量直径。SAP的用量从0.025 mg DW cm^{-2}增加到0.1 mg DW cm^{-2}时，形成的土壤团聚体平均重量直径值相似，因此我们建议人工藻类结皮培养时，SAP的最小用量为0.025 mg DW cm^{-2}。

ACSAP处理形成的藻类结皮的抗破坏性、固沙粒径、平均重量直径和藻类结皮生物量等指标均显著高于单独使用AC处理形成的结皮。同时，添加了SAP的蓝藻生长良好，表现为丝状体长度的增长和胶结能力的增强。Schulten（1985）的研究认为，蓝藻的生物量和胶结能力呈显著的正相关关系。Xie等（2007）的研究同样发现，蓝藻丝状体的长度和养分的捕

获能力呈显著的正相关。SAP的使用增强了土壤的物理属性，因为SAP增加了土壤的有效水分含量，增强了土壤的通透性和渗透率，同时降低了土壤的紧实度、阻止了土壤侵蚀和径流（Ekebafe *et al.*，2011）。藻类结皮的抗压程度与其生物量显著相关，抗压程度随着生物量的增加而增加（Xie *et al.*，2007）。和AC相比，ACSAP处理显著增加了念珠藻形成结皮的生物量，表现在菌丝体的增长及其对沙粒的胶结。这是因为SAP最明显的特征是增强了土壤的稳定性。显然，ACSAP处理中土壤稳定性显著高于AC处理。在自然条件下，雨滴冲击对BSC的破坏大于机械破坏。这一现象也证明了ACSAP比AC的处理更有利于人工藻类结皮的形成。

AC产生的胞外多糖含量高于ACSAP处理。胞外多糖能够使沙粒物理胶结。然而，我们的研究发现胞外多糖含量高低与蓝藻生物量并没有直接的关系。但是胞外多糖的产生与微生物的分泌物显著相关（Costerton *et al.*，1981；Whitfield，1988）。SAP为蓝藻细胞提供了较多的水分，ACSAP处理中念珠藻细胞的失水速度低于AC处理。由于SAP较高的持水能力，在干燥条件下蓝藻细胞产生了较少的胞外多糖。前人的研究表明，胞外多糖的增加增强了土壤颗粒的胶结能力，因为胞外多糖改变了土壤的微形态特征（Belnap and Gardner，1993；Malam Issa *et al.*，2007）。尽管AC处理和ACSAP处理中土壤平均重量直径间的差异不显著，但是SAP促进了土壤颗粒的胶结作用得到了证实。而不同念珠藻用量条件下（10 mg FW cm^{-2}、15 mg FW cm^{-2}、20 mg FW cm^{-2}和25 mg FW cm^{-2}），胞外多糖的产量没有显著差异。因此，我们建议人工藻类结皮的培养，藻细胞的用量为10 mg FW cm^{-2}。

由于高分子聚合物降解困难，它的使用产生了严重的环境问题（Azahari *et al.*，2011）。出于这些原因，近些年，大量的无毒、生态友好型和可降解的高分子聚合物产品应运而生。一些学者研究了这些材料在防沙、固沙方面的应用（Yang *et al.*，2007a；Gokcen，2010；Azahari *et al.*，2011；Liu *et al.*，2012）。研究结果表明，这些新型化学固沙剂可以短时间固定沙面，而若干年后固定的沙面仍然会退化到原来的状态。本研究结果表明，通过ACSAP可以促进人工藻类结皮的形成，并能保持沙面稳定。因此，ACSAP方法可以作为一种新型固沙方法应用于沙漠治理（Park *et al.*，2015）。

5.1.4 人工藻类结皮在沙区的应用

为探索人工藻类结皮在沙区防沙固沙的应用，在实验室研究的基础上，我们将研究成果在野外条件下进行了实践应用。研究区设在腾格里沙漠东南缘沙坡头站试验区流动沙区。

试验共设置了6个处理，T1：对照（流沙）；T2：藻混合液（藻液使用量为5 g DW m^{-2}）；T3：固沙剂（浓度2%）+藻混合液（藻液使用量为5 g DW m^{-2}）；T4：固沙剂（浓度2%）；T5：锯末（锯末使用量为300 mL m^{-2}）+藻混合液（藻液使用量为5 g DW m^{-2}）；T6：固沙剂（浓度2%）+锯末（锯末使用量为300 mL m^{-2}）+藻混合液（藻液使用量为5 g DW m^{-2}）；N：自然条件下发育4年的藻类结皮。试验小区面积为4 m^2（2 m×2 m），每处理设置3次重复。试验所用藻液为蓝藻藻液，蓝藻在沙坡头藻培养基地培养12天。试验所用蓝藻为念珠藻、席藻、鱼腥藻、单歧藻和伪枝藻按照1：1：1：1：1的比例混合后的混合藻液。所用固沙剂为W—OH高新复合固化材料（可降解材料，无污染，降解物不会造成二次污染）。

图5-8为不同处理5个月后的照片，从图中可以很直观地看到不同处理条件下人工藻类结皮形成的状况。为了更直观地描述人工藻类结皮的形成状况和发育特征，我们选取了盖度、叶绿素含量、厚度、抗风蚀程度、土壤理化属性以及碳固定和释放特征等指标对人工藻类结皮的特征进行了全面描述。

（1）不同处理条件对人工藻类结皮盖度的影响

从图5-9a可以看出，试验开始后2个月和5个月的时间，对照处理（T1）和固沙剂处理（T4）条件下没有藻类结皮形成；T2处理藻类结皮盖度分别为4.1%和6.0%；T3处理的盖度分别为12.4%和14.3%；T5处理的盖度分别为3.4%和4.3%；T6处理的盖度分别为5.5%和6.2%；自然发育4年的藻类结皮（N）盖度为8.2%。上述结果表明，固沙剂+藻液的处理（T3）更有利于人工藻类结皮的形成，其盖度高于其他处理和自然发育4年的藻类结皮的。同时，接种后5个月的人工藻类结皮的盖度均高于接种2个月时的盖度，说明了沙区通过人工添加藻种的方法能够促进藻类结皮的形成。

（2）不同处理条件对人工藻类结皮叶绿素含量的影响

从图5-9b可以看出，试验开始后2个月和5个月的时间，对照处理（T1）和固沙剂处理（T4）条件下由于没有藻类结皮形成，所以叶绿素含量为0；T2处理的藻类结皮叶绿素含量分别为8.8 mg g^{-1}和10.4 mg g^{-1}；T3处理的叶绿素含量分别为10.2 mg g^{-1}和13.4 mg g^{-1}；T5处理的叶绿素含量分别为4.2 mg g^{-1}和5.6 mg g^{-1}；T6处理的叶绿素含量分别为5.2 mg g^{-1}和6.3 mg g^{-1}；自然发育4年的藻类结皮叶绿素含量为32.3 mg g^{-1}。其中T3处理（固沙剂+藻混合液）叶绿素含量高于其他处理的，但是显著低于自然发育4年的藻类结皮的。同时，接种后5个月的人工藻类结皮叶绿素含量均高于接种2个月时的叶绿素含量，说明通过水溶液繁殖的藻类，能够较好地适应土壤环境。

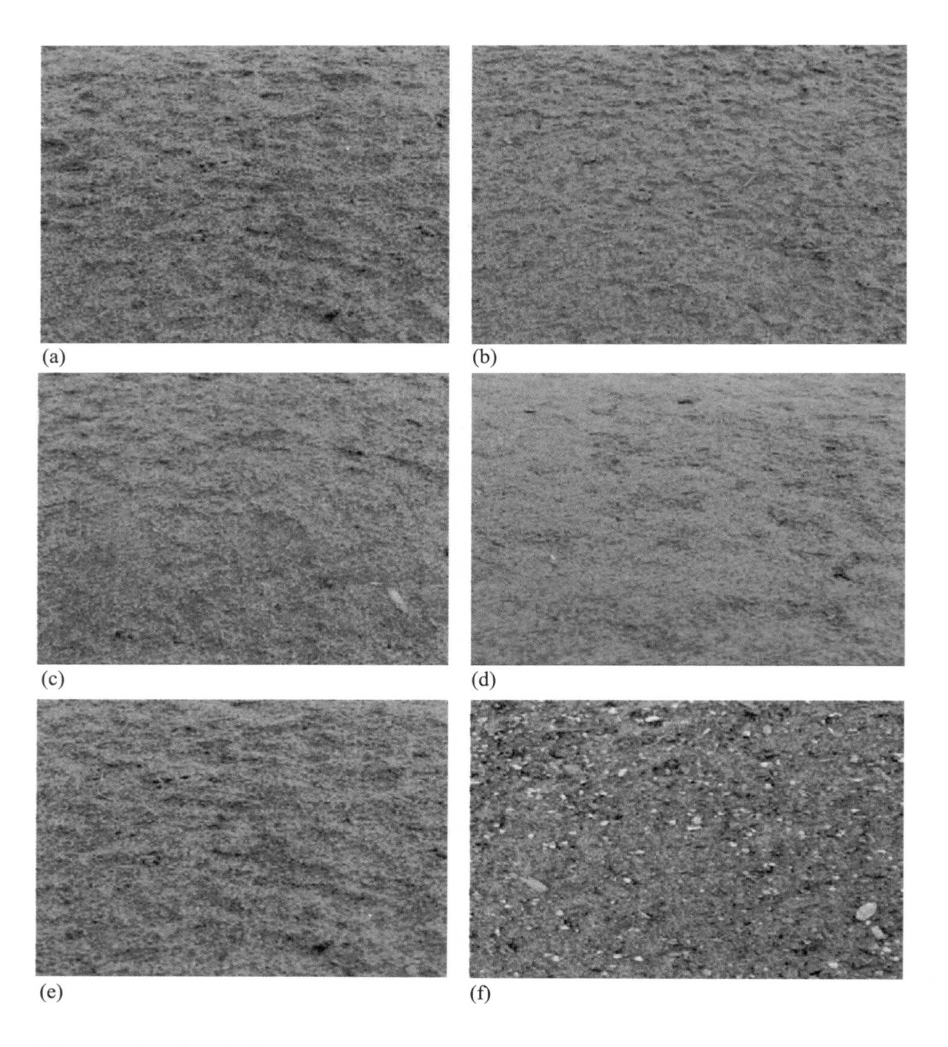

图 5-8　不同处理条件下形成的人工藻类结皮：(a) 对照（流沙）(T1)；(b) 藻混合液（T2）；(c) 固沙剂 + 藻混合液（T3）；(d) 固沙剂（T4）；(e) 锯末 + 藻混合液（T5）；(f) 固沙剂 + 锯末 + 藻混合液（T6）

Figure 5-8　Artificial algae crust at different culture treatment : (a) control (sand) (T1); (b) mixed liquor of algae (T2); (c) sand solidification agent+mixed liquor of algae (T3); (d) sand solidification agent (T4); (e) saw powder+mixed liquor of algae (T5); (f) sand solidification agent+saw powder+mixed liquor of algae (T6)

（3）不同处理条件对人工藻类结皮厚度的影响

图5-9c描述了接种5个月后人工藻类结皮的厚度。对照处理（T1）的藻类结皮厚度

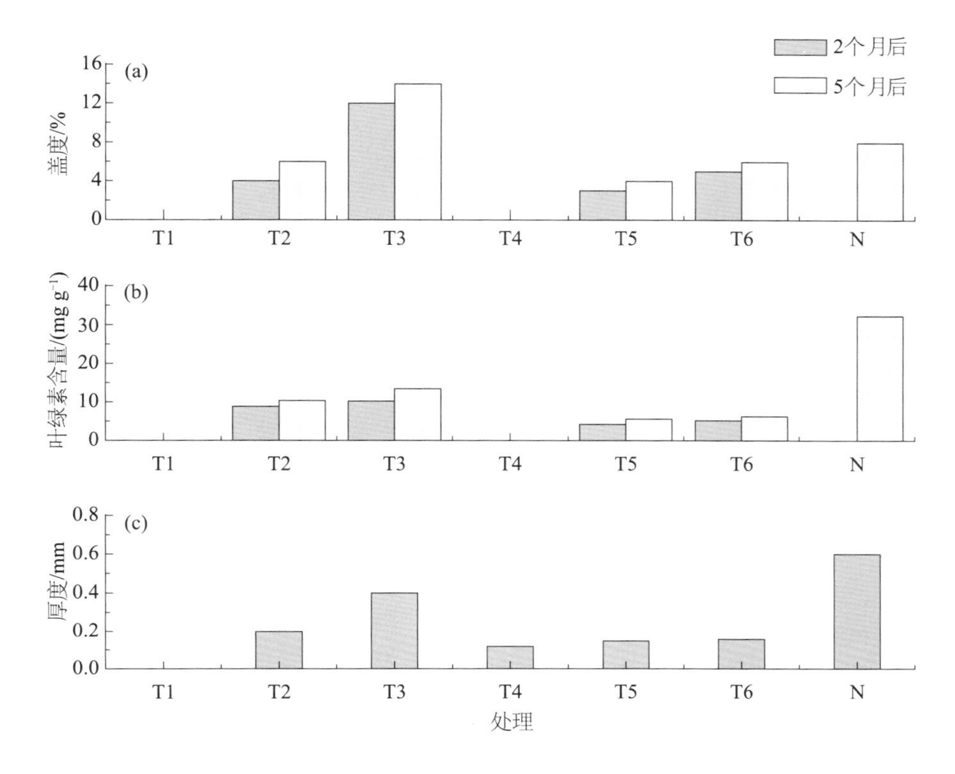

图 5-9　不同处理条件下人工藻类结皮和自然发育 4 年藻类结皮的盖度（a）、叶绿素含量（b）和厚度（c）
Figure 5-9　Compare with coverage（a）, chlorophyll contents（b）and thickness（c）of artificial algae crust and 4 years natural algae crust

为0；T2处理的藻类结皮厚度为0.22 mm；T3处理的藻类结皮厚度为0.39 mm；T4处理中没有藻类结皮形成，但是由于固沙剂对沙粒的固定作用，沙面表层形成了化学结皮，厚度为0.12 mm；T5处理的藻类结皮厚度为0.15 mm；T6处理的藻类结皮厚度为0.16 mm。人工培养的藻类结皮中T3处理培养的藻类结皮厚度最大。自然发育4年的藻类结皮厚度为0.6 mm，显著高于人工培养的藻类结皮。

（4）不同处理条件形成的人工藻类结皮抗风蚀程度比较

使用便携式风洞对人工藻类结皮覆盖土壤的起沙风速进行了测定，研究结果表明，自然发育4年的藻类结皮起沙风速最大，为10 m s^{-1}；其次为T3处理，起沙风速为8 m s^{-1}，对照处理的起沙风速最低，为4.5 m s^{-1}。T2、T5和T6的起沙风速分别为5 m s^{-1}、6 m s^{-1}和7 m s^{-1}（图5-10）。

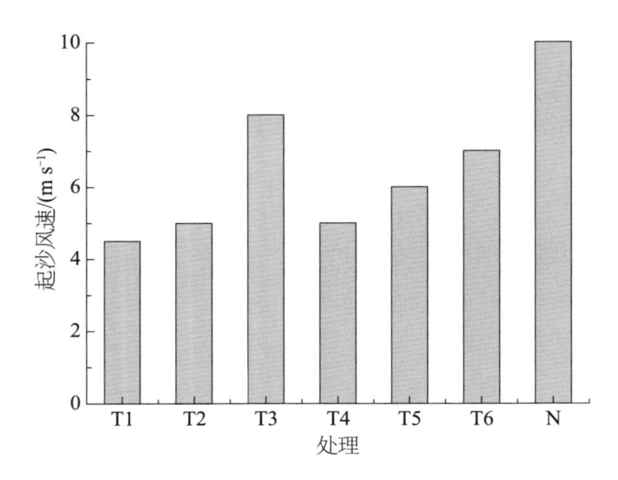

图 5-10　不同处理条件下人工藻类结皮和自然发育 4 年藻类结皮的起沙风速

Figure 5-10　Compare with threshold velocity for transportation of sand of artificial algae crust and 4 years natural algae crust

（5）不同处理条件人工藻类结皮土壤理化属性

对设置的6个处理和自然发育4年的藻类结皮表层0.5 cm土壤理化属性进行了分析。研究结果表明，对照T1的土壤有机质含量为1.79 g kg^{-1}，全氮含量为0.017 g kg^{-1}，全磷含量为0.30 g kg^{-1}；pH为8.98；电导率为48.4 μS cm^{-1}。T2 ~ T6的土壤有机质含量为1.89 ~ 1.99 g kg^{-1}，全氮含量为0.021 ~ 0.024 g kg^{-1}，全磷含量为0.30 ~ 0.31g kg^{-1}；pH为8.67 ~ 8.98；电导率为41.6 ~ 50.3 μS cm^{-1}；自然发育4年的藻类结皮的土壤有机质含量、全氮和全磷含量分别为2.1 g kg^{-1}、0.028 g kg^{-1}和0.35 g kg^{-1};pH为8.85；电导率为59.3 μS cm^{-1}（图5-11）。从上述结果可以很明显地看出，人工藻类结皮的土壤有机质含量、全氮和全磷含量高于流沙，但是低于自然发育4年的藻类结皮，而pH和电导率值与流沙和自然发育4年的藻类结皮一致。

（6）不同处理条件人工藻类结皮的碳释放特征

对人工藻类结皮和自然发育4年的藻类结皮碳释放速率和光合速率进行了测定。研究发现，无论是碳释放速率还是光合速率，人工藻类结皮均低于自然发育4年的藻类结皮，但是高于对照和固沙剂处理。人工藻类结皮的碳释放速率和光合速率分别为0.34 ~ 0.59 μmol m^{-2} s^{-1}和0.28 ~ 0.42 μmol m^{-2} s^{-1}；自然发育4年的藻类结皮碳释放速率和光合速率分别为0.78 μmol m^{-2} s^{-1}和0.52 μmol m^{-2} s^{-1}；对照和固沙处理的碳释放速率分别为

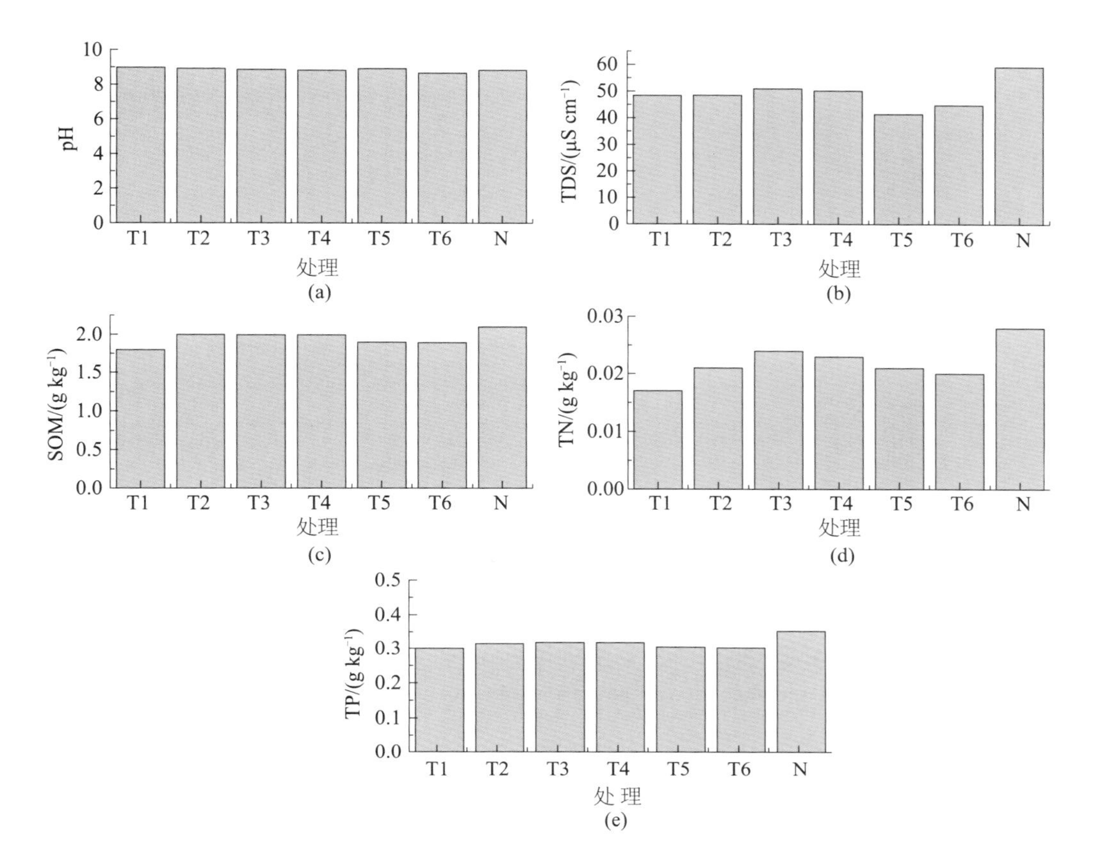

图 5-11 不同处理条件下人工藻类结皮和自然发育 4 年藻类结皮的土壤属性。TN：全氮；TDS：电导率；TP：全磷；SOM：土壤有机质

Figure 5-11 Compare with soil properties of artificial algae crust and 4 years natural algae crust. TN:total nitrogen；TDS: total dissdned solids；TP: total phosphorus；SOM:soil organic matter

0.18 μmol m^{-2} s^{-1}和0.21 μmol m^{-2} s^{-1}（图5-12）。综上所述，无论在实验室条件下还是在野外实践中，固沙剂+混合藻液的处理方式更有利于人工藻类结皮的形成（图5-13为沙坡头站试验区应用人工藻类结皮进行防沙固沙5个月后的治理效果）。

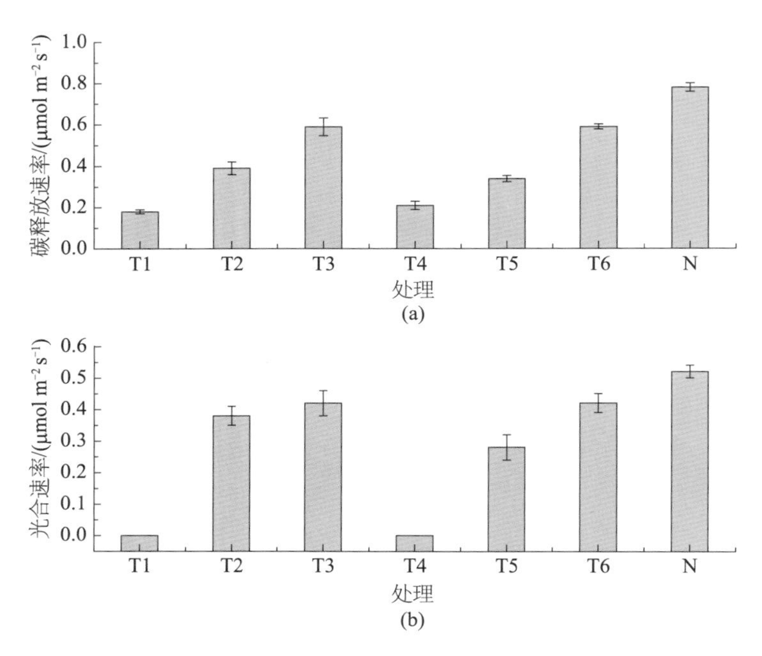

图 5-12　不同处理条件下人工藻类结皮和自然发育4年藻类结皮的土壤碳释放速率（a）和光合速率（b）
Figure 5-12　Compare with respiration (a) and photosynthetic (b) rates of artificial algae crust and 4 years natural algae crust

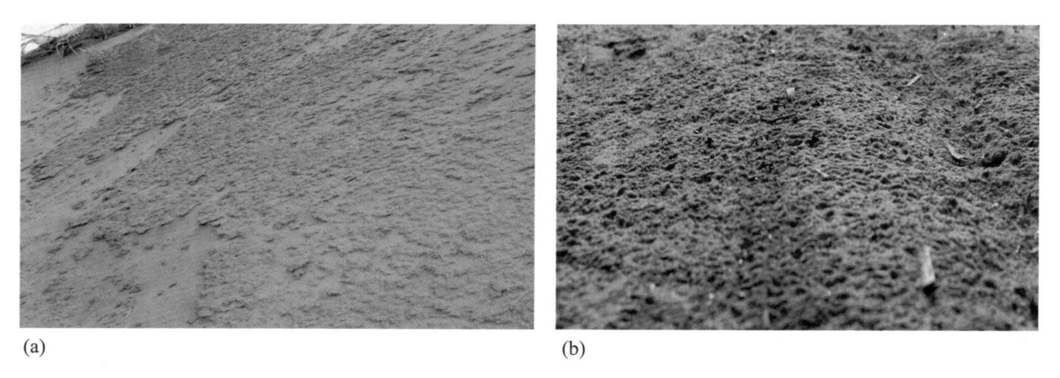

图 5-13　沙坡头站培育的人工藻类结皮及其在防沙治沙中的应用
Figure 5-13　Artificial algae crust cultured by Shapotou station and its application in prevention and control of desertification

经过2014年和2015年两年的野外试验，我们发现人工藻类结皮的固沙、治沙作用主要表现在：① 增强沙面的抗风蚀能力。因为藻种接入沙面后，其丝状体同沙粒颗粒结合，所分泌的有机凝胶体和多聚糖将松散的土粒黏结在一起，形成了一个致密的抗蚀层，直接增强沙土表面的稳定性和抗风蚀的能力。② 改善土壤理化属性。沙面接入藻种后，藻类通过其生长和代谢作用，带动土壤异养微生物的生长，增加了沙漠地表中的生物多样性，从而促进沙质的矿化过程和土壤物质循环及流动，并有利于改变土壤的理化性质，增加了土壤有机质、全氮和全磷的含量，促进了土壤的演化。③增加土壤水分含量。接入藻种后形成的固沙藻类结皮在发育过程中，藻类结皮中的水稳性土壤团聚体和有机质含量大大增加，土壤的吸湿性、可塑性明显提高，这些都有利于保持沙漠表层土壤的含水量和降低土壤水分的挥发速率；此外，藻接种使用的固沙剂自身具有良好的保水作用，同样增加了土壤水分含量。

尽管自然条件下形成的藻类结皮叶绿素含量、厚度、抗风蚀程度、土壤理化属性以及碳固定（光合速率）和释放（碳释放速率）特征等指标均高于或显著高于人工藻类结皮，但是人工藻类结皮具有形成速度快的特点，这将为人工藻类结皮防沙固沙的实践变成现实提供了条件。高新复合固化材料与沙、土颗粒可形成黏结性能良好的弹性固化体，具有高度的抗紫外线性、抗压（拉）性、耐风蚀性以及保水性等。高新复合固化材料为藻类的拓殖提供了稳定的沙面环境，促进了其定居与发育。然而，可持续沙漠化防治的理念是基于化学固沙和生物固沙技术的高新结合，初期的化学固沙为后期的可持续生物固沙提供基础，而后期的生物固沙是实现可持续固沙和恢复、改善荒漠化生态环境的保障。

5.1.5 藻类结皮人工培养存在的问题

（1）水分

2014年和2015年，我们的野外研究结果发现，接种后15～20天内，每隔2天对人工藻类结皮进行水分补充后和雨养条件下（接种后不进行水分补充，水分来源是自然降水或吸湿凝结水）人工藻类结皮的盖度存在显著差异，进行补充水分的人工藻类结皮盖度极显著高于雨养条件下人工藻类结皮的盖度。这一野外研究结果与张丙昌等（2013）在实验室条件下研究结果一致，后者认为在裸沙上不施加水分，具鞘微鞘藻不能形成藻类结皮，藻类生物量极低；随着施水量的增加，藻类生物量、结皮厚度和抗压强度均显著增加，从初期形成的藻类结皮发育成为稳定的藻类结皮；而且施水量在3～4 L m^{-2} d^{-1}，连续补充水分10～15天即可形成稳定的藻类结皮。

在无水或少水条件下，不能形成藻类结皮的主要原因是：在干旱区无论碳固定还是碳释放，降水量与频次都会对胞外多糖的合成产生重要影响，土壤水分有助于蓝藻胞外多糖的合成；蓝藻代谢的活化取决于可利用水分的量，土壤水分的获得和吸收将启动蓝藻不同的代谢过程（如光合作用、呼吸和固氮过程）(Mazor *et al.*,1996)。无论是室内研究还是野外试验，我们都可以发现，接种后早期的人工补水措施有助于藻类胞外多糖的合成和代谢活动的正常进行，而胞外多糖的积累有助于恢复其耐旱能力，适应外界干旱环境，从而增加藻类生物量，提高其耐旱能力，促进人工藻类结皮的形成。简言之，早期的水分获得是成功形成人工藻类结皮的关键因素。然而，水分是干旱区最主要的生态限制因子，如果依靠前期不断地进行水分补充，在大范围的防沙治沙实践中无法实现。因此，如何培养出抗旱性更强、更适合人工藻类结皮形成的藻类，是人工藻类结皮构建需要突破的关键技术。

（2）沙埋

2014年和2015年春季，对上一年度进行的野外条件下培养的人工藻类结皮进行了调查，我们发现各个试验小区的人工结皮均有沙埋发生，沙埋厚度从0.5 ~ 5 mm不等。沙埋厚度较高的区域，人工藻类结皮盖度退化明显。王伟波等（2007）在实验室条件下，研究了干燥沙子不同掩埋时间（0天、5天、10天、15天、20天、30天）和深度（0 cm、0.2 cm、0.5 cm、1 cm、2cm）对人工藻类结皮生物量、叶绿素荧光活性和胞外多糖的影响。结果表明，随着沙埋时间的延长和深度的增加，人工藻类结皮的F_v/F_m值和胞外多糖含量均呈现出逐渐降低的趋势，但是在20天和30天沙埋处理之间，两者在不同沙埋深度均不存在显著性差异；生物量的降低出现在沙埋处理的20天和30天，在不同的沙埋深度这两种处理之间差异亦不显著。F_v/F_m值和胞外多糖含量的协同降低说明两者之间或许存在着一定的关联。上述研究表明，沙埋显著影响人工藻类结皮的生长和存活。而沙埋去除后将对BSC产生哪些影响？贾荣亮等（2010）对沙埋干扰解除后腾格里沙漠人工植被区四种典型BSC的光合作用、暗呼吸作用与叶绿素荧光参数进行跟踪测定，研究了沙埋干扰解除后BSC光合生理恢复机制。结果表明，沙埋干扰解除后四种BSC净光合速率增加，暗呼吸速率先降低后增加。沙埋干扰去除后净光合速率、暗呼吸速率受沙埋深度和沙埋时施水量的影响，分别与沙埋深度和施水量呈反比和正比关系。沙埋干扰解除后四种BSC PS Ⅱ光化学效率随时间逐渐增加，证实了BSC沙埋干扰解除后积极自行修复的内在生理机制。因此，沙埋去除后BSC将会自行恢复。

沙埋是沙区，尤其是风沙活动频繁的干旱荒漠生态系统最常见的干扰因素之一。沙埋通过改变BSC生境的光照、温度和土壤理化性质影响BSC的生长和存活。Campbell（1979）研究发现，当土壤潮湿时，被掩埋的丝状蓝藻以5 mm $(24\ h)^{-1}$的速度向上移动；而当土壤干燥时，丝状蓝藻因不能移动而被掩埋致死。人工藻类结皮构建初期，藻的生存能力比较弱，而沙埋的存在严重威胁着藻类结皮的进一步发育和形成。人工结皮比较脆弱，而且自身的抗逆能力也较弱。因此，人工藻类结皮构建完成后，如何防治沙埋对其的危害，是构建人工藻类结皮需要面对的问题。

藻类结皮人工培育作为一种新型的防沙治沙手段，在防止和治理沙漠化的实践中也取得了实质性的进展和突破。然而，在技术的关键环节方面尚需进一步的优化和创新，仍然需要通过大量的科学研究和生产实践不断进行改进和完善。

5.2 藓类结皮人工培养及其应用

藓类结皮是BSC演替的高级阶段（Belnap and Lange，2003；李新荣，2012）。藓类植物的脱水复苏机制使其具有较强的生理耐旱能力。与维管植物相比，藓类具有较低的水势。环境变湿润后，植株利用水势梯度迅速吸收周围环境中的水分，进行生理活动。藓类具有很强的修复作用，它能将因干旱损伤的膜系统全部修复（Bewley，1979）。干旱环境促进了藓类植株的抗旱能力，表现出了植物对环境的适应性。

藓类植物具有强大的无性繁殖能力和耐旱能力，可以利用藓类植物的生物学特性，从环境较适宜、藓类植物群落面积较大的区域采样，进行机械粉碎，使用藓类植物的茎叶碎片人工培养出一定面积的人工藓类结皮，然后移植到野外环境。通过人工方法加速其定居、扩繁，与周围土壤复合形成具有较强后期维持能力的藓类结皮，从而对荒漠化防治、植被重建和生态恢复起到促进作用。此外，藓类结皮覆盖土壤的稳定性较藻类结皮和地衣结皮高。因此，通过人工培养的方法实现藓类结皮的快速拓殖和定居，也将进一步丰富人工结皮防沙治沙的手段和方法。

5.2.1 藓类结皮人工培养的流程和方法

藓类结皮人工培养，主要包括4个方面：① 藓类配子体的采集；② 藓类人工培养；③ 野外接种；④ 管理与维护。在我国腾格里沙漠东南缘、古尔班通古特沙漠和毛乌素沙地等地区，构成藓类结皮的优势种和先锋种各异（李新荣，2012）。因此，BSC中藓类的人工培养，应以所在区域的优势种或广布种为主。藓类结皮人工培养的主要过程和步骤如下（图5-14）。

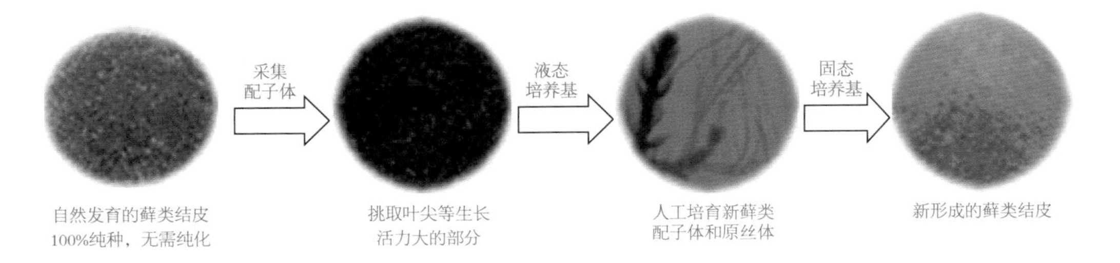

图 5-14　藓类结皮分离、纯化培养流程
Figure 5-14　Flow diagram of artificial moss crust cultivation

① 采取自然条件下发育良好的藓类结皮。样品采集，选择单一藓种组成的藓类结皮。单一藓种形成的藓类结皮培养时不需要进行纯化处理，节约了培养时间，简化了培养步骤。采样时首先将采样框用力压入土壤中，至采样框完全进入土壤中，用75%酒精消毒过的铲子轻轻取出不锈钢框内约50 cm^2 藓类结皮，装入信封中待用。

② 藓类配子体的采集。用75%酒精消毒过的剪刀，将藓类上部发育良好的叶尖剪下，待用。

③ 藓类人工培养。将采集好的藓类配子体放入液态培养基中。称取剪切下的藓类配子体1 g（干重），加入装有300 mL Knop营养液（Knop营养液组分和用量详见表5-2）的三角烧瓶中。将三角烧瓶置于转速为140 r min^{-1} 摇床上进行初步培养，室内温度控制在25 ~ 30 ℃，光强控制在600 lx，室内培养时间20天左右。培养完成后将培养好的藓类置于消毒后的医用纱布上，放置在阴凉通风处，自然风干后待用。

表 5-2　人工藓类培养 Knop 营养液组分和用量
Table 5-2　Components and dosage of Knop nutrient solution for artificial moss cultivation

组分	$Ca(NO_3)_2 \cdot 4H_2O$	KNO_3	$MgSO_4 \cdot 7H_2O$	KH_2PO_4	$ZnSO_4 \cdot 7H_2O$
用量 / （$mg\ L^{-1}$）	1000	250	250	250	3

④ 接种。取步骤③中干燥的藓，均匀撒播在沙面（经多种方法尝试，发现手工播撒方法最好）。撒播后每隔 2 天对土壤水分进行补充，补充水分的标准为土壤表面湿润即可。

5.2.2 不同浓度培养液对真藓原丝体生长的影响

本研究以腾格里沙漠东南缘沙坡头地区发育真藓结皮为研究对象。在我国腾格里沙漠东南缘，真藓是组成藓类结皮的先锋种，同时也是藓类结皮形成的建群种和优势种（李新荣，2012）。因此，BSC中藓类的人工培养选择了真藓为研究对象。分别使用了浓度为25%、50%、75%和100%的Knop营养液对真藓进行了21天的培养，研究结果如图5-15所示。随着培养时间的增加，4种浓度下真藓的原丝体均呈现出逐渐增加的趋势。培养21天后，营养液浓度25%、50%、75%和100%的真藓的原丝体长度分别为1286.9 mm、1062.7 mm、962.5 mm和1009.3 mm。25%浓度营养液培养的真藓原丝体生长最快最长；其次是50%浓度、100%浓度和75%浓度（图5-15a）。

对4种培养浓度下真藓的原丝体长度的增加量进行了计算。结果表明，100%浓度下日均增长量为85.2 mm，最大增长量出现在第14～17天，期间日均增长量为218.3 mm；75%浓度下日均增长量为83.4 mm，最大增长量出现在第14～17天，期间日均增长量为162.9 mm；50%浓度下日均增长量为92.55 mm，最大增长量出现在第14～17天，期间日均增长量为233.69 mm；25%浓度下日均增长量为111.4 mm，最大增长量出现在第14～17天，期间日均增长量为252.7 mm。从上述研究结果可以看出，4种培养浓度条件下，真藓的原丝体长度的最大日均增加量均出现在第14～17天，也就是说，最大增长量出现在培养的中后期。总体来看，4种培养浓度下真藓的原丝体长度增加量呈现出单峰曲线。从日均最大生长量看，25%浓度营养液培养的真藓原丝体生长最快最长；其次是50%浓度、100%浓度和75%浓度。日均增长量呈现出和日均最大生长量相同的趋势（图5-15b）。

对4种培养浓度下真藓的原丝体长度的增长率进行了计算，结果表明，100%浓度的日均增长率为55.0%，最大增长率分别出现在第2～9天和第14～17天，期间日均增长率分别为147%和81.8%；75%浓度的日均增长率为58.6%，最大增长率分别出现在第2～9天和第14～17天，期间日均增长率分别为176.3%和42.89%；50%浓度的日均增长率为66.5%，最大增长率分别出现在第2～9天和第14～17天，期间日均增长率分别为182.8%和89.1%；25%浓度的日均增长率为68.66%，最大增长率分别出现在第2～9天和第14～17天，期间日均增长率分别为182.72%和85.2%。从上述研究结果可以看出，4种培养浓度条件下，真藓的原丝体长度的最大日均增加率均出现在第2～9天和第14～17天。也就是说，最大增长率出现在培养的初期和后期，而中期增长速率较慢。总体来看，4种浓度下培养的真藓，原丝体长度的增加率呈双峰曲线，日均增长率呈现出快－慢－快的生长趋势。从日均增长率看，4种培养浓度下增长率大小排序为25%>50%>75%>100%浓度（图5-15c）。

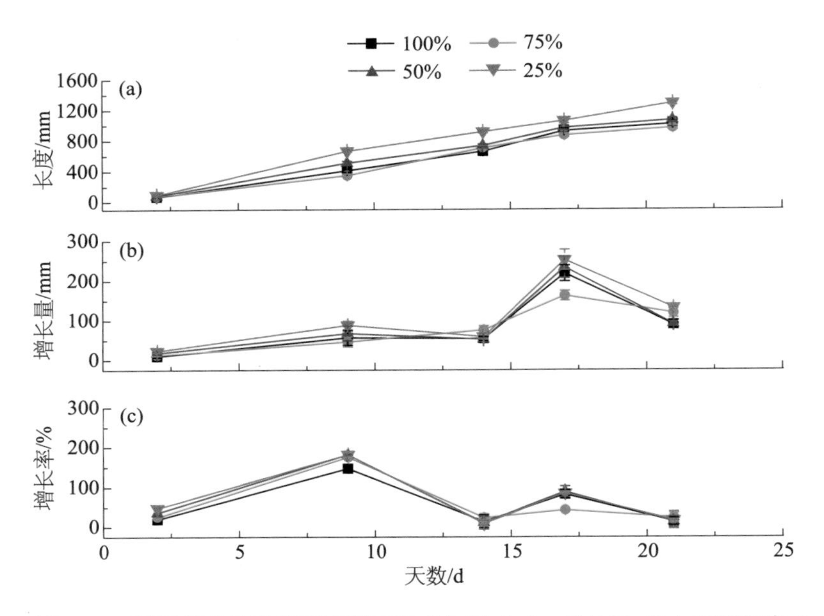

图 5-15　不同营养液浓度下真藓原丝体长度（a）、增长量（b）和增长率（c）的日变化
Figure 5-15　Diurnal variation of length（a）, increment（b）and increment rate（c）of *B. argenteum* protonema at different concentration nutrient solution

本研究结果与前人的研究结果相似，即营养液对藓类的发育表现为低浓度促进、高浓度抑制。王显蓉（2014）的研究认为，不同的藓种具有不同的适宜浓度。30 mg L^{-1} 腐殖酸溶液和40 mg L^{-1} 土壤浸出液有利于真藓和短叶扭口藓的生长发育，低于10 mg L^{-1} 的糖粉溶液有利于真藓的生长发育，30 mg L^{-1} 时表现为抑制作用。30 g L^{-1} 的葡萄糖显著抑制银叶真藓的形成发育。腐殖酸和土壤浸出液中仅有40 mg L^{-1} 腐殖酸对葫芦藓结皮的形成有影响，其他物质无显著影响。牛粪浸出液中大量的细菌病原体会污染土壤基质，引起霉菌污染，对三种藓类结皮的生长具有很明显的抑制作用。

综上所述，无论是绝对增长量还是相对增长量，25%浓度下真藓原丝体生长速度均最快，随着营养液浓度的增加真藓原丝体生长速度受到了抑制，说明低浓度的营养液更适宜真藓原丝体生长。因此，使用Knop营养液进行真藓原丝体液体培养时，营养液浓度以25%为宜。

5.2.3　不同基质对藓类结皮形成的影响

将25%、50%、75%和100% Knop营养液浓度培养的真藓分别接种到蛭石、沙子、黄土和锯末4种基质上，进行了30天的培养，研究结果如图5-16所示。

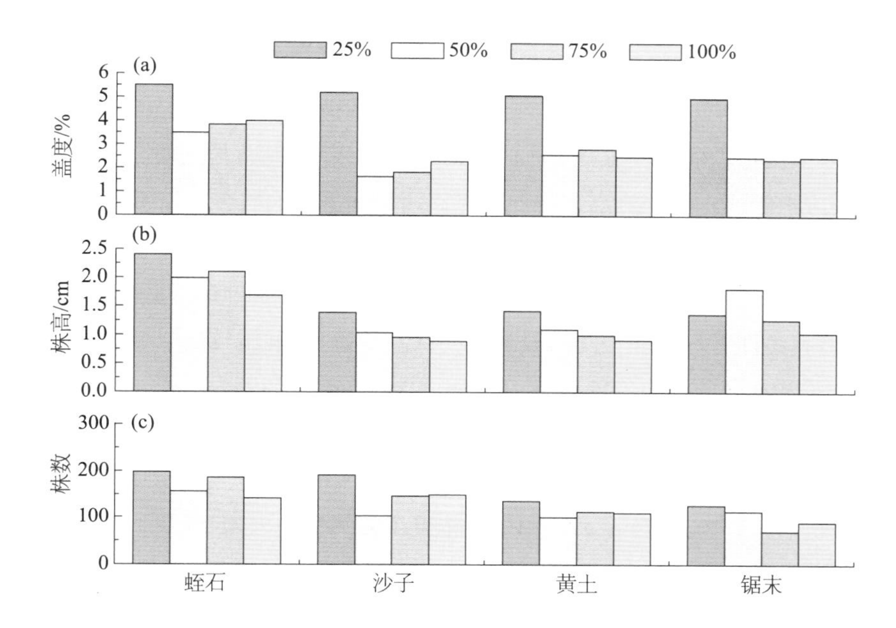

图5-16　不同基质（蛭石、沙子、黄土和锯末）下人工真藓盖度、株高和株数

Figure 5-16　Coverage，height and number of artificial *Bryum argenteum* crust at different culture substrates

图5-16a描述了不同Knop营养液浓度培养真藓接种到蛭石、沙子、黄土和锯末4种基质上的盖度。在蛭石基质上，25%、50%、75%和100%浓度下的盖度分别为5.5%、3.5%、3.8%和4%，25%浓度培养的真藓盖度显著高于其他浓度，其他浓度间差异不显著；在沙子基质上，4种浓度下的盖度分别为5.2%、1.6%、1.8%和2.3%，25%浓度培养的真藓盖度显著高于其他浓度，其他浓度间差异不显著；在黄土基质上，4种浓度下的盖度分别为5.1%、2.6%、2.8%和2.5%，25%浓度培养的真藓盖度显著高于其他浓度，其他浓度间差异不显著；在锯末基质上，4种浓度下的盖度分别为5.0%、2.5%、2.4%和2.5%，25%浓度培养的真藓盖度显著高于其他浓度，其他浓度间差异不显著。对相同浓度培养条件下真藓在不同培养基质的盖度进行平均值计算，结果表明，25%、50%、75%和100%平均盖度分别为5.2%、2.6%、2.5%和2.8%，25%浓度培养的真藓盖度显著高于其他浓度，其他浓度间差异不显著；对不同浓度培养条件下真藓在相同培养基质的盖度进行平均值计算，结果表明，蛭石、沙子、黄

土和锯末平均盖度分别为4.2%、2.7%、3.2%和3.1%，蛭石培养的真藓盖度显著高于其他基质，其他基质间差异不显著。

图5-16b描述了不同Knop营养液浓度培养下，接种到蛭石、沙子、黄土和锯末4种基质上的真藓高度。在蛭石基质上， 25%、50%、75% 和100% 浓度下的高度分别为2.4 cm、1.9 cm、2.1 cm和1.7 cm，25%浓度培养的真藓株高高于其他浓度；在沙子基质上，4种浓度下的高度分别为1.4 cm、1.0 cm、0.96 cm和0.89 cm，25%浓度培养的真藓株高高于其他浓度；在黄土基质上，4种浓度下的高度分别为1.4 cm、 1.1 cm 、 1.0 cm和0.91 cm，25%浓度培养的真藓株高高于其他浓度；在锯末基质上，4种浓度下的高度分别为1.3 cm、1.8 cm、1.2 cm和1.0 cm，50%浓度培养的真藓株高高于其他浓度。对相同浓度培养条件下真藓在不同培养基质的株高进行平均值计算，结果表明，25%、50%、75%和100%浓度下平均盖度分别为1.6 cm、1.5 cm、1.3 cm和1.1 cm，25%浓度培养的真藓株高高于其他浓度，100%浓度株高最低；不同浓度培养条件下真藓在相同培养基质的株高进行平均值计算，结果表明，蛭石、沙子、黄土和锯末平均高度分别为2.05 cm、1.07 cm、1.11 cm和1.38 cm，蛭石培养的真藓株高显著高于其他浓度，其次为锯末基质，沙子和黄土基质培养的真藓株高最低。

图5-16c描述了不同Knop营养液浓度培养下，接种到蛭石、沙子、黄土和锯末4种基质上的真藓株数。在蛭石基质上，25%、50%、75%和100%浓度下的株数分别为198株、156株、186株和141株，25%浓度培养的真藓株数最多，其次为75%浓度；在沙子基质上，4种浓度下的株数分别为191株、103株、146株和150株，25%浓度培养的真藓株数最多；在黄土基质上，4种浓度下的株数分别为136株、100株、112株和110株，25%浓度培养的真藓株数最多，其次为75%浓度；在锯末基质上，4种浓度下的株数分别为126株、113株、70株和90株，25%浓度培养的真藓株数最多，其次为50%浓度。对相同浓度培养条件下真藓在不同培养基质的株数进行平均值计算，结果表明，25%、50%、75%和100%平均株数分别为163株、118株、128株和122株，25%浓度培养的真藓株数高于其他浓度；不同浓度培养条件下真藓在相同培养基质的株数进行平均值计算，结果表明，蛭石、沙子、黄土和锯末平均株数为170株、147株、114株和100株，蛭石培养的真藓株数显著高于其他浓度，其次为沙子基质，黄土和锯末基质真藓株数最低。

BSC是由隐花植物及其分泌物等与土壤形成的复合体，而藓类结皮的培养基质是否是制约藓类结皮形成的因素，一直备受争议。一部分研究者认为，基质并不构成制约因素。因为藓类是只有假根的非维管植物，基质中丰富的养分物质并不能通过假根大量转移到藓类体内，满足其生长发育的需求（孙守琴等，2009）。另外一部分研究学者则认为，藓类植物体内所需要的元素除了来自于大气外，主要来自于生长基质，如土壤、岩石、枯枝落叶层及树冠淋溶等（Rambo and Muir，1998）。赵小艳等（2011）在室内培育双色真藓、真藓、土生对齿藓结皮，发现双色真藓在黏粒与有机质含量较高、保水性能较强的基质上发育良好；粗粒含量增加、表层透气性强的基质上真藓发育良好；而有机质含量过高反而会显著制约土生对齿藓的生长发育。由于藓类种类的不同，其生活特性、对环境的依赖程度、对营养元素的选择等方面各有差异；同时，不同的基质其空隙度、保水、保温能力亦有所不同，进而可能会影响藓类植物的代谢活动。

土壤（基质）水分含量、相对空气湿度是藓类培育的首要限制因子（Proctor，1972）。胡人亮（1987）研究认为，大部分藓类植物其生长发育最适宜的空气相对湿度应大于32%，最适空气温度为10～21℃。陈彦芹等（2009）在室内人工培养藓类结皮时发现，当土壤含水量低于60%相对含水量时藓类的萌发受到抑制。赵小艳等（2011）通过失水复水过程发现，双色真藓、真藓和土生对齿藓最适光合含水量均大于80%，分别为82.14%～87.10%、80.04%～83.19%和85.90%～89.19%。由此可见，水分是显著影响人工藓类结皮的形成和发育的重要因素。本研究中使用的蛭石、沙子、黄土和锯末4种基质（图5-17），对其保水性比较发现蛭石>锯末>黄土>沙子，而4种基质培养的藓类结皮的盖度和高度的变化趋势与基质保水性的顺序一致。

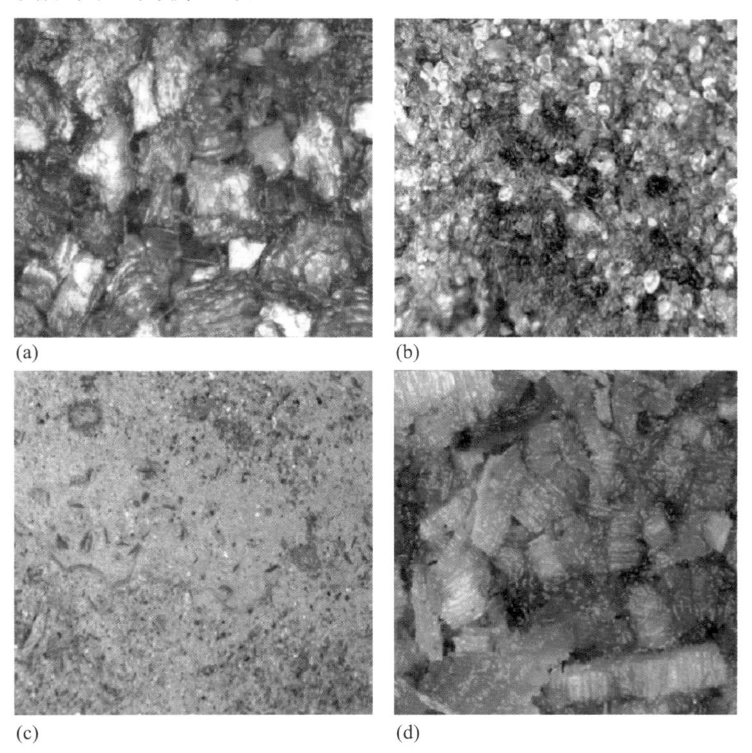

图5-17　不同基质培养20天后的真藓（放大20倍）：（a）蛭石；（b）沙子；（c）黄土；（d）锯末
Figure 5-17　Artificial *B. argenteum* at different culture substrate after 20 days（magnify 20 times）:（a）vermiculite roseite；（b）sand；（c）loess；（d）saw dust

综上所述，对盖度、高度和株数进行综合分析发现，蛭石、沙子、黄土和锯末4种基质上25%浓度的Knop营养液培养的真藓盖度、株高和株数均为最高；而4种浓度培养的真藓在蛭石基质上生长状况最好。此外，室内接种试验进一步论证了25%浓度Knop营养液培养的真藓活性最好。

5.2.4 藓类结皮人工培养存在的问题

（1）人工培养藓类结皮的自我维持和更新

判断藓类结皮人工培养是否成功，判断标准应至少包括两个方面：① 成功构建出一定盖度的人工结皮；② 人工结皮具有较强的后期维持能力和自我更新能力，能在较长时间内正常生长和发育，并发挥其防治土壤侵蚀和固定沙面的功能。

通过我们的室内研究以及其他研究人员的野外研究可以发现，短期内成功构建出一定盖度的人工藓类结皮已经可以实现。室内研究方面，陈彦芹等（2011）的研究发现，在40天的培养周期内，藓类结皮的盖度和密度持续上升。许书军（2007）的研究发现，30天左右的培养周期可形成15 cm^2的刺叶墙藓结皮。野外研究方面，孙俊峰等（2005）的研究表明，以河沙为基质人工培养的真藓结皮成活率保持在80%以上，培养180天后盖度保持在60%以上。在半固定沙丘上用双色真藓为优势种构建藓类结皮，研究结果表明，双色真藓的成活率高达90%，两年内藓类结皮盖度呈现出持续增加的趋势。然而，随着培养时间的延长，两年后藓类结皮盖度呈明显下降趋势（贾艳等，2012）。田桂泉等（2005）在沙坡头地区成功培养出了以真藓为优势种的藓类结皮，但从第3个月开始，真藓结皮盖度开始衰退，而第11个月时真藓株全部死亡，藓类结皮完全退化。

上述研究结果表明，无论在室内条件还是野外环境下，均可以通过人工培养的方式培养出盖度和密度较高的藓类结皮，但这种人工培养藓类结皮的后期维持能力较差。如何构建出稳定的藓类结皮，如何维持或管理成功构建的藓类结皮，使其能够进行自我维持和自我更新，是藓类结皮人工培养亟待解决的问题。

（2）藓类的工厂化生产

人工培养藓类结皮的基础是接种材料。自然条件下，藓类主要以其茎叶碎片进行无性繁殖进而发育形成藓类结皮，若直接以自然条件下发育的藓类结皮作为接种材料，由于破坏原生藓类结皮而导致其发展的不可持续，该方法不能大范围推广使用。但是，可以通过藓类植物的组织培养获取接种材料。近些年，有关藓类植物组织培养的研究已有不少成果。高永超等（2002）总结了藓类植物组织培养的影响因素，包括培养基组成、植物激素的添加、温度、pH、光照等。藓类植物组织培养常用的培养基有MS、Knop、Beneche、Nitsh，其中MS培养基和Knop培养基适合于大多数的藓类植物。潘一廷等（2005）在诱导小立碗藓的愈伤组织时发现，葡萄糖是重要的影响因子之一，葡萄糖含量超过4%不利于愈伤组织的诱导。藓类植物组织培养适宜的温度通常在25 ℃，光照强度为40 $\mu mol\ m^{-2}\ s^{-1}$。一般有关藓类结皮接种量的研究认为，人工藓类结皮构建时，藓类用量约为500 ~ 750 g m^{-2}（陈彦芹等，2009；王显蓉，2014）。尽管上述方法在室内和野外条件下均能培养出藓类结皮，但是由于藓类产量的限制，构建藓类结皮的面积仍然很小。也就是说，人工繁殖藓类的产量将直接决定着人工藓类结皮的覆盖面积。目前，尚未见到藓类工厂化生产技术和生产体系的报道。因此，如何将藓类进行工厂化生产是将人工藓类结皮大范围应用到实践中必须解决的问题。

5.3 地衣结皮人工培养及其应用

地衣结皮是BSC的主要类型之一，处于BSC演替的过渡阶段。地衣是地衣专化型真菌与一些低等光合共生物如藻类及菌类紧密结合成的体内胞外互利共生型生态系统。藻类进行光合作用制造营养被菌类利用，而菌类供给藻类水分及矿物质，形成一种互利共生的关系。由于地衣在沙表结皮形成中起着重要的作用，它可以利用菌丝和假根黏合沙粒，有效地束缚沙粒的流动，从而起到固沙的作用，进而减少荒漠地表的风蚀和水土流失。因此，通过人工培养形成人工地衣结皮对防沙治沙具有重要意义。

5.3.1 地衣人工培养的流程和方法

地衣的人工培养，主要包括4个方面：① 地衣的采集；② 共生菌的分离和培养；③ 共生藻的分离和培养；④ 接种。在我国腾格里沙漠东南缘、古尔班通古特沙漠和毛乌素沙地等地区，构成地衣结皮的优势种和先锋种各异。因此，BSC中地衣的人工培养应以所在区域的优势种或广布种为主。地衣培养的主要过程和步骤如下（图5-18）。

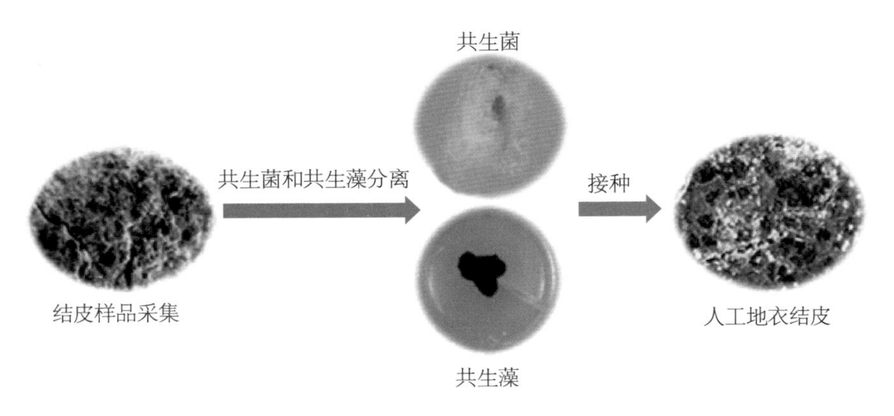

图 5-18　地衣结皮的分离、纯化培养流程
Figure 5-18　Flow diagram of artificial lichen crust cultivation

（1）地衣的采集

采取自然条件下发育良好的地衣结皮。样品采集选择所在区域发育良好的优势结皮。采样时首先将采样框用力按入土壤中，至采样框完全进入土壤中，用75%酒精消毒过的铲子轻轻取出不锈钢框内约50 cm^2 地衣结皮，装入信封中待用。

（2）共生菌的分离和培养

共生菌可以从子囊孢子、分生孢子、裂芽、粉芽和菌体碎片中分离出来，使用固体培养基进行培养。培养基为LB（Lilly and Barnett）培养基，主要组分和用量详见表5-3。

表 5-3　地衣培养基组分和用量（LB 培养基）
Table 5-3　Components and dosage of LB nutrient solution for lichen cultivation

组分	用量	组分	用量
葡萄糖	10.0 g	天冬酰胺酸	2.0 g
KH_2PO_4	1.0 g	$MgSO_4 \cdot 7H_2O$	0.5 g
$Fe(NO_3)_3 \cdot 9H_2O$	0.2 mg	$MnSO_4 \cdot 4H_2O$	0.1 mg
$ZnSO_4 \cdot 7H_2O$	0.2 mg	琼脂	15～20 g
维生素 B1	0.1 mg	蒸馏水	1000 mL
维生素 H	5 μg		

① 孢子释放。把野外采集的地衣体洗净，放置几天使其与周围环境达到水分平衡。从地衣体上取下子囊盘或子囊壳，并将其放在有蒸馏水的培养皿中浸泡4 h或在流动水中清洗。通过挤压去除多余的水分。将其用凡士林固定在培养皿底部，然后用含水4%的琼脂培养基盖在培养皿上部，将培养基放在培养皿的上面盖上，并防止琼脂污染。在适当的间隔（依据释放时间，通常为24 h）多次旋转琼脂层。另外在潮湿的环境中也可以释放到玻璃片上或无菌的薄膜上，然后用蒸馏水将孢子冲下，立即转移到培养基上。用超薄膜将培养皿封口，在培养箱中进行培养，条件为无光，温度15 ~ 20 ℃。孢子萌发情况的检验可以在高倍电子显微镜下进行，观察孢子的释放情况。

② 从地衣体中分离出共生菌。用消毒的手术刀将新鲜的地衣切成薄片，然后存放于装有蒸馏水的小试管（25 mL）中或放在15 ℃环境下的潮湿的滤纸上。然后将冲洗后的地衣体碎片在研钵中磨碎并加水混匀成悬浮液。经过滤后的滤液再经过第二次过滤。从第二次过滤的滤纸上取出少部分接种至斜面培养基上。新的髓状菌丝通常会在2 ~ 3周后延长。采用无菌技术切取一部分新长出的菌丝，转移到试管培养基上。此步骤应进行多次重复，以保证生长的真菌最有可能为共生菌而不是生长于地衣体表面或内部的其他真菌。

③ 共生菌的保存。共生菌可以存放很长时间（1年左右）。利用解剖刀将培养的共生菌分为几部分（通常为5 mg）。将各个部分放在培养皿中的LB培养基中，在15 ℃无光条件下培养2 ~ 3个月。每2 ~ 3个月重复一次该步骤。通常共生菌在15 ~ 20 ℃时生长速率最快。在共生菌的培养中，培养基的pH对群体生长有着重要的影响。每种共生菌均有其最适生长pH，通常为5 ~ 6，太高或太低均会阻碍其生长。在共生菌群体保存培养中，光照强度没有明确的要求。地衣共生菌可以在液氮中保存很长时间，仍具有活性。

（3）共生藻的分离和培养

① 采样方法详见共生菌分离和培养中地衣采样方法。

② 共生藻的分离和培养。

地衣体的清洗：对于大型地衣（叶状和枝状），从地衣体顶端切下面积约1 cm^2的地衣体，然后放在蒸馏水中浸泡5 ~ 10 min，用软毛刷（画笔刷或毛笔等）在流动的蒸馏水中清洗地衣体表面，然后用无菌水冲洗。对于小型地衣，将其放入装有1 ~ 2 mL的无菌水和一滴聚氧乙烯失水山梨醇单月桂酸酯（吐温-20）的小试管中。超声波粉碎3 min。离心（离心机转速2000 rpm）去除地衣体表面的杂质和附属物。

地衣体匀浆液的制备：清洗之后，将地衣体放在一个无菌的载玻片上。在解剖镜下，利用针锉平的小刀仔细刮去地衣皮层。在显微镜下，取下藻层并转移到新的无菌载玻片上。在无菌玻片上滴一滴无菌水，用另一个玻片把藻层盖上，轻轻压玻片，将地衣体磨成较小的碎片。这样共生菌和共生藻就分开了。两种共生体均悬浮在液体中。除了清洗步骤，所有的操作都在超净工作台或无菌条件下进行。所有仪器均应高压灭菌（15 ~ 20 min，121 ℃，1 个标准大气压）或干热灭菌（30 min，180 ℃）。用酸或清洁剂清洗玻片，然后用蒸馏水冲洗。

共生藻的分离：在皮氏培养皿的固体琼脂培养基上滴2 ~ 3滴含有共生菌藻的悬浊液。绿藻的培养用1 × 的NBBM培养基，而蓝藻则需要用BG11培养基（主要组分和用量详见表5-1）。另外，也可以将含有共生菌藻的悬浊液喷到皮氏培养皿的琼脂培养基上，在15 ℃的培养箱中培养15天。通常培养应保持在15 ~ 20 ℃，光照强度为10 ~ 27 μmol m^{-2} s^{-1}的条件下。培养基最初应放于低光强条件下，培养约30天后（时间长短取决于共生藻的种类），琼脂培养基上就会出现较少的共生藻群。

无菌培养群体的获得：无菌共生藻的培养群体的获得主要有以下两种方法。

直接法：在高倍电子显微镜下将未受污染的群体选出来，并移植到适宜的固体培养基上。在试管或皮氏培养皿中的*Trebouxia*有机营养培养基用于绿藻，BG11培养基用于蓝藻。如果未受污染就可获得无菌培养的蓝藻或绿藻群体。

喷雾法：该技术适用于绿藻单细胞的分离。对于单细胞绿藻或无菌群体获得很有用。具体步骤：从单细胞中获得群体，然后从生长在琼脂上的群体中选择污染低（或未污染）的群体，并转移到试管中1 × NBBM的斜面培养基上，培养若干周；在10 mL的离心管中加入1 mL蒸馏水和一滴吐温-20，将共生藻悬浮液移入，超声波粉碎。这样会使共生藻群同附着在其细胞壁表面上的细菌和其他共生菌及污染物分开；移去上清液，在离心剩下的共生藻细胞中加入1 mL无菌水和吐温-20，重复以上操作约10次；在离心管底部插入一个毛细管，并保持直立。通过毛细管的小开口引出压缩空气，压缩空气穿过伸出离心管的毛细管，藻培养液就会被冲出毛细管形成喷雾；快速将含有培养基（通常为*Trebouxia*有机营养培养基）的皮氏皿通过喷雾，培养皿上就会附着一层藻细胞的悬浊液；1周或2周后，将未污染的藻群体移到合适的培养基上。

（4）接种

首先将培养的共生藻藻液用喷雾器均匀地接入沙地表面，藻液用量为干重1 g m^{-2}，然后将培养的共生菌喷洒到沙地表面。接种后，前20天每隔2天用喷雾器进行浇水（表面湿润即可），下雨期间停止浇水。1 ~ 2个月后便会有地衣形成。图5-19为人工培养的地衣——石果衣（*Endocarpon pusillum*）。

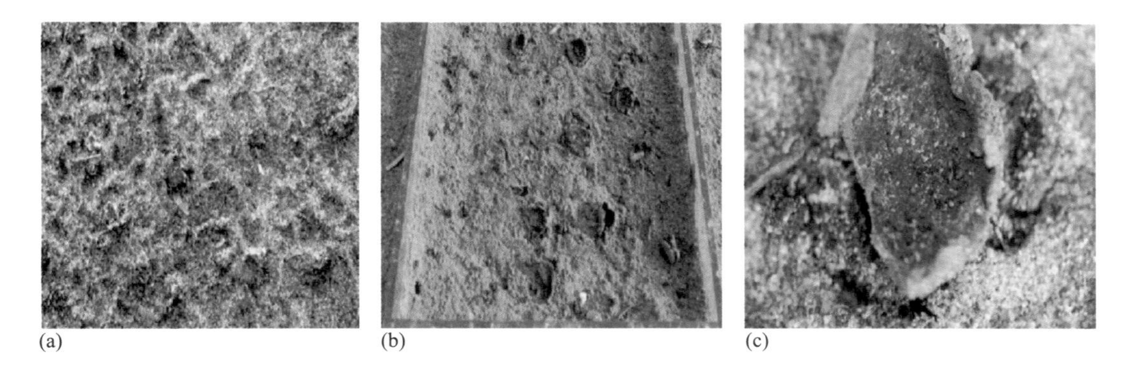

图 5-19　野外培养的人工地衣（石果衣）
Figure 5-19　Artificial lichen (*Endocarpon pusillum*) cultivation under field condition

5.3.2　人工地衣结皮的研究现状

地衣是共生藻和共生真菌以紧密而特殊的共生关系形成的复合真核生物体，藻类和真菌细胞被胶质蛋白连接在一起，藻细胞完全被菌丝所包围而形成原植体。地衣通常是由1种真菌和1种藻组合的复合有机体，真菌是主要成员。地衣是真菌和光合生物之间稳定而又互利的共生体（李冬雪和丁雨龙，2001；Trembley *et al.*，2002）。因此，理论上将地衣结皮中的共生真菌和共生藻进行人工培养，便能够进行人工地衣结皮培养。然而，地衣共生体并非真菌与藻类的简单加和，而是经过长期的相互作用演化而成的既不同于真菌又不同于藻类的生物体。此外，研究表明，人工分离培养的共生菌和共生藻与地衣共生体菌藻相比，其结构、生理、化学和遗传均有较大区别（魏江春，1998）。虽然地衣中的光合共生生物合成的糖类被共生真菌所利用这一假设已被同位素标记实验所证实，而且来自不同光合共生生物的不同糖类及其在地衣体内的转移途径及速度的研究也取得了明显进展，但要解决地衣的人工培养和提高生长速度的问题，还需走相当遥远的路程。人工地衣结皮的培养仍需克服许多困难。

5.4　人工结皮在实践中的应用

在我国温带荒漠区，沙漠和沙漠化土地面积约152×10^4 km^2，约占国土总面积的16%，有十大沙漠和四大沙地。这些沙漠和沙地从我国东部一直延伸到西部，穿越了湿润地区、半湿润地区、半干旱地区和干旱地区。然而，在我国广袤的沙区使用人工结皮固沙技术进行防沙治沙，如何选择人工结皮的类型？应当遵循什么样的规律或原则？

土壤质地和化学性质强烈地影响着BSC类型和群落种的组成。与稳定性差、质地较粗的土壤相比，在较稳定、质地较细的土壤（如含石膏的土壤和细黏土）上发育的蓝藻、地衣和藓类拥有更高的盖度和更多样的种群，而质地较粗的土壤仅能支持移动性强、长菌丝体的蓝藻（如具鞘微鞘藻）分布（Kleiner and Harper，1977）。如果不考虑土壤质地这一因素，相对于邻近土层较深的土壤，土层较浅的土壤上的蓝藻、地衣和藓类的多样性较高。土壤化学性质也能影响BSC的形成、盖度和组成。例如，碱性土壤容易形成蓝藻占优势的BSC，而酸性土壤则容易形成绿藻占优势的BSC（Grondin and Johansen，1995）。碳酸钙含量高的土壤则能形成地衣结皮，在高钙质土壤上BSC的盖度和地衣的多样性较高；在大多荒漠区由石灰石或石膏衍生的高钙质土壤支撑的地衣盖度能达到80%，而毗邻的低钙质土壤仅能支撑10%的地衣盖度；高等植物在这类土壤上分布稀疏且物种多样性也受到限制。因此，可以根据土壤质地和土壤化学性质选择人工结皮的类型。

在塔克拉玛干沙漠、库布齐沙漠、巴丹吉林沙漠以及浑善达克沙地，流沙占沙漠面积的90%以上，流沙限制了藓类结皮的发育。因此，需要选择能够在恶劣的环境条件下生长和繁殖的藻类结皮。在黄土高原与沙地或沙漠的过渡区，如鄂尔多斯高原南部、腾格里沙漠南部以及一些固定较好的沙地，如准噶尔盆地和科尔沁沙地，土壤质地相对较细，有利于支撑藓类结皮和地衣结皮的形成。而在降水量较大的黄土高原地区，高等植物冠层之间的“裸地”因其更细的土壤质地支撑了地衣和藓类结皮的形成。黄土高原地区大面积高等植物植被的破坏，为地衣结皮和藓类结皮的形成提供了机会。因此，人工藓类结皮和地衣结皮可以用于该区域治理。此外，在黄土高原地区植被盖度较高的区域或新建立的人工植被区进行BSC恢复，如露天煤矿排土场（Zhao *et al.*，2016），在植被灌丛下或灌丛间进行人工藓类结皮的接种。在东部沙区，降水条件较好，土壤质地也较细，适宜藓类结皮的形成和发育，因此，可以先通过人工藻类结皮的接种，待土壤环境改善后再进行人工藓类结皮的接种。在各个沙区的BSC退化区，如活化斑，若活化斑面积较大（BSC重度退化区），首先使用人工藻类结皮进行沙面的固定，然后选择与活化斑周边发育一致的BSC进行人工培养和接种；若活化斑面积较小（BSC轻度退化区），直接选择与活化斑周边发育一致的BSC进行人工培养和接种。综上所述，在防沙和治沙的初期以及土壤质地较粗的区域，应选择环境适应性较强的人工藻类结皮；在土壤质地较细和降水量较高的区域，首先使用人工藻类结皮进行土壤环境的改善，再选择对环境要求较高的人工藓类结皮或人工地衣结皮。

无论在室内研究，还是在野外实践中，藻类结皮、藓类结皮和地衣结皮等BSC人工培养和野外接种技术都取得了突破性的进展。人工结皮固沙技术也日趋成熟，将逐渐为我国防沙治沙服务。然而，由于每个防沙治沙区域都有自己的特点和特殊性，人工结皮固沙技术的应用需要因地制宜。人工结皮固沙技术的广泛应用也面临着严峻的挑战，还有很长的路要走。

参考文献

陈兰周, 刘永定, 宋立荣. 微鞘藻胞外多糖在沙漠土壤成土中的作用. 水生生物学报, 2002, 26(2): 155−159.

陈彦芹, 赵允格, 冉茂勇. 黄土丘陵区藓结皮人工培养方法试验研究. 西北植物学报, 2009, 29(3): 586−592.

陈彦芹, 赵允格, 冉茂勇. 4种营养物质对藓结皮形成发育的影响. 西北农林科技大学学报（自然科学版）, 2011, 39(5): 44−50.

高永超, 沙伟, 张晗. 苔藓植物的组织培养. 植物生理学通讯, 2002, 38(6): 607−610.

胡春香, 刘永定. 土壤藻生物量及其在荒漠结皮的影响因子. 生态学报, 2003, 23(2): 284−291.

胡春香, 刘永定, 宋立荣, 黄泽波. 半荒漠藻结皮中藻类的种类组成和分布. 应用生态学报, 2000, 11(1): 61−65.

胡人亮. 苔藓植物学. 北京: 高等教育出版社, 1987.

贾荣亮, 李新荣, 谭会娟, 贺郝钰, 苏洁琼. 沙埋干扰去除后生物土壤结皮光合生理恢复机制. 中国沙漠, 2010, 30(6): 1299−1304.

贾艳, 白学良, 单飞彪, 白少伟, 詹洪瑞. 藓类结皮层人工培养试验和维持机制研究. 中国沙漠, 2012, 32(1): 54−59.

李冬雪, 丁雨龙. 地衣协同进化研究进展. 南京林业大学学报, 2001, 25(1): 56−60.

李新荣. 荒漠生物土壤结皮生态与水文学研究. 北京: 高等教育出版社, 2012.

潘一廷, 施定基, 杨明丽, 吴鹏程, 杜桂森. 小立碗藓愈伤组织诱导和培养. 植物生理学通讯, 2005, 4(1): 293−296.

孙俊峰, 陈其兵, 王怡, 杨江山. 苔藓植物联合乡土草种应用于植被恢复工程的初步研究. 四川草原, 2005, 26(3): 16−18.

田桂泉, 白学良, 徐杰, 王先道. 腾格里沙漠固定沙丘藓类植物结皮层的自然恢复及人工培养试验研究. 植物生态学报, 2005, 29(1): 164−169.

王伟波, 杨翠云, 唐东山, 李敦海, 刘永定, 胡春香. 实验室条件下沙埋对人工藻结皮生物量、光合活性和胞外多糖的影响. 中国科学C辑（生命科学）, 2007, 37(2): 241−245.

王显蓉. 培养基对耐旱藓结皮生长发育的影响. 西北农林科技大学硕士学位论文, 2014.

魏江春. 地衣、真菌和菌物的研究进展. 生物学通报, 1998, 34(12): 1−5.

魏江春. 沙漠生物地毯工程——干旱沙漠治理的新途径. 干旱区研究, 2005, 22(3): 287−288.

许书军. 典型荒漠苔藓人工繁殖特征与抗御干热环境胁迫的生理生化机制研究.上海交通大学博士学位论文, 2007.

张丙昌, 王敬竹, 张元明, 华邵. 水分对具鞘微鞘藻构建人工藻结皮的作用. 应用生态学报, 2013, 24(2): 535−540.

张元明, 陈晋, 王雪芹, 潘慧霞, 辜智慧, 潘伯荣. 古尔班通古特沙漠生物结皮的分布特征. 地理学报, 2005, 60(1): 53−60.

赵小艳, 田桂泉, 王铁娟. 沙地与黄土丘陵区生物结皮层五种藓类的植物荧光日变化比较研究. 内蒙古师范大学学报（自然科学版）, 2011, 40(1): 77−81.

Azahari NA, Othman N, Ismail H. Biodegradation studies of polyvinyl alcohol/corn starch blend films in solid and solution media. *Journal of Physical Science*, 2011, 22: 15−31.

Belnap J, Gardner JS. Soil microstructure in soils of the Colorado Plateau: The role of cyanobacterium *Microcoleus vaginatus*. *The Great Basin Naturalist*, 1993, 53: 40−47.

Belnap J, Lange OL. *Disturbance and Recovery of Biological Soil Crusts: Structure, Function, and Management Ecological Studies*. Berlin: Springer−Verlag, 2003.

Bewley JD. Physiological aspects of desiccation tolerance. *Annual Review of Plant Physiology*, 1979, 30: 195−238.

Bowker MA. Biological soil crust rehabilitation in theory and practice: An underexploited opportunity. *Restoration Ecology*, 2007, 15: 13−23.

Campbell SE. Soil stabilization by a prokaryotic desert crust: Implications for Precambrian land biota. *Origins of Life*, 1979, 9: 335−348.

Corti A, Cinelli P, D' Antone S, Kenawy ER, Solaro R. Biodegradation of poly (vinyl alcohol) in soil environment: Influence of natural organic fillers and structural parameters. *Macromolecular Chemistry and Physics*, 2002, 203: 1526−1531.

Costerton JW, Irvin RT, Cheng KJ. The bacterial glycocalyx in nature and disease. *Annual Review of Microbiology*, 1981, 35: 299−324.

Ekebafe LO, Ogbeifun DE, Okieimen FE. Polymer applications in agriculture. *Biokemistri*, 2011, 23: 81−89.

Elbert W, Weber B, Burrows S, Steinkamp J, Büdel B, Andreae MO. Contribution of cryptogamic covers to the global cycles of carbon and nitrogen. *Nature Geoscience*, 2012, 5: 459−462.

Gokcen Y. Effects of polyvinylalcohol (PVA) and polyacrylamide (PAM) as soil conditioners on erosion by runoff and by splash under laboratory conditions. *Ekoloji*, 2010, 19: 35−41.

Grondin AE, Johansen JR. Seasonal succession in a soil algal community associated with a beech−maple forest in Northeastern Ohio, USA. *Nova Hedwigia*, 1995, 60: 1−12.

Kleiner EF, Harper KT. Soil properties in relation to cryptogamic groundcover in Canyonlands National Park. *Journal of Range Management Archives*, 1977, 30: 202−205.

Liu J, Shi B, Lu Y, Jiang H, Huang H, Wang G, Kamai T. Effectiveness of a new organic polymer sand−fixing agent on sand fixation. *Environmental Earth Science*, 2012, 65: 589−595.

Malam Issa O, Défarge C, Le Bissonnais Y, Marin B, Duval O, Bruand A, D'Acqui L, Nordenberg S, Annerman M. Effects of the inoculation of cyanobacteria on the microstructure and the structural stability of a tropical soil. *Plant and Soil*, 2007, 290: 209−219.

Mazor G, Kidron GJ, Vonshak A, Abeliovich A. The role of cyanobacterial exopolysaccharides in structuring desert microbial crusts. *FEMS Microbiology Ecology*, 1996, 21: 121−130.

Park CH, Li XR, Jia RL, Hur JS. Effects of superabsorbent polymer on cyanobacterial biological soil crust formation in laboratory. *Arid Land Research and Management*, 2015, 29: 55−71.

Proctor MCF. An experiment on intermittent desiccation with *Anomodon viticulosus* (Hedw.) Hook. & Tayl. *Journal of Bryology*, 1972, 7: 181−186.

Rambo TR, Muir PS. Bryophyte species associations with coarse woody debris and stand ages in Oregon. *Bryologist*, 1998: 366−376.

Schulten JA. Soil aggregation by cryptogams of a sand prairie. *American Journal of Botany*, 1985, 72: 1657−1661.

Trembley ML, Ringli C, Honegger R. Morphological and molecular analysis of early stages in the resynthesis of the lichen *Baeomyces rufus*. *Mycological Research*, 2002, 106: 768−776.

Uysal H, Taysun A, Kose C. Toprak özellikleriyle birlikte kümeleþmeyi saðlayan bazý polimerlerin laboratuvar koþullarý altýnda su erozyonu üzerine etkileri, İ: Mansuz N, Unver Ⅰ Cayci G (eds.). *Ý lhan Akalan Toprak ve Ç evre Sempozyumu, 7Kasým 1995*. Ankara, 1995,101−111.

Whitfield C. Bacterial extracellular polysaccharides. *Canadian Journal of Microbiology*, 1988, 34: 415−420.

Xie ZM, Liu YD, Hu CX, Chen LZ, Li DH. Relationships between the biomass of algal crusts in fields and their compressive strength. *Soil Biology and Biochemistry*, 2007, 39: 567−572.

Yang J, Cao H, Wang F, Tan T. Application and appreciation of chemical sand fixing agent−poly (aspartic acid) and its composites. *Environmental Pollution*, 2007a, 150: 381−384.

Yang J, Wang F, Fang L, Tan T. Synthesis, characterization and application of a novel chemical sand−fixing agent−poly (aspartic acid) and its composites. *Environmental Pollution*, 2007b, 149: 125−130.

Zhao Y, Zhang P, Hu YG, Huang L. Effects of re−vegetation on herbaceous species composition and biological soil crusts development in a coal mine dumping site. *Environmental Management*, 2016, 57: 298−307.

索引

J

K

L

M

N

Q

R

S

T

U

W

X

Y

Z

#

反盗版举报电话　（010）58581999　58582371　58582488

反盗版举报传真　（010）82086060

反盗版举报邮箱　dd@hep.com.cn

通信地址　北京市西城区德外大街4号

高等教育出版社法律事务与版权管理部

邮政编码　100120